Surveying and Geomatics Engineering

Other Titles of Interest

Applied GPS for Engineers and Project Managers, by Clement A. Ogaja. (ASCE Press 2011). This book introduces civil engineers—especially those who are not already licensed surveyors—to the fundamental principles of global positioning technology (ISBN 978-0-7844-1150-6).

CORS and OPUS for Engineers: Tools for Surveying and Mapping Applications, edited by Tomás Soler (ASCE/Geomatics Division 2011). This collection of 22 articles assembles the latest thinking on the use of two advanced services—CORS and OPUS—for obtaining accurate positional coordinates to use in high-accuracy surveying (ISBN 978-0-7844-1164-3).

NAVSTAR Global Positioning System Surveying, by the U.S. Army Corps of Engineers (ASCE Press 2000). The U.S. Army Corps of Engineers Technical Engineering and Design Guide No. 28 provides technical specifications and procedural guidance for surveying with the NAVSTAR GPS (ISBN 978-0-7844-0353-2).

ASCE Manuals and Reports on Engineering Practice No. 152

Surveying and Geomatics Engineering

Principles, Technologies, and Applications

Sponsored by the
Surveying Committee of the
Surveying and Geomatics Division of the
Utility Engineering and Surveying Institute of the
American Society of Civil Engineers and the

National Geodetic Survey

Edited by
Daniel T. Gillins, Ph.D., P.L.S.
Michael L. Dennis, Ph.D., P.E., P.L.S.
Allan Y. Ng, P.L.S.

ASCE

UESI
UTILITY ENGINEERING
& SURVEYING
INSTITUTE

Published by the American Society of Civil Engineers

Library of Congress Cataloging-in-Publication Data

Names: Gillins, Daniel T., editor. | Dennis, Michael L., editor | Allan Y. Ng, editor.
Title: Surveying and geomatics engineering : principles, technologies, and applications / edited by Daniel T. Gillins, Michael L. Dennis, Allan Y. Ng.
Description: Reston, Virginia : American Society of Civil Engineers, [2022] | Series: ASCE manuals and reports on engineering practice ; no. 152 |"Sponsored by the Surveying Committee of the Surveying and Geomatics Division of the Utility Engineering and Surveying Institute of the American Society of Civil Engineers and the National Geodetic Survey." | Includes bibliographical references and index. | Summary: "MOP 152 provides a comprehensive yet general overview to help support education and inform practicing engineers on the important role of the surveying engineer"-- Provided by publisher.
Identifiers: LCCN 2021059081 | ISBN 9780784416037 (paperback) | ISBN 9780784484005 (pdf)
Subjects: LCSH: Surveying. | Geomatics.
Classification: LCC TA545 .S8535 2022 | DDC 526.9--dc23/eng/20220114
LC record available at https://lccn.loc.gov/2021059081

Published by American Society of Civil Engineers
1801 Alexander Bell Drive
Reston, Virginia 20191-4382
www.asce.org/bookstore | ascelibrary.org

***Errata:* Errata, if any, can be found at https://doi.org/10.1061/9780784416037.**

ISBN 978-0-7844-1603-7 (print)
ISBN 978-0-7844-8400-5 (PDF)
ISBN 978-0-7844-8422-7 (ePub)

Manufactured in the United States of America.
27 26 25 24 23 22 1 2 3 4 5

MANUALS AND REPORTS ON ENGINEERING PRACTICE

(As developed by the ASCE Technical Procedures Committee, July 1930, and revised March 1935, February 1962, and April 1982)

A manual or report in this series consists of an orderly presentation of facts on a particular subject, supplemented by an analysis of limitations and applications of these facts. It contains information useful to the average engineer in his or her everyday work, rather than findings that may be useful only occasionally or rarely. It is not in any sense a "standard," however, nor is it so elementary or so conclusive as to provide a "rule of thumb" for nonengineers.

Furthermore, material in this series, in distinction from a paper (which expresses only one person's observations or opinions), is the work of a committee or group selected to assemble and express information on a specific topic. As often as practicable the committee is under the direction of one or more of the Technical Divisions and Councils, and the product evolved has been subjected to review by the Executive Committee of the Division or Council. As a step in the process of this review, proposed manuscripts are often brought before the members of the Technical Divisions and Councils for comment, which may serve as the basis for improvement. When published, each manual shows the names of the committees by which it was compiled and indicates clearly the several processes through which it has passed in review, so that its merit may be definitely understood.

In February 1962 (and revised in April 1982), the Board of Direction voted to establish a series titled "Manuals and Reports on Engineering Practice" to include the manuals published and authorized to date, future Manuals of Professional Practice, and Reports on Engineering Practice. All such manual or report material of the Society would have been refereed in a manner approved by the Board Committee on Publications and would be bound, with applicable discussion, in books similar to past manuals. Numbering would be consecutive and would be a continuation of present manual numbers. In some cases of joint committee reports, bypassing of journal publications may be authorized.

A list of available Manuals of Practice can be found at http://www.asce.org/bookstore.

CONTENTS

FOREWORD

Surveying engineering has been at the heart of the National Geodetic Survey (NGS) since our founding as the Survey of the Coast in 1807 by President Thomas Jefferson. The purpose, as stated by Congress, was "…for completing an accurate chart of every part of the coasts…." In 1878, under the name of the United States Coast and Geodetic Survey (USC&GS), our mission expanded to include the entire country. We continued under that name for nearly 100 years, until the creation of the National Oceanic and Atmospheric Administration (NOAA) in 1970. Today, the geodetic work of the USC&GS continues under the name NGS, as an office within NOAA's National Ocean Service.

With this long and distinguished history, NGS is often touted as our nation's first civilian *science* agency. Although our name has changed and the breadth of our work has grown, our core mission remains the same: to support surveying and mapping in the United States. We accomplish this mission through the creative application of scientific principles and mathematics to solve practical problems for the benefit of society—the very definition of *engineering*.

The NGS engineering ethos is an enduring connection not just to civil engineering in general but also to the American Society of Civil Engineers (ASCE) in particular. This connection is best exemplified by Major William Bowie, a geodetic engineer and Chief of the USC&GS Division of Geodesy from 1915 until his retirement in 1936. Bowie joined ASCE in 1917 and founded the ASCE Surveying and Mapping Division in 1926, serving as Chair until his death in 1940. The name was changed to Surveying Engineering Division in 1986, and then to Geomatics Division in 1997. It was renamed Surveying and Geomatics Division (SGD) in 2015, along with the creation of the ASCE Utility Engineering and Surveying Institute (UESI). As a UESI division, SGD carries forward the spirit of what Bowie started, by ensuring that surveying engineering remains an integral part of civil engineering.

This manual of practice is an extension of Major Bowie's foresight and passion. Founded on science, rooted in mathematics, and focused on modern practical applications, it serves as a general but comprehensive reference for practicing engineers, surveyors, and other geospatial professionals. Within its pages, topics range from the fundamentals of geodesy, coordinate systems, least-squares adjustments, and error propagation through the latest innovations in satellite, terrestrial, mobile, and airborne technologies. It provides best practices for these technologies, as well as for other topics, such as establishing geodetic control, performing construction surveys, and employing geographic and building information systems in civil engineering. Two of the manual's three editors are NGS employees who authored three of the 15 chapters herein. All other chapters are written by authors who are experts in their respective fields. This manual is a unique compilation of professional and practical material—all located in this one magnificent source.

A common thread running through the manual is the importance of the National Spatial Reference System (NSRS) to modern geospatial applications. The NSRS serves as the foundation for civilian surveying and mapping in the United States and its territories. It forms an essential link between the geospatial professions and NGS, whose mission is to define, maintain, and provide access to the NSRS.

The NSRS has never been more important than it is today for ensuring that the work products of surveying engineering are correctly georeferenced and meet accuracy requirements. The vital role of the NSRS continues to grow as society seeks reliable integration and sharing of an ever-expanding array of geospatial data. NGS is currently in the process of modernizing the NSRS, and portions of this manual highlight these changes. As positioning technologies and methodologies evolve, the NSRS modernization will ensure that NGS remains a key partner in surveying and geomatics engineering. This is consistent with what Major Bowie started at ASCE in 1926; and there is no doubt that he would heartily endorse this manual were he alive today. Nearly a century later, this manual is a fitting legacy to Bowie's vision and the ongoing mission of NGS.

Juliana P. Blackwell
Director National Geodetic Survey

PREFACE

In 1985, ASCE published a manual of practice on engineering surveying. It has not been updated since, and the world we are living in today is a very different place when compared with that of 1985. This new reference manual provides a sorely needed and complete update on the modern practice of surveying and geomatics engineering.

In the early 1980s, the digital revolution had just begun with the introduction of the IBM PC. The internet was not widely available, nor was email. In the surveying world, the electronic total station was being introduced, and the US government was just starting to build the Global Positioning System (GPS) satellite constellation.

Today, as some predicted, the use of GPS and other Global Navigation Satellite Systems (GNSS) has become ubiquitous. High-resolution commercial satellites and mobile lidar systems are creating consumer map databases that support vehicle navigation systems in most countries around the world.

As this new Manual of Practice on surveying engineering was being written, mobile computing, autonomous vehicles, augmented and virtual reality, and—perhaps of the greatest potential impact—unmanned aircraft system (UAS) technology were all poised to create the next revolution in surveying and mapping.

None of this would be possible without the surveying engineer who performs work at the intersection of the two fields of surveying and engineering.

For those readers who might be familiar with the 1985 engineering surveying manual of practice, the prologue is an incredibly detailed history of surveying, complete with diagrams and references. I will not even attempt to duplicate that effort. Rather than looking back, it is hoped that this new manual will help establish a vision of the current and near-future role of the surveying engineer.

As stated in the prologue of the previous manual,

> Lasers are being tested for acquisition of vertical control in photogrammetry and for providing line and grade in construction related surveying. Similar to radar-equipped Terrain Profile Recorder systems already in use, the laser beam achieves greater precision and resolution in air-to-ground measurement.

This was the early vision for what would become airborne lidar.

It continued,

> As a result of the technological breakthrough in surveying and mapping, the survey engineer of the 1980's must be better trained in a much broader field of science than the surveyor of even a decade ago. A background in higher mathematics, computer technology, photogrammetry, geodetic science, and electronics is necessary for today's survey engineer to compete in this rapidly expanding discipline. But like his Nineteenth Century counterpart, the survey engineer should possess those admirable traits of early American character, energy, perseverance and close observation that led some young men who began as axemen and rodmen, to rise in their profession until they became distinguished engineers.

These admonitions in the 1980s are all the more true today. The modern surveying engineer is now faced with an all-encompassing digital suite of sensors and computer software that strain the limits of any practicing professional to keep pace with the changes that are taking place on the technology front. Surveying engineers should seek higher education opportunities and continue learning after graduation to stay abreast with all the latest advancements. With the multitude of changes, many have begun referring to the profession as "geomatics engineering" as it now encompasses a broader field.

Perhaps the greatest challenge facing today's professionals, from a technical point of view, lies in the shift from 2D, paper-based documentation to 3D digital models that range in scale from an individual project to the entire globe. Satellite navigation systems such as GPS require that the surveying engineer understand the principles of geodesy to judge the accuracy of their 3D positions.

Despite the strong need for more education to keep up with the technology and to support engineering projects, it has been documented that civil engineering university programs are offering fewer courses in surveying and geomatics. Most states allow licensed professional engineers to perform activities associated with surveying engineering; however, many engineers do not necessarily receive the specialized skills, education, and training.

So, this manual is part of a call to action for the engineering community. There are exciting new opportunities for the surveying engineering profession to step up and assume a leadership role as the world struggles with profound changes, including urbanization, climate change, energy demand, decaying infrastructure, and the Internet of Things. These are just a few of the mega trends that require the knowledge and expertise of the surveying engineer.

This manual is intended to provide a foundation for the practice of surveying engineering, but it cannot replace the need for fundamental, continuous education and training because the technology and the profession are not going to stand still. The manual provides background information on the basics of geodesy, least-squares adjustments, error propagation, coordinate systems and transformations, surveying and remote sensing equipment, identification and establishment of control, construction surveying, best practices, and more. Because entire books could be devoted to each of these topics, this manual only provides a summary, and it points the readers to important references where they can find more details. The manual is meant to provide a comprehensive but general overview to help support education and inform practicing engineers on the important role of the surveying engineer. It is too important for this not to occur.

Gene V. Roe

Blue-Ribbon Review Panel

Kevin M. Kelly, Ron Singh, Dave Zilkoski

ACKNOWLEDGMENTS

This Manual of Practice was prepared over a period of several years under the direction of the Surveying Committee of the Surveying and Geomatics Division of the Utility Engineering and Surveying Institute (UESI) of ASCE. Numerous people contributed to the writing, review, and editing of the manual. Thanks to everyone for their service; without their help, this manual would not have come to fruition.

The manual was created following ASCE procedures for creating manuals of practice. First, authors with expertise and experience in a given subject area were invited to draft a chapter. Then, each chapter was sent to a minimum of two experts for peer review. After responding to the review comments and making revisions, the entire draft manual was submitted to three individuals on a Blue-Ribbon Panel to review its overall content. Afterward, the editors made revisions and edits in response to the Blue-Ribbon Panel's review comments. Finally, the manual was sent to the Executive Committee of the Surveying and Geomatics Division and approved for publication. I believe that the expertise of the many authors and the multiple rounds of review have greatly improved the overall quality of the manual.

I would like to thank the following individuals for serving as authors for chapters of the manual; many of these people also helped serve as peer reviewers of other chapters. Special thanks to Dimitrios Bolkas, Michael Bunn, Earl F. Burkholder, Erzhuo Che, Michael L. Dennis, Charles D. Ghilani, N. W. J. Hazelton, Lisa D. Herzog, Jaehoon Jung, Andrew C. Kellie, Matthew S. O'banion, Michael J. Olsen, Chris Parrish, Gene V. Roe, Michael J. Starek, Yelda Turkan, Jan Van Sickle, Marlee A. Walton, Benjamin E. Wilkinson, and David E. Woolley.

I also would like to thank the following people who served as peer reviewers: Steve Breidenbach, Jacky C. K. Chow, Georgia Fotopoulos, David Garber, Chris Glantz, Alec Grkovic, John Hamilton, Steve Hennessee, Brent Jones, Thomas Meyer, Norman Miller, Carol Morman, Jie Shan, Dru Smith, Tomas Soler, Tom Taylor, David Totman, Greg Wear, David Widmer, and Mark Woodson. I am also grateful for the excellent reviews by the three members of the Blue-Ribbon Review Panel: Kevin M. Kelly, Ron Singh, and Dave Zilkoski.

UESI staff provided valuable support and guidance throughout the entire process of creating the manual. Thanks to John Segna, who helped start the process, Diane Swecker, who continued and encouraged its completion, and Susan Reid, who was our primary and most helpful staff support.

Finally, I would like to thank my spirited coeditors who spent countless hours of service meeting, reviewing, revising, and responding to the many important needs required to edit and prepare the final draft of the manual. My most sincere gratitude to Michael L. Dennis and Allan Y. Ng. Your help was invaluable.

Daniel T. Gillins, Ph.D., P.L.S., Chief Editor
The National Geodetic Survey

CHAPTER 1
ENGINEERING SURVEYING WITHIN ASCE

Gene V. Roe

INTRODUCTION

ASCE defines engineering surveying as

> those activities involved in the planning and execution of surveys for the location, design, construction, operation, and maintenance of civil and other engineered projects. Engineering surveying may be regarded as a specialty within the broader professional practice of engineering and includes all surveying and mapping activities required to support the sound conception, planning, design, construction, maintenance, and operation of engineered projects. Engineering surveying excludes the surveying of real property for the establishment of land boundaries, rights-of-way, easements, and the dependent or independent surveys or resurveys of the public land survey system. ASCE believes that this definition should be adopted by state engineering licensing boards.

These two paragraphs, taken from ASCE Policy Statement 333 (ASCE 2018), define the profession of engineering surveying on which this manual has been written and provided to the public. In addition, we believe that it is important to call the reader's attention to a subtle but critically important distinction in the use of the terms *engineering surveying* and *surveying engineer*. In the first case, we are referring to a type of surveying—an activity. In the second case, it is a type of engineer—an individual. The order in which the terms are used does matter within the profession and particularly when using these terms with the general public. Hence, in this manual, the terms *engineering surveying* and *surveying engineering* should be considered synonymously because they are used interchangeably in many publications. As noted in the following, there are other common phrases for describing this type of surveying activity.

GEOMATICS AND GEOSPATIAL ENGINEERING

The term surveying engineering represents the intersection of the two fields. Many other terms or labels are used to describe this combination of professions, both within and outside the United States. Some of them are discussed as follows.

Since 1926, the Surveying and Geomatics Division has been the lead group within ASCE concerning all topics related to surveying engineering and, more broadly, "mathematics of the Earth," as the term geomatics implies. At the time of writing this manual, the Surveying and Geomatics Division had recently joined a new ASCE institute. The Utility Engineering and Surveying Institute will now be the home of the surveying engineer within ASCE. ASCE also maintains the prestigious *Journal of Surveying Engineering*, which is one of the leading surveying journals worldwide.

Since the twentieth century, the technology for practicing surveying engineers has changed dramatically. The space and digital age has created new tools, equipment, software, workflows, and opportunities. Technologies such as electronic distance measurement, digital levels and theodolites, Global Navigation Satellite Systems (GNSS), laser scanning, digital photogrammetry, unmanned aircraft systems (UASs), geographic information systems (GISs), building information modeling (BIM), and more have emerged and become common tools in practice. Because of these revolutionary changes, many have begun using "geomatics engineering" as a better name to describe the many different fields within the profession than "surveying engineering." As a compromise, this manual is titled *Surveying and Geomatics Engineering* to encompass the many advancements in the profession over the years while keeping the more widely known word "surveying."

A search beyond the boundaries of ASCE finds that the US Army Corps of Engineers (USACE) prefers the use of the term *geospatial engineering* to describe the work that this group performs. Also known as Geospatial Information and Services (GI&S) and formerly called Mapping Charting and Geodesy (MC&G), geospatial engineering remains a core mission of USACE, providing the foundation layers for geospatial intelligence (GEOINT).

Outside the United States, the surveying engineer, in some cases, will adopt generic ISO 9001 standards to manage overall quality. In the United Kingdom, university courses are offered in "civil engineering surveying" and "geospatial surveying engineering."

In the final analysis, it is not the nomenclature that should be the focus; rather, it is the education, training, skills, and most important, the professional role that the surveying engineer can be relied on by industry and the public to provide.

In the case of this manual, it is ASCE, an engineering organization, that is seeking to clearly define the role of a professional who is specialized in the field of surveying (geomatics) engineering. This manual is meant to serve as a reference targeted for the professional, practicing engineer who wants to know more about surveying engineering. Numerous important topics on the profession of surveying and geomatics engineering are summarized, and each chapter provides references where interested readers can find more details. It is hoped that this manual will serve as a useful reference for professionals engaged in surveying engineering.

SURVEYING ENGINEER TODAY

Today, the surveying engineer is being relied on to provide an increasing level of sophisticated mapping, positioning, control, monitoring, and in some cases, navigation information. With the world finally experiencing the digital transformation that has been promised for many decades, the 2D, paper-based paradigm is being replaced by 3D digital models that demand higher levels of mathematical training and understanding.

As noted in the Introduction to ASCE's (1985) engineering surveying manual of practice,

> No important civil project can be properly designed or built without the assistance of the survey engineer. Yet today, the public, including many architects and engineers, has little awareness and knowledge of modern surveying and its relationship to the advancement of civilization.

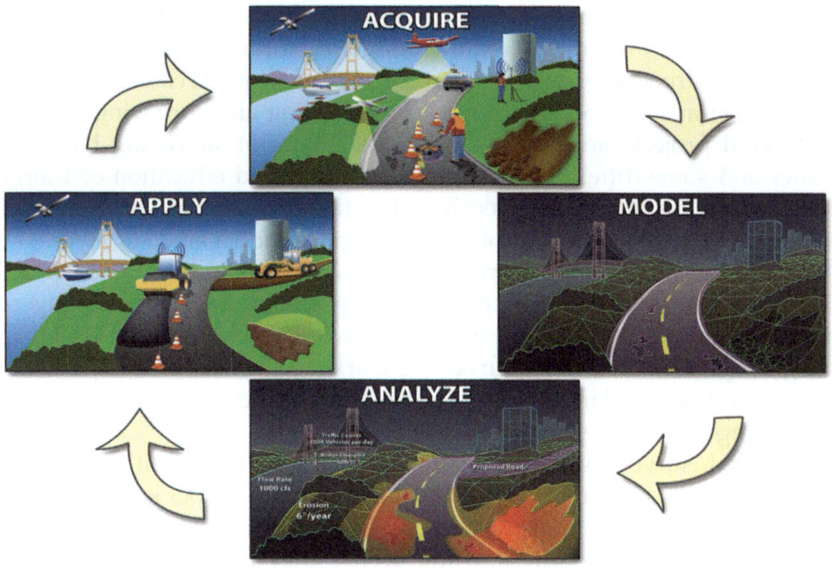

Figure 1-1. Life cycle of the asset management process.
Source: NCHRP (2016), reproduced with permission from the National Academies Press, Washington, DC.

This manual is intended to clearly define the areas of expertise where the surveying engineer can contribute and add value to an engineering project. It is meant to provide knowledge on current surveying engineering practice and explain how modern tools can be used to assist civil projects so as to accomplish them more efficiently and accurately.

Looking forward, as the civil engineering profession transitions to the use of 3D models throughout the life cycle of built environment assets, the role of the surveying engineer will require becoming managers of geospatial data. The surveying engineer is encouraged to engage with the owners of engineered assets to become a trusted adviser in not just the design and construction phases but the entire life cycle of the asset management process (Figure 1-1). In a 3D world, accurately determining and maintaining a current location requires the expertise of the surveying engineer.

The one constant in the world in which we live is change, and the rate of change seems to be accelerating. There is no shortage of issues facing society that will benefit from the expertise of the surveying engineer. The list is a long one—from decaying infrastructure to urbanization, resilience, energy efficiency, and autonomous vehicles. The opportunity for the surveying engineer to contribute to the overall improvement of our planet is extensive and growing.

PROFESSIONAL LICENSING AND CERTIFICATION

As defined previously, ASCE excludes the specialty of boundary surveying from the definition of surveying engineering. Boundary surveying requires specific education and training to research boundary and property rights, identify boundary control, study boundary law, find controlling property corners, restore lost and obliterated corners, understand legal descriptions, protect bona fide rights, among others. It could be said that boundary surveying is a special type of surveying, and it is recognized that many of the tools, equipment, and workflows are similar to those used in surveying engineering for making measurements, determining coordinates on marks, and developing models. In addition, many engineering projects require boundary surveys to determine the limits or boundaries for a construction project, location of rights-of-way and easements, and existence of any gaps or overlaps. Hence, both boundary surveying and surveying engineering are important to civil projects and must coexist. Each type of surveying has overlapping characteristics and some differences that require specialized education or training. Civil engineering students do not typically receive education in boundary surveying unless they are seeking dual licensure as both a professional engineer and a land surveyor.

In the United States, each state's licensing board is free to determine, in conjunction with their legislature, what the definition of surveying entails. Because property rights are important to the public, all states require individuals to hold a professional land surveying license to legally perform boundary surveying work. Most states also require individuals to hold either a professional land surveying license or a professional engineering license to perform surveying engineering.

Thus, in general, surveying engineering can be performed by either a licensed land surveyor or a professional engineer, but only if they have the proper knowledge, skills, and training to do the work. Individuals who do not have this competency should not engage in surveying engineering, regardless of the license they hold. Some professional land surveyors may have expertise in surveying engineering, but this is not necessarily true. The same can be said about professional engineers.

The transition to a 3D digital model–based paradigm of the built environment may provide the opportunity to recognize the importance of clearly defining the roles, specialties, overlap, and differences of boundary surveying and surveying engineering. The public, as well as other professionals such as architects and asset owners, are entitled to a clear understanding of the roles and duties of the boundary surveyor and the surveying engineer.

CHAPTER TOPICS

It is important to note that the intent of this manual is to offer the reader an overview of a number of key topics that require proficiency by the surveying engineer. It is not meant to be an exhaustive treatment of any one of these topics, and the reader is encouraged to seek additional references and training when faced with the need for additional understanding. The manual does not serve as a treatise on surveying and geomatics engineering. Instead, it provides a summary of key topics and then provides the reader with a list of recommended references at the end of each chapter where it is possible to find more details. Hence, this manual should be considered as a reference manual, targeted at professional and practicing engineers and surveyors who want to know more about surveying engineering.

The manual can be divided into three sections. Chapters 2 to 5 provide the fundamentals on surveying engineering, Chapters 6 to 10 introduce equipment for surveying, and Chapters 11 to 15 provide recommendations on best practices.

In Chapter 2, the basics of geodesy are introduced as the foundation for working in a 3D world. Chapter 3 discusses the principles of coordinate systems and map projections to accommodate the need for surveying in the real world. Chapter 4 provides fundamentals on local, regional, and global coordinate transformations, addressing a common task in surveying engineering. Chapter 5 discusses least-squares adjustments and the all-important issue of error management.

Chapter 6 provides an overview of GNSS and their common use in surveying. Chapter 7 presents the fundamentals on leveling and total stations. Chapter 8 introduces terrestrial (static) laser scanning, and Chapter 9 discusses mobile laser scanning and mapping. Chapter 10 provides a summary of aerial laser scanning, aerial photogrammetry, and UASs.

The importance of establishing an accurate control network is discussed in Chapter 11. Chapter 12 looks into the specific issues associated with supporting construction activities, and Chapter 13 lays out the requirements for proper documentation of survey work. Chapter 14 introduces GIS and BIM techniques and summarizes the different applications for which they are used in surveying engineering. Finally, Chapter 15 discusses the aspects of professional services and design professionals' agreements.

Note that this manual assumes that the reader has a basic knowledge of fundamental surveying principles covered in an introductory surveying course offered in a civil engineering curriculum, such as angular and distance measurements and computations, horizontal and vertical curves, azimuth-bearing computations, traversing, area and earthwork computations, topographic mapping, and so forth. Should this not be the case, the following publications are useful references to accompany this manual in providing that important information.

- Anderson, J. M., and E. M. Mikhail. 1997. *Surveying: Theory and Practice*. 7th ed. Columbus, OH: McGraw-Hill.
- Ghilani, C. D. 2021. *Elementary Surveying*. 16th ed. New York: Pearson.
- Moffitt, F. H., and J. D. Bossler. 1997. *Surveying*. 10th ed. New York: Pearson.

Finally, in addition to this manual, the surveying engineer should stay up to date with the recent advances available in the *Journal of Surveying Engineering*, among other sources.

REFERENCES

ASCE. 1985. *Engineering surveying manual, MOP 64*. Reston, VA: ASCE.

ASCE. 2018. *Engineering surveying definition*. Policy statement 333. Reston, VA: ASCE.

NCHRP (National Cooperative Highway Research Program). 2016. *Use of advanced geospatial data, tools, technologies, and information in Department of Transportation projects*. NCHRP Synthesis 446. Washington, DC: Transportation Research Board.

CHAPTER 2
GEODESY AND GEODETIC COMPUTATIONS

Earl F. Burkholder

INTRODUCTION

Geodesy is both a science and an art that includes activities devoted to determining the size and shape of the Earth, defining (in a global sense) the location of objects in/on/near the Earth, and understanding the forces of nature governing the movement of those objects. From an engineering perspective, concepts from geodesy form the foundation for mapping the location of engineered facilities/objects and for monitoring the changing location of those objects within a variety of complex systems. Location relies on geometrical geodesy for its definition. Concepts of physical geodesy are focused on the Earth's gravity field and are used to explain why objects are where they are, to address issues of how and why objects move as they do, and to provide the basis for physical height systems (i.e., with respect to Earth's gravity). Vanícek and Krakiwsky (1986) and Meyer (2010) provide a comprehensive description of the scope of geodesy.

What is the difference between *spatial data* and *geospatial data*? The generic term spatial data includes elements of geometry relating to the size, shape, and location of objects and may include geodetic considerations, whereas the term geospatial data refers specifically to those spatial data referenced to the Earth and presumes inclusion of geodetic concepts as needed. Given the existing ambiguity, it could be argued either way that spatial data are a subcomponent of geospatial data or geospatial data are a subcomponent of spatial data.

BRIEF HISTORY OF GEODESY

Questions relating to the size and shape of the Earth go back in time to when humankind first contemplated the reason for a curved shadow of the Earth on the moon during an eclipse or why the mast of an approaching off-shore vessel was visible before the hull—both physical evidence of a curved Earth. Eratosthenes (276 to 195 BCE) is credited for first accurately documenting the size of the Earth and used the meridian arc length of 5000 stadia from Syene (on the Tropic of Cancer near the Nile River) to the port city of Alexander (where the Nile River meets the Mediterranean Sea) and the angle subtended at the center of the Earth (1/50th of a circle) as determined by the angular difference of the rays of the Sun arriving at noon on the longest day of the year at Syene and at Alexander. Although there were incidental defects in Eratosthenes' determination and some uncertainty exists in the conversion of units (185 m per stadium), the value of 250,000 stadia that Eratosthenes obtained (5000 stadia per sector × 50 sectors) is within about 16% of the currently accepted value (Goldstein 1984).

Eratosthenes assumed a spherical Earth—it is almost but not quite. Subsequent observations over the centuries led one group of people (the French) to believe that Earth is prolate, that is, the distance from the center of the Earth to the North Pole is greater than the equatorial radius. Another group of people (the British) were convinced that the Earth is oblate, that is, the polar distance from the center of the Earth to the poles is shorter than the equatorial radius. Two geodetic surveys sponsored by the French Academy of Science from 1635 to 1641 CE showed conclusively that Isaac Newton and his British followers were correct—the Earth is flattened at the poles (Smith 1986).

Non-uniformity of weights and measures was one cause of unrest preceding the French Revolution in the 1790s (Adler 2002). Visionaries of the time decided that a decimally divided, physically defined standard of length was justified. Realization of that vision included using a precise geodetic survey of the Earth's meridian quadrant lying within France to compute the arc distance from the equator to the pole as accurately as possible and setting this distance equal to 10,000,000 m. This physical dimension was etched into 16 iron bars, one of which was brought to the United States in 1805 by Ferdinand R. Hassler, who later became the first superintendent of the Survey of the Coast, a predecessor of today's National Geodetic Survey (NGS). That bar was subsequently replaced by the National Prototype Meter in 1889 or 1890 (NGS 1986). Since then, the length of the meter has not changed, but the definition of its reproduction has evolved from a platinum bar to being 1,650,763.73 wavelengths of krypton-86 gas to being 9,192,631,770 oscillations of a cesium-133 atom to being the distance light will travel in a vacuum in 1/299,792,458 of a second. The speed of light has been accepted as a constant by the scientific community since 1975.

The establishment and maintenance of the National Spatial Reference System (NSRS) in the United States is the responsibility of NGS of the National Oceanic and Atmospheric Administration (NOAA), Department of Commerce. From its modest origins when established as the "Survey of the Coast" in 1805 by then-President Thomas Jefferson, the agency has evolved over the last 200 years to its present comprehensive mission, which is to define, maintain, and provide access to the NSRS to meet the nation's economic, social, and environmental needs. The NSRS is a consistent coordinate system that defines the latitude, longitude, height, scale, gravity, and orientation throughout the United States. The NSRS includes a network of physical control marks, an up-to-date national shoreline, and a set of models for describing dynamic and geophysical processes that affect geospatial measurements. The horizontal and vertical control marks are referenced to various NGS-defined datums, which is discussed in greater detail toward the end of this chapter.

Development of the physical infrastructure in the United States has relied on the integrity of the geodetic survey control networks since the early 1800s. Examples include the location and gradients for canals, the network of railroads, navigation related to national coastal and inland waterways, the interstate system of highways, the location of pipelines and electric transmission lines, the location of airports and navigation aids for air traffic, and others. Although the US Public Land Survey System—defining the ownership boundaries of many tracts—was laid out with respect to a system of initial points throughout the country, the subsequent location of most cadastral surveys in the United States is also referenced to geodetic control set by NGS.

The benefits that society realizes from geodesy as a discipline are based on two distinct concepts—measurements and datums—also described loosely as technology and models. The evolution of measurement science is recognized as part of the digital revolution with the development of a plethora of sensors that provide digital measurements such as voltages, temperatures, time, and distances. These outputs are processed via tested models to obtain, among others, spatial data components used by geodetic professionals and practitioners alike in the context of the defined datums.

There are records that humankind has known since the days of Pythagoras that the Earth is not flat. However, as humans, we stand erect and walk on what appears to be a flat Earth. From this perspective, the concepts of plane Euclidean geometry are appropriate for making maps and describing where we are. Plane surveying has been an important part of infrastructure development throughout the United States, but when the scope of any project goes beyond legitimate assumptions of the flat-Earth model, geodetic considerations of the Earth's size and shape provide the logical foundation and tools for handling geospatial data—whether analog or digital. Modern geodesy defines the framework that acknowledges and accommodates the characteristics of 3D digital spatial (and geospatial) data.

GEOMETRICAL ELEMENTS OF GEODESY

As stated in the previous section, the Earth is not flat, but neither is it perfectly spherical. Earth's polar axis is shorter than its equatorial axis, and the equator is well-modeled by a circle. So, an oblate ellipsoid is commonly used as a mathematical object to model Earth's macroscopic geometric shape. An oblate ellipsoid can be produced by rotating an ellipse about its minor axis, forming a figure of revolution, symmetrical about the minor axis, known as an *ellipsoid of revolution*. Its name is frequently shortened and referred to as an *ellipsoid*, and that name will be used in the remainder of this chapter. The United States has long adopted the convention of referring to its Earth models as *reference ellipsoids*.

Oblate ellipsoids can be defined by the length of the major axis and either the length of the minor axis or a parameter specifying the shortening of the minor axis in ratio with the major axis. Reference ellipsoids are defined both ways, and various *best fitting* ellipsoids have been used in the United States over the years. The mathematics of ellipsoids is more conveniently written in terms of the semimajor axis (a) and the semiminor axis (b), being the distances from the center of the ellipsoid to the furthest and closest points on the ellipsoid, respectively.

The defining parameters are interrelated according to the following equations:

f denotes flattening and is defined as

$$f = \frac{a-b}{a} = 1 - \frac{b}{a} \tag{2-1}$$

e^2 denotes flattening and is defined as

$$e^2 = \frac{a^2-b^2}{a^2} = 1 - \frac{b^2}{a^2} \tag{2-2}$$

According to these definitions, $0 < b \leq a$, $0 \leq f < 1$, and $0 \leq e^2 < 1$. If $a = b$, then the ellipsoid would, in fact, be a sphere, and both f and e^2 would equal 0. As b gets smaller and smaller relative to a, f and e^2 both approach 1. Realistic reference ellipsoids for the Earth have a flattening value around 1/298, and so the flattening value is often more conveniently reported as a reciprocal.

Several convenient relationships derived from the definitions include

$$e^2 = 2f - f^2 \quad (2\text{-}3)$$

$$b = a(1 - f) \quad (2\text{-}4)$$

$$\frac{b^2}{a^2} = 1 - e^2 \quad (2\text{-}5)$$

$$b^2 = a^2(1 - e^2) \quad (2\text{-}6)$$

Table 2-1 lists three ellipsoids commonly encountered in practice in the United States, giving both defining parameters and derived values. The first ellipsoid is regional and is specific to North America, and the other two are global and used throughout the world. Other ellipsoids are used in other parts of the world for regional datums, but all modern global reference frames recommend the reference ellipsoid of the Geodetic Reference System 1980 (GRS 80) or the World Geodetic System 1984 (WGS 84), which are nearly identical, as shown in Table 2-1.

Modern reference ellipsoids such as GRS 80 and WGS 84 are low-order models of Earth's gravity field. In fact, these ellipsoids are actually defined by four parameters: a, Earth's geocentric gravitational constant (including the atmosphere), a dynamical form factor of the Earth, and Earth's angular velocity (Hofmann-Wellenhof and Moritz 2006).

Any position or point on the surface of the ellipsoid can be identified with two independent coordinates: latitude and longitude. Traditionally, latitude and longitude are reported in angular units of degrees, minutes, and seconds. Latitude is the angle formed between the intersection of a line that is projected downward and normal (perpendicular) to the ellipsoid surface from the point of interest and the ellipsoid's equatorial plane. Latitude is, therefore, the angular distance north or south of the ellipsoid's equator—north is positive, and south is negative. Values range from −90° (South Pole) to +90° (North Pole). Geodetic longitude is measured in the east–west direction, and it is defined as the angle in the plane of the equator from the prime meridian plane to the meridian plane passing through the point of interest. Meridians are great circles on the circumference of the ellipsoid and pass through its north and south poles. Historically, by convention, the

Table 2-1. Selected Reference Ellipsoids.

Ellipsoid	Defined by	Derived values
Clarke Ellipsoid of 1866 (used for NAD 27)	$a = 6{,}378{,}206.4$ m	$1/f = 294.978\,698\,214\ldots$
	$b = 6{,}356.583.8$ m	$e^2 = 0.006\,768\,657\,997\,29\ldots$
Geodetic Reference System 1980 (used for NAD 83 and ITRFs)	$a = 6{,}378{,}137$ m	$b = 6{,}356{,}752.3141\ldots$ m
	$1/f = 298.257\,222\,101$	$e^2 = 0.006\,694\,380\,022\,90\ldots$
World Geodetic System 1984 (defined by DoD for GPS)	$a = 6{,}378{,}137$ m	$b = 6{,}356{,}752.3142\ldots$ m
	$1/f = 298.257\,223\,563$	$e^2 = 0.006\,694\,379\,990\,14\ldots$

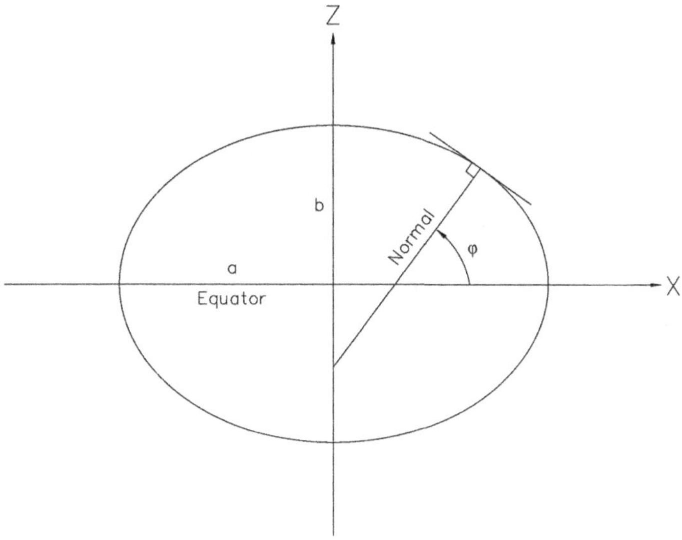

Figure 2-1. Two-dimensional meridian section of the ellipsoid (i.e., ellipse) through the prime meridian.

prime meridian was assigned to the Greenwich Meridian that passes through the center of the transit instrument of the Royal Observatory Greenwich; however, modern reference frames have shifted the prime meridian 102 m eastward (Malys et al. 2015). Mathematically, longitude is counted as positive eastward from 0° at the prime meridian for a full circle of 360°. However, by international agreement in 1884, "longitude is to be counted in two directions up to 180° from the Greenwich Meridian, east longitude being positive and west longitude minus" (House 1980). Nonetheless, many scientific organizations only use positive longitudes east of the prime meridian.

Figure 2-1 is a two-dimensional meridian section through the prime meridian. (Note that the section plane includes the X-axis that informs the reader that this is the prime meridian. However, all meridians are identical in size and shape, so the figure exemplifies all meridians.)

The generic equation of a two-dimensional ellipse is

$$\frac{X^2}{a^2}+\frac{Z^2}{b^2}=1 \tag{2-7}$$

where X and Z are the Cartesian coordinates of positions on the ellipse; and a and b are the semimajor and semiminor axes, respectively. Note that if $a = b$, then Equation (2-7) becomes the equation for a circle.

A sphere has a constant radius, but the radius of curvature in the meridian section of an ellipse varies continuously, with the radius of curvature at the pole being larger than the radius of curvature at the equator. Given the position on an ellipse in terms of geodetic latitude (φ), the radius of curvature in the meridian at that point is given by

$$M=\frac{a(1-e^2)}{(1-e^2\sin^2\varphi)^{3/2}} \tag{2-8}$$

Lying in a meridional plane, M is the ellipsoid's north–south radius of curvature. The length of the normal (N) as shown in Figure 2-1 is also known as the radius of curvature in the prime vertical (a plane containing the normal and perpendicular to the meridian at that location) and is given by

$$N = \frac{a}{\sqrt{1-e^2 \sin^2 \varphi}} \tag{2-9}$$

Radii of curvature are associated with circles, and the plane of the circle for N is perpendicular to the meridian; therefore, N is the east–west radius of curvature. When at the poles ($\varphi = \pm 90°$), it can be readily seen that $M = N$, which confirms the ellipsoid's rotational symmetry about the minor axis.

When reducing horizontal distances to the ellipsoid (Hofmann-Wellenhof and Moritz 2006), it is commonplace to use the Gauss mean radius of curvature at a specified geodetic latitude, which is computed as

$$R_{\text{mean}} = \sqrt{M \times N} = \frac{a\sqrt{1-e^2}}{(1-e^2 \sin^2 \varphi)} \tag{2-10}$$

The radius of curvature in the meridian (M) and the prime vertical (N) can also be combined to compute the radius of curvature in a specific azimuth, as given in Equation (3-5) of Chapter 3.

GEODETIC COORDINATE SYSTEMS

The previous section of this chapter provided definitional aspects of reference ellipsoids. As discussed, the position of any point on the surface of a reference ellipsoid can be identified in terms of latitude (φ) and longitude (λ). In addition, ellipsoids are 3D objects geometrically, so they can also serve as a 3D datum, allowing a third coordinate, called a *geodetic height* or *ellipsoid height*. An ellipsoid height (h) is defined as the shortest linear offset from the ellipsoid to any point in 3D space. Positive values are above the reference ellipsoid, and negative values are below it.

Right-handed, Cartesian coordinate systems (X, Y, Z) are also commonly used in geodesy for expressing the 3D coordinates of a point. By placing the center of the reference ellipsoid at the origin of the Cartesian coordinate system, and by aligning its minor (rotational) axis with the Z-axis, equations can be readily derived to convert between coordinates of a point expressed in (φ, λ, h) with coordinates of the same point in (X, Y, Z). In other words, the two coordinate systems can become interchangeable.

With the coordinate systems defined, the next step is to "realize" them into a "datum" or "reference frame" for practical use. Today, space geodesy methods are used to first produce celestial reference frames that are afterward transformed into terrestrial reference frames that rotate with the Earth's diurnal motion, thus becoming "Earth-fixed" in some sense. In addition, when realized in accordance with the conventions of the International Terrestrial Reference System (ITRS), the origin of the coordinate system is set at the center of mass of Earth (as best as possible), thus becoming "Earth-centered." These modern realizations are, therefore, known as Earth-centered Earth-fixed (ECEF) reference frames. The Z-axis is oriented parallel with a conventional polar axis that is close to Earth's rotational axis, and the X–Y-plane is set perpendicular to the Z-axis and contains the Earth's center of mass. The meridian section formed by the X–Z-plane defines the Prime Meridian (Figure 2-2). The X–Y-plane also defines the plane of the Equator, which is a circle whose radius equals a.

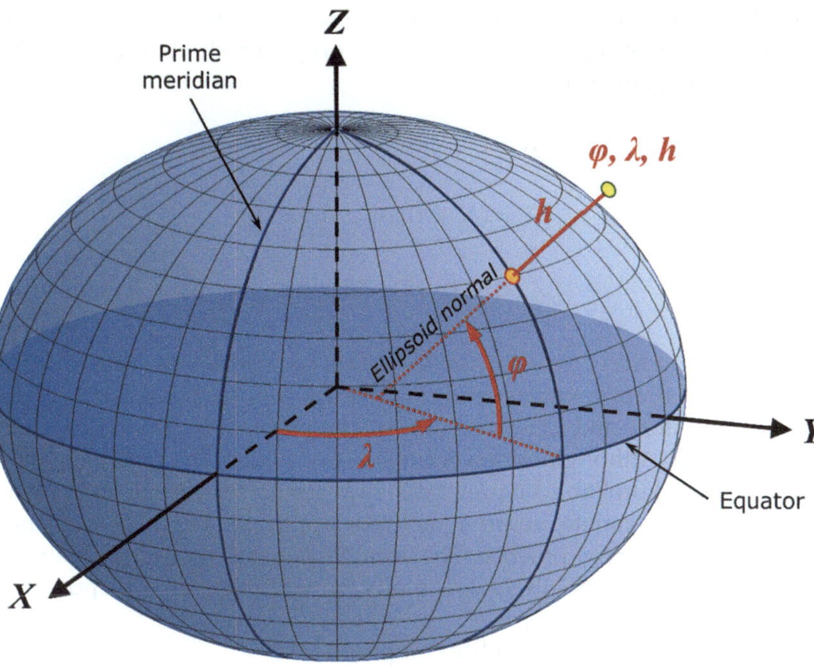

Figure 2-2. Three-dimensional diagram showing geodetic and geocentric coordinate systems.

The resulting coordinate system embodies the rules of solid geometry. In the ECEF system, the (X, Y, Z) components are known as "geocentric Cartesian coordinates," because they represent the linear distances from the Earth's center of mass along three mutually perpendicular axes. When realized according to the ITRS, the linear scale will be in meters. The (φ, λ, h) components are known as the "geodetic coordinates," based on the chosen reference ellipsoid.

In summary, Figure 2-2 shows the geocentric coordinate system having the following characteristics:

- Origin is located at the Earth's center of mass, hence geocentric.
- Conventional right-handed rectangular coordinate system is used.
- X–Y-plane is coincident with the equator, and the Z-axis is closely aligned with a conventional polar axis. [Note that the rotation axis for Earth wobbles and moves in a roughly circular pattern around the poles, up to 8 to 10 cm per day. This motion is removed from Global Navigation Satellite Systems (GNSS) orbits so that the resulting errors are not propagated to the coordinates derived from GNSS receivers.]
- Meter is most commonly the unit of length.
- Rules of solid geometry (including linear algebra and vectors) are applicable throughout.

Figure 2-2 also shows the traditional geodetic coordinate system having the following characteristics:

- Latitude is measured north and south of the equator—north is positive, south is negative.
- Longitude is measured east and west 0° to 180° from the prime meridian. (Many scientific organizations only use positive longitudes eastward 0° to 360°.)
- Ellipsoid height is measured up (positive) and down (negative), normal from the ellipsoid.
- Latitude and longitude are commonly reported in degrees, minutes, and seconds.

Converting between geocentric and geodetic coordinates is well studied. The conversion of (φ, λ, h) to (X, Y, Z) is straightforward, with the only wrinkle coming from the fact that N does not pass through the origin.

$$X = (N + h)\cos\varphi \cos\lambda \tag{2-11}$$

$$Y = (N + h)\cos\varphi \sin\lambda \tag{2-12}$$

$$Z = [N(1 - e^2) + h)] \sin\varphi \tag{2-13}$$

Perhaps surprisingly, the conversion of (X, Y, Z) to (φ, λ, h) is quite difficult. This has led to a number of iterative, approximate, and closed-form solutions. For a review of available methods, see Featherstone and Claessens (2008). The problem is, fundamentally, a quartic in the sine of latitude. Although closed-form solutions exist, the formulas are relatively complicated and would probably be preferable only for places very deep in the interior of the Earth. Instead, approximate, iterative methods applicable to places near the ellipsoid surface are preferred in survey practice. The iterative algorithms are relatively simple and computationally efficient, and there are many competitors. One popular method is presented as follows:

First, geodetic longitude can be readily computed by

$$\lambda = \tan^{-1}\left(\frac{Y}{X}\right) \tag{2-14}$$

In implementation, it is better to use the atan2 function rather than the generic atan function to account for the quadrant of the X and Y values.

Geodetic latitude follows from inverting previous equations and using an iterative technique because φ is in both sides of the equation:

$$\tan\varphi = \frac{Z}{P}\left(1 + \frac{e^2 N \sin\varphi}{Z}\right) \quad \text{where } P = \sqrt{X^2 + Y^2} \tag{2-15}$$

An initial value for latitude and N must first be approximated. This can be done using

$$\varphi_{\text{initial}} = \tan^{-1}\left(\frac{Z}{P(1 - e^2)}\right) \tag{2-16}$$

$$N_{\text{initial}} = \frac{a}{\sqrt{1 - e^2 \sin^2 \varphi_{\text{initial}}}} \tag{2-17}$$

These values from Equations (2-16) and (2-17) are iterated in Equation (2-15), and the steps are repeated until the results converge to some user-specified tolerance.

Once iteration stops, the final values of φ and N are inserted in the following to find the ellipsoid height:

$$h = \frac{P}{\cos\varphi} - N \tag{2-18}$$

MODELS USED IN GEOMETRICAL GEODESY

Models are used to provide a connection between the physical world and the abstraction of "location" in the human mind, on a map, or in a computer database. Some models are quite simple—for example, the two-dimensional flat-Earth model used for plane surveying—but the Earth is not flat, and more complex models are needed to preserve curved-Earth integrity. Geometrical models used in geodesy are embodied in various projected coordinate systems (covered in more detail in Chapter 3). Briefly, a spherical Earth model utilizes equations of spherical trigonometry and finds application in disciplines such as navigation and geography. Because the Earth is not perfectly spherical but rather flattened at the poles, a better ellipsoidal model is used for those engineering applications not adequately served by flat-Earth or spherical-Earth models.

The geometrical geodesy model uses the ellipsoid as the basis for computations. Over the years, latitude and longitude positions on the ellipsoid have been determined by various methods such as astronomical observations, triangulation, precise traverse, and aerotriangulation. More recently, methods such as GNSS, lidar, and photogrammetry are also used. The common denominator is that observations, measurements, and data are reduced to their geometrical equivalent of distances and azimuths on the ellipsoid. The equations are proven and well known but not necessarily simple. In geometrical geodesy, the third dimension is the distance above or below the ellipsoid, the ellipsoid height.

Map projections were invented many years ago to "flatten the Earth" and to facilitate the use of plane Euclidean geometry equations on many kinds of maps. This simpler 2D model provides many of the benefits associated with the use of latitude and longitude positions. A driving factor is that the plane geometry equations are less onerous than the associated geodesy equations. However, simplicity does not come without a price. It is impossible to flatten the curved Earth without distorting angles, distances, shape and/or area. However, cartographers have designed special conformal projections having the characteristic that angles on the reference ellipsoid can be portrayed without distortion on a map. Nonetheless, depending on the chosen map projection, certain curved-Earth distances and areas are still distorted when shown on a flat map (see Chapter 3 for more discussion). The trade-off is that distance distortion can be limited to some predetermined tolerance and the geographical extent of the map projection is limited accordingly. Examples of mapping systems that make use of conformal map projections used for engineering applications include the US State Plane Coordinate System and the Universal Transverse Mercator coordinate system (see Chapter 3 for more details). In addition, low-distortion map projections are gaining popularity for many engineering/surveying applications and are described in Chapter 3.

Elevation is the traditional model for the third dimension. Additional details on elevation are included later in this chapter, following a discussion of physical geodesy concepts. The point here is that modern measurement systems collect and process 3D spatial data and that the conventional computational approach is to use (φ, λ, h) or to use 2D coordinates from map projection equations along with 1D elevations (the difference between ellipsoid heights and elevations is discussed later). However, additional efficiencies are available as practice evolves to embrace the use of an integrated 3D model for spatial data applications. One approach includes the 3D global spatial data model (GSDM), which is a 3D solution based on the geocentric Cartesian coordinates and solid geometry equations implemented in 3D space (Burkholder 2018). The GSDM does not use a map projection and short distances are estimated directly from the underlying geocentric coordinates (see Chapter 3 for more details). The geodetic azimuth of a line is also estimated from a pair of geocentric coordinate values. See Burkholder (2016a) for a comparison of computational models and a simple numerical example of the GSDM. Also, see Chapter 3 for the advantages and limitations of using coordinate systems based on a local geodetic horizon frame.

GEODETIC FORWARD AND INVERSE COMPUTATIONS

The geodetic forward computation consists of beginning at some known latitude/longitude position and computing the latitude/longitude position of an unknown location based on a specified reference ellipsoid and using a geodetic distance (s) on the ellipsoid (or "ellipsoid distance") along with the geodetic azimuth (α) at the starting point. A geodetic inverse computation consists of finding the direction and distance from one known point to another based on a specified reference ellipsoid. Both the forward and the inverse computations are performed on the ellipsoid surface and often require complicated geodetic equations, depending in part on the distance involved. This being the case, methods have been developed that simplify the algorithms and make the computations much easier for routine applications.

Several options for geodetic forward and inverse computations are described. Figure 2-3 is applicable to various methods. Regardless of which option is selected, it is incumbent on the user to verify that acceptable results are obtained. Of course, getting identical results from several methods is also a great confidence builder. In some cases, only the equations for the geodetic forward and inverse are given, whereas others provide only a computational software tool. Some provide both.

Highly regarded geodetic forward and inverse solutions can be found in Vincenty (1975) and Karney (2012). Software programs for performing both forward and inverse solutions based on the Vincenty algorithms are available on the NGS Geodetic Toolkit (NGS 2022a) website, which includes tools for computing 3D geodetic forward and inverse solutions. The 3D geodetic forward tool will provide the geodetic coordinates (φ, λ, h) of the forepoint given the geodetic coordinates of the beginning point along with one of the following: (1) the delta-north, east, and up components from the beginning point; (2) the delta-X, Y, and Z components from the beginning point; (3) the delta-height, forward azimuth, and geodetic distance; or (4) the delta-height, forward azimuth, and mark-to-mark (3D) distance. The 3D geodetic inverse tool (INVERSE3D) provides all four of these outputs given the geodetic or geocentric coordinates of two points. Vincenty's algorithms are a standard to which the results of other methods are often compared.

Another method for performing geodetic direct and inverse computations is by map projection. Equations for the actual direct and inverse computations involve plane Euclidean geometry and are, in general, less complicated depending on the map projection than using geodesy equations on the ellipsoid (see Chapter 3 for more details). The catch is that map projections provide grid distances and azimuths directly, which must be converted to geodetic values [Equations (3-1) and (3-2)]. In addition, geodetic coordinates must be converted to plane coordinates (and vice versa), and software is available for making the conversions. Even so, prudent practice involves checking the results of the process.

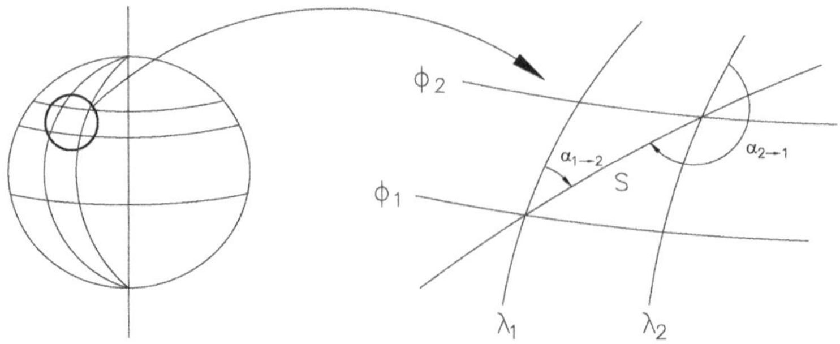

Figure 2-3. Geodetic forward and geodetic inverse computations.

The GSDM allows estimating geodetic forward and inverse computations in 3D space using measured spatial components for the forward computation and geocentric coordinate values for the inverse computation [see Burkholder (2018) for more details]. An example, including a comparison of computed results, is included in Burkholder (2016b).

Finally, for most "small" engineering projects, forward and inverse computations can be greatly simplified by assuming a local geodetic coordinate system and by using plane surveying techniques rather than using the geodetic methods discussed herein or a map projection [Equations (3-9) and (3-10)]. See Chapter 3 for more on local geodetic coordinate systems and their limitations.

PHYSICAL GEODESY

Although geometrical geodesy finds many applications in engineering and utilizes geometry to describe (map) where things are, physical geodesy is that branch of science that investigates the gravity field as related to the internal distribution of mass within the Earth (geophysical exploration), as related to the size and shape of the Earth (elevations and level surfaces), and as related to understanding how/why objects move as they do (in a force field). Gravity is the weakest of the fundamental physical forces, but because the force of gravity is accumulative and because gravity acts over very long distances, it has an enormous impact on many aspects of human activities, including engineering.

Physical geodesy is a particular area of interest for both engineers and scientists. Engineers need to understand how physical forces affect the design, construction, and operation of constructed facilities while scientists strive to gain a better understanding of the origin, nature, and impact of all physical forces. The study of physical geodesy can be highly theoretical and accommodating the impact of physical geodesy in engineering applications can become quite challenging. This summary recognizes both.

Newton recognized that the force of gravity experienced on the surface of the Earth is really the vector sum of gravitational attraction (between objects) and centrifugal force caused by the rotating Earth. As a vector, gravity has both magnitude and direction. Both components have a direct impact on the location and movement of objects. A level surface is always perpendicular to the direction of gravity and the magnitude of gravity relates to equilibrium in the third dimension—elevation and leveling.

Latitude and longitude are references for location in geometrical geodesy, but the reference for the third dimension is nominally mean sea level—governed by gravity (physical geodesy). Herein lies the challenge for engineers. In a geometric 3D coordinate system, Earth's center of mass is the reference for the third dimension as related to height above or below the reference ellipsoid. With GNSS and other 3D measurement systems, 3D location is essentially a solved problem. However, traditionally, horizontal and vertical data have separate origins and separate datums (datums are discussed in the next section). It is helpful to be aware of the following concepts and definitions:

- Geopotential surface is a continuous surface on which the Earth's gravity potential is constant at every point. It is any surface along which a body can be moved without doing any work against the Earth's gravitational and centrifugal forces, and it is also called a "level surface." When bodies of the same mass are moved from one geopotential surface to another, the same amount of work is developed or expended regardless of the route followed.
- Geoid is the geopotential surface of the Earth's gravity field coinciding with the global mean sea level of the oceans.

- Elevation (as a generic term) is the distance measured along the (curved) plumb line from a geopotential surface to the point of interest. Most commonly, the reference geopotential surface is the geoid, and in such cases, the elevation is known as an "orthometric height." Differences in orthometric heights are more useful than differences in ellipsoid heights in engineering design because they show in which direction water will flow (although only approximately, because technically, differences in geopotential numbers show the flow of water).
- Ellipsoid height is the distance of a point of interest above or below the reference ellipsoid as measured along the ellipsoid normal.
- Geoid height (or undulation) is the distance from the ellipsoid to the geoid, taken along a line normal to the reference ellipsoid. Negative geoid heights mean that the geoid is beneath the ellipsoid, and positive geoid heights mean that the geoid is above the ellipsoid. Except for rigorous theoretical considerations (because orthometric height is actually measured along the curved plumb line, but the difference is less than a millimeter nearly everywhere on Earth), the simple equation used worldwide is

  $$h = H + N \tag{2-19}$$

  where

  h = Ellipsoid height,
  H = Orthometric height, and
  N = Geoid height (or undulation).

  Figure 4-2 illustrates the relationship among these three types of heights.
- A geopotential number is the potential energy difference between two points. It has units of work (energy) per unit mass (e.g., m^2/s^2) and is unchanged for a given geopotential (level) surface. Because work is the product of gravity and distance, the distance between geopotential surfaces changes as gravity changes. If the geopotential numbers for two points are equal, then they are on the same geopotential surface and water will not flow between them. NGS currently converts leveling observations to geopotential numbers prior to running leveling network adjustments. Geopotential number differences are approximated by multiplying the leveled-height difference between two points with the mean of the surface gravity values at the two points. Leveling is a path-dependent process (i.e., the results depend on the route taken), resulting in nonunique heights. Converting leveled height differences to geopotential number differences removes the path dependence. In addition, geopotential numbers can be converted to a variety of other types of heights such as orthometric and dynamic heights (see Chapter 7 for more details).
- A dynamic height is found by dividing the geopotential number at a point by the value of normal gravity on the reference ellipsoid at 45° latitude (9.806 199 m/s^2). As such, they are simply scaled geopotential numbers that have units of length, because $(m^2/s^2)/(m/s^2) = m$. However, they have no geometric meaning (e.g., two points with the same dynamic height, in general, will not have equal orthometric heights). Dynamic heights are a specialty application and are used to compute precise hydraulic grade lines over areas of large extent. An example is computing hydraulic head (and grade lines) in the Great Lake System in the United States as part of the International Great Lakes Datum of 1985.

Geometric measurements are typically made with total stations (as angles, distances), GNSS equipment, lidar instruments, and cameras. In addition to these, physical geodesy also utilizes measurements from gravimeters, tide gauges, automatic levels, and clocks (Kleppner 2006).

Physical geodesy has a significant impact on engineering projects in the area of geoid modeling. As noted from Equation (2-19), the geoid height at a point can be determined by subtracting an orthometric height from the ellipsoid height. A common approach in practice is to derive orthometric heights by leveling from bench marks with known orthometric heights (Chapters 7 and 11) and measure ellipsoid heights directly using GNSS (Chapter 6).

Regretfully, the sparseness of reliable bench marks creates a challenge of logistics and efficiency for many projects. Orthometric heights for a project can be determined efficiently from GNSS data if the geoid height and ellipsoid heights in the project area are reliably known. Although the geoid is an irregular surface, it is continuous and is devoid of sharp changes. This means that, in the simplest case, a single value of geoid height can be used for small "local" projects. However, a better assumption is that the ellipsoid and the geoid surfaces in an area have an irregular separation, and geoid heights are modeled with techniques such as those employed by NGS. NGS provides raster grids of geoid models that can be used to find geoid heights at a given latitude and longitude. Hybrid geoid models are recommended for use when working in the current NSRS, such as to allow transformation between North American Datum of 1983 (NAD 83) ellipsoid heights and North American Vertical Datum of 1988 (NAVD 88) orthometric heights.

Hybrid geoid modeling is the process of using available orthometric and ellipsoid height data to modify geoid heights of an existing purely gravimetric geoid model to find an acceptable "best fit" for geoid heights in a given area. It is important to understand that a hybrid geoid model is not really a model of the geoid. A hybrid geoid is instead a surface that best matches an existing vertical datum (such as NAVD 88) created exclusively for the purpose of serving as a tool to transform between ellipsoid and orthometric heights (see Chapter 4 for more information). Given the importance of estimating reliable orthometric heights (elevations) from GNSS-derived ellipsoid heights, NGS has, for years, worked diligently to provide reliable solutions for the user community. NGS has developed hybrid geoid models for the United States by taking a gravimetric geoid model and fitting it in a least-squares sense to bench marks distributed across the United States that have both "known" orthometric heights from differential leveling referenced to the nation's vertical datum (e.g., NAVD 88) and "known" ellipsoid heights measured with GNSS referenced to the horizontal datum or reference frame (e.g., NAD 83). Note that these datums are discussed in greater detail in the following paragraphs. Thus, hybrid geoid models provide the separation between the ellipsoid in the realized reference frame and the vertical datum for the NSRS. Over the years, NGS has continued to refine its hybrid geoid models and made use of the increasing amount of GNSS data collected on bench marks. Examples of the various hybrid geoid models developed by NGS include GEOID96, GEOID99, GEOID03, GEOID09, GEOID12B, and GEOID18. Additional information on geoids, geoid heights, and geoid modeling can be found on the NGS website (NGS 2022b). Raster grids of the hybrid geoid models can be downloaded for use, and online tools are available, allowing entry of a latitude/longitude and receiving the geoid height at the position from a given geoid model (with uncertainty estimated). NGS also provides purely gravimetric "scientific" and "experimental" geoid models, but these should not be used to transform between NAD 83 ellipsoid heights and NAVD 88 orthometric heights.

The geoid modeled values provided by NGS software give geoid heights to the nearest millimeter; however, this should not imply that the geoid heights are known to the nearest millimeter. For GEOID18, network errors in the geoid heights are in the range of about 5 to 10 cm, and they are poorly understood in rugged terrain. However, the *relative difference* between geoid heights of adjacent points is estimated to be about 2 cm (68% confidence) (Ahlgren et al. 2020). This is because the shape (slope) of the geoid is known better than the geoid heights themselves. This means a user can obtain better orthometric heights if geoid height differences are used with ellipsoid height differences to obtain an orthometric height

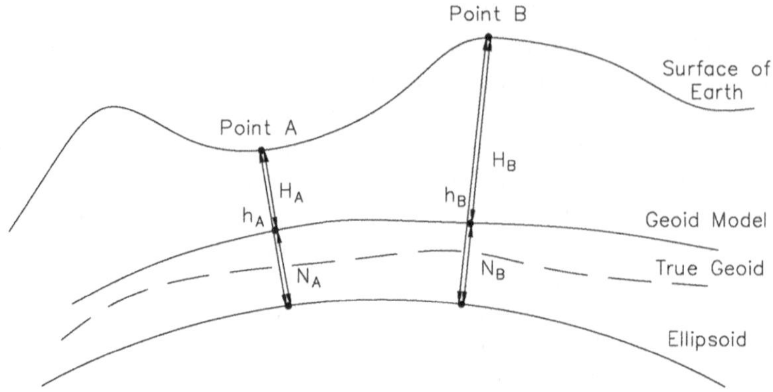

Figure 2-4. Determining orthometric height using GNSSs and geoid modeling.

difference. With a reliable orthometric height difference in hand, the orthometric height of Point B in Figure 2-4 can be determined by attaching that known difference to the known elevation at Point A.

For Figure 2-4, Equation (2-20) is used to compute the orthometric height of Point B using the geoid height difference and the ellipsoid height difference between Points A and B. Note that all geoid heights in the conterminous United States (CONUS) are negative, and the geoid heights are positive for most of Alaska. Figure 2-4 depicts a positive geoid height because the geoid is above the ellipsoid. In other words, the negative sign for a negative geoid height must be included when using Equations (2-19) and (2-20).

$$\Delta h = h_B - h_A \quad \text{(typically from GNSSs)}$$

$$\Delta N = N_B - N_A \quad \text{(from a geoid model like GEOID18)}$$

$$\Delta H = \Delta h - \Delta N$$

$$H_B = H_A + \Delta H = H_A + (h_B - h_A) - (N_B - N_A) \qquad (2\text{-}20)$$

Gravity is enormously important when studying the geoid. In the past, the location and slope of the geoid were primarily estimated from differential leveling and astronomical observations for deflection of the vertical. Equations (2-19) and (2-20) are critical and highly useful when relating orthometric heights to ellipsoid heights. However, a purely gravimetric geoid determined from terrestrial, airborne, and satellite gravity data and models of topographic gravitation can provide a better estimate of the location and shape of the geoid more efficiently and over larger areas. NGS is in the process of developing a purely gravimetric geoid model known as GEOID2022 as part of its modernization of the vertical component of the NSRS.

Investigations into the feasibility of GEOID2022 started in 2014 when NGS began publishing a series of annual experimental geoid (xGEOID) models. The models contain the gravity data from the latest satellite gravity models, the terrestrial gravity, and most important, the airborne gravity from the Gravity for the Redefinition of the American Vertical Datum (GRAV-D) project (Smith 2007). The purpose of the models is to demonstrate the geoid improvement over areas where GRAV-D data are available. These geoid models are categorized into two types: Type A without airborne gravity and Type B with airborne gravity. The differences between the two are the contribution of the airborne gravity. The xGEOIDs provide a preliminary but increasingly accurate view of the changes expected for the release of a new geopotential (vertical) datum, the North American-Pacific Geopotential Datum of 2022 (NAPGD2022). See NGS 2022c for more information about the experimental geoid models, including interactive tools that use the models to compute geoid and orthometric heights.

In this future system, surveying engineers will be able to derive orthometric heights at points of interest by first measuring them directly with GNSS to obtain ellipsoid heights and then by differencing the ellipsoid heights with geoid heights from the GEOID2022 model to obtain NAPGD2022 orthometric heights. Unlike a hybrid geoid model, GEOID2022 does not attempt to convert to legacy vertical datums based on differential leveling. The stated goal is for orthometric heights to be derived with GNSS and GEOID2022 [using Equation (2-19)] to less than 2 cm at 68% confidence (NGS 2021b).

DATUMS

A generic definition for a datum is a reference to which other values are referenced. More specifically, a survey datum is the basis of coordinates and elevations. A plane survey datum in an arbitrary coordinate system could be a specified point of beginning with assumed coordinate values. Or a local vertical datum could be the elevation used for a locally described/recognizable feature—for instance, the elevation of the top step of the local courthouse or the top of a rebar driven in the ground at a construction site. Of course, such arbitrary or local systems are suggested only for small engineering projects.

It is much better practice to use a geodetic datum when executing an engineering survey. A geodetic datum, as discussed in this manual, is rigorously defined on a regional or global scale and provides a common reference for geospatial data in the area covered by the datum. Over the last 200 years, NGS has provided a network of physical marks with published coordinates referenced to a geodetic datum that can be used as control (Chapter 11). An understanding of various issues related to geodetic datums can enhance the quality of professional practice in many circumstances. These issues include the following:

- Difference between horizontal and vertical datums.
- Difference between regional and global datums.
- Knowing the authority responsible for published data—agency/organization/individual.
- Knowing the legacy of published values, whether current or historical.
- Knowing how to transform data from one datum to another.
- Understanding how a 3D datum is (or is not) compatible with traditional 1D/2D datums.
- Being able to contribute to the discussion of whether the satellite orbits do (or will ever) constitute an acceptable substitute to the physical ground-based control monument.

A brief summary of these issues is included herein, but additional information sources include the following:

- NGS website (NGS 2022d)
- Glossary of Mapping Sciences (ASCE, ACSM, ASPRS 1994).
- Geodetic Glossary (NGS 1986).

Horizontal Datums

A horizontal geodetic datum is called a 2D datum and provides the basis for latitude and longitude positions published for surveyed points. However, with the advent of GNSS, modern horizontal datums are actually in 3D because they also provide the basis for published ellipsoid heights. Today, horizontal datums are more frequently referred to as "reference frames."

In the past, horizontal datums were realized by first setting physical marks (see passive marks, Chapter 11) in the ground and then measuring the angles (predominantly by triangulation) or distances (less commonly) between them using conventional techniques. Great horizontal geodetic networks were surveyed and adjusted to fit certain adopted sets of coordinates.

Regional horizontal datums were established by adopting latitude and longitude values for given point(s) on a specified reference ellipsoid whose geometric surface as closely as possible approximated the geoid in the region. Additional specifications for network orientation, geoid height and deflection of the vertical at an initial point, and unit of length were also required.

In 1927, the horizontal geodetic survey networks in the United States were adjusted by least squares. The resulting network of passive marks along with their adjusted geodetic latitudes and longitudes was referred to as the North American Datum of 1927 (NAD 27), which is a regional horizontal datum. The adjustment realized the Clark Ellipsoid of 1866 and was fixed to an adopted latitude and longitude position at the triangulation station "Meades Ranch" in north central Kansas. Network orientation was defined by holding fixed a south azimuth to station "Waldo." The geoid height and the deflection of the vertical were both assumed to be zero at Meades Ranch. Until replacement, geodetic surveys were then rigidly constrained to the passive marks with published NAD 27 coordinates. Although becoming less commonplace, NAD 27 is still used in some localities and by some organizations.

Beginning in 1974, a major effort (that turned out to take more than 10 years) was launched to update and readjust the horizontal geodetic network in the United States (Schwarz 1989). The newly adjusted network was named the North American Datum of 1983 (NAD 83). NAD 83 used the Geodetic Reference System of 1980 (GRS 80) reference ellipsoid. In addition, unlike NAD 27, the datum origin was taken to be the center of mass of Earth as nearly as could be determined at the time, and NAD 83 was meant to be a global horizontal datum. Scale and orientation for NAD 83 were controlled by a combination of space geodesy technologies such as laser ranging, very long baseline interferometry, Doppler data, and astronomical observations. The NAD 83 project was completed in 1986 and, in general, represents more than an order of magnitude improvement over NAD 27. NAD 83 was published before the impact of GNSS positioning could be fully realized.

By the time NAD 83 was published in 1986, GNSS became a viable positioning technology embraced by the user community. One justification for readjusting the horizontal network in NAD 27 was that, using existing instruments (EDM and theodolites), the user community could routinely perform surveys that were more accurate than the underlying NAD 27 network. Ironically, with the advent of GNSS, the user community again acquired the ability to survey more accurately (in some cases) than the underlying NAD 83 network. NGS addressed this challenge by performing GNSS control network surveys on the passive marks on an essentially state-by-state basis. These surveys became known as high precision geodetic networks (HPGNs) and later were called high accuracy reference networks (HARNs). As a way to keep track of the various upgrades, NGS adopted the practice of adding a qualifying year to the datum name, referred to as a "datum tag." For example, the original HPGN for Oregon became known as NAD 83(1991), and a later adjustment in Oregon was referred to as NAD 83(1998). Other states were assigned similar designators and ultimately the original NAD 83 adjustment became known as NAD 83(1986). Typically, the datum tag indicates the year the adjustment was performed for determining station coordinates and constitutes new (regional) realizations of NAD 83. Because of GNSS technology, it also became possible to directly measure and report ellipsoid heights. Thus, NAD 83 began to function as a 3D geometric datum for latitude, longitude, and ellipsoid heights rather than only as a 2D horizontal datum; however, the phrase "horizontal datum" for NAD 83 is still commonly used today.

As technology and practice continued to evolve, the compatibility of one state HPGN/HARN with that of an adjoining state completed at a different time became a challenge. In addition, another wave of technology was incompatible with the HPGN/HARN networks, namely, what is known today as the NOAA Continuously Operating Reference Station (CORS) network. A station in this network, or a CORS, continuously collects GNSS data. In the 1990s, NGS embarked on an ambitious program to establish a well-distributed network of CORSs wherein GNSS data were logged and uploaded to NGS for central processing. NGS owns and operates some stations, but by far, most of the stations are owned by others who establish and operate the installation in accordance with NGS specifications. Coordinates derived from the stations' data published by NGS are considered highly accurate and well monitored. These coordinates serve as excellent control and access to NAD 83. See Chapter 11 for more on the NOAA CORS network.

Eventually, the incompatibilities between state HPGN/HARN networks and the NOAA CORS network were eliminated by a comprehensive readjustment of the HPGN/HARN surveys in 2007 constrained to the derived coordinates of the CORSs in NAD 83 (Pursell and Potterfield 2008). The resulting seamless network became the NSRS2007 realization of NAD 83, and "feathering" at state boundaries became a moot issue. Unlike the HPGN/HARN surveys, the 2007 adjustment consisted only of GNSS observations (more specifically, GPS only); stations connected only by terrestrial observations were omitted. Another national adjustment of the GNSS survey networks on passive marks was completed in 2012, constrained to the latest available coordinates for stations in the NOAA CORS network determined in 2011 (Dennis 2020). This resulted in the current realization of NAD 83 referred to as NAD 83(2011) epoch 2010.00. For the first time, an epoch date was formally used as part of the realization name in recognition of the importance of time in determining positions. Actually, the (2011) datum tag refers specifically to coordinates referenced to the North America tectonic plate (i.e., the stations are intended to move with the plate). A datum tag of (PA11) is used for stations referenced to the Pacific plate, such as those on Hawaii and American Samoa. Likewise, the (MA11) datum tag is for stations referenced to the Mariana plate, such as those on Guam and the Commonwealth of the Northern Mariana Islands. Together, these three datum tags refer to the latest realization of NAD 83, all at epoch 2010.00. Incidentally, such plate-fixed frames cannot be considered truly Earth-fixed since the plates move with respect to the overall Earth. Therefore, it is not appropriate to refer to their (X, Y, Z) coordinates as ECEF but rather as "geocentric Cartesian" (or simply "geocentric").

During the life of NAD 83, the published values on many control marks have changed slightly because of better realizations (as explained previously) and being driven by a collection of more accurate data with improved instrumentation. Although some users are frustrated by changes in the control coordinates, these changes are typically cataloged by the year in which the new values are realized. It is still the same NAD 83 datum, but the values use a datum tag that specifies a different (better) realization. As mentioned previously, the latest realization of NAD 83 also includes an explicit epoch date for all stations. Previously, epoch dates were only used for stations in tectonically active areas (such as California).

The World Geodetic System of 1984 (WGS 84) is the latest datum used by the US Department of Defense in a series of realizations known as WGS 60, WGS 66, and WGS 72 (NGA 2014). WGS 84 is very similar to, but distinctly different from, NAD 83. They are similar in that both are 3D global geodetic datums with their origins located at Earth's center of mass—well, not quite. Currently, the location of the Earth's center of mass is known to an accuracy of about 2 cm, whereas in the 1980s (when NAD 83 was originally realized), the uncertainty in the location of the Earth's center of mass was about 2 m. Since being originally published, the NAD 83 positions have been refined based on better data and models, but the origin and alignment of NAD 83 has not been changed in light of better knowledge of the location of the center of mass. In contrast, the origin and alignment of the WGS 84 realization is updated periodically by NGA based on an analysis of signals from satellites that orbit the Earth's center of mass.

The reference ellipsoids for WGS 84 and NAD 83 are examples of similarity. NAD 83 utilizes the GRS 80 ellipsoid, whereas WGS 84 uses the WGS 84 ellipsoid. The semimajor axis, a, for both reference ellipsoids is the same, and the flattening values are very slightly different. (See the defining values in Table 2-1 and compare the small difference of 0.0001 m in the derived value of b, the semiminor axis.) This means that one could use the GRS 80 and the WGS 84 ellipsoids interchangeably; however, because of different origins in their realizations, coordinates referenced from NAD 83 differ substantially (>1 to 2 m) with coordinates referenced from the latest realization of WGS 84.

Datum differences and similarities foster the possibility of incompatible uses. GPS satellite orbits are defined in WGS 84, whereas other satellite systems (e.g., GLONASS and Galileo; see Chapter 6) use different datums. Equipment manufacturers are, in general, good at processing data consistent with the design assumptions of the various systems and at providing output in specified format/datums. Standard practice is for default output values to be those that the customer expects. However, sometimes settings can be changed (maybe unwittingly), and the system delivers results not compatible with client preferences. There is an unconfirmed story about a person who, over a period of years, unwittingly delivered WGS 84 results to clients with the understanding the data being provided were in NAD 83. A less clear-cut example is that NAD 83 vector components are nearly identical with WGS 84 vector components, at least over short distances. In that case, mixing datum components may have little or no detrimental consequences. Prudent practice is to studiously avoid mixing NAD 83 and WGS 84 datum components inappropriately.

The ITRS is very similar to WGS 84 with the exception that, instead of being a product of the US Department of Defense (DoD), the ITRS is observed, computed, and published by the International Earth Rotation and Reference System Services (IGN 2021). The ITRS provides International Terrestrial Reference Frames (ITRFs) that are a product of the international scientific community and serve as an international standard. The ITRFs are based on the "best" technology available irrespective of the source, and the latest realizations of WGS 84 aim to be aligned with the latest ITRF. Comparisons are routinely made, and although small differences exist, they are, as a rule, statistically insignificant. The ITRFs are based on multiple space geodesy techniques, including very long baseline interferometry, satellite laser ranging, Doppler ranging integrated on satellite, and GNSS. The International GNSS Service (IGS) makes use of only the GNSS data in the solution to align with an ITRF. Similarly, WGS 84 only makes use of GPS data when aligning with an ITRF (but from a different set of ground tracking stations).

A brief summary of horizontal datums/reference frames in the United States includes the following:

- NAD 27 was based on the Clarke Ellipsoid of 1866 and is a regional geodetic datum.
- NAD 83 replaces NAD 27, is a global geodetic datum, and uses the GRS 80 ellipsoid. Although it is commonly called a horizontal datum, it is actually a 3D reference frame for latitudes, longitudes, and ellipsoid heights.
- WGS 84 is a global reference frame developed by the DoD in support of military activities. WGS 84 is also the reference frame on which the constellation of GPS satellites is operated.

- ITRFs serve as the international standard for all global reference frames and are a product of the international scientific community. The latest realizations of modern reference frames (e.g., for WGS 84 and from IGS) aim to align with the latest ITRF.

Vertical Datums

A vertical datum is a 1D datum and provides the basis for elevations. Orthometric heights are meant to be referenced from the geopotential surface known as the geoid, and many efforts have been taken to develop datums that approximate this surface locally, regionally, and nationally.

In the past, vertical datums were realized by setting physical marks or bench marks in the ground and then surveying and leveling between them (Chapter 7). Significant leveling work has been done throughout the United States

In the 1920s, all leveling/vertical networks in the United States were combined, and a 1D least-squares adjustment was performed constrained to 26 tide gauge stations distributed along the coasts of North America. Twenty-one of the tide stations were in the United States and five of them were in Canada. Zero elevation, as determined by averaging readings at each tide gauge over long periods of time, was held in the adjustment of the leveling network. The resulting adjusted elevations were published on the bench marks and referred to as being in the Sea Level Datum of 1929. In the decades following 1929, it became apparent that the internal consistency of the interconnected level loops was better than the means of the tide gauges, meaning that zero elevation was not identical with mean sea level. Therefore, the name of the datum was changed in 1973 from the Sea Level Datum of 1929 to the National Geodetic Vertical Datum of 1929 (NGVD 29) (Berry 1976). None of the published elevations on the bench marks were changed, only the name of the datum was changed. Until replacement, subsequent vertical surveys were rigidly constrained to the bench marks with published NGVD 29 heights.

In the process of computing the hydraulic head for water stored in the Great Lakes system, scientists realized that a system of work units (geopotential numbers) for the water surface could more accurately portray available potential energy than using elevations. The International Great Lakes Datum of 1955 (IGLD 55) was defined in terms of geopotential numbers, and IGLD 55 dynamic heights were computed as the geopotential number divided by normal gravity. The Hudson Bay region of North American is still rebounding owing to the removal of the glacial burden of 10,000 years ago, with the consequence that periodic updates to the IGLD are required. IGLD 85 was later published as an updated IGLD (Zilkoski 1992). Another update called IGLD (2020) is under development and planned for release in 2025. Unlike previous IGLDs that were based on leveling, IGLD (2020) will use a geoid-based vertical datum (i.e., NAPGD2022) that will be accessible using modern GNSS technology (Heck and Craymer 2021).

Following the publication of NAD 83, NGS undertook a readjustment of the vertical network and published the results as the North American Vertical Datum of 1988 (NAVD 88). In addition to legacy data, the adjustment was based on interconnecting loops of approximately 81,500 km of new first-order leveling and anchored to a single bench mark at Father Point/Rimouski, Quebec, Canada.

The new NAVD 88 datum and the IGLD 85 datum have identical geopotential numbers. The main difference between them is that IGLD 85 dynamic heights are obtained by dividing the geopotential numbers by normal gravity at 45° latitude, whereas the NAVD 88 Helmert orthometric heights (elevations) are obtained by dividing the same geopotential number by mean gravity on the plumbline with modeled NAVD 88 surface gravity. An additional difference is that a small "hydraulic correction" is added to the computed dynamic height so that the IGLD85 heights match actual water levels on the Great Lakes.

Chapters 4 and 7 give additional information on dynamic heights and hydraulic correctors. Extensive information on both IGLD 85 and NAVD 88 is provided by Zilkoski (1992) and Zilkoski et al. (1992).

A brief summary of vertical datums in the United States includes the following:

- The National Geodetic Vertical Datum of 1929 (NGVD 29) was constrained to 26 tide gauges.
- The International Great Lake Datums of 1955 and 1985 use geopotential numbers and dynamic heights (as well as hydraulic correctors).
- The North American Vertical Datum of 1988 (NAVD 88) is based on an adjustment of the vertical network constrained to one bench mark. (IGLD 85 and NAVD 88 share a common origin and common geopotential numbers.) Other vertical datums have also been realized for islands in the United States. These datums are based on leveling networks constrained to a tide station on the island.

New Datums—The Modernized National Spatial Reference System

Soon, NAD 83 and NAVD 88 will be replaced by the modernized NSRS. NGS has drafted three "blueprint" documents that discuss the intended development of the modernized NSRS and how to work in it (NGS 2021a, b, c). One of the major features of the new system is that it enables better modeling and accounting for time-dependent coordinates. It could be said that the modernized NSRS will be 4D. NGS estimated that the work would be accomplished sometime around 2022, but it will more likely be finished sometime between 2025 and 2027.

The first blueprint (NGS 2021a) discusses plans for replacing NAD 83 with four new, plate-fixed terrestrial reference frames (TRFs). These new reference frames will be named as the following:

- North American Terrestrial Reference Frame of 2022 (NATRF2022),
- Pacific Terrestrial Reference Frame of 2022 (PATRF2022),
- Mariana Terrestrial Reference Frame of 2022 (MATRF2022), and
- Caribbean Terrestrial Reference Frame of 2022 (CATRF2022).

All four of these new reference frames will first be aligned with the future ITRF2020; then, they will rotate according to a plate rotation model for each tectonic plate for each frame. Thus, the latitudes and longitudes will appear stable in areas where overall crustal motion is well-modeled by plate rotation. The remaining horizontal and vertical motions will be modeled by the future Intra-Frame Deformation Model of 2022. All these efforts allow a careful treatment of coordinates that are time-dependent.

The second blueprint (NGS 2021b) addresses the geopotential aspects of the modernized NSRS, including the vertical datum, the geoid, gravity, deflections of the vertical, and other quantities related to the Earth's gravity field. All these related items will be defined in an internally consistent geopotential datum known as the North American-Pacific Geopotential Datum of 2022 (NAPGD2022). Time-dependent models will be developed for the following:

- Global model of the geopotential field of Earth;
- Regional, purely gravimetric geoid model known as GEOID2022, which will be defined to best fit the global mean sea level at the reference epoch of NAPGD2022;
- Regional deflection of the vertical model (DEFLEC2022); and
- Regional surface gravity model (GRAV2022).

The three aforementioned regional models cover CONUS, Alaska, Hawaii, the Caribbean, Canada, Mexico, Central America, Greenland, American Samoa, and Guam/Commonwealth of Northern Mariana Islands. Orthometric heights referenced in NAPGD2022 will be defined as equal to the ITRF2020 ellipsoid heights minus geoid heights from GEOID2022 (note that ITRF2020 ellipsoid heights will be identical to ellipsoid heights in the four 2022 TRFs). Thus, orthometric heights will be primarily accessed through GNSS technology, although leveling can still be done to determine high-accuracy orthometric height differences.

The third blueprint (NGS 2021c) presents ideas and recommendations on how to conduct work in the modernized NSRS. One of the major themes of this document is how to deal with time-dependent coordinates. It lays out ideas for how often published coordinates will be updated and defines and discusses "survey epoch" and "reference epoch" coordinates. It also discusses the increasing role of the Online Positioning User Service and the NOAA CORS network in the new system (see Chapter 11 for more on both topics).

As can be seen from this chapter, geometry is not the entire scope of the spatial data. The physical laws of gravity impact the measurement environment of many spatial data and monitoring the behavior of objects within the Earth's gravity field cannot be avoided. Ideally, the implementation of the modernized NSRS will enable all spatial data users to accommodate both geometry and gravity in routine geodetic computations.

REFERENCES

Adler, K. 2002. *The measure of all things: The seven-year odyssey and hidden error that transformed the world*. New York: Free Press.

Ahlgren, K., G. Scott, D. Zilkoski, B. Shaw, and N. Paudel. 2020. *GEOID18*. NOAA Tech. Rep. NOS NGS 72. Silver Spring, MD: National Oceanic and Atmospheric Administration, National Geodetic Survey.

ASCE, ACSM (American Congress on Surveying and Mapping), and ASPRS (American Society of Photogrammetry and Remote Sensing). 1994. *Glossary of the mapping sciences*. New York: ASCE, ACSM, ASPRS.

Berry, R. M. 1976. "History of geodetic leveling in the United States." *Surv. Mapp.* 36 (2): 137–153.

Burkholder, E. F. 2016a. "Comparison of geodetic, state plane, and geocentric computational models." *Surv. Land Inf. Sci.* 74 (2): 53–59.

Burkholder, E. F. 2016b. "3D geodetic inverse." *Surv. Land Inf. Sci.* 75 (1): 17–21.

Burkholder, E. F. 2018. *The 3-D global spatial data model: Foundation of the spatial infrastructure*. 2nd ed. Boca Raton, FL: CRC Press.

Dennis, M. L. 2020. *The National Adjustment of 2011: Alignment of passive GNSS control with the three frames of the North American Datum of 1983 at epoch 2010.00: NAD 83 (2011), NAD 83 (PA11), and NAD 83 (MA11)*. NOAA Tech. Rep. NOS NGS 65. Silver Spring, MD: National Oceanic and Atmospheric Administration, National Geodetic Survey.

Featherstone, W. E., and S. J. Claessens. 2008. "Closed-form transformation between geodetic and ellipsoidal coordinates." *Stud. Geophys. Geod.* 52 (1): 1–18.

Goldstein, B. R. 1984. "Eratosthenes on the "measurement" of the earth." *Hist. Math.* 11 (4): 411–416.

Heck, J., and M. Craymer. 2021. "Updating the international Great Lakes datum: Enabling the integration of water and land management in the Great Lakes region." In *Proc., FIG e-Working Week 2021: Smart Surveyors for Land and Water Management—Challenges in a New Reality. Netherlands (virtual): International Federation of Surveyors.*

Hofmann-Wellenhoff, B., and H. Moritz. 2006. *Physical geodesy*. 2nd ed. New York: Springer.

House, D. 1980. *Greenwich time and the discovery of the longitude*. Oxford, UK: Oxford University Press.

IGN (Institut Géographique National). 2021. "International terrestrial reference frame." *ITRF solutions*. Accessed February 8, 2022. Itrf.ign.fr/ITRF_solutions/index.php.

Karney, C. F. F. 2012. "Algorithms for geodesics." *J. Geod.* 87: 43–55.

Kleppner, D. 2006. "Time too good to be true." *Physics Today*, March 10–11.

Malys, S., J. H. Seago, N. K. Pavlis, P. K. Seidelmann, and G. H. Kaplan. 2015. "Why the Greenwich meridian moved." *J. Geod.* 89: 1263–1272.

Meyer, T. H. 2010. *Introduction to geometrical and physical geodesy: Foundations of geomatics*. Redlands, CA: ESRI Press.

NGA (National Geospatial-Intelligence Agency). 2014. *Department of defense world geodetic system 1984: Its definition and relationships with local geodetic systems*. Version 1.0.0. NGA-STND.0036_1.0.0_WGS84. Springfield, VA: NGA.

NGS (National Geodetic Survey). 1986. *Geodetic glossary*. Rockville, MD: NGS, National Oceanic and Atmospheric Administration.

NGS. 2021a. *Blueprint for the modernized NSRS, part 1: Geometric coordinates and terrestrial reference frames*. NOAA Tech. Rep. NOS-NGS 62. Silver Spring, MD: NGS.

NGS. 2021b. *Blueprint for the modernized NSRS, part 2: Geopotential coordinates and geopotential datum*. NOAA Tech. Rep. NOS-NGS 64. Silver Spring, MD: NGS.

NGS. 2021c. *Blueprint for 2022, part 3: Working in the modernized NSRS*. NOAA Technical Rep. NOS-NGS 67. Silver Spring, MD: NGS.

NGS. 2022a. "INVERSE/FORWARD/INVERS3D/FORWARD3D Computation Utilities." *NGS Geodetic Tool Kit*. Accessed February 7, 2022. https://geodesy.noaa.gov/TOOLS/Inv_Fwd/Inv_Fwd.html.

NGS. 2022b. "Geoid models." Accessed February 8, 2022. https://geodesy.noaa.gov/GEOID/.

NGS. 2022c. "Experimental GEOID modeling (xGEOID)." Accessed February 8, 2022. https://beta.ngs.noaa.gov/GEOID/xGEOID/.

NGS. 2022d. "National Geodetic Survey: Positioning America for the future." Accessed February 8, 2022. https://geodesy.noaa.gov/.

Pursell, D. G., and M. Potterfield. 2008. *National readjustment final report*. NAD 83(NSRS2007). NOAA Tech. Rep. NOS-NGS 60. Silver Spring, MD: National Oceanic and Atmospheric Administration, National Geodetic Survey.

Schwarz, C. R. 1989. *North American datum of 1983*. NOAA Professional Paper NOS 2. Silver Spring, MD: National Oceanic and Atmospheric Administration, National Geodetic Survey.

Smith, D. 2007. *The GRAV-D project gravity for the redefinition of the American vertical datum*. NOAA Special Publication NOS-NGS 7. Silver Spring, MD: National Oceanic and Atmospheric Administration, National Geodetic Survey.

Smith, J. R. 1986. *From plane to spheroid: Determining the figure of the Earth from 3000 B.C. to the 18th century Lapland and Peruvian survey expeditions*. Rancho Cordova, CA: Landmark Enterprises.

Vanícek, P., and E. J. Krakiwsky. 1986. *Geodesy: The concepts*. 2nd ed. Amsterdam, Netherlands: North-Holland.

Vincenty, T. 1975. "Direct and inverse solutions of geodesics on the ellipsoid with applications of nested equations." *Surv. Rev.* 23 (176): 88–93. https://doi.org/10.1179/sre.1975.23.176.88.

Zilkoski, D. 1992. "North American datum and international great lakes datum: They are now one and the same." In *Proc., US Hydrographic Conference '92*. Baltimore, MD: The Hydrographic Society of America.

Zilkoski, D., J. Richards, and G. Young. 1992. "Special report: Results of the general adjustment of the North American vertical datum of 1988." *Surv. Land Inf. Syst.* 52 (3): 133–149.

CHAPTER 3

MAP PROJECTIONS AND LOCAL COORDINATE SYSTEMS

Michael L. Dennis

INTRODUCTION

Coordinate systems are an integral part of modern surveying and geomatics. For our purposes, *coordinates* are ordered sequences of real numbers that uniquely define a location with respect to a coordinate system origin. If there are two numbers, they define a two-dimensional (2D) location on a surface such as a plane or sphere. If there are three numbers, they define three-dimensional (3D) location in space. A coordinate can also be one number (1D), which in surveying engineering is usually an elevation or height, although it can also be a location along a linear feature, such as stationing on a highway alignment (note that stationing can also be a considered a quasi-3D coordinate system if it includes offsets and elevations). Time is sometimes considered as another dimension in addition to spatial coordinates. Although the role of time is not considered in this chapter, it is discussed in Chapter 4 in the context of time-dependent transformations.

Within the context of surveying engineering, coordinate systems can be divided into two broad categories, those that are georeferenced and those that are not. *Georeferenced* means the coordinate system is defined with respect to a geometric reference system or frame (also known as "geodetic datum" or "geographic coordinate system"), as described in Chapter 2. The term *coordinate reference system* (CRS) or *spatial reference system* (SRS) is often used for such georeferenced systems, although the term *reference* is not always included [e.g., the State Plane Coordinate System (SPCS)].

This chapter deals almost entirely with georeferenced coordinate systems. The reasons are both practical and philosophical. Prior to the 1990s, georeferencing a survey was usually only practical for large projects, but this is no longer the case. A number of factors have made georeferencing simple and sometimes even automatic: dense systems of geodetic control; readily available sophisticated geospatial software; and modern positioning technology such as robotic total stations and Global Navigation Satellite Systems (GNSSs). Because in most cases georeferencing is now so simple, there is rarely a justifiable reason not to do it. Georeferenced spatial products are more valuable and more useful because they can be combined with other geospatial datasets. This is becoming increasingly important as ever more accurate and high-resolution datasets become available, such as imagery and digital elevation models. Effectively consuming such datasets require the use of accurately georeferenced products. Nonetheless, situations still occur where an engineering survey cannot be easily georeferenced, and in such cases, an arbitrary coordinate system must be used. Because of this, the salient characteristics of such systems are addressed at the end of this chapter.

Georeferenced coordinate or spatial reference systems range in extent from global to local. Global systems can be extended outward from the Earth to include satellite orbits, so that the satellites are in the same coordinate system as the survey on the ground (an essential characteristic of GNSS). The type of coordinates used for the entire Earth are usually expressed as latitude and longitude (φ, λ)—plus ellipsoid height (h) for modern 3D systems—or as geocentric Cartesian coordinates (X, Y, Z). Both these types of coordinates are described in Chapter 2. Such global coordinates can be used for small local areas as well. But a vast majority of engineering applications use different types of Cartesian coordinates, divided into two main types in this chapter: *projected coordinate systems* and *nonprojected local coordinate systems*.

The topics addressed in this chapter are intended as introductory for surveying engineering in a practical and applied context, and so, they cannot be covered in much depth. The interested reader is encouraged to use the references at the end of the chapter to pursue the topics in greater detail (e.g., Iliffe and Lott 2008, Snyder 1987, Stem 1990, Van Sickle 2017).

MAP PROJECTIONS

Map projections are ubiquitous. We use them to represent the curved Earth on something "flat," such as a sheet of paper or a computer screen. Even a photograph of Earth from space is a specific type of projection (perspective) that can be completely defined mathematically. The word "projection" itself provides an intuitive description of how a projection works—like an image projected onto a screen. Indeed, some projections can be constructed entirely geometrically by placing an imaginary light source at a specific location and tracing rays from the source, through the Earth, and projecting features from the Earth's surface onto a flat surface. Figure 3-1 illustrates this for a spherical Earth with a ray source at the center (gnomonic projection), at the opposite side of the globe (stereographic projection), and at infinity (orthographic projection).

Imagining light rays that "project" the curved surface of the Earth onto a flat surface is conceptually useful, and it will be employed in this chapter for explaining some projection concepts. However, such geometric construction cannot be done for most map projections, and it cannot be done for any projections based on a reference ellipsoid. More formally, a map projection is a functional (one-to-one) conversion between geodetic coordinates of latitude and longitude (φ, λ) and projected or "grid" coordinates. The geodetic coordinates

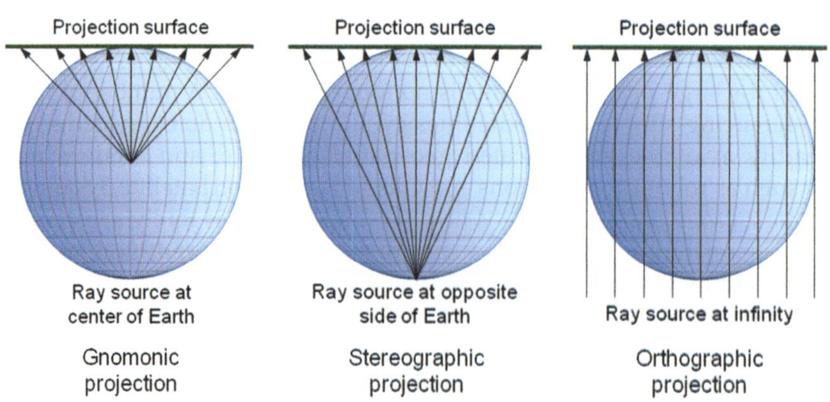

Figure 3-1. Projections constructed geometrically (based on a spherical model of the Earth). Source: Courtesy of NOAA, reproduced with permission.

are in angular units, and the projected coordinates are in linear units. The projected coordinates can be expressed as x, y or easting, northing (E, N). For surveying in the United States, the convention for grid coordinates is typically as northing, easting (N, E), and this convention will be followed in this chapter. However, it is important to realize that $(x, y) = (E, N)$ is also common, especially in certain applications, and so one should always be aware of the convention used. Great confusion and costly mistakes can arise when misinterpreting coordinates as being in (E, N) convention when, in fact, they were in (N, E) or vice versa.

Projecting from (φ, λ) to (N, E) is often called the *direct* (or *forward*) *solution*, and "de-" projecting from (N, E) to (φ, λ) the *inverse solution*. Either way, there is a unique relationship between (φ, λ) and (N, E) coordinate pairs. It is also possible to create grid-type (N, E) or (E, N) systems from geodetic (φ, λ) and geocentric (X, Y, Z) coordinates without using a formal map projection. These systems are called *local horizon* systems and are discussed later in this chapter.

The "flat" surface on which coordinates are projected is called the *developable surface*. There are three types—plane, cylinder, and cone—as shown in Figure 3-2. Each of these is "flat" in the sense that it can be represented as a plane without distortion because it has an infinite radius of curvature in at least one direction. Conceptually, the cylinder and cone can be "cut" parallel to their central axis (which is the direction of infinite radius of curvature) and laid flat without changing the relationship between the projected coordinates. Another way to think of a developable surface is that there is only one kind, the cone. A cone of infinite height is a cylinder, and a cone of zero height is a plane.

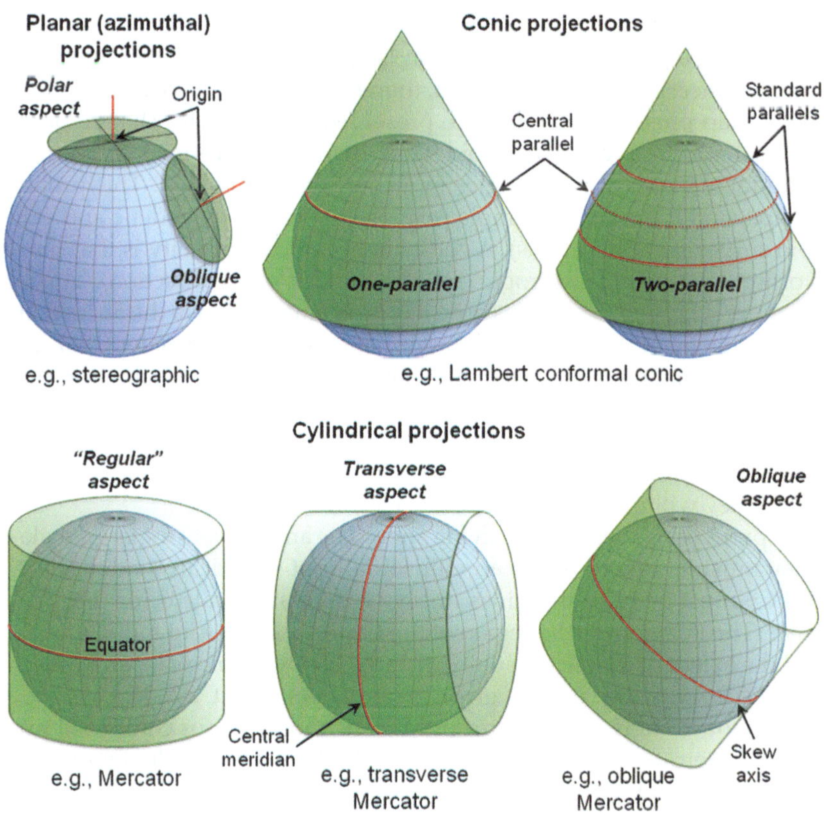

Figure 3-2. Map projection developable surfaces and their projection axes.
Source: Courtesy of NOAA, reproduced with permission.

Map Projections Appropriate for Surveying and Engineering

When a map projection is associated with a specific geodetic datum (i.e., geometric reference frame or geographic coordinate system), it is called a *projected coordinate system* (PCS). A PCS must always include a projection type, geodetic datum, and linear unit.

Thousands of map projection types have been developed, and about a hundred are commonly used for a wide range of geospatial applications. Fortunately, the list of projections appropriate for surveying engineering is much shorter, because they should be

1. Appropriate for large-scale mapping (for covering fairly small portions of the Earth),
2. Widely available and well-defined in commercial geospatial software packages, and
3. Conformal.

For the purposes of this chapter, large-scale mapping is at a scale and accuracy suitable for surveying and engineering design and analysis. The term "map scale" is typically applied to printed maps, plats, and plans. However, the term is ambiguous for digital products because they can be printed or viewed at any scale. To resolve this ambiguity, map projection scale error can be used as a criterion rather than the scale at which the map is displayed. "Scale error" is the error in distance (linear distortion) on a map, and, in general, it increases as the size of the mapped area increases. Here, "large scale" is considered mapping with a scale error of less than about 1 part in 2000. Later in this chapter, the meaning of such linear distortion is explored in depth.

Outside this chapter, there is no general agreement as to what constitutes "large"- and "small"-scale mapping, and it is context dependent. Here, small-scale maps are considered those that cover entire states, countries, continents, and even most or all parts of the Earth, but they have a much greater scale error than large-scale maps. They are not appropriate for surveying engineering applications and, therefore, will not be covered in this chapter.

In *conformal* map projections, angles are preserved at all points. That is, projected lines intersect at the same angle as on the ellipsoid. Thus, meridians (lines of longitude) and parallels (lines of latitude) intersect at right angles in a conformal projection. This property locally preserves shape, and it provides a simple relationship between geodetic and grid azimuth (important in surveying applications). A consequence of conformality is that the scale error (linear distortion) at a point is isotropic (i.e., the same in every direction). This property is important for surveying engineering because it is problematic to have distances that vary with direction. The mathematical definition of conformality is given by Snyder (1987, pp. 27–28).

For all nonconformal projections (such as equal area projections), lines on the ellipsoid, in general, do not intersect at the same angle on the mapping plane, and the scale error varies with direction, so there is no unique linear distortion at a point. These characteristics make nonconformal projections inappropriate for most surveying engineering applications, and, therefore, they will not be discussed further in this chapter.

Conformal projections used for surveying should be appropriate for large-scale mapping. As such, they must be referenced to an ellipsoidal (not spherical) model of the Earth. In addition, a PCS must be used that does not have excessive scale error (linear distortion) within its *zone*. The PCS zone is the region of the Earth where it is intended that a given PCS be used, and the extent of a zone is usually dictated by allowable scale error. As mentioned previously, a PCS zone with a scale error greater than about 1:2000, or ± 500 parts per million (ppm), would likely not be appropriate. The most famous conformal projection of all, the "regular" Mercator (cylindrical with the cylinder axis coincident with the polar axis of the Earth, as shown in Figure 3-2), is not appropriate for large-scale mapping at latitudes beyond 2° of the equator because the scale error is much too great. At this latitude, the scale error is $1{:}1652 = +605$ ppm for a Mercator projection tangent at the equator. At 40° latitude, it increases to about $+304{,}000$ ppm $= +30.4\%$. This means that 10 km on the surface of the Earth is represented as 13.04 km by the projection.

Working outside the intended bounds of a projection zone can cause problems other than just large-scale error. Many projection equations for the ellipsoid include summation series containing an infinite number of terms. When developing projection algorithms, infinite series must be truncated; this requires decisions as to the number of terms to include. Such decisions are based on the desired accuracy with respect to the zone extents. Using a map projection far outside its zone can cause coordinate computation errors that cannot be corrected if the original geodetic coordinates are unavailable. It can also cause coordinate errors that grow with repeated direct and inverse calculations.

Finally, a conformal projection suitable for large-scale mapping should also be widely available in commercial geospatial software. It is usually counterproductive to create a custom projection that cannot be used by others. However, there are cases where the benefits of a custom projection type are sufficient to warrant their use, although such solutions are often proprietary and may be problematic for data sharing or implementation within geospatial software.

Based on these three criteria, the number of conformal map projections appropriate for surveying engineering applications reduces to the four listed in Table 3-1 and shown in Figure 3-2: Transverse Mercator (TM), Lambert Conformal Conic (LCC), Oblique Mercator (OM), and stereographic. Table 3-1 also indicates which of the projections are used in the

Table 3-1. Conformal Projections used for Large-Scale Surveying and Engineering Applications.

Projection	Type	Usage	Comments
Transverse Mercator (TM)	Cylindrical	SPCS, UTM	Often used for areas elongate in the north–south direction. Perhaps the most widely used projection for large-scale mapping. Also called the Gauss–Krüger projection
Lambert conformal conic (LCC)	Conical	SPCS	Often used for areas elongate in the east–west direction. Also widely used for both large- and small-scale mapping. Includes both the one-parallel and the two-parallel versions (which are mathematically identical)
Oblique Mercator (OM)	Cylindrical	SPCS	Often used for areas elongate in an oblique direction. Not used as often as the TM and LCC projections, but widely available in commercial software. A common implementation is the Hotine OM (also called rectified skew orthomorphic)
Stereographic (oblique and polar aspects)	Planar (azimuthal)	UPS	Suitable for small areas, but for large areas, the scale error is greater than TM, LCC, or OM because it does not curve with the Earth in any direction. Polar aspect (origin at the Earth's poles) used for polar regions. Can be implemented as "ordinary" or "double" stereographic, but the resulting coordinates are different

SPCS = State Plane Coordinate System, UTM = Universal Transverse Mercator, UPS = Universal Polar Stereographic.

Source: Courtesy of NOAA, reproduced with permission.

following well-known PCSs: the State Plane Coordinate System (SPCS), Universal Transverse Mercator (UTM) system, and Universal Polar Stereographic (UPS) system. All discussion of the OM projection in this chapter will be restricted to the Hotine version, which is the type used for SPCS.

Defining Parameters of Projected Coordinate Systems

Each of the projection types listed in Table 3-1 has a specific set of five or six defining parameters. One is k_0, the projection scale (factor) on the *projection axis*. The projection axis is the line along which the projection scale error is minimum and (usually) constant with respect to the reference ellipsoid. As shown in Figure 3-2, it is the *central meridian* (λ_C) for the TM, the *central parallel* (ϕ_C) for the LCC, and the *skew axis* for the OM. The scale is not quite constant along the OM skew axis but is minimum at a single point (the local origin) and increases slowly along the axis with distance from the origin. The stereographic projection does not have a projection axis per se but rather a single point of minimum scale at its origin.

The k_0 value defines the scale relationship between the ellipsoid and developable surfaces, as listed subsequently and shown in Figure 3-3. (Note that with the descriptions of this surface being "inside," "below," "outside," "above," "intersecting," and so on, the ellipsoid surface is provided for conceptual purposes only; these projections cannot be constructed geometrically, and so, these descriptions should not be taken literally.)

- $k_0 < 1$. The developable surface is inside ("below") the ellipsoid and intersects the ellipsoid along two curves on either side of the projection axis, beyond which the developable surface is outside ("above") the ellipsoid. In this case, the projection is called *secant* because it cuts through the ellipsoid. Secant is the most common projection configuration for published PCSs (such as SPCS, UTM, and UPS) because it covers the largest region with the least absolute scale error with respect to the ellipsoid. Positive and negative scale errors are balanced for the secant projection zone as shown in Figure 3-3, with approximately the middle 71% of the developable surface below the ellipsoid and the outer 14.5% on either side above the ellipsoid.

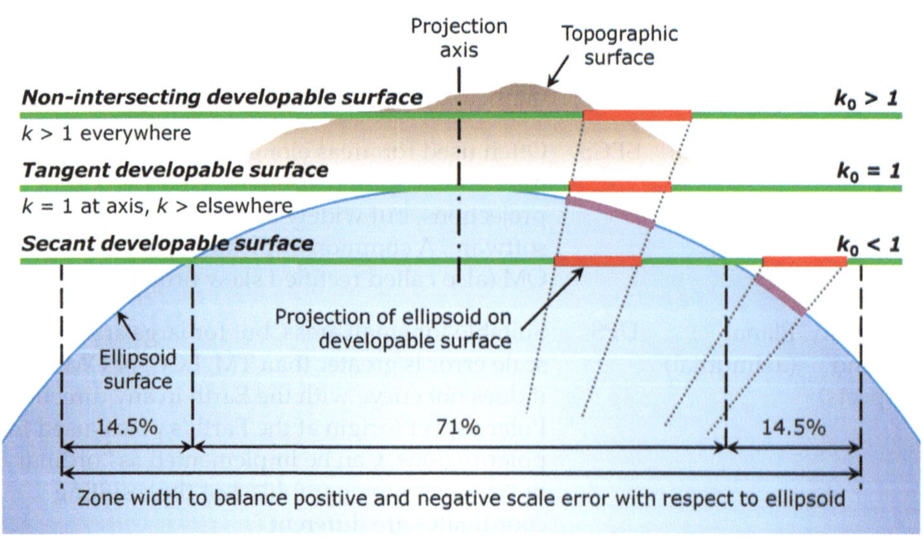

Figure 3-3. Secant, tangent, and nonintersecting map projection developable surfaces. Source: Courtesy of NOAA, reproduced with permission.

Positive scale error (linear distortion) means that the projected distance is longer than the (curved) distance on the ellipsoid, and negative scale error means the projected distance is shorter than the ellipsoid distance. A common value for SPCS is $k_0 = 0.9999$, which corresponds to negative 1:10,000 or −100 ppm.
- $k_0 = 1$. The developable surface is tangent to the ellipsoid. That is, it touches the ellipsoid along its projection axis (or at a single point for the Stereographic projection).
- $k_0 > 1$. The developable surface is above the ellipsoid and the two surfaces do not intersect anywhere. The motivation for this case is usually to place the developable surface near the topographic surface, which is typically above the ellipsoid. This configuration is used to decrease the linear distortion of the projected coordinates at the ground surface and is discussed in the section on conformal projection distortion.

In addition to the projection axis scale, at least four other parameters are needed to fully define the projections listed in Table 3-1. Two of these are the latitude and longitude of its *geodetic origin* (φ_0, λ_0). The geodetic origin may or may not be located on the projection axis. It is always on the central meridian of the TM ($\lambda_0 = \lambda_C$) but often is not on the central parallel of the LCC projection ($\varphi_0 \neq \varphi_C$), in which case at least six parameters are required to define an LCC. The other two parameters are the projected coordinate values of the geodetic origin, often called the *grid origin* and specified as *false northing* (N_0) and *false easting* (E_0) in this chapter. Grid origin values are usually selected such that projected coordinates are positive within the zone. An additional (sixth) parameter called the *skew axis azimuth* (α_0) is required for the OM projection to specify the orientation of its skew axis (α_0 can also be defined implicitly by using two points on the desired geodesic curve passing through the origin).

Although k_0 must be specified for a projection, it can be defined implicitly for the LCC projection if $k_0 < 1$. In this case, two standard parallels are specified where $k_0 = 1$ (i.e., the two latitudes where the LCC cone intersects the ellipsoid), as shown in Figure 3-2. One standard parallel is north (φ_N) and one is south (φ_S) of the central parallel, that is, $\varphi_N > \varphi_C > \varphi_S$, with the central parallel approximately halfway between. The further apart the standard parallels, the smaller k_0 on φ_C. For $\varphi_C = 40°$, a separation between the north and south parallel of 1° (111 km) gives $k_0 \approx 0.999\ 962$; a separation of 2° (222 km) gives $k_0 \approx 0.999\ 848$. It is important to understand that the one- and two-parallel LCC versions are mathematically identical. For both types, the central parallel is the projection axis, and for every two-parallel LCC, an exactly equivalent one-parallel LCC can be specified. The only difference is that the projection axis scale is defined implicitly for the two-parallel LCC. This is merely an arbitrary convention that uses two standard parallels to replace a single k_0 value. Such LCC definitions also have $\varphi_0 \neq \varphi_C$, which means that six parameters are required to define the projection. In the most general case, an additional scale can be assigned to φ_C for a two-parallel LCC, for a total of seven defining parameters. However, such an approach is not recommended because it creates confusion and defeats the purpose of using φ_N and φ_S to implicitly define k_0; it is tantamount to specifying k_0 twice.

Despite the mathematical equivalence of the one- and two-parallel LCC, some software implementations allow the use of only the two-parallel version, with no ability to explicitly define k_0. In these cases, setting the standard parallel to the same values causes a division by zero error. Such poorly written LCC algorithms are becoming less common, but users should be aware of this potential problem.

The typical required defining parameters for the four conformal projections appropriate for surveying are summarized in Table 3-2. The OM projection is a more complex case, in part because it can be defined in more than one way. The convention in Table 3-2 uses a local origin at the location of interest for the Hotine OM. However, the origin is actually on the equator of an intermediate surface of constant curvature (called the *aposphere*) used in the OM calculations. In any case, if an OM is used, the type and parameter convention must

Table 3-2. Typical Defining Parameters for Large-Scale Conformal Map Projections.

Projection	Symbol	Parameter	Comments
All types	N_0	False northing	Specifies grid coordinates at the geodetic origin of all projection types
	E_0	False easting	
Transverse Mercator (TM)	φ_0	Latitude of origin	Geodetic origin is usually selected far enough south that $N_0 = 0$ and still has positive coordinates within the zone
	λ_C	Central meridian	
	k_0	Scale on central meridian	
Lambert conformal conic (LCC)	φ_0	Latitude of origin	The two LCC versions are mathematically identical (i.e., every two-parallel LCC has an exact one-parallel equivalent). An additional scale can be applied to φ_C of the two-parallel LCC, but this is not common or recommended. Standard parallels are often specified as 1 and 2 rather than N and S, and the order is irrelevant in computations
	λ_C	Central meridian	
	One-parallel LCC		
	φ_C	Central parallel	
	k_0	Scale on central parallel	
	Two-parallel LCC		
	φ_N	North standard parallel	
	φ_S	South standard parallel	
Oblique Mercator (OM); specifically Hotine OM with N_0 and E_0 defined at local origin	φ_0	Local latitude of origin	The true origin for N_0 and E_0 is at λ on the equator of the intermediate surface used in the calculations (not on the equator of the ellipsoid). The scale is not constant on the skew axis but increases slowly away from the local origin. The convergence angle is zero on the central meridian at the local origin, and it is minimum (but not zero) on the central meridian north and south of origin
	λ_0	Local longitude of origin	
	k_0	Scale on skew axis at local origin	
	α_0	Azimuth of skew axis at local origin	
Stereographic	φ_0	Latitude of origin	The scale error increases in all directions from the geodetic origin. For the polar aspect, the geodetic origin is at the poles
	λ_0	Longitude of origin	
	k_0	Scale at the geodetic origin	

Source: Courtesy of NOAA, reproduced with permission.

be clearly identified. In addition, the OM-projected coordinate axes are oriented with respect to the skew axis, rather than with respect to north. To provide coordinates where the projected (grid) north direction coincides with geodetic north at the local origin, projected OM coordinates are almost always rotated ("rectified"). This rotation is why one of the common alternate names for this projection is rectified skew orthomorphic.

In addition to the projection parameters discussed previously, a PCS must specify two parameters for the reference ellipsoid. One is the ellipsoid semimajor axis (equatorial radius), a. The second describes its departure from sphericity, which is usually provided as one of the following values: inverse flattening, $1/f$; semiminor axis (polar radius), b; or second eccentricity squared, e^2. The relationship between these parameters is $f = (a - b)/a$ and $e^2 = f(2 - f)$. Additional information on reference ellipsoids is given in Chapter 2.

The PCS definition must also specify a linear unit, which is often the meter. However, in the United States, output coordinates are usually provided in either US survey feet (sft) or international feet (ift). By definition, 1 sft = (1200/3937) m and 1 ift = 0.3048 m (exact) and confusing these two types of feet is a common problem. Although the sft is longer than the ift by only 2 ppm (about 0.01 ft per mile), explicitly specifying the type of foot is critically important because projected coordinate values are almost always very large. Even in a PCS where output-projected coordinates are used as sft or ift, to avoid confusion, the meter is often used for the defining parameters. This approach is used for the State Plane Coordinate System of 1983 (SPCS 83).

Although the sft has been widely used for SPCS 83 and for surveying practice in general, this will not be the case in the future. Having two nearly identical definitions of the foot in current use creates confusion that causes mistakes. In addition, when the sft was defined in 1959, it was intended for only limited temporary use to allow time for transition to the ift. This was supposed to occur in 1989, when SPCS 83 was officially adopted. To resolve these problems, the National Institute of Standards and Technology (NIST) and the National Geodetic Survey (NGS) jointly made a final determination that the sft will be deprecated on December 31, 2022, as published in the *Federal Register* (2020). After that date, the sft will be used only in legacy applications (such as SPCS 83 and SPCS 27) and will not be supported by NGS for any future version of State Plane or for any other component of the National Spatial Reference System (NSRS).

Equations for computing coordinates and distortion from the ellipsoidal form of the four conformal projections discussed previously are complex and beyond the scope of this chapter. For additional information on performing the calculations, the interested reader is referred to Stem (1990) and Snyder (1987).

Map Projection Distortion

Map projection distortion is an unavoidable consequence of representing a curved surface on a flat surface. It is an error in size, shape, or orientation of a feature on the Earth as represented on the developable surface. Distortion cannot be eliminated; the best that can be done is to account for or minimize its effect. Because it is the difference between actual conditions and their representation, distortion must be clearly understood for surveying engineering applications.

Distortion can be split into two categories: angular and linear. Angular distortion is the error in projected direction or orientation. Linear distortion is the error in projected scale or distance. Details for both types are described subsequently for conformal projections.

Angular Distortion. For conformal projections, angular distortion at a point is the *convergence (mapping) angle*, γ. The convergence angle is the difference between grid (map) north and geodetic ("true") north, as shown in Figure 3-4. On the central meridian (λ_C), γ is zero, and it is positive east and negative west of λ_C. The magnitude of γ increases with the distance east or west from λ_C, and the rate of change increases (nearly) linearly with increasing latitude, as shown in Figure 3-5. The LCC projection was used to create Figure 3-5 for points on the central parallel set to the latitude values shown. Figure 3-5 can also be used for the TM projection. For points within 200 km of λ_C, the maximum difference with the LCC is only 12 arcseconds (at 80° latitude).

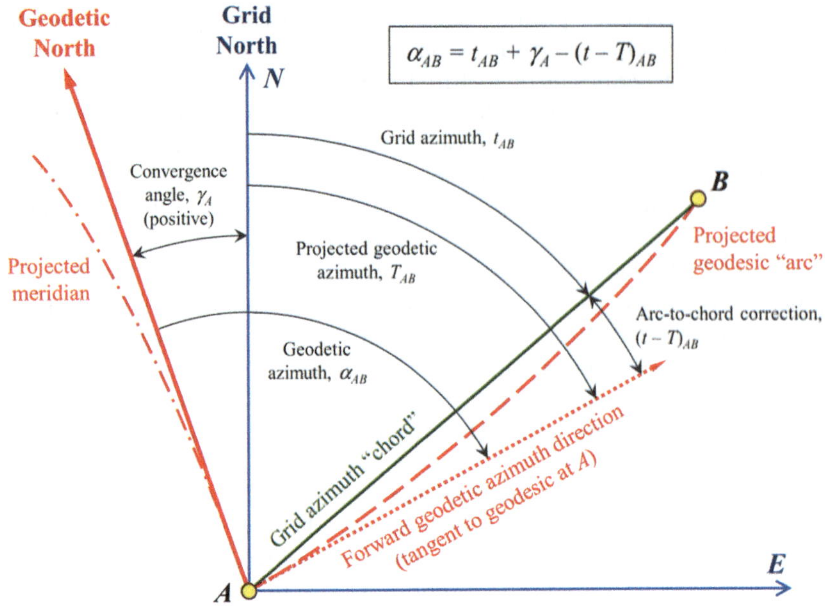

Figure 3-4. Geodetic and grid azimuths, convergence angle, and the arc-to-chord correction.

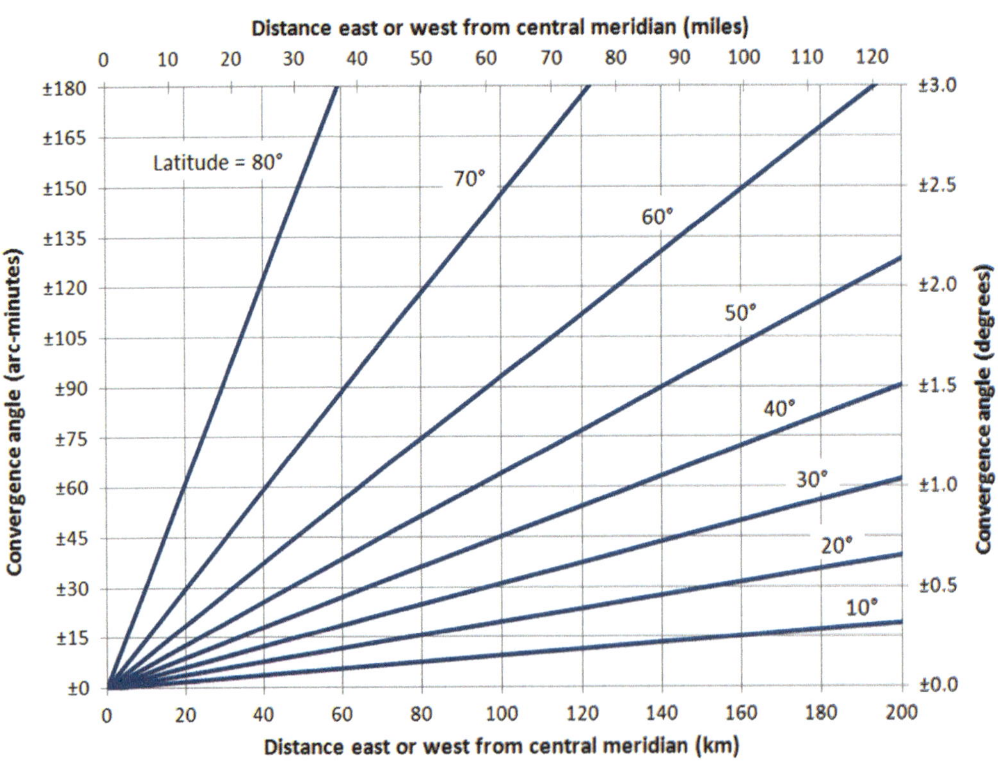

Figure 3-5. Convergence angles as a function of distance from the central meridian at various latitudes.

For the stereographic and Hotine OM projections, the rate of change of γ is about the same as shown in Figure 3-5 with respect to their projection origins. For example, 200 km east or west of a local origin of $\varphi_0 = 40°$ for a stereographic and OM (with skew axis $\alpha_0 = 45°$), $\gamma = \pm 1.499°$ and $\pm 1.520°$, respectively. These values are close to $\gamma = \pm 1.505°$ for an LCC and $\gamma = \pm 1.504°$ for a TM (Figure 3-5). OM convergence is fairly insensitive to α_0; even for $\alpha_0 = 88°$ in the example here, $\gamma = \pm 1.506°$. At $\alpha_0 = 0°$ the OM reduces (approximately) to a TM. At the other extreme, for $\alpha_0 = 90°$ and $\phi_0 = 0°$, a regular Mercator is used in place of an OM and γ is zero everywhere (despite zero convergence, the regular Mercator is not suitable for surveying engineering applications because of extreme scale error, as discussed previously).

Another characteristic of conformal projections that is useful for surveying applications is the relatively simple relationship between grid and geodetic azimuth. The geodetic azimuth from point A to B (α_{AB}) is related to the grid azimuth t_{AB} as

$$\alpha_{AB} = t_{AB} + \gamma_A - (t - T)_{AB} \tag{3-1}$$

where $(t - T)_{AB}$ is the *arc-to-chord* correction. This correction is referred to as $(t - T)$, because it is the difference between the projected geodetic azimuth, T, and the grid azimuth, t. As shown in Figure 3-4, the "straight line" grid azimuth forms a "chord" of the projected geodesic "arc" between points, hence the name. It is a function of the distance of the midpoint of line AB from the projection axis and the length of the line parallel to the projection axis, and its sign depends on its location relative to the projection axis as well as its azimuth. Sometimes it is also called the *second term* (or *difference*) correction because it can be considered a second-order correction to γ. In many situations, $(t - T)$ is negligible and can be ignored, but not always. For example, a line with its midpoint 200 km from an LCC or TM projection axis has a $(t - T)$ magnitude of 5.1″ (arc-seconds) for a 10 km-length parallel to the projection axis and 0.5″ for a 1 km length (for latitudes within about 55° of the equator). These corrections decrease to about 1.3″ and 0.1″, respectively, if the midpoint is 50 km from the projection axis. For all UTM and even the largest SPCS zones, $(t - T)$ will never exceed about 0.8″ for lines less than 1 km long, and it usually will be much less. Additional details on the arc-to-chord correction are given in Stem (1990).

Convergence angles are usually not as much of a concern as linear distortion because little can be done to reduce convergence. It can be minimized only by staying close to the projection central meridian or by limiting surveying and mapping activities to equatorial regions of the Earth. Although convergence cannot be reduced, it can be dealt with on survey plats in various ways such as by giving both the grid and the geodetic azimuth on a line or by giving the convergence angle at the line endpoints.

Linear Distortion. Map projection linear distortion is manifested as a difference in distance between a pair of projected (map grid) coordinates and the true horizontal distance at the surface of the Earth. Although differences in finite distances are often used to describe linear distortion, it is formally defined at a point using infinitesimal (differential) distances. In this chapter, linear distortion is defined as

$$\delta = k\left(\frac{R_G}{R_G + h}\right) - 1 \tag{3-2}$$

where

k = Grid point (scale) factor (or error),

h = Ellipsoid height, and
R_G = Geometric mean (or Gaussian) radius of curvature

$$R_G = \frac{a\sqrt{1-e^2}}{1-e^2\sin^2\varphi} \tag{3-3}$$

where, for the reference ellipsoid used,

φ = Geodetic latitude,
a = Semimajor axis, and
e^2 = First eccentricity squared.

The variation in R_G with latitude is shown in Figure 3-6, based on the GRS 80 ellipsoid with $a = 6{,}378{,}137$ m (exact) and $e^2 = 0.006\ 694\ 380\ 022\ 90\ldots$. Geometric mean radius is used to represent the "average" curvature at a point, because curvature varies with direction.

The value of k can be considered the part of distortion caused by Earth curvature. It is calculated from the projection equations as a function of position only and can be computed using many geospatial software packages. Algorithms for computing k can also be obtained from a variety of sources such as Stem (1990) and Snyder (1987).

The quantity in parentheses in Equation (3-2) is the *elevation* or *height (scale) factor* and, as the name suggests, is the part of distortion caused by variation in ellipsoid height. Despite often being called an "elevation" factor, it is actually the height above the reference ellipsoid, specifically the ellipsoid used in the projection equations. The term *elevation* usually refers to orthometric height, H, which is related to ellipsoid height as $h = H + N_g$, where N_g is the geoid height (more information on these and other heights is given in Chapter 2). Worldwide h can differ from H by more than 100 m, so it is good practice to use h to calculate δ. Note that a height error of 100 m causes an error in δ of 16 parts per million (ppm).

The product of k and the height factor is often called the *combined (scale) factor*, which represents total linear distortion. For surveying applications, linear distortion is typically small, and so the combined factor is usually very close to 1. To facilitate evaluation of distortion, 1 is subtracted from the combined factor in Equation (3-2) to give a value that can be conveniently represented in ppm (e.g., a combined factor of 0.999 98 is -20 ppm distortion). Linear distortion can be positive or negative. Negative distortion means that the projected (map grid) length is shorter than the horizontal ground length. Positive distortion means the projected (map grid) length is longer than the ground length.

One can think of linear distortion as the unavoidable consequence of the projection developable surface departing from the reference ellipsoid. Although no ellipsoidal forms of conformal projections can be created geometrically, it can be useful to think of linear distortion increasing as the "distance" between the developable surface and the ellipsoid increases, as shown in Figure 3-7. This diagram is similar to Figure 3-3 and is for a secant map projection, such as used for SPCS and UTM. It also shows the topographic surface above both the ellipsoid and projection surface, which is the typical case and is why linear distortion with respect to the ground surface is usually of greater magnitude than with respect to the ellipsoid.

Although total linear distortion is always a combination of distortion due to curvature and ellipsoid height, it is instructive to consider the two components separately. Distortion values for each component are given in Tables 3-3 and 3-4. Table 3-3 gives the zone width

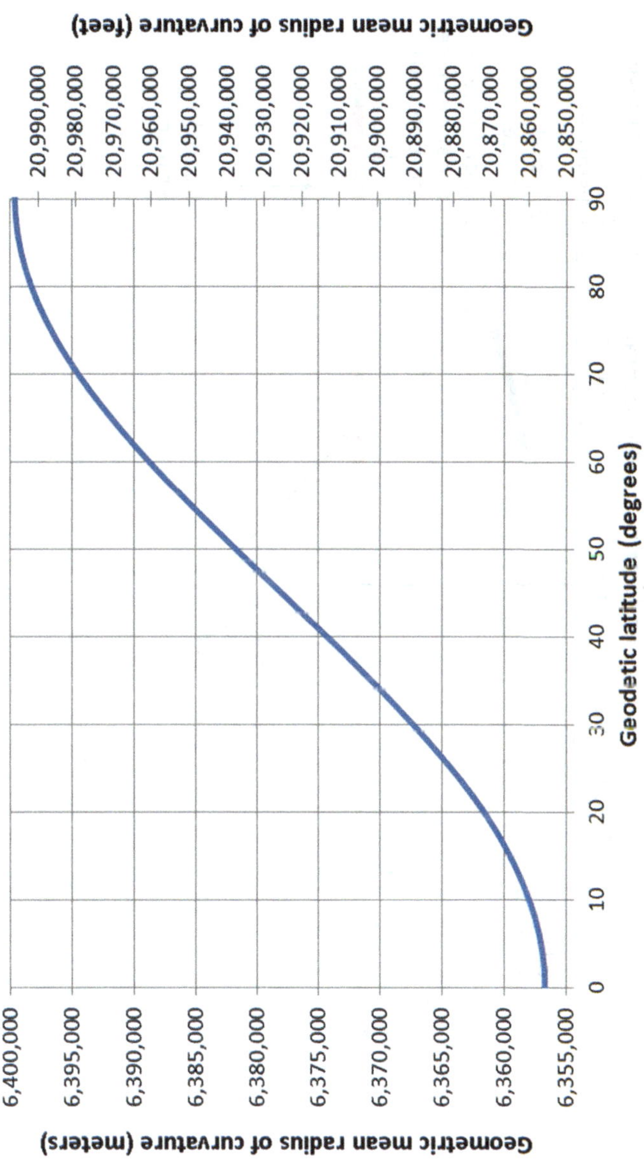

Figure 3-6. Geometric mean radius of curvature as a function of latitude for the GRS 80 ellipsoid.

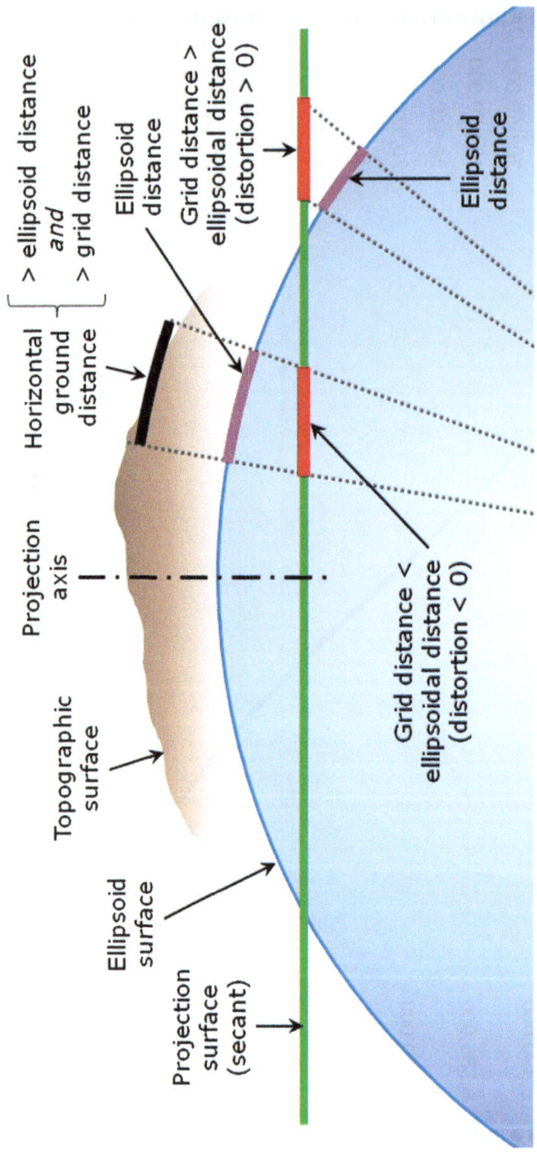

Figure 3-7. Linear distortion of secant map projection with respect to ellipsoid and topography.
Source: Courtesy of NOAA, reproduced with permission.

Table 3-3. Linear Distortion of Projected Coordinates Caused by Earth Curvature Only (k).

	Maximum linear distortion due to curvature, $\delta + 1 = k$		
Maximum zone width for secant projections[a]	Parts per million (mm/km)	Feet per mile	Ratio (absolute value)
25 km (16 mi)	±1	±0.005	1:1,000,000
57 km (35 mi)	±5	±0.026	1:200,000
81 km (50 mi)	±10	±0.053	1:100,000
114 km (71 mi)	±20	±0.11	1:50,000
180 km (112 mi)	±50	±0.26	1:20,000
255 km (158 mi) (e.g., SPCS)	±100	±0.53	1:10,000
510 km (317 mi) (e.g., UTM)	±400	±2.11	1:2,500

[a]Computed for TM projection at 40° latitude. If latitude ≤70°, the listed zone widths agree with TM and other conformal projections to within ±2 km for $\delta \leq 100$ ppm and within ±10 km for $\delta = 400$ ppm.

Source: Dennis (2016), © ASCE.

corresponding to various maximum magnitudes of linear distortion. Distortion is expressed in ppm (mm/km), in ft/mi, and as a unitless ratio. The second to last row in the table gives a nominal zone width of 255 km (158 mi) for SPCS, corresponding to its well-known scale error of 1:10,000 (±100 ppm). The last row is for UTM, which can have a zone 510 km (317 mi) wide for its allowable distortion of 1:2500 (±400 ppm).

Table 3-4 gives linear distortion for various ranges of ellipsoid heights with respect to the projection surface. In Table 3-3, the row with ±20 ppm distortion corresponds to a zone width of 114 km for distortion caused by curvature alone, yet it is limited to a height range of 250 m (±125 m). This is a modest variation in topographic height for many areas, such as most of the western United States. In such situations, topographic relief is often the main source of distortion. However, the location relative to the projection axis (which controls distortion curvature) is also important and can dominate distortion if the projection axis is far away from a point of interest.

Distortion is defined at a point, but often it must be computed over a line or finite distance. The part caused by height, $R_G/(R_G + h)$, can be estimated as the mean h of the line endpoints and R_G at the midpoint. The part of distortion caused by curvature, k, increases with approximately the square of distance from the projection axis. Because the rate of change in k is the first derivative of the squared distance, it is nearly linear, with a magnitude of 0.024 ppm/km^2 perpendicular to the projection axis. At 250 km from the projection axis (about the maximum distance for SPCS zones), k changes at a rate of about 0.024 ppm/km^2 × 250 km = 6 ppm/km. This means that at 250 km from the axis, k will increase (or decrease) by 6 ppm if the distance from the axis increases (or decreases) by an additional 1 km. The rate decreases to 2.4 ppm/km at 100 km from the projection axis, which covers the vast majority of SPCS zones.

Table 3-4. Linear Distortion of Projected Coordinates Caused by Ellipsoid Height Only.

	Maximum distortion due to height, $\delta + 1 = R_G/(R_G + h)$		
Height below (−) and above (+) projection surface	Parts per million (mm/km)	Feet per mile	Ratio (absolute value)
±30 m (±98 ft)	±4.7	±0.025	~1:213,000
±125 m (±410 ft)	±20	±0.10	~1:51,000
±300 m (±980 ft)	±47	±0.25	~1:21,000
+600 m (+2,000 ft)[a]	−94	−0.50	~1:10,600
+1,000 m (+3,300 ft)[b]	−157	−0.83	~1:6,400
+4,400 m (+14,400 ft)[c]	−690	−3.6	~1:1,450

[a]Approximate mean topographic height of North America.

[b]Approximate mean topographic height in CONUS west of 100°W longitude.

[c]Approximate maximum topographic height in conterminous United States.

Source: Dennis (2016), © ASCE.

Strictly speaking, k for a projected line should be determined by integrating along the line. The integration to estimate k_{AB} for a line from point A to B can be approximated as

$$k_{AB} = \frac{1}{6}(k_A + 4k_m + k_B) \tag{3-4}$$

where k_A and k_B are the values at the line endpoint and k_m is the value at the midpoint. But Equation (3-4) need not be used for all cases. For a line 10 km long (perpendicular to the projection axis), using the mean k of the endpoints will agree with Equation (3-4) to within 0.2 ppm, and using k at the midpoint will agree to within 0.1 ppm. For a 20 km line, these differences increase to 0.8 and 0.4 ppm, respectively. These relationships hold everywhere within SPCS and UTM zones. Additional details on distance computation methods are given later in this chapter.

PROJECTED COORDINATE SYSTEMS

State Plane Coordinate System

Existing and Historic Versions of the State Plane Coordinate System. Large-scale conformal projected coordinate systems for surveying engineering became common in the United States after the creation of the SPCS in the 1930s by NGS (then the US Coast and Geodetic Survey). The first version is now called SPCS 27 because it was referenced to the North American Datum of 1927 (NAD 27) in most areas. The second (current) version is called SPCS 83 because all zones are now referenced to the North American Datum of 1983 (NAD 83). See Dennis (2018a) for additional details on the history and development of both versions of the SPCS.

The main purpose for establishing SPCS was to make it possible to perform georeferenced geodetic surveys using plane trigonometry, without the complex mathematics of geodesy. This characteristic was important prior to widespread availability of microcomputers (and associated software). Originally, SPCS was particularly useful for surveys covering large areas such as highway projects. This is still true today, but the widespread use of computers and GNSSs have led to SPCS being commonly used for a broad range of applications, from engineering plans to GIS datasets.

SPCS is a multizone system of conformal PCSs where each zone was designed using the TM, LCC, or Hotine OM projections. The area of coverage for a zone was typically selected such that scale error (linear distortion) with respect to the ellipsoid would not exceed 1:10,000 (±100 ppm). To achieve this requirement, the zone width (perpendicular to the projection axis) cannot exceed 255 km (158 mi) at mid-latitudes. Therefore, the TM projection was usually used for zones elongate predominantly in a north–south direction and the LCC projection for zones elongate east–west. Only a single zone uses the OM projection (for the panhandle of Alaska), which is elongate in an oblique direction.

The entire extent of SPCS 27 and 83 is shown in Figure 3-8, which also indicates changes that occurred from SPCS 27 to 83. Current conterminous US (CONUS) SPCS 83 zones are shown in Figure 3-9, and the Alaska zones are shown in Figure 3-10. Zones in CONUS consist of TM or LCC projections. For most zones, the projection type can be discerned from the zone shape and orientation in Figures 3-9 and 3-10. The same projection type is used for all zones in a state, with the exception of Florida and New York (which use both the TM and LCC) and Alaska (which uses all three SPCS projection types).

Nearly all SPCS zones are defined by state or county boundaries; the exceptions are the zones in Alaska, the Louisiana Offshore Zone, and one county in Washington state (Grant) split across its two SPCS 83 zones. Zone boundaries are usually selected to stay within the nominal maximum zone width of 255 km to the extent possible. However, several zones exceed this width by a considerable amount, especially in SPCS 83. The widest is the Montana SPCS 83 LCC zone (about 516 km wide), which covers the entire state with a single zone (versus three zones for SPCS 27). Both versions of the Louisiana Offshore Zone (also an LCC) are of comparable width, but its boundaries are not as well defined (it covers essentially the entire Gulf of Mexico).

For SPCS 27, a total of 134 zones were designed covering all 50 states plus one for each of four territories (American Samoa, Guam, Puerto Rico, and the US Virgin Islands). Three of these zones (based on a scaled version of the Clarke 1866 ellipsoid) replaced the original three for Michigan in 1964. The final maximum of 131 SPCS 27 zones were achieved in 1968, when the Louisiana Offshore Zone was added. Although the name "SPCS 27" implies that NAD 27 was used for all areas, different datums were used for island areas beyond visible range of the North America mainland: Hawaii, American Samoa, Guam, Puerto Rico/US Virgin Islands, and four island datums in the Bering Sea (as shown in Figure 3-10).

When NAD 27 was officially replaced by NAD 83 in 1989, SPCS 27 was updated to SPCS 83. It currently consists of 125 zones covering the same areas as SPCS 27 (except for American Samoa, which has no SPCS 83 zone). The last change occurred in 2001, when the Kentucky statewide LCC zone was added (Figure 3-9). This is an interesting change, because the existing two LCC zones still remain, resulting in two "layers" of zones covering the same geographic area. The concept of layered zones has been adopted for the future SPCS, as discussed in the next section.

Most SPCS 83 zone definitions remained the same as those used for SPCS 27 in terms of projection type, geodetic parameters, and area of coverage. In these cases, the only change was in the false northings and/or false eastings, so that SPCS 83 coordinates would differ substantially from SPCS 27 values (usually by several to hundreds of miles). The defining linear unit was also changed from US survey feet to meters. As shown in Figures 3-8 and 3-9,

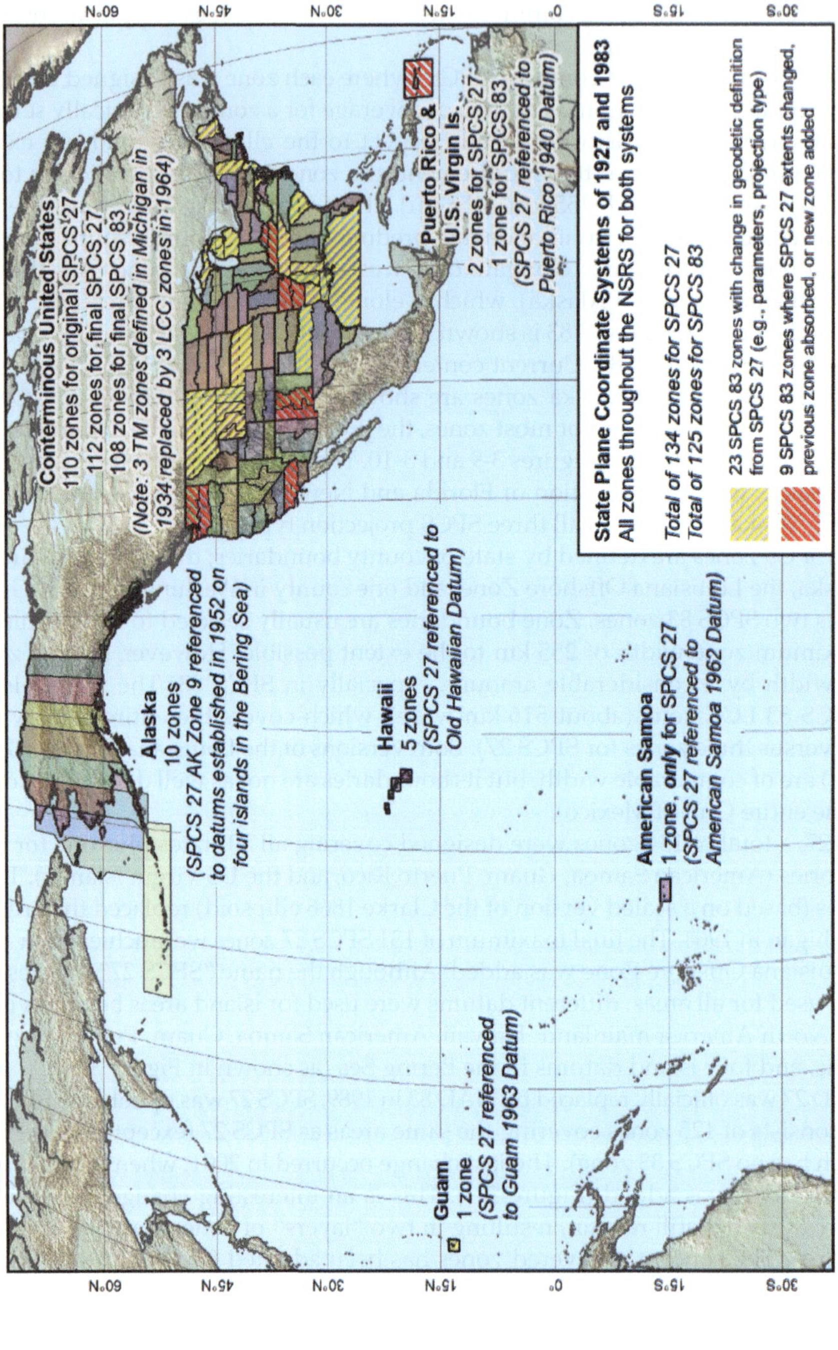

Figure 3-8. State Plane Coordinate Systems of 1927 and 1983, showing changes from SPCS 27.
Source: Courtesy of NOAA, reproduced with permission.

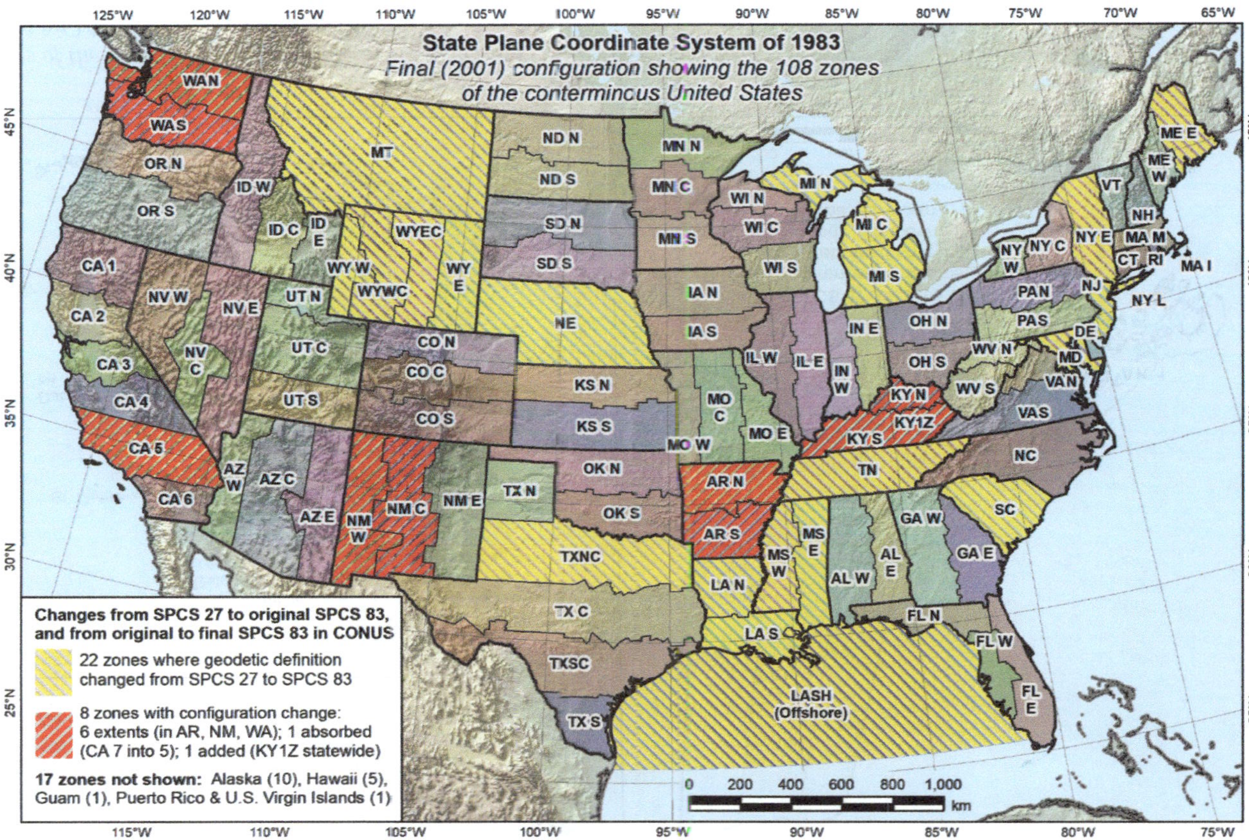

Figure 3-9. Current zones of the State Plane Coordinate System of 1983 (SPCS 83) in CONUS, showing changes from SPCS 27. Source: Courtesy of NOAA, reproduced with permission.

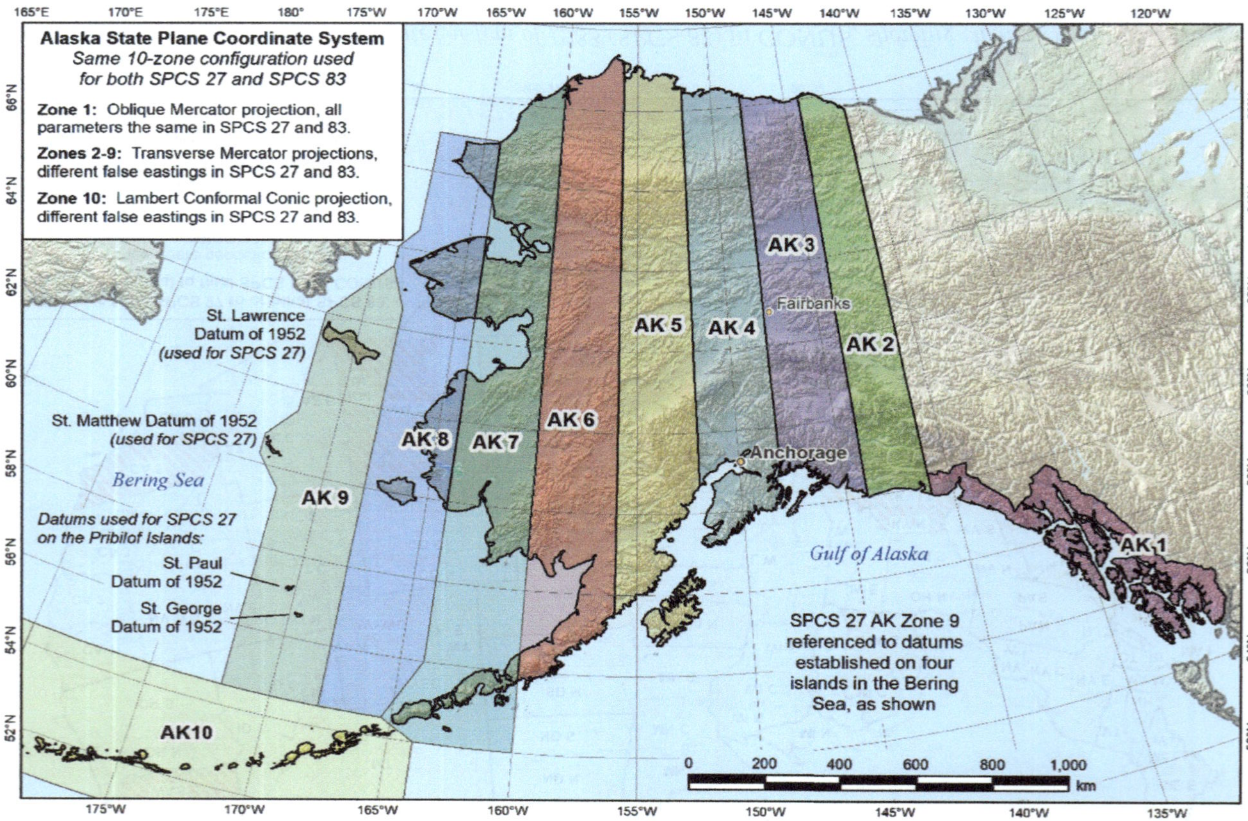

Figure 3-10. Zones of the State Plane Coordinate System of 1983 (SPCS 83) in Alaska.
Source: Courtesy of NOAA, reproduced with permission.

more significant changes were also made. Zone extents were modified in Arkansas, New Mexico, and Washington, and two zones were absorbed into other existing zones: the California Los Angeles zone (CA Zone 7) became part of CA Zone 5, and the US Virgin Islands zone was combined with the Puerto Rico zone. Various geodetic characteristics were also modified for 23 zones by switching from a nonconformal projection type (Guam); changing parameters of latitude, longitude, and/or projection scale (Louisiana, Maine, Maryland, Mississippi, New Jersey, New York, Tennessee, Texas, and Wyoming); ending the use of a scaled ellipsoid (Michigan); and combining multiple zones into a single statewide zone (Montana, Nebraska, and South Carolina). Replacing multiple zones with a single zone increased the scale error from the nominal target of 1:10,000 (±100 ppm). This is particularly evident for the single zone of Montana, which has a scale error on the central parallel of 1:1646 (−607 ppm) with respect to the ellipsoid, as compared with 1:11,207 (−89 ppm) for the largest of its three SPCS 27 zones.

The history and defining parameters for all SPCS 83 and 27 zones are available in Appendix A and B of Dennis (2018a), respectively. Stem (1990) includes SPCS 83 projection algorithms and information on angular and linear distortions.

State Plane Coordinate System of 2022. As part of modernizing the NSRS, NGS will replace NAD 83 with four new terrestrial reference frames (TRFs) in the near future, probably between 2025 and 2027. Because the original intent was to adopt them in 2022, they are called the TRFs of 2022. A new SPCS is currently being created as part of this change, called the State Plane Coordinate System of 2022 (SPCS2022), and all its zones will be referenced to the 2022 TRFs. Although the geodetic coordinates (latitude, longitude, and ellipsoid height) will differ from NAD 83 by up to a few meters, the same reference ellipsoid used for SPCS 83 (GRS 80) will also be used for SPCS2022, and the same three projection types will be used (LCC, TM, and Hotine OM). This means that much of the existing projection mathematics can be used for SPCS2022, but there will be some differences, as discussed subsequently.

Despite using the same reference ellipsoid and projection types, SPCS2022 will differ significantly from SPCS 83 in most states. The main differences are organized into six categories, as listed in the following numbered sections. At the time of this writing (March 2022), SPCS2022 is still under development, so some of the characters may change slightly once SPCS2022 is finalized.

Minimize Linear Distortion at the Topographic Surface Unlike SPCS 27 and 83 zones, SPCS2022 zones are being designed to minimize linear distortion at the topographic rather than ellipsoid surface. This design approach is used because surveying and mapping work usually occurs on the ground and not on the ellipsoid. Recall from Table 3-4 that the mean topographic ellipsoid heights of CONUS west of the 100th meridian exceed 1000 m. A height increase of this magnitude causes −157 ppm of distortion, which by itself exceeds the nominal design criterion of ±100 ppm used for SPCS 27 and 83 with respect to the ellipsoid.

The basic idea is to increase the projection axis scale such that the total linear distortion [given by Equation (3-2)] is minimized for a zone. This is shown schematically in the left diagram of Figure 3-11. However, only modifying the scale will not minimize distortion in cases where there is systematic slope across a zone. In such bases, the projection axis scale and location (and orientation for the OM) are also modified to more uniformly minimize distortion over the entire extent of a zone, as shown in the right diagram of Figure 3-11. This is the general approach used by NGS for designing SPCS2022 zones.

In designing SPCS2022 zones, NGS also takes into account population distribution, so that distortion can be further reduced in populated areas. This is done by using population density and the location of cities and towns. It has a particularly significant effect on design in the western United States, where population tends to be irregularly distributed and

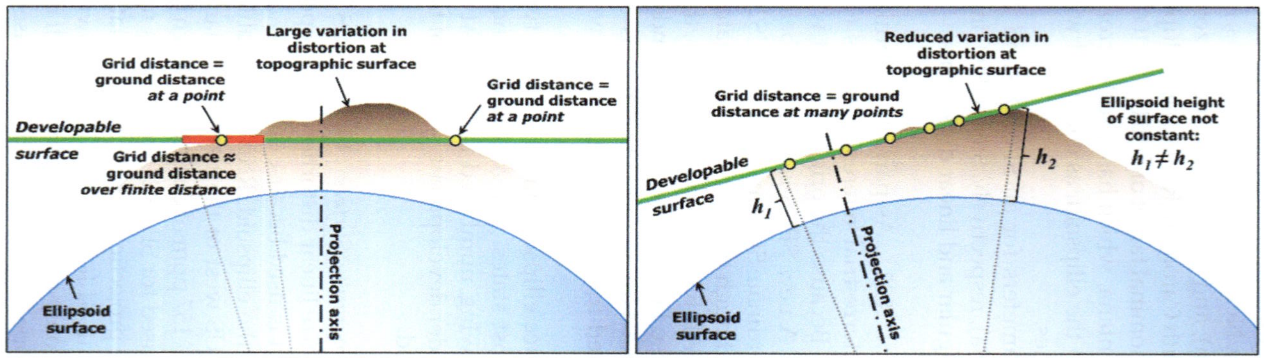

Figure 3-11. Schematic representing how linear distortion can be reduced by changing projection axis location.
Source: Courtesy of NOAA, reproduced with permission.

concentrated in lower elevation areas. The main performance objective is to achieve a "distortion design criterion" selected from specific distortion ranges of ±5, ±10, ±20, ±30, ±40, ±50, ±75, ±100, ±150, ±200, ±300, or ±400 ppm. The final zone design is assigned the lowest distortion design criterion that includes at least 90% of the population, 75% of cities and towns (by location only), and 50% of the total zone area. Details of the design approach are described by Dennis (2018b) (Chapter 3).

Stakeholder Involvement in the Design Process Groups identified as SPCS2022 stakeholders were an integral part of the design process. Stakeholder groups consist of one or more of the following state organizations: departments of transportation, state GIS or cartographer offices, professional surveying, engineering, and other geospatial organizations; and universities or colleges with a geospatial curriculum. A process was established for stakeholders to make formal requests and proposals for SPCS2022. Requests were made for zone designs by NGS, and proposals were for designs by stakeholders themselves. A total of 41 states submitted 36 requests and 31 proposals. These helped define certain zone characteristics and determine how many zone "layers" were needed and, thus, the total number of zones (zone layers and the number of zones are discussed in the following two items).

Not all states made requests and proposals, but most did, and many that did not had already indicated that they were satisfied with the NGS approach (so they felt no need to make formal requests). In this way, state stakeholders were able to get the type of SPCS2022 that best suited their diverse needs.

Zone "Layers" In SPCS2022, a state can have up to three zone "layers." Zones in a layer will overlap some or all the zones in another layer within a state, as is the case for the existing SPCS 83 zones in Kentucky (which consist of a statewide zone layer and a two-zone layer).

Every state and territory will have a single statewide zone designed by NGS. Two of these will each cover two territories, one for Puerto Rico and the US Virgin Islands (as already done for SPCS 83), and one for Guam and the Commonwealth of the Northern Mariana Islands, for a total of 54 statewide zones.

States may also have one or two multiple-zone layers. If a state has two multiple-zone layers, one must completely cover the entire area of the state, and the other must cover only part of the state. Partial-coverage layers were intended for mountainous states, so that they could design zones only for limited areas where low distortion was desired (such as in valleys where most of the population is located).

Figure 3-12 shows the preliminary number of zone layers in each state. Most states (28) will have two layers, a statewide zone and a complete-coverage multiple-zone layer. Of states with multiple-zone layers, Montana is the only one that will have a partial-coverage layer without a complete-coverage layer. Most states with three layers are in mountainous western states, as intended, with some exceptions (most notably in Florida). Hawaii will have two layers on only its two largest islands, Hawaii and Maui.

Large Increase in the Number of Zones SPCS2022 will have far more zones than either SPCS 83 (125) and SPCS 27 (131), but the number of zones varies greatly among states, because each state is able to request or propose zones. The preliminary number of zones by state is shown in Figure 3-13, which varies from one statewide zone only (18 states) to a maximum of 90 in three layers (Utah), for a total of 971 zones. Of these, 54 are statewide zones designed by NGS, as well as NGS designs of 105 zones in multiple-zone layers and two "special-use" zones (described subsequently). The remaining 810 zones are designed by stakeholders, nearly all of which consist of low-distortion projections (LDPs), as discussed in the next item. The total number of zones in some of the multiple-zone layers may decrease to some extent before SPCS2022 is finalized, but it will not increase.

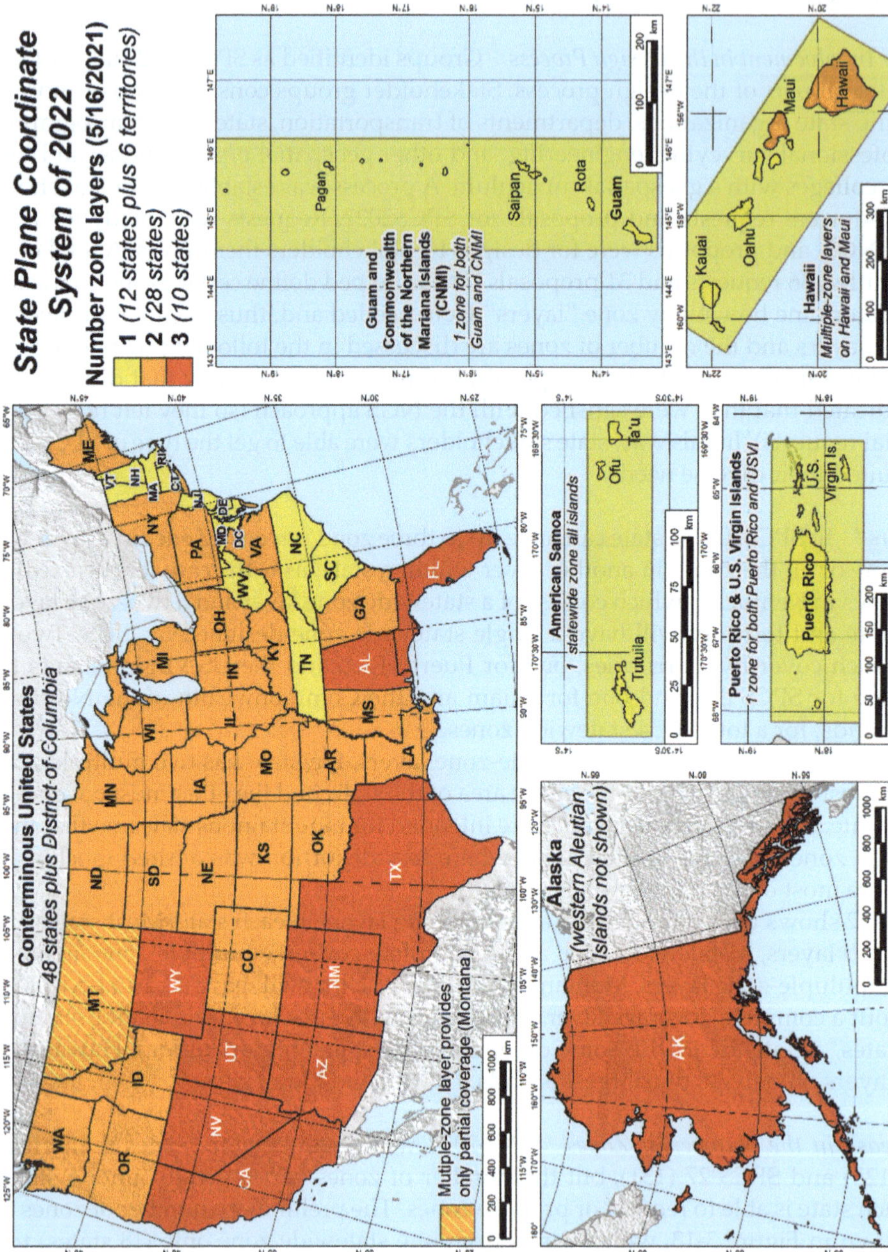

Figure 3-12. Number of zone layers for the State Plane Coordinate System of 2022 (preliminary). Source: Courtesy of NOAA, reproduced with permission.

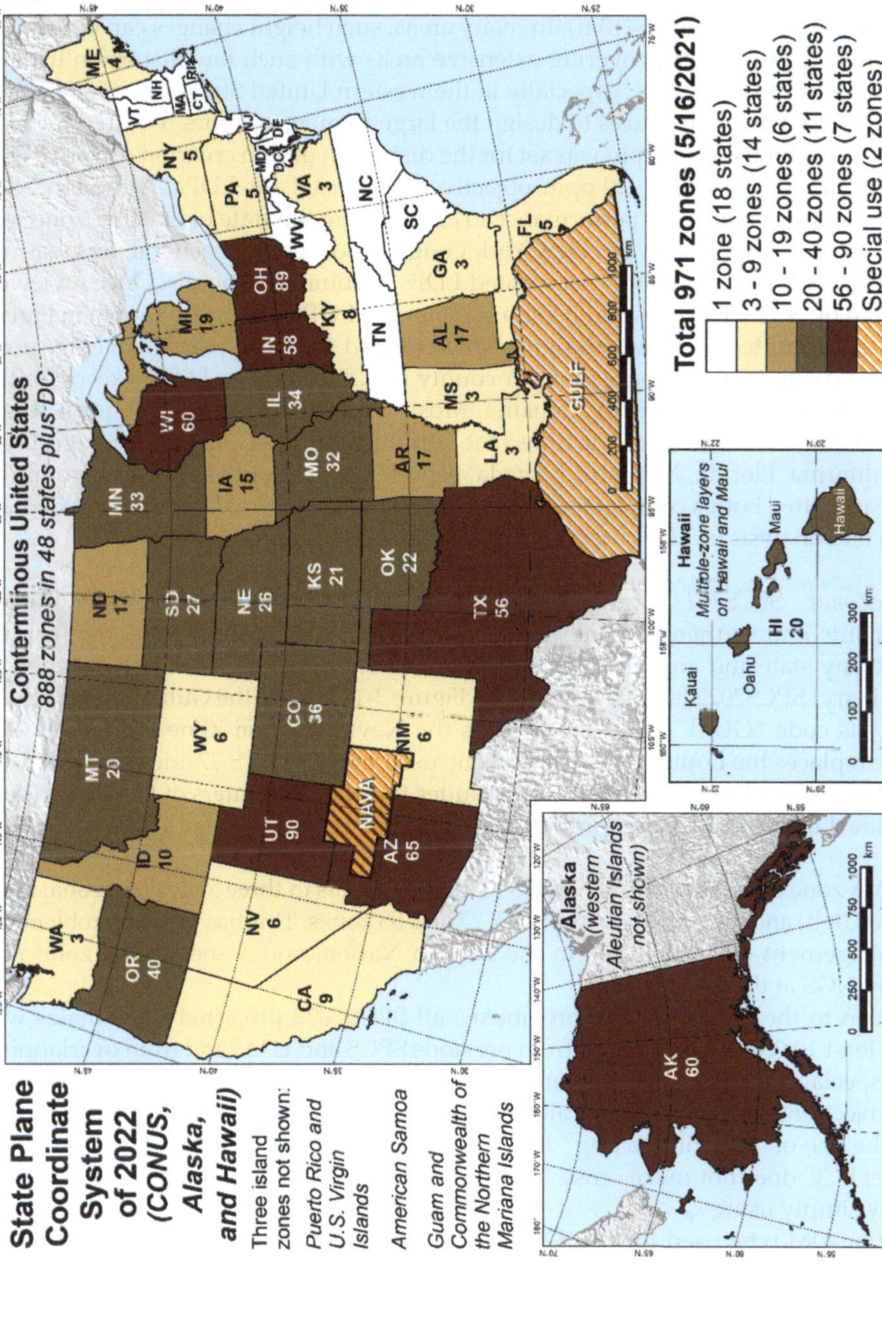

Figure 3-13. Number of zones in each state for the State Plane Coordinate System of 2022 (preliminary). Source: Courtesy of NOAA, reproduced with permission.

Low-Distortion Projections Designed by Stakeholders *Low-distortion projection* (LDP) coordinate systems are designed to achieve distortion so low that projected grid distances are nearly equal to actual horizontal ground distances. A typical design objective is ±20 ppm (±0.1 ft/mi) for an area of interest. As shown in Table 3-3, this level of distortion corresponds to a maximum zone width of 114 km (71 mi). But this only applies to areas with no topographic relief, which is usually not the case. Table 3-4 shows that ±20 ppm also occurs for height changes of ±125 m (±410 ft). In many areas, such height changes can occur over short distances. Consequently, covering extensive areas with such low distortion usually requires a large number of zones, especially in the western United States.

NGS does not have the resources to design the large number of zones required for LDP coverage. A lower limit of ±50 ppm was set for the distortion design criterion used by NGS, substantially higher than the ±20 ppm objective often used for LDPs. Although NGS designed some zones with LDP performance in cases where the state or existing zone was small (e.g., Rhode Island, Delaware, New York Long Island Zone), in general, this was not done. Instead, NGS allowed states that wanted LDPs to submit proposals. Once reviewed and approved by NGS, the states could submit their own LDP designs. As shown in Figure 3-14, 28 states submitted their own designs, which ranged from one partial-coverage zone (Florida and Wyoming) to 88 zones (one per county for Ohio). All are LDPs, except for the zone in Florida and three zones in Alabama. Most state-submitted designs for a single complete-coverage layer, except for nine that submitted only a partial-coverage layer (Alaska, California, Florida, Montana, Nevada, New Mexico, Texas, Utah, and Wyoming). Two states submitted both a complete- and a partial-coverage layer (Alabama and Arizona). The design and analysis of LDPs are covered in detail later in this chapter.

Special-Use Zones SPCS2022 allows the establishment of zones for well-defined geographic regions that are in more than one state. Because they are not in a single state, they cannot "belong" to any state and are called *special-use* zones. Two of these zones are included in the (preliminary) SPCS2022 layout, as shown in Figure 3-13. One is the Gulf of Mexico Zone (labeled by its code "GULF"), and the other is the Navajo Nation Zone ("NAVA"). The GULF zone replaces the Louisiana Offshore Zone used in both SPCS 27 and 83. It has been reclassified as a special-use zone because it includes the offshore waters of five states (Alabama, Florida, Louisiana, Mississippi, and Texas) and most of its coverage area is not within any state.

The NAVA zone covers the entire Navajo Nation, which is in three states (Arizona, New Mexico, and Utah) and is covered by five SPCS 27 and 83 zones. This has proven problematic for the management of spatial data in the Navajo Nation, and a special-use zone was designed by NGS at their request.

In addition to the changes listed previously, all SPCS2022 projected coordinates will differ by at least 10,000 m horizontally from previous SPCS and UTM and from overlapping layers and special-use zones within a state. Another change worth noting is the type of LCC and OM projections. Only the one-parallel version of the LCC can be used in SPCS2022 to allow for the use of "nonintersecting" LCCs (i.e., with a central parallel $k_0 > 1$) where a two-parallel LCC does not make sense. The performance of two-parallel (secant) LCCs is achieved by simply using $k_0 < 1$.

The Hotine OM type used for SPCS2022 is the "center" (or "local") version. It differs from the "natural" version used for SPCS 27 and 83 only by where the grid origin is defined (false northing and easting). For SPCS2022, the grid origin is defined at the local origin, which usually falls within the zone. This is a more intuitively obvious definition than the natural OM, which places the grid origin where the skew axis intersects the aposphere equator. This location does not correspond to the equator of the reference ellipsoid and is only of academic interest in deriving the projection.

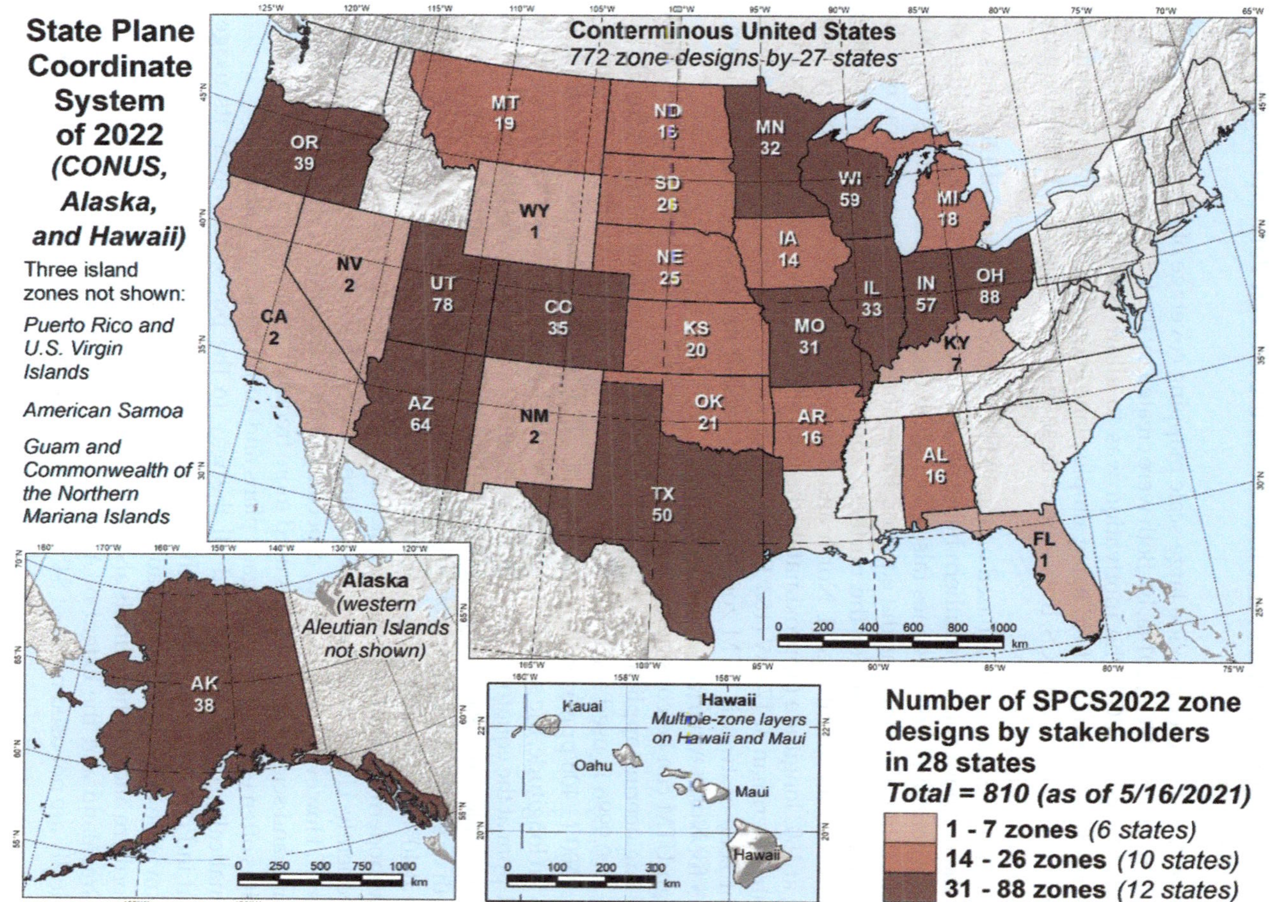

Figure 3-14. Number of zones designed by states for the State Plane Coordinate System of 2022 (preliminary). Source: Courtesy of NOAA, reproduced with permission.

Details about the characteristics of SPCS2022 are given in the SPCS2022 Policy and Procedures (NGS 2019a, b). The policy allows for changes to the system after its official rollout, and so even the "final" layout and designs adopted by NGS may change in the future.

Universal Grids: Universal Transverse Mercator and Universal Polar Stereographic Coordinate Systems

Two other PCSs together provide large-scale mapping coverage of the entire Earth: The UTM and UPS systems (NGA 2014a, b). Both were created by the US Department of Defense in the 1940s. As with SPCS, the motivation was to allow computations of positions, distances, and directions in the field using simple plane geometry rather than complicated geodetic methods. Also, like SPCS, the usage of these systems (especially UTM) has grown well beyond their original intent.

As the name indicates, UTM is based on the TM projection. It consists of 120 zones, with 60 in both the northern and the southern hemispheres. Each zone is 6° wide (3° either side of its central meridian). For all zones false easting (on the central meridian) is 500,000 m, and false northing is defined at the equator. False northing where the central meridian intersects the equator is zero for the northern zones and 10,000,000 m for the southern zones. The zones are numbered 1 North and 1 South through 60 North and 60 South beginning at 180° longitude (approximately the International Date Line). Zones 1N and 1S have a central meridian of 177°W (183°E), and the central meridian increases by 6° for each subsequent zone eastward. UTM zones and their central meridian values are shown in Figure 3-15 for North America.

Although UTM can be considered a large-scale map projection nominally suitable for surveying engineering applications, the scale error range is greater than that for SPCS. A value of $k_0 = 0.9996$ is used for the central meridian, corresponding to a maximum scale error of 1:2500 (±400 ppm) with respect to the ellipsoid. Because the zone width is a constant 6° in longitude, the scale error at the edges of the zone varies with latitude, ranging from +981 ppm at the equator, to +407 ppm at 40° latitude, and to −359 ppm at 80° latitude (note that the scale error is entirely negative within a UTM zone at latitudes beyond about 57°20′ north or south). The corresponding linear width of a zone is 666 km at the equator, to 512 km at 40° latitude, and 116 km at 80° latitude.

In many modern civilian surveying applications, UTM scale error is considered excessive, and consequently, UTM is used much less often than SPCS for engineering projects in the United States. However, it still is used, especially by military organizations, as well as GIS professionals and scientists. UTM was used more frequently for civilian engineering in the past, and it is often encountered in legacy plans and corresponding datasets.

Mathematically, UTM can be used all the way to the poles. However, its usage is impractical in Polar Regions because the zones become very narrow (e.g., a zone is only about 12 km wide at 89° latitude). Because of this, UTM is normally not used in such cases, with the northern and southern limits also based in part on geography (such as the northern coast of Greenland and the extent of Antarctica). Based on these limits, northern UTM zones typically are not used north of 84°N latitude and southern UTM zones south of 80°S latitude.

The UPS system is used in place of UTM to provide coverage of the Polar Regions with one zone for the North Pole and one zone for the South Pole. As with the UTM, these zones can be used well beyond the polar areas, but typically their usage is restricted to the region north of 84°N latitude for North UPS and south of 80°S latitude for South UPS.

To balance negative and positive scale errors over such large regions, a value of $k_0 = 0.994$ (smaller than for UTM) is used for the projection origin (at the poles). This corresponds to

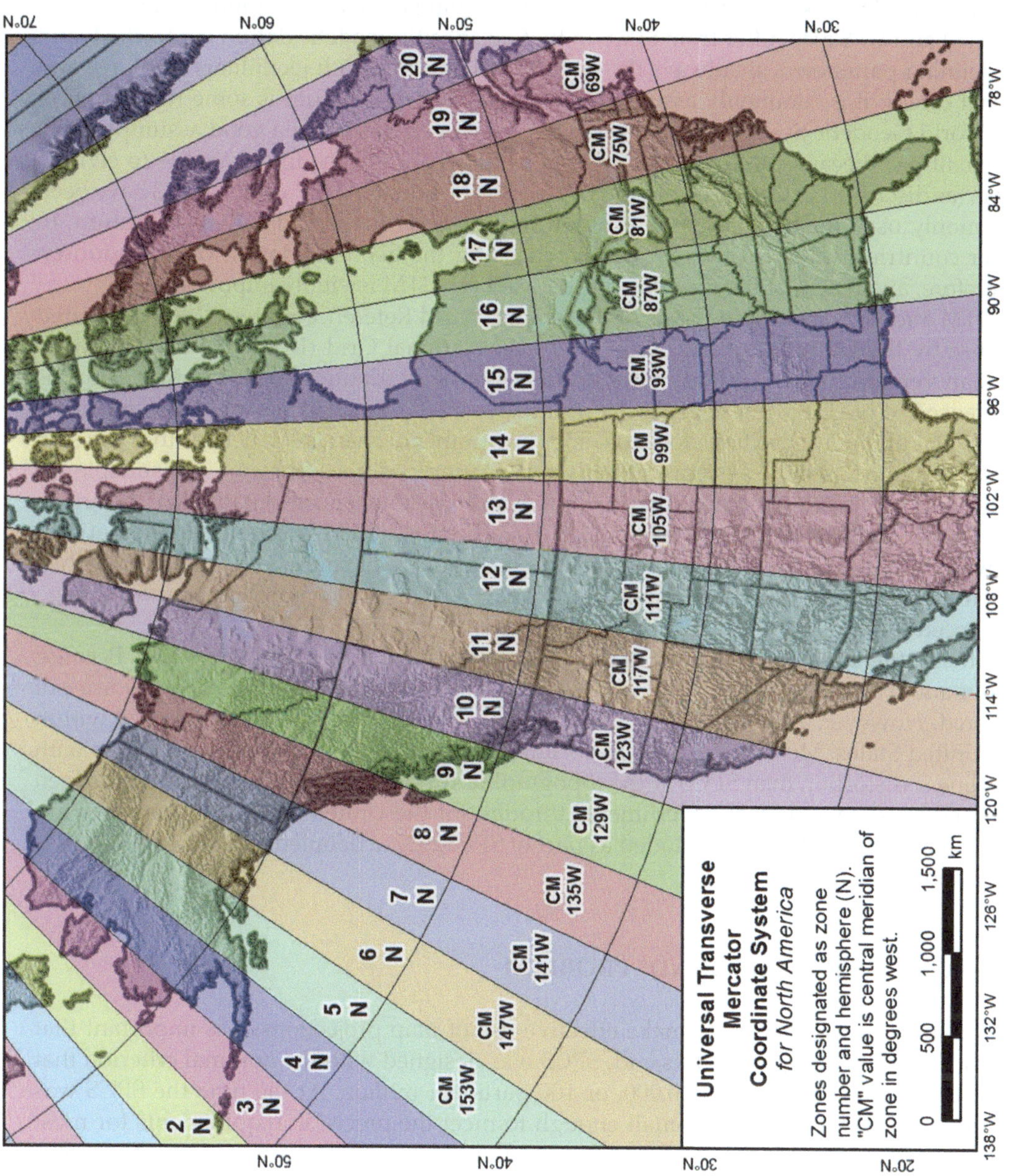

Figure 3-15. Zones of the Universal Transverse Mercator coordinate system in North America.

a negative scale error of 1:167 (−6000 ppm) at the poles for both zones. The scale error becomes less negative with increasing distance from the poles, reaching a value of zero ("true scale" with respect to the ellipsoid) at 81°06′52.3″ north and south latitudes. At the southern limit of North UPS (84°N latitude), the scale error is negative 1:306 (−3270 ppm). The northern limit South UPS is far enough from the pole (80°S latitude) that the scale error is positive 1:622 (+1608 ppm). Such large linear distortion over a majority of the area covered by UPS makes it unsuitable for most surveying engineering applications.

UTM and UPS also differ from SPCS in that no specific geodetic datum is specified, and its defining parameters are always the same for all datums. With its military pedigree and global extent, it is commonly assumed that the appropriate datum is some realization of the World Geodetic System of 1984 (WGS 84). However, this is not a good assumption, in no small part because accurate WGS 84 coordinates are not readily available to civilian users (as discussed in Chapter 4). In the United States, both NAD 27 and NAD 83 have been commonly used for UTM, as well as other datums such as the Old Hawaiian Datum. In other countries, UTM is often implemented using the geodetic datum of that country. Therefore, awareness of the geodetic datum used for UTM is vitally important.

UTM and UPS are also the basis of the Military Grid Reference System (MGRS) created for use by the US military and NATO. The US National Grid (USNG) is a UTM-based civilian version for the United States and is essentially identical with the MGRS but is referenced to NAD 83. These grid systems are designed to provide a single alphanumeric string that uniquely specifies a location, and the number of characters dictates the precision of the position. As these systems are only for locating positions, they are not suitable for surveying applications (because they are not readily useable for computations of distances, directions, areas, etc.). They are included here because UTM positions are sometimes provided in MGRS/USNG format (which can be a source of confusion) and because they are used in other geospatial applications such as GIS and emergency services (e.g., Enhanced 911).

Other large-scale PCSs are used in the United States and throughout the world. Because this manual is intended for a US audience, systems outside the United States are not covered. However, a large number of PCSs in addition to SPCS and UTM are used within the United States. Most of these are LDPs intended to provide projected coordinates with less linear distortion than SPCS at the topographic surface. Many LDP systems have been adopted by state and local governments throughout the United States. Because they are commonly used and will be included as part of SPCS2022, a detailed discussion of LDPs is provided in the next section.

THE "GRID VERSUS GROUND" PROBLEM

For surveying engineers to make effective use of map projections, it is important that linear distortion be small. To this end, SPCS was designed with the nominal criterion that scale error does not exceed 1:10,000, or 100 parts per million. At the time the SPCS was created, 1:10,000 was deemed small enough to meet the precision requirements for most survey work. However, the SPCS design criterion considered distortion only with respect to the reference ellipsoid. As shown previously, the actual distortion is usually greater at the topographic surface. In engineering and surveying applications, it is often important to represent the actual horizontal distance on the ground. However, projected (grid) distances do not equal ground distances, and in most cases, the grid distances are shorter. This difference is frequently referred to as the "grid versus ground" problem.

Although the term "ground distance" is often used by surveyors, there is no universal agreement on its definition. In this chapter, *horizontal ground distance* is defined, as shown in Figure 3-7, as the (curved) distance parallel to the ellipsoid at the mean ellipsoid height of the endpoints. It can be computed as the ellipsoid distance (geodesic) multiplied by a scale factor to bring it to the topographic surface (ellipsoid distances are distances on the ellipsoid surface and were discussed in Chapter 2). The scale factor can be computed as $1+\bar{h}/\bar{R}_G$, where $\bar{h}$ is the mean ellipsoid height of the endpoints and $\bar{R}_G$ is the geometric mean radius of curvature at the mean latitude of the endpoints. An alternative to $\bar{R}_G$ is to instead use the radius of curvature in the azimuth at the midpoint of the line between the endpoints, $\bar{R}_\alpha$. Using $\bar{R}_\alpha$ can give somewhat more accurate results because, in general, it better represents the radius of curvature along the path between the endpoints than does the geometric mean. It is calculated at the midpoint as

$$\bar{R}_\alpha = \frac{MN}{M\sin^2\alpha + N\cos^2\alpha} \tag{3-5}$$

where

α = Azimuth,
M = Meridian radius of curvature (north–south), and
N = Prime vertical radius of curvature (east–west).

M and N are defined in Chapter 2 in Equations (2-8) and (2-9), respectively. Note that $R_G = \sqrt{MN}$.

An alternative to scaling the ellipsoidal distance is to use the geocentric Cartesian coordinates to compute a ground distance as

$$D_g = \sqrt{\Delta X^2 + \Delta Y^2 + \Delta Z^2 - \Delta h^2} \tag{3-6}$$

where ΔX, ΔY, and ΔZ are the change in geocentric coordinates, and Δh is the change in ellipsoid height at the endpoints. Equation (3-6) does not account for curvature, but for distances of less than about 30 km (20 mi), the increase resulting from curvature is less than 1 ppm. To account for curvature, D_g can be multiplied by the following curvature correction factor:

$$C_C = \frac{2\bar{R}_G \sin^{-1}\left(\frac{D_g}{2\bar{R}_G}\right)}{D_g} \tag{3-7}$$

where all variables are as defined previously. When Equation (3-6) is corrected using Equation (3-7), the value for D_g agrees with that based on scaled ellipsoidal distance to within 2 mm for distances up to 200 km (i.e., within 0.01 ppm).

Perhaps the most common method for computing ground distances is to scale projected (grid) distances, rather than use the geodetic approaches mentioned previously. Scaling "to ground" is normally accomplished by dividing the grid distance by the combined scale factor [$\delta + 1$ according to Equation (3-2)]. However, using the combined scale factor with projected distances requires computing distortion of the line, as described at the end of the previous section. If the ground distance for projected line AB is computed using the mean h of the endpoints, R_G of the midpoint, and Equation (3-4) for k_{AB}, the accuracy is about 0.01 m for a 100-km line (0.1 ppm). If the mean k values of the line endpoints are used instead of Equation (3-4), the accuracy reduces to nearly the same as that of Equation (3-6) without curvature correction, to about 3 cm for a 30 km line (1 ppm).

Approximate accuracies of various ground distance calculation methods are summarized in Table 3-5. The error of each method is given as calculated distance minus the Vincenty ellipsoid distance (described in Chapter 2); comparisons were computed at 40° latitude and azimuth of about 40°. Although the computations were done on the ellipsoid (i.e., with $h = 0$), the accuracy estimates (at the significance given) are applicable to both ground and ellipsoid distances.

The method in the rightmost column of Table 3-5, local geodetic horizon, is discussed later in this chapter. All other methods are accurate to within 1 mm for distances of less than 10 km (about 6 mi). For distances greater than 10 km, the concept of "horizontal ground distance" is not clear, except in regions of very mild topographic relief. Even in such regions, there are likely no practical surveying or engineering use cases for "ground" distances of more than 50 km (but for completeness, Table 3-5 includes distances of up to

Table 3-5. Approximate Accuracies of Various Ground and Ellipsoid Distance Calculation Methods.

	Ground distance error as calculated value minus ellipsoid distance (m and ppm)						
Distance (km)		Scaled grid distance using mean *k* at endpoints	Scaled grid distance using integrated *k* [Equation (3-4)]	Delta geocentric without curvature correction [Equation (3-6)]	Delta geocentric with curvature correction [Equations (3-6) and (3-7)]	Local geodetic horizon [Equation (3-8)]	
10	m	−0.001	0.0000	−0.001	0.0000	−0.004	m
	ppm	−0.1	0.000	−0.1	0.000	−0.4	ppm
20	m	−0.01	0.0000	−0.01	0.0000	−0.03	m
	ppm	−0.4	0.000	−0.4	0.000	−1.5	ppm
30	m	−0.03	−0.0001	−0.03	0.0000	−0.12	m
	ppm	−1	−0.003	−1	0.000	−4	ppm
50	m	−0.12	−0.0006	−0.13	0.0000	−0.5	m
	ppm	−2.4	−0.012	−2.6	0.000	−10	ppm
100	m	−0.8	−0.01	−1.1	0.0001	−4.5	m
	ppm	−8	−0.1	−11	0.001	−45	ppm
200	m	−6	−0.14	−8	0.0012	−30	m
	ppm	−30	−0.7	−40	0.006	−150	ppm
500	m	−110	−6	−130	0.1	−550	m
	ppm	−220	−12	−260	0.2	−1100	ppm

500 km). For distances between 10 and 50 km, either grid distance scaling using Equation (3-4) or curvature-corrected delta geocentric Cartesian ($\Delta X, \Delta Y, \Delta Z$) coordinates [Equations (3-6) and (3-7)] should be used. Only curvature-corrected ($\Delta X, \Delta Y, \Delta Z$) coordinates should be used for distances of 100 to 200 km, and above 200 km, none of the methods in Table 3-5 should be used. Instead, Vincenty ellipsoidal distances (Chapter 2) should be used (either on the ellipsoid or scaled to ground, as appropriate).

There are two fundamental ways to deal with the grid versus ground problem. One is to provide the actual (computed) ground distance per the preceding discussion. When this approach is used, the ground distances are often provided implicitly by giving the projected (grid) distances (such as SPCS distances) along with linear distortion values usually as the combined scale factor $\delta + 1$ [Equation (3-2)]. This can be done as a single $\delta + 1$ value for a project if it is small, for parts of larger projects, or for individual distances. The grid distances are divided by $\delta + 1$ to determine their corresponding ground distances within the limitations given in Table 3-5.

The other way to handle the "grid versus ground" problem is to minimize PCS linear distortion to a level where it is negligible (with respect to the accuracy of the product). When this approach is used, it is sometimes referred to as providing "ground coordinates." But it is important to realize that there is no such thing as "ground coordinates." Projected coordinates are always with respect to the developable surface; they can never be "at ground." Instead, the linear distortion with respect to the topographic surface is reduced such that grid distances are effectively the same as ground distances, within a specific tolerance and area of interest.

Methods for Reducing Map Projection Distortion

Minimizing linear distortion only makes sense for conformal map projections. With this fact in mind, there is essentially only one way to minimize linear distortion, and that is to place the conformal projection developable surface at the topographic surface. Three common methods for reducing linear distortion are discussed subsequently.

Scale an Existing Map Projection to the Topographic Surface. Soon after creation of the SPCS, surveyors and engineers began applying a scale factor to SPCS coordinates to bring "grid to ground," so that the projected distances were (nearly) equal to the actual horizontal distance at the topographic surface. This approach has existed for several decades and is often called "modified" or "scaled" State Plane, or State Plane "at ground," because it is usually applied to SPCS coordinates (although it is sometimes used with UTM and other conformal PCSs). It was adopted largely because the projection calculations themselves are complex and difficult. Prior to widely available personal computers, it simply was not practical to create custom projections. So, scaling existing projected coordinates was the only viable approach in most cases.

Scaling "to ground" is accomplished simply by dividing all coordinates by a single $\delta + 1$ value (i.e., combined factor) for an area. Equation (3-2) is used to compute δ by selecting a location (which fixes the values of k and R_G) and choosing an appropriate representative ellipsoid height (h_0) for the area of interest. A slight refinement involves computing $\delta + 1$ at several locations over an area and taking the mean as the value for coordinate scaling.

Consider a project in SPCS coordinates scaled to ground using a single point. Usually, the point is selected near the center of a project area. In this example, the point is at 40°N latitude, so $R_G = 6{,}374{,}384$ m, and a value of $k = 0.999\,930\,767$ is computed from the SPCS projection equations. For a representative project area ellipsoid height of $h_0 = 1600$ m (5250 ft), the combined scale factor is computed using Equation (3-2) as $\delta + 1 = 0.999\,679\,843$. All SPCS coordinates for the project can then be divided by that value so that the distance between scaled coordinate pairs will be close to the horizontal ground distances. For example, a distance of 1000.00 m (3,280.84 ft) based on SPCS coordinates becomes a scaled ("ground") distance of 1000.32 m (3,281.89 ft)—a difference of 0.32 m (1.05 ft). Whether this adequately represents a horizontal distance on the ground depends on both the actual topographic height of the points and where they are located with respect to the point used to compute $\delta + 1$. A more detailed discussion on the combined effects that both height and location have on distortion is given in the following section on LDPs.

Despite the popularity of "modified" State Plane for reducing map projection distortion, it has the following liabilities:

- Generates coordinates with values similar to "true" SPCS, which can cause confusion. However, this problem can be eliminated by translating grid coordinates to get different (usually smaller) values distinct from SPCS.
- More difficult to implement in GIS and other geospatial software because of the scaling process creating an incomplete or ambiguously defined coordinate system. Scaling essentially defines a new projection, and its parameters can be back-calculated from the scaled values. The resulting projection definition can then be used to georeference the product in geospatial software. However, a projection can only be determined if the coordinates of the point about which scaling is performed are provided [usually scaling is about the grid origin (0, 0) but not always]. In addition, a back-calculated projection, in general, has "messy" parameters (i.e., with nonterminating trailing decimal places), which can be problematic for portability and implementation, especially if repeated projections and deprojections are performed.
- Does *not* reduce convergence angles (they are the same as the original SPCS). This means that the arc-to-chord correction is also the same as SPCS, which can be large if the design area is distant from the projection axis.
- Most important, it usually does not minimize distortion over as large an area as other methods. Applying a single combined scale factor is limited both by the size of the area and by its location with respect to the axis of the projection being scaled. In general, the extent of low-distortion coverage decreases as distance increases from the projection axis. In most cases, the SPCS axis does not pass through the project area and in addition may be oriented in a direction that decreases the area of low distortion coverage.

Modify the Reference Ellipsoid to Place Its Surface near the Topographic Surface. Once personal computers became readily available, creating and using custom coordinate system definitions became practical. One such custom approach for minimizing linear distortion is to modify the reference ellipsoid so that its surface is near the topographic surface and then define a tangent (or slightly secant) conformal map projection so that it is also near the topographic surface.

This method requires five or six projection parameters based on the projection type used (as listed in Table 3-2) plus two parameters for the new ellipsoid. The typical approach is to place the geodetic origin of the projection near the center of the design area. Because the modified ellipsoid surface is already near the ground surface, the projection axis scale is set to $k_0 = 1$ (tangent) or slightly secant, such as $k_0 = 0.999\,99$ (10 ppm less than 1). Because the ellipsoid has been scaled to the topographic surface, the ellipsoid heights with respect to the scaled ellipsoid are near zero. Consequently, the height ("elevation") factor part of the distortion is near 1 and, thus, $\delta + 1 \approx k$.

There are different ways to modify the ellipsoid. One is to add a value of h_0 for the area of interest to the ellipsoid semimajor axis, a (or by using h_0 to scale a), while holding the flattening constant. This gives a new semiminor axis of $b' = (a + h_0)(1 - f)$. An alternate approach is to add h_0 to both axes, which gives a new flattening $f' = (a - b + 2\,h_0)/(a + h_0)$. For positive h_0 (the typical case), this decreases the flattening. The significance is that such changes to the ellipsoid also change the latitude of a point. As an example, consider $h_0 = 1600$ m used in the previous "modified" SPCS example. For the GRS 80 ellipsoid, adding 1600 m to a with a constant f causes a maximum change in latitude of 0.17373″ (5.363 m) at 45 degrees latitude. In contrast, adding 1600 m to both a and b causes a maximum change in the latitude of 0.00015″ (0.005 m) at 22.5° and 67.5° latitude and nearly zero (<0.02 mm) at 45° latitude (these apply to both north and south latitudes).

With either approach, the modified ellipsoid parameters are used in the mapping equations to generate projected coordinates. This is an important point, because the change to the ellipsoid is essentially a datum transformation. So, to use a PCS based on a modified ellipsoid requires not only that the new ellipsoid parameters be in the projection equations, but also that the geodetic coordinates (latitude, longitude, and ellipsoid height) of a point being projected first be transformed to the new ellipsoid. The new geodetic coordinates will have ellipsoid heights near zero, and so when they are projected, their distortion will be low (mostly owing to k alone). Datum transformations are described in Chapter 4, and for this case, the transformation consists of computing the geocentric Cartesian (X, Y, Z) coordinates of the points and then computing new transformed latitude and ellipsoid height (φ' and h') using the modified ellipsoid parameters (changing the ellipsoid parameters has no effect on longitude). Consider the same example point at $\varphi = 40°00'0.00000''$N and $h = 1600.000$ m with 1600 m added to a and holding the flattening constant. Its transformed coordinates are $\varphi' = 40°00'00.17119''$N and $h' = 2.214$ m. Thus, the height component of its linear distortion is nearly zero: $R_G/(R_G + h) - 1 = 6{,}374{,}384/(6{,}374{,}384 + 2.214) - 1 = 0.999\,999\,653 - 1 = -0.347$ ppm. If the method of adding 1600 m to both a and b is used instead, then $\varphi' = 40°00'00.00005''$ and $h' = 0.002$, and the linear distortion caused by height is -0.0003 ppm.

This method of reducing map projection distortion has been fairly widely used. However, there are complications and liabilities with this approach that make it problematic:

- In addition to a custom projection definition, it requires a new ellipsoid for every PCS. Therefore, five or six projection parameters plus two ellipsoid parameters are required for a total of seven or eight parameters to define each system.
- Using a modified ellipsoid means that coordinates must be transformed to the new ellipsoidal system prior to being projected. Therefore, the projection algorithm must include a datum transformation, and this can make these systems more difficult to implement.
- Transformed latitudes of points can differ substantially from the original values. The maximum change occurs for the method where the flattening is held constant, and the change in latitude can exceed a meter for topographic heights greater than 300 m. Although this is not a problem in and of itself, it can cause confusion over which latitude values are correct. It can also cause an implementation problem resulting in incorrect projected coordinates if the original geodetic coordinates are not transformed prior to projecting.
- Modifying the ellipsoid by changing the semimajor and semiminor axes by the same amount results in a much smaller change in latitude. However, this causes the ellipsoid flattening to change, and the resulting projected coordinates will be conformal to the modified ellipsoid rather than the original reference ellipsoid. This departure from conformality is small at the topographic heights used for LDPs (typically <3,000 m). But it is nonetheless problematic, because the intent of a conformal PCS is to be conformal with respect to the reference ellipsoid that is used as part of the reference frame definition not some other arbitrary ellipsoid.

- Modifying an ellipsoid by changing the length of its axes (regardless of whether flattening is held constant) by itself cannot be used to reduce linear distortion in most situations where there is a systematic change in ellipsoid height in the design area (i.e., areas with an overall nonzero topographic slope). This severely limits the applicability of this method to optimally minimize linear distortion in situations frequently encountered in LDP design (including the one used as an example later in this chapter).

As mentioned previously in this chapter, a modified ellipsoid was used to reduce linear distortion for the three Michigan SPCS 27 zones. These new definitions were adopted in 1964, along with a change from TM to LCC for all three zones. In this case, a was scaled while holding flattening constant. It is likely that this approach was used to ensure that the resulting projections were also conformal with respect to the unscaled Clarke 1866 ellipsoid.

Beginning in the 1980s, Wisconsin and Minnesota designed NAD 83–based county LDP coordinate systems using various modified reference ellipsoid approaches. Unlike what was done in Michigan for SPCS 27, in Wisconsin, the length of both ellipsoid axes was increased by the same amount. This minimized the change in latitude, but the change in ellipsoid flattening made the projections (slightly) nonconformal with respect the GRS 80 ellipsoid. This problem was unknown to the designers; the larger difficulty was getting the system to work in commercial software packages. This became such a problem that Wisconsin abandoned their modified ellipsoid approach and redesigned all zones with respect to GRS 80 by changing only the projection parameters; that same system will be used for SPCS2022. Minnesota also had difficulty with usage of their system, and it has been completely redesigned for SPCS2022 based on GRS 80 and with a much smaller number of zones.

In summary, the modified reference ellipsoid approach to designing LDPs is more complex and difficult to implement and yet does not result in improved performance (i.e., lower distortion). In fact, in cases where topography is systematically sloped, modifying the ellipsoid will not achieve distortion as low as the LDP design approach described next. This limitation can be overcome by translating the ellipsoid in addition to changing its size and/or shape. However, this is an even more complex approach that would probably result in greater departures in conformality with respect to the reference ellipsoid.

Because of the problems with using a modified reference ellipsoid for LDPs, this approach is not recommended. It is also not permitted for design of LDPs used in SPCS2022, according to NGS policy (NGS 2019a).

Define a Projection Developable Surface near the Topographic Surface. Rather than modifying the reference ellipsoid and then defining a tangent (or near tangent) projection, the projection developable surface can instead be placed near the topographic surface. The original reference ellipsoid is used, and so no additional ellipsoid parameters (and corresponding geodetic transformation of points) are required, and the projection remains conformal with respect to the ellipsoid used for the reference frame (datum). Defining such a projection is done in a manner similar to the two other methods, by using a representative ellipsoid height, h_0. In this case, the projection axis scale is computed as $k_0 = 1 + h_0/R_G$. The design point from the previous examples with $\varphi = 40°\text{N}$ and $h_0 = 1600\ \text{m}$ gives $k_0 = 1 + 1600/6{,}374{,}384 = 1.000\,251$. This k_0 value becomes one of the defining parameters of the projection with respect to a standard reference ellipsoid, such as GRS 80 (used for SPCS 83 and SPCS2022).

As with the modified ellipsoid approach, designing a custom projection requires that five or six projection parameters be defined, as listed in Table 3-2: projection axis scale, geodetic origin, and grid origin (and projection axis orientation for the OM). The type of projection must also be selected, and it is not always obvious what type will result in the lowest linear distortion for a given design area. In addition to the projection type, projection axis location (and orientation for an OM projection) also effect distortion. All these considerations for LDP design are discussed in the next section and closely follow the development given by Dennis (2016).

Low-Distortion Projection Coordinate Systems

Understanding Linear Distortion Behavior. What exactly happens when a PCS is developed that is intended to reduce linear distortion? To help answer this question a hypothetical situation is presented, which will be followed with an example of a similar actual LDP design. Consider a project site where low-distortion projected coordinates are desired. As discussed previously, a common method is to scale an existing projection (usually SPCS) "to ground" to create a "modified" State Plane system. Although this approach can work well for certain applications, it has several drawbacks (as listed previously). The most important is that the projection type and its axis location cannot be changed. This can severely reduce the extent of the area where low distortion can be achieved. Performance can be significantly compromised when the projection axis is distant from the design area and the projection axis orientation or topographic slope is unfavorable.

Figure 3-16 illustrates this situation for scaling an existing LCC projection and compares it with an LCC LDP design. Figure 3-16(a) is a secant LCC as used for existing SPCS 83 and SPCS 27, with an implicitly defined central parallel based on the spacing between north and south standard parallels (where the conic developable surface intersects the ellipsoid). Figure 3-16(b) is the same LCC scaled to ground at the indicated design location. Scaling to ground is performed by dividing the coordinates by a selected single value of $\delta + 1$ (i.e., the combined factor). Where the developable surface intersects topography (at the design location in Figure 3-16), linear distortion is zero. But elsewhere, the distortion is not optimally minimized. This is particularly the case for the situation illustrated, with the topography, in general, sloping down to the north. Figure 3-16(c) shows an LDP design that takes advantage of the topographic slope to increase the area of low-distortion coverage. This is accomplished by moving the projection axis (central parallel) north and adjusting its scale until the projection surface coincides with topography. The same process can be used with other conformal projections, depending on the direction of the overall topographic slope (e.g., a TM can be used for areas of east–west slope and an OM for areas that slope in an oblique direction).

The key point of the situation in Figure 3-16(c) is that linear distortion is always a function of three properties of the projection axis: its scale, location, and orientation. When modified State Plane is used, only one of these properties (scale) is modified, which can significantly limit options for reducing distortion. This limitation with modified State Plane has contributed to the common but mistaken idea that the projection axis scale is the only property that matters in designing an LDP and that the LDP scale is determined solely by a consideration of topographic height (i.e., by using the height or "elevation" factor). Figure 3-16 shows that this is not true, and this point will be further illustrated in the next section by an analogous example of an actual LDP design using the LCC projection.

Figure 3-16 does not explicitly show one of the design methods discussed previously: modifying the ellipsoid so that its surface is at the topographic surface. This method can be visualized by simply considering a larger ellipsoid, concentric with the one shown, with its surface intersecting the design point. A projection can then be specified for Case (b) or (c), or some other projection entirely. Such a projection—based on a modified ellipsoid—will perform in the same way as a scaled projection based on the unaltered reference ellipsoid.

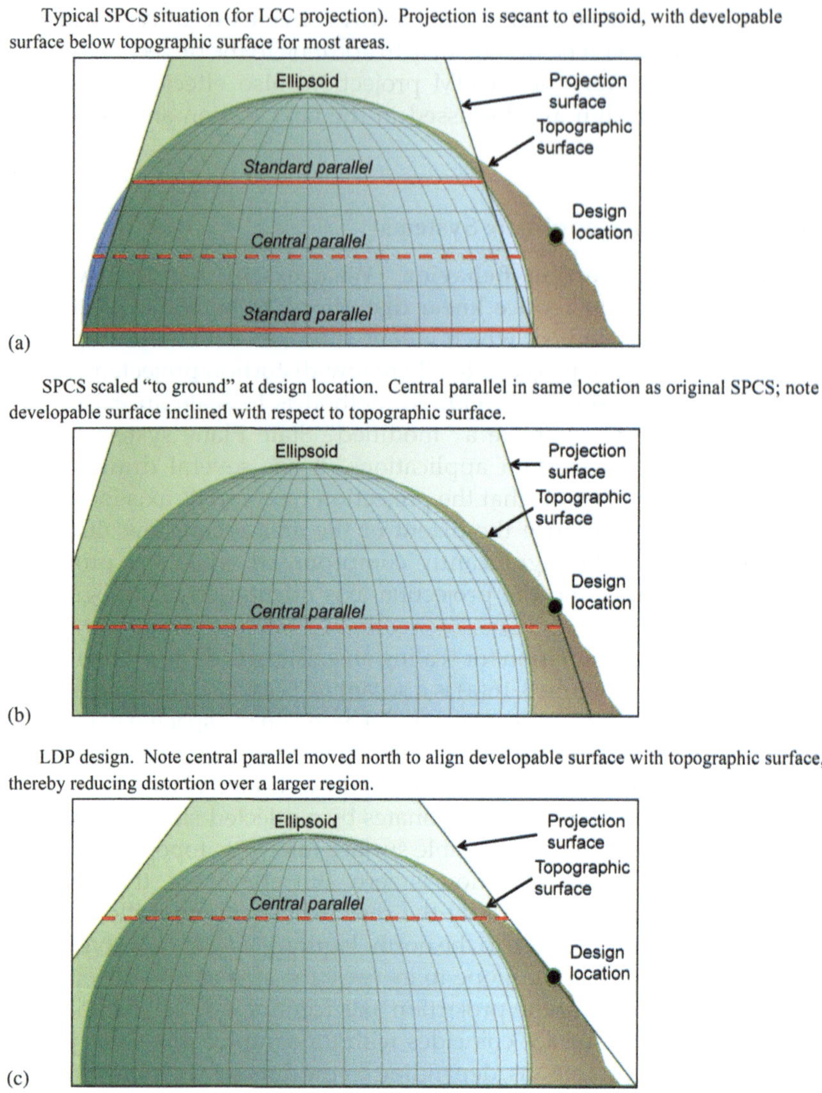

Figure 3-16. Comparison of (a) SPCS, (b) "modified" SPCS, (c) LDP. (a) Typical SPCS situation (for LCC projection). Projection is secant to the ellipsoid, with the developable surface below the topographic surface for most areas. (b) SPCS scaled "to ground" at the design location. The central parallel is in the same location as in the original SPCS; note that the developable surface is inclined with respect to the topographic surface. (c) LDP design. Note that the central parallel is moved north to align the developable surface with the topographic surface, thereby reducing distortion over a larger region.
Source: Dennis (2016), © ASCE.

With regard to performance, the two methods can be considered similar (assuming that the topographic slope is also taken into account when selecting the projection axis location). However, as indicated previously, the scaled ellipsoid approach is more complex, and it can result in a departure from conformality, depending on how the ellipsoid is modified. The use of ellipsoid height for modifying the ellipsoid is also awkward when iteratively changing the projection axis location as part of the design process, as done in Figure 3-16(c) and in the following design example.

Low-Distortion Projection Performance Example. An example of actual LDP performance like that shown schematically in Figure 3-16(c) is provided for the Bend–Redmond–Prineville zone of the Oregon Coordinate Reference System (OCRS), a statewide system of LDPs (this same zone will also be part of SPCS2022). The region of this OCRS zone is shown in Figure 3-17. The overall design objective was ±20 ppm for the region and ±10 ppm within the three largest towns (Bend, Redmond, and Prineville). In this example, distortion is also evaluated in three additional towns (Sisters, Culver, and Madras) to represent performance in the overall region.

NAD 83 ellipsoid heights for an arbitrary location in all six towns are shown in Figure 3-17, and their mean height was used as a "representative" value for an initial LDP design, $h_0 = 871$ m (2860 ft). The design area is about 70 km (approximately 45 mi) north–south and about 60 km (approximately 35 mi) east–west. Based on Table 3-3, distortion can be limited to ±10 ppm for a zone width of 80 km, so it appears that the distortion criterion can be achieved, at least with respect to Earth curvature. The height range is 422 m, that is, ±211 m. Based on the ±4.7 ppm per ±30 m in Table 3-4, this range corresponds to ±33 ppm (not an encouraging observation, considering the design objective of ±20 ppm overall).

Because the design area is somewhat longer north–south than east–west, the TM projection might seem the most appropriate for LDP design. However, topographic height overall decreases from south to north, and such slope tends to favor the LCC projection, as illustrated schematically in Figure 3-16(c). In the actual design for this OCRS zone, both projection types were evaluated. But the LCC provided low distortion over a substantially larger area, and in the interest of brevity, only the LCC is evaluated in this example. For the initial design, the projection axis (central parallel) was placed near the center of the region, at $\varphi_c = 44°20'00''$N. The projection developable surface was brought to the topographic surface by computing an initial projection axis scale as $k_0 = 1 + h_0/R_G$. Using the initial central parallel, φ_c, in Equation (3-3) gives $R_G = 6{,}377{,}600$ m (20,923,900 ft). Thus, the projection axis scale factor is $k_0 = 1.00014$ (rounded to five decimal places).

Equation (3-2) is used to compute total linear distortion at specific points for a given LDP. Distortion values for the initial LDP design are given in the left column of Table 3-6. The distortion for Bend (the largest town in the region) is −18.5 ppm. Despite using the mean topographic height for determining the projection scale, the distortion magnitude exceeds the ±10 ppm criterion for Bend. This can be fixed in Bend by increasing the projection scale by 20 ppm, to $k_0 = 1.00016$, which will change its distortion to +1.5 ppm. However, this will also increase distortion at all other points by 20 ppm, yielding a maximum of +69.9 ppm in Madras, which is much greater than the target maximum of 20 ppm for the overall area.

Changing projection scale has essentially no effect on distortion variability—the range and standard deviation will be about the same regardless of the scale. For a given projection, variability can only be changed by changing the location of the projection axis (and skew axis azimuth for the OM). The result of doing this is shown in Table 3-6 (the axis scale was also changed so that mean distortion was within ±10 ppm for all cases). The range and standard deviation decrease from 68.4 and ±24.6 ppm, respectively, to minimums of 19.5 and ±7.6 ppm for $\varphi_c = 44°45'$N. The design with $\varphi_c = 44°40'$N has slightly greater variability, but it has $\delta < 10$ ppm in Bend, Redmond, and Prineville, and the variability is also less in these towns.

Evaluating distortion values at several discrete points is usually not sufficient for optimizing an LDP design. A more comprehensive evaluation can be done by computing distortion on a regular high-resolution grid. Distortion can then be visualized and analyzed everywhere, as shown in the maps in Figure 3-18 for both the initial and final LDP designs. All areas within ±20 ppm are shaded green, and the zero distortion contour is shown.

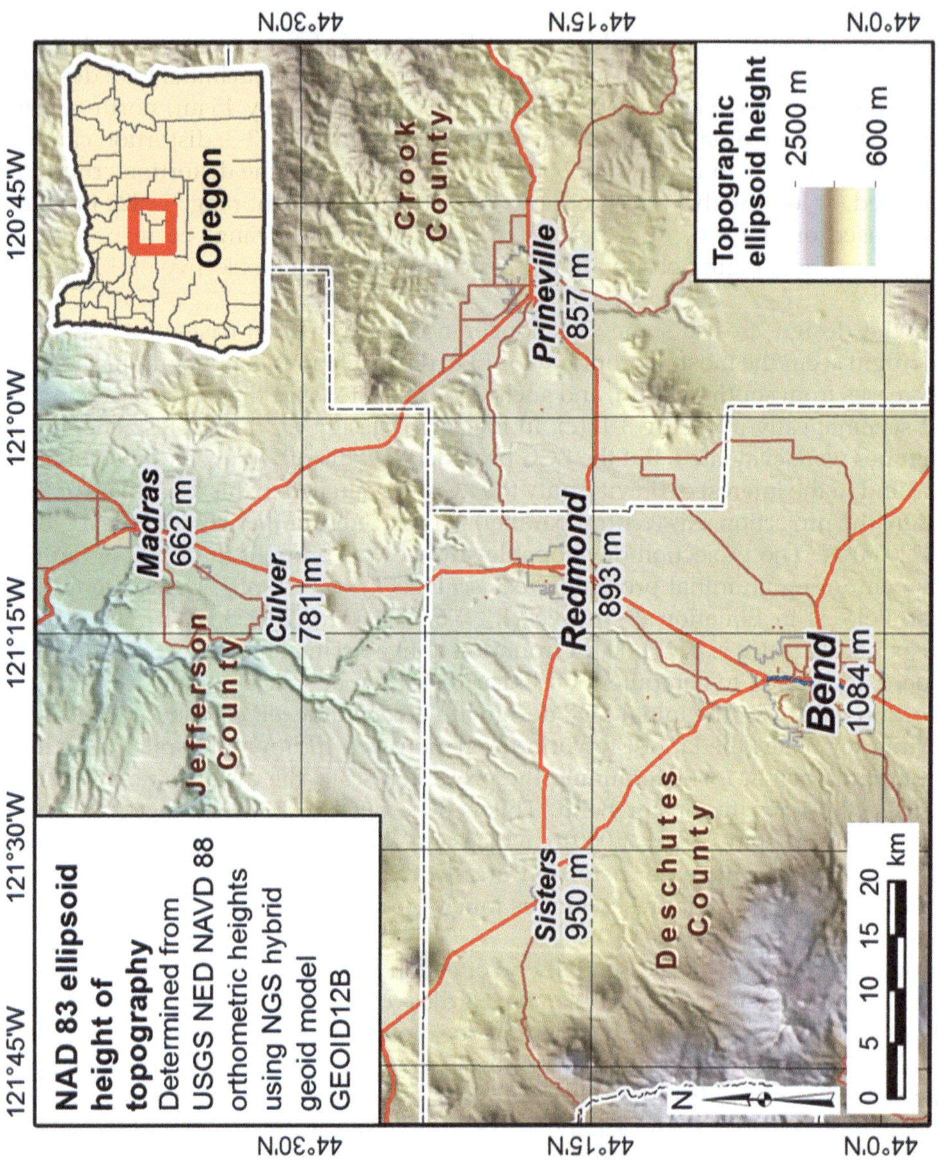

Figure 3-17. LDP example design area, showing the topographic ellipsoid heights of towns.
Source: Dennis (2016), © ASCE.

Table 3-6. Distortion Performance for Six Different LCC Projection Design Alternatives.

LCC axis scale	Initial 1.00014	1.00013	1.00012	Final 1.00012	1.00011	1.00010
Axis latitude	44°20′N	44°30′N	44°35′N	**44°40′N**	44°45′N	44°50′N
Location	Linear distortion (parts per million)					
Bend	−18.5	−10.4	−8.2	**6.1**	12.4	20.9
Redmond	0.5	−2.2	−5.4	**3.5**	4.5	7.6
Prineville	5.8	1.7	−2.2	**6.0**	6.3	8.7
Sisters	−8.7	−12.3	−16.0	**−7.6**	−7.0	−4.4
Culver	23.2	7.6	−2.0	**0.6**	−4.8	−8.1
Madras	49.9	28.9	16.6	**16.4**	8.3	2.3
Mean	8.7	2.2	−2.9	**4.2**	3.3	4.5
Range	68.4	41.2	32.6	**23.9**	19.5	29.0
Std deviation	±24.6	±15.0	±10.8	**±7.8**	±7.6	±10.4

Source: Dennis (2016), © ASCE.

Substantial improvement in performance is achieved by moving φ_c north by 20 arcminutes from the initial mean value. After inspecting the performance in other areas of the design region, a design with $\varphi_c = 44°40'$N and $k_0 = 1.00012$ was selected for the final design (bold column in Table 3-6).

As mentioned previously, despite the popularity of "modified" SPCS, the performance is almost always inferior to that of a carefully designed LDP. To illustrate, distortion maps are shown in Figure 3-19 for the SPCS 83 Oregon South Zone, both original and "modified" by scaling "to ground" such that it gives the same distortion in Bend as the final LDP design (+6 ppm). The difference in performance with LDPs is striking, although all are based on the LCC projection. Even the initial LDP design covers a far larger area with low distortion than does SPCS scaled to ground.

For both the original and scaled SPCS, low distortion (±20 ppm) is only achieved in a narrow band that is more or less parallel to the projection axis (located 100 km south of Bend). Scaling SPCS has essentially no effect on the width of the band; it merely shifts it so that it is centered on Bend. This is a vivid example of how changing the projection scale has virtually no impact on variability. Indeed, the range and standard deviation for both the original and scaled SPCS are 274 and ±97 ppm, respectively (versus a range of 24 ppm and standard deviation of ±8 ppm for the final LDP).

Selecting Final Design Parameters for Low-Distortion Projections. When designing LDPs (or any projections), it is good practice to use simple and "clean" values for the defining parameters. This is consistent with how SPCS and UTM are defined. The only values that affect distortion are the projection axis scale and location (and axis orientation for the OM projection). Although the other parameters for geodetic origin and false northing and easting have no effect on distortion, they still must be specified. In addition, to serve

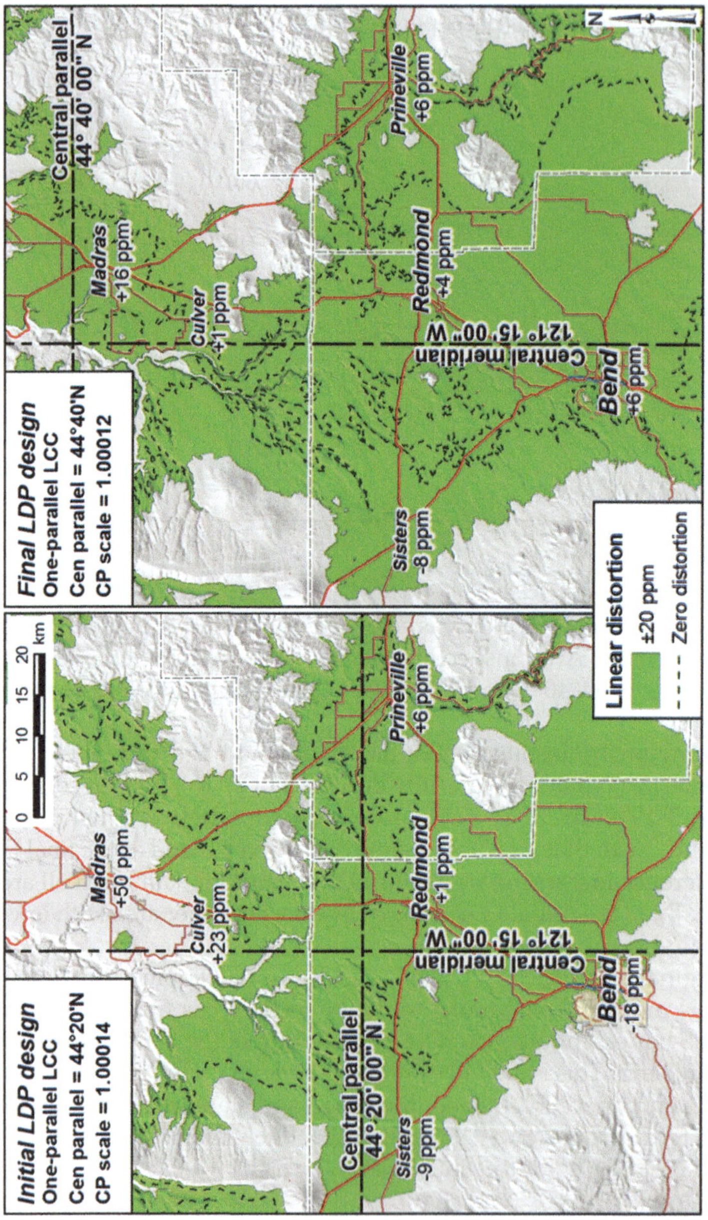

Figure 3-18. Areas with ±20 ppm linear distortion in an example for initial and final LDP designs.
Source: Dennis (2016), © ASCE.

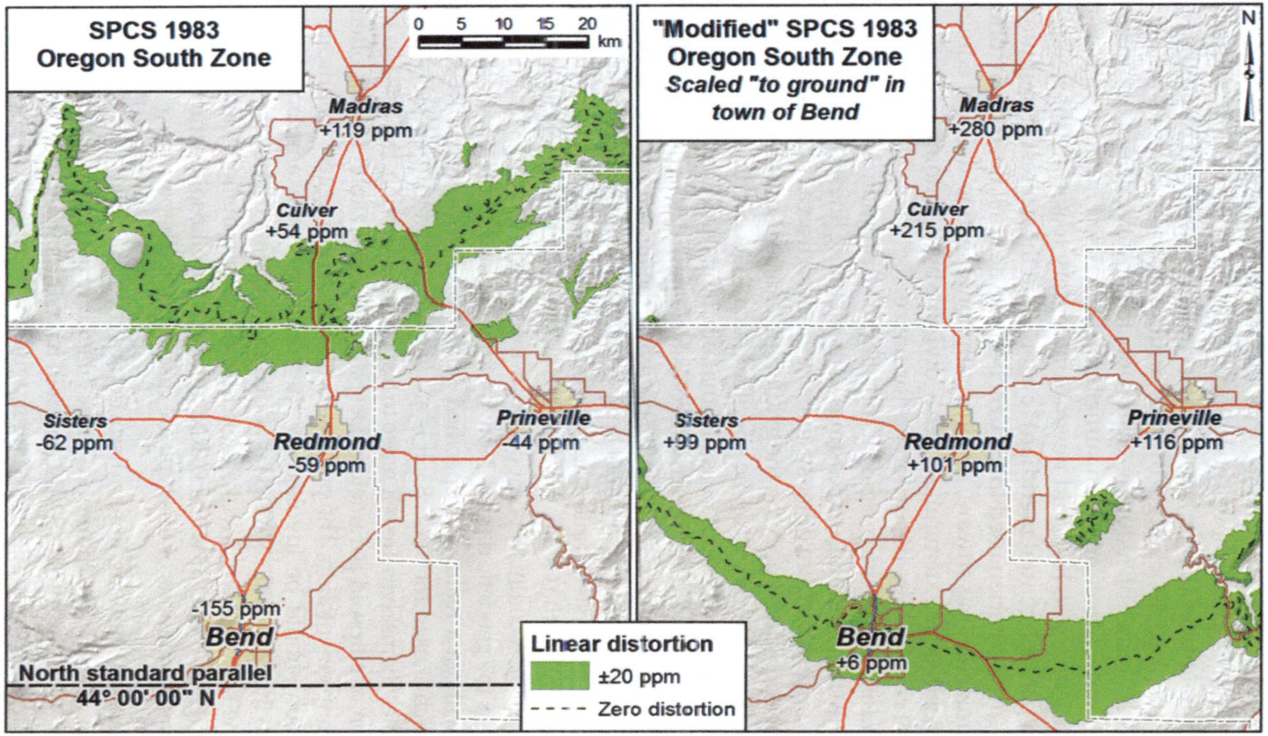

Figure 3-19. Areas with ±20 ppm linear distortion in a design example for an original and "modified" SPCS 83 Oregon South Zone. Source: Dennis (2016), © ASCE.

as a PCS, the definition must include a specific geodetic datum and linear unit. The following are recommendations for defining these values:

- *Projection type.* A conformal projection type that is widely available in commercial software packages should be used, as listed in Table 3-1 (TM, LCC, OM). The oblique stereographic can also be used, but it is highly unlikely that it will perform better than the other three, because it does not curve with the Earth in any direction.
- *Projection scale.* It is almost never necessary to specify a defining k_0 value with more than six decimal places (1 ppm); note that a change in height of 6 m (20 ft) causes the sixth decimal place to change by one unit. Additional digits rarely improve performance in terms of reducing distortion for an overall project area. For large areas, five decimal places are often sufficient, as was used in the design example.
- *Projection geodetic origin.* It should never be necessary to specify the geodetic origins (e.g., central parallel, central meridian) to more than the nearest arcminute or nearest 5 arcminutes (the criterion used for SPCS). Because decimal degrees are often used as defining parameters in software, origins with nonrepeating decimal equivalents are also recommended (i.e., arcminutes evenly divisible by 3), if doing so for the projection axis does not compromise distortion performance.
- *Projection grid origin.* The grid origin has no effect on distortion. Use whole numbers for the grid origin (false northing and easting) in the defining linear unit such that projected coordinates are distinct from other systems in the design area (such as SPCS and UTM). Many other options for specifying the grid origin can be used, based on preference and convenience. For example, it can be specified such that coordinates in the design area are the smallest positive values possible; that northings and eastings within a zone are never equal; or that coordinates of multizone systems are not equal in different zones.
- *Projection geodetic datum and linear unit.* Specifying the linear unit is essential, and if feet are used, it is vital that the type of foot be specified (international or US survey). An LDP is not a coordinate system unless it is associated with a geometric reference system (geodetic datum). For the OCRS, the datum was specified as the North American Datum of 1983 (NAD 83). A common point of confusion exists regarding what actually should be specified as the geodetic datum. Note that the geometric reference system definition is NAD 83 without a realization specified, such as "(2011) epoch 2010.00." Exactly the same approach is used for SPCS 83; it is always referenced to "generic" NAD 83. Only the coordinates themselves are reference to a specific realization, but this has no effect on the projection or ellipsoid parameters. Despite this, a realization label is often included in software implementation. The purpose is to serve as a "trigger" to indicate whether a transformation should be invoked by the software. For example, if a GIS dataset referenced to the "original" NAD 83 is added to a GIS data frame referenced to NAD 83 "HARN," the software will invoke a grid-based transformation (called NADCON) to align the "original" NAD 83 with HARN, although the ellipsoid parameters for both are identical. Transformations of this type are discussed in Chapter 4.

Figure 3-20 is a map showing linear distortion of the final design for the current (NAD 83) version of the OCRS Bend–Redmond–Prineville Zone used in the LDP design example. This zone will also be a part of SPCS2022 and will have the same parameters shown on the map, except for the false northing and easting, which will differ by those shown in Figure 3-19 by at least 10,000 m.

Low-Distortion Projected Coordinate Systems Adopted in the United States. Many low-distortion PCSs have been created and are in use throughout the United States, and they are too many to list here. Most of these are single-zone LDP systems intended for only one particular area such as a city or county. But there are others consisting of multiple zones designed to cover an entire state or even multiple states as an integrated coordinate system

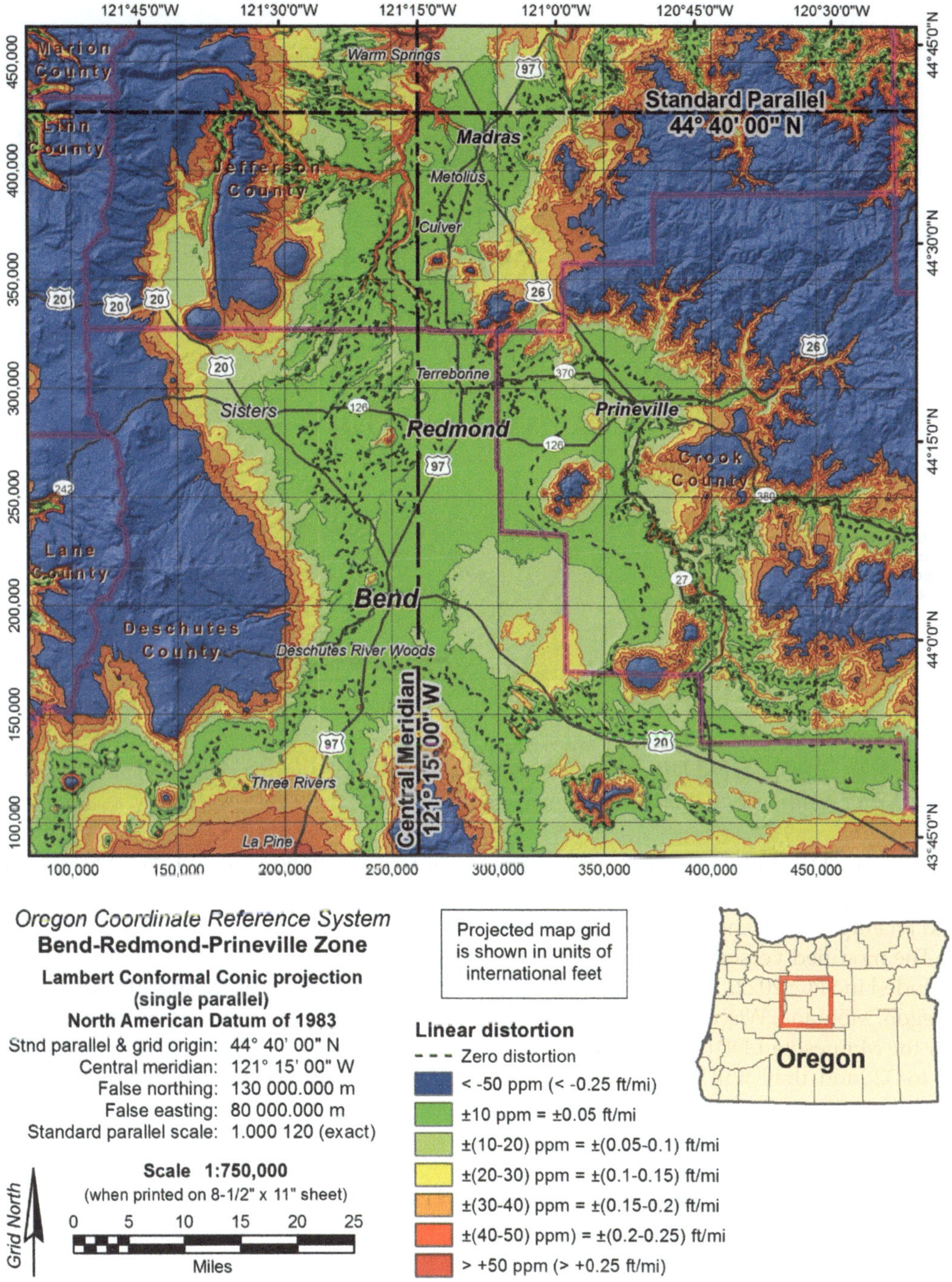

Figure 3-20. OCRS Bend–Redmond–Prineville Zone Linear distortion and defining parameters. Source: Courtesy of NOAA, reproduced with permission.

(e.g., as described by Armstrong et al. 2017, Badger 2016, Wisconsin SCO 2015). Examples of both single- and multiple-zone LDP systems adopted by various government agencies in the United States are shown in Figure 3-21 and listed in Table 3-7. More LDP systems are available than the ones listed here, which provides an indication as to the extent and variety of systems. Note that LDP usage is not restricted to high-elevation areas. In addition, with the exception of the Minnesota County Coordinate System, all the systems given here are

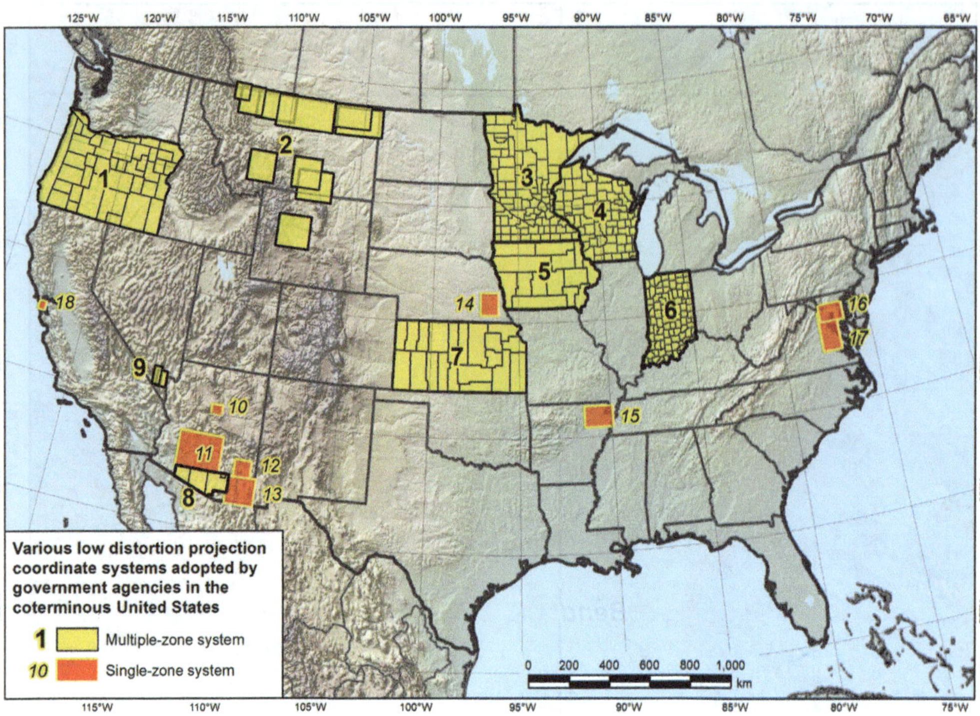

Figure 3-21. Examples of LDP coordinate systems currently used in the conterminous U.S. (See Table 3-7 for additional information.)

LDPs created by defining a projection with respect to the GRS 80 ellipsoid. The Minnesota system was created by scaling the reference ellipsoid, resulting in a projection definition and an ellipsoid for each of its 97 zones.

The modified versions of many of the LDP systems in Figure 3-21 and Table 3-7 will be included in SPCS2022. These include those shown in Indiana, Kansas, Minnesota, Montana, Oregon, Montana, Wisconsin, and Wyoming as well as some of the zones shown in Arizona. For the Minnesota LDP designs used in SPCS2022, the number of zones will decrease from 97 to 32, and they will be based on projections referenced to GRS 80, rather than the modified ellipsoid approach used currently.

NONPROJECTED LOCAL COORDINATE SYSTEMS

Local Geodetic Horizon Systems

Map projections are not the only way to convert between geodetic and georeferenced local Cartesian coordinates. Another way is to define a *local horizon* system. Such systems are created by establishing a locally "horizontal" plane that is perpendicular to the local "up" direction at the coordinate origin point of the plane. The two axes within the plane are oriented such that one axis (passing through the origin) points to geodetic (or astronomic) north, and the other axis points in a perpendicular direction to the right (east). The three axes together create a 3D local north-east-up (*NEU*) system.

The *NEU* system described is set up to provide coordinates in the manner normally encountered for surveying in the United States, in the sequence north, east, up. Local horizon

Table 3-7. Examples of LDP Coordinate Systems Currently used in CONUS (Figure 3-21).

Map num	System name	Number of zones	References and additional information
1.	Oregon Coordinate Reference System (OCRS)	39	See Armstrong et al. (2017)
2.	Rocky Mountain Tribal Coordinate Reference System (RMTCRS)	10	See Gladstone (2014)
3.	Minnesota County Coordinate System	97	Initially established in the 1980s; see Minnesota Department of Transportation (2009)
4.	Wisconsin Coordinate Reference Systems (WISCRS)	72	Initially established in 1995 as the Wisconsin County Coordinate System and later modified in 2006 to create WISCRS; see Wisconsin SCO (2015)
5.	Iowa Regional Coordinate System (IaRCS)	14	See Dennis et al. (2014)
6.	Indiana Geospatial Coordinate System (InGCS)	92	See Badger (2016)
7.	Kansas Regional Coordinate System (KRCS)	20	See Dennis (2017a)
8.	Pima County Coordinate System (PCCS)	4	See Dennis (2017b)
9.	Nevada Coordinate Reference System (NCRS)	2	Developed in cooperation with city of Las Vegas, NV, for engineering, surveying, and GIS; see Las Vegas (n.d.)
10.	City of Flagstaff Low-Distortion Grid Coordinate System	1	Developed in cooperation with City of Flagstaff, AZ, for engineering, surveying, and GIS

(*Continued*)

Table 3-7. (*Continued*).

Map num	System name	Number of zones	References and additional information
11.	Maricopa County Low Distortion Projection	1	Created by city of Surprise, AZ, to provide coverage for Maricopa County, AZ; see Surprise (n.d.)
12.	Gila Valley Low-Distortion Projection (GVLDP)	1	Created by the city of Safford, AZ, to provide coverage for most of the Gila River valley in Graham County
13.	Cochise County Low-Distortion Projection	1	Developed in cooperation with Cochise County, AZ, engineering, surveying, and GIS
14.	Lancaster County Grid Low-Distortion Projection	1	One of the oldest LDPs established by a local government agency in the 1980s (with assistance from NGS)
15.	Craighead County Low-Distortion Projection Coordinate System	1	Created by the city of Jonesboro, AR, to provide coverage for the entire county
16.	WMATA Low-Distortion Projection Coordinate System	1	For the entire rail system of the Washington Metropolitan Area Transit Authority (WMATA)
17.	SEHSR-RAPS Low-Distortion Coordinate System	1	For Southeast High Speed Rail (SEHSR) Richmond Area to Potomac Segment (RAPS) project, VA
18.	City and County of San Francisco 2013 Coordinate System (SFCS13)	1	See San Francisco (2016)

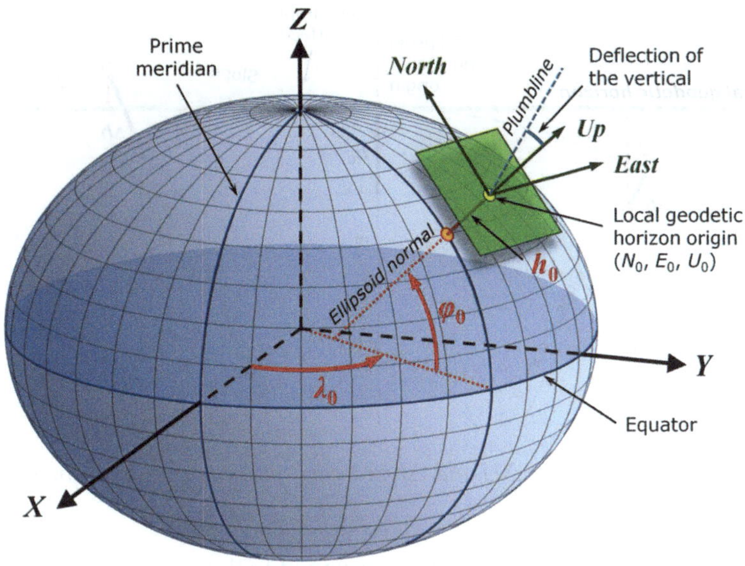

Figure 3-22. Local geodetic horizon (LGH) coordinate system at height h_0 above the ellipsoid.

systems are also set up as right-handed systems, as east, north, up (*ENU*) or north, east, down (*NED*). Right-handed local horizon coordinate systems are commonly employed as an intermediate step in determining coordinates for multiple-component positioning systems. Examples of such applications are aerial and terrestrial surveying/mapping systems that integrate coordinates from multiple sensors, such as those combining input from lidar, cameras, GNSS, and inertial measurement units (IMUs). Local horizon systems are also used to perform geodetic computations in a local frame, which is particularly useful when combining observations from different technologies such as GNSS and optical total stations.

The "up" direction of an *NEU* system can coincide either with the gravity vector (plumbline) at the origin point or with the geodetic reference frame ellipsoid normal. For an IMU, "up" coincides with the plumbline, because orientation is referenced to local gravity (as sensed by the IMU accelerometers). If "up" coincides with the ellipsoid normal, the system is a LGH system, as shown in Figure 3-22. The angle between the plumbline and the ellipsoid normal is the deflection of the vertical (DoV), which is less than 1 arcminute at most locations on the Earth. The mean DoV for the conterminous United States is about 6 arcseconds. LGH systems are sometimes called *local tangent planes*, because the plane is perpendicular to the ellipsoid normal. However, the plane is only truly tangent to the ellipsoid if the ellipsoid height of the plane origin is zero (i.e., the plane touches the ellipsoid at its origin), which is often not the case.

An *NEU* system can be the basis of coordinates that are used directly for creating surveying engineering products such as engineering plans. Such a system based on GNSS-derived coordinates will be an LGH system, because the up component of GNSS is along the ellipsoid normal. To create such a coordinate system, the latitude and longitude of the origin point (φ_0, λ_0) are specified. Then, the 3D *NEU* coordinates of any point can be computed with respect to that origin as

$$\begin{aligned} N &= N_0 - \Delta X \sin\varphi_0 \cos\lambda_0 - \Delta Y \sin\varphi_0 \sin\lambda_0 + \Delta Z \cos\varphi_0 \\ E &= E_0 - \Delta X \sin\lambda_0 - \Delta Y \cos\lambda_0 \\ U &= U_0 + \Delta X \cos\varphi_0 \cos\lambda_0 + \Delta Y \cos\varphi_0 \sin\lambda_0 + \Delta Z \sin\varphi_0 \end{aligned} \tag{3-8}$$

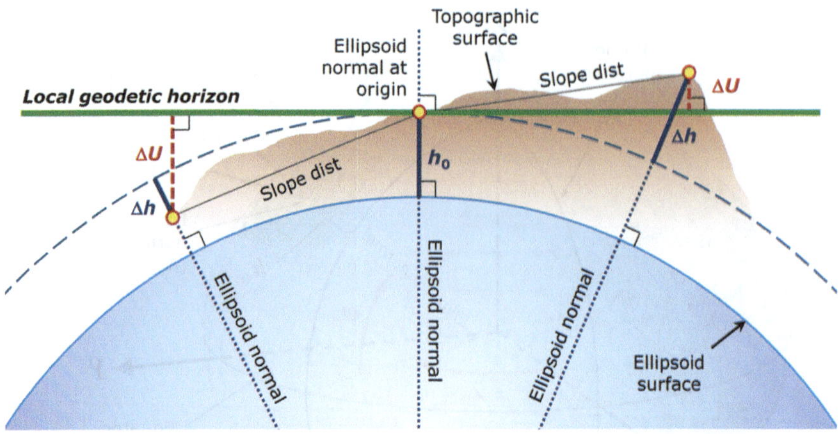

Figure 3-23. Relationship between ΔU and Δh in local geodetic horizon coordinate systems.

where (N_0, E_0, U_0) are (arbitrary) coordinates assigned to the origin, and (ΔX, ΔY, ΔZ) are the geocentric Cartesian coordinate differences (deltas) from the origin to the point of interest. LGH systems are also useful for computing coordinate differences using the geodetic coordinates of two points, without having to use map projection coordinates. The delta north, delta east, and delta up (ΔN, ΔE, ΔU) differences are computed using Equation (3-8) with one point of the pair at the origin and $N_0 = E_0 = U_0 = 0$.

An LGH *NEU* system is truly 3D. In contrast, PCSs are strictly 2D. When using projected coordinates where 3D is required (such as engineering projects), the third dimension is created simply by assigning a height ("elevation") at the appropriate horizontal location, which is geometrically perpendicular to the plane represented by the projection. However, heights are not referenced to a projection "mapping plane" but rather some type of "vertical datum," such as the geoid for orthometric heights or the reference ellipsoid for ellipsoid heights. Combining heights with 2D projections results in coordinate systems that are not truly 3D and are sometimes instead called "2.5D". The reason is that a map projection gives only a 2D horizontal location, and height is a 1D value at that location referenced to a completely different system (such as a vertical datum), and the two are, in general, not mutually perpendicular. Combining them creates a point in space that, although nominally 3D, does not correctly represent the actual 3D geometry on the Earth. Despite this problem, such 2.5D systems are successfully used for 3D projects because the relationship between the horizontal and the vertical is well-defined and can be handled mathematically without loss of rigor. Indeed, even when LGH systems are used, they are often used as hybrid *NEh* systems that are also 2.5D. An *NEh* system uses the N and E equations from Equation (3-7) but uses delta ellipsoid height Δh (or some other delta height) in place of ΔU. The problem with ΔU is that it gives heights with respect to the LGH plane rather than a geodetic datum. Such ΔU heights, although truly 3D, are not directly useful as "heights" in the conventional sense of the word. The relationship between Δh and ΔU is shown in Figure 3-23 and is further discussed subsequently.

An LGH provides a way to create a georeferenced local Cartesian coordinate system—and to compute distances—without using a map projection. However, the LGH approach may not be appropriate for either coordinates or distances in all situations, especially for large areas. Users should be fully aware of the following LGH characteristics and limitations:

- *Delta up versus delta height.* As discussed previously and shown in Figure 3-22, $\Delta U < \Delta h$. In most surveying engineering situations, Δh (or some other delta "elevation") is of greater interest and utility than ΔU, which is why 2.5D *NEh* systems are often used rather than truly 3D *NEU* systems. For small areas, ΔU and Δh can be considered essentially equal: they differ by less than 1 mm for distances of less than 100 m. However, the difference increases rapidly to 0.08 m at a distance of 1 km and about 7.8 m at 10 km.

- *Distances*. The 3D (ΔN, ΔE, ΔU) distance is always exactly equal to the 3D (ΔX, ΔY, ΔZ) distance. However, "horizontal" (ΔN, ΔE) distances (in the LGH plane) do not necessarily represent horizontal as the term is normally used. When the LGH is tangent to the ellipsoid, an LGH distance will be shorter than an ellipsoidal distance by less than 1 mm for distances of less than about 6 km. This difference increases rapidly, as shown in Table 3-5, to −0.004 m in 10 km and −0.5 m in 50 km. These differences are significantly larger than the other distance computation methods in Table 3-5.
  In addition, for computing horizontal distances between points *A* and *B*, difference distances will be obtained depending on whether *A* or *B* is at the LGH origin, and this horizontal difference increases as the height difference between *A* and *B* increases. For zero height difference, the horizontal distance between *A* and *B* differs by less than 0.5 mm for distances of less than 100 km if *A* and *B* are swapped for the LGH origin. However, for a height difference of 100 m, the difference in horizontal distance is about 10 cm when *A* and *B* are only about 6 km apart.
- *"Up" direction*. As stated previously, the LGH plane is perpendicular to the ellipsoid normal at the plane origin. This situation works well for GNSS applications, because GNSS are consistent with ellipsoid heights. However, care should be used when combining LGH systems with local horizon systems referenced vertically to gravity, such as from total stations, laser scanners, or IMUs. The vertical (and, thus, the horizontal) will be inclined with respect to that of the LGH by the deflection of the vertical (Chapter 7).
- *Nonconformality*. LGH systems are not conformal. This means that the linear distortion at a point varies with direction, shapes are not locally preserved, lines intersect at an angle different from that on the ellipsoid, and there is no simple relationship between an LGH azimuth and a geodetic azimuth. The one exception is azimuths from the origin of an LGH, which differ from geodetic by less than 0.1 arcsecond for distances under 200 km. However, azimuths between all other point pairs in an LGH system will differ from the geodetic azimuth, usually by a substantial amount (depending on the length of the line, its orientation, and its distance from the LGH origin).
- *Software support*. At the time of writing this chapter (September 2021), LGH coordinate systems are apparently supported in few (if any) commercial geospatial software packages used in surveying, engineering, and GIS. Although this situation may change in the future, it currently makes the adoption of such coordinate systems difficult, especially if working or sharing data across platforms. This limitation should be taken into account before adopting LGH systems for production work.

Nongeoreferenced Local Coordinate Systems

Using georeferenced coordinate systems is highly recommended whenever possible. Nonetheless, situations exist where it is not practical to georeference survey work. In such situations, it is still necessary to use a coordinate system, even if the system cannot be connected to geodetic reference frames. Because GNSSs are by default georeferenced (even if the user is not aware of it), this discussion will focus on coordinate systems established using conventional optical instruments such as total stations and terrestrial laser scanners.

Nongeoreferenced local coordinate systems are similar to the local horizon systems discussed in the previous section. When established using terrestrial instruments, the "up"

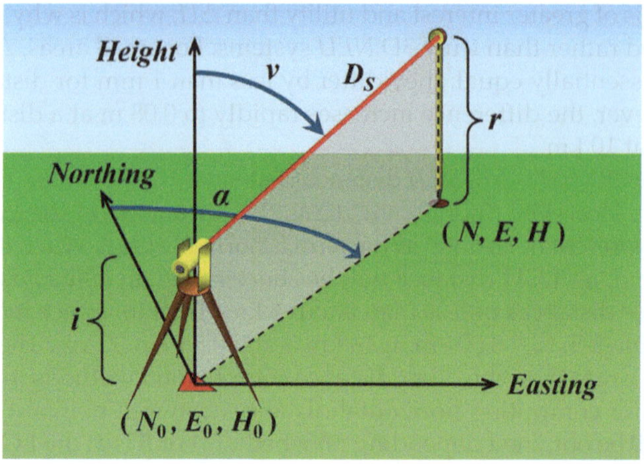

Figure 3-24. Establishing an arbitrary coordinate system using a terrestrial instrument. Source: Courtesy of NOAA, reproduced with permission.

component of the coordinate system will be aligned with the local gravity vector (plumbline). Often, this will be the plumbline passing through the vertical axis of the initial instrument setup. Nongeoreferenced projects are typically not large (less than a few hundred meters in extent), and so the plumbline orientation (deflection of the vertical) will likely be nearly the same everywhere.

To provide some insight into establishing a local arbitrary coordinate system using a terrestrial instrument, a simplified example is provided here for a total station, as shown in Figure 3-24. For this situation, northing, easting, and height (*N*, *E*, *H*) coordinates can be determined as

$$\begin{aligned} N &= N_0 + D_S \cos\alpha \sin\nu \\ E &= E_0 + D_S \sin\alpha \sin\nu \\ H &= H_0 + D_S \cos\nu + i - r \end{aligned} \tag{3-9}$$

where

N_0, E_0, and H_0 = (Arbitrary) northing, easting, and height at the instrument setup;
D_S = Observed slope distance (range);
α = Observed horizontal angle (azimuth);
ν = Observed zenith angle; and
i and r = Instrument and the prism rod heights, respectively.

From Figure 3-24, the inverse relationships can also be developed as

$$D_S = \sqrt{(N - N_0)^2 + (E - E_0)^2 + (H - H_0 - i + r)^2}$$

$$\alpha = \tan^{-1}\left(\frac{E - E_0}{N - N_0}\right) \tag{3-10}$$

$$v = \cos^{-1}\left(\frac{H - H_0 - i + r}{D_S}\right)$$

Sometimes in practice, engineers will set $N_0 = E_0 = 3000$ m (10,000 ft) and set $H_0 = 300$ m (1,000 ft). These false northings and eastings are primarily done to prevent negative coordinates. Although this practice is common, it is better practice to make N_0 and E_0 significantly different so that it is more obvious which coordinate is a northing and which coordinate is an easting. As discussed previously, northings and eastings can be reported in (N, E) or (E, N) conventions; thus, they are easily mixed up when their values are similar. For this reason, the northings and eastings in the SPCS are designed to be significantly different to reduce the confusion.

Because the vertical component of the coordinate system is parallel to the plumbline, the computed height differences will nominally be orthometric heights (assuming more or less constant DoV in the work area) (see Chapter 7). Because the coordinate system is not georeferenced, the axes, in general, will not be oriented with the actual geodetic cardinal directions, and, thus, the horizontal angle will not be the azimuth. In such situations, north can be assumed or approximated (say from a magnetic compass). If points that have a known azimuth are observed, they can be used to orient the survey with respect to north.

This example of an arbitrary coordinate system is simplified. To obtain accurate results, other factors should be taken into account, such as curvature and refraction, atmospheric propagation delays caused by variable temperature and pressure, and horizontal offsets of retroreflective prisms. In addition, the instrument as well as the prism rod should be calibrated and in good working order. A discussion of these factors is beyond the scope of this chapter, but they are nonetheless important considerations for determining accurate coordinates. To learn more about the use of total stations and terrestrial laser scanners, see Chapters 7 and 8, respectively.

SUMMARY

This chapter gave an overview of the types of local Cartesian coordinate systems most often used in modern surveying engineering practice. Given the importance of integrating spatial data from a variety of sources, an emphasis was placed on georeferenced coordinate systems, although nongeoreferenced systems were briefly discussed. The Cartesian systems provide "grid" coordinates referenced to two (or three) mutually perpendicular axes. Two axes form a nominally horizontal plane, with the axes in the north (y) and east (x) directions for georeferenced systems. The third axis is perpendicular to this plane and defines an "up" direction for elevations or heights.

The most common local (horizontal) coordinate systems in use today are projected coordinate systems (PCSs). Only four widely used, large-scale, conformal projections are appropriate for surveying and engineering: Transverse Mercator, Lambert Conformal Conic, Oblique Mercator, and Stereographic (although the first three are usually preferred over the oblique aspect of the Stereographic). For high-accuracy work, the scale error should be small enough to avoid excessive linear distortion (i.e., the difference between projected distances and actual ground distances). A large-scale conformal PCS commonly used in surveying and engineering is SPCS.

Although SPCS was designed to support surveying engineering, in many cases, its linear distortion is considered excessive. One method for reducing distortion is to scale SPCS to the topographic surface to bring "grid to ground" (often called *modified* State Plane). However, this approach is not readily compatible with digital spatial data (such as in GIS) and is too inflexible to optimally reduce distortion. Another approach is to create a custom LDP coordinate system that minimizes distortion for a specific area. Such LDPs can be created in two ways: (1) by modifying the reference ellipsoid so that its surface is near the topographic surface and then defining a projection referenced to the modified ellipsoid, or

(2) by designing a projection such that its developable surface is near the topographic surface without modifying the ellipsoid. The second method, in general, produces lower distortion in design areas with an overall nonzero topographic slope. However, even in situations where both methods produce similar results, modifying the reference ellipsoid is not recommended because it is more a complex approach that can be difficult to implement correctly. It can also create a PCS that is not conformal with respect to the ellipsoid used for the reference frame, depending on how the ellipsoid is modified.

Engineering projects typically require accurate positions in 3D space. However, in most cases, 2D (planar) systems are used that are given some type of elevation or height perpendicular to the plane to create a quasi-3D system, sometimes called *2.5D*. The only truly 3D system discussed in this chapter was the LGH, which provides 3D north, east, and up coordinates. Although LGH systems are useful in many geospatial computations and applications, they have not been widely adopted (and are not recommended) for surveying engineering coordinate systems, for the following reasons: they are not conformal; the up component differs from the height component by a significant amount for horizontal distances of more than a few hundred meters; and, in general, they are not supported in commercial geospatial software.

Some of the limitations in handling 3D data sets and coordinate systems may diminish in the future as software becomes more geodetically sophisticated and as the traditional approaches to manipulating spatial data and creating engineering plans evolve. Fortunately, the 2.5D systems—despite not being truly 3D—are used successfully in the 3D engineering environment. The important part is not so much the specific type of coordinate system used (they all have benefits and liabilities) but to have a solid working understanding of the mathematical and physical relationships between them. Such knowledge is an essential part of modern surveying engineering practice.

REFERENCES

Armstrong, M. L., J. Thomas, K. Bays, and M. Dennis. 2017. *Oregon coordinate reference system handbook and map set*. Version 3.01. Salem, OR: Oregon Dept. of Transportation, Geometronics Unit.

Badger, M. G. 2016. *Indiana geospatial coordinate system handbook and user guide*. Version 1.05. Indianapolis, IN: Indiana Dept. of Transportation and Land & Aerial Survey Office.

Dennis, M. L. 2016. "Ground truth: Low distortion map projections for engineering, surveying, and GIS." In *Proc., Pipelines 2016: Out of Sight, Out of Mind, Not Out of Risk*, edited by B. L. Livingston, C. Cate III, A. Pridmore, J. W. Heidrick, and J. Geisbush, 857–869. Kansas City, MO: ASCE.

Dennis, M. L. 2017a. *The Kansas regional coordinate system: A statewide multiple-zone low-distortion projection coordinate system for the state of Kansas*. Topeka, KS: Kansas Dept. of Transportation.

Dennis, M. L. 2017b. *Pima County coordinate system*. Tucson, AZ: Pima County, Information Technology Dept., GIS Division.

Dennis, M. L. 2018a. *The state plane coordinate system: History, policy, and future directions*. NOAA Special Publication NOS NGS 13. Silver Spring, MD: National Oceanic and Atmospheric Administration, National Geodetic Survey.

Dennis, M. L. 2018b. "Of planes and plumblines: Map projections and differential leveling in a GNSS-based 3D geodetic framework." Ph.D. thesis, Oregon State University. Accessed February 8, 2022. https://ir.library.oregonstate.edu/concern/graduate_thesis_or_dissertations/0k225h38j.

Dennis, M. L., N. Miller, and G. Brown, 2014. *Iowa regional coordinate system handbook and user guide*. Version 2.10. Ames, IA: Iowa Dept. of Transportation.

Federal Register. 2020. "Deprecation of the United States (U.S.) survey foot." 85 FR 62698. Vol. 85, No. 193, pp. 62698–62708. Accessed February 8, 2022. https://www.federalregister.gov/documents/2020/10/05/2020-21902/deprecation-of-the-united-states-us-survey-foot.

Gladstone, W., ed. 2014. *Rocky Mountain tribal coordinate reference system*. Version 1.0. Billings, MT: Northern Engineering and Consulting, Inc.

Iliffe, J. C., and R. Lott. 2008. *Datums and map projections: For remote sensing, GIS and surveying*. 2nd ed. Dunbeath Mill, UK: Whittles Publishing.

Las Vegas. n.d. "Nevada coordinate reference system (NCRS) overview." City of Las Vegas, NV. Accessed February 8, 2022. https://files.lasvegasnevada.gov/public-works/Nevada-Coordinate-Reference-System-PDF.pdf.

Minnesota Department of Transportation. 2009. *Map projections and parameters*. Saint Paul, MN: Minnesota Dept. of Transportation, Office of Land Management.

NGA (National Geospatial-Intelligence Agency). 2014a. *The universal grids and the transverse mercator and polar stereographic map projections*." Version 2.0.0. NGA.SIG.0012_2.0.0_UTMUPS. Springfield, VA: NGA.

NGA. 2014b. *Universal grids and grid reference systems*. Version 2.0.0. NGA.STND.0037_2.0.0_GRIDS. Springfield, VA: NGA.

NGS (National Geodetic Survey). 2019a. *State plane coordinate system of 2022 policy*. NGS 2019-1214-01. Silver Spring, MD: National Oceanic and Atmospheric Administration, National Geodetic Survey.

NGS. 2019b. *Procedures for design and modification of the state plane coordinate system of 2022*. NGS 2019-1214-01-A1. Silver Spring, MD: National Oceanic and Atmospheric Administration, National Geodetic Survey.

San Francisco. 2016. *City and County of San Francisco datums, coordinates systems, reference frames and acronyms*. Version 3.9. San Francisco, CA: City and County of San Francisco, Dept. of Public Works.

Snyder, J. P. 1987. *Map projections—A working manual*. US Geological Survey Professional Paper 1395. Washington, DC: United States Geological Survey.

Stem, J. E. 1990. *State plane coordinate systems of 1983*. NOAA Manual NOS NGS 5. Silver Spring, MD: National Oceanic and Atmospheric Administration, National Geodetic Survey.

Surprise. n.d. "Maricopa County, Arizona low distortion projection." City of Surprise, AZ. Accessed February 8, 2022. https://www.surpriseaz.gov/DocumentCenter/View/3906/Maricopa-County-Low-Distortion-Projection-20ppm?bidId=.

Van Sickle, J. 2017. *Basic GIS coordinates*. 3rd ed. Boca Raton, FL: CRC Press.

Wisconsin SCO (State Cartographer's Office). 2015. *Wisconsin coordinate reference systems*. 2nd ed. Madison, WI: Wisconsin SCO.

CHAPTER 4

LOCAL, REGIONAL, AND GLOBAL COORDINATE TRANSFORMATIONS

Michael L. Dennis

INTRODUCTION

A common task in surveying engineering is transformation of coordinates, which can range from modifications of a local planar system for a project site to complex geodetic operations at a continental or global scale. The latter are often referred to as datum, reference frame, or geographic coordinate system transformations, but in this chapter, all are considered as various types of coordinate transformations. Many types of transformations are available, but those covered in this chapter are ones typically encountered in surveying engineering practice. They are broadly divided into two main groups, *equation-based* and *grid-based*.

Equation-based transformations are, as the name implies, performed using an equation or a set of equations, and all the ones considered here are of closed form. The equations can range in complexity from a simple single vertical shift value for a local site to a global 3D rotation, translation, and scale where all parameters vary with time.

Grid-based transformations typically consist of an array of points at specific locations where transformation values are defined, usually on grids of uniform spacing. Points at arbitrary locations are transformed by interpolating between the grid points, and usually, the type of interpolation is specified. These transformations are typically for large regions, such as a state or country, and can be global in extent. They can be horizontal, vertical, or 3D.

Equation-based and grid-based transformations can also be combined. This type of transformation is used by the National Geodetic Survey (NGS) for its Horizontal Time-Dependent Positioning (HTDP) software to model plate tectonic motion and localized crustal motions as well as sporadic episodic motions such as earthquakes (Pearson and Snay 2012, NGS 2021a).

Map projections and local geodetic horizon (LGH) systems (covered in Chapter 3) are considered coordinate *conversions* (rather than *transformations*), as are conversions between curvilinear geodetic (latitude, longitude, and ellipsoid height) and geocentric X, Y, Z Cartesian coordinates (covered in Chapter 2). The difference with the transformations in this chapter and the conversions in Chapters 2 and 3 is that the latter ones convert angular coordinates (latitude and longitude) to linear coordinates such as northings and eastings or—when ellipsoid height is included—to geocentric X, Y, Z coordinates or LGH north, east, up systems. In contrast, for the transformations, the input and output *type* of coordinates are the same, but the coordinate *values* are different. For example, projected coordinates can be rotated, translated, and scaled, but both the original and final coordinates are planar. Likewise, a datum transformation from one geodetic reference frame to another can be based on different ellipsoids oriented differently with respect to the Earth, but both the input and output coordinates are geodetic.

EQUATION-BASED TRANSFORMATIONS

Local Horizontal and Vertical Transformations

Modern surveying is usually done by 3D data collection techniques with equipment such as total stations, Global Navigation Satellite Systems (GNSSs), and laser scanners. Three-dimensional transformations can be performed within this context, but the most common approach in surveying is to perform the horizontal and vertical transformations separately. A widely used application of such transformations is for generating local coordinates in GNSS surveys. This process is built into many commercial software packages and is variously called *calibration* or *localization*. This topic is discussed later in this section.

Horizontal and 3D Transformations. A mathematical operation often performed by surveying engineers is the transformation of horizontal coordinates, often done so that one set of coordinates can be made to match another set. Most surveying engineering applications use coordinate systems that are conformal. In such systems, the linear distortion (or scale error) is the same in all directions from a point, and angles (thus shapes) are locally preserved, which is why conformal map projections such as State Plane are so widely used, as discussed in Chapter 3. Because a 2D similarity transformation is a conformal operation (i.e., a change in scale is the same in all directions), it is often the most appropriate for transforming horizontal coordinates.

In the United States, surveying engineers usually work with coordinate pairs given as northing and easting (N, E) with angles reckoned positive clockwise. This convention forms a left-handed coordinate system, and the similarity transformation to compute (N', E') target (output) coordinates from (N, E) source (input) coordinates is given by the following pair of equations:

$$\begin{aligned} N' &= (1+S)\left[(N-N_0)\cos R-(E-E_0)\sin R\right]+(T_N+N_0) \\ E' &= (1+S)\left[(N-N_0)\sin R+(E-E_0)\cos R\right]+(T_E+E_0) \end{aligned} \tag{4-1a}$$

where

T_N and T_E = North and east translations, respectively;
S = Fractional (differential) part of the scale change (unitless);
R = Clockwise positive horizontal rotation angle; and
N_0 and E_0 = Source coordinates about which the transformation is performed.

If the transformation is done with respect to the source coordinate system origin, then $N_0 = E_0 = 0$. This simplifies Equation (4-1a), which can also be expressed in matrix form as

$$\begin{bmatrix} N' \\ E' \end{bmatrix} = (1+S)\begin{bmatrix} \cos R & -\sin R \\ \sin R & \cos R \end{bmatrix}\begin{bmatrix} N \\ E \end{bmatrix}+\begin{bmatrix} T_N \\ T_E \end{bmatrix} \tag{4-1b}$$

Both Equations (4-1a) and (4-1b) are also called a *4-parameter transformation*. The 2×2 matrix of sine and cosine functions in Equation (4-1b) is the *rotation matrix*, with rotation of coordinates about the input coordinate origin (N_0, E_0) = (0, 0). The geometry of this transformation is illustrated in Figure 4-1 for a pair of input points P_1 and P_2, for all

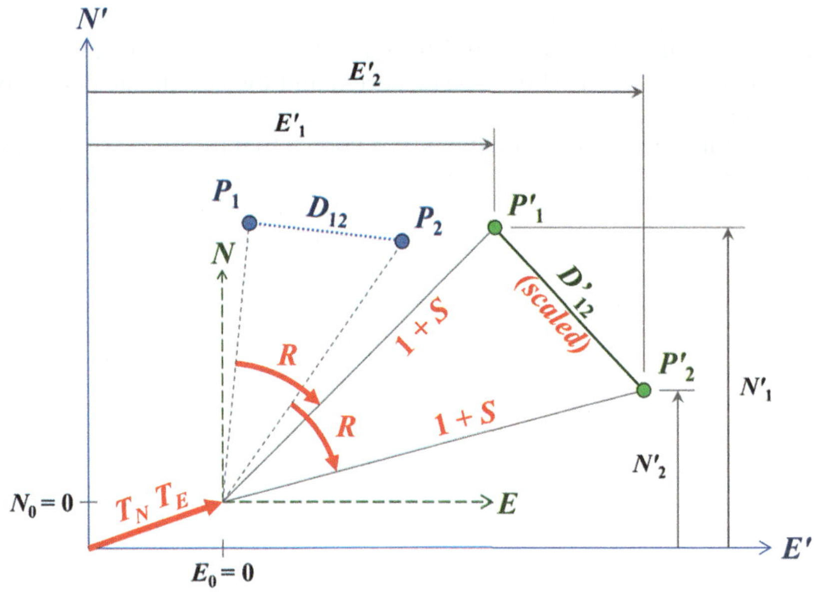

Figure 4-1. Horizontal (2D) similarity coordinate transformation about origin (0, 0) for a left-handed (N, E) coordinate system (rotation positive clockwise).

transformation parameters as positive values. It shows a (dashed) vector formed from the origin to each of the two input points. These vectors are rotated clockwise by positive R, scaled by a factor of $(1 + S)$ that is greater than 1, and then translated in positive north (T_N) and east (T_E) directions. The distance between the transformed points, D'_{12}, is greater than the original distance D_{12} by $(1 + S)$, and the azimuth between the points is increased by the rotation angle R. The output coordinate values are with respect to the new (translated) coordinate system origin.

Although surveying engineers in the United States often use the left-handed convention of (N, E) and positive clockwise rotation, the right-handed conventions is frequently used in other geospatial applications and software, and it is the typical mathematical convention. In this case, the coordinate order is (E, N), the rotation is positive counterclockwise, and Equation (4-1b) becomes

$$\begin{bmatrix} E' \\ N' \end{bmatrix} = (1+S) \begin{bmatrix} \cos R & -\sin R \\ \sin R & \cos R \end{bmatrix} \begin{bmatrix} E \\ N \end{bmatrix} + \begin{bmatrix} T_E \\ T_N \end{bmatrix} \quad (4\text{-}2)$$

Note that Equation (4-2) will give the same results as (4-1b) if the sign of its rotation is changed, and vice versa.

When performing the transformation, the order of computations matters: Different answers are obtained if the coordinates are first rotated and then translated versus first translated and then rotated. Both approaches work if done correctly and consistently, but typically translation is done after rotation and scaling, and this is the convention used in this chapter.

Usually, there is a set of common points with coordinates in both systems, and Equation (4-1a, b) or (4-2) are rearranged to solve for the four parameters. A minimum of two 2D points are required to determine a solution. Typically, more than two common points are known, and the most likely parameters are determined using least squares (Chapter 5). The quality of the least-squares solution is often evaluated by inspecting the coordinate residuals at each point. All points should have residuals of similar magnitude, any that differ substantially from others should be inspected and the coordinate pairs removed or added as necessary.

In many surveying engineering situations, it is known that the two systems should have the same scale, for example, if both are intended to accurately represent horizontal "ground" distances (as discussed in Chapter 3). In such cases, the departure of the solved scale factor from 1 can also be used to evaluate the transformation. However, for the final transformation, the scale factor should be fixed to exactly 1 if the intent is to preserve the distances of the system being transformed (although, in general, this will increase the residuals).

No specific guidelines are available on the magnitude of residuals for evaluating the quality of a 4-parameter similarity transformation, because it depends on the situation. In general, the error of the residuals should not substantially exceed the error of the coordinates used for the transformation (such as by more than a factor of 2). If one coordinate set has larger errors than the other, then the larger errors should be used as the criterion for evaluating residuals. For example, consider a set of old project coordinates with an expected overall accuracy of 10 cm that is transformed to align with the new coordinates of 2 cm accuracy. Then the root-mean-square error (RMSE) of the residuals should be less than about 10 cm, and the largest residual should be less than about twice that (20 cm). Note that this is but a crude guideline; any given situation may have its own idiosyncrasies.

Having a transformation tool does not mean it can be used for all situations. A limitation with planar (2D) similarity transformations is that they usually will not work well for datasets covering large areas. Excessively large residuals will likely result for areas greater than about 10 km (6 mi) in extent if the two point sets are from different coordinate systems (or if either of the coordinate systems is unknown). The reason for this limitation is simply that the Earth is not flat, and so a part of the error budget is the difference in the distribution of coordinate system linear distortion. The four available parameters cannot adequately account for the variation in distortion, and so it is manifested by increased residuals. Probably, the most common example of such transformation abuse is the use of "calibrations/localizations" over large areas, which is discussed later in this chapter.

The 4-parameter 2D transformation can be generalized to allow scale to vary with direction, as well as to allow skewing of the coordinate axes, to create a horizontal 6-parameter *affine transformation*. A minimum of three coordinate pairs are needed to define an affine transformation. The affine transformation has many applications, but it should be used with caution for survey coordinate transformations. If this transformation is selected, it should be for a specific reason, because it assumes that scale varies with direction and allows for skewing (shearing) of the axes. In other words, in general, the affine transformation does not preserve the angular relationship between points. An example where this is appropriate is the transformation of aerial photo coordinates obtained from films to account for differential shrinkage. However, most surveying coordinate systems are conformal, so it is usually not appropriate to use an affine transformation. Because both the source and target coordinates used for a transformation will also contain error, a least-squares solution using the affine transformation will almost always yield smaller residuals than a conformal transformation. The smaller residuals do not mean that the affine transformation is better; it merely means that it has more degrees of freedom to better fit the data. However, the resulting transformed coordinates will be warped in a manner inconsistent with conformal systems. Spatial relationships can change when an affine transformation is used (e.g., a tangent curve will likely become nontangent). It is for this reason that most surveying engineering software performs only conformal transformations, which are a special case of affine transformations. However, other geospatial software may default to the general affine, so care should be exercised in performing transformations.

A further generalization of the affine transformation is the 8-parameter *projective transformation*. A minimum of four coordinate pairs are needed to define a projective transformation. This transformation has all the properties of the affine, but parallelism is not preserved because the projective transformation allows change of shape (e.g., squares can be transformed to trapezoids). An example application is to transform oblique aerial imagery directly to a mapping coordinate system, without first orthorectifying the imagery. As with the affine transformation, it is usually not used in surveying engineering for transformations among conformal systems.

Equation (4-2) can be extended to 3D conformal coordinate transformations, also known as 7-parameter similarity transformations [Equation (4-2) is used rather than (4-1b) because right-handed coordinate systems are more common for 3D transformations, and this convention is used for all 3D transformations in this chapter.]. In this case, the local coordinate order is east, north, up (E, N, U), and rotations of vectors are positive in the counterclockwise direction when looking down each coordinate axis toward the origin. Explicitly, this transformation of vectors is given by

$$\begin{bmatrix} E' \\ N' \\ U' \end{bmatrix} = (1+S) \begin{bmatrix} \cos R_U & -\sin R_U & 0 \\ \sin R_U & \cos R_U & 0 \\ 0 & 0 & 1 \end{bmatrix} \begin{bmatrix} \cos R_N & 0 & \sin R_N \\ 0 & 1 & 0 \\ -\sin R_N & 0 & \cos R_N \end{bmatrix} \begin{bmatrix} 1 & 0 & 0 \\ 0 & \cos R_E & -\sin R_E \\ 0 & \sin R_E & \cos R_E \end{bmatrix} \begin{bmatrix} E \\ N \\ U \end{bmatrix} + \begin{bmatrix} T_E \\ T_N \\ T_U \end{bmatrix} \tag{4-3}$$

where the rotations R_E, R_N, and R_U rotate the 3D vector (E, N, U) counterclockwise about the east, north, and up axes to obtain the transformed vector (E', N', U'). Note the presence of three rotation matrixes, as compared to just one in Equation (4-2). Also, note that the upper left 2×2 part of the left-most rotation matrix (about the U-axis) is the same as the 2D (horizonal) rotation matrix in Equation (4-2). This type of right-handed transformation is often used for terrestrial laser scanner data and to transform among various sensors in mobile and aerial survey applications (such as between camera and inertial measurement unit body frames). In many types of software implementation, the three rotation matrices in Equation (4-3) are replaced by quaternions, because they are more numerically stable and are computationally efficient (see Chapter 8 for a description of quaternions and their use in terrestrial laser scanning).

Vertical Transformations. Vertical transformations are probably the most common type of transformation used for surveying data. Such transformations can range from a simple single vertical shift to "inclined plane" corrections as well as to complex grid-based models that take into account various physical or other phenomena that effect heights, such as topographic gravitation, network distortion, sea surface topography, and hydraulic correctors (discussed later in this chapter). NGS provides several grid-based vertical transformation models, which again is discussed later in this chapter.

The widespread adoption of GNSS technology is the main reason vertical transformations are so commonly used. GNSS equipment provides purely geometric coordinates, and the "vertical" component is ellipsoid heights. However, in most engineering applications, orthometric heights ("elevations") are preferred because they more accurately indicate where water will go under the influence of gravity. In fact, the only type of height that gives true change in hydraulic head is *dynamic height*, as discussed in Chapters 2 and 7. A detailed discussion of dynamic heights is beyond the scope of this chapter, but they are mentioned again in the section on grid-based vertical transformations in the context of Great Lakes hydraulic correctors (and are covered in more detail in Chapters 2 and 7). Although orthometric heights technically give only an approximate change in hydraulic head, they often provide sufficiently accurate estimates for many engineering design and analysis applications.

An additional advantage of orthometric heights is that their relationship with ellipsoid heights is simple. As discussed in Chapter 2 (and shown in Figure 4-2), the relationship between the two is given by $h = H + N_g$, where h is the ellipsoid height, H is the orthometric height, and N_g is the geoid height (sometimes called *geoid undulation* or *separation*). N_g is negative everywhere in CONUS, and so $h < H$ (but $h > H$ in most of Alaska). Strictly speaking, the relationship is actually approximate ($h \approx H + N_g$), because the plumbline for H is curved (as shown in Figure 4-2) and so it intersects the geoid at a different location from h. However, in most locations on Earth, this approximation usually causes $\ll 1$ mm of error, and so for practical applications, it can be considered negligible.

As discussed in Chapter 2, NGS has created a series of "hybrid" geoid models that convert between the North American Datum of 1983 (NAD 83) ellipsoid heights and North American Vertical Datum of 1988 (NAVD 88) orthometric heights. The latest of these models is GEOID18, which is specifically intended for use with NAD 83 (2011) epoch 2010.00 ellipsoid heights (Ahlgren et al. 2020). So, if such an ellipsoid height is known, simply using GEOID18 will give an NAVD 88 orthometric height. In CONUS, the estimated overall accuracy of the model is 1.39 cm root-mean-square error (RMSE), although it varies with location. A map of the estimated accuracy of GEOID18 is shown in Figure 4-3.

Using a hybrid geoid model alone with a NAD 83 ellipsoid height will rarely equal the exact NAVD 88 height of a leveled bench mark, because all three values contain error. A common way to resolve this discrepancy is to use GNSS to measure the h of a bench mark with a known H and compute a vertical shift, ΔV, so that

$$H = h - N_g + \Delta V \tag{4-4}$$

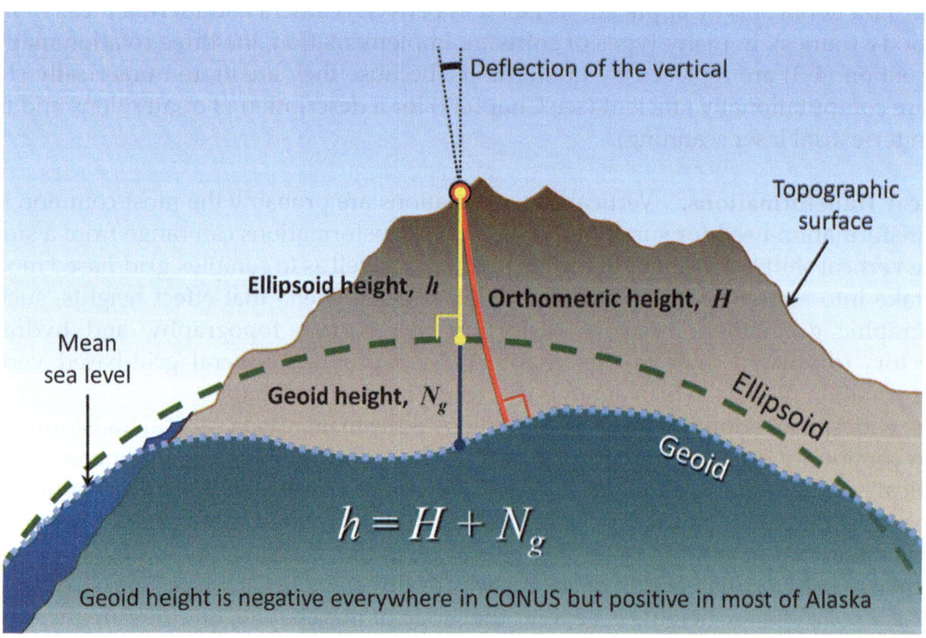

Figure 4-2. Relationship among orthometric, ellipsoid, and geoid heights.
Source: Courtesy of NOAA, reproduced with permission.

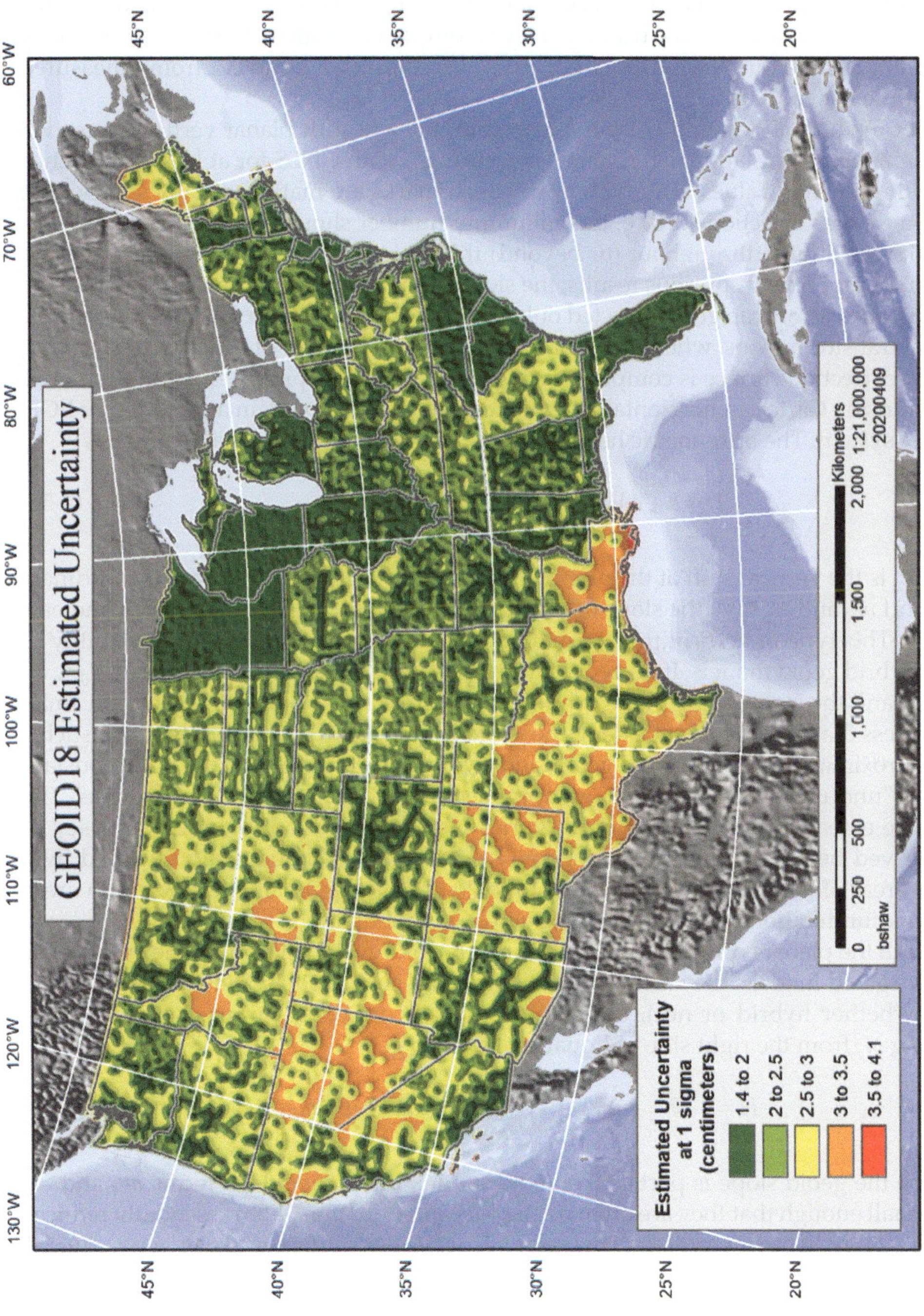

Figure 4-3. Estimated error of the GEOID18 model with respect to NAVD 88 (at 68% confidence).
Source: Courtesy of NOAA, reproduced with permission.

equals the published value for a specific bench mark (e.g., one selected as vertical control for an engineering project). The relative accuracy of the hybrid geoid model is then used to carry orthometric heights to other locations. Because the relative accuracy of the model is typically better than about 2 ppm, constraining to a single bench mark should give GNSS-derived orthometric heights to an accuracy of 2 cm (or better) within 10 km of the constrained bench mark. Such vertical accuracy is comparable with that of GNSS technology itself. A refinement of this process for larger areas is to compute a mean ΔV from a measured h for multiple bench marks.

Another method of deriving H from h is to compute a best-fit planar vertical correction surface for an area. This approach requires measuring h with GNSS for at least three bench marks with known H values. Although three bench marks are mathematically sufficient to compute a planar correction surface, this minimum number should never be used. At least four marks distributed throughout (or beyond) the project area should be used to ensure redundancy and prevent spurious results; the slope of the plane is very sensitive to errors in height, and such errors can go undetected only if three bench marks are used. As with the 2D conformal transformations, when more than the minimum number of bench marks is used, the planar correction surface is computed using least squares. Typically, the computation is performed using Cartesian horizontal coordinates, which could be from a projected or LGH coordinate system. The orthometric height, H, at point (N, E) with ellipsoid height h is

$$H = h + \Delta V_0 + (N - N_0)m_N + (E - E_0)m_E \tag{4-5a}$$

where ΔV_0 is the vertical shift at the origin coordinates (N_0, E_0) of the projected coordinate system, and m_N and m_E give the slope of the planar correction surface in the north and east directions. This type of vertical transformation was developed prior to the availability of accurate hybrid geoid models. If no geoid model is used with this approach, then $\Delta V_0 \approx N_g$ at (N_0, E_0), and the slope of the plane is intended to match the slope of the geoid. In small areas, say less than several km in extent, the geoid can be adequately modeled with such a planar approximation. However, the geoid slope with respect to the ellipsoid is not constant (the geoid "undulates"), and so, such an approach is not appropriate for large areas. The appropriate extent for such an approach varies with location and intended accuracy of GNSS-derived heights. In general, the geoid slope cannot be adequately modeled by a plane for areas of extent larger than about 10 to 20 km (and as small as a few km in areas where geoid undulations are more severe, such as in mountainous regions).

Although the planar correction surface approach was developed prior to the availability of accurate geoid models, they are still used along with such models. When a geoid model is used (whether hybrid or not), the sloping planar correction surface is modified by subtracting N_g from the right side of Equation (4-5a) to get

$$H = h - N_g + \Delta V_0 + (N - N_0)m_N + (E - E_0)m_E. \tag{4-5b}$$

Because the geoid slope is part of the geoid model, the planar slope terms m_N and m_E become small enough that they are often negligible, and Equation (4-5b) essentially reduces to Equation (4-4). In other words, a planar sloping correction surface is usually not required (and is not recommended) if an accurate high-resolution hybrid geoid model is used, such as GEOID18.

The vertical transformations in Equations (4-4) and (4-5a, b) in this section, together with the 2D conformal transformation of Equation (4-1a, b) of the previous section, constitute a process known as "calibration" or "localization" in most commercial GNSS surveying software. This process is explored in the following section.

On the Use of "Calibrations/Localizations" for GNSS-Derived Local Coordinates. GNSS positioning is based on a global (geodetic) coordinate system. Once the geocentric X, Y, Z coordinates have been determined, the GNSS part of the process is done. All subsequent operations used to obtain the final coordinate values are coordinate transformations or conversions of some type. The general overall sequence is as follows:

1. Transform geocentric X, Y, Z coordinates from the GNSS frame to a "local" or "regional" datum/frame, for example from the World Geodetic System 1984 (WGS 84) to the North American Datum of 1983 (NAD 83). Note that this is sometimes a "null" transformation that does nothing, for example by treating WGS 84 and NAD 83 as essentially the same, although they differ substantially (as discussed later in this chapter).
2. Convert local/regional X, Y, Z coordinates to local/regional latitude, longitude, and ellipsoid height (φ, λ, h).
3. Compute N, E "grid" coordinates from φ, λ using a map projection (or LGH) conversion.
4. Apply a geoid model to compute orthometric heights H ("elevations") from h.
5. *Optional:* Perform a "calibration/localization" to compute final coordinates.

"Calibration" and "localization" are terms used by various commercial software vendors. For this discussion, they can be considered synonymous, and "calibration" is usually used for both. They are transformations that generate final N, E, H coordinates from GNSS devices after Steps 1 through 4 have been completed. Although there is some minor variation in options and details among software vendors, all do essentially the same thing: use Equation (4-1a, b) for horizontal calibrations, and Equations (4-4) or (4-5a, b) for single- or multiple-point vertical calibrations, respectively.

Site calibrations are commonly used by surveying engineers and yet considerable confusion exists about their purpose, when they should be used, how they should be used, and even their mathematical form. On this latter point, calibrations are often generically (and misleadingly) described as a method for converting "WGS 84" to "local/regional" coordinates. Sometimes, these descriptions go so far as to describe a calibration as a 3D geodetic datum transformation. However, in reality, a calibration instead decomposes into two separate planar horizontal and vertical operations, and either can be performed without the other.

Horizontal Calibration/Localization

A horizontal calibration is the conformal transformation given by Equation (4-1a, b) or by Equation (4-2) if the software uses a right-handed convention for computations). It is performed on projected (grid) coordinates and not on geodetic coordinates. Thus, it is a purely planar (nongeodetic) operation. Before the calibration can be performed, the geodetic coordinates must first be projected to a mapping plane (Step 3 in the aforementioned sequence). Even if the user does not first select a projection, the software will automatically create one before performing a calibration. For example, it could be a Transverse Mercator projection with its central meridian through the first point in the calibration list (which may or may not be scaled "to ground"; see Chapter 3 for a description of "grid versus ground"). Then, the projected northings and eastings are transformed to best match the northings and eastings of the calibration points. This is usually done via a best-fit unweighted least-squares planar rotation, translation, and scale. Usually, an option exists to set the scale factor to exactly 1, if desired (often done to preserve the distances of the system being transformed).

A persistent perception seems to exist among users of high-accuracy GNSS—especially real-time GNSS—that a horizontal calibration must be performed to generate "grid" coordinates, particularly "local" project coordinates. However, this is not true. A map projection (or LGH) alone can generate grid coordinates. A horizontal calibration merely makes these grid coordinates match other grid coordinates from some other source. Doing so increases the complexity of a coordinate system by adding an additional four parameters to its definition. These parameters often cannot be used across different software platforms and are usually not compatible with GIS. These disadvantages limit data transferability and compromise georeferencing.

When creating a coordinate system for a project, rather than using a horizontal calibration to create a coordinate system that matches, say, a previous survey, the reverse can be done: use Equation (4-1a, b) to transform an existing survey to a rigorously defined projected coordinate system. This process might be called *anticalibration* or *antilocalization*, and it is particularly useful when combining multiple surveys of record (such as boundary surveyors for a large engineering project). Importantly, it allows creation of a coordinate system that is based on a standardized map projection definition, without the need for additional translation, rotation, and/or scale parameters obtained from a calibration. Using standardized definitions facilitates georeferencing and results in data that are fully compatible with other geospatial software such as GIS (Chapter 14).

Although a horizontal calibration usually is not needed for *defining* a coordinate system, it is nonetheless a useful and important tool for certain surveying applications. Three categories are listed as follows:

1. *When matching coordinates for undefined (or poorly defined) systems*, especially in real-time applications. For example, when matching point coordinates on engineering plans for construction staking with no coordinate system definition.
2. *When aligning real-time surveys with different datum realizations*, such as when using a real-time network (RTN) referenced to, say, SPCS 83 (2011) epoch 2010.00 for a project that requires using a previous realization of NAD 83 (although the best approach will be to perform a least-squares network adjustment, this option is not available for real-time applications). In this case, calibrating to stations in the previous realization of NAD 83 is a viable means for real-time alignment of a survey with a controlling set of coordinates over an area of limited extent (such as for a project). This is especially appropriate if the same projected coordinate system is used (which is usually the case).
3. *As a tool to facilitate searching for points in the field using real-time GNSS*, such as boundary corners generated by coordinate geometry from a survey plat. Such calibrations are not intended for defining final coordinate systems but rather are "disposable" because they can be refined repeatedly while searching and are discarded after the field searches are completed. Nonetheless, it is a tremendously useful application of horizontal calibrations.

Even in cases where a horizontal calibration is necessary (and is not disposable), it is best to first define a coordinate system that provides a good match with the target coordinates even before performing a calibration. This way, when the calibration is performed, the translation and rotations are small, and the scale is close to 1 (and can be forced exactly to 1). The guiding idea is that the calibration should be a minor adjustment of the coordinates (as would occur for Case 2 listed previously), rather than a major transformation. This is important because calibration is a planar (nongeodetic) operation, so large adjustments become problematic. As discussed previously, this problem becomes particularly acute as the size of the area covered increases, and when the coordinates are from different or undefined coordinate systems. The latter is often the case, especially for coordinates generated from records of survey, engineering plans, or legal descriptions. Because linear distortion is a function of both location and height, there is no simple rule as to the maximum size area appropriate for including in a horizontal calibration as part of a coordinate system definition. A rough guideline for keeping maximum residuals less than about 10 cm (0.3 ft) is to limit the maximum extent of calibration areas to within about 10 km (6 mi). For variation in topographic height, a 60 m (200 ft) change in height causes about 10 ppm of horizontal distortion (10 cm in a 10 km distance).

To illustrate the errors that can occur when performing horizontal calibrations for large areas, consider a 2D conformal transformation from one well-defined conformal system to another. The example here is a transformation from the Arizona Central zone of the State Plane Coordinate System of 1983 (SPCS 83 AZ C) to Zone 12 North of the Universal Transverse Mercator coordinate system (UTM 12N). The appropriate strategy to follow in this situation is to reproject rather than use a planar transformation. However, using a planar transformation shows how it is impacted by the size of an area, even when the coordinates are known perfectly and there is no measurement error. In this example, five points in common between the two systems are used to compute best-fit 2D conformal transformations (horizontal calibrations). The area considered is square, with four of the points at the corners and one in the center, as shown on the left side of Figure 4-4. Horizontal calibrations were performed (with the scale factor allowed to float) for boxes of increasing size, as shown in the plot on the right side of Figure 4-4. The plot gives the root-mean-square error (RMSE) in cm of the calibration as a function of the dimensions (in km) of the square box. For a square box with dimensions of 10 km, the RMSE is 4.7 cm. The RMSE increases rapidly as the size of the box increases, reaching 19 cm for a box with dimensions of 20 km. Note that the RMSE is the minimum possible for this case, because for an actual calibration, there will also be measurement errors.

The important point in this example is that significant calibration error can occur *only* because of differences in the spatial distribution of coordinate system linear distortion. For this case, both systems use the Transverse Mercator projection, and the difference in distortion is entirely because of the 55 arcminute separation between their central meridians (the difference in the central meridian scale is a nonvarying constant). As stated previously, the correct approach here is to reproject rather than calibrate. However, in many cases, this is not possible because often the target coordinate system is not well defined (and may not be known at all). In such situations, calibrations of large areas should be avoided. The only time that the size of an area does not contribute to calibration error is when both coordinate

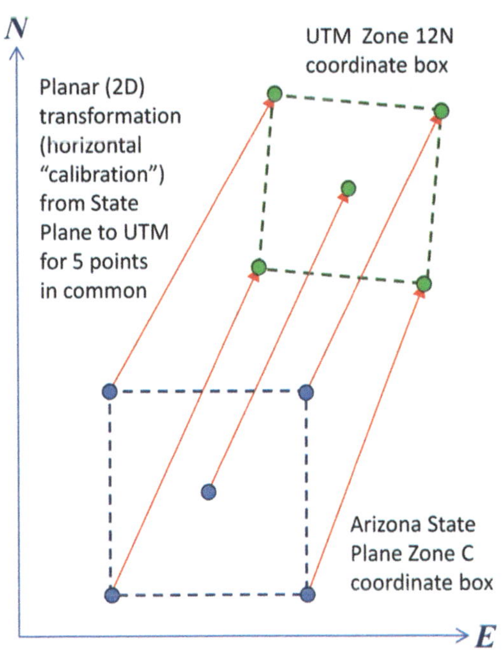

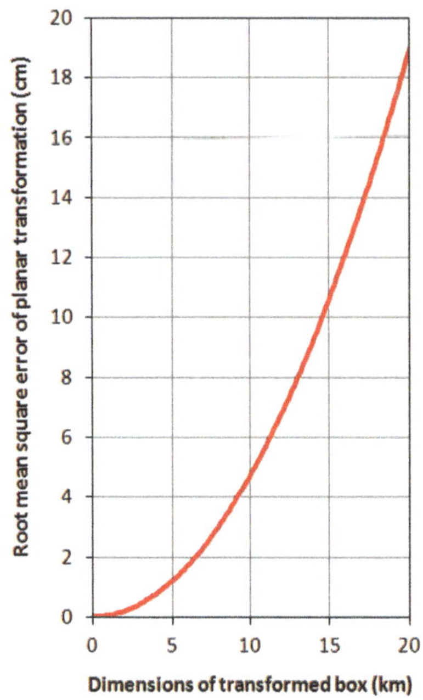

Figure 4-4. Error of planar transformation (horizontal calibration) as a function of the size of the area.

sets are based on projections with *exactly* the same parameters that contribute to linear distortion (not just the same projection *type*).

Vertical Calibration/Localization

A high-resolution hybrid geoid model, such as GEOID18, combined with an accurate NAD 83 (2011) epoch 2010.00 ellipsoid height, h, will almost never match the published NAVD 88 orthometric height, H, of a bench mark, whether it was determined from differential leveling or GNSS equipment in an NGS Height Modernization Survey. Matching a specific bench mark orthometric height is often a requirement for engineering projects (although measurement error means that such heights can never be perfectly known). To accomplish this, a vertical calibration can be performed that consists of a simple vertical shift, ΔV, to match a project bench mark, often combined with a hybrid geoid model, according to Equation (4-4). The relative accuracy of the hybrid geoid model then allows the determination of GNSS-derived H at other locations, usually within an accuracy of 2 ppm with respect to the constrained bench mark. Whenever possible, the GNSS-derived H should be checked against other bench marks of accuracy comparable to the constrained bench mark.

When multiple bench marks are available for calibration, the mean vertical offset for all bench marks can be used to determine the needed vertical shift. This also provides redundancy and confirmation of the relative accuracy of the hybrid geoid model. However, commercial software is not designed to use a single vertical offset with multiple points. Instead, it automatically uses all vertical calibration points with unweighted least squares to compute the planar correction surface given by Equation (4-5a, b). It is important to note that this is a tilted planar *correction* surface and not a plane per se. That is, it is a model that applies vertical correction values that vary linearly with respect to the projected coordinates. So, like a horizontal calibration, a vertical calibration requires grid coordinates *before* the calibration can be performed.

If no hybrid geoid model is used, the multipoint vertical calibration correction surface of Equation (4-5a) provides a planar model of the local geoid. If a geoid model is used, the correction surface is computed after the geoid heights are applied, according to Equation (4-5b). However, using an inclined planar correction surface along with a hybrid geoid model is usually unnecessary and can even be counterproductive. The maximum curvature (rate of change of slope) of GEOID18 in CONUS is about ± 1 ppm/km, and about half of CONUS has a magnitude >0.03 ppm/km, which means the slope changes by more than 0.6 ppm in 20 km. So for large areas (long dimensions of more than about 20 km), it makes no sense to use a planar model because the geoid surface "undulates," and modeling it with a plane will usually *increase* error. However, even in cases where the geoid slope itself is essentially planar, the slope of a planar correction surface is usually at the "noise" level, that is, less than can be reliably measured with GNSS. In such situations, an inclined planar correction surface will not hurt anything, but it does something else: it creates a coordinate system definition that increases complexity without improving performance. When using vertical calibration points that have accurate leveled heights over a large enough area (say more than a few kilometers in extent), the slope of the correction surface is usually less than 1 ppm (0.2 arcsecond). It is somewhat absurd to use such a correction, because the "improvement" to GNSS-derived orthometric heights is not even detectable (1 ppm = 1 cm in 10 km). This leads to some rules of thumb for evaluating the slope of multipoint vertical calibrations done with a hybrid geoid model:

- If the slope >2 ppm (0.4 arcsecond), there may be a problem, and both h and H of the points used should be checked.
- If the slope >5 ppm (1 arcsecond), then something is probably wrong and at least one of the calibration points is likely in error.

- If the slope >10 ppm (2 arcseconds), something is definitely wrong (unless all calibration points are in a very small area, say less than 1 km in extent).

If a geoid model is not used, then the slope of the correction surface should match the direction and magnitude of the geoid model slope to the ppm levels cited previously. The slope magnitude of GEOID18 in CONUS is shown in Figure 4-5. The average geoid slope in CONUS is about 30 ppm (6 arcseconds), which means that, on average, the change in orthometric height differs from the change in ellipsoid height by 30 cm (1 ft) in 10 km—a large difference when attempting to determine orthometric height differences with GNSS equipment. Less than 0.8% of CONUS has a geoid slope >100 ppm, mostly (but not entirely) in areas of rugged topography in the western United States. So, if the slope of the correction surface (without a hybrid geoid model) is greater than 100 ppm, there is likely an error in the vertical calibration. The maximum geoid slope in CONUS is approximately 250 ppm.

These guidelines likely work well for most places in CONUS. In other areas (such as Alaska), the situation can be quite different because of the lack of bench marks for creating hybrid geoid models. Still they should provide reasonable guidance even in such areas.

The residuals of a multipoint vertical calibration should also be used to evaluate the quality of the calibration. As with horizontal calibration, the RMSE of the residuals should be roughly the same or better than the accuracy of the h or H values used (whichever has greater error). However, small residuals do not guarantee good results, which is why the slope of the planar correction surface should also be evaluated. Obviously, if three vertical calibration points are used, the residuals will be zero, and it is recommended to *never* use only three points. However, even when more are used, it is still possible to get small residuals even with erroneous calibration results, depending on the number and distribution of vertical calibration points. In summary, the best approach for most situations is to use a hybrid geoid model and *not* use an inclined planar correction surface (i.e., to not perform a multipoint vertical calibration).

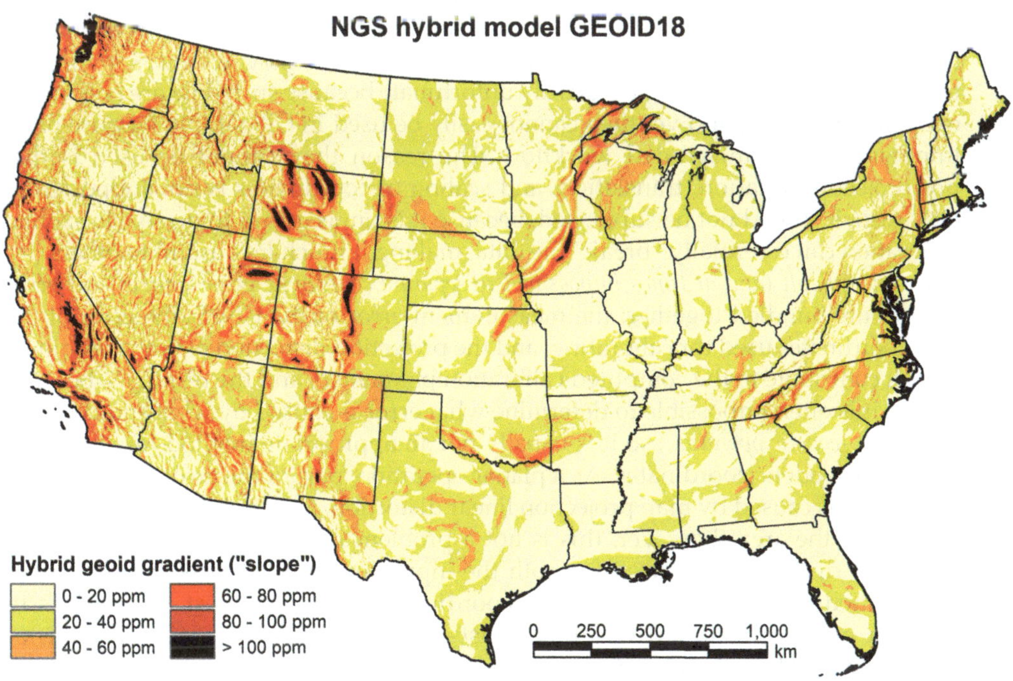

Figure 4-5. GEOID18 slope magnitude with respect to the NAD 83 GRS 80 ellipsoid. Source: Courtesy of NOAA, reproduced with permission.

Given the complications posed by determining orthometric heights with GNSS, it is worth noting that NGS will adopt a new geopotential ("vertical") datum as part of modernizing the National Spatial Reference System (NGS 2021b) in the 2025 to 2027 time frame. This geopotential datum will be defined entirely by a purely gravimetric geoid model, and so access to orthometric heights will be via GNSS-derived ellipsoid heights. The geopotential datum should do away with most of the vertical calibration issues discussed here. However, situations will still exist where it will be necessary to optimally combine GNSS with leveling, and so familiarity with these technologies and the relationship between ellipsoid and orthometric heights will still be required.

Summary Discussion on Calibration/Localization

To summarize, horizontal calibrations are usually unnecessary for defining coordinate systems, and vertical calibrations are often necessary but can usually be done without resorting to an inclined planar correction surface based on multiple points (by using a one-point vertical calibration or applying a constant vertical shift). Calibration is nonetheless popular because of an important advantage: it is easy. Although easy is laudable, it comes with two disadvantages: (1) it is also easy to accidentally generate erroneous results, and (2) it is only easy for the person doing the calibration. Other data users downstream suffer, because the resulting coordinate system definitions are often more complicated than needed, and the approach is not standardized. Except for cases where a calibration is used to best-fit an existing well-defined coordinate system, a system defined by horizontal calibration is not compatible with GIS. The disadvantages with calibrations are summarized as follows:

- *Increases the complexity of a coordinate system definition without improving performance.* In a vast majority of applications, the horizontal calibration is completely unnecessary for defining a coordinate system, local or otherwise. Yet, a common mistaken belief persists that a calibration must be performed to get local coordinates.
- *Decreases data transferability.* The use of local coordinate systems based on calibrations (especially horizontal calibrations) is probably the main reason why it is so difficult to get survey data into GIS. This is truly unfortunate because such data are potentially extremely valuable, and yet they cannot easily be used in GIS simply because of an unnecessary and misguided workflow. The reason they cannot be used is that postprojection rotation, translation, and scaling of coordinates is not supported in GIS. Also, it should not be supported, because it is unnecessary to do so. Horizontal calibration can essentially "break" the data, making it much less useful to others.
- *Makes it difficult (and at times impossible) to separate coordinate error sources.* All error sources are bundled together: the measurement error of the GNSS, the error of the grid coordinates used for calibrating, and the projection distortion of both coordinate sets all contribute to the computed calibration parameters and the residuals. Because of this, it is often impossible to determine what error is caused by what source.
- *It is not a geodetic operation,* so it is usually not an appropriate method for aligning a survey to geodetic coordinates. As a planar operation, part of the error budget in the calibration is caused by map projection linear distortion, an effect that increases with the size of the area (however, this is not a problem if a projection with the same distortion distribution is used for both the input and output). It is in addition an unweighted operation based only on coordinate values. The best approach is to perform an appropriately weighted least-squares adjustment of a network of redundant GNSS vectors (see Chapters 5 and 11).
- *Too much emphasis is placed on inspecting only the residuals for assessing quality.* Although residuals are important for evaluating a calibration, it is also important to be aware

that small residuals do not guarantee good results. Other characteristics should also be evaluated, such as the scale of a horizontal calibration and the slope of a multipoint vertical calibration.

Global Equation–Based Coordinate Transformations

Helmert Transformations. The transformations in this section are based on equations applied to global coordinate systems, even when they are only intended for specific regions. Often they go by the name *datum*, *reference frame*, or *geographic coordinate* transformations. Although in principle they can be defined for small areas, in most cases, they are used for countries, continents, or the entire Earth. This can be partly attributed to the equations being global in scope, so determining reliable transformation parameters from standard least-squares procedures using coordinates at common points should be done over large areas (say more than several hundred kilometers in extent). Because of the global scale of the equations, for the remainder of this section, they will be referred to as datum transformations.

Equation-based datum transformations are usually 3D and applied to the entire Earth coordinate system. Perhaps the most common type is the Helmert (similarity) transformation (but other 2D types are addressed later in this chapter). It is a conformal transformation, so it preserves shape because the scale is constant at a point (i.e., does not vary with direction). In general, three translations, three rotations, and a scale are applied to the coordinate system. Common current practice is to first convert latitude, longitude, and ellipsoid height to geocentric Cartesian (X, Y, Z) coordinates and then perform a Helmert transformation. The transformation is typically described, as listed subsequently, by the number of parameters used (although the first two are special cases of the general linearized Helmert 14-parameter transformation):

- *3-Parameter*. Three-dimensional translation of origin as $T_X = \Delta X$, $T_Y = \Delta Y$, and $T_Z = \Delta Z$
- *7-Parameter*. Three translations *plus* three rotations (one about each of the axes, R_X, R_Y, R_Z), *plus* a scale factor, S.
- *14-Parameter*. A 7-parameter transformation where each of the seven parameter varies with time, that is, each has a *velocity* (a linear rate of change with time).

After the transformation is performed, the new X', Y', Z' values are converted to new latitude, longitude, and ellipsoid height values. This process is illustrated schematically in Figure 4-6. Note that the dimensions of the reference ellipsoid (a and b axes) will only change if the input and output ellipsoids are different. The 3-, 7-, and 14-parameter transformations are discussed in more detail subsequently.

A "solid body" transformation is also possible where all parameters are zero and the only change is in the ellipsoid parameters (i.e., the size and shape of the ellipsoid). Such a transformation will only change latitude and ellipsoid height and not longitude. An example of this was given in Chapter 3 for scaling the ellipsoid to the topographic surface to create a low-distortion map projection. This type of transformation is also commonly used in geospatial software as a "null" transformation to enforce equivalence between frames, for example the commonly assumed equivalence between NAD 83 and generic "WGS 84" (this is further discussed subsequently).

The Helmert transformation was given previously in Equation (4-3) for 3D transformation of local coordinates. The only difference for datum transformations is that local east, north, up (E, N, U) coordinates are replaced with global geocentric (X, Y, Z) coordinates and rotations about the (X, Y, Z) axes, which gives

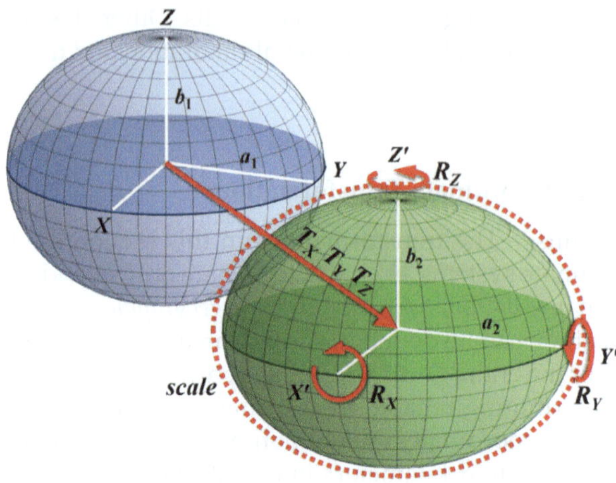

Figure 4-6. Helmert 7-parameter datum transformation and reference ellipsoids.

$$\begin{bmatrix} X' \\ Y' \\ Z' \end{bmatrix} = (1+S) \begin{bmatrix} \cos R_Z & \sin R_Z & 0 \\ -\sin R_Z & \cos R_Z & 0 \\ 0 & 0 & 1 \end{bmatrix} \begin{bmatrix} \cos R_Y & 0 & -\sin R_Y \\ 0 & 1 & 0 \\ \sin R_Y & 0 & \cos R_Y \end{bmatrix} \begin{bmatrix} 1 & 0 & 0 \\ 0 & \cos R_X & \sin R_X \\ 0 & -\sin R_X & \cos R_X \end{bmatrix} \begin{bmatrix} X \\ Y \\ Z \end{bmatrix} + \begin{bmatrix} T_X \\ T_Y \\ T_Z \end{bmatrix} \tag{4-6}$$

Equation (4-6) can be simplified considerably because the rotations for datum transformations are typically very small (usually much less than 1 arcsecond), so small-angle approximations of $\sin R \approx R$ and $\cos R \approx 1$ can be used (for R in radians). Using these approximations, the three rotation matrices are multiplied together to give a single 3×3 rotation matrix. The rotation matrix can be further simplified by assuming that the product of rotations is negligible (because the rotations are small). For rotations of less than 1 arcsecond, the error of these combined approximations for 3D coordinates at the surface of the Earth is less than 1 mm. These simplifications give the following transformation equation (which must be evaluated using rotations in radians):

$$\begin{bmatrix} X' \\ Y' \\ Z' \end{bmatrix} = \begin{bmatrix} T_X \\ T_Y \\ T_Z \end{bmatrix} + \begin{bmatrix} 1+S & R_Z & -R_Y \\ -R_Z & 1+S & R_X \\ R_Y & -R_X & 1+S \end{bmatrix} \begin{bmatrix} X \\ Y \\ Z \end{bmatrix} \tag{4-7a}$$

Equation (4-7a) can be expanded into three individual equations:

$$\begin{aligned} X' &= T_X + (1+S)X + R_Z Y - R_Y Z \\ Y' &= T_Y - R_Z X + (1+S)Y + R_X Z \\ Z' &= T_Z + R_Y X - R_X Y + (1+S)Z \end{aligned} \tag{4-7b}$$

The simplification represented by Equation (4-7a, b) is often called the *Bursa–Wolf transformation*, although it will continue to be referred to as the (simplified) *Helmert transformation* throughout this chapter. It is typically evaluated with counterclockwise positive rotations in the United States, Canada, and Australia, whereas clockwise positive rotations are commonly used in Europe, by NATO, and by the International Earth Rotation

Service (IERS). In commercial geospatial software, using counterclockwise positive rotations is sometimes called the *coordinate frame* convention, and clockwise positive rotations are called the *position vector* convention. The two are usually treated as identical, apart from the rotation sense, but they actually differ. For the coordinate frame convention, the points remain fixed, and the frame axes rotate counterclockwise. For the position vector convention, the frame remains fixed and the points (as vectors from the origin) are rotated clockwise. Apart from the rotation sense, the two methods yield essentially identical results, and in this chapter, the coordinate frame method is used for Helmert transformations (with counterclockwise positive rotations).

As indicated previously, a Helmert transformation can be reduced to a 3-parameter transformation by removing the rotation and scale parameters so that there are only three translations, simply $X' = X + T_X$, $Y' = Y + T_Y$, and $Z' = Z + T_Z$. Although less accurate than 7-parameter transformations, they are commonly used, especially for relating older local datums to the global reference frame. The US Defense Mapping Agency (now the National Geospatial-intelligence Agency, NGA) defined the relationship between the WGS 84 datum and about 125 local/regional datums throughout the world using 3-parameter transformations (NGA 2014). These transformations were widely adopted in commercial software and are still in use today, even in situations where they are no longer appropriate.

The Helmert transformation equations can be extended so that each of the seven parameters varies with time, to create a 14-parameter transformation. The change with time is a *velocity* (rate of change with time) and is denoted by placing a dot over the symbol, for example, $\dot{R}_X$, $\dot{T}_X$, $\dot{S}$. At time t for a time span of Δt with respect to some reference time t_0 (i.e., $\Delta t = t - t_0$), the time-varying Helmert transformation parameters are as follows:

$$
\begin{array}{lll}
\text{Translation} & \text{Rotation} & \text{Scale} \\
T_X(t)=T_X(t_0)+\dot{T}_X\Delta t & R_X(t)=R_X(t_0)+\dot{R}_X\Delta t & S(t)=S(t_0)+\dot{S}\Delta t \\
T_Y(t)=T_Y(t_0)+\dot{T}_Y\Delta t & R_Y(t)=R_Y(t_0)+\dot{R}_Y\Delta t & \\
T_Z(t)=T_Z(t_0)+\dot{T}_Z\Delta t & R_Z(t)=R_Z(t_0)+\dot{R}_Z\Delta t &
\end{array}
\tag{4-8}
$$

Equations (4-7a, b), and (4-8) together represent the general 14-parameter transformation. If Δt is zero or none of the parameters vary with time, it reduces to a 7-parameter transformation. Any 14-parameter transformations can be recast as a 7-parameter transformation by replacing the translation, rotation, and scale parameters at time t_0 with the values at the time of interest, $t = t_0 + \Delta t$, according to Equation (4-8).

If the parameters are desired at an epoch t that differs from t_0, the following equation can be used:

$$P(t) = P(t_0) + \dot{P} \times (t - t_0) \tag{4-9}$$

where

P = Any of the transformation parameters in Equations (4-7a, b) or (4-8),
t = Desired epoch,
t_0 = Given epoch, and
$\dot{p}$ = Rate (velocity) of the parameter.

Transformations among Commonly Used Reference Frames. Helmert transformations are often used to convert geodetic coordinates among reference frames. Commercial geospatial software can contain hundreds of parameter sets for different versions and combinations of transformations, far too many to list here. For example, the European Petroleum Survey Group (EPSG) Geodetic Parameter Dataset (EPSG 2022) is widely used by software vendors to obtain transformation parameters; as of February 2022, it contained more than 2200 transformations (although these include transformations other than Helmert). Another commonly used source for geodetic transformation definitions is the International Standards Organization (ISO) Geodetic Registry (ISO 2022).

The most common global frame in such software is generic "WGS 84," but it usually lacks any clear identification as to which WGS 84 realization is used for a particular parameter set. In addition, parameters for this and other frames often lack epoch dates and there can be multiple forms of what appears to be the "same" transformation. This has led to considerable confusion of which transformation parameters are "best" for various applications. This section will focus on two commonly used global reference systems and one regional system. These are the International Terrestrial Reference System, ITRS (IGN 2021); the World Geodetic System of 1984, WGS 84 (NGA 2014); and the three frames of the North American Datum of 1983, NAD 83 (Schwarz 1989, Snay 2012).

International Terrestrial Reference System

The International Terrestrial Reference System (ITRS) is important because most modern reference frames (including WGS 84) have been aligned to one or more of its various realizations, called International Terrestrial Reference Frames (ITRFs) (IGN 2021). The ITRFs have changed over time as positioning abilities have improved and to accommodate the relative movement of stations within the frame (such as caused by plate tectonics and earthquakes). Each ITRF is based on multiple geodetic techniques, usually GNSS, very long baseline Interferometry (VLBI), satellite laser ranging (SLR), and Doppler orbitography and radiopositioning integrated by satellite (DORIS). There have been 13 ITRFs, from ITRF88 to ITRF2014 (Altamimi et al. 2016, 2017). Transformation parameters for each are listed in Table 4-1, from ITRF2014 to all prior ITRFs at epoch 2010.00, in reverse chronological order, as defined by IERS, but with the opposite rotation convention (IGN 2020).

The ITRF transformation parameters in Table 4-1 also apply to realizations of the International GNSS Service (IGS), which are closely associated with the ITRS (IGS 2017). These are GNSS-only realizations aligned to a specific existing ITRF. For example, both IGS14 and IGb14 are GNSS-only realizations determined after—and aligned with—ITRF2014. A lower-case letter replacing "S" indicates that the IGS solution has been updated (often because of changes, removals, or additions of a few specific GNSS stations), but it is still aligned with the same ITRF. The resulting coordinate differences at individual existing stations are usually minor (a few cm or less) and may not affect all stations, but the overall (global) change is negligible, and so they are all considered identical. That is, the transformation parameters between ITRF2014, IGS14, and IGb14 are defined as zero (although small coordinate differences can exist at individual stations). The same is true for the other IGS/IGb solutions, as provided in Table 4-1, in which they are grouped with their parent ITRFs.

The trend in Table 4-1 for the ITRF translation and scale parameters show that the change from one ITRF to the next usually decreases with time. The trend for rotations is not quite so clear, although the change for all three rotation components is zero for the last three ITRFs. Figure 4-7 illustrates the maximum change in parameters with respect to ITRF2014 at epoch 2010.0. Translations and rotations are given as 3D values, with rotations given at a radius of 6137 km to represent horizontal change at the surface of the Earth and in the same linear units as the 3D translation (mm). Likewise, scale change is given at the same radius to represent change in height at the Earth surface. Change in both translation and scale shows a fairly consistent decline with time as the ITRS becomes more refined. A large

Table 4-1. Transformation Parameters from ITRF2014 to Past ITRFs, WGS 84, and the Three NAD 83 Frames, all at Epoch 2010.0 (Rotations Counterclockwise Positive).

Parameters	T_X mm	T_Y mm	T_Z mm	R_X mas	R_Y Mas	R_Z mas	S ppb
Parameter rates (velocities)	mm/year	mm/year	mm/year	mas/year	mas/year	mas/year	ppb/year
ITRF2008 [IGS08; IGb08; WGS 84 (G1674, G1762)]	1.6	1.9	2.4	0.00	0.00	0.00	−0.02
	0.0	*0.0*	*−0.1*	*0.00*	*0.00*	*0.00*	*0.03*
ITRF2005 [IGS05]	2.6	1.0	−2.3	0.00	0.00	0.00	0.92
	0.3	*0.0*	*−0.1*	*0.00*	*0.00*	*0.00*	*0.03*
ITRF2000 [IGS00; IGb00; WGS 84 (G1150)]	0.7	1.2	−26.1	0.00	0.00	0.00	2.12
	0.1	*0.1*	*−1.9*	*0.00*	*0.00*	*0.00*	*0.11*
ITRF97 = ITRF96 = ITRF94 [~WGS 84 (G873)[a]]	7.4	−0.5	−62.8	0.00	0.00	−0.26	3.80
	0.1	*−0.5*	*−3.3*	*0.00*	*0.00*	*−0.02*	*0.12*
ITRF93	−50.4	3.3	−60.2	2.81	3.38	−0.40	4.29
	−2.8	*−0.1*	*−2.5*	*0.11*	*0.19*	*−0.07*	*0.12*
ITRF92	15.4	1.5	−70.8	0.00	0.00	−0.26	3.09
	0.1	*−0.5*	*−3.3*	*0.00*	*0.00*	*−0.02*	*0.12*
ITRF91 [~WGS 84 (G730)[a]]	27.4	15.5	−76.3	0.00	0.00	−0.26	4.49
	0.1	*−0.5*	*−3.3*	*0.00*	*0.00*	*−0.02*	*0.12*
ITRF90	25.4	11.5	−92.8	0.00	0.00	−0.26	4.79
	0.1	*−0.5*	*−3.3*	*0.00*	*0.00*	*−0.02*	*0.12*

(Continued)

Table 4-1. (*Continued*)

Parameters	T_X mm	T_Y mm	T_Z mm	R_X mas	R_Y Mas	R_Z mas	S ppb
Parameter rates (velocities)	mm/year	mm/year	mm/year	mas/year	mas/year	mas/year	ppb/year
ITRF89	30.4	35.5	−130.8	0.00	0.00	−0.26	8.19
	0.1	*−0.5*	*−3.3*	*0.00*	*0.00*	*−0.02*	*0.12*
ITRF88	25.4	−0.5	−154.8	−0.10	0.00	−0.26	11.29
	0.1	*−0.5*	*−3.3*	*0.00*	*0.00*	*−0.02*	*0.12*
WGS 84 original (Transit)	85.4	−505.5	−315.8	−18.30	0.30	−7.26	−6.21
	0.1	*−0.5*	*−3.3*	*0.00*	*0.00*	*−0.02*	*0.12*
NAD 83 (North America plate)[b]	1005.30	−1909.21	−541.57	26.78138	−0.42027	10.93206	0.36891
	0.79	*−0.60*	*−1.44*	*0.06667*	*−0.75744*	*−0.05133*	*−0.07201*
NAD 83 (Pacific plate)	910.9	−2012.9	−586.3	22.749	26.560	−25.706	2.12
	0.1	*0.1*	*−1.9*	*−0.384*	*1.007*	*−2.186*	*0.11*
NAD 83 (Mariana plate)	910.9	−2012.9	−586.3	28.711	11.785	4.417	2.12
	0.1	*0.1*	*−1.9*	*−0.02*	*0.105*	*−0.347*	*0.11*

[a]*WGS 84 (G873) and (G730) aligned with the ITRF to within approximately ±5 cm and ±10 cm, respectively.*

[b]*Based on IGS-adopted relationship where ITRF96 ≠ ITRF97; NAD 83 transformations for the Pacific and Mariana plates are based on the IERS-adopted relationship where ITRF96 = ITRF97.*

Source: Courtesy of NOAA, reproduced with permission.

Note: IGS14, IGb14, and WGS 84 (G2139) considered equivalent to (i.e., aligned with) ITRF2014.

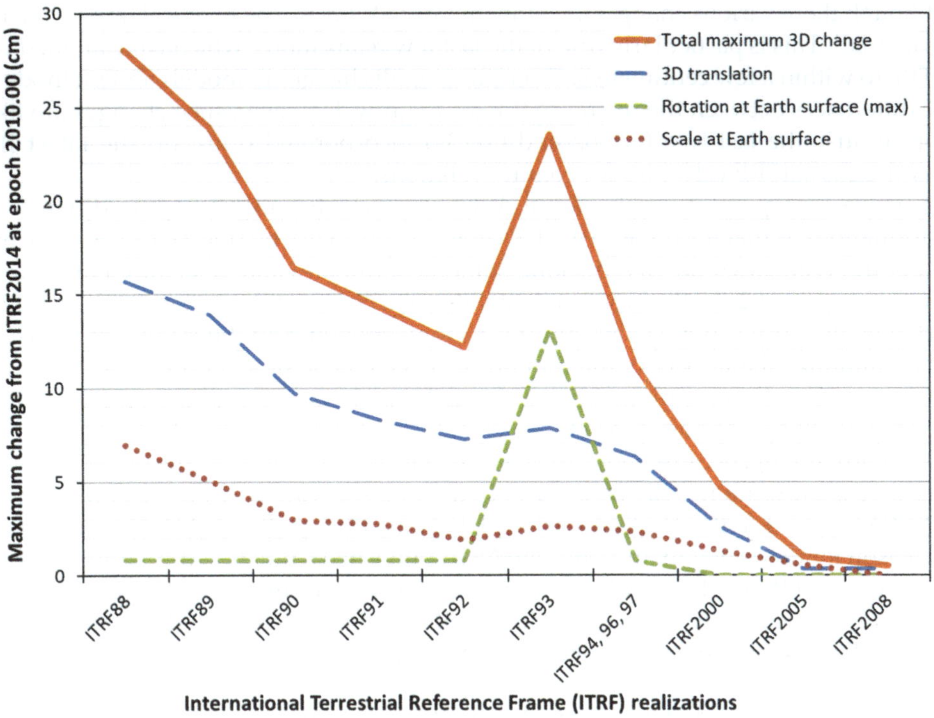

Figure 4-7. Magnitude of translation, rotation, and scale components at the nominal earth surface for transformation from ITRF2014 to past ITRFs (at epoch 2010).

spike in the change of rotation occurs for ITRF93 because it was artificially rotated. The following ITRF94 realization was reset to enforce the condition that the sum of all tectonic plate rotations should be zero (called *no net rotation*); all ITRFs other than ITRF93 conform to that condition. This figure shows that the ITRS has become very stable in the last three realizations; the maximum 3D change since ITRF2005 (released in 2006) has been less than about 1 cm.

World Geodetic System of 1984

The situation for WGS 84 is similar to that for ITRS, and there have been a total of seven realizations. The first has no specific identifier and is called WGS 84 *original* or *Transit* (because of its initial relationship with the previous Transit Doppler system). The subsequent six realizations each have an identifier of the form "G*nnnn*," where *nnnn* is the GPS week that the realization was adopted (numbered sequentially beginning the week of January 6, 1980). Each is listed in Table 4-1 associated with the ITRF to which it was aligned. The latest is WGS 84 (G2139), which corresponds to January 2021 and is aligned with ITRF2014 (NGA 2021). The prior two realizations, WGS 84 (G1674) and (G1762), were both aligned with ITRF2008 (NGA 2014). The reason for the two WGS 84 realizations aligned with ITRF2008 is that the second one (G1762) conformed better to IERS 2010 conventions (Petit and Luzum 2010), and so it is better aligned with ITRF2008. Although G1674 and G1762 are shown as identical in Table 4-1, there is actually a systematic difference between them. The largest component of the difference is manifested as a decrease in ellipsoid height of about 4 to5 cm at the surface of the Earth. This was largely causes by NGA changing how it modeled GPS antennas in the G1762 solution.

Although these various changes are refinements to WGS 84, the practical impact on most users is small. This is particularly true of the last few realizations, which were aligned with the ITRF to within a few centimeters. Over time, small changes in models and methods can cause differences between the reference frames to grow. However, such changes are likely minor because the last two (G1762 and G2139) incorporated (most of) the latest IERS conventions and ITRF velocities for common stations.

An important point to note about WGS 84, which can affect coordinates and transformations, is that the broadcast GPS satellite coordinates broadcast are set to the mid-epoch of the year, based on GPS tracking station coordinates and velocities. For example, for the entire year of 2022, the broadcast satellite coordinates will be at an epoch of 2022.50. The change in epoch is done to account for tectonic motion, and the middle of the year is used to minimize the impact of the motion over the course of an entire year. Horizontal motion caused by tectonic plate rotation varies by plate (and location on a plate), but at some NGA tracking stations, it can exceed 7 cm/yr with respect to the WGS 84 frame. This is a fact worth noting, because most users access the WGS 84 frame via the broadcast orbits.

Another complication with WGS 84 is that some software assumes it is the same as NAD 83 referenced to the North America plate (sometimes called NAD 83 *CONUS*). The main reason for this is that many years ago NGA published a null (identity) transformation between original WGS 84 and NAD 83 CONUS. This transformation was widely adopted, but it was only accurate to ±2 m (according to NGA). Since then, IERS determined a more accurate relationship between original WGS 84 (Transit) and ITRF90, and this is the one given in Table 4-1.

The WGS 84 transformation parameters in Table 4-1 are based on the best available information and are consistent with the parameters used in the NGS Horizontal Time-Dependent Positioning (HTDP) software current at this time of this writing (March 2022). However, readers should be aware that it is not possible to provide authoritative WGS 84 transformations, because it is a product of NGA within the Department of Defense. As such, nonauthorized individuals and organizations do not have access to the information used to define the WGS 84 realizations or their actual relationships with the various ITRFs. Therefore, the parameters involving WGS 84 in Table 4-1 should be considered approximate and used with caution. For a more detailed discussion of WGS 84 transformations, see Kelly and Dennis (2022).

Three Frames of the North American Datum of 1983

The three NAD 83 frames are defined with respect to three tectonic plates: North America, Pacific, and Mariana. The latest realization of these frames is at an epoch of 2010.00 and is based on a nationwide adjustment of NGS control begun in 2011 (Dennis 2020). NGS identifies NAD 83 realizations by so-called datum tags in parentheses after "NAD 83," which is usually the year the initial adjustment was done for a particular realization. In this case, this would be 2011, but to distinguish each of the frames from one another, they are instead identified by four characters "*xx*11" specified as follows:

(2011) = North America plate (e.g., CONUS, Alaska),
(PA11) = Pacific plate (e.g., Hawaii, American Samoa), and
(MA11) = Mariana plate (e.g., Guam, Commonwealth of Northern Mariana Islands).

Transformation parameters for the NAD 83 frames are given at the bottom of Table 4-1. Both the translation and rotation parameters are large in comparison with ITRF values. For the translations, this is because of the nongeocentricity of NAD 83; its origin is offset from the geocentric ITRS origin by the magnitude of the translations (approximately 2.2 m). The large rotations are in part because of the NAD 83 frames being fixed to specific tectonic plates, each of which rotates horizontally with respect to the ITRS. Note that the translation and scale parameters for the PA11 and MA11 realizations are identical; only their rotations differ. This is because both were aligned with ITRF2000 such that the two were identical with each other at epoch 1993.62 and with NAD 83 referenced to the North America plate (Snay 2003). However, they are each referenced to different tectonic plates, which are parameterized as rotations and are manifested as horizontal change only.

For each of the three NAD 83 frames, the transformation parameters given in Table 4-1 remain the same for all realizations of that frame; they do not change with a new NAD 83 realization. This is because they were defined by their relationship with a specific ITRF at a specific epoch, and this relationship remains fixed (although the ITRF itself has been updated several times). NAD 83 referenced to the North America tectonic plate was jointly defined by the United States and Canada in 1999, and this definition has not changed (Craymer et al. 2000, NGS 2000). A set of parameters was adopted relating NAD 83 to ITRF96 at epoch 1997.0. However, the equivalence ITRF96 = ITRF97 defined by IERS (as given in Table 4-1) was *not* used for NAD 83. Instead, the relationship determined by IGS with ITRF96 ≠ ITRF97 (based on GPS data only) was used instead (Soler and Snay 2004). Normally, all parameters between IGS and ITRF realizations are the same, but they were not in this one case. The United States and Canada decided to adopt the IGS convention of ITRF96 ≠ ITRF97 for NAD 83 because GPS was the dominant technology used for precise geodetic work. This decision has created a somewhat complicated relationship between NAD 83 and ITRS, because the relationship of NAD 83 with ITRF96 is not the same as with ITRF97 (as it is for the IERS convention of ITRF96 = ITRF97). This difference is accounted for in the transformation parameters for NAD 83 in Table 4-1. This difference amounts to a maximum of about 8 cm at the surface of the Earth at epoch 2010.00 (and about 2 cm at the defining epoch of 1997.0).

Because the relationship between each of the NAD 83 frames and the ITRS is fixed, Helmert transformation parameters like those in Table 4-1 cannot be used to transform among the various NAD 83 realizations. Likewise, HTDP also cannot be used for this purpose, as addressed in more detail at the end of this chapter.

Helmert Transformation Limitations and Other Equation-Based Methods. Although Helmert transformations are commonly used, it is important to realize that a transformation using only 14 Helmert parameters will, in general, *not* give correct results if the input (source) and output (target) epochs are not the same. The reason is that the regions corresponding to where the reference frames are used do not behave like perfectly rigid bodies because of both continuous and episodic deformations (e.g., earthquakes). The difference is especially pronounced in areas such as the western United States, but it is also manifested elsewhere (as shown near the end of this chapter where HTDP is discussed). Yet, there is a common misconception that the seven parameter rates in a 14-parameter transformation account for differential crustal motion (e.g., between different tectonic plates), but this is not true. The parameter rates account only for the change in coordinates from one reference frame to another over time, with respect to a specific reference epoch. Because the parameters are global, their rates cannot represent the rigid rotation of more than a single tectonic plate. For example, the parameter rates between NAD 83 "fixed" to the North America tectonic plate and a global frame (such as an ITRF) include the rotation of the North America plate. However, only the North America plate (and no other plate) is represented in the rates. Because only a single, simple rigid rotation can be represented by the parameter rates, it is also impossible for them to account for motion in the deforming parts of the plate.

Because of this limitation, whenever a transformation between reference frames is performed where the input and output epochs differ, crustal motion that occurs between the two epochs must be accounted for, in addition to the Helmert transformation. HTDP is an example of a program that does both, and it is discussed in the section on combined methods. This discussion will show the errors that occur when 14-parameter transformations are used with unequal input and output epochs, as well as when the seven parameter rates are ignored.

Equation-based transformations other than Helmert have often been used, especially in the past. One is the Molodensky method, which can convert directly between latitude and longitude without the need to convert to and from geocentric Cartesian (X, Y, Z) coordinates. The ΔX, ΔY, ΔZ values are incorporated directly into the transformation equations, along with the change in ellipsoid semimajor axis (Δa) and flattening (Δf), if applicable. There are two versions of the Molodensky method, the "standard" and the abridged. As the name implies, the abridged Molodensky method is a simplified method, and it is less accurate. The Molodensky methods were developed before digital computers, yet they are still included in many geospatial software packages.

Another equation-based datum method that is fairly common is the Molodensky–Badekas transformation. It is sometimes called a *7 + 3* or *10-parameter transformation*, because it has three more parameters than the Helmert transformation. The additional parameters allow accounting for rotation and scaling in a region of interest rather than at the geocenter to help reduce coupling between rotations and translations. This is accomplished by adding an (X_0, Y_0, Z_0) rotation and scale origin of the area of interest. A disadvantage of this method is that it is not fully reversible, because the rotation origin is associated with the original datum being transformed.

The final type of equation-based datum transformations mentioned in this chapter are those created through development of empirical multiple regression equations (MREs). These transformations consist of polynomials that directly convert the change in latitude and longitude, (and ellipsoid height if applicable) without the need to use intermediate geocentric (X, Y, Z) coordinates. MRE transformations were developed to better model the change in coordinates for continental-sized regions than could be achieved using other transformations. The approach is particularly useful for modeling irregular network distortion. MREs of up to ninth degree were adopted by NGA for transformations between several local datums and WGS 84, and they achieved accuracies of about 2 m. However, such methods are sensitive to the points used to develop the MREs, and large errors can occur if they are used outside the polygon defined by those points.

GRID-BASED COORDINATE TRANSFORMATIONS

Equation-based transformations are limited to situations where the difference between input and output coordinates is fairly uniform. For situations where the coordinate differences are highly variable, there simply are not enough terms in the equations to accurately resolve these differences. MRE transformations were developed to address this problem, but even they are limited; there is only so much irregularity that can be represented reliably by polynomials.

Consider the difference in coordinates between the NAD 27 and (original) NAD 83 datums. A large part of the difference in coordinates is because of lower accuracies and network distortions of NAD 27 compared with those of NAD 83. Nonetheless NGA developed simple 3-parameter transformations between NAD 27 and WGS 84 (which can be considered equivalent to NAD 83 in this low-accuracy context). Twenty of these transformations were developed for various regions throughout North and Central America and the Caribbean. The one for all of CONUS has standard deviations of 5 to 6 m for each of the geocentric (X, Y, Z) axes, resulting in roughly 20 m of horizontal error at 95% confidence. The corresponding MREs developed by NGA for CONUS is accurate to within about 5 m at 95% confidence, a considerable improvement. In contrast, the grid-based transformation developed by NGS for this case—NADCON—is accurate to within about 0.4 m at 95% confidence, more than an order of magnitude better than the NGA MREs (Dewhurst 1990).

Grid-based transformations consist of regular grids of coordinate differences where the differences at locations between grid points are interpolated. They are used when the differences are too irregular for standard transformation equations or MREs. These transformations are typically for large regions, such as a state or country, and can be global in extent. To ensure consistent results, the interpolation method is specified. In most cases, a single grid of uniform spacing is used for particular regions. However, there are implementations that use multiple grids of different spacings that overlie one another. This approach allows more accurate interpolations in areas where a higher density of points was used to create the grid. The National Transformation version 2 (NTv2) uses this method, which was adopted by Canada, Australia, and New Zealand to perform various datum transformations.

NGS has developed several types of grid-based transformations. The ones covered in this section are NADCON, VERTCON, VDatum, and geoid models (in the context of this chapter, the geoid models are considered transformation models because they are used to transform between ellipsoid heights and orthometric heights). All these transformations use latitude and longitude grids (not projected coordinate grids). Each of these grid-based transformations is discussed subsequently, along with a consideration of their accuracies. For some, the estimated accuracies can be quite good, as low as a few cm or even less. Such accuracies approach that of modern surveying. Nonetheless, it should be emphasized that these are transformation *models*. It cannot be expected that any of these models will maintain the full accuracy of the surveys used to determine the original coordinates. However, these models are suitable for various mapping and charting purposes that do not require the accuracies associated with most surveying engineering applications. No transformation can improve the accuracy of spatial data, and in many cases, they can degrade accuracy. Users are cautioned to be aware of the context in which they use these transformations, be cognizant of their accuracies, and to use them conservatively.

Although NADCON and VERTCON are discussed as separate products, both are included as part of the NGS Coordinate Conversion and Transformation Tool, NCAT (NGS 2022). NCAT is a web-based application that allows seamless use of these transformation models through a graphical user interface. It also performs coordinate conversions between geodetic and projected coordinates (State Plane and UTM). At the time of this writing (March 2022), NCAT does not yet include geoid models. However, these will be added in the future, as will Helmert transformations and crustal motion models to perform "combined" transformations of the type currently performed by HTDP (discussed previously and in greater detail near the end of this chapter).

NADCON (Geometric Coordinate Transformations)

Evolution of 2D NADCON (Versions 1.0 through 4.2). NADCON and its transformation grids have evolved significantly since its initial release in 1990 as version 1.0 (Dewhurst 1990). This and other early purely 2D versions of NADCON (through 4.2) have been superseded by version 5.0 (discussed subsequently). However, an awareness of these earlier versions of NADCON is important, because they still exist in commercial software, and they have been used extensively to transform datasets that are still in use today. NADCON has a complicated history that is covered only briefly here. It includes incorporation of the 3D GEOCON grids (discussed subsequently) and rebuilding all 2D and 3D grids in NADCON 5.0. For more details, refer to the official NGS technical report on NADCON 5.0 (Smith and Bilich 2017).

NADCON 1.0 was originally developed by NGS to transform geodetic coordinates (latitude and longitude) between NAD 27 and original NAD 83 (1986). A map of the NADCON-derived magnitude of horizontal differences between NAD 27 and NAD 83 for CONUS is shown in Figure 4-8. The differences range from nearly zero in northern Michigan to more than 100 m in the Pacific Northwest.

Only CONUS and Alaska were actually referenced to NAD 27, and transformation grids were created for each of these two regions. An additional five grids were also included in NADCON 1.0 for islands too far from the North American mainland to be tied to NAD 27 using classical (optical) methods, and each had its own datum. These grids are for the Hawaiian Islands (several datums collectively known as the "Old Hawaiian Datum"), the Puerto Rico Datum of 1940, and three datums realized in the 1950s for the Alaskan islands of St. George, St. Paul, and St. Lawrence in the Bering Sea.

The overall horizontal accuracy of the original NADCON 1.0 grids for CONUS was about 0.4 m (at 95% confidence). As mentioned previously, NGA has also developed 3-parameter

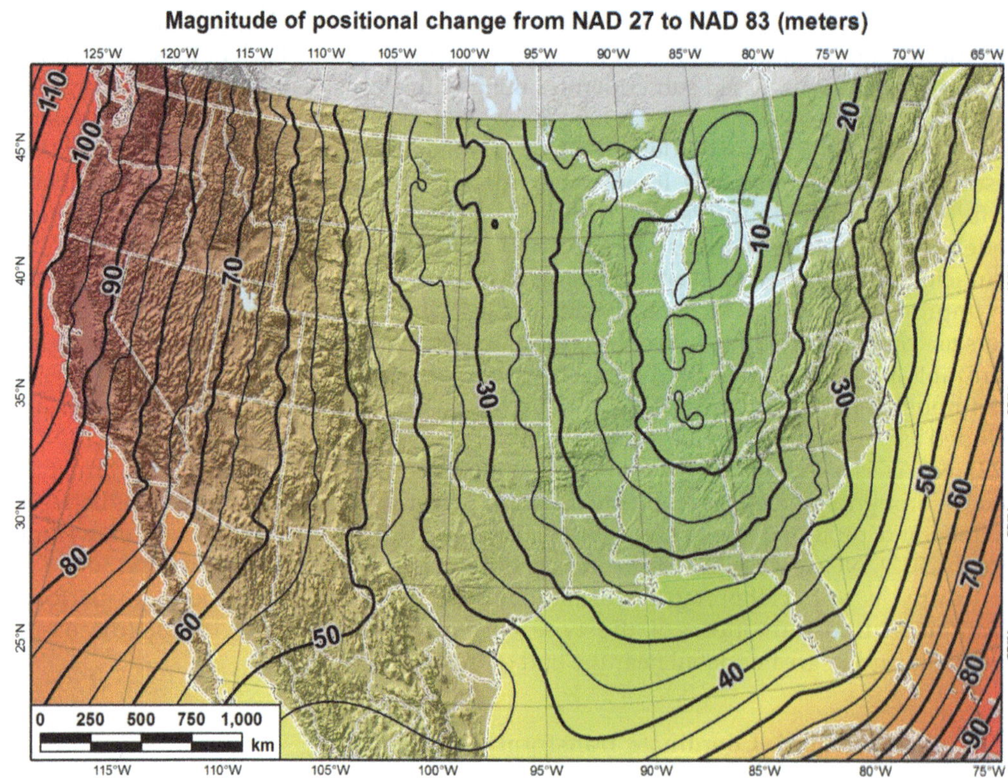

Figure 4-8. NADCON-derived horizontal difference between NAD 27 and NAD 83 (meters). Source: Courtesy of NOAA, reproduced with permission.

and MRE transformations between NAD 27 and WGS 84 (equivalent to NAD 83 in this context). The 3-parameter NGA NAD 27 transformations, especially those for CONUS, have been incorporated into geospatial software and many GNSS receivers. The CONUS NAD 27 NGA transformations do nominally the "same" transformation as NADCON, yet they can differ by more than 20 m. This discrepancy is a potential problem, because a user might invoke the transformation without knowing which method was used. Such possible errors show the importance of knowing exactly what datum transformation methods are used in software. Moreover, this problem is not limited to NADCON, as is shown later.

Beginning in 1992, NADCON was updated to version 2.1, which included additional grids to perform transformations between NAD 83 (1986) and newer realizations of NAD 83, based on control surveys done with GPS. These refinements of NAD 83 control were mostly done in the 1990s, for the most part on a state-by-state basis. Initially, these control surveys were called High Precision Geodetic Networks (HPGNs), which were later changed to High Accuracy Reference Networks (HARNs). Multiple essentially state-specific NADCON 2.1 grids were created to transforms between NAD 83 (1986) and the NAD 83 HPGN/HARN realizations. As with the original NADCON, only 2D (horizontal) transformations were performed (because original NAD 83 does not have well-defined ellipsoid heights). The estimated accuracy of the HPGN/HARN transformations is about 0.1 m at 95% confidence. Figure 4-9 is a map of the horizontal differences in Oregon for the

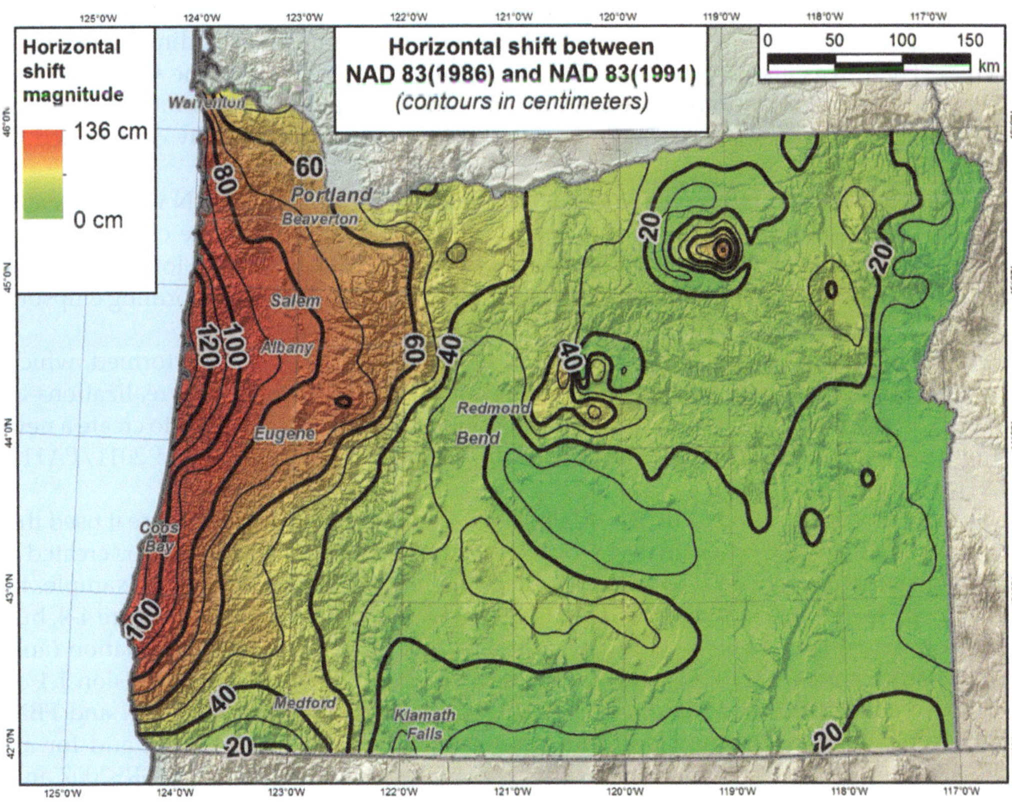

Figure 4-9. NADCON HPGN horizontal shift in Oregon from NAD 83 (1986) to NAD 83 (1991). Source: Courtesy of NOAA, reproduced with permission.

transformation between NAD 83 (1986) and NAD 83 (1991) based on the Oregon–Washington NADCON HPGN/HARN grid. Note the large magnitude of shifts along the coast as compared to those inland. These larger shifts are because of tectonic motion associated with the Juan de Fuca plate subducting under the North America plate off the Oregon coast.

In 2011, NADCON was updated to version 4.2, which included the creation of a graphical user interface. For this version, grids were added to convert between original NAD 83 (1986) and the American Samoa Datum of 1962, the Guam Datum of 1963, and the United States Standard Datum (USSD); USSD predates NAD 27 and was used by NGS in the early 1900s. This was the last update to the program and grids that went by the name "NADCON" prior to current version 5.0.

Creation of 3D Transformation Grids (GEOCON). After the HARN surveys, additional regional GPS surveys were performed, which are collectively called Federal Base Networks (FBNs), again on a largely state-by-state basis. However, no consistent naming convention was used for these regional surveys; their existence could only be inferred by the presence of more than one datum tag (a year in parentheses following "NAD 83"). Some states had two datum tags and, thus, had both "HARN" and "FBN" realizations, whereas some had only one.

In 2007, NGS performed a national adjustment of all GNSS observations in its database which superseded HARN and FBN control for all marks with GNSS-derived coordinates; all classically determined marks were unchanged (Pursell and Potterfield 2008). This resulted in the NSRS2007 realization of NAD 83 for passive marks (excluding those in the Pacific). There was also a concurrent CORS96 realization defined for the Continuously Operating Reference Stations (CORSs), but this can be considered functionally identical with NSRS2007 because the 2007 adjustment of passive marks was rigidly constrained to CORSs.

As with the HARN/HPGN networks, a grid-based tool called GEOCON was created to perform transformations between the NSRS2007 and the latest available coordinates of prior regional surveys (which was a mix of HARN and FBN surveys). However, in this case, the transformation was 3D, because they also included grids for transforming ellipsoid heights.

In 2012, another national adjustment of all GNSS observations was performed, which was constrained to the CORS-based 2011, PA11, and MA11 epoch 2010.00 realizations of NAD 83 (Dennis 2020). The results of this nationwide adjustment were used to create a new set of grids, called GEOCON 11, to transform between the NSRS2007 and 2011/PA11/MA11 realizations.

There were complications with the original GEOCON transformation because it used the latest available pre-2007 coordinates, which were a mix of HARN and FBN. This created a confusing situation for NAD 83 between the 1986 and 2007 realizations. For example, in Oregon, the HARN realization is NAD 83 (1991), which is the one shown in Figure 4-9, but there was also a later FBN NAD 83 (1998) realization. To help clarify this situation (and address other issues), GEOCON was updated to version 2.0 that combined version 1.1 of GEOCON and GEOCON 11 into a single tool. For states that had both HARN and FBN realizations, version 2.0 could perform a transformation between the two, and for all states, it transformed between HARN/FBN and NSRS2007, and between NSRS2007 and 2011/PA11/MA11. Because there were no NSRS2007 coordinates for the Pacific, transformation grids were created for the most recent pre-2011 realization of Pacific marks to the PA11 and MA11 frames of the 2011 national adjustment.

For HARN and FBN, GEOCON continued to use state-specific grids. Despite the effort to fix problems with defining state-specific realizations as HARN or FBN in GEOCON 2.0, inconsistencies remained both within CONUS and elsewhere in the NSRS.

Combining all Transformation Grids: NADCON 5.0. In 2016, all the transformations supported by NADCON and GEOCON were combined into a single product called NADCON 5.0. This was a comprehensive rebuild and replacement with three main objectives: (1) to improve how users interface with NGS datum transformation software, (2) to correct known issues with the existing grids and software, and (3) to prepare for the replacement of NAD 83 in as part of NSRS Modernization. NADCON 5.0 includes 44 separate transformations (30 latitude/longitude, 14 ellipsoid height) computed in 9 different regions. NADCON 5.0 is available online from NGS as an online interface through NCAT, as a web service, and as a downloadable program that includes the transformation grids.

NADCON 5.0 nominally performs the same transformations as NADCON 4.2 and GEOCON 2.0, and a guiding principle was to keep it as consistent as possible with existing products. Nonetheless, it differs from both previous products on a number of ways, summarized as follows:

- *Grid extents.* An entire new set of consistent grid extents was defined, and the state-specific HARN/FBN grids were eliminated. Ten regions are defined: CONUS, Alaska, and Puerto Rico/US Virgin Islands; three regions in the Pacific (Hawaii, American Samoa, and Guam/Mariana Islands); and four regions in the Bering Sea (St. Paul, St. George, St. Mathews, and St. Lawrence islands).
- *Gridding method.* For all grids, splines in tension were used to create grids for NADCON 5.0 (as was done for all versions of GEOCON). Grids for all prior versions of NADCON were created using "minimum curvature." A spline tension parameter of 0.4 was used for all NADCON 5.0 grids based on experience from GEOCON and additional experiments.
- *Interpolation method.* Biquadratic interpolation is used for all grids in NADCON 5.0, the same as done for all versions of GEOCON. All prior versions of NADCON used bilinear interpolation.
- *Generate new grids.* All transformation grids were re-created from scratch based on new data pulls from the NGS Integrated Database (NGSIDB). A guiding principle was that the grids must be based on information that was actually in the NGSIDB and nothing else. The grid spacing for NADCON ranges from 1 arcminute for the grids involving NSRS2007 and 2011/PA11/MA11 (the same as in GEOCON 2.0) to 30 arcminutes for the NAD 27 to NAD 83 (1986) grid for Alaska.
- *Produce error grids.* Grids were created to give estimates of the transformation error for all grids. These values were provided as "total error." It was computed as square root of the sum of the squared "method noise" (caused by data gaps) and "data noise" (variability within a transformation grid cell).
- *Use only sequential transformations.* All transformations are sequential in chronological order, and none can be "skipped" to go from one to another. For example, a transformation from USSD to NAD 83 (1986) must step through NAD 27, rather than go to NAD 83 (1986) directly. This is in contrast to NADCON 4.2, which went directly from USSD to NAD 83 (1986).
- *Clarify "HARN" and "FBN" transformations.* As indicated previously, HARN and FBN realizations (and their use in previous NGS transformations) are extremely complicated. For NADCON 5.0, the following terminology is used: HARN means the

first post-1986 GPS-survey-based realization in a state/region, and FBN means the second post-1986 GPS-survey-based realization in a state (if it existed). This means that there is a HARN realization for all CONUS states (plus DC), but an FBN realization for only a subset of 19 states. The HARN/FBN terminology is also used only for CONUS. Elsewhere, the various realizations are identified by the year of their datum tag.

Figure 4-10 shows a simplified schematic of the sequential transformation path used in NADCON 5.0 for 2D and 3D transformations. Note that there are three NAD 83 realizations between NAD 83 (1986) and NAD 83 (2007) for only the Puerto Rico/US Virgin Islands region. Elsewhere, there are at most two of these types of NAD 83 realizations supported in NADCON 5.0.

Figures 4-11 and 4-12 show NADCON 5.0 horizontal and ellipsoid height transformations, respectively, for CONUS from NAD 83 (2007) to NAD 83 (2011). In both figures, the upper map shows the horizontal or vertical shifts, and the lower map gives the estimated error of the transformations, with both shifts and errors given in centimeters.

In Figure 4-11, the maximum horizontal shift is 48 cm, the mean is 2.3 cm, and the standard deviation is ±1.3 cm. Overall, the largest shifts occur in the west mainly because of tectonic motion that occurred between the epochs used for the 2007 and 2011 adjustments. Epoch 2010.00 was used for the 2011 adjustment, but the epoch of the 2007 adjustment was not entirely straightforward. For stations in the western states of California, Oregon, Washington, Nevada, Arizona, and Alaska, an epoch of 2007.00 was used. Elsewhere in CONUS, the nominal epoch was 2002.00 (that of the constrained CORS). This disparity in epochs makes it somewhat difficult to state the epoch of the coordinates, but in any event, the two adjustments have an epoch difference of 3 to 8 years. The minimum error estimate of the horizontal transformation is 1.8 cm, the maximum is 36 cm, the mean is 0.2 cm, and the standard deviation is ±0.4 cm. The largest transformation error estimates are in California, probably because of both continuous tectonic motion and earthquakes.

In Figure 4-12, the minimum and maximum vertical shifts are −34 and +32 cm, the mean is −1.8 cm, and the standard deviation is ±1.3 cm. The reason for the mean not being closer to zero is that a systematic decrease in NAD 83 ellipsoid heights occurred because the

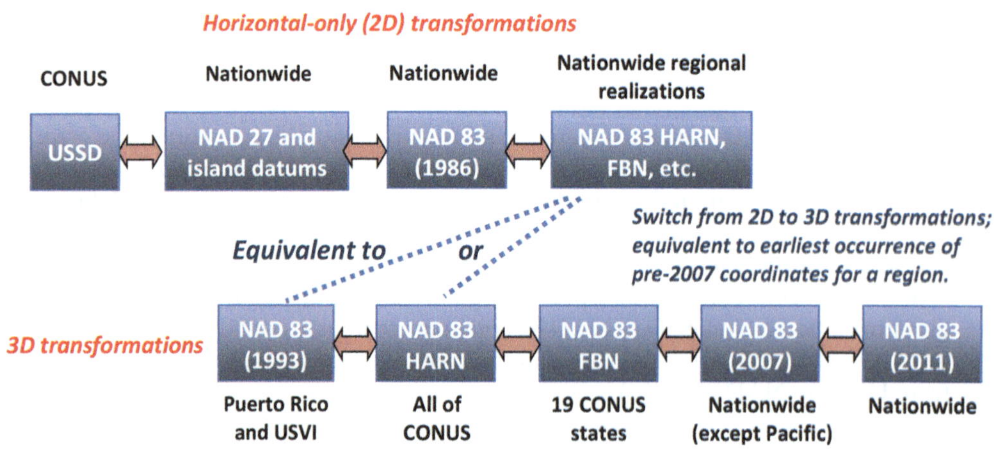

Figure 4-10. Simplified schematic of the 2D and 3D transformations in NADCON 5.0. Note that HARN and FBN terminology is used only in CONUS (datum tag years are used outside of CONUS).

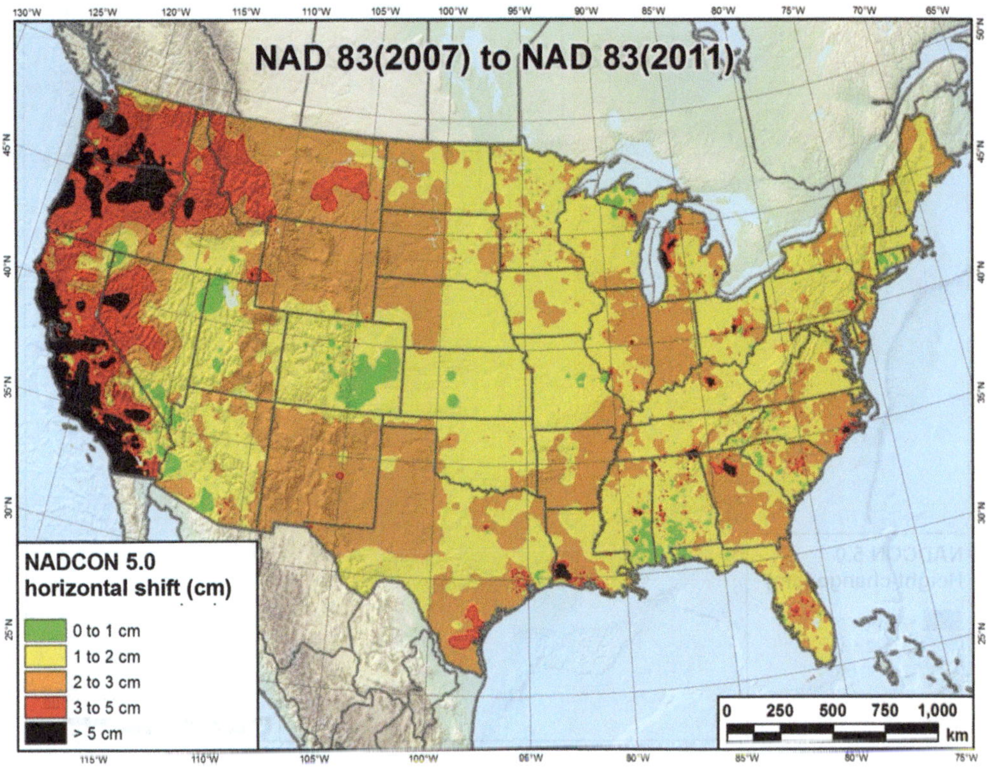

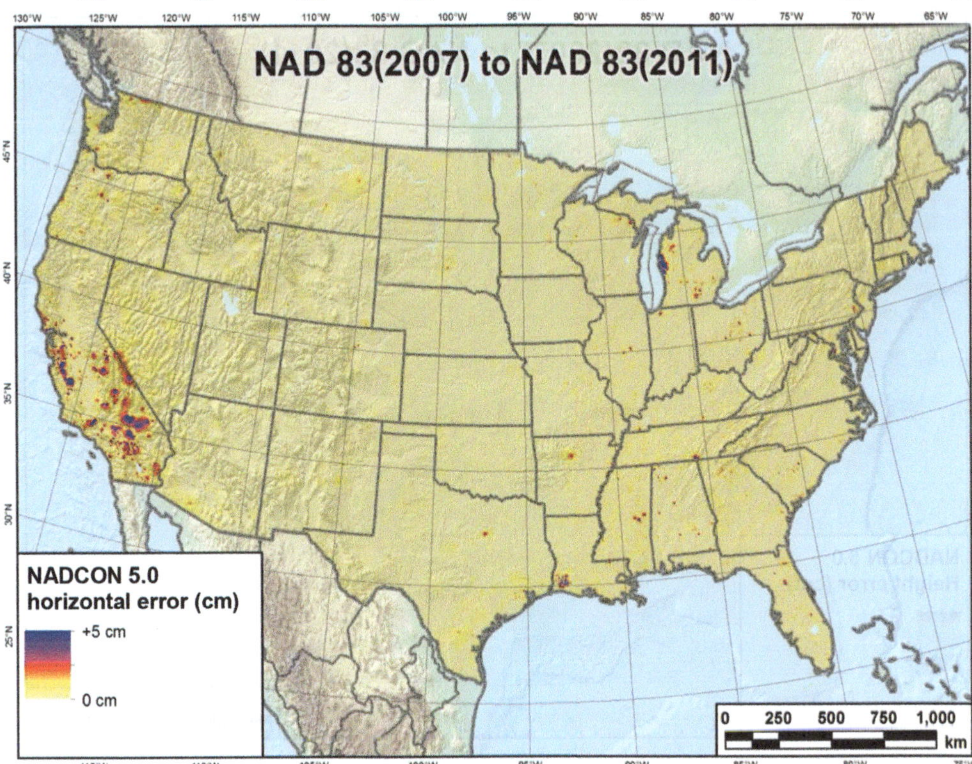

Figure 4-11. NADCON 5.0 horizontal shift and errors between NAD 83 (2007) and (2011). Source: Courtesy of NOAA, reproduced with permission.

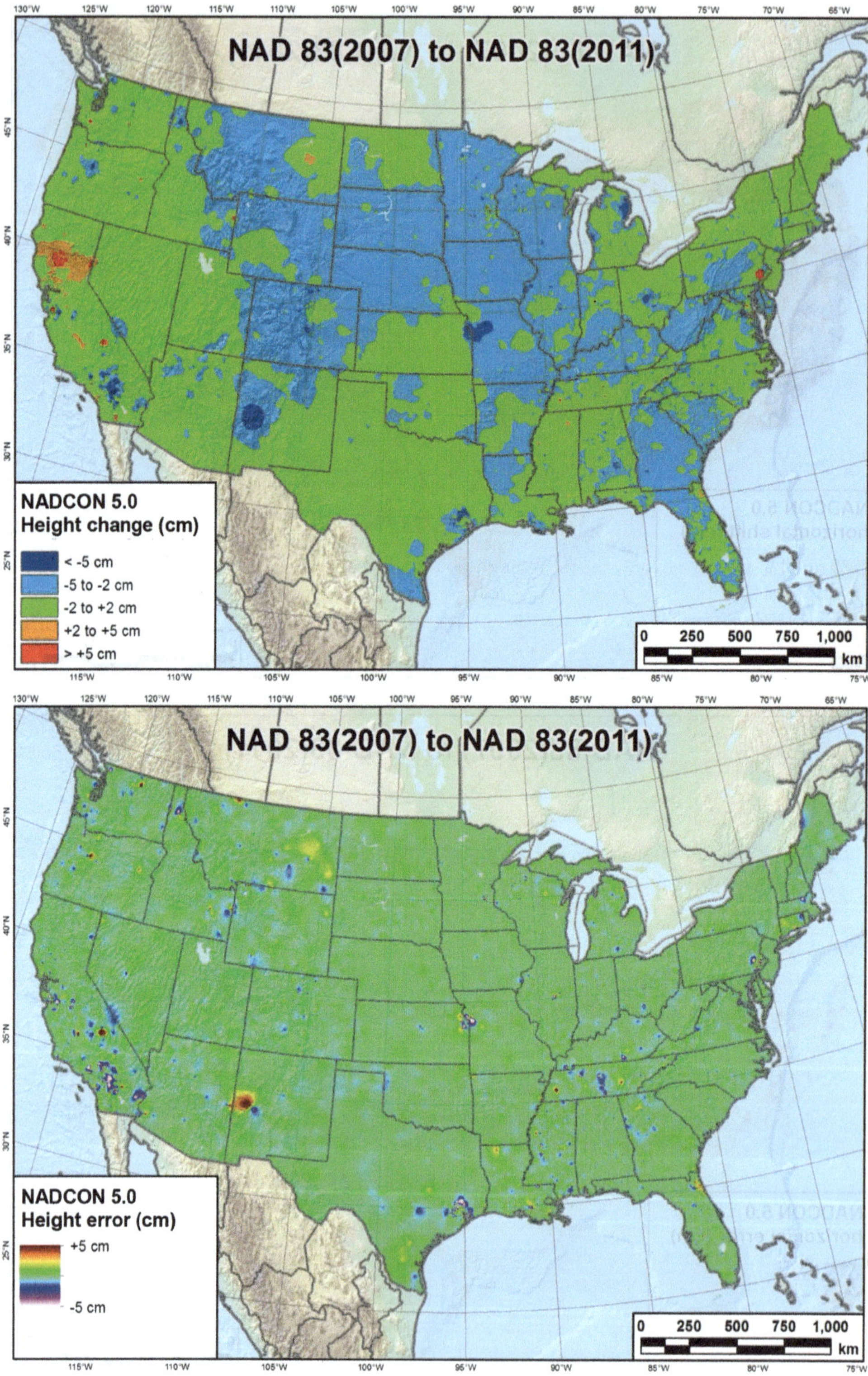

Figure 4-12. NADCON 5.0 vertical shift and errors from NAD 83 (2007) to (2011). Source: Courtesy of NOAA, reproduced with permission.

CORS used as constraints were reprocessed using absolute antenna models. The minimum and maximum error estimates of the vertical transformation are −33 and +24 cm, the mean is 0.0 cm, and the standard deviation is ±0.5 cm.

VERTCON (Vertical Transformations). VERTCON performs vertical transformations between the National Geodetic Vertical Datum of 1929 (NGVD 29) and the North American Vertical Datum of 1988 (NAVD 88). Grid coverage is only available for CONUS (none in Alaska) as a model of NAVD 88 minus NGVD 29 orthometric height ("elevation") differences. Figure 4-13 is a map showing the vertical change in CONUS from NGVD 29 to NAVD 88.

Similarly to NADCON, a new version of VERTCON (3.0) was created to replace the previous 2003 version (2.1), which used exactly the same transformation grids and methods as the original 1994 version (2.0). As with NADCON 5.0, VERTCON 3.0 is primarily accessed through NCAT. A description of the development of VERTCON 3.0 is not given here. Interested readers are referred to the official NGS technical report (Smith and Bilich 2019).

An important part of developing the original version of VERTCON was the modeling of predictable, physical components of the differences between the NAVD 88 and NGVD 29 datums. Systematic differences existed between NAVD 88 and NGVD 29 in part because of how gravity was used to correct the leveling observations. NAVD 88 utilized a surface gravity model (derived from measured gravity values) and an estimate of the gravitational effect of topography on leveling, whereas NGVD 29 used only the theoretical (normal) gravity, which only applies a small correction to north–south component of leveled lines.

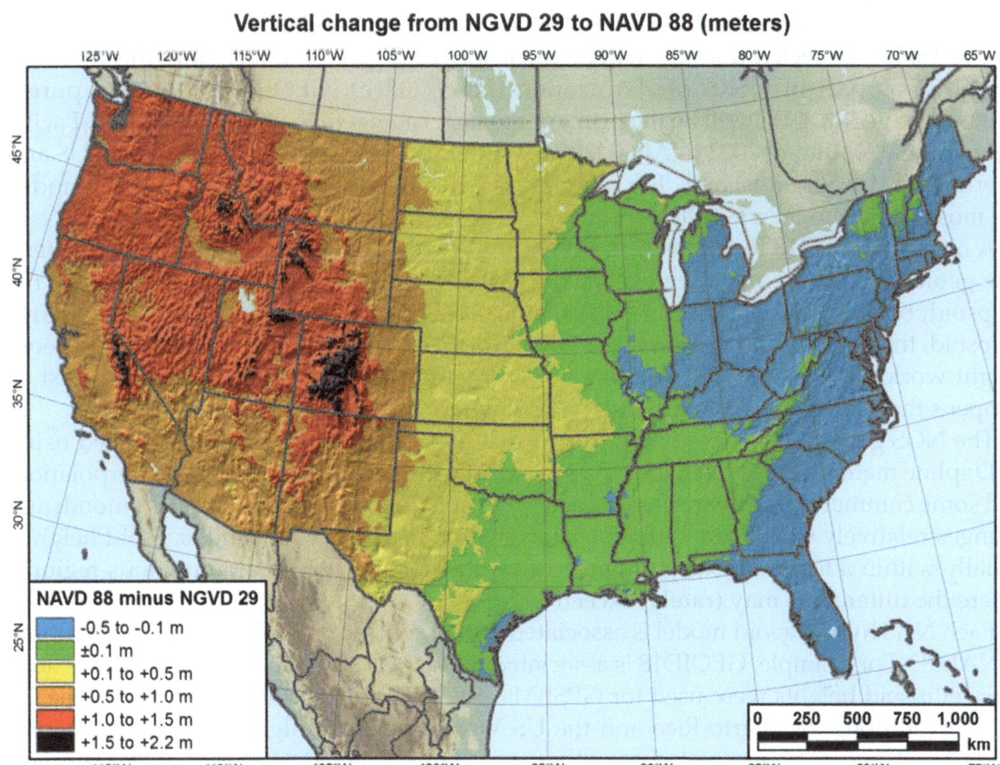

Figure 4-13. VERTCON transformation from NGVD 29 to NAVD 88 for CONUS. Source: Courtesy of NOAA, reproduced with permission.

Consequently, the two datums are represented by somewhat different types of orthometric heights: NAVD 88 by Helmert orthometric heights and NGVD 29 by normal orthometric heights. The effect of refraction on geodetic leveling was also modeled.

Based on a comparison of 381,833 leveled bench marks with both NAVD 88 and NGVD 29 heights, the estimated overall accuracy of the physical models used for VERTCON was 4 cm at 95% confidence. The 381,833 known height differences were then used to make corrections to the physical models and create the final VERTCON transformation grids. In areas where a high density of bench marks were used to correct the physical models (such as the eastern United States), VERTCON may yield conversions of better accuracy than 4 cm.

National Geodetic Survey Hybrid and Gravimetric Geoid Models (Vertical Transformations)

NGS has created several so-called hybrid geoid models to perform transformations between NAD 83 ellipsoid heights and NAVD 88 orthometric heights. These models were discussed previously in this chapter in the section on "calibrations/localizations" because they are fundamental to the process of determining orthometric heights with GNSS (the relationship among orthometric, ellipsoid, and geoid heights is shown in Figure 4-2). The models are called *hybrid* because they are created by modifying purely gravimetric geoids with differences between NAVD 88 leveled heights and GNSS-derived NAD 83 ellipsoid heights, observed at "GPS Bench Marks" (GPSBMs). This essentially converts the gravimetric geoid into a model of the NAVD 88 datum surface relative to the NAD 83 ellipsoid, with the gravimetric geoid providing the information to "fill" the model between the discrete GPS BM points.

There have been nominally seven NGS hybrid geoid models, with the latest being GEOID18. It is expected that this will be the last hybrid geoid model before the switch to GEOID22 as part of NSRS Modernization (NGS 2021b). GEOID22 will be a purely gravimetric geoid, and it will be the primary means of access to the North American-Pacific Geopotential Datum of 2022 (NAPGD2022). GEOID18 provides coverage for CONUS and Puerto Rico/US Virgin Islands. For other areas (such as Alaska and some Pacific islands), the most recent model is GEOID12B.

A map of GEOID18 for CONUS is shown in Figure 4-14, with geoid heights in meters. The geoid heights are negative everywhere in CONUS because the geoid is below the ellipsoid, but they are positive in most of Alaska. For a geoid model of the entire Earth ellipsoid, the positive and negative geoid heights tend to balance out, so the average geoid height worldwide should be close to zero (and varies between about ± 100 m). It just so happens that CONUS is in a part of the world where geoid heights are negative.

The NGS geoid models are 1 arcminute grids and geoid heights are interpolated using a 1D spline method. Earlier versions of NGS geoid software used biquadratic interpolation, and some commercial software uses bilinear interpolation. Because the geoid is smooth and changes relatively slowly, the different interpolation methods yield similar geoid heights, usually within a few mm except in areas of rapid change (such as mountainous regions) where the differences may (rarely) exceed 1 cm.

Each NGS hybrid geoid model is associated with a specific realization of a specific frame of NAD 83. For example, GEOID18 is associated with NAD 83 (2011) epoch 2010.00 because these ellipsoid heights were used for GPSBMs. Two GEOID18 grids were created, one for CONUS and one for Puerto Rico and the US Virgin Islands (Ahlgren et al. 2020). Earlier geoid height grids were created for other regions as well: Alaska, Hawaii, Guam and Northern Mariana Islands, and American Samoa. The model for Hawaii is not a hybrid because there is no defined vertical datum in Hawaii, although the Coast and Geodetic Survey (now NGS) published heights referenced above a Hawaiian Datum in 1930

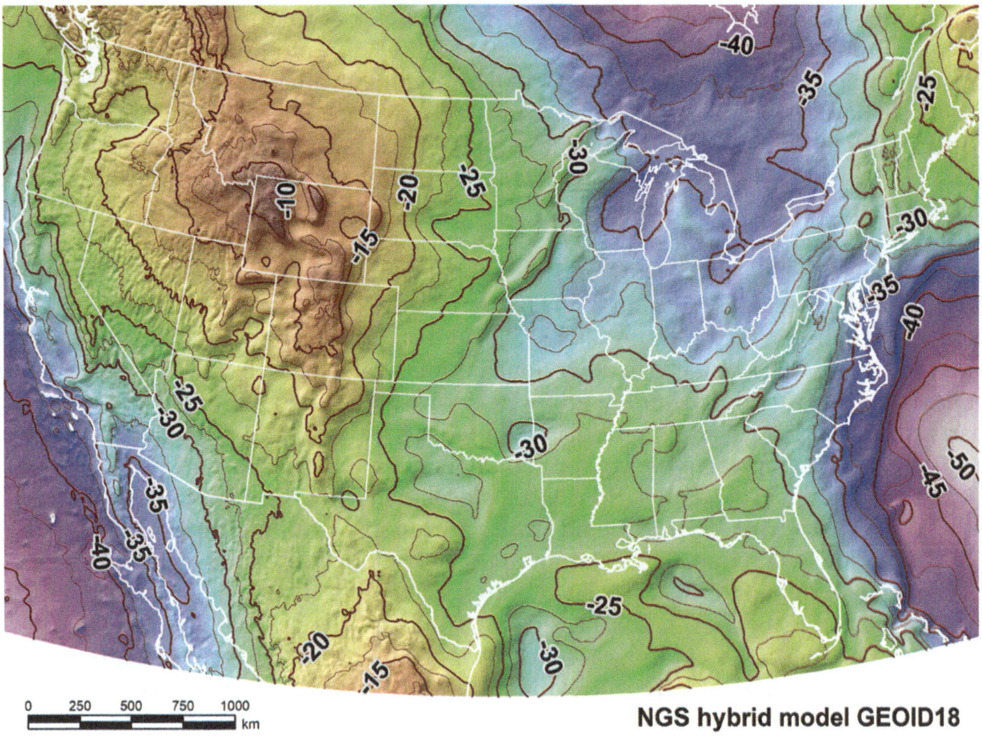

Figure 4-14. Hybrid geoid model GEOID18 for CONUS (geoid heights in meters). Source: Courtesy of NOAA, reproduced with permission.

(Rappleye 1929). A final set of 32,357 GPSBMs was used for the CONUS GEOID18 grid, and the resulting estimated overall accuracy of the model was 1.39 cm root-mean-square error (RMSE), although it varied with location, as shown in Figure 4-3. A map of GEOID18 slope for CONUS is shown in Figure 4-5.

The first step in creating a hybrid geoid model is the determination of a purely gravimetric geoid model. These models are also available from NGS, although most are referenced to the global reference frame (for specific ITRS realizations). Thus, they are with respect to ITRF/IGS (not NAD 83) ellipsoid heights (but using the same GRS 80 ellipsoid), and so they cannot be used for NAVD 88 orthometric heights. NGS also currently creates experimental gravimetric geoid models as part of its preparations for transition to the new vertical (geopotential) datum, NAPGD2022, as part of NSRS Modernization. These experimental models are also referenced to the ITRF/IGS frame but cover a larger portion (nearly one quarter) of the Earth than the gravimetric geoids used for hybrid geoid models.

A gravimetric geoid model of the entire Earth is available from NGA based on the Earth Gravitation Model 2008, EGM2008 (Pavlis et al. 2012). This geoid model is used by NGS as part of creating its geoid models and is also used by NGS products and services such as VDatum. EGM2008 is referenced to WGS 84 and so is also consistent with the ITRS, but it is not consistent with NAD 83 or NAVD 88.

VDatum (Vertical Transformations)

VDatum performs transformations among a variety of tidal, orthometric, and ellipsoidal vertical datums, as well as the International Great Lakes Datum of 1985 (IGLD 85), which is a dynamic height datum (NGS 2021c). It was developed jointly by NOAA's National Geodetic Survey (NGS), Office of Coast Survey (OCS), and Center for Operational Oceanographic Products and Services (CO-OPS), and it is available for download as a local Java installation. The VDatum website provides extensive documentation on the uncertainty of its transformations, by considering the cumulative contributions of each transformation source. This includes estimates of the accuracy of the sea surface topography and each of the local tidal datums.

VDatum can perform some of the NADCON horizontal transformations as well as some transformations between NAD 83 and the global frame. However, its main purpose is vertical transformations between NAVD 88 (or other geodetic vertical datums) and tidal datums, and between NAVD 88 and IGLD 85. Tidal datums are referenced not only to local mean sea level (LMSL), but also to other tidal datums, which are selected depending on the application. For example, mean lower low water (MLLW) is used as the NOAA nautical Chart Datum reference for chart products and predicted tides, and mean high water (MHW) is typically used to represent the coastal shoreline in mapping products.

VDatum includes some of the grids for transformations discussed previously: NADCON grids, VERTCON grids, hybrid geoid grids, and global gravimetric geoid grids (e.g., EGM2008). However, what sets VDatum apart from the other transformations is that it provides transformation grids associated with specific local tidal and Great Lake datums, which are available for specific coastal regions. These grids include the IGLD 85 hydraulic corrector grids and inverse topography of sea surface grids, which give location of the vertical datum (such as NAVD 88) relative to LMSL. It also includes grids for six tidal models relative to LMSL, listed in order of increasing height: (1) mean lower low water (MLLW), (2) mean low water (MLW), (3) diurnal tide level (DTL), (4) mean tide level (MTL), (5) mean high water MHW, and (6) mean higher high water (MHHW). Each of these tidal datums has specific practical and legal applications.

Figure 4-15 is a map of the IGLD 85 hydraulic correctors. These correctors are added to NAVD 88 dynamic heights (*not* to orthometric heights) to transform them to IGLD 85 dynamic heights. NGS datasheets for all stations with adjusted leveled NAVD 88 heights give the dynamic height, in addition to the Helmert orthometric height. A change in dynamic height gives the change in hydraulic head, which is why they are used for the Great Lakes. NAVD 88 and IGLD 85 were actually codefined, so theoretically they should be the same, and no correctors should be required (Zilkoski 1992). However, the correctors became necessary to ensure that the water levels at the Great Lakes gauging stations matched the dynamic heights determined from leveling (see Zilkoski et al. 1992 for more details).

COMBINED EQUATION- AND GRID-BASED TRANSFORMATIONS

The previous sections of this chapter considered equation- and grid-based coordinate transformations separately. This section considers a specific transformation tool that combines these methods to produce more accurate results than either method alone—the NGS program Horizontal Time-Dependent Positioning (HTDP).

HTDP transforms coordinates across time and between spatial reference frames. It can be used to estimate horizontal velocities and displacements between two epoch dates, update positions and geodetic observations (such as GNSS vectors) to a specific reference frame and epoch and transform positions and velocities between reference frames. The transformation between reference frames also includes the ability to transform between epochs.

When transforming between reference frames, HTDP uses the 14-parameter transformation in Equations (4-7a, b) and (4-8). An important thing to understand about

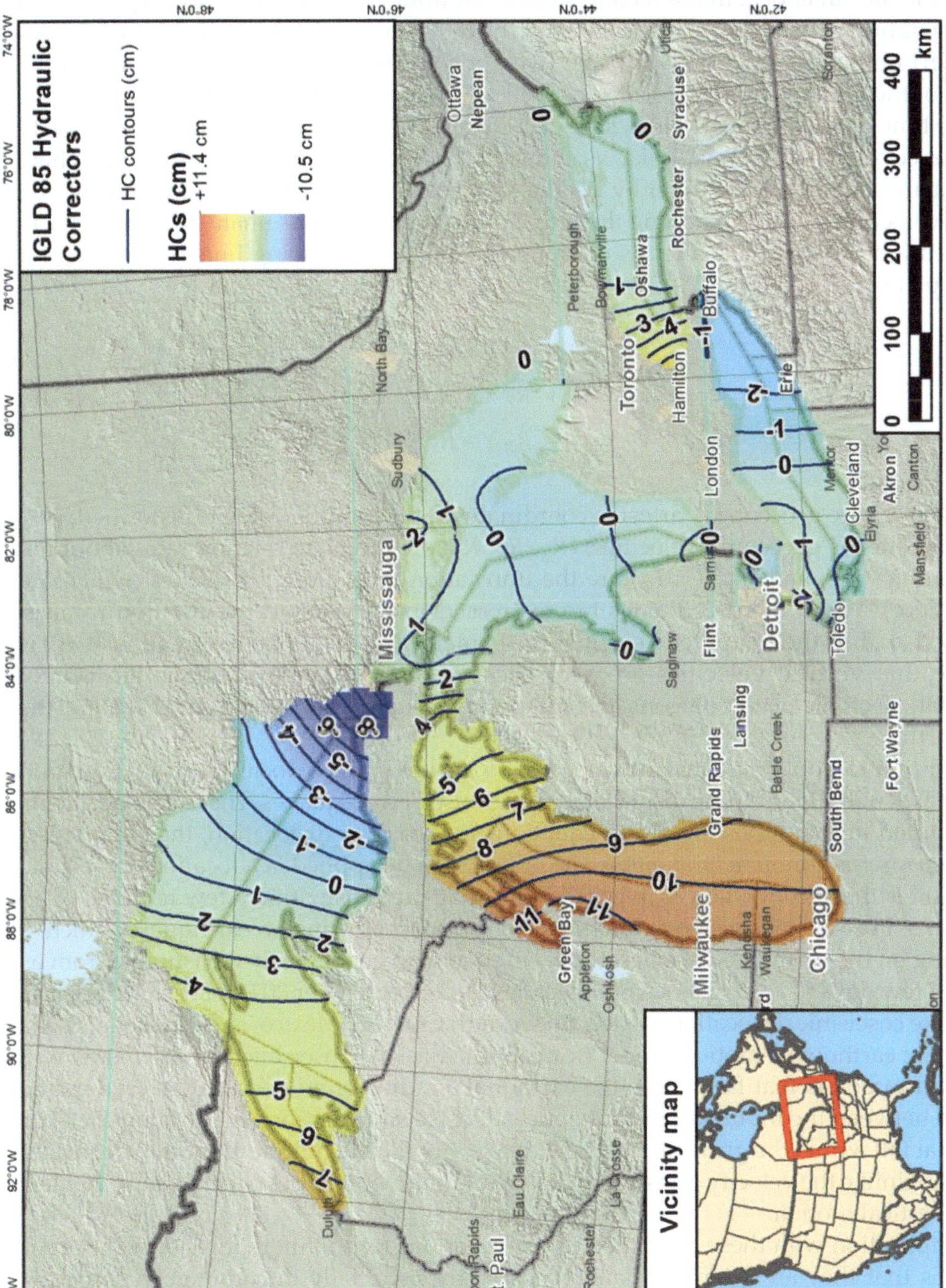

Figure 4-15. VDatum hydraulic correctors to convert NAVD 88 dynamic heights to IGLD 85 dynamic heights. Source: Courtesy of NOAA, reproduced with permission.

such transformations is that the 14-parameter transformation is sufficient by itself *only* for the case where the input (source) and the output (target) epochs are the same. If the input epochs are different, then the transformation is done in two steps. In the first step, velocities are used to compute the displacement that occurs over the time difference Δt from the input to the output epoch. In the second step, a 14-parameter transformation is performed using the output epoch. The velocities used in the first step are determined using equation- and/or grid-based methods between the two epochs. The method used depends on the location of the point being transformed as well as the input and output epochs.

At the time of this writing (March 2022), HTDP includes ten constant horizontal velocity grids that cover CONUS and Alaska, as shown in Figure 4-16. For transformations where the input and output epochs differ by Δt, the displacements are computed by multiplying the grid velocities by Δt and adding them to the 14-parameter coordinate difference. At all other locations, the displacement is computed using the following linear velocities for tectonic plates:

$$\begin{aligned} V_X &= \dot{T}_X + \dot{R}_Y Z - \dot{R}_Z Y \\ V_Y &= \dot{T}_Y + \dot{R}_Z X - \dot{R}_X Z \\ V_Z &= \dot{T}_Z + \dot{R}_X Y - \dot{R}_Y X \end{aligned} \qquad (4\text{-}10)$$

where (X, Y, Z) are geocentric Cartesian coordinates at the location being transformed; (V_X, V_Y, V_Z) are the 3D geocentric velocities; $(\dot{R}_X, \dot{R}_Y, \dot{R}_Z)$ are the plate rotation rates about the X-, Y-, and Z-axes; and $(\dot{T}_X, \dot{T}_Y, \dot{T}_Z)$ are the translation rates along these axes (which are usually zero). When the coordinates are expressed in meters and the rotation rates in radians/year, then the computed velocities are in meters/year [although Equation (4-10) is 3D, note that vertical velocities are zero because tectonic plate rotations are purely horizontal]. Tectonic plate polygons and rotation parameters are defined in HTDP for the entire Earth.

In addition to applying constant velocities to Δt using the velocity grids or Equation (4-10), HTDP will superimpose displacements because of earthquakes. These displacements are computed in two ways, one using equations for *coseismic* dislocation, the other using grids for *postseismic* motion, and either one or both can be applied at a location. Coseismic dislocation is the abrupt changes in coordinates that happens within a few minutes of an earthquake. Postseismic motion is the transient motion following an earthquake which—depending on the earthquake's magnitude—may remain geodetically measurable from as short as a few days to as long as several decades. As of September 2021, HTDP includes 32 earthquake coseismic dislocation models (in western CONUS, Mexico, and Alaska) and one postseismic earthquake motion model (for central Alaska). Coseismic displacements will be applied to points that fall within the earthquake radius of influence if the earthquake occurred between the input and output epochs. Postseismic motion will be computed for points that fall within the grid after the earthquake, but the magnitude of motion decreases with time. Unlike the velocity grids and tectonic rotations, both types of earthquake models include vertical motion.

A complication with these combined transformations (with different input and output epochs) is that they are often not handled correctly, especially in commercial software. Although some commercial software has incorporated the HTDP grids, tectonic plate rotations, and earthquake models, most do not even perform 14-parameter transformations. A common example of this is the transformations between NAD 83 (2011/PA11/MA11) and the global reference frame, usually identified as generic WGS 84 (i.e., without the realization specified). Here, WGS 84 is represented by ITRF2008/IGS08 at epoch 2005.0 (the epoch if the realization). Most commercial software uses only the seven (nontime varying)

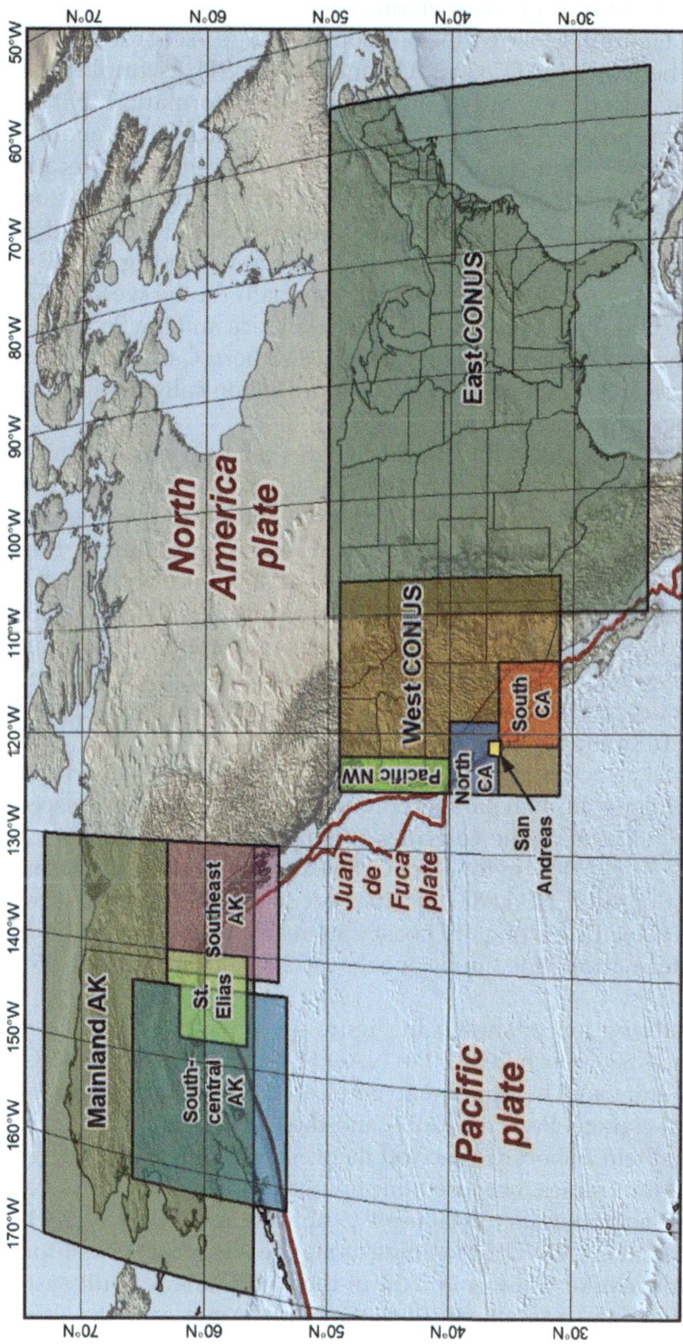

Figure 4-16. Velocity grids in HTDP (as of September 2021), along with adjacent tectonic plates (used for points that fall outside the velocity grids).
Source: Courtesy of NOAA, reproduced with permission.

parameters in Table 4-1 and ignore the parameter rates (velocities). In other words, the software performs a Helmert 7-parameter transformation at the reference epoch, which is $t_0 = 1997.00$ for the NGS-published transformation from IGS08 to NAD 83. This is not correct, because the epochs for NAD 83(2011/PA11/MA11) are all 2010.00. An illustration of the magnitude of this error is given in Figure 4-17.

The upper map in Figure 4-17 shows the horizontal change in coordinates in CONUS for the transformation between NAD 83 (2011) and "WGS 84" as implemented in most commercial software. Specifically, it is a 7-parameter transformation that uses only the seven constant Helmert parameters. The coordinate shifts range from about 60 to 130 cm from southeast to northwest but change gradually and smoothly with location, which is characteristic of such equation-based transformations. The error of this transformation with respect to the transformation between NAD 83 (2011) epoch 2010.00 and IGS08 epoch 2005.00 is shown in the lower map of Figure 4-17 (determined using HTDP v3.3.0). Note that the error contours are much more irregular, especially in the western United States, particularly along the boundary between the North America and the Pacific tectonic plates (as represented by the San Andreas Fault system in southern California). Along the west coast, the error can reach 30 cm, but even in the more tectonically benign eastern United States, the errors range from 10 to 17 cm.

Other choices of equal input and output epoch dates can be used, and some will reduce the errors shown in the lower map of Figure 4-17. The least error that can be achieved using a 14-parameter transformation is one with equal input and output epochs of 2005.00. The error is least for this epoch because the Helmert transformation aligns the frames at a common epoch of 2005.00 for NAD 83, and so the 5-year difference with epoch 2010.00 uses velocities in the NAD 83 frame. The NAD 83 velocities are less than the IGS08 velocities because NAD 83 is nominally fixed to the North America tectonic plate, and, thus, the reference frame approximately rotates with the plate. However, the plate model used has a slightly greater rotation rate than actual, and so there are residual velocities that increase north (with distance from the rotation point).

The highest NAD 83 velocities for CONUS on the stable part of the North America plate are at mid-longitudes along the 49th parallel border with Canada (e.g., can exceed 0.3 cm/yr in northernmost North Dakota). The velocities are low, so the resulting horizontal error is less than $0.3 \text{ cm/yr} \times 5 \text{ yr} = 1.5 \text{ cm}$ for all of CONUS except for the west coast states (and Alaska). The greatest CONUS NAD 83 velocities occur in California, as shown in Figure 4-18. The NAD 83 velocities along its coast can reach 5 cm/yr, so the error of this 14-parameter transformation (for equal epochs of 2005.00) can reach 25 cm at some locations.

The correct transformation is shown in Figure 4-19, with "correct" defined as the transformation from IGS08 epoch 2005.00 to NAD 83 epoch 2010.00 (the NGS-published epochs for these two frames). The upper map gives horizontal change, and the lower map ellipsoid gives height change. Both the (2011) and the (PA11) frames are shown, with the "boundary" between them a short distance off the west coast of CONUS. This is not really a boundary between the frames, because both frames can actually be used anywhere on Earth. It is represented as a boundary in Figure 4-19 only because NAD 83 (2011) is normally used everywhere on land in CONUS, including along the west coast of California. Because of this, parts of southwestern California west of the San Andreas Fault system (that are actually on the Pacific plate) are referenced to the North America plate—hence the large horizontal NAD 83 (2011) velocities in southern California. A similar phenomenon occurs for NSRS stations in the Caribbean because they are referenced to the North America plate, although they are on the Caribbean plate. For example, in Puerto Rico, the NAD 83 horizontal velocities are about 1.9 cm/yr.

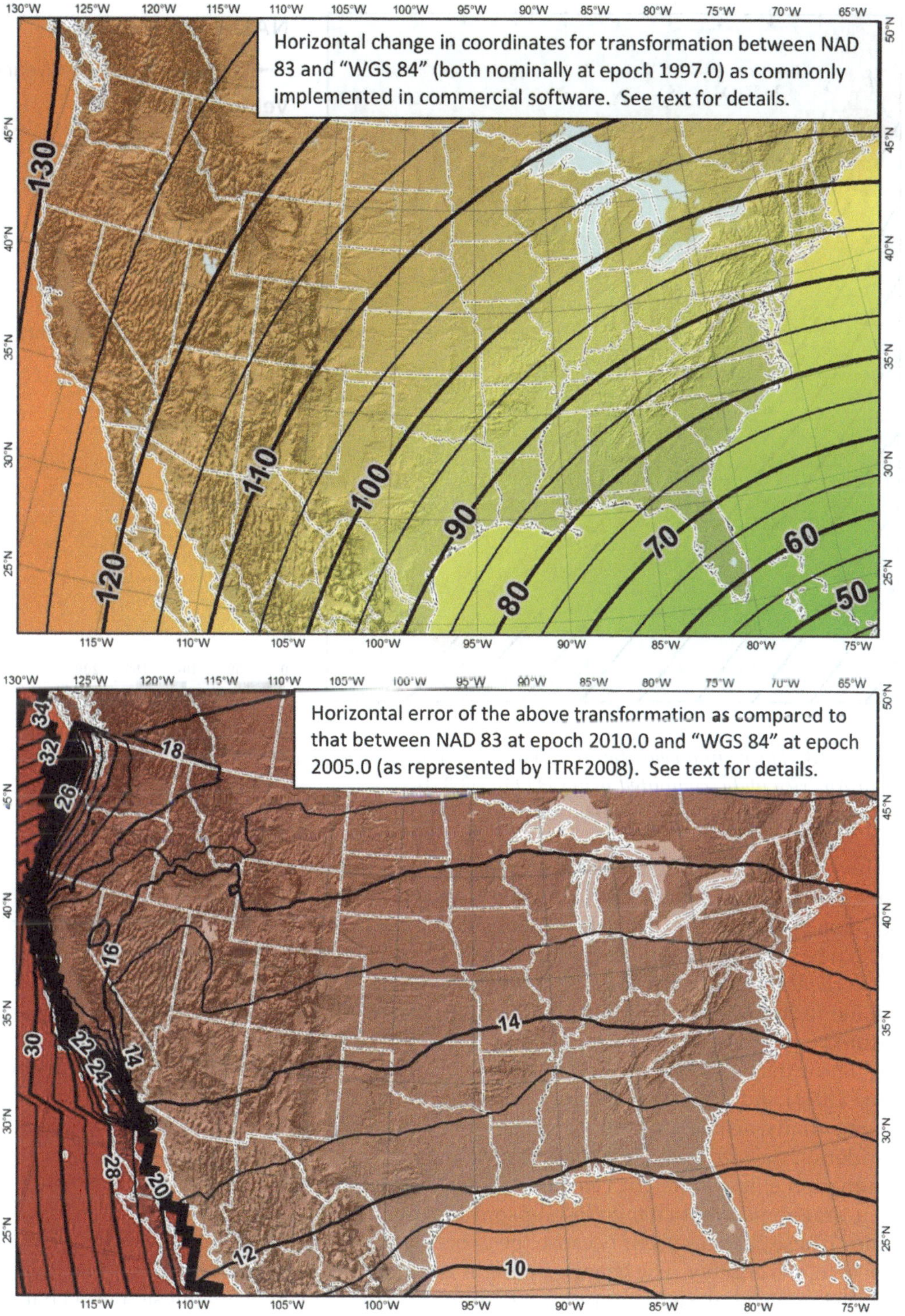

Figure 4-17. Horizontal shift and error of "WGS 84" to NAD 83 (2011) transformation as implemented in most commercial software (contours in cm).
Source: Courtesy of NOAA, reproduced with permission.

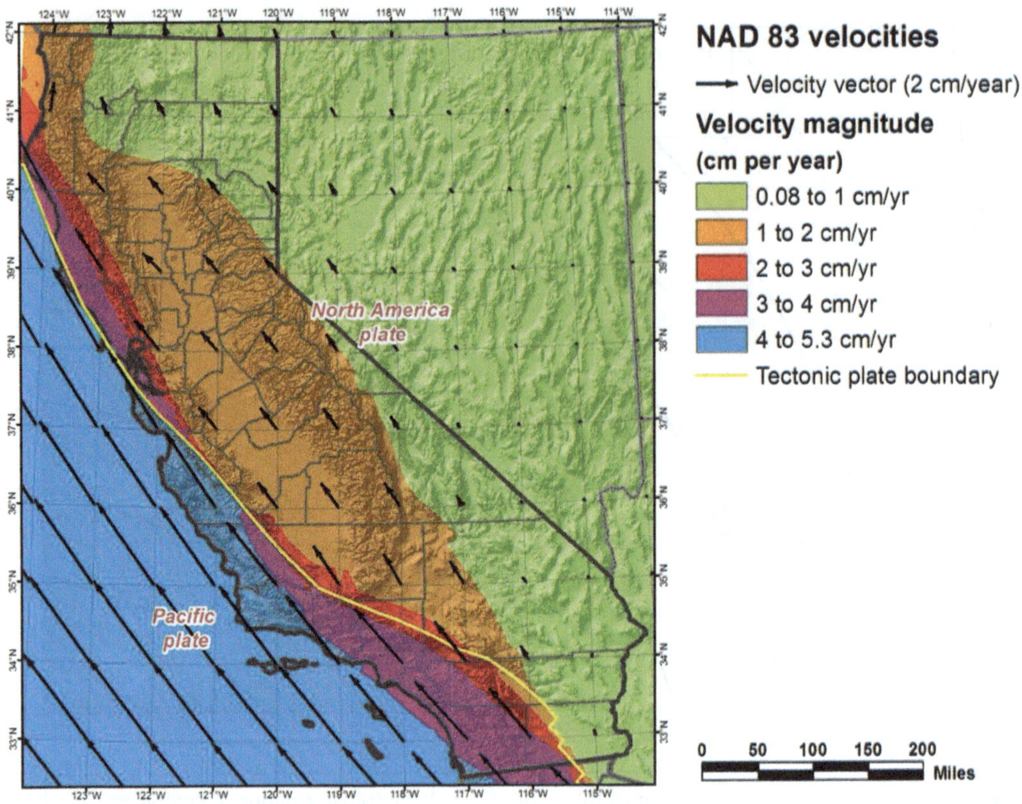

Figure 4-18. Horizontal velocities in CA and NV with respect to NAD 83 (2011) referenced to the North America plate (computed using HTDP v3.3.0).
Source: Courtesy of NOAA, reproduced with permission.

The ellipsoid height changes in Figure 4-19 are much smoother than the horizontal change because at this time HTDP mainly models horizontal tectonic motion, with the exception of earthquake dislocations (as can be discerned by the contours in central Alaska). This can be partly attributed to the fact that large-scale tectonic motion is mostly horizontal, and vertical motion is much more local, varied, and difficult to model. Because vertical tectonic motion is (mostly) not modeled, virtually all the ellipsoid height changes is captured by the 14-parameter transformation. The discontinuity in the change of ellipsoid height near the boundary between the North American and Pacific plates is a consequence of the (2011) and (PAa11) transformation models being developed separately. Thus, they are essentially artifacts of model development and are actually quite small, especially considering the size of the regions covered.

Although ellipsoid height velocities are (so far) usually not modeled by HTDP, there is still a time-dependent change in ellipsoid heights between the NAD 83 and the global frames. These changes occur because the geocenter (origin) of the frames are moving with respect to one another. The change in origin is represented by the translation velocities in Table 4-1, which have a magnitude of 1.7 mm/yr. This velocity is manifested as a change in ellipsoid height with time, but this change varies with location, from zero to the full magnitude of 1.7 mm/yr. The rate of change in ellipsoid height is about 1.1 mm/yr at the geometric center of CONUS.

The application of datum transformations that have unequal epoch dates goes beyond giving the relationship between NGS-published IGS08 and NAD 83 coordinates. An

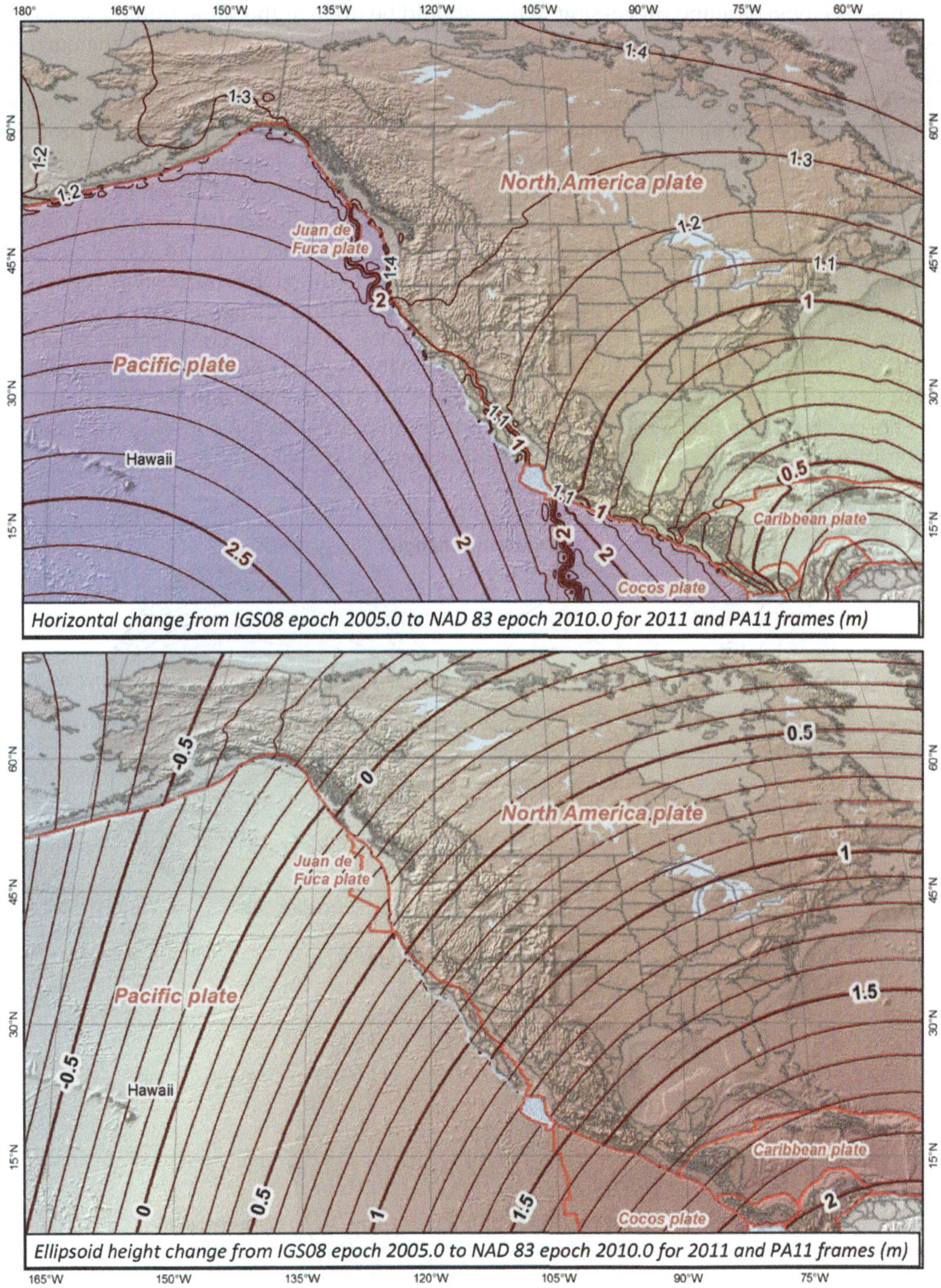

Figure 4-19. Horizontal and ellipsoid height change from IGS08 epoch 2005.0 to NAD 83 epoch 2010.0 for both the (2011) and (PA11) frames.
Source: Courtesy of NOAA, reproduced with permission.

example is GNSS baseline processing, which should be done in the same frame and epoch as the satellite orbits. Precise satellite orbits are provided in the IGS frame at the current epoch. So, to get NAD 83 (2011), epoch 2010.00 requires that the GNSS results be transformed to a different epoch [this is what is done with the NGS Online Positioning User Service (OPUS)]. Another example is Satellite-Based Augmentation Systems (SBAS), which provide real-time corrections. As with GNSS baseline processing, these systems provide corrections referenced to the current ITRF/IGS frame, also usually at the current epoch of the GNSS data. The same is true for Precise Point Positioning (PPP) and other online GNSS processing services. So, if such a service provides coordinates that are "accurate" to, say, 2 cm, one should be aware that the accuracy is likely with respect to the global frame at the epoch of the data. As has been shown previously, such positions may disagree with the desired datum by a significant amount. For example, consider a point in San Diego, California (32°42′N, 117°12′W), a tectonically active area. Its NAD 83 (2011) epoch 2010.00 coordinates differ from its ITRF2014 epoch 2020.00 coordinates by 1.79 m horizontally. Yet, 0.42 m of this horizontal difference is because of 10 years of differential tectonic motion that cannot be captured by a 14-parameter datum transformation.

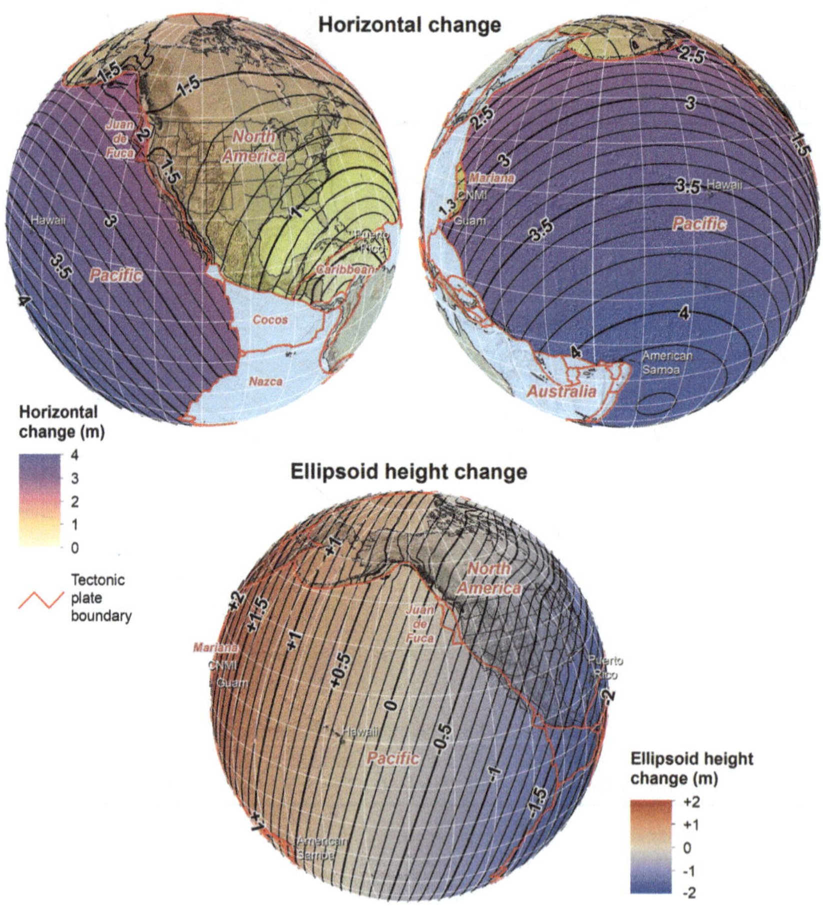

Figure 4-20. Estimated horizontal and ellipsoid height change from NAD 83 epoch 2010.0 to the new 2022 terrestrial reference frames at epoch 2020.0 with tectonic plates labeled (contours in meters).
Source: Courtesy of NOAA, reproduced with permission.

Another application of unequal epoch datum transformations is for estimating the change in coordinates from NAD 83 (2011) epoch 2010.00 to the future 2022 terrestrial reference frames at a different epoch. This transformation is shown in Figure 4-20 for both the expected horizontal and ellipsoid height change at an epoch of 2020.0. Note that the horizontal shifts are different for different tectonic plates, as is the case in Figure 4-19, but the shifts are not quite the same as in Figure 4-19 because of the different epochs. The ellipsoid height changes provide a visualization of how the entire reference ellipsoid shifts, with minimum and maximum shifts of ± 2.2 m and a contour where the change in height will be zero. The ellipsoid height change is shown everywhere on Earth without regard to tectonic plates because the plate motion is almost entirely horizontal. However, the effect of vertical crustal motion on ellipsoid height over the 10-year time span is discernible in some areas. This is most evident for the post-glacial uplift centered on Hudson Bay in Canada and for the tectonically active areas in central and southern Alaska.

Similarly, datum transformations can be used to illustrate the expected change from NAVD 88 to NAPGD2022, which will be based on a purely gravimetric geoid model (the best current estimate of this geoid model is the NGS experimental model xGEOID20B). The change is shown in Figure 4-21, which was created by transforming xGEOID20B from ITRF2014 to NAD 83, and then subtracting it from hybrid geoid model GEOID18. The transformation puts the two models in the same frame, which allows comparison. The most prominent feature of the difference is the approximately 1 m "tilt" from the southeast to northwest of NAVD 88 with respect to NAPGD2022.

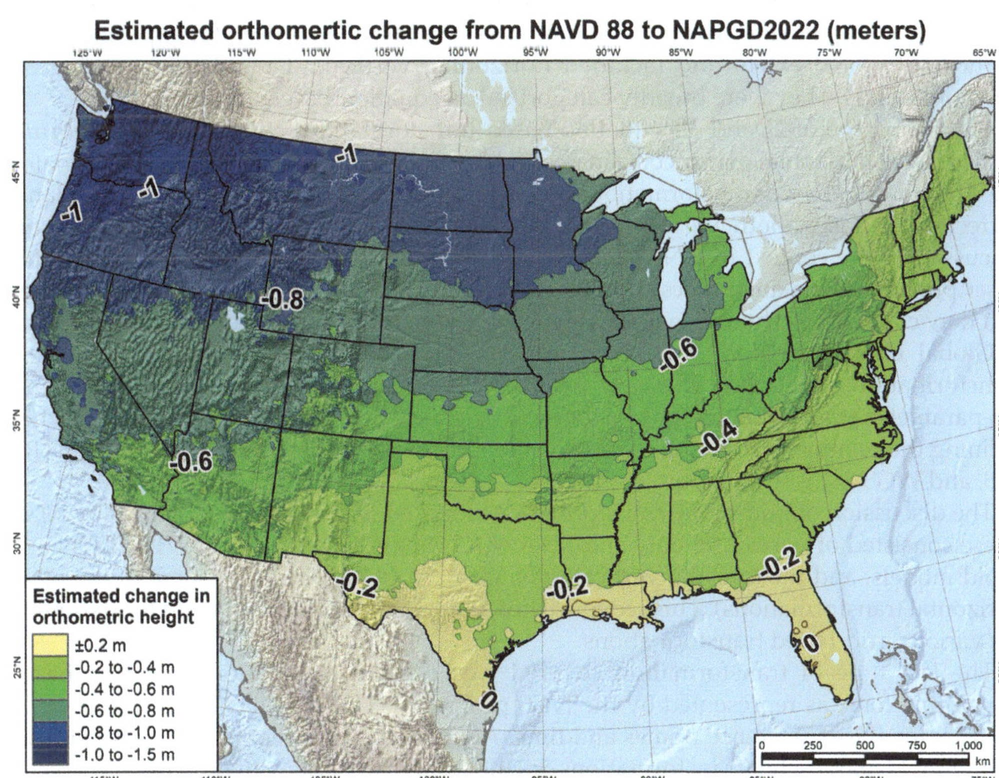

Figure 4-21. Estimated orthometric height change from NAVD 88 to NAPGD2022. Source: Courtesy of NOAA, reproduced with permission.

HTDP is a powerful and useful tool, but one should keep in mind that the crustal motion grids are updated as new data are acquired and after events (such as large earthquakes) occur. Thus, different versions of HTDP may not give the same results.

It is also important to understand that HTDP should not be used to transform among the various NAD 83 realizations. Only a single definition for NAD 83 referenced to the North America plate exists in HTDP, as described by Soler and Snay (2004). Because of this, it is impossible for HTDP to correctly transform among NAD 83 realizations (e.g., 2011, NSRS2007, CORS96, HARN, FBN, etc.). However, a common misconception persists that HTDP can be used to transform among various realizations of NAD 83 referenced to the same tectonic plate, for example between NAD 83 HARN and NAD 83 (2011). HARN coordinates are identified on NGS datasheets by a year in parentheses after the NAD 83 designation, for example "NAD 83 (1992)." This year is called a *datum tag* and typically refers to the year a defining adjustment was performed—it is *not* an epoch. So, it would be completely incorrect to use HTDP with an input epoch of 1992.00 and an output epoch of 2010.00 to transform NAD 83 (1992) to NAD 83 (2011) epoch 2010.00. To perform such transformations among NAD 83 realizations, NCAT should be used instead (NGS 2022). The same is true for the NAD 83 frames referenced to the Pacific and Mariana plates. Only a single definition for each exists in HTDP, and so NCAT should be used for transforming among NAD 83 realizations in these areas as well.

SUMMARY

This chapter provided an overview of a wide range of coordinate transformations used in surveying engineering. The transformations ranged from ones used for small local sites to those of global extent and included horizontal, vertical, 3D, and time-varying 3D transformations. They were broadly categorized as equation-based and grid-based.

The main type of 2Dand 3D equation-based transformations covered were conformal both for local horizontal planar coordinates and for global geocentric Cartesian coordinates. The 1D equation-based transformations included simple vertical shifts and inclined planar correction surfaces. Both the conformal horizontal and vertical transformations were discussed in the context of GNSS "calibrations/localizations." The discussion included descriptions of the mathematics of calibrations and provided recommendations on when and how calibrations should be used as well as when and how they should *not* be used.

Global equation–based transformations focused mainly on 7-parameter Helmert transformations, including the case where each parameter varies linearly with time (14-parameter transformations). The time-varying aspect is particularly important when defining the relationship among the various realizations of global reference frames (ITRF, IGS, and WGS 84).

The discussion of grid-based transformations focused mainly on those created by NGS. These consisted of horizontal-only and 3D NADCON grids, vertical-only VERTCON and geoid models, and the (mostly) vertical-only VDatum (which includes some of the other horizontal transformations). This section also discussed the applicability and limitations of the various grid-based transformations.

The last type of transformation covered was combined equation- and grid-based transformations, as represented by the NGS program HTDP and its ability to transform coordinates across reference frames and time. An important concept in this section is that although 14-parameter transformations include time variation, they cannot correctly transform coordinates when the input and output epoch dates are different. The reason for this is that the Earth's crust does not behave like a rigid body, often even within defined tectonic plates. This section concluded by showing how combined equation- and grid-based transformations can be used to estimate the change of the current US geometric and vertical datums to the new datums that will be released in the near future as part of the NSRS Modernization. Additional details about modernizing the NSRS are given in the three NGS "blueprint" technical reports (NGS 2021d, e, f).

REFERENCES

Ahlgren, K., G. Scott, D. Zilkoski, B. Shaw, and N. Paudel. 2020. *GEOID 18*. NOAA Technical Rep. NOS-NGS 72. Washington, DC: National Oceanic and Atmospheric Administration, National Geodetic Survey.

Altamimi, Z., P. Rebischung, L. Métivier, and X. Collilieux. 2016. "ITRF2014: A new release of the international terrestrial reference frame modeling nonlinear station motions." *J. Geophys. Res. Solid Earth* 121 (8): 6109–6131.

Altamimi, Z., P. Rebischung, L. Métivier, and X. Collilieux. 2017. *Analysis and results of ITRF2014*. IERS Tech. Note 38. Frankfurt am Main, Germany: Verlag des Bundesamts für Kartographie und Geodäsie.

Craymer, M., R. Ferland, and R. S. Snay. 2000. "Realization and unification of NAD 83 in Canada and the U.S. via the ITRF." In Vol. 120 of *Towards an Integrated Global Geodetic Observing System*, edited by R. Rummel, H. Drewes, W. Bosch, and H. Hornik, 118–121. Berlin: Springer.

Dennis, M. L. 2020. *The national adjustment of 2011: Alignment of passive GNSS control with the three frames of the North American Datum of 1983 at epoch 2010.00: NAD83 (2011), NAD83 (PA11), and NAD83 (MA11)*. NOAA Tech. Rep. NOS-NGS 65. Washington, DC: National Oceanic and Atmospheric Administration, National Geodetic Survey.

Dewhurst, W. T. 1990. *NADCON: The application of minimum-curvature-derived surfaces in the transformation of positional data from the North American Datum of 1927 to the North American Datum of 1983*. NOAA Technical Memorandum NOS-NGS-50. Washington, DC: National Oceanic and Atmospheric Administration, National Geodetic Survey.

EPSG (European Petroleum Survey Group). 2022. "Geodetic parameter dataset." *International Association of Oil and Gas Producers*. Accessed February 8, 2022. https://epsg.org/home.html.

IGN (Institut Géographique National). 2020. "Transformation parameters." *International Terrestrial Reference Frame*. Accessed February 8, 2022. https://itrf.ign.fr/trans_para.php; ITRS and WGS84 parameters on this page available in "WGS84.TXT" file at ftp://itrf-ftp.ign.fr/pub/itrf/.

IGN. 2021. "International terrestrial reference frame." *ITRF solutions*. Accessed February 8, 2022. itrf.ign.fr/ITRF_solutions/index.php.

IGS (International GNSS Service). 2017. "Chronology of IGS reference frame usage." Accessed February 8, 2022. http://acc.igs.org/igs-frames.html.

ISO (International Standards Organization). 2022. "ISO geodetic registry." *Coordinate Operations: Transformation*. Accessed February 8, 2022. https://geodetic.isotc211.org/register/geodetic/Transformation.

Kelly, K. M., and M. L. Dennis. 2022. "Transforming between WGS84 realizations." *J. Surv. Eng*. 148 (2): 04021031.

NGA (National Geospatial-Intelligence Agency). 2014. "World Geodetic System 1984: its definition and relationships with local geodetic systems." NGA-STND.0036_1.0.0_WGS84, Version 1.0.0. Accessed February 8, 2022. https://earth-info.nga.mil/index.php?dir=wgs84&action=wgs84/.

NGA. 2021. "(U) recent update to WGS 84 reference frame and NGA transition to IGS ANTEX, office of geomatics/GNSS division." Approved for Public Release #21-520. Accessed February 8, 2022. https://earth-info.nga.mil/php/download.php?file=(U)WGS%2084(G2139).pdf.

NGS (National Geodetic Survey). 2000. "NGS adopts transformation." ITRF97 to NAD83 Transformation. Accessed February 8, 2022. https://geodesy.noaa.gov/PUBS_LIB/transformationITRF97toNAD83.html.

NGS. 2021a. *HTDP—Horizontal Time-Dependent Positioning*. Includes software, user guide, and revision log. Accessed February 8, 2022. https://geodesy.noaa.gov/TOOLS/Htdp/Htdp.shtml.

NGS. 2021b. "New Datums: Replacing NAVD 88 and NAD 83." Accessed February 8, 2022. https://geodesy.noaa.gov/datums/newdatums/index.shtml.

NGS. 2021c. "Vertical datum transformation: Integrating America's elevation data." *VDatum*. Accessed February 8, 2022. https:// vdatum.noaa.gov/welcome.html.

NGS. 2021d. *Blueprint for the modernized NSRS, part 1: Geometric coordinates and terrestrial reference frames*. NOAA Tech. Rep. NOS NGS 62. Silver Spring, MD: National Oceanic and Atmospheric Administration, National Geodetic Survey.

NGS. 2021e. *Blueprint for the modernized NSRS, part 2: Geopotential coordinates and geopotential datum*. NOAA Tech. Rep. NOS NGS 64. Silver Spring, MD: National Oceanic and Atmospheric Administration, National Geodetic Survey.

NGS. 2021f. *Blueprint for the modernized NSRS, part 3: Working in the modernized NSRS*. NOAA Tech. Rep. NOS-NGS 67. Silver Spring, MD: National Oceanic and Atmospheric Administration, National Geodetic Survey.

NGS. 2022. "NGS Coordinate Conversion and Transformation Tool (NCAT)." Accessed February 8, 2022. https://geodesy.noaa.gov/NCAT/.

Pavlis, N. K., S. Holmes, S. Kenyon, and J. K. Factor. 2012. "The development and evaluation of the Earth Gravitational Model 2008 (EGM2008)." *J. Geophys. Res.* 117 (B4): B04406.

Pearson, C., and R. Snay. 2012. "Introducing HTDP 3.1 to transform coordinates across time and spatial reference frames." *GPS Solutions* 17 (1): 1–15.

Petit, G., and B. Luzum. 2010. "IERS conventions 2010." (IERS Tech. Note No. 36) International Earth Rotation and Reference Systems Service. Accessed February 8, 2022. https://www.iers.org/SharedDocs/Publikationen/EN/IERS/Publications/tn/TechnNote36/tn36.pdf?__blob=publicationFile&v=1.

Pursell, D. G., and M. Potterfield. 2008. *National readjustment final report*. NAD 83(NSRS2007). NOAA Technical Rep. NOS NGS 60. Washington, DC: National Oceanic and Atmospheric Administration, National Geodetic Survey.

Rappleye, H. S. 1929. *First-order leveling in Hawaii*. Special Publication 161. Washington, DC: US Department of Commerce, Coast and Geodetic Survey.

Schwarz, C. R. 1989. *North American datum of 1983*. NOAA Professional Paper NOS 2. Washington, DC: National Oceanic and Atmospheric Administration, National Geodetic Survey.

Smith, D., and A. Bilich. 2017. *NADCON 5.0: Geometric transformation tool for points in the national spatial reference system*. NOAA Tech. Rep. NOS NGS 63. Washington, DC: National Oceanic and Atmospheric Administration, National Geodetic Survey.

Smith, D., and A. Bilich. 2019. *The VERTCON 3.0 project: Creating vertical transformations for points in the national spatial reference system, including VERTCON 3.0 release 20190601*. NOAA Technical Rep. NOS NGS 68. Washington, DC: National Oceanic and Atmospheric Administration, National Geodetic Survey.

Snay, R. 2003. "Introducing two spatial reference frames for regions of the Pacific Ocean." *Surv. Land Inf. Sys.* 63 (1): 5–12.

Snay, R. 2012. "Evolution of NAD 83 in the United States: Journey from 2D toward 4D." *J. Surv. Eng.* 138 (4): 161–171.

Soler, T. and Snay, R. 2004. "Transforming positions and velocities between the international terrestrial reference frame of 2000 and North American Datum of 1983." *J. Surv. Eng.* 130 (2): 49–55.

Zilkoski, D. B. 1992. "North American vertical datum and international great lakes datum: They are now one and the same." In *Proc., U.S. Hydrographic Conference '92*. Baltimore, MD: The Hydrographic Society of America.

Zilkoski, D. B., J. H. Richards, and G. M. Young. 1992. "Special report: Results of the general adjustment of the North American vertical datum of 1998." *Surv. Land Inf. Sys.* 52 (3): 133–149.

CHAPTER 5

ANALYSIS AND ADJUSTMENT OF OBSERVATIONAL ERRORS

Charles D. Ghilani

INTRODUCTION

Because of instrumental and personal limitations as well as variations in environmental conditions, every observation contains error. In engineering surveys, this is often evidenced by geometric closure conditions not being met. For example, when all the angles are measured about a point to surrounding stations, as shown in Figure 5-1, the sum of these angles should be 360°. However, this sum is seldom achieved because of errors in the observations. Thus, the proper planning and execution of surveys to meet certain specifications or standards requires a knowledge of these error sources and how they affect the final accuracy of the survey. This chapter investigates these error sources. It discusses how to adjust the observations so that the final adjusted unknowns are the most probable values for the given set of observations and explore statistical procedures that can be used to identify errors after an adjustment. Although this chapter does not discuss every possible error source that may be found in a measurement nor every measurement type, it discusses the types of errors that occur commonly in engineering survey observations.

The accuracy of surveying results should never be judged solely on the basis of personal judgment or experience. Commonly accepted statistical methods must be used to provide an unbiased evaluation. Some of the fundamental principles and methods associated with the evaluation of surveying observations are discussed in this chapter. More detailed treatments of these subjects can be found in the references at the end of this chapter.

TYPES OF ERRORS

Mistakes, which are also called blunders, can occur because of carelessness, fatigue, poor communication, confusion in the observer, or confusion in the observation process. For example, if the observer is unsure about the name of a sighted station, an incorrect station name may be entered into the field notes. Mistakes are not considered in the following discussions on errors. However, they must be detected by carefully checking one's work through repeated observations and by performing additional observational checks. Thus, it is important to have more observations than are required to simply get the solution to the problem. These additional observations are known as repeat or *redundant observations* because they are over and above what is required to obtain a solution, such as the elevation of all the bench marks in the project. The number of redundant observations in the

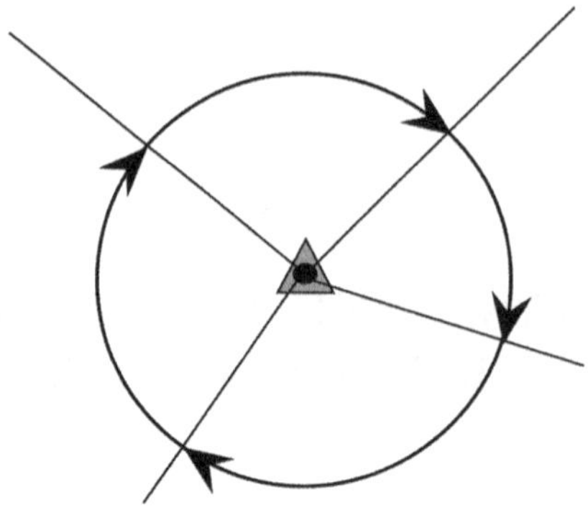

Figure 5-1. Angular horizon closure at the station.

adjustment is also known as the adjustment's *degrees of freedom*. Large mistakes such as station misidentifications must be removed before any further analysis of a survey is undertaken. Failure to catch large mistakes will often result in a nonsensical adjustment or no adjustment at all.

There are two types of errors. Systematic errors, which are also called biases, follow physical laws that can be modeled mathematically and are often removed through computations or field procedures. Random errors are those errors that remain in the observations after mistakes and systematics errors are removed. Random errors occur because of instrumental and personal limitations as well as small variations in environmental conditions.

Although systematic errors may be small, they can accumulate throughout the measurement process. For example, a steel tape will expand with increasing temperature. A 10°C difference in temperature from a standard temperature of 20°C will result in a measurement error of 3.5 mm for a 30 m steel tape because of its expansion. Although this systematic error is small, it will result in a cumulative error of 35 mm if this same tape is used to measure a 300 m line under the same conditions. This type of error can be removed by applying a well-known correction to taping observations, which in this case is $0.0000116(T - 20°)L$, where T is the temperature, in Celsius, of the tape at the time of the observation and L is the length being measured with the tape.

Most systematic errors can also be removed by instrument calibration and field procedures. For example, ideally, the horizontal line of sight of a theodolite should be perpendicular to its tilting axis. If not, the angle observed in one face of the instrument will be too small, whereas the angle measured in the other face will be too large. Owing to use, abuse, wear, or simply being out of calibration, this condition is sometimes not satisfied. To compensate for this error, the observer must measure an angle in both faces of the instrument and average the angular values. Because the error is equal in size but opposite in sign for each observation, the average will remove this systematic error from the observation. If a quality survey is to be performed, the observer must be aware of the presence of systematic errors in their observations and follow proper procedures to remove these errors from the observations before an adjustment is attempted.

As previously stated, random errors are those that remain in measured values after mistakes and systematic errors are removed from the observations. They are caused by factors beyond the control of the observer and thus cannot be eliminated by computation or field procedures. However, random errors in surveying tend to be normally distributed, can be examined using statistics, and can be adjusted by a procedure known as least squares.

POPULATION VERSUS SAMPLE

In theory, it is possible to continue re-measuring the length of one distance during one's entire lifespan, each time obtaining a similar but sometimes slightly different value for the length. Assuming this same person had a family, an offspring of this person could continue to measure the length of the same distance, and their offspring could repeat this measurement process. As the measurements continue, the total number of observations would approach a population of measurements for this length but never be the entire population because the population contains an infinite number of observations. Obviously, this entire process would be counterproductive. Thus, survey engineers collect samples of data to estimate population values of central tendency and data variation. Statistics is the study of these estimates.

Measures of central tendency include the mean, median, and mode. The median and mode are seldom used with engineering survey data. However, the mean is often used. The mean for a population of data is known as the true value and is typically denoted by μ. Because the true value for a population is not discernable, we often estimate its value with the sample mean $\bar{y}$, which is determined from n observations of y and is computed as

$$\bar{y} = \frac{\sum_{i=1}^{n} y_i}{n} \tag{5-1}$$

In addition, the data in a population of engineering survey measurements are assumed to be normally distributed, which means that individual observations can and do vary. Using the true value, the error for an individual observation in a population is defined as

$$\varepsilon_i = y_i - \mu \tag{5-2}$$

where ε_i is the error for observation y_i.

A measure for data variation in a population is known as its variance, σ^2, which is computed as

$$\sigma^2 = \frac{\sum \varepsilon_i^2}{n} \tag{5-3}$$

The square root for the variance in Equation (5-3) is known as the *standard error*. However, because μ is never known, the population's variance and standard error are also unknowns. Thus, we can only estimate these values. That is, because only samples of data from a population are collected, an estimate for the population variance is given by the *sample variance*, which is computed as

$$S^2 = \frac{\sum_{i=1}^{n} (\bar{y} - y_i)^2}{n-1} = \frac{\sum_{i=1}^{n} v_i^2}{n-1} \tag{5-4}$$

where S^2 is the sample variance, v is the residuals of the observations computed as $\bar{y} - y_i$, and all other terms as previously defined. The square root of the value in Equation (5-4) is known as the *standard deviation for the sample*. The standard deviation, S, is an estimate for the standard error σ, as is the sample variance an estimate for the population's variance, and the sample mean $\bar{y}$ an estimate for μ.

Equations (5-3) and (5-4) clearly show that population statistics differ from sample statistics, and it is important to understand such dissimilarities. This chapter carefully uses Greek symbols for population statistics and alphamerical characters for sample statistics. However, it is common in practice for the symbol "σ" to be used for a measurement's standard deviation. Therefore, it is important to examine the context of the writing to determine if σ is referring to a sample or population statistic.

Example 5-1. Twelve repetitions of the same angle are observed. The second's portion of angles is 24″, 22″, 27″, 12″, 36″, 23″, 12″, 24″, 34″, 30″, 34″, and 20″. What is the mean and standard deviation for this sample of data?

Solution

$$\text{Mean: } \bar{y} = \frac{298''}{12} = 24.8''$$

Data	v	v^2	Data	v	v^2	Data	v	v^2
24″	0.8″	0.694	36″	−11.2″	124.694	34″	−9.2″	84.028
22″	2.8″	8.028	23″	1.8″	3.361	30″	−5.2″	26.694
27″	−2.2″	4.694	12″	12.8″	164.694	34″	−9.2″	84.028
12″	12.8″	164.7	24″	0.8″	0.694	20″	4.8″	23.361
					Σ	298″		689.667

Note: Standard deviation: $S = \sqrt{689.667/(12-1)} = \pm 7.9''$

Now assume that the data in Example 5-1 represent a population of normally distributed data, with a mean of 24.8″ and a standard error of ±7.6″. If we were to randomly select any two observations and compute a mean and standard deviation for these observations, we might see results similar to that provided in Table 5-1.

Note in Table 5-1 that no sample has a mean of 24.8″ nor a standard deviation of ±7.6″, which are the population mean and standard error. Although the sample mean and standard deviation are estimates for the population mean and standard error, there is considerable variation in these samples' values. In fact, Sample 4, which has a sample mean closest to the population mean, has a standard deviation of ±17.0″, which is considerably

Table 5-1. Mean and Standard Deviation from Two Randomly Selected Values in Example 5-1.

Sample	Mean	S	Sample	Mean	S	Sample	Mean	S
1	30.0″	±8.5″	4	24.0″	±17.0″	7	23.5″	±0.7″
2	21.0″	±1.4″	5	17.5″	±7.8″	8	28.5″	±2.1″
3	28.5″	±2.1″	6	28.0″	±8.5″	9	29.0″	±7.1″

different from the standard error of ±7.6″. Statisticians noted this at the turn of the twentieth century and derived sampling statistics. From their studies, they created the t and χ^2 distributions to create confidence intervals for the population mean and standard error based on the sample's mean, standard deviation, number of elements sampled, and desired level of confidence.

The t distribution can be used to create a confidence interval for the population mean or an individual data element in a sample. The confidence interval for the population mean is computed as

$$\bar{y} - t_{\alpha/2,r}\frac{S}{\sqrt{n}} < \mu < \bar{y} + t_{\alpha/2,r}\frac{S}{\sqrt{n}} \tag{5-5}$$

where the critical t value is determined from the t distribution at α level of confidence so that the probability statement in Equation (5-5) is with n observations of which r are redundant. It should be noted that α is divided by 2 so that the test is distributed from the center for the distribution. Also, note that $S/\sqrt{n}$ in Equation (5-5) is defined as the *standard deviation of the mean*, which will be derived in Example 5-2. Thus, in Example 5-1, the 95% confidence interval for the population mean, μ, is $(19.8 < \mu < 29.9)$ where $t_{0.05/2,11}$, which equals 2.20, is obtained from the t distribution table, which can be found in references at the end of this chapter, using 0.025 (0.05/2) level of confidence and 11 redundant observations.

Using the same critical value for $t_{0.05/2,11}$, a similar confidence interval for a single observation, such as $y_i = 24.8$, is determined as

$$\bar{y} - t_{\alpha/2,r}S < y_i < \bar{y} + t_{\alpha/2,r}S \tag{5-6}$$

$$24.8 - 2.20(7.9) = 7.41 < y_i < 42.3 = 24.8 + 2.20(7.9)$$

Note that the 95% range for an observation using Equation (5-6) with a t-distribution multiplier is 12% larger than that obtained if the critical value of 1.96 from the normal distribution was used. This increase in confidence interval size is due to the uncertainty in the estimates for both the sample mean and standard deviation because only 12 observations were obtained, of which 11 were redundant.

Using critical values from the χ^2 distribution allows us to determine a confidence interval for the population variance based on the number of redundant observations and sample variance. Because, as shown in Figure 5-2, the χ^2 distribution is not symmetric like the normal and t distributions, the confidence interval for the population variance is dependent on two different critical values from the χ^2 distribution. Using these two critical values, the

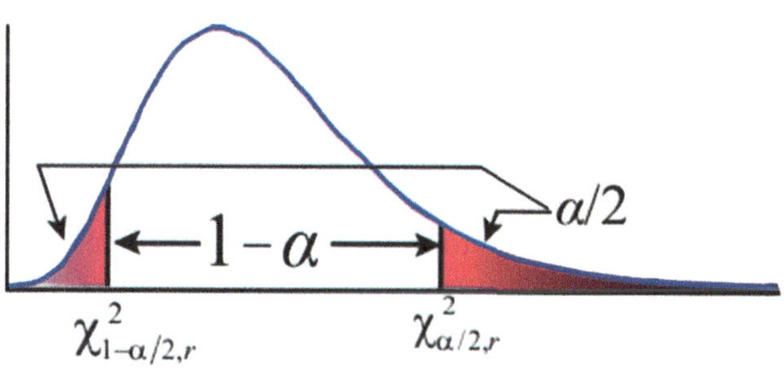

Figure 5-2. Chi-square distribution showing a 1 − α confidence interval.

$(1-\alpha)\%$ confidence interval for the standard error based on the sample's standard deviation S and number of redundant observations r is

$$\frac{rS^2}{\chi^2_{\alpha/2,r}} < \sigma^2 < \frac{rS^2}{\chi^2_{1-\alpha/2,r}} \tag{5-7}$$

Thus, the 95% confidence interval for the population's variance based on the standard deviation of ±7.9″ and 11 redundant observations from Example 5-1 is

$$\frac{11(7.9^2)}{\chi^2_{0.025,11}} = \frac{11(7.9^2)}{21.92} = 31.46 < \sigma^2 < 180.74 = \frac{11(7.9^2)}{3.82} = \frac{11(7.9^2)}{\chi^2_{0.975,11}}$$

This 95% range of 31.46 to 180.74 shows the uncertainty in the true value for the population's variance based on a sample of data with limited observations. In fact, 95% of the time, the standard error will lie between ±5.6″ and ±13.4″. If these same sample values were based on having only four angle observations as is typically observed in practice with three redundant observations, the range for the population variance would be between 20.0 and 867.6 yielding a 95% range for the standard error between ±4.5″ to ±29.5″. As is discussed later, this fact is important to remember when it comes to weighting a least-squares adjustment.

Another sampling distribution that is often used with least-squares adjustments is the F distribution. This distribution determines the range at a specified level of confidence for the ratio of two sample variances. Its use often comes when a least-squares adjustment has more than the minimal number of controls to fix it in space. For example, the least-squares adjustment of a horizontal survey on a map projection (see Chapter 3) is fixed positionally in space with known or assumed coordinates for one point and a known or assumed direction for one course. A Global Navigation Satellite System (GNSS) baseline vector adjustment is fixed by having one station's geocentric XYZ coordinates fixed, as is a differential leveling adjustment fixed by having one bench mark fixed in height. These adjustments are said to be *minimally constrained*. When a survey has more than the minimum amount of control, it is said to be fully or *overconstrained*. However, these additional controls help identify both blunders in control and systematic errors in a survey and thus are considered important to a survey.

Once the blunders are removed, systematic errors may still be present in the data. For example, if the observed horizontal distances are not reduced to the map surface as discussed in Chapter 3, the survey will result in coordinates for points that are shifted from their true positions by systematic errors. A technique that is used to identify this problem is to first run a least-squares adjustment using only the minimum amount of control required to fix the survey in space. One statistic computed from a minimally constrained adjustment is its reference variance, which indicates the fit of the observations with respect to themselves. Following the minimally constrained adjustment, the adjustment is performed with a data set that includes the additional control. This results in an overconstrained adjustment where the observations must also agree with the additional control. Again, this overconstrained adjustment has a reference variance. If there are no systematic errors in the observations and no blunders in the control, the ratio of these two reference variances should statistically equal one. Critical values from the F distribution are used to determine the acceptable range for the ratio of the two sample variances based on α level of confidence. This range is computed as

$$\frac{1}{F_{\alpha/2,r1,r2}}\frac{S_1^2}{S_2^2} < \frac{\sigma_1^2}{\sigma_2^2} < \frac{S_1^2}{S_2^2}F_{\alpha/2,r2,r1} \tag{5-8}$$

where S_1^2 and S_2^2 are the reference variances from the two adjustments, and $F_{\alpha/2}$ are the two critical values from the $F_{\alpha/2}$ distribution. Note that in Equation (5-8), the lower-tail critical value from the $F_{\alpha/2}$ distribution is determined using the degrees of freedom (redundant observations) for the first and second adjustments ($r1$ and $r2$, respectively), whereas the upper-tail is determined from the $F_{\alpha/2}$ distribution using $r2$ and $r1$ degrees of freedom. In other words, these are two different critical values from the $F_{\alpha/2}$ distribution because it, like the χ^2 distribution, is not symmetric. An example in the use of this distribution and the techniques used in the detection and removal of blunders in observations are discussed later.

LEAST-SQUARES ADJUSTMENTS

Least-squares adjustments are simply the simultaneous solution of a set of equations that represent the observations in terms of a set of unknown parameters and renders the sum of the squared residuals a minimum. In most least-squares adjustments, these unknown parameters are the coordinate values of the stations. The minimized sum of the squared residuals produces the most probable value for the unknowns (Ghilani 2017, Chapter 10). However, it is imperative that the blunders be removed from the observations; that is, flawed data will result in flawed adjusted parameters.

A least-squares adjustment contains two components both of which must be correct to generate the most probable values from the adjustment. These two components are the functional and stochastic models. The functional model is the set of well-known equations that represent the observations in terms of the unknown parameters. For example, the horizontal length of a distance between two stations I and J can be expressed thus:

$$\sqrt{(n_j - n_i)^2 + (e_j - e_i)^2} - l_{ij} \tag{5-9}$$

where (n_i, e_i) and (n_j, e_j) are the map projection northing and easting coordinates for Stations I and J, respectively; and l_{ij} is the map projection-reduced horizontal distance between Stations I and J. (See Chapter 3 for a discussion on the reduction of horizontal distances to a map projection surface.) Because this equation is nonlinear, it is linearized using a first-order Taylor series approximation during the adjustment process, which is

$$\left(\frac{e_i - e_j}{IJ}\right)_0 de_i + \left(\frac{n_i - n_j}{IJ}\right)_0 dn_i + \left(\frac{e_j - e_i}{IJ}\right)_0 de_j + \left(\frac{n_j - n_i}{IJ}\right)_0 dn_j = l_{ij} - IJ_0 + v_l \tag{5-10}$$

where

de_i, dn_i, de_j, and dn_j = Corrections to approximate (n, e) station coordinates;
IJ_0 = Horizontal distance as computed in Equation (5-9) with the approximate coordinates; and
v_l = Residual remainder.

Equations like Equation (5-10) are programmed into the least-squares adjustment software. Further exploration of equations used in surveying engineering adjustments can be found in Ghilani (2017), as well as other references at the end of this chapter.

The stochastic model is the weighting model for each observation. It controls the distribution of errors during the adjustment process. A correctly weighted least-squares adjustment will closely reproduce (or estimate) the observational errors. The least-squares solution can be represented using the matrix expression of

$$X = \left(A^{T}WA\right)^{-1} A^{T}WL \tag{5-11}$$

where

X = Solution for the unknown parameters, which for a horizontal adjustment are de_i, dn_i, de_j, and dn_j in Equation (5-10);
A = Matrix of coefficients for each observation as shown by the values in the parenthetical portions $()_0$ of Equation (5-10); and
W = Matrix of weights for each observation, which comprises the stochastic model for the adjustment.

ERROR PROPAGATION

It can be easily shown that errors from observations propagate throughout the computational process. For example, a check on the manufacturer's specifications for an electronic distance measurement (EDM) instrument lists the uncertainty in the distance measurement in terms of a constant error *a* and a scalar error *b*, which is expressed in terms of parts-per-million (ppm). Thus, an EDM may have a specified observational error of ±(3 mm + 3 ppm). Assuming fixed starting coordinates and direction in the computations, the resulting coordinates for the second station will contain the errors from the distance observation. This can be seen in Figure 5-3. When computing the third station using an

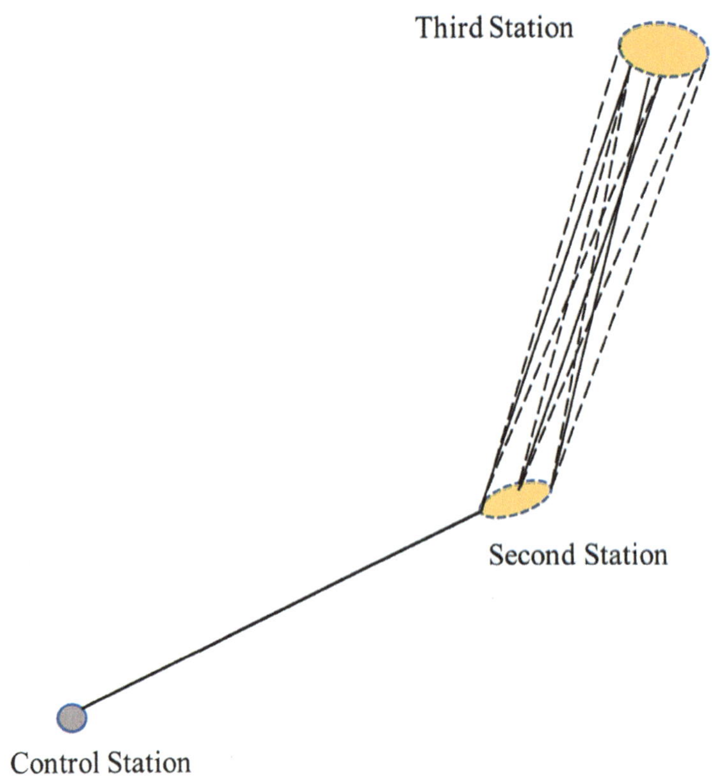

Figure 5-3. Error propagation in traverse computations.

angle and distance observation, its resultant coordinates will contain the errors in the second station's coordinates as well as those in the observed distance and angle at the second station. The result is that the coordinates for the third station will contain more errors than either of the previous stations. The errors continue to propagate as the survey proceeds along the traverse, and thus the station farthest from the control will always contain the most errors.

The propagation of errors from their source to a computed quantity can be summarized in two general expressions. The first, the General Law of the Propagation of Variances (GLOPOV), determines how errors propagate in a system of equations. As discussed later, GLOPOV is used for determining the uncertainties in computed quantities after an adjustment. The propagation of random errors through functions using GLOPOV is determined as

$$\Sigma_{zz} = A^{\mathrm{T}}\Sigma A \tag{5-12}$$

where Σ_{zz} is the covariance matrix for the computed quantities using function Z (such as shown in Equation (5-10); and A is the coefficient matrix of the computing function, as was the case in Equation (5-11).

Σ is the covariance matrix for the observations, expressed as

$$\Sigma = \begin{bmatrix} \sigma_{x1}^2 & \sigma_{x1x2} & \cdots & \sigma_{x1xn} \\ \sigma_{x2x1} & \sigma_{x2}^2 & & \sigma_{x2xn} \\ \vdots & & \ddots & \\ \sigma_{xnx1} & \sigma_{xnx2} & \cdots & \sigma_{xn}^2 \end{bmatrix} \tag{5-13}$$

where the diagonal elements are the variances for the observations $x1, x2, \ldots, xn$; and the off-diagonal elements are the covariances between the indicated pairs of observations.

When observations are mathematically independent (i.e., the covariances equal zero), the Special Law of the Propagation of Variances (SLOPOV) can be used. SLOPOV can be mathematically expressed as

$$\sigma_z = \sqrt{(a_{x1}\sigma_{x1})^2 + (a_{x2}\sigma_{x2})^2 + \cdots + (a_{xn}\sigma_{xn})^2} \tag{5-14}$$

where

σ_z = Derived standard error for the computed quantity using function Z,
$a_{x1}, a_{x2}, \ldots, a_{xn}$ = Coefficients of the function Z, and
$\sigma_{x1}, \sigma_{x2}, \ldots, \sigma_{xn}$ = Standard errors for the observations $x1, x2, \ldots, xn$.

SLOPOV is used in many elementary surveying computations because surveying engineering observations are often mathematically independent of each other.

Some common error propagation equations that are derived from SLOPOV are

$$\text{Error of a sum:}\quad \sigma_z = \sqrt{\sigma_{x1}^2 + \sigma_{x2}^2 + \cdots + \sigma_{xn}^2} \tag{5-15a}$$

where σ_z is the standard error in function that sums $x1, x2, \ldots, xn$ observations with standard errors of $\sigma_{x1}, \sigma_{x2}, \ldots, \sigma_{xn}$, respectively.

$$\text{Error of a series:}\quad \sigma_z = \sigma_x\sqrt{n} \tag{5-15b}$$

where $x1, x2, \ldots, xn$ observations have the same standard error of σ_x.

Example 5-2. The mean of observations $y_1, y_2, \ldots, y_n$ is computed using Equation (5-1) as

$$\bar{y} = \frac{y_1 + y_2 + \cdots + y_n}{n}$$

Assuming the standard error of each observation is σ_y, what is the error of the mean?

Solution

Using Equation (5-14), we see that the coefficient for each observation in Equation (5-1) is $1/n$. Thus, by Equation (5-14), the error of the mean is

$$\sigma_{\bar{y}} = \sqrt{\left(\frac{1}{n}\sigma_y\right)^2 + \left(\frac{1}{n}\sigma_y\right)^2 + \cdots + \left(\frac{1}{n}\sigma_y\right)^2} = \sqrt{\frac{n\sigma_y{}^2}{n^2}} = \frac{\sigma_y}{\sqrt{n}}$$

Note in Example 5-2 that if we substitute in the standard deviation for each observation as S_y, we obtain the standard deviation of the mean, $S_{\bar{y}} = S_y/\sqrt{n}$. The standard deviation of the mean was used in Equation (5-5) to compute a confidence interval for the population mean based on critical values from the *t* distribution, sample mean, and standard deviation.

WEIGHTS

As previously discussed, the equations that represent surveying observations are well known and programmed into least-squares software packages. Although the functional model is included in the surveying adjustment software, the stochastic or weighting model is not. That is, only its application, as shown in Equation (5-11), is included in the software. Weights for mathematically dependent observations, such as those in a GNSS survey, are computed as

$$W = \sigma_0^2 \, \Sigma^{-1} \tag{5-16}$$

where

W = Weight matrix;
σ_0 = Scalar that is typically assigned a value of one before the adjustment; and
Σ = Covariance matrix of the observations, which in a GNSS survey are the baseline vector components of (ΔX, ΔY, ΔZ).

Unfortunately, GNSS baseline processing software is notoriously optimistic when estimating the variances in Σ. As a result, it is common to iterate a factor to scale up the variances when developing the stochastic model for the GNSS survey.

For a set of independent observations, which is the case in many engineering surveys, the weight for an individual observation is computed as

$$w_i = \sigma_0^2/\sigma_i^2 \tag{5-17}$$

where σ_i is the standard error of the observation. Because σ_0^2 is usually assigned a value of one, independent observations are typically weighted in a least-squares adjustment as

$$w_i = 1/\sigma_i^2 \tag{5-18}$$

Note that Equations (5-17) and (5-18) use the standard error of the observation, which is a population value that is not known. Thus, it is common to substitute the sample standard deviation for the standard error in Equation (5-18) because this is an estimate for the standard error. However, as was shown before using Equation (5-7), when the size of the sample is small, as is typically the case in surveying engineering practice, there can be a large range for the standard error. Thus, an alternative method of weighting is to determine estimates for the standard errors of observations using error propagation principles.

For example, the error in an angle observation has the major error sources of pointing and reading, instrument-centering, and target-centering. Because each of these error sources can be estimated, the overall error in an observed angle can also be estimated. For example, the standard error in pointing and reading is determined using the ISO 17123-3 standard, which is published as part of the instrument's specifications. According to Ghilani (2017, Chapter 7), the estimated standard error in pointing and reading of an angle using a digital theodolite is

$$\sigma_{\alpha_{pr}} = \pm\frac{2(\text{ISO})}{\sqrt{n}} \tag{5-19}$$

where

$\sigma_{\alpha_{pr}}$ = Estimated standard error in pointing and reading an angle,
ISO = Digital theodolite's published ISO 17123-3 uncertainty, and
n = Number of repetitions of the angle.

According to Ghilani (2017, Chapter 7), the estimated standard error in target-centering is

$$\sigma''_{\alpha_t} = \pm\frac{\sqrt{D_{BS}^2 + D_{FS}^2}}{D_{BS}D_{FS}}\sigma_t\rho \tag{5-20}$$

where

σ''_{α_t} = Estimated standard error in the angle owing to the target-centering error σ_t in units of seconds;
D_{BS} and D_{FS} = Sight distances to the backsight and foresight targets, respectively; and
ρ = Conversion factor of 206,264.8″/rad.

Finally, the standard error caused by the instrument-centering is

$$\sigma''_{\alpha_i} = \pm\frac{D_3}{D_{BS}D_{FS}}\frac{\sigma_i}{\sqrt{2}}\rho \tag{5-21}$$

where σ''_{α_i} is the estimated standard error caused by the instrument-centering error of σ_i in units of seconds, D_3 is the distance between the backsight and foresight stations, and all other variables as defined for Equation (5-20).

Example 5-3. What is the estimated standard error for an angle of 53° with a backsight distance of 233 ft (71.0 m) and a foresight distance of 385 ft (117 m) if the ISO 17123-3 standard for the instrument is ±3″, estimated target-centering error is ±0.01 ft (±3 mm), and estimated instrument-centering error is ±0.005 ft (±1.5 mm). The angle observation was repeated two times.

Solution

By Equation (5-19):

$$\sigma_{\alpha_{pr}} = \pm\frac{2(3'')}{\sqrt{2}} = \pm 4.2''$$

By Equation (5-20):

$$\sigma''_{\alpha_t} = \pm\frac{\sqrt{233^2 + 385^2}}{233(385)}0.01(206,264.8) = \pm 10.3''$$

Using the law of cosines, the distance between the targets is

$$D_3 = \sqrt{233^2 + 385^2 - 2(233)385 \cos 53^\circ} = 385 \text{ ft}$$

By Equation (5-21):

$$\sigma''_{\alpha_i} = \pm\frac{385}{233(385)}\frac{0.005}{\sqrt{2}}(206,264.8) = \pm 3.1''$$

Each of these error sources is independent, and thus the error of a sum as given by Equation (5-14) provides the overall estimate for the standard error in the angle, which is

$$\sigma''_{\alpha} = \sqrt{4.2^2 + 10.3^2 + 3.1^2} = \pm 11.5''$$

As previously stated, the manufacturers of EDMs list the precision of their instruments in terms of a constant error *a* and a scalar error *b*. However, these error sources do not account for the centering errors of the instrument and target during the measurement process. Thus, using Equation (5-14), the estimated error in an electronically measured distance is computed as

$$\sigma_l = \sqrt{\sigma_t^2 + \sigma_i^2 + a^2 + \left(\frac{b}{1,000,000}D\right)^2} \tag{5-22}$$

where σ_l is the estimated standard error in the observed distance D, and all other variables are as previously defined.

Example 5-4. A distance of 125.768 m is measured using an EDM with a manufacturer's specified accuracy of 3 mm + 3 ppm. The estimated instrument-centering error is ±2 mm and the estimated target-centering error is ±5 mm. What is the estimated standard error in the observed distance?

Solution

By Equation (5-22):

$$\sigma_l = \sqrt{5^2 + 2^2 + 3^2 + \left(\frac{3}{1,000,000}125,678\right)^2} = \pm 6.2 \text{ mm}$$

Similarly, the errors in a differential leveling survey can be estimated. The two major random error sources in differential leveling are the ability to level the instrument and the ability to read the level rod. With automatic levels, whether optical or digital, the manufacturer's specifications for an instrument typically list the compensator's setting accuracy in terms of arc units such as $\pm 0.5''$ and a minimum reading capability in terms of a minimum reading per sight length such as ± 1 mm/30 m. However, this second uncertainty can be estimated by the user. For example, one may estimate that their ability to read an English rod graduated to the nearest 0.01 ft (3 mm) is about ± 0.005 ft (± 1.5 mm) per 100 ft (30.5 m). Using these two values, the standard error in a leveled height difference between two points can be estimated (Ghilani 2017, Chapter 9) as

$$\sigma_{\Delta H} = D\sqrt{2n(\sigma_{r/D}^2 + \sigma_\alpha^2)} \tag{5-23}$$

where

$\sigma_{\Delta H}$ = Estimated standard error in the leveled height difference between two bench marks,
D = Average sight distance for each rod reading,
$\sigma_{r/D}$ = Estimated or specified error in reading the leveling rod, and
σ_α = Compensator's ability to set a horizontal line.

Example 5-5. An automatic level with a compensator setting accuracy of $\pm 0.5''$ is used to determine the leveled height difference between two bench marks that are connected by a leveling line that is 600 m long. The average sight distance during the leveling process was 50 m, and the minimum reading accuracy for the instrument was ± 1.0 mm per 30 m. What is the estimated standard error in a leveled height difference between two bench marks?

Solution

Each setup of the instrument resulted in an advance in a distance of 100 m. Thus, there were six [i.e., 600/[2(50)]] setups between the two bench marks. Using Equation (5-23), the estimated standard error for the leveled height difference is

$$\sigma_{\Delta H} = 50\sqrt{2(6)\left[\left(\frac{1.0}{30}\right)^2 + \left(\frac{0.5}{206{,}264.8}\right)^2\right]} = \pm 5.8 \text{ mm}$$

Because the error for each setup in a differential leveling line often has the same amount of random error, it is not uncommon in practice to assign weights for differential leveling networks based on the distance between the bench marks or the number of setups between bench marks. Using Equation (5-23), it has been shown (Ghilani 2017, Section 10.6) that the weights for differential leveling networks can be assigned as

$$w_i = 1/l_i \tag{5-24a}$$

$$w_i = 1/n \tag{5-24b}$$

where

w_i = Assigned weight for each leveling height difference between bench marks,
l_i = Leveled distance between the bench marks, and
n = Number of setups between bench marks.

However, if these methods of weighting are used, the postadjustment χ^2 test, as discussed later, cannot be performed because σ_0 is unknown and thus can no longer be assumed to be 1. In fact, the postadjustment, a posteriori, reference variance will tend to approach zero if Equations (5-24a) or (5-24b) are used to weight a differential leveling adjustment.

PREPARING DATA FOR AN ADJUSTMENT

Depending on the type of adjustment being attempted, the observational data usually need to be corrected for systematic errors so that only random errors are left in the adjustment. For example, when performing a horizontal least-squares adjustment, all the horizontal distances must be reduced to the mapping surface. Failure to reduce the horizontal lengths to map lengths will result in systematic errors being present in the data. It is also important to transform any geodetic azimuths to grid azimuths. Failure to account for the difference between map directions and geodetic directions will result in a rotated network of stations that, if connected to more than one set of control coordinates or control directions, will result in an incorrect solution or possibly no solution. In GNSS adjustment software, the corrections to observations must take into account the differences between observations made with respect to the ellipsoid rather than the geoid (Chapter 2). For leveling data, corrections must be applied to convert leveled height differences to orthometric height differences. Observed vertical angles must be corrected for deflection of the vertical. Thus, it is important to be aware of the sources for systematic errors, and corrections must be applied to achieve a proper adjustment.

In addition to the removal of systematic errors, mistakes must also be uncovered and removed. For example, failure to uncover misidentified stations or a transcription error may result in a nonsensical solution or no solution when the equations involved in the functional model are nonlinear as in a horizontal adjustment.

In simple traverse situations, mistakes in angle and distance observations can often be uncovered using graphical techniques. For example, a blunder in an angle observation will result in rotating the remaining courses in a traverse about the station that has the angular blunder. As shown in Figure 5-4(a), when this happens, the perpendicular bisector of the traverse misclosure line will point near the station with the blunder. In Figure 5-4(a), an angular blunder at Station *E* rotates the remaining courses of the traverse such that the

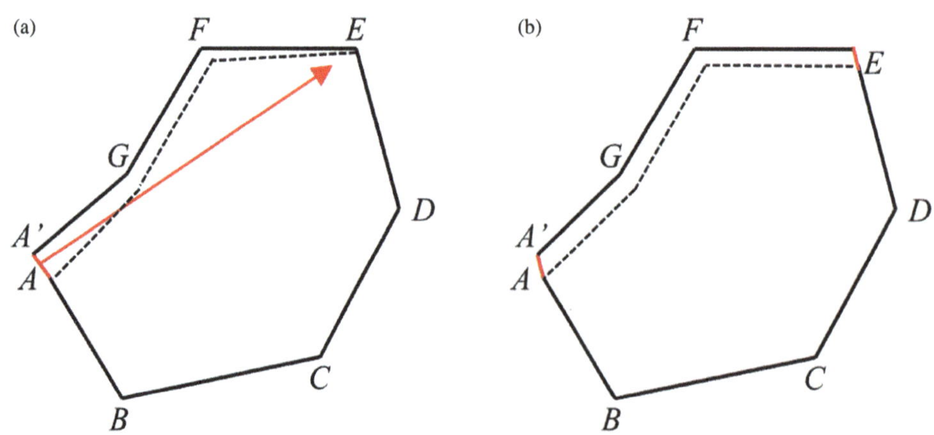

Figure 5-4. (a) A simple closed traverse with an angular blunder at station E, (b) a simple closed traverse with a distance blunder in course DE.

misclosure line of the traverse between the Starting Station A and Closing Station A' is a chord for the circle that is centered at Station E. This error can be verified by assuming control coordinates for the station with the assumed angular error, E, in Figure 5-4(a), and assuming a direction for the connected course, which would be course EF. The computations to determine the closing station's coordinates, which would be E', never use the angle at station E. Thus, the closing error will be acceptable if the angle at E contains the blunder.

As shown in Figure 5-4(b), when a distance blunder occurs as in course DE, the remaining courses are shifted by the amount of the blunder. In this case, the azimuth or back azimuth of the misclosure line for the traverse will nearly match the azimuth of the course with the blunder. However, because there are random errors in all the observations, the closure line's azimuth and azimuth of the suspected course with the blunder will differ slightly. Only re-observing the distance with the suspected blunder, which is DE in Figure 5-4(b), will resolve the situation.

Mistakes such as station misidentifications can sometimes result in many hours of searching to uncover. This is why it is always recommended that field crews create site log sheets when occupying stations, whether with GNSS receivers or optical instruments. Although it is inevitable that field crews will sometimes have entry errors in a survey controller, they seldom will make the same mistake on a handwritten site log sheet. By having two different sources of the crew's field procedures, station misidentifications and transcription errors can often be uncovered. This technique is often used in static GNSS surveys. However, it may be beneficial in some optical surveys also.

POSTADJUSTMENT STATISTICS

Following a least-squares adjustment, error propagation principles are used to develop postadjustment statistics. Some of the primary postadjustment statistics are the reference variance or its square root known as the *standard deviation of unit weight*, standard deviations for the unknown parameters and adjusted observations, and error ellipses. Except for the reference variance, these statistical values are all derived from the cofactor matrix for the unknowns, which is computed in the least-squares solution as $(A^TWA)^{-1}$ and is represented as Q_{xx}.

The adjustment's reference variance, a sample statistic, is computed as

$$S_0^2 = \frac{V^TWV}{r} \tag{5-25}$$

where

S_0^2 = Sample reference variance for the adjustment;
V = Vector of all the observational residuals in the adjustment;
W = Matrix of the observational weights; and
r = Number of redundant observations, which is the number of additional observations that are necessary to solve the matrix system shown in Equation (5-11).

Using the cofactor matrix Q_{xx} for the unknowns and the standard deviation of unit weight S_0, the standard deviations for the unknowns are computed as

$$S_{x_i} = S_0\sqrt{q_{ii}} \tag{5-26}$$

where

S_{x_i} = Standard deviation for the unknown parameter x_i,
S_0 = Sample reference variance for the adjustment, and
q_{ii} = Diagonal element of the Q_{xx} matrix corresponding to the unknown parameter x_i.

Using GLOPOV [Equation (5-12)], the standard deviations in the adjusted observations can be similarly computed with the cofactor matrix Q_{ll}, which is defined as

$$Q_{ll} = AQ_{xx}A^{T} \tag{5-27}$$

Although the standard deviations for the unknown coordinates are informative, they are not oriented in the direction of the largest and smallest errors at the station. This can be seen in Figure 5-3 at the second station. To determine orientation of the largest and smallest errors at a station, the submatrix of the Q_{xx} matrix that corresponds to the station's unknown parameters must be orthogonalized. The eigenvalues of the orthogonalized submatrix provide the values for the computation of the semiaxes of the error ellipse or ellipsoid. However, often only the two-dimensional error ellipse is determined, and the error in the third dimension of height is represented by its standard deviation. The components of the error ellipse are shown in Figure 5-5. Thus, the 2-by-2 submatrix for the horizontal component of a station is rotated by t to obtain a maximum value for the error of S_u and a minimum value for the error of S_v for the station.

Using a 2-by-2 submatrix from Q_{xx} that corresponds to the unknowns for a station with xy coordinates, the three error ellipse components are computed as

$$\tan 2t = \frac{2q_{xy}}{q_{yy} - q_{xx}} \tag{5-28}$$

$$S_u = S_0\sqrt{q_{xx}\sin^2 t + 2q_{xy}\cos t \sin t + q_{yy}\cos^2 t} \tag{5-29}$$

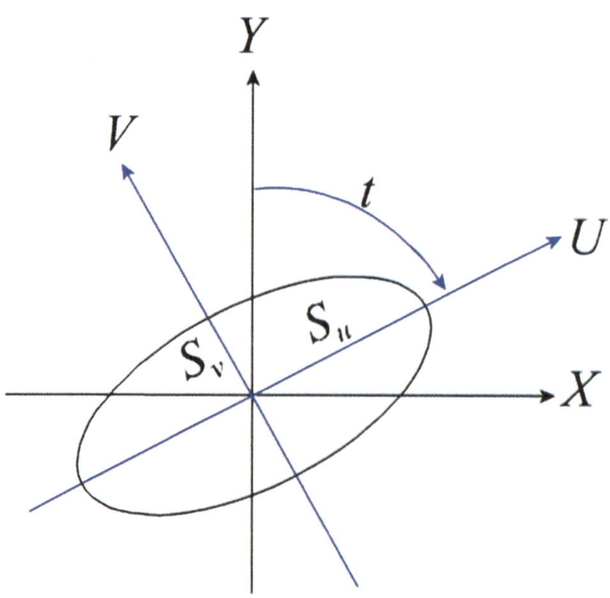

Figure 5-5. Components of an error ellipse and its relation to the X–Y axes.

$$S_v = S_0\sqrt{q_{xx}\cos^2 t - 2q_{xy}\cos t \sin t + q_{yy}\sin^2 t} \tag{5-30}$$

where

t = Rotation angle for the semimajor axis of the ellipse from the y-axis,
S_u = Length of the semimajor axis of the ellipse,
S_v = Length of the semiminor axis,
q_{xx} = Diagonal element in the 2-by-2 submatrix of the Q_{xx} matrix corresponding to the easting (x) coordinate of the station,
q_{yy} = Diagonal element of the submatrix corresponding to the northing (y) coordinate of the station, and
q_{xy} = Off-diagonal or covariance element in the submatrix corresponding to the station.

The error ellipse computed using Equations (5-28) through (5-30), which is known as the *standard error ellipse*, has a probability of only 35% to 39%. However, most standards call for an error ellipse of much higher probability, such as 95%. To compute a 95% error ellipse for a station, a critical value from an F_α distribution must be determined. The multiplier to obtain an error ellipse at a $(1-\alpha)$% is computed using

$$c = \sqrt{2F_{\alpha,2,r}} \tag{5-31}$$

where

c = Multiplier to apply to Equations (5-29) and (5-30),
$F_{\alpha,2,r}$ = Critical value from an F_α distribution, and
r = Number of redundant observations in the adjustment.

For example, the multiplier for a 95% error ellipse from an adjustment having 15 redundant observations would be computed as $c = \sqrt{2F_{0.05,2,15}} = 2.71$, where $F_{0.05,2,15}$ is 3.6823. Thus the 95% error ellipse semiaxes would be computed as

$$S_{u95\%} = 2.71S_0\sqrt{q_{xx}\sin^2 t + 2q_{xy}\cos t \sin t + q_{yy}\cos^2 t}$$

$$S_{v95\%} = 2.71S_0\sqrt{q_{xx}\cos^2 t - 2q_{xy}\cos t \sin t + q_{yy}\sin^2 t}$$

It should be pointed out that some standards require the user present an equivalent circular error rather than error ellipse data for the stations in the adjustment. A positional circular error denotes the probability within a radius of position and thus provides a single error estimate for a point. The probability required in standards is typically 95%. To compute the 95% error circle for a point, a constant C is computed using the semimajor and semiminor axes from the least-squares adjustment as

$$C = \frac{S_v}{S_u} \tag{5-32}$$

Then, using a polynomial from Leenhouts (1985), the 95% radial error of a point is approximated as

$$r_{95\%} = Su(1.960790 + 0.004071C + 0.114276C^2 + 0.371625C^3) \tag{5-33}$$

POSTADJUSTMENT BLUNDER DETECTION

For any postadjustment statistics to be valid, the adjustment must be free from blunders and correctly weighted. The first statistic to check is the adjustment's reference variance S_0^2. The adjustment's reference variance represents a sampling statistic of its a priori value σ_0^2, which was assigned a value of 1 in Equations (5-16) and (5-18) during the weighting of the observations. Thus, a χ^2 test is often performed after adjustment comparing its population value of σ_0^2 against the postadjustment value of S_0^2. This is known as the *goodness-of-fit* test. If S_0^2 is statistically equal to σ_0^2, the adjustment is said to pass the test, which means that σ_0^2 is in the range as determined by Equation (5-7).

However, this test can fail for multiple reasons. First, it is possible that one or more observations contain a blunder. If this is the case, the residuals for these observations will be too large. It is also possible that the weights are incorrect for the observations. That is, they could be too large if the assigned standard errors in Equations (5-16) or (5-18) are too small for the errors in the observations or too small if the assigned standard errors are too large. Finally, all statistical tests are performed at a particular probability level, which is known as the level of confidence. Thus, α% of the time, the test will fail with observations containing only random errors and having valid weights. For example, a t-test performed at a probability of 95% will fail 5% of the time when nothing is incorrect. Because the possibility of incorrect weighting or the presence of observational blunders is possible and even more likely, the surveyor is expected to use due diligence to check the observations and stochastic (weighting) model when the test fails. Thus, failure to pass the goodness-of-fit test should only be thought of as a warning flag that requires further investigation.

An F distribution is used to check observational data for systematic and/or possible control errors when an overconstrained adjustment is possible. For example, assume a minimally constrained adjustment, which is performed using map projection coordinates, passes the goodness-of-fit test. This means that the data agree with themselves. However, let us further assume that there is additional control that can be added to the adjustment to create an overconstrained adjustment. When this is done, the overconstrained adjustment produces a second reference variance for the same data. If there are no systematic errors or control errors present in the data, the ratio of the two reference variances should statistically equal 1.

For example, assume that a minimally constrained adjustment's reference variance is 1.089 with 16 degrees of freedom. However, when the additional control is added to the adjustment, the overconstrained adjustment has a reference variance of 4.639 with 14 degrees of freedom. Using Equation (5-8), the 99% confidence interval ($\alpha = 0.01$) is

$$\frac{1}{4.20}\frac{1.089^2}{4.639^2} = 0.01 < \frac{\sigma_1^2}{\sigma_2^2} < 0.22 = \frac{1.089^2}{4.639^2}3.97$$

Because the 99% confidence interval does not contain 1, which is the ratio of the two population reference variances, it fails the F-test. Thus, the observational data contain a systematic error, which means they do not fit the control or the control contains errors. However, like the goodness-of-fit test, this test does not identify the specific problem. Thus, the survey engineer must practice due diligence in examining both the data and their procedures. It is wise to be certain that all possible systematic errors have been eliminated from the observations before suspecting problems with the control.

A quick check for blunders in the data can be performed by reviewing all the residuals of the observations at the completion of the adjustment. Three common methods can be used to do this. First, the assigned standard errors for the observations can be compared against their residuals using a confidence interval for the residuals that is derived from Equation (5-6) as $\pm t_{\alpha/2,r}S$. Using this technique, the number of redundant observations r is taken from the degrees of freedom of the adjustment, which is computed as the number of independent observation equations minus the number of unknowns. Thus, a single critical value $t_{\alpha/2,r}$ is multiplied by the observation's assigned standard error and checked against its residual. Often the assigned standard errors for a particular observational type are very similar, and a single computation of the largest assigned standard error by the critical t value will uncover any large observational blunders.

Another method, which was developed by Willem Baarda in the 1960s, involves a technique known as *data snooping*. This method computes a standard deviation for each residual using error propagation techniques. These values are then used to determine what is known as the *standardized residuals* for the observations. The standardized residuals are then compared against a rejection criterion derived from the t distribution. In practice, researchers found that a rejection criterion of 3.29 worked well.

In 1976, Alan Pope modified the data snooping method by introducing the *tau* distribution. This method recognized that the reference variance from an adjustment with a blunder was flawed and thus suggested a critical value τ from the tau distribution be used as a rejection criterion rather than a critical value from the t distribution. See Ghilani (2017) for more on data snooping and the τ criterion.

All three methods share one fact in common, which is that only one observation should be removed as a suspected blunder from the adjustment at a time and that this observation should be the one that is farthest from the rejection criteria. This is due to the fact that any adjustment will spread a blunder to the observation's nearest neighbors, sometimes resulting in multiple observations being apparently flagged as blunders. Thus, the adjustment will need to be repeated several times to remove all the blunders from an observational data set. Because this is nothing more than a very short computer process, this is repetition should not be thought of as being onerous.

STATISTICAL METHODS OF DETERMINING MAP ACCURACY

Modern map accuracy standards are based on a statistical analysis of a check survey's coordinates versus the map coordinate values. These standards can also be applied to data from lidar surveys. The importance of these standards is that it makes the user of the map or data aware of its accuracy, and thus the user can determine its suitability for a particular project. The standards often require a certain number of checkpoints distributed throughout the mapping region. In addition, the check survey must be of a higher quality than the map coordinates. Often this means that an optical or GNSS survey is used as the check survey for the photogrammetric mapping and lidar or laser scanning data. Because the check survey coordinates are of a higher quality than the map coordinates, the check survey coordinates are assumed to be without error. Discrepancies, which are thought of as residuals in this instance, are computed between the check survey coordinates and map coordinates, and from these, standard deviations and root-mean-square errors (RMSE) are computed for each coordinate type. The mapping accuracy is then based on the RMSE values for both the horizontal and vertical components of the map. RMSE is defined as the square root of the average of squared discrepancies for points tested and is computed as

$$\text{RMSE} = \sqrt{\frac{\Sigma_{i=1}^{n}\left[x_i - f(x_i)\right]^2}{n}} \tag{5-34}$$

Table 5-2. Map Coordinates versus Surveyed Check Point Coordinates.

	Surveyed check point coordinates			Map coordinates			Discrepancies		
Pt	(1) x (m)	(2) y (m)	(3) h (m)	(4) E (m)	(5) N (m)	(6) H (m)	(7) Δx	(8) Δy	(9) Δz
1	14,891.153	35,874.351	24.589	14,891.114	35,874.320	24.704	0.031	0.039	−0.115
2	13,480.214	35,789.318	38.674	13,480.233	35,789.263	38.595	0.055	−0.019	0.079
3	28,924.086	36,182.489	56.483	28,924.051	36,182.497	56.394	−0.008	0.035	0.089
4	29,716.158	28,315.168	5.641	29,716.176	28,315.123	5.677	0.045	−0.018	−0.036
5	15,314.604	43,648.084	43.891	15,314.661	43,648.148	45.945	−0.064	−0.057	−0.054
				Number of checkpoints			5	5	5
				Mean error (m)			0.0118	−0.004	−0.0074
				Standard deviation (m)			0.049	0.041	0.088
				RMSE (m)			0.045	0.037	0.079
				$RMSE_r$			0.058		
				95% RMSE			0.100		0.155

where x_i are the check survey's coordinates for the n sample of points, and $f(x_i)$ are the coordinates measured from the map or from a lidar or laser scanning project.

For example, consider the coordinates listed in Table 5-2. Columns 1 to 3 list the coordinates of points as determined from a check survey, Columns 4 to 6 list the coordinates of the same points as determined from the map, and Columns 7 to 8 list the discrepancies between the coordinates. As an example, the discrepancies for point (Pt) 1 are computed as

$$\begin{aligned} \text{In } x: &\quad 14{,}891.153 - 14{,}891.114 = 0.031\text{ m} \\ \text{In } y: &\quad 35{,}874.351 - 35{,}874.320 = 0.039\text{ m} \\ \text{In } z: &\quad 24.589 - 24.704 = -0.115\text{ m} \end{aligned}$$

After all the discrepancies are determined, the means and standard deviations for the discrepancies are computed using Equations (5-1) and (5-4), respectively. Once these values are determined, the RMSE values for each coordinate type are computed using Equation (5-34). For example, the RMSE for the x-coordinate discrepancies shown in Table 5-2 is computed as

$$\text{RMSE}_x = \sqrt{\frac{0.031^2 + 0.055^2 + (-0.008)^2 + 0.045^2 + (-0.064)^2}{5}} = \pm 0.045\text{ m}$$

Similar computations are performed for the Δy and Δz components, the results of which are shown in Columns 8 and 9. Finally, the horizontal RMSE_r is computed as

$$\text{RMSE}_r = \sqrt{\text{RMSE}_{\Delta x}^2 + \text{RMSE}_{\Delta y}^2} \tag{5-35}$$

Typically, the accuracy standards are listed at 95%. When this is the case, the 95% accuracies are given as

Horizontal:

$$\text{RMSE}_{@95\%} = 2.447\left(\frac{\text{RMSE}_r}{\sqrt{2}}\right) = 1.7038(\text{RMSE}_r) \tag{5-36}$$

Vertical:

$$\text{RMSE}_V = 1.9600(\text{RMSE}_{\Delta z}) \tag{5-37}$$

CONCLUSIONS

The importance of understanding the sources of errors in observations, their detection, and adjustment cannot be overstated. Many deeds, plats, and engineering drawings have been used with blunders undiscovered. When this happens, the user of the flawed document, whether a professional engineer, surveyor, or citizen, is then misled by the information contained in the document until errors are uncovered. When this happens, the correction for an undiscovered blunder or systematic error can often lead to unforeseen additional costs. Thus, due diligence should always be practiced when performing an engineering survey.

This chapter gives a brief overview of the use of statistics and least-squares adjustments to ensure that blunders and systematic errors are uncovered before the data are released for additional use. In addition, the reader must investigate the proper procedures in collecting data and the sources of systematic errors that must be removed before any adjustment of data is further attempted. For further investigation on these topics, the reader should refer to the books listed in the reference section of this and other chapters in this book.

REFERENCES

ASPRS (American Society for Photogrammetry and Remote Sensing). 2015. "ASPRS positional accuracy standards for digital geospatial data." *Photogramm. Eng. Remote Sens.* 81 (3): A1–A26.

Buckner, R. B. 1983. *Surveying measurements and their analysis.* Cordova, CA: Landmark Enterprises.

FGDC (Federal Geodetic Data Committee). 1998a. "National standard for spatial data accuracy." *Geospatial positioning accuracy standards*, Part 2. FGDC-STD-007.2-1998. Accessed February 9, 2022. http://www.fgdc.gov/standards/projects/FGDC-standards-projects/accuracy/part2/chapter2.

FGDC. 1998b. "National standard for spatial data accuracy." *Geospatial positioning accuracy standards,* Part 3. FGDC-STD-007.3-1998. Accessed February 9, 2022. http://www.fgdc.gov/standards/projects/FGDC-standards-projects/accuracy/part3/chapter3.

Ghilani, C. 2017. *Adjustment computations: Spatial data analysis.* 6th ed. Hoboken, NJ: Wiley.

Ghilani, C. 2022. *Elementary surveying: An introduction to geomatics.* 16th ed. Upper Saddle River, NJ: Pearson.

Leenhouts, P. 1985. "On the computation of bi-normal radial error." *Navigation* 32 (1): 16–28.

Mikhail, E. 1976. *Observations and least squares.* New York: Harper & Row.

Mikhail, E., and G. Gracie. 1981. *Analysis and adjustment of survey measurements.* New York: Van Nostrand Reinhold.

CHAPTER 6
SATELLITE-BASED SURVEYING TECHNOLOGY

Jan Van Sickle

BRIEF OVERVIEW OF GLOBAL POSITIONING SYSTEM

Global Navigation Satellite Systems (GNSSs) is a term that describes any satellite system that provides positioning, navigation, and timing on a regional or global basis. One of the completed global systems is the United States' global positioning system (GPS). Other complete, global system are the Russian Federation's Globalnaya Navigatsionnaya Sputnikovaya Sistema (GLONASS) and the Chinese's BeiDou Navigation Satellite System (BDS). Satellite systems in development that will likely reach similar capabilities soon will also take their place within the GNSS definition. Galileo, the system administered by the European Union (EU), and regional systems such as the Indian Regional Navigation Satellite System (IRNSS) and the Japanese Quasi-Zenith Satellite System (QZSS) are often included in GNSS as well. Much of the following information applies to both the GPS and the larger GNSS context.

This chapter is intended to provide an overview of GNSS for the surveying engineer. However, a single chapter cannot cover this topic in much depth. More detailed introductory information on GNSS can be found in see Kaplan 1996, NovAtel 2015, and Van Sickle 2015. For references with more technical detail, see Hofmann-Wellenhof et al. 2008, Leick and Tatarnikov 2015, and Teunissen and Montenbruck 2017. Information about the geodetic principles and mathematics that underlie GNSS can be found in Strang and Borre 1997 and Teunissen and Kleusberg 1998.

GLOBAL POSITIONING SYSTEM SEGMENTS

Space Segment

The current GPS constellation under full operational capability consists of 24 satellites (Figure 6-1). There are more satellites than those in orbit because the constellation includes several orbital spares. It is necessary to keep four satellites, one in each of the four slots in the six orbital planes.

Each is inclined to the equator by 55 degrees in a symmetrical arrangement that covers the globe completely. That coverage is somewhat weaker in higher latitudes. For the middle latitudes, it provides the best coverage of Earth with the fewest number of satellites, reduces the influence of the Earth's gravitational pull, and provides uploading and monitoring sufficiency for the control segment.

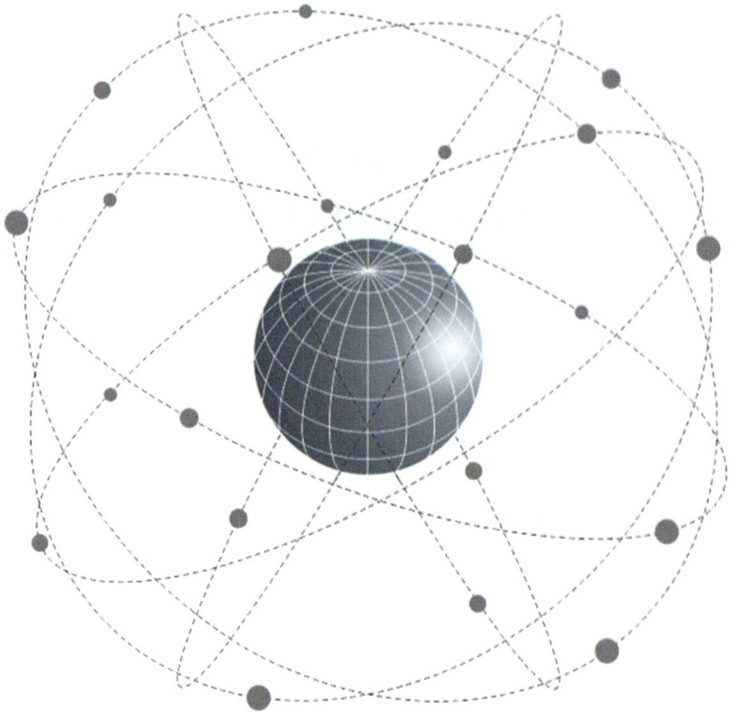

Figure 6-1. GPS constellation.
Source: Van Sickle (2015), reproduced with permission from Taylor & Francis Group.

Control Segment

The control segment is composed of government tracking and uploading facilities distributed around the world (Figure 6-2). These facilities monitor the L-band signals from the GPS satellites, update their navigation messages, track the satellite's health and maneuvers, and many other undertakings, including even battery recharging.

The Master Control Station (MCS) resides at the Consolidated Space Operations Center at Schriever (formerly Falcon) Air Force Base near Colorado Springs, Colorado. It has been manned by the 2nd Space Operations Squadron (2SOPS) since 1992.

2SOPS controls the satellite's orbits. The group monitors the state of each satellite's onboard battery, solar panels, and propellant systems, among others, and keeps the information within limits, because the older it gets, the more its veracity deteriorates. The required two-way communication with the constellation relies on a network of ground antennas and monitoring stations (MS). These stations track the entire GPS constellation. The stations added to the control segment in 2005 augmented the information forwarded to the MCS with data from Washington, DC; England; Argentina; Ecuador; Bahrain; and Australia. Since then, even more have come online, including 6 Air Force and 11 National Geospatial-Intelligence Agency (NGA) MS. Now, every GPS satellite is tracked by at least three of these stations all the time. Testing has shown that the augmented control segment contributes to an increase in the accuracy of real-time GPS of 15% or more. Another modernization program known as Next Generation Operational Control System, or OCX, is on its way. It will facilitate the full control of the new GPS signals such as L5, L2C, and L1C and the latest ongoing GPS III program.

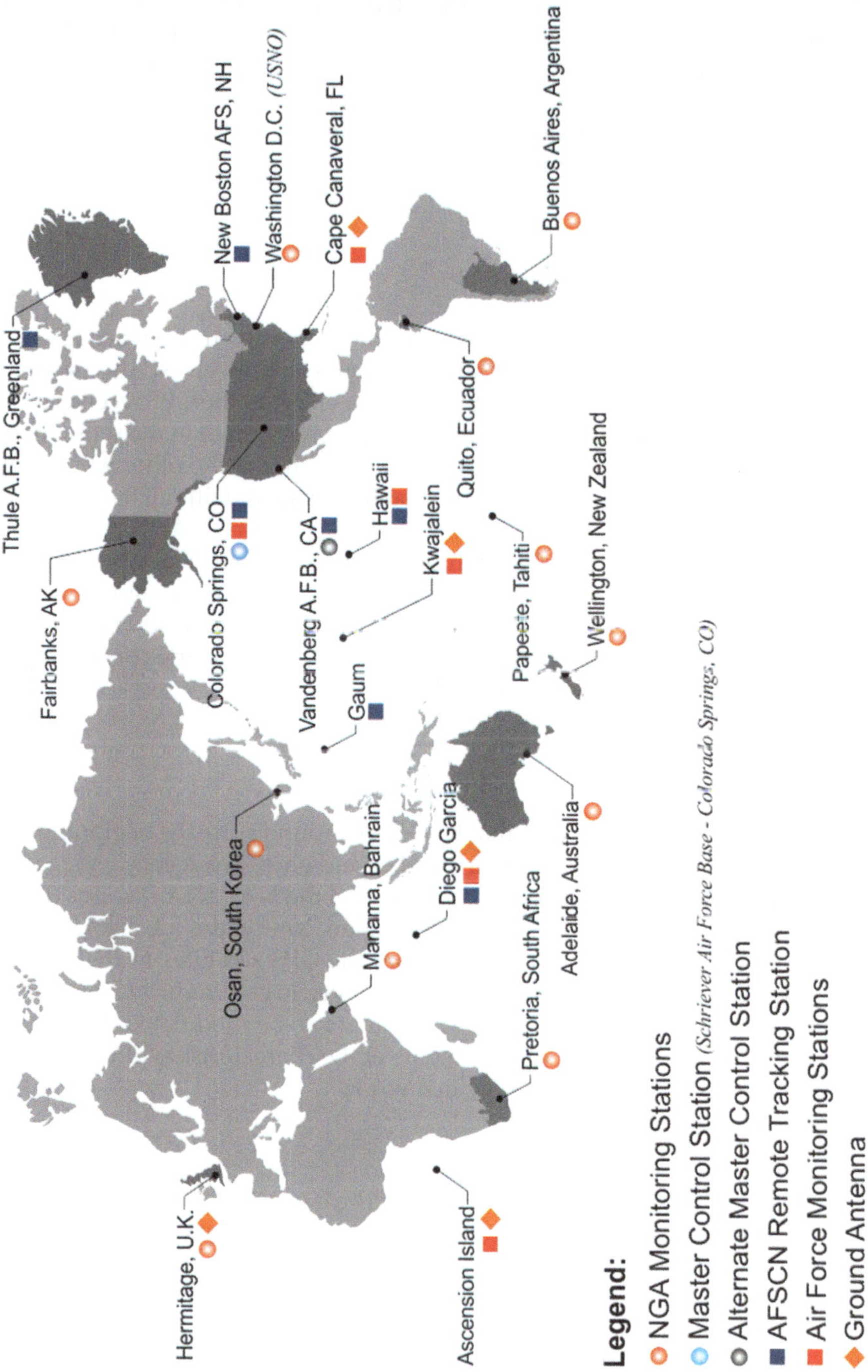

Figure 6-2. Control segment.
Source: Van Sickle (2015), reproduced with permission from Taylor & Francis Group.

User Segment

The US military planned, built, and continues to maintain the GPS. A large component of the user segment of the system is the military. Surveyors and geodesists were among the first civilians to use the GPS. They are sophisticated users, but their number is limited. The explosion of applications for the technology is among the general public. The uses of the GPS in positioning, navigation, and timing (PNT) applications in precision agriculture, machine control, aviation, railroads, and the marine industry are just the beginning. It is no exaggeration to say that GPS technology has transformed civilian businesses and lifestyles around the world.

GLOBAL POSITIONING SYSTEM SIGNALS

GPS positions are based on the measurement of the ranges from the receivers to the satellites. The control points for GPS work are the satellites themselves. The ranges are measured with signals that are broadcast from the GPS satellites to the GPS receivers in the microwave part of the electromagnetic spectrum. It is a passive system in the sense that the satellites transmit signals; the user's equipment receives them, not the other way around. This allows millions of GPS receivers to monitor the satellite's signals without overburdening the system.

The GPS signals must carry a great deal of information. A GPS receiver must be able to gather all the information it needs to determine its own position from the signals it collects from the satellites (Langley 1991). A GPS signal uses codes to communicate: (1) the time on the satellite, (2) the instantaneous position of the satellite, (3) atmospheric corrections, and (4) a satellite identification system. It is done with codes.

Codes

There are three legacy codes in the GPS that have been around since the beginning of the system, and there are new codes. All are binary, zeroes and ones. The legacy codes are the precise P(Y) code, the coarse/acquisition (C/A) code, and the NAV (NAV) code. The new codes include the M code, the L2C code, and the L1C code. New codes are also present on the new carrier, L5. All these codes must travel from the GPS satellites to the receivers. Carrier waves provide the conveyance for the trip. However, to communicate information, a carrier wave has to have at least one characteristic that can be modulated. GPS carriers use phase modulation to do the job. Looking at some of the characteristics of waves makes a productive start toward understanding how that works.

Wavelength and Frequency

A wavelength with a duration of 1 s, known as *1 cps*, has a frequency of 1 Hz in the International System of Units (SI). A frequency of 1 Hz is rather low.

The three fundamental frequencies assigned to GPS carriers come from a part of the electromagnetic spectrum known as the *L-band*. The L-band includes high frequencies from approximately 390 MHz to 1,550 MHz with wavelengths from 15 to 30 cm. The two legacy GPS carriers are L1 at 1,575.42 MHz, which is just a little above the L-band range, and L2 at 1,227.60 MHz. A third carrier is also being broadcast by some GPS satellites. It is known as *L5*, and its frequency is 1,176.45 MHz.

NAV Messages

One of the most important legacy codes is known as the *NAV code* (Van Dierendock 1978). Today, there are several NAV messages being broadcast by GPS satellites; nevertheless, the legacy NAV message continues to be one of the mainstays on which GPS positioning relies. The NAV code is broadcast at a low frequency of 50 Hz on both the L1 and the L2 GPS carriers. It carries information that has been uploaded by the control segment to the satellite. It carries information about the location of the GPS satellites, called the *ephemeris*, as well as other data used in both time conversions and offsets called *clock corrections*. Both GPS satellites and receivers have clocks on board. The NAV code also communicates the health of the satellites in orbit and information about the ionosphere. The ionosphere is a layer of atmosphere through which the GPS signals must travel to reach the user. This code includes data called *almanacs* that provide a GPS receiver with enough little snippets of ephemeris information to calculate the coordinates of all the satellites in the constellation with an approximate accuracy of a couple of kilometers.

There is a new, improved navigation code carrier on L2C called CNAV. CNAV is more compact than the legacy NAV, in which a receiver can get the first fix faster. CNAV can support 32 satellites using only 75% of its bandwidth with a fraction of its packet types and can grow to accommodate 63 satellites. CNAV is also more flexible. For example, there could be a packet with differential correction data like from a Satellite-Based Augmentation System (SBAS) (e.g., Wide Area Augmentation System or WAAS in the United States). In fact, a packet is assigned to the time offset between the GPS and the GNSS; more about that later. An unhealthy satellite can be flagged; that is just the sort of quick access needed to support safety-of-life (SOL) applications. CNAV has forward error correction (FEC) capability in which data are sent, giving the receiver the ability to detect errors at the single bit level and correct them without the need to have the data retransmitted. In fact, four new NAV messages are coming into play: L2-CNAV, CNAV-2, L5-CNAV, and one military message, MNAV.

Pseudorandom Noise Codes

The P and C/A codes also carry information from the GPS satellites to the receivers. They are modulated onto carrier waves, as are the new M, L2C, and L1C codes. However, unlike the NAV code, these codes are not vehicles for broadcasting information that has been uploaded by the control segment; instead, they carry the raw data from which GPS receivers derive their position and time measurements. The P and C/A codes are complicated, so complicated that they appear to be noise. They are known as *pseudorandom noise* or PRN codes. Actually, they are carefully designed. They have to be, and they must be capable of repetition and replication.

P(Y) Code

The P(Y) code is a 37 week-long military code. It is generated at a rate of 10.23 mbps and carried on both L1 and L2. Each GPS satellite is assigned a 1-week-long part of the P-code all its own and repeats its portion every 7 days. This is one way GPS receivers distinguish one satellite's transmission from another.

C/A Code

The C/A code's generation is 10 times slower than the P(Y) code. The C/A code rate is 1.023 mbps. Each GPS satellite broadcasts its own completely unique 1,023 bit C/A code and repeats its C/A code every millisecond making satellite identification quite straightforward. The C/A code is broadcast on L1 only and used to be the only civilian GPS code, but it no longer is. It has been joined by a new civilian signal known as *L2C* that is carried on L2.

ERROR SOURCES

The understanding and management of errors is indispensable for finding the true geometric range ρ from either a pseudorange (p) or carrier phase observation (Φ)

$$p = \rho + d_\rho + c(dt - dT) + d_{ion} + d_{trop} + \varepsilon_{m_p} + \varepsilon_p \text{ (pseudorange)} \tag{6-1}$$

$$\Phi = \rho + d_\rho + c(dt - dT) + \lambda N - d_{ion} + d_{trop} + \varepsilon_{m_\Phi} + \varepsilon_\Phi \text{ (carrier phase)} \tag{6-2}$$

Both equations include environmental and physical limitations called *range biases*.

Atmospheric errors are among the biases; the ionospheric effect, d_{ion}, and the tropospheric effect, d_{trop}. Other biases are clock errors symbolized by ($dt - dT$) and receiver noise, ε_p and ε_φ, multipath, ε_{mp} and $\varepsilon_{m\varphi}$, and orbital errors, d_ρ. Each of these biases comes from a different source and is independent of one another; they combine to obscure the true geometric range.

Ionospheric Effect (d_{ion})

One of the largest errors in GPS positioning is attributable to the atmosphere. The GPS signal is refracted and diffracted as it passes through the Earth's atmosphere. The atmosphere alters the apparent speed and, to a lesser extent, the direction of the signal. This causes an apparent delay in the signal's transit from the satellite to the receiver. A consequence of the dispersive nature of the ionosphere is that the apparent time delay for a higher-frequency carrier wave is less than it is for a lower-frequency wave. The word dispersive in this context means that the apparent time delay contributed by the ionosphere depends on the frequency of the signal. That means that L1, 1,575.42 MHz, is not affected as much as L2, 1,227.60 MHz, and L2 is not affected as much as L5, 1,176.45 MHz.

Satellite Clock Bias (dt)

The satellite clock bias can be quite large, especially if the broadcast clock correction available in the NAV message is not used by the receiver. As time is a critical component in the functioning of the GPS, it is important to correct this bias.

Receiver Clock Bias (dT)

The third largest error in GPS positioning can be caused by the receiver clock, its oscillator. Both a receiver's measurement of phase differences and its generation of replica codes depend on the reliability of this internal frequency standard.

Orbital Bias (d_ρ)

Orbital bias has the potential to be the fourth largest error. It is addressed in the broadcast ephemeris.

Tropospheric Effect (d_{trop})

The fifth largest error can be attributed to the effect of the troposphere (Brunner and Welsch 1993). The troposphere is that part of the atmosphere closest to Earth. It is part of the electrically neutral layer of the Earth's atmosphere and is not ionized. The troposphere and the ionosphere are by no means alike in their effect on the satellite's signal. Although the troposphere is refractive, its refraction of a GPS satellite's signal is not related to its frequency. The refraction is tantamount to a delay in the arrival of a GPS satellite's signal. This means that the range between a receiver and a satellite will be shown to be a bit longer than it actually is. The troposphere is also nondispersive for frequencies below 30 GHz or so. Therefore, L1, L2, and L5 are equally refracted.

Multipath

Multipath is a range delay symbolized by ε_{mp} in the pseudorange equation and $\varepsilon_{m\varphi}$ in the carrier phase equation. As the name implies, the reception of the GPS signal is via multiple paths rather than from a direct line of sight (Figure 6-3). Multipath differs from both the apparent slowing of the signal through the ionosphere and troposphere and the discrepancies caused by clock offsets. The range delay in multipath is the result of the reflection of the GPS signal. Repeat satellite geometries produce similar multipath signals; thus, multipath can be temporally correlated, but it is spatially uncorrelated. Its temporal correlation is periodic, and it may or may not introduce bias in a GPS measurement. Its lack of spatial correlation means it is different at every spatial location.

Multipath occurs when part of the signal from the satellite reaches the receiver after one or more reflections or scattering from the ground, a building, or another object. Multipath in carrier phase observations is harder to mitigate than multipath in pseudoranges.

Receiver Noise

Receiver noise is directly related to thermal noise, dynamic stress, and so on in the GPS receiver itself. Receiver noise is also a spatially uncorrelated error source. The effects of receiver noise on carrier phase measurements symbolized by ε_{φ}, like multipath, are small when compared with their effects on pseudorange measurements, ε_{p}.

DIFFERENTIAL GLOBAL POSITIONING SYSTEM VERSUS RELATIVE POSITIONING

Solutions

There are a variety of ways to limit the effects of the biases in GPS work, but whether the techniques involve the methods of data collection or ways of processing the data, the objective is the management of errors.

GPS work is sometimes divided into three categories: positioning, navigation, and timing (PNT). Most often, GPS surveying is concerned with the first of these, which is positioning. In general, there are two techniques used in surveying, kinematic and static. In static GPS surveying sessions, the receivers are motionless during the observation. Because static work most often provides higher accuracy and more redundancy than kinematic work, it is usually done to establish control (Chapter 11). The results of static GPS surveying are processed after the session is completed. In other words, the data are *post-processed*. In kinematic GPS surveying, the receivers are either in periodic or in continuous motion. Kinematic GPS is done when real time, or near real time, results are needed. When

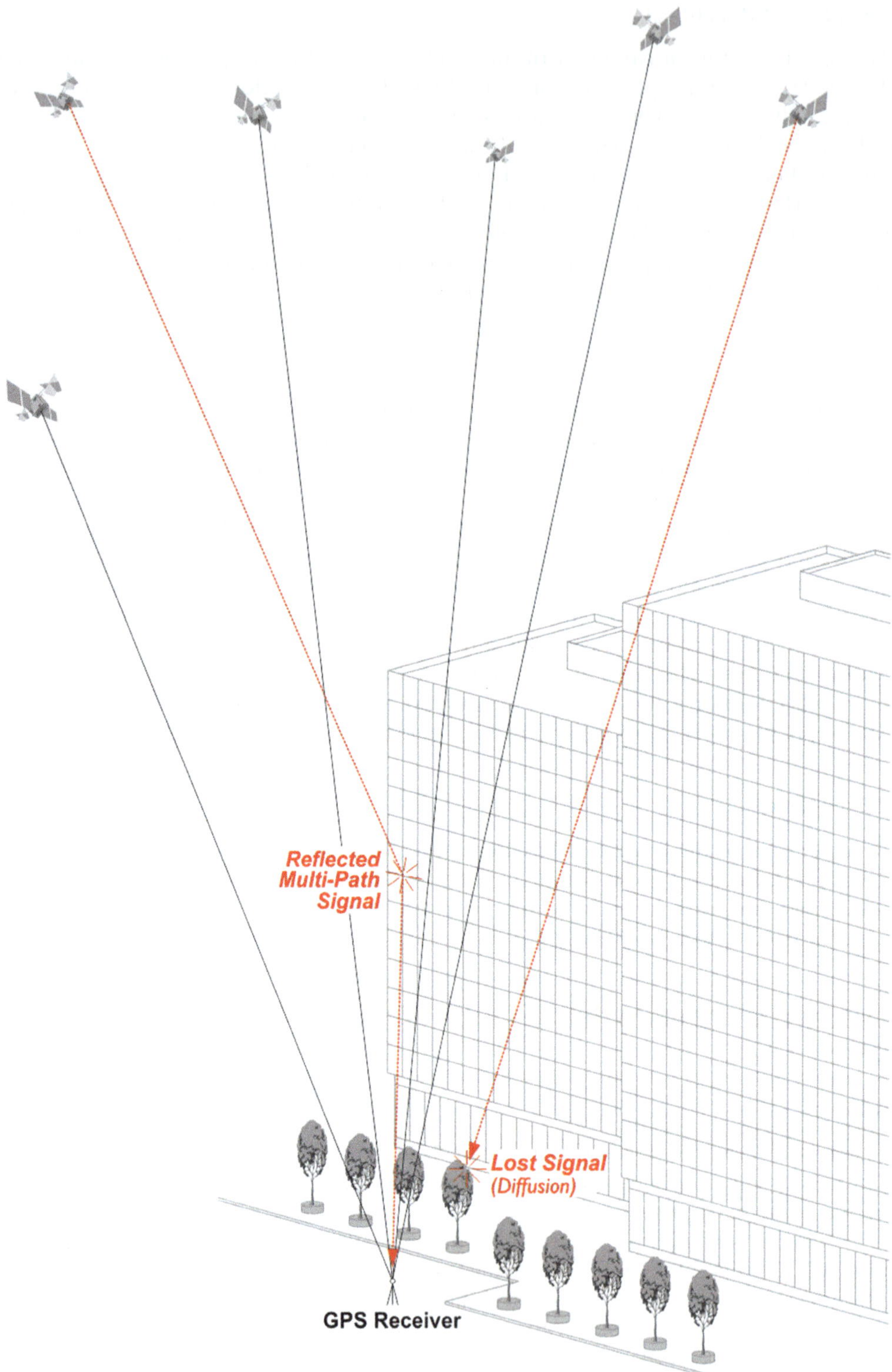

Figure 6-3. Multipath.
Source: Van Sickle (2015), reproduced with permission from Taylor & Francis Group.

the singular objective of kinematic work is positioning, the receivers move periodically using the start and stop methodology. When the receivers are in continuous motion, the objective may be the acquisition of the location, attitude, and velocity of a moving platform (i.e., navigation) or positioning.

Single Point

Single-point GPS, also known as the *navigation solution* or *absolute positioning*, is the most familiar and ubiquitous application of the technology. It is a pseudorange code-based solution rather than a carrier phase solution. It is the solution used by cell phones, GPS-enabled cameras, car navigation systems, and many more devices. As shown in Figure 6-4, it is characterized by a single receiver measuring pseudoranges to a minimum of four satellites simultaneously in which the receiver must also rely on the information it receives from the satellite's navigation messages to learn the positions of the satellites, the satellite clock offset, the ionospheric correction, and so on. Even if all the data in the navigation message contained no errors, and they surely do, four unknowns remain, namely the position of the receiver in three Cartesian coordinates, u_x, u_y, and u_z; and the receiver's clock error dT_u. Three pseudoranges provide enough data to solve for u_x, u_y, and u_z. In addition, the fourth pseudorange provides the information for the solution of the receiver's clock offset.

A unique solution is found here because the number of unknowns is not greater than the number of observations. The receiver tracks a minimum of four satellites simultaneously; therefore, these four equations can be solved simultaneously for every epoch of the observation. An *epoch* in the GPS is a very short period of observation time and is usually just a small part of a longer measurement. Single-point positioning with its reliance on the navigation message is in a sense the fulfillment of the original idea of the GPS.

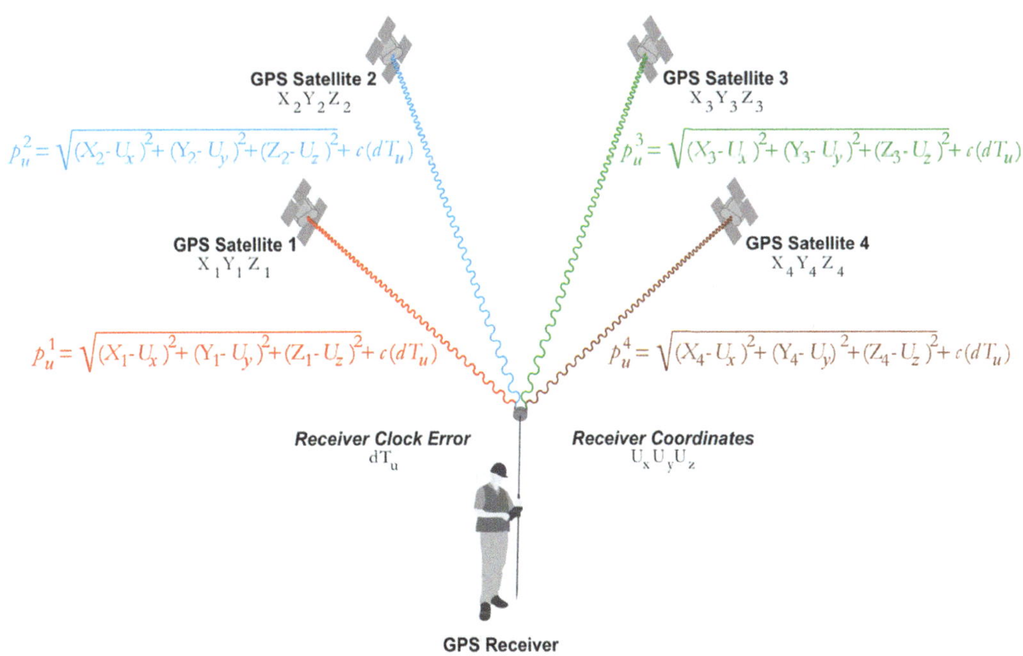

Figure 6-4. Single-point solution.
Source: Van Sickle (2015), reproduced with permission from Taylor & Francis Group.

Relative Positioning

A minimum of two receivers are involved in relative or differential positioning. The term *differential GPS*, or DGPS, sometimes indicates the application of this technique with coded pseudorange measurements, whereas relative GPS indicates the application of this technique with carrier phase measurements. These definitions are by no means universal, and the use of the terms relative and differential GPS has become virtually interchangeable.

In relative positioning, one of the two receivers involved occupies a "known" position during the session. It is the base. The objective of the work is the determination of the position of the other, the rover, relative to the base. Both receivers observe the same constellation of satellites at the same time, and because, in typical applications, the vector between the base and the rover, known as a *baseline*, is so short compared with the average 20,200 km altitude of the GPS satellites, there is extensive correlation between observations at the base and the rover. In other words, the two receivers record very similar errors and because the base's position is known, corrections can be generated there that can be used to improve the solution at the rover.

Postprocessing

In many ways, postprocessing is the heart of a GPS control operation. GPS measurements are composed of fundamentally biased ranges. Therefore, GPS data-processing procedures are really a series of interconnected computerized operations designed to minimize the more difficult biases and extract the true ranges. As previously discussed, the biases originate from a number of sources: imperfect clocks, atmospheric delays, cycle ambiguities in carrier phase observations, and orbital errors. If a bias has a stable, well-understood structure, it can be estimated. In other cases, multifrequency observations can be used to measure the bias directly, as in the ionospheric delay, or a model may be used to predict an effect, as in tropospheric delay.

Correlation of Biases

When two or more receivers observe the same satellite constellation simultaneously, a set of correlated vectors are created between the co-observing stations. Most GPS practitioners use more than two receivers in static control work. Therefore, even when several receivers are set up on widely spaced stations, as long as the data are collected by the receivers simultaneously from the same constellation of satellites, they will record very similar errors. It is the simultaneity of observation and the correlation of the carrier phase *observables* (Langley 1993) that make the extraordinary GPS accuracies possible because biases that are linearly correlated can be virtually eliminated by differencing the data sets of a session.

GLOBAL POSITIONING SYSTEM SURVEY PLANNING

If a static GPS control survey is carefully planned, it usually progresses smoothly. However, there is seldom anything more expensive in a GPS project than a surprise. The first task is determining the observation windows.

Before proceeding, it is important to note that this section discusses planning and executing surveys using only the GPS. As previously mentioned, and as presented in more detail later in the chapter, other GNSSs are also now available (e.g., GLONASS, Galileo, BeiDou). Most modern receivers and antennae are capable of collecting multi-GNSS data.

Although the planning only mentions the GPS as an example, the reader should note that the methods are generic and can be extended to planning a multi-GNSS survey. GPS is only one of several viable and available types of GNSS, and surveying engineers with capable equipment can plan to collect data from additional space vehicles (SVs) in other GNSSs using the techniques discussed previously.

Figure 6-5 is a GPS satellite azimuth and elevation table showing that there are changes in the available constellation from eight GPS SVs between 0:00 and 0:50, six between 1:00 and 1:30, seven from 1:40 to 2:00, and back to eight at 2:10. The constellation never dipped below the minimum of four satellites, and the position dilution of precision (PDOP), a measure of the relative geometry of the satellites to the receiver, was low throughout. When visible satellites are close together in the sky, the geometry is weak and the PDOP is high; when the satellites are spread apart, the geometry is strong and the PDOP value is low (Figure 6-6). In Figure 6-5, the PDOP varied between a low of 1.7 and a high of 3.0. Over the interval covered by the table, the PDOP never reached the unsatisfactory level of 5 or 6, which is when a planner should avoid observation. Using this chart, the GPS survey designer might well have concluded that the best available window was the first. There was nearly an hour of 8-satellite data with a PDOP below 2. However, the data indicated that accurate observations could be made at any time covered here, except for one challenge: it was the middle of the night. It is worth noting that the ionospheric error is usually smaller after sundown.

Figure 6-7 shows data from later in the day. It covers a period of 2 h when a constellation of five and six satellites was always available. However, through the first hour, from 6:30 to 7:30, the PDOP hovered around 5 and 6. For the first half of that hour, four of the satellites—PRN 9, PRN 12, PRN 13, and PRN 24—were all near the same elevation. During the same period, PRN 9 and PRN 12 were only approximately 50 degrees apart in azimuth, as well. Although a sufficient constellation of satellites was constantly available, the survey designer may well have considered only the last 30 to 50 min of the time covered by this chart as suitable for observation.

Supposing that the period from 7:40 to 8:30 was found to be a window with low PDOP, the planner may have regarded it as a single 50 min session or divided it into shorter sessions. One aspect of that decision was probably the length of the baseline in question. In a static GPS, a long line of 30 km may require 50 min of 6-satellite data, but a short line of 3 km may not. Another consideration was probably the approximation of the time necessary to move from one station to another. In any case, it is not only the number of satellites available that matters, it is also their configuration, which is why the designer must also consider their elevations and azimuths in the planning. Looking at those parameters, fortunately, the window under consideration stands up.

The next step in the static GPS survey design is determining the number of sessions needed in the observation schedule. For illustration, suppose that the project includes 14 points in a planned static GPS network. They will be symbolized by m. Presume that there are four multifrequency GPS receivers available for this project. They will be symbolized by r. There will be five observation sessions each day during the project. They will be symbolized by d. The initial plan calls for 40 occupations by the four receivers used on the project, and because there are 14 points to be positioned, this means each one will be visited an average of 2.857 times. This value shows the redundancy contributed by the plan. Clearly, the repetition of baselines increases the reliability of the results. That redundancy will be symbolized by n. Equation (6-3) is a formula for determining the actual number of sessions necessary for completing the project.

$$s = \frac{(m \cdot n)}{r} + \frac{(m \cdot n)(p-1)}{r} + k \cdot m \tag{6-3}$$

Satellites Azimuth and Elevation Table I

Point: *Morant* Lat *36:45:0 N* Lon *121:45:0W* Ephemeris: *9/24/2014*

Date: *Wed., Sept. 29, 2014* Mask Angle: *15 (deg)* Zone: *Time Pacific Day (-7)*

24 Satellites: *1 2 3 7 9 12 13 14 15 16 17 18 19 20 21 22 23 24 25 26 27 28 29 31*

Sampling Rate: *10 minutes*

Time	El	Az	El	Az	El	Az	El	Az	El	Az	El	Az	El	Az	El	Az	PDOP
								constellation of 8 SVs									
SV	2		16		18		19		27		28		29		31		
0:00	16	219	15	317	77	121	66	330	41	287	23	65	36	129	30	109	1.7
0:10	20	221	18	314	73	131	67	341	44	292	22	60	32	132	33	104	1.8
0:20	24	223	20	310	68	137	68	353	47	297	21	56	28	135	35	99	1.8
0:30	28	226	22	306	64	142	68	5	50	302	20	51	24	138	36	93	1.9
0:40	32	229	23	302	59	146	67	17	52	308	18	48	20	140	37	88	1.8
0:50	36	232	24	297	54	148	66	28	55	314	16	44	16	142	38	82	1.8
								constellation of 6 SVs									
SV	2		16		18		19		27		31						
1:00	40	235	24	293	49	151	65	39	58	320	38	76					3.0
1:10	43	239	24	288	44	153	63	49	61	328	37	70					3.0
1:20	47	244	24	283	40	155	61	57	64	336	36	64					2.8
1:30	51	249	23	278	35	156	59	65	66	345	34	60					2.6
								constellation of 7 SVs									
SV	2		7		16		18		19		27		31				
1:40	54	254	16	186	22	273	30	157	56	73	68	356	32	55			2.3
1:50	57	260	21	186	20	269	26	158	53	79	70	9	29	51			2.2
2:00	60	268	25	186	19	264	22	159	50	85	71	23	26	48			2.0
								constellation of 8 SVs									
SV	2		7		16		18		19		26		27		31		
2:10	66	276	30	185	16	260	17	160	47	91	15	319	71	38	23	45	1.7

Figure 6-5. Satellite azimuth and elevation table 1.
Source: Van Sickle (2015), reproduced with permission from Taylor & Francis Group.

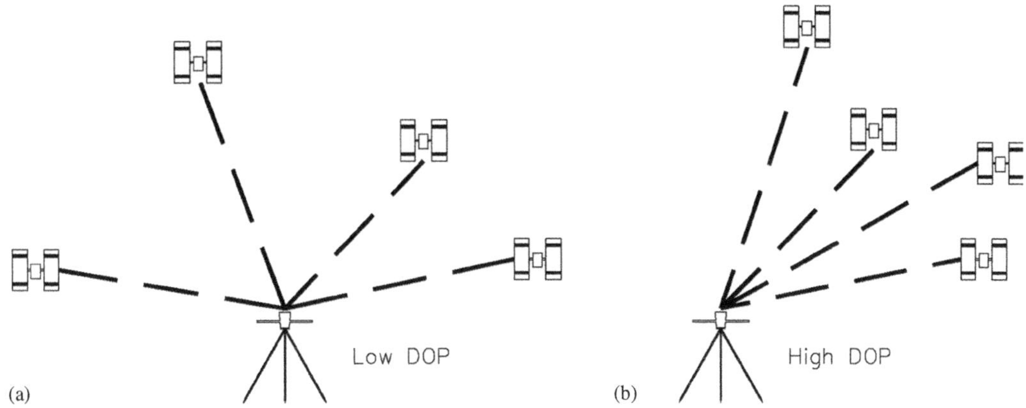

Figure 6-6. (a) Low dilution of precision and strong geometry, (b) high dilution of precision and weak geometry.

where

m = Total number of stations (existing and new) = 14,
d = Number of possible observing sessions per observing day = 5,
r = Number of receivers = 4 dual frequency,
n = Redundancy = 2.857,
p = Production factor is typically 1.1, and
k = Safety factor of 0.1 is recommended for GPS projects within 100 km of a company's home base. Beyond that radius, an increase to 0.2 is advised.

The substitution of the appropriate quantities for the illustrated project increases the prediction of the number of observation sessions required for its completion:

$$s = \frac{(m \cdot n)}{r} + \frac{(m \cdot n)(p-1)}{r} + k \cdot m$$

$$s = \frac{(14)(2.857)}{4} + \frac{(14)(2.857)(1.1-1)}{4} + (0.2)(14)$$

$$s = \frac{(40)}{4} + \frac{(4)}{4} + 2.8$$

$$s = 10 + 1 + 2.8$$

$$s = 14 \text{ sessions (rounded to the nearest integer)} \qquad (6\text{-}4)$$

Independent Lines

Presuming a project is done with four receivers, only three of the possible six baseline vectors that are produced by each simultaneous occupation of those receivers are independent. They are also known as *nontrivial lines*. The other three of the six lines are called trivial and are known as *dependent lines* (Figure 6-8). In building the resulting network, it is best to restrict the use of the lines created in each observing session to the absolute minimum needed to produce a unique solution.

Satellites Azimuth and Elevation Table

Point: *Morant* Lat *36:45:0 N* Lon *121:45:0W* Ephemeris: *9/24/2014*

Date: *Wednesday, Sept. 29, 2014* Mask Angle: *15 (deg)* Zone: *Time Pacific Day (-7)*

24 Satellites: *1 2 3 7 9 12 13 14 15 16 17 18 19 20 21 22 23 24 25 26 27 28 29 31*

Sampling Rate: *10 minutes*

Time	El Az	El Az	El Az	El Az	El Az	El Az	El Az	El Az	PDOP
				constellation of 5 SVs					
SV	7	9	12	13	24				
6:30	28 54	60 271	61 319	62 15	48 177				6.3
6:40	24 57	60 261	66 314	57 19	53 176				6.0
6:50	21 60	59 252	70 305	53 22	58 175				5.3
7:00	18 62	57 243	73 292	49 25	63 172				4.6
				constellation of 5 SVs					
SV	9	12	13	20	24				
7:10	54 235	74 274	44 28	16 308	68 169				4.8
7:20	51 229	74 255	40 32	20 310	72 163				5.7
7:30	47 224	72 238	37 35	23 311	77 153				5.1
7:40	43 219	68 226	33 38	27 313	80 134				4.0
				constellation of 6 SVs					
SV	9	12	13	16	20	24			
7:50	39 215	64 218	29 41	16 149	31 314	81 102			2.1
8:00	35 212	59 213	26 45	19 146	36 314	80 73			2.3
8:10	31 209	54 209	23 48	23 143	40 315	76 57			2.4
8:20	27 207	49 206	19 52	27 140	44 314	72 49			2.5
8:30	23 204	44 204	16 55	30 137	48 314	67 45			2.5

Figure 6-7. Satellite azimuth and elevation table 2.
Source: Van Sickle (2015), reproduced with permission from Taylor & Francis Group.

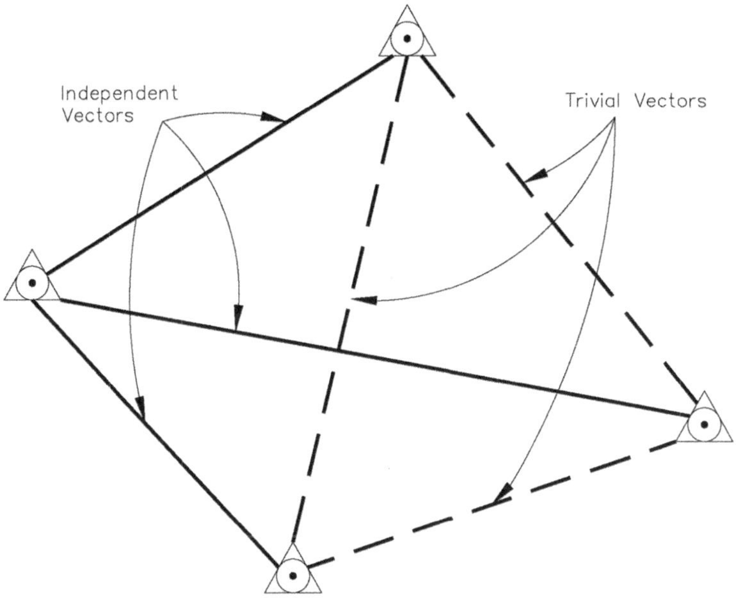

Figure 6-8. Independent and trivial vectors.

As mentioned, whenever four receivers are used, six lines are created by each session. However, any three of those lines will fully define the position of each occupied station in relation to the others in the session. Therefore, the user can consider any three of the six lines as independent. A common approach is to designate the shorter baseline vectors as independent and discard the longer baseline vectors as dependent. However, once the decision is made, only those three baselines are included in the network. The remaining baselines are then considered trivial and discarded. In practice, the three shortest lines in a four-receiver session are usually deemed the independent vectors, and the three longest lines are eliminated as trivial (dependent).

Another aspect of the distinction between independent and trivial lines involves the concept of error of closure or loop closure. A series of baseline vector components from more than one GPS session, forming a loop or closed figure, are added together. The closure error is the ratio of the length of the line representing the combined errors of all the vector's components to the length of the perimeter of the figure. Any loop closures that only use baselines derived from a single common GPS session will yield unrealistically small closure errors. Also, the inclusion of any of the trivial lines adds essentially no redundancy or geometric strength to the lines of the session, because they are all derived from the same data.

The number of independent occupations is an important source of redundancy. Redundancy is then partly defined by the number of independent baselines that are measured more than once, as well as by the percentage of stations that are occupied more than once. Although it is not possible to repeat a baseline without reoccupying its endpoints, it is possible to reoccupy a large percentage of the stations in a project without repeating a single baseline. These two aspects of redundancy in the GPS—the repetition of independent baselines and the reoccupation of stations—are somewhat separate.

Station Data Sheet

The principles of high quality field notes have a long tradition in surveying engineering, and they will continue to have validity for some time to come. In GPS surveying, the ensuing paper trail will not only fill subsequent archives; it has immediate utility. For

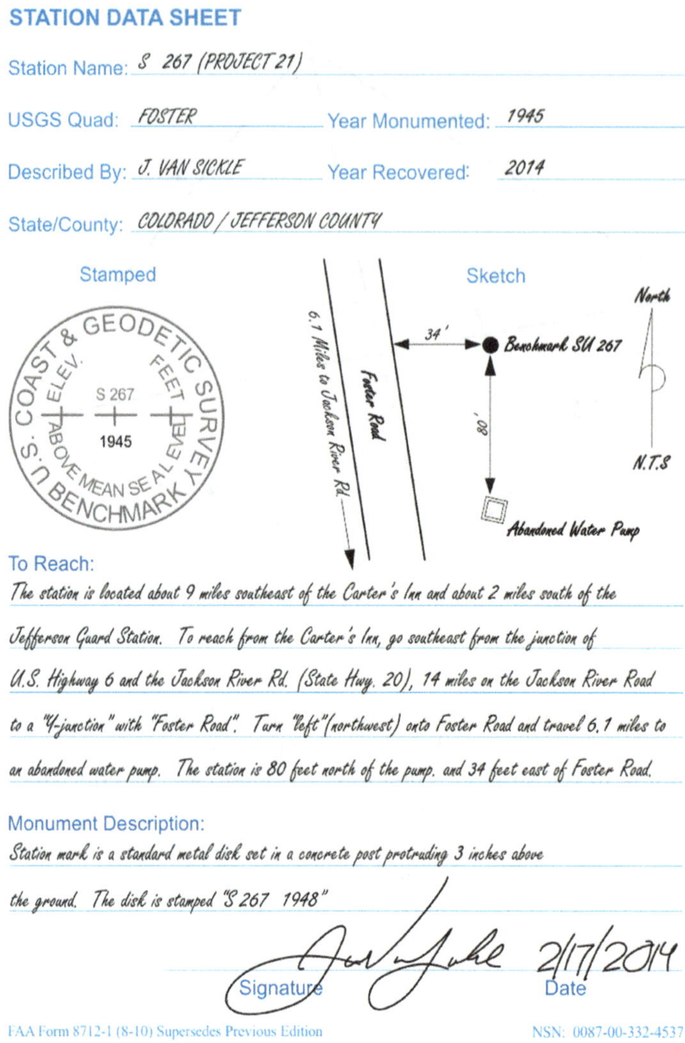

STATION DATA SHEET

Station Name: S 267 (PROJECT 21)

USGS Quad: FOSTER Year Monumented: 1945

Described By: J. VAN SICKLE Year Recovered: 2014

State/County: COLORADO / JEFFERSON COUNTY

Stamped

Sketch

To Reach:

The station is located about 9 miles southeast of the Carter's Inn and about 2 miles south of the Jefferson Guard Station. To reach from the Carter's Inn, go southeast from the junction of U.S. Highway 6 and the Jackson River Rd. (State Hwy. 20), 14 miles on the Jackson River Road to a "Y-junction" with "Foster Road". Turn "left" (northwest) onto Foster Road and travel 6.1 miles to an abandoned water pump. The station is 80 feet north of the pump. and 34 feet east of Foster Road.

Monument Description:

Station mark is a standard metal disk set in a concrete post protruding 3 inches above the ground. The disk is stamped "S 267 1948"

Signature Date 2/17/2014

FAA Form 8712-1 (8-10) Supersedes Previous Edition NSN: 0087-00-332-4537

Figure 6-9. Station data sheet.
Source: Van Sickle (2015), reproduced with permission from Taylor & Francis Group.

example, the station data sheet is often an important bridge between on-site reconnaissance and the actual occupation of a monument.

Although every organization develops its own unique system of handling its field records, most have some form of the station data sheet. The datasheet illustrated in Figure 6-9 is merely one possible arrangement of the information needed to recover the station.

Observation Logs

Most static GPS control operations require their receiver operators to keep a careful log of each observation. Usually written on a standard form, these field notes provide a written record of the measurements, times, equipment, and other data that explains what actually occurred during the observation itself. The log illustrated in Figure 6-10 is one method that includes some of the information that might be used to document one session at one station.

OBSERVATION LOG

JOB NUMBER: UL42396

OBSERVER	STATION	JULIAN DATE	DATE
J. VAN SICKLE	S 267 (POINT 14)	50	2/17/2014

LATITUDE	LONGITUDE	HEIGHT
47°52′39″	115°00′48″	3,241.09 feet

PLANNED OBSERVATION SESSION	SESSION NAME	ACTUAL OBSERVATION SESSION
START TIME: 9:10am STOP TIME: 10:10am	0014 050 2	START TIME: 9:10am STOP TIME: 10:10am

ANTENNA TYPE	ANTENNA HEIGHT ABOVE STATION MONUMENT	
ON-BOARD	BEFORE OBSERVATION	AFTER OBSERVATION
	METERS: 1.585	METERS: 1.585
MASK ANGLE: 15°	FEET: 5.20	FEET: 5.20

METEOROLOGICAL DATA

TIME	RELATIVE HUMIDITY	BAROMETER	THERMOMETER (D)	THERMOMETER (W)
9:30am	30%	29.94	37°F	35°F

VISIBILITY DIAGRAM COMPLETED? (Y) N	TOP OF MONUMENT ABOVE THE SURFACE: 3 IN.
STATION DATA SHEET COMPLETED? (Y) N	TOP OF MONUMENT BELOW THE SURFACE:

SV PRN TRACKED	
3	16
12	20
13	24

COMMENTS:
DATA FROM SV 16 APPEARS HEALTHY
DESPITE WATER PUMP OBSERVATION

FAA Form 8710-1 (4-00) Supersedes Previous Edition — NSN: 0082-00-582-5507

Figure 6-10. Observation log.
Source: Van Sickle (2015), reproduced with permission from Taylor & Francis Group.

GLOBAL NAVIGATION SATELLITE SYSTEM SURVEYING TECHNIQUES

Static

Static GPS surveying (or static GNSS surveying, more commonly speaking, as more satellite systems are available) is used on control surveys. It involves several stationary receivers simultaneously collecting data from at least four satellites during observation sessions that usually last from 30 min to 2 h. It requires planning. The requisites include more than one receiver, four or more satellites, and a mostly unobstructed sky above the stations to be occupied.

The selection of satellites to track, start and stop times, mask elevation angle, assignment of data file names, reference position, bandwidth, and sampling rate are some options useful in the static mode. In general, in GNSS networks, the more well-chosen horizontal control stations that are available, the better. Refer to Chapter 11 for more details on identifying and tying into existing control.

Real-Time Kinematic

Most GNSS surveying relies on the idea of relative positioning. Today, the most commonly used method is real-time kinematic (RTK). It utilizes receivers on reference stations that provide correction signals to the end user via a data link, sometimes over the Internet, radio signal, or cell phone, and often in real time (Figure 6-11). As previously discussed, errors in satellite clocks, imperfect orbits, the trip through the layers of the atmosphere, and many other sources contribute inaccuracies to GNSS signals by the time they reach a receiver. These errors are variable, so the best way to correct them is to monitor them as they happen. An effective way to do this is to set up a GNSS receiver on a station whose position is known exactly, a base station. This base station receiver's computer can calculate its position from satellite data, compare that position with its actual known position, and find the difference. The resulting error corrections can be communicated from the base to the rover. While this is happening, another GNSS receiver or rover moves from place to place, collecting the positions relative to the base station. Real-time positioning is built on the foundation of the idea that, with the important exceptions of multipath and receiver noise, GNSS error sources are spatially correlated. This spatial correlation is one method of reducing known systematic errors in GNSS measurements. In other words, the closer the rover is to the base, the more the errors at the ends of the baseline match. The shorter the baseline, the more the errors are correlated. The longer the baseline, the less the errors are correlated.

The agreed on protocol for communication between base stations and rovers was first designed for use in marine navigation by an organization known as the Radio Technical Commission for Maritime Services (RTCM). RTCM is an independent not-for-profit organization that is supported by an international membership that includes both governmental and nongovernmental institutions. Its goals are educational, scientific, and professional.

In 1985, the RTCM Special Committee (SC-104) created a standard that is still more used than any proprietary formats that have come along since. RTCM is open. The most recent versions are RTCM 3.1 and RTCM 3.2; version 3.1 adds a network correction message and version 3.2, announced in 2013, introduces a feature known as *multiple signal messages* (MSMs). MSMs include the capability to handle the European Galileo and the Chinese BeiDou GNSS systems in the RTCM protocol.

RTK requires a real-time wireless connection be maintained between the base station and the rover. This is commonly done with radios, although other wireless communication

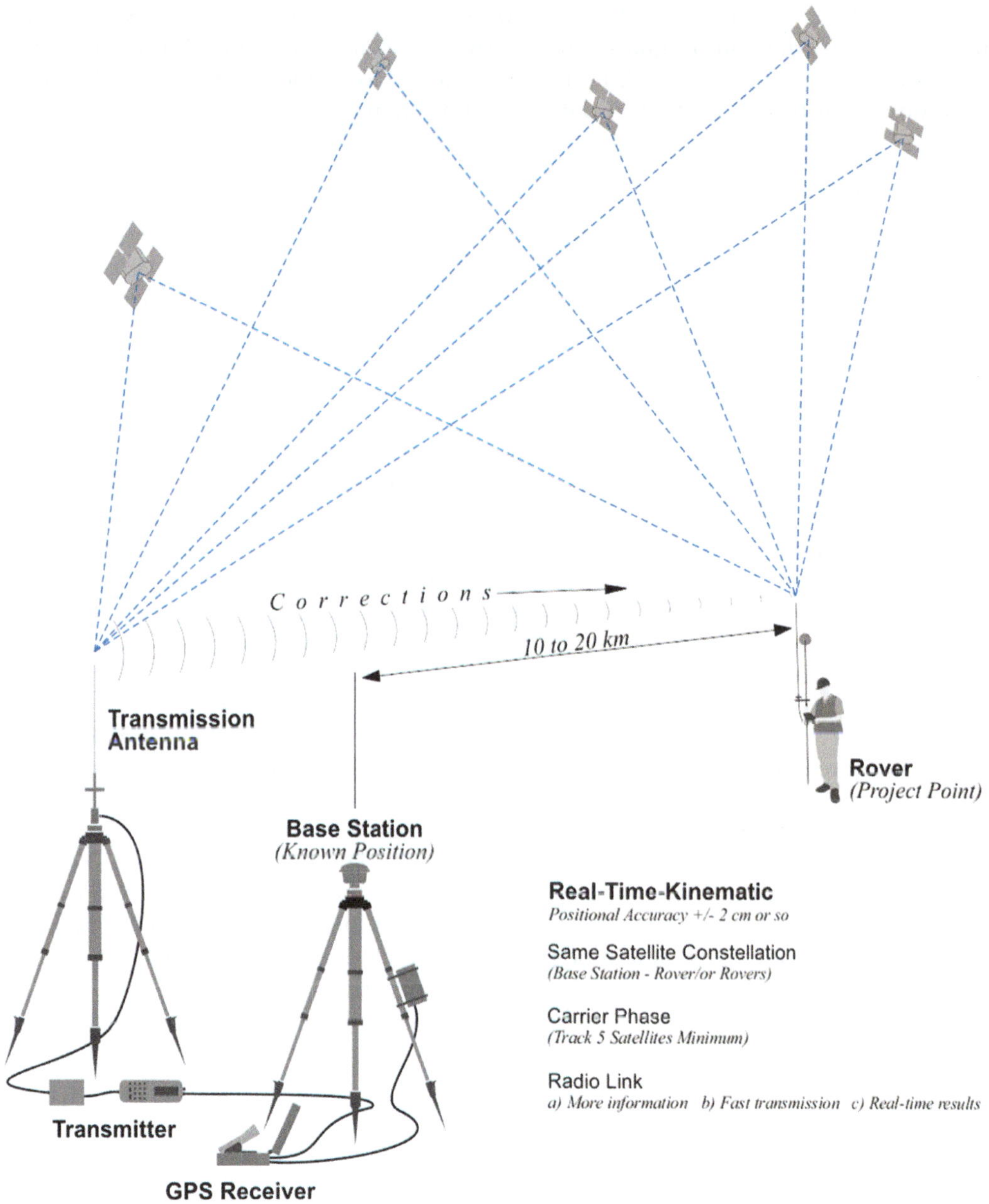

Figure 6-11. Real-time kinematic.
Source: Van Sickle (2015), reproduced with permission from Taylor & Francis Group.

options are available. The radio receiving antennas for the rovers will either be built into the GNSS antenna or be present as separate units. Usually, the radio antenna for the data transmitter and the rover is an omnidirectional whip antenna; however, at the base, it is usually on a separate mast and has a higher gain than that at the rover. It is best when the distance between the base station and the rovers is less than 10 km (6 mi). However, the length of the baselines may be further limited by the effective range of the radio data link. In areas with high radio traffic, it can be difficult to find an open channel. The Federal

Communications Commission (FCC) is concerned with some RTK GNSS operations interfering with other radio signals, particularly voice communications. It is important for GNSS surveyors to know that FCC licensure is required to perform RTK and also that voice communications have priority over data communications.

Real-Time Networks

There can be no doubt that RTK dominates the GNSS surveying applications. However, the requirements of setting up a GNSS reference station on a known position, the establishment of a radio frequency transmitter, and all attendant components before a single measurement can be made are both awkward and expensive. This, along with the limitation of short baselines for centimeter-level work, makes RTK cumbersome. Therefore, services have arisen around the world to provide RTCM real-time corrections to surveyors by a different means. Real-time networks (RTNs) have been implemented by both governments and commercial interests. The services are sometimes free and sometimes require a subscription or the payment of a fee before the surveyor can access the broadcast corrections over a data link via a modem such as a cell phone or some other device.

These benefits are accomplished by the services gleaning corrections from a whole network of permanent, continuous GNSS base stations rather than just a single base. In this way, quality control is facilitated by the ability to check corrections from one continuous GNSS base station with those generated from another, and should a station go off-line or give incorrect values, other stations in the network can take up the slack with minimal accuracy loss.

The central idea underlying RTN differential corrections is the combination of observations from several base stations at known positions used to derive a model of an entire region (Figure 6-12). So rather than being considered as isolated beacons with each covering its own segregated area, the base stations are united into a network. The data from the network can then be used to produce a virtual model of the area of interest. From this model, distance-dependent biases such as ionospheric, tropospheric, and orbit errors can be calculated. Once the roving receiver's place within that network is established, it is possible to predict the errors at that position with a high degree of certainty. Not only can the network be used to model errors in a region more correctly, but the multibase solution also can improve redundancy. Solving several baselines that converge on a project point simultaneously rather than relying on just one from a typical single-base RTK setup adds more certainty to the resulting coordinates.

Coordinates for the RTN base stations are used as control, and they are typically computed by RTN managers. Any errors in these coordinates are propagated to the measurements taken by the rovers. See Chapter 11 for discussion and references on checking the coordinates of RTN base stations.

Precise Point Positioning

Many networks of continuous GNSS stations are available around the world to support differential processing. There are also many regional networks such as the Australian Fiducial Network administered by the Australian Surveying and Land Information Group (AUSLIG), EUREF with its EUREF Permanent Network, the NOAA Continuously Operating Reference Station Network administered by NGS, AFREF, NAREF, SIRGAS, as well as many commercial networks. The convenience of these networks has contributed substantially to the extraordinary expansion of relatively high accuracy GNSS applications that rely on differential processing. However, there is an alternative in both real-time and post-processed work. It is known as *precise point positioning* (PPP); see Kouba and Héroux (2001).

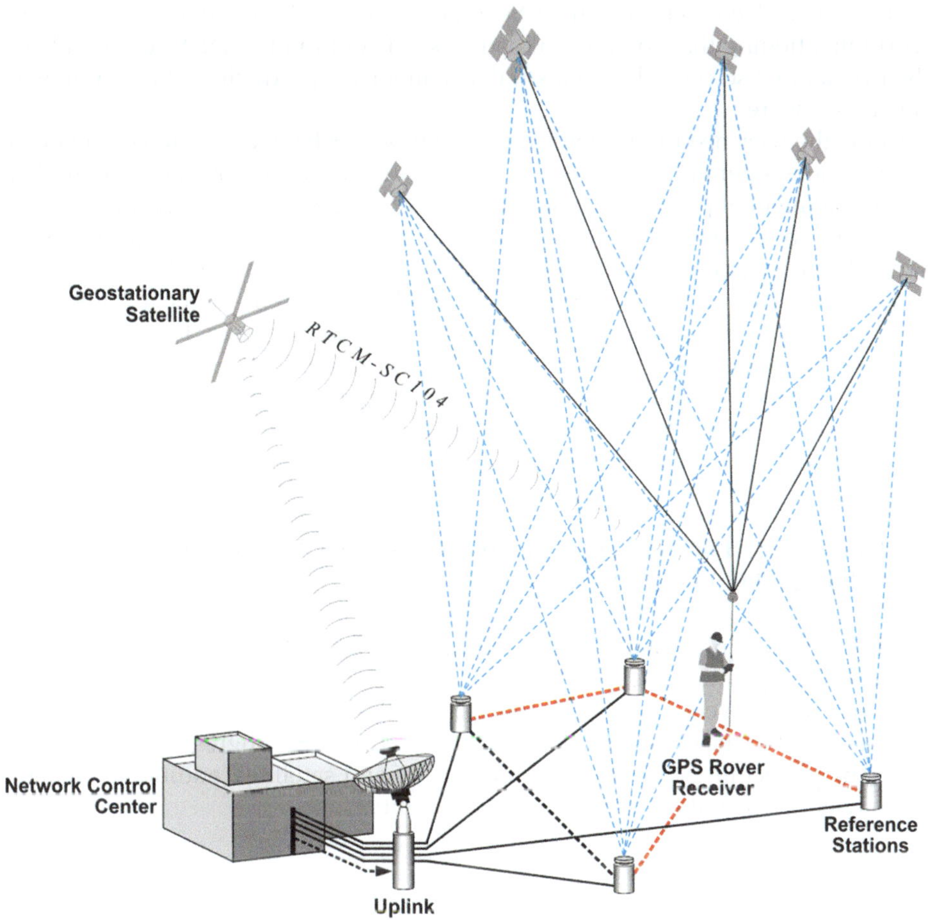

Figure 6-12. Real-time network RTN.
Source: Van Sickle (2015), reproduced with permission from Taylor & Francis Group.

It was mentioned previously that SPP could be a real-time solution using a single receiver measuring to a minimum of four satellites simultaneously. Undoubtedly, this is the most common GNSS solution outside of geodesy and surveying. Its weakness is that the receiver must rely on the information it collects from the satellite's navigation message to learn the positions of the satellites, the satellite clock offset, the ionospheric correction, and so on. These data contain substantial errors. However, there is a source that contains much more accurate data about the satellite orbits and the clocks, and that enables SPP to achieve higher accuracy and thereby presents some advantages over differential methods. One source is the International GNSS Service (IGS), a collaboration among more than 200 organizations in more than 80 countries. As a public service, it collects and archives GNSS data from a worldwide network of more than 300 continuous GNSS stations. It formulates precise satellite ephemerides and clock solutions from these data. Up to eight IGS analysis centers are involved in the processing, and then IGS freely distributes the results. In other words, these data allow users to process their observations using the positions of the satellites and the state of the clocks derived from the period of time the satellites were being tracked. That period of time includes the moment the user actually made the observations. Because these data reflect the precise position of the satellites and clock offsets during the actual measurements, it stands to reason that they are more precise than the broadcast ephemeris and clock corrections can be. Several categories of these data are available. It is also worth mentioning that commercial networks of continuous GNSS stations also exist, and their data are also provided in a similar manner for producing PPP solutions using commercial software.

With PPP, the user need not establish base stations or have access to corrections from reference stations operated by others. It ameliorates the limits on the baseline lengths imposed by differential processing, and the solution is global. It can work anywhere in the world. The PPP corrections are expressed in a global reference frame, such as the International Terrestrial Reference Frame 2014, which offers better overall consistency than does a local or regional solution. However, PPP currently has disadvantages. One of the most persistent is the time necessary to resolve the cycle ambiguity. The time necessary to move from a float to a fixed solution is extended because the ambiguity cannot be assumed to be an integer as it is in a differenced solution. Currently, the convergence can take 20 min or more. The situation may be improved with the increase in satellite observables that will come with the inclusion of signals other than the GPS (e.g., GLONASS, Galileo) in the IGS PPP solution.

GLOBAL POSITIONING SYSTEM MODERNIZATION AND GLOBAL NAVIGATION SATELLITE SYSTEM

The GPS was put in place with amazing speed considering the technological hurdles and reached its Fully Operational Capability (FOC) on July 17, 1995. The oldest satellites in the current constellation were launched in the early 1990s. It is not surprising that there are plans in place to alter the system substantially. In 2000, the US Congress authorized the GPS III effort. The project involves new ground stations and satellites, additional civilian and military navigation signals, and improved availability.

GPS Satellite Blocks

Figure 6-13 is an illustration that summarizes the improvements made in the satellite blocks that have made up the GPS constellation over the years.

L2C. Two new codes will be broadcast on the carrier, L2, which previously carried only one military signal exclusively, the P(Y) code. Now, L2 will carry a new military signal, the M-code, and a new civil signal as well. This is a code that was first announced in March 1998. It is transmitted by all block IIR-M satellites and subsequent blocks and is known as *L2C*. The "C" is for civil. L2C is actually composed of two PRN signals: the civil-moderate length code, CM, and the civil-long length code, CL. L2C has better autocorrelation and cross-correlation protection than the C/A code because both the CM and CL codes are longer than the C/A code. Longer codes are easier to keep separate from the background noise. In practice, this means that these signals can be acquired with more certainty by a receiver that can maintain a lock on them more surely in marginal situations where the sky is obstructed. There is also another characteristic of L2C that pays dividends when the signal is weak. Although CM carries newly formatted navigation data and is, therefore, known as the *data channel*, CL does not. It is without data and is known as a *pilot channel*. A pilot channel can support longer integration when the signal received from the satellite is weak.

L5. L5 is the new carrier being broadcast on the Block IIF satellites. It is centered on 1,176.45 MHz. The basic structure of L5 looks similar to that of L1. Two PRN codes are present on this carrier. Both L5 codes have a 10.23 MHz chipping rate. This is the same rate that has been available on the P(Y) code from the beginning of the system. However, this is

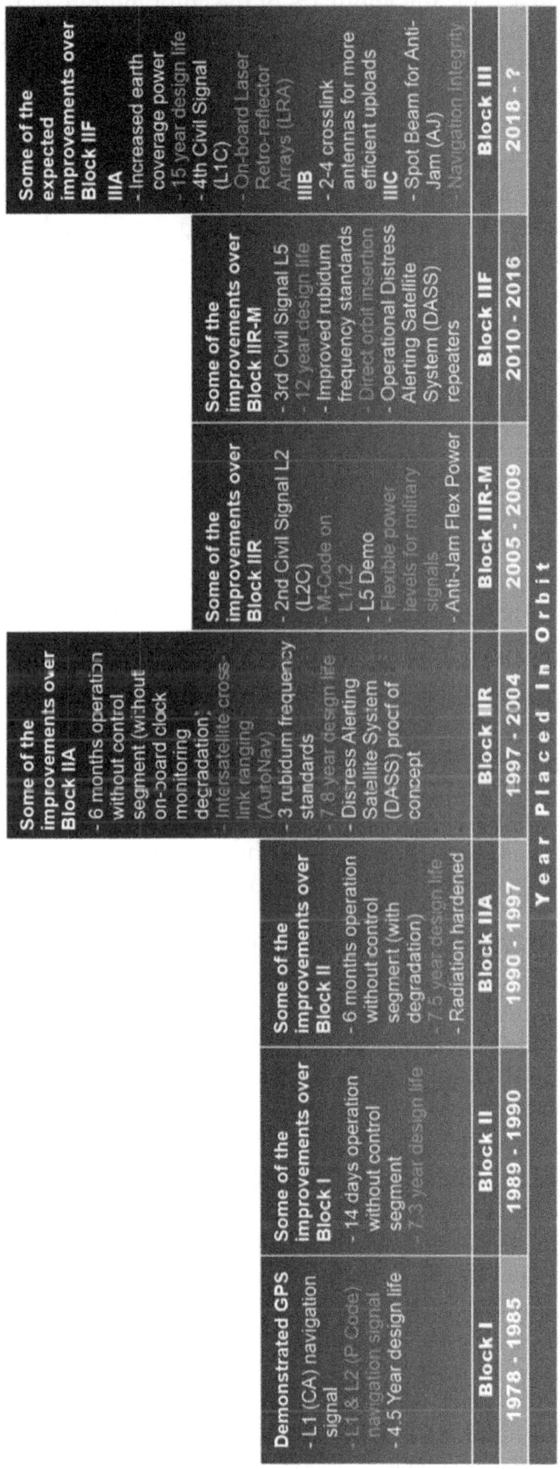

Figure 6-13. GPS modernization.
Source: Van Sickle (2015), reproduced with permission from Taylor & Francis Group.

the fastest chipping rate available in any civilian code. L5 has the only civilian codes that are both 10 times longer and ten times faster than the C/A code. Because the maximum resolution available in a pseudorange is typically about 1% of the chipping rate of the code used, the faster the chipping rate, the better the resolution. L5 has about twice as much power as L1, and because L5 does not carry military signals, it achieves an equal power split between its two signals. In this way, L5 lowers the risk of interference and improves multipath protection. It also makes the data-less signal easier to acquire in unfavorable and obstructed conditions. Unlike L2C, L5 users will benefit from its place in a band designated by the International Telecommunication Union for the Aeronautical Radionavigation Navigation Service worldwide. Therefore, it is not prone to interference with ground-based navigation aids and is available for aviation applications. Although no other GPS signal occupies this band, L5 does share space with one of the Galileo signals, E5. L5 also incorporates FEC.

L1C. Another civil signal will be broadcast by the Block III satellites. It is known as *L1C*. As a result of an agreement between the United States and the European Union (EU) reached in June 2004, this signal will be broadcast by both GPS and Galileo's L1 Open Service signals. It will also be broadcast by the Japanese QZSS, the Indian Regional Navigation Satellite System (NAVIC aka IRNSS), and China's BeiDou system. The L1C design shares some of the M code characteristics. The navigation message carried by L1C will be known as *CNAV-2*. Both CNAV and CNAV-2 have 18% more satellite ephemeris information than NAV. The structure of CNAV-2 will virtually ensure that once a receiver acquires the satellite, the time to first fix will not exceed 18 seconds.

Global Navigation Satellite Systems

The GPS is one component of the worldwide effort known as the GNSS. Another component of the GNSS is the GLONASS system of the Russian Federation, and a third is the Galileo system administered by the EU. And it is likely that more constellations will be included in the GNSS, such as the Japanese QZSS, the Indian Regional Navigation Satellite System (NAVIC aka IRNSS), and the Chinese BeiDou Satellite Navigation and Positioning System. They will be augmented by both ground-based augmentation systems and space-based augmentation systems (SBASs) deployed by the United States, Europe, Japan, China, and Australia. One immediate effect of the GNSS is the substantial growth of the available constellation of satellites, and the more signals that are available for positioning and navigation, the better. The concept is that these networks of satellites and others will begin to work together to provide PNT solutions to users around the world. The objectives of this cooperation are interoperability and compatibility.

Globalnaya Navigationnaya Sputnikovaya Sistema. Russia's Globalnaya Navigationnaya Sputnikovaya Sistema (Global Orbiting Navigation Satellite System), known as *GLONASS*, did not reach full operational status before the collapse of the Soviet Union. It is the first generation of the Uragan satellites that reached orbit in October 1982, a little more than 4 years after the GPS constellation began. A total of 87 Uragan satellites were launched, and a nearly full constellation of 24 made up of 21 satellites in three orbital planes with three on-orbit spares was achieved in 1996.

Galileo. At the time of this writing, the European Union's civilian-controlled Galileo system is expected to reach full operational status in 2022. The satellites are on orbit at a nominal height of about 23,222 km above the Earth. The full constellation will include 30 satellites in three planes, 10 in each plane higher than the GPS constellation. Galileo's ground segment has two centers, one at the Fucino Control Centre in Italy and one at the Oberpfaffenhofen Control Centre in Germany. There is also a network of sensor stations that provide data to the centers through the dedicated data dissemination network, and the uplink stations communicate the calculated integrity, time, and other information back up to the satellites. There are also two telemetry, tracking, and command stations, one at Kourou, French Guiana, and one at Kiruna, Sweden. The stations are all located in areas controlled by European nations. There are four Galileo signals; E5a, E5b, E6, and E2-L1-E1. E5a and E2-L1-E1 overlap the existing L1 and L5 GPS signals (Issler et al. 2003). Galileo has defined five levels of service that will be provided by the system. They include the Open Service (OS), which uses the basic signals and is quite similar to GPS and GLONASS. OS is free and available for timing and positioning applications. The SOL Service is along the same line but provides increased guarantees, including integrity monitoring, meaning that users are warned if there are signal problems. SOL will be a global service and will include both a critical level service and a less accurate noncritical level. The Public Regulated Service (PRS) is encrypted and is meant to assist public security and civil authorities. PRS is under government control and provides significant jamming protection. PRS will be provided on E6 and E2-L1-E1. Likely applications will include emergency services, law enforcement, intelligence services, and customs. The Search and Rescue Service (SAR) is intended to enhance space-based services and improve response time to distress beacons and alert messages. The availability of the Open Service, Search and Rescue, and PRS will coincide with Initial Operational Capability. Other services, such as the encrypted custom solutions and unique applications of the Commercial Service (CS), will follow as Full Operational Capability (FOC) is achieved.

BeiDou Navigation Satellite System (BDS). A fourth GNSS system, joining those undertaken by the United States (GPS), Russia (GLONASS), and Europe (Galileo) is the Chinese BeiDou Naviation Satellite System. The system is named after the Big Dipper. The first generation of the system is known as *BeiDou-1.* It is a regional satellite navigation system that services a portion of the Earth from 70°E longitude to 140°E longitude and from latitude 5°N to 55°N. There are three orbital categories, namely, geostationary (GEO), medium earth orbit (MEO), and the inclined geosynchronous orbit (IGSO). As of January 2022, the BeiDou constellation consisted of 44 operational satellites; 7 are in GEO, 27 are in MEO, and 10 are in the IGSO.

There are two service types. Open Service is available to the public and offers an autonomous (not differentially corrected) positional accuracy of 10 m, 0.2 m/s velocity accuracy, and timing accuracy within 20 ns. Authorized Service is not available to the public. It is available to the military, specifically the militaries of China and Pakistan.

BeiDou is supports two services: Radio Navigation Satellite Service (RNSS) and Radio Determination Satellite Service (RDSS). RDSS includes a two-way short message service capable of handling10 million messages per hour using three GEO satellites across mainland China. There is a global message service using inter-satellite cross-links with a capacity of 200,000 messages per hour using 14 MEO satellites. RNSS also includes a Satellite-Based Augmentation Service (SBAS). The network includes 150 reference stations, information from these stations is processed in the BeiDou control/ground segment and the resulting corrections are sent to the BeiDou GEO satellites, which broadcast them via the D2 NAV message to the BeiDou user's receivers. It also includes ionospheric corrections and clock corrections to other GNSSs.

The BeiDou control/ground segment is comprised of an MCS, two upload stations and a network of 30 widely distributed MSs. The MCS is responsible for the operational control of the system, including orbit determination, navigation messages, and ephemerides that are based on the China Geodetic Coordinate System 2000 (CGCS2000), which is within a few centimeters of the ITRF.

Quazi-Zenith Satellite System. The first demonstration satellite of the Japanese, Quazi-Zenith Satellite System (QZSS), named QZS-1, was launched in 2010 by the Japan Aerospace Exploration Agency (JAXA) from the Tanegashima Space Center. It is expected to have a design life of 10 years. The system is nicknamed Michibiki, meaning *guide*. Two additional quasi-zenith satellites, QZS-2 and QZS-3, and one geostationary satellite, QZS-4, are in orbit at the equator. The four-satellite constellation is planned to increase to seven satellites in the future.

The QZSS is intended to provide satellites in highly elliptical orbits that are inclined and geosynchronous. The orbits are designed so the satellites will always be available at high elevation angles that are almost directly overhead in Japan, Oceania, and East Asia. This is the origin of the term *quasi-zenith*.

The QZSS is primarily a multisatellite regional augmentation system (SBAS). The objective is to improve satellite positioning, navigation and timing services in urban canyons and mountainous areas in the region. The system is designed to improve the positioning of GPS and Galileo receivers.

Indian Regional Navigation Satellite System. The building of the Indian Regional Navigation Satellite System (NAVIC aka IRNSS) was authorized by the Indian government in 2006. It provides position, navigation, and timing service in a region from 30°S latitude to 50°N latitude and from 30°E longitude to 130°E longitude. The region embraced is approximately 1,500 km around India.

The space segment of the IRNSS is comprised of seven satellites. All the satellites are continuously visible to the extensive Indian control segment's 21 stations located across the country, including the Master Control Center at Hassan, Karnataka. They are named with the prefix IRNSS-1. The IRNSS will provide two levels of service, a public Standard Positioning Service and an encrypted Restricted Service (RS).

FUTURE

As of early 2022, 118 PNT satellites are aloft. The systems provide users with a variety of signals and codes. Probably there may be many more satellites that will enable new applications in areas where their scarcity has been a hindrance. For civilian users, new signals will provide more protection from interference, the ability to compensate for ionospheric delays with pseudoranges, and wide-laning or even tri-laning capabilities. For everybody, there will be an improvement in accuracy, wider availability, higher integrity, and greater reliability. So, it is no surprise that there is great anticipation.

REFERENCES

Brunner, F. K., and W. M. Welsch. 1993. "Effect of the troposphere on GPS measurements." *GPS World* 4 (1): 42–51.

Hofmann-Wellenhof, B., H. Lichtenegger, and E. Wasle. 2008. *GNSS-Global navigation satellite systems*. Wien, Austria: Springer.

Issler, J.-L., G. W. Hein, J. Godet, J. C. Martin, P. Erhard, R. Lucas-Rodriguez, and T. Pratt. 2003. "Galileo frequency & signal design." *GPS World* 14 (6): 30–37.

Kaplan, E. D. 1996. *Understanding GPS, principles and applications*. Boston: Artech House.

Kouba, J., and P. Héroux. 2001. "Precise point positioning using IGS orbit and clock products." *GPS Solutions* 5 (2): 12–28.

Langley, R. B. 1991. "The GPS receiver: An introduction." *GPS World* 2 (1): 50–53.

Langley, R. B. 1993. "The GPS observables." *GPS World* 4 (4): 52–59.

Leick, A., L. Rapoport, and D. Tatarnikov. 2015. *GPS satellite surveying*. 4th ed. Hoboken, NJ: Wiley.

NovAtel. 2015. *An introduction to GNSS: GPS, GLONASS, BeiDou, Galileo and other global navigation satellite systems*. 2nd ed. Calgary, Alberta: NovAtel.

Strang, G., and K. Borre. 1997. *Linear Algebra, geodesy, and GPS*. Wellesley, MA: Wellesley-Cambridge Press.

Teunissen, P., and O. Montenbruck, eds. 2017. *Springer handbook of global navigation satellite systems*. Cham, Switzerland: Springer.

Van Dierendock, A. J., S. S. Russell, E. R. Kopitzke, and M. Birnbaum. 1978. "The GPS navigation message." *Navigation* 25 (2): 147–165.

Van Sickle, J. 2015. *GPS for land surveyors*. 4th ed. Boca Raton, FL: CRC Press.

CHAPTER 7
LEVELING AND TOTAL STATIONS

N. W. J. Hazelton

INTRODUCTION TO LEVELS AND LEVELING

Differential leveling is the process by which the elevation difference between points is measured directly. This allows the determination of the vertical location of individual points. Levels and leveling rods are the instruments used for this process. These instruments are described in the first section of this part of the chapter.

With differential leveling, it is important to realize that elevations themselves are not measured. Elevation differences are measured, and elevations are derived from these measurements. The procedures for using a level and leveling rod, and the means of deriving elevations from the measurements, are described in the second section.

All measurements are subject to error, and errors and methods of finding and correcting them are discussed in the third section.

As measurements always include some random error, various means of adjusting the measurements to allow consistent derived information are discussed in the "Errors in Leveling" section.

This discussion of leveling is limited to differential leveling using levels and leveling rods. It does not cover the use of photogrammetry, total stations and trigonometric leveling, barometers, tacheometric methods, water levels, Global Navigation Satellite System (GNSS)-based leveling, plane table methods, laser leveling, and other methods of vertical location of points, nor does it cover precise leveling. A deeper discussion of these and other methods can be found in Bossler and Hazelton (2015).

Differential leveling is discussed in all introductory surveying textbooks such as Cole (2014), Ghilani (2022), Kavanagh and Mastin (2013), McCormac et al. (2012), Nathanson et al. (2017), Uren and Price (2010), Bhavikatti (2009), Bannister et al. (1998), Anderson and Mikhail (1997), Moffitt and Bossler (1997), and Whyte and Paul (1997).

NGS also provides several publications on precise leveling and error sources, such as NOS 73-NGS 8 (Whalen 1979), NOS-NGS 3 (NGS 2020a), and NOS-NGS 34 (NGS 1982). These publications are freely available and provide a detailed description on levels, leveling procedures, and errors associated with leveling instruments and observations.

LEVELING INSTRUMENTS AND EQUIPMENT

Levels

Levels are designed to project a horizontal plane by aligning a telescope at right angles to the vertical at the location of the instrument. In days gone by, a physical bubble tube was used to align the telescope (Figure 7-1), but modern levels have an automatic system to fine-tune the leveling of the instrument (Figure 7-2).

Modern levels are classified as either: automatic levels (because of the automatic compensation system), where an optical reading on the leveling rod is made by the operator reading where the level telescope's crosshair intersects the leveling rod's scale (Figure 7-3); or digital levels (built on an automatic level's basic design), where an electronic measurement is made by the instrument with a special leveling rod with a barcode pattern on its face (Figure 7-4). Most modern digital levels are capable of making electronic readings on the barcode pattern and optical readings on a standard leveling rod, but, of course, electronic readings are more efficient.

Levels are secured to a tripod, which allows an approximate leveling of the instrument. The instrument's foot screws are then used to center a small circular bubble, which brings the compensator in the telescope into its operating range. The telescope is then turned about the vertical axis and pointed to the leveling rods in turn, and the value on the rod beneath the horizontal crosshair is observed and recorded. With automatic levels, this is done

Figure 7-1. Wye level with a bubble tube below the telescope.

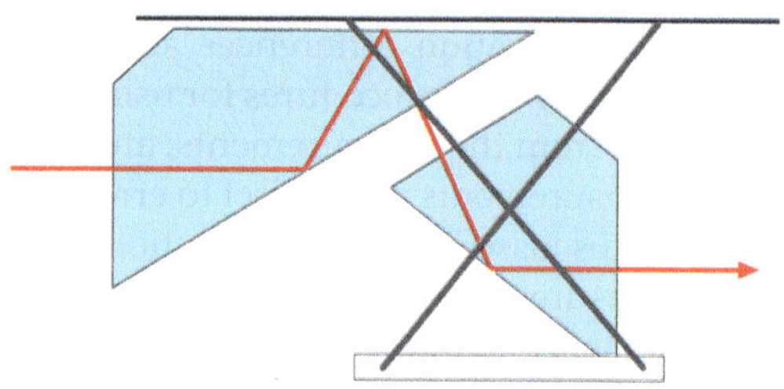

Figure 7-2. Suspended prism system to compensate for small tilts in the telescope.

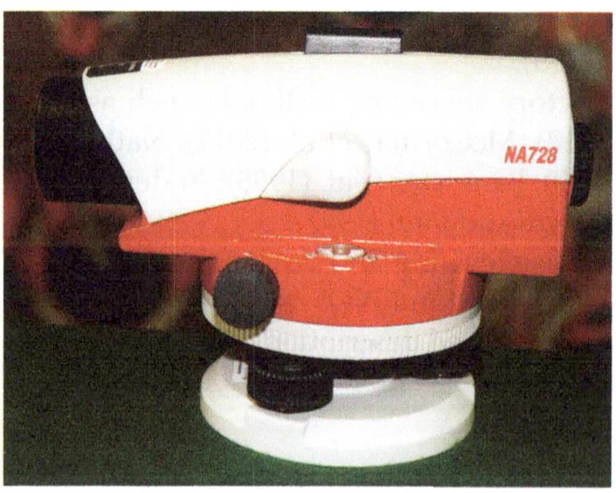

Figure 7-3. Automatic level. A small circular bubble can be seen in the middle of the red area. The level has three foot screws.

Figure 7-4. Digital level and barcode rod. The keyboard allows a more extensive data entry with observations. The eyepiece is in the middle of the panel.

by the operator; with digital levels, it is handled by the instrument, under operator control. The plane projected through the telescope from the horizontal crosshair is termed the *line of collimation*.

The telescope and the crosshairs must be in sharp focus for observations. To do this, first, blur the image through the telescope, then get the crosshairs into sharp focus using the ring around the eyepiece lens. Then, focus the telescope on the object to be observed. If this has been done correctly, the eye moving behind the eyepiece will not cause the crosshairs to move with respect to the object being sighted. This means all parallax has been removed, but only for one eye. Different operators and using the other eye may require adjusting the crosshairs' focus.

Adjusting the telescope to proper focus is critical for quality observations, especially when using digital levels. Having the level properly focused also means the sensor is properly focused.

Before each day's leveling begins, the instrument should be checked using a two-peg test, described subsequently, and adjusted if necessary. All automatic levels can be checked and adjusted in much the same way, but digital levels have included software that is specific to the manufacturer and the model, which must be used for proper adjustment.

Tripods

Tripods (Figure 7-5) allow surveying equipment to be held in a stable location during operation, possibly vertically above a ground mark. For the instrument to be kept stable,

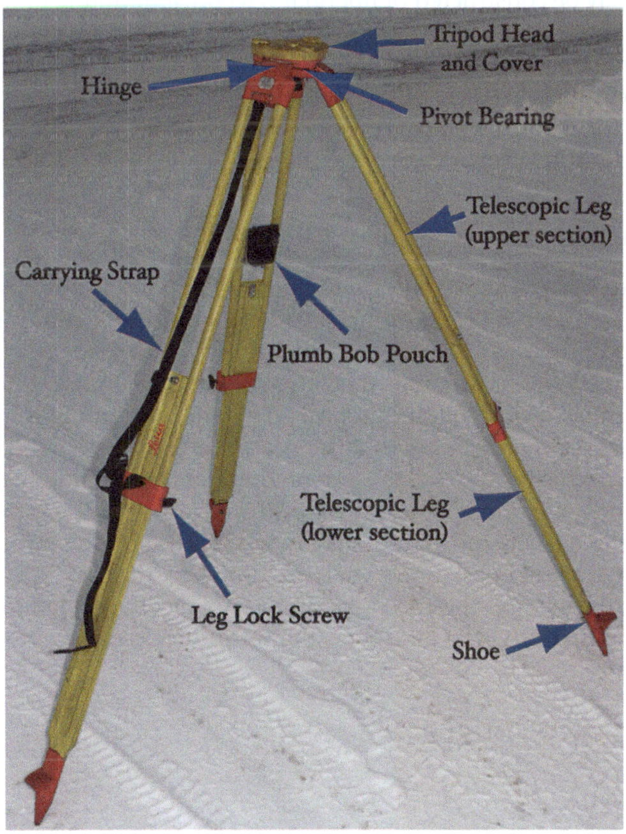

Figure 7-5. Tripod and its parts.

the tripod needs to be stable. This means that the various joints and locking screws need to be tight, the feet pushed well into the ground (where possible) or otherwise braced, and the tripod in good condition. Fixed-leg tripods increase stability and reduce maintenance needs because their legs are of a fixed length, and they do not telescope, as depicted in Figure 7-5. A tripod with telescoping legs can be less stable because the legs can slip if a leg lock screw is loose. Poorly maintained tripods can degrade results. All tripods need to be checked and maintained for high-quality work.

Leveling Rods

Leveling rods are also known as "leveling staves" (singular "staff") in some parts of the world. They consist of a system for holding a graduated scale vertical and rigid. This allows measurements above or below the level's line of collimation in the vertical direction. Examples of leveling rods are shown in Figure 7-6.

For automatic levels, the pattern on the rod is designed to be easily seen and read by the operator through the level's telescope, and the graduations may be in feet or meters. For digital levels, a barcode pattern is used. The barcodes are manufacturer-specific, so levels and rods from the same manufacturer must be used together.

The leveling rod must be held vertically for the observation, and this is accomplished by using a small circular bubble, either attached to the leveling rod or held against it by the operator. This allows the correct reading to be made immediately and is absolutely required for digital levels. Leveling rods need to be checked frequently for rigidity and consistency, especially across joints.

Leveling rods may be constructed of fiberglass, wood, aluminum, or invar. For the most precise work, it is recommended to use an invar rod because it hardly undergoes thermal expansion or contraction. It is also recommended to use a rod without any joints, as worn or loose joints can introduce errors in the readings.

A common rod for construction surveying is called a *Lenker rod*. These rods can be set to directly read the elevation of marks as opposed to only the height or scale on the rod.

Leveling Rod Bubbles

Some leveling rods do not have built-in bubbles. These simple devices, as in Figure 7-7, provide a small circular bubble to be used to help hold the leveling rod vertical. The bubble is attached to the rod or to a small bracket that is held against the rod. Most rod bubbles can be adjusted to ensure that they are truly vertical in use. Rod bubbles should be checked and adjusted at frequent intervals. Leveling rod bubbles are recommended over other practices for making sure that the rod is vertical, such as a procedure known as *waving the rod*. For this procedure, the person holding the rod slowly rocks it away and toward the instrument while keeping the tip of the rod over the occupied point. The instrument person reads the rod while it is being waved, and the minimum reading on the rod is recorded, as it is assumed that this is when the rod is vertical. Although frequently taught, the waving-the-rod procedure is not recommended for precise leveling because the bottom tip of the rod has edges that introduce eccentricities in the readings.

CORE LEVELING PROCEDURES

Setting up the Level

The tripod is first opened and the legs extended. The tripod head should be about chest high when set up to allow easy reading of instruments. The legs are spread to allow stability.

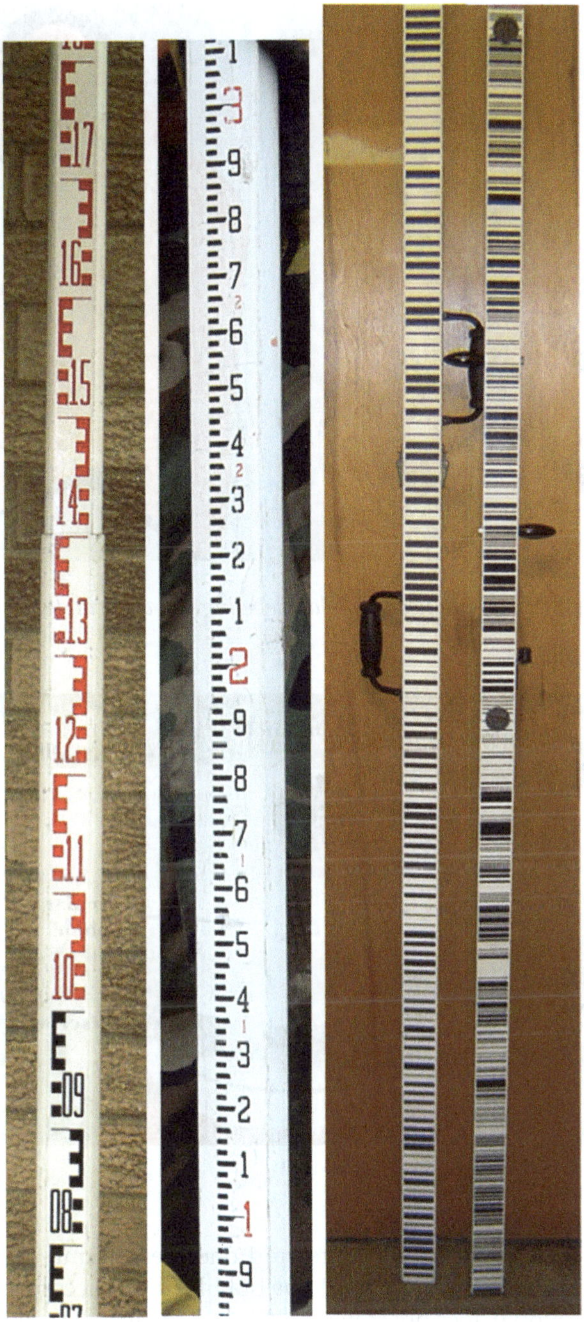

Figure 7-6. (Left to right) Metric E-pattern rod, Foot rod, Sokkia barcode rod, Leica barcode rod.

The level is removed from its case and attached to the tripod using the central screw on the tripod head. NEVER let go of the instrument until it is securely attached to the tripod. Make sure that the foot screws are in about the middle of their travel. Adjust the legs, mainly moving them from side to side, to get the small circular bubble (on the level) somewhere near the center ring. Once this is done, press the tripod legs' shoes well into the ground to stabilize the tripod. Turn the level so that it is at right angles to the line of travel, if base leveling (see later in this section) and center the small circular bubble using the foot screws

Figure 7-7. (Left) Rod bubble being used to hold a leveling rod vertical. (right) Adjustment screws for bubble.

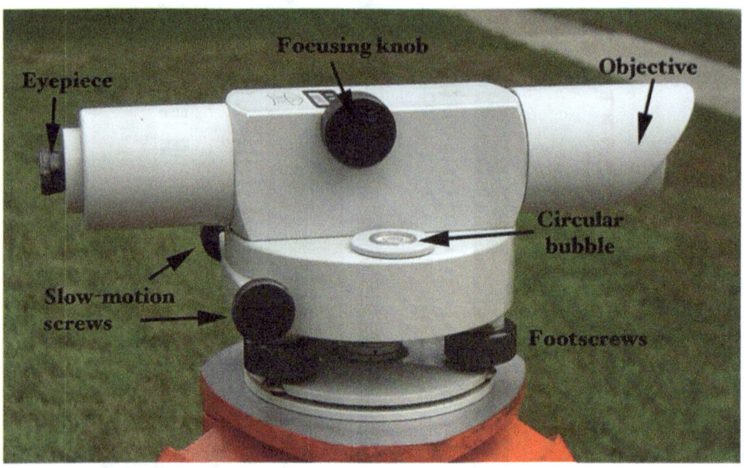

Figure 7-8. Parts of an automatic level.

(Figure 7-8). Note that the bubble moves toward one's left thumb as the foot screws are turned. Turn the instrument 180° and check the bubble. If it is acceptable, the rod is ready to be observed. If the bubble does not stay central when the instrument is rotated 180°, then it is out of adjustment and should be adjusted.

Reading the Rod

With the leveling rod held vertical, its scale is observed through the level's telescope, and the value on the scale where the horizontal crosshair appears to sit on the rod is recorded. Figure 7-9 shows how the two primary rod scales are read. For digital levels, a button is pressed on the instrument to scan the barcode on the rod and take a reading. The measurements can be stored in the onboard data collector of the instrument.

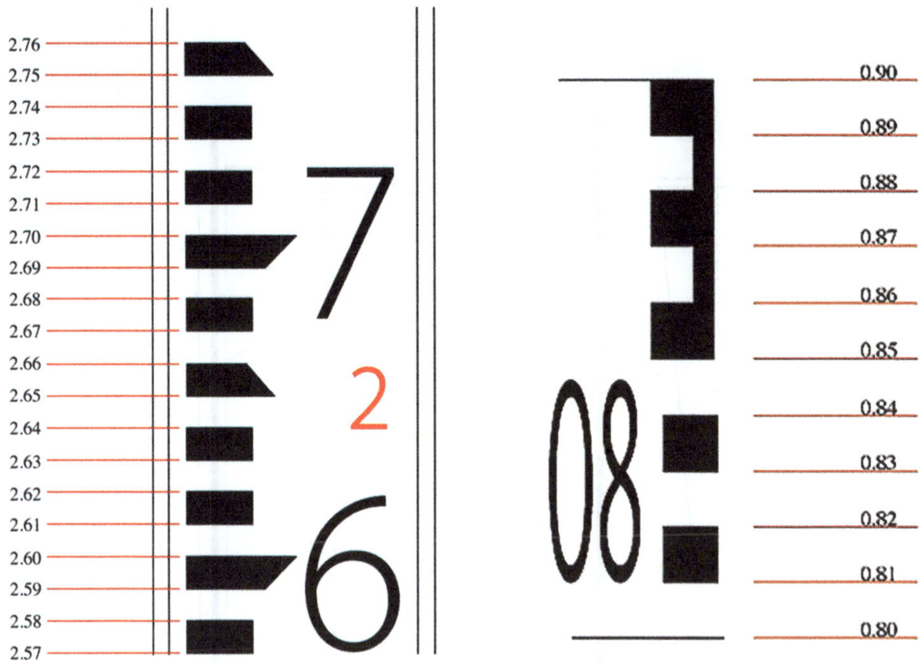

Figure 7-9. Reading the leveling rod's scale. (Left) A foot scale, read directly to 0.01 ft, then estimated to 0.001 ft, if necessary. (Right) A metric scale, read directly to 0.01 m, then estimated to 0.001 m, if necessary.

Two-Peg Test

To check that the level is operating correctly, a two-peg test is used. This is also the basis for adjusting the level. If the level is out of adjustment, then the telescope is tilted, and when the instrument is rotated, the line of collimation will sweep out a shallow cone, rather than a horizontal plane. However, observations made at an equal distance from the telescope will have equal errors caused by this tilt, whereas observations made at different distances will have different amounts of error. Two marks ("pegs") are placed in the ground, usually about 30 m (100 ft) apart, and two setups are made to check the level, as shown in Figure 7-10.

Setting up the level midway between two leveling rods, the rods are read, and the difference between the rod readings is calculated. This difference will also be the elevation difference, without any error from a tilted telescope.

The level is then set up close to one of the leveling rods, usually less than 2 m (6 ft) away, and the leveling rods are observed again. The difference in the readings should be the same as from the first setup, within the reading tolerance. If this is the case, the level is considered to be in adjustment. If the two differences do not agree, the instrument needs to be adjusted before any work can be done.

Through normal use and wear, all levels will become maladjusted at some time and perform poorly during two-peg tests. Levels vary by manufacturer, and practitioners should consult manuals for a given level prior to making any adjustments. Experienced operators may know how to adjust the level immediately while in the field; inexperienced operators should send the instrument away for adjustment by qualified technicians.

It is recommended to perform adjustments in good atmospheric conditions while in the shade and with tripods firmly set up and highly stable. Leveling manuals will prescribe the

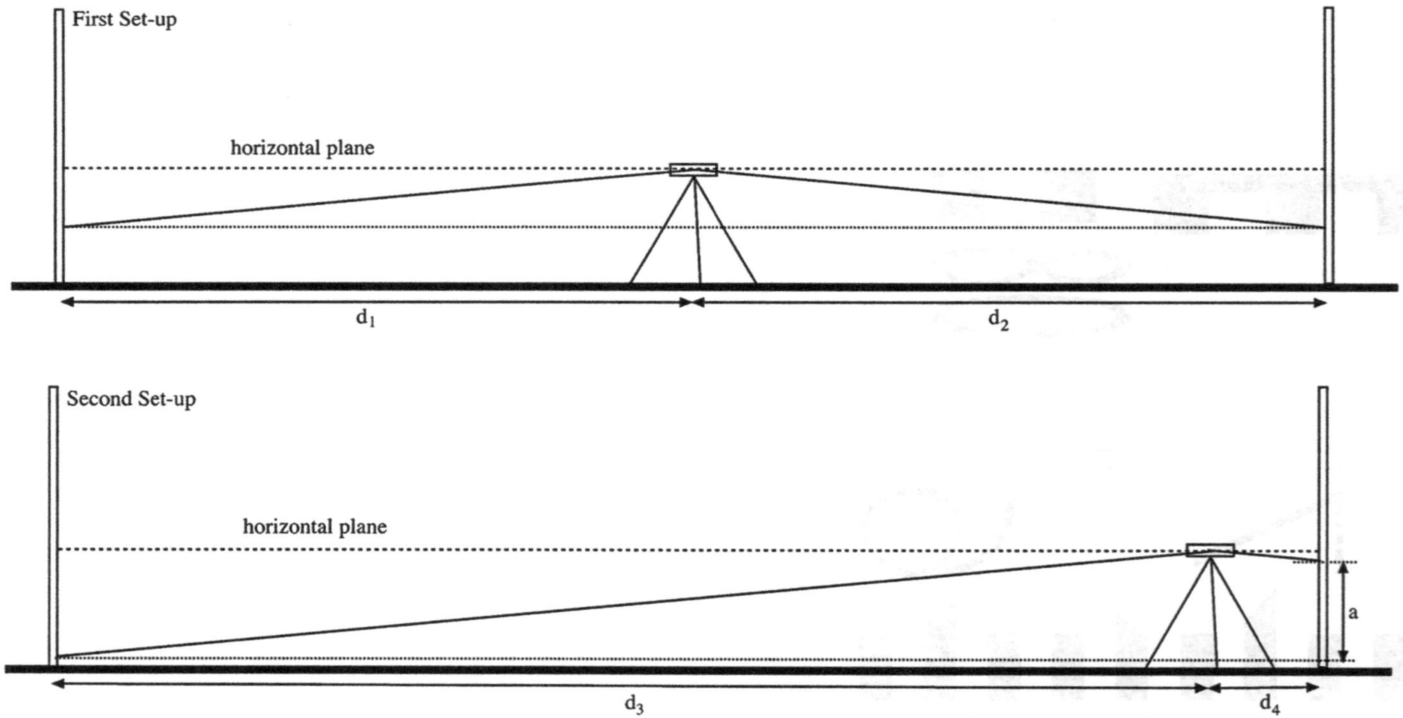

Figure 7-10. Two-peg test. (Top) The first setup is midway between the leveling rods and the difference in readings is calculated. (Bottom) The second setup is very close to one rod, and the difference between the readings is again calculated. Any difference between the differences indicates that the level is not in adjustment.

order and methods for adjusting nuts and screws with specific tools and pins. A final check of all adjustments should be made to ensure that everything is satisfactory. Digital levels often provide step-by-step instructions and digital workflows for checking and adjusting the level.

Even while properly adjusted, it is important to understand that no level is perfect. All levels will have some systematic errors in the line of collimation, and it remains important to follow proper field procedures (e.g., balancing the distances to the foresights and backsights as discussed later) to reduce these types of systematic errors even with calibrated equipment. For example, the crosshair reticle will move post adjustment to find its equilibrium point, and the level should, therefore, be checked and adjusted frequently.

Base Leveling

The basic process of leveling is measuring a set of connected elevation differences, using intermediate points, to determine the overall elevation difference between two points of interest. At each setup, the first observation is made to a known point, or the point previously observed, and is termed the *backsight*. The final observation of a setup is to the point to be used to transfer the leveling forward to the next setup, and this is termed the *foresight*.

Leveling used to transfer elevation between points will require only a single backsight and a single foresight at each setup. This type of leveling is termed *base leveling*. The process at a single setup is shown in Figure 7-11.

Once a setup has been completed, the person holding the backsight rod moves forward to establish the foresight point for the next setup. The person operating the level moves forward to midway between the rods. The person holding the foresight rod remains at the same point and becomes the backsight for the next setup, as shown in Figure 7-12.

To minimize any errors caused by the level being out of adjustment, as well as the effects of Earth curvature and atmospheric refraction, the level is placed as close as is reasonable to midway between the two rods. Similarly, the locations used for backsight and foresight rod placement usually have marks placed in the ground to provide a stable location for the rod between setups. These points may be nails or pegs in the ground or suitably stable and pointy objects at the appropriate place. They are known as *turning points* or *change points*. The critical characteristics of these points are stability and an unambiguous point for the rod's baseplate.

The values that are observed on the rod at each location are immediately recorded in the field book or in a digital level's storage. The recording and reducing methods are discussed subsequently. The values recorded should be checked by the operator, as it is very easy to make transcription errors between reading the rod and what is written in the level book.

General Leveling

Although base leveling provides a foundation for subsequent leveling, in general, leveling the elevation of a number of points is required, such as for a topographic or engineering survey. To achieve this, additional observations may be made to a leveling rod located at points of interest. Note that the backsight and foresight are required, and they bracket these additional observations, allowing them to be connected to the rest of the leveling work, as shown in Figure 7-13. These additional observations are termed *side shots* or *intermediate sights*.

It is important to note that making backsight and foresight observations to every turning point means that these points have been measured twice, and so, include a basic form of independent check; side shots are measured once and, therefore, have no independent

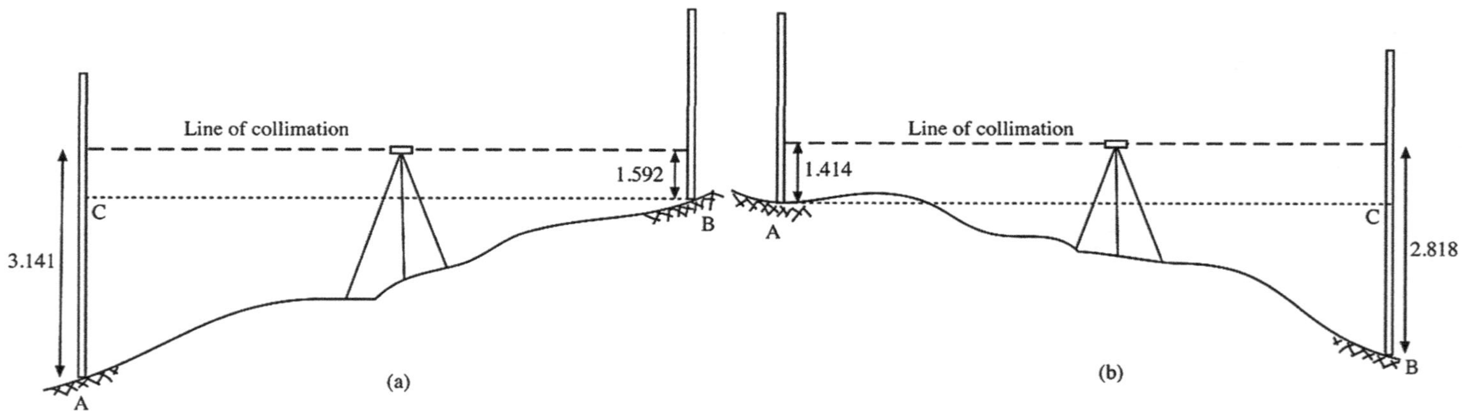

Figure 7-11. *Elevation difference measurements. Assuming that the leveling is from point A to point B, (a) represents a rise, as the foresight reading (to B) is less than the backsight reading (to A). The diagram in (b) represents a fall, as the foresight reading (to B) is greater than the backsight reading (to A). The difference in the rod readings in (a) is AC, and in (b) is CB, and this is also the elevation difference in each case.*

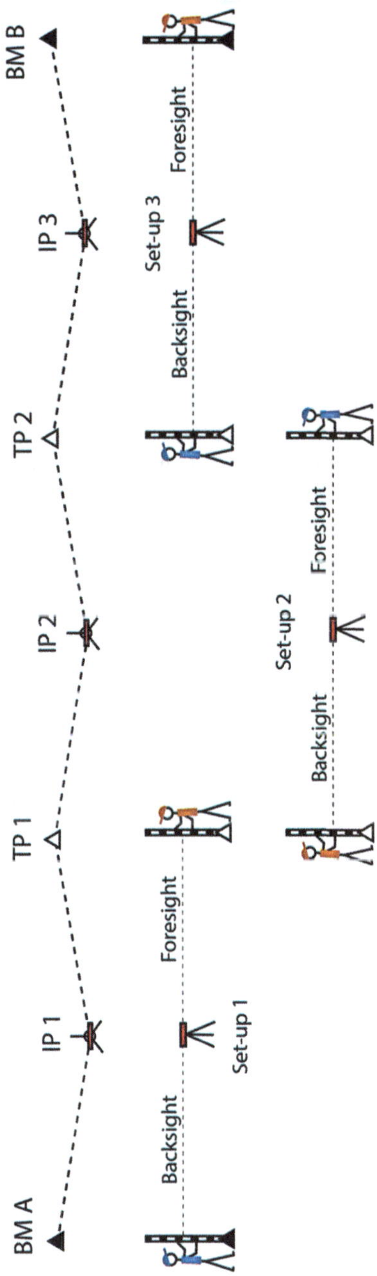

Figure 7-12. Base leveling from Bench mark A (BM A) to Bench mark B (BM B). Three setups are needed, using two turning points (TP1 and TP2) and three instrument points (IP1, IP2, IP3). The people (depicted in blue and orange) holding the leveling rods indicate the sequence of movements. The backsight is read first, followed by the foresight.

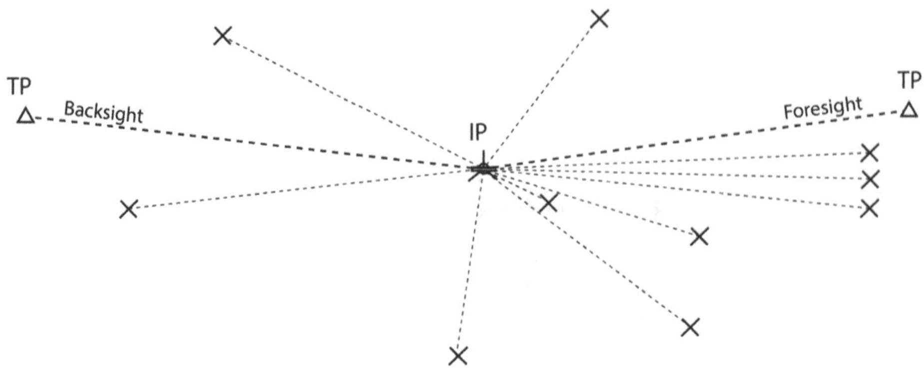

Figure 7-13. A single setup where the backsight and foresight (made to turning points, TP) bracket a number of side shots (made to the points marked with crosses). The sequence of observations starts with the backsight, collects all the side shots, then finishes with the foresight.

check. Side shots should not be used for points that are of importance in the survey, or they should be checked by reobserving the same point from a second setup. It is best practice to treat important points as turning points (foresights/backsights) in the base leveling.

Although it is desirable to keep the backsight and foresight approximately equal in length while leveling, the distance from the instrument to side shot points will often be highly variable. It is, therefore, critical that the instrument be in adjustment, so a daily two-peg test is required.

Recording Methods

There are three basic methods for recording leveling observations. The method used will depend on the requirements of the project. Base leveling and leveling where the elevation of points is not required at the time of observation should be done using the rise and fall method. Leveling for setting out points for construction is better done using the height of collimation method (also known as the height of instrument method). Topographic leveling on a regular grid should be done using the grid leveling method.

Height of Collimation Method. This is the method that is widely known and used in the United States. As the elevation is carried through the entire method, the elevation of the starting bench mark should be known before leveling begins. The method focuses on the elevation of the points, rather than the elevation differences being measured, and although this has advantages for applications where elevations are required during the leveling process (such as for setting out points), it makes checking and error detection in the field more difficult.

Stepping through the process of observing and recording a series of setups (also known as a *flight* of levels), the observation process is discussed, and the measurements are entered in the appropriate places in a booking form, as shown in Figure 7-14. Note that this is only one of several different designs for a booking form, and some people enter data from the bottom to the top.

Starting from a bench mark (TBM A), which is used as the backsight of the first setup, the bench mark's elevation is entered in the RL column at ①, with the suitable annotation in the Comments column. The backsight observation is entered in the Backsight column at ②, and the backsight value is added to the elevation of the backsight point to obtain the height of the collimation line of the instrument, and this value is noted in the Height of Instrument column at ③. Note that all values associated with this point are recorded on the same line.

BS	HI	FS	RL	Comments
② 8.42	③ 828.65		① 820.23	TBM A RL 820.23
⑥ 11.58	⑦ 839.03	④ 1.20	⑤ 827.45	TP
		⑧ 4.01	⑨ 835.02	Cap
		①⓪ 3.52	①① 835.51	Cover
		①② 3.03	①③ 836.00	Spike
①⑥ 3.68	①⑦ 830.27	①④ 12.44	①⑤ 826.59	TP
		①⑧ 10.05	①⑨ 820.22	TBM A RL 820.23
②⓪ 23.68		②① 23.69	②③ 820.23	
		23.68	②④ 820.22	
		②② 0.01	②⑤ 0.01	Misclosure

Figure 7-14. Booking form for the height of collimation method. The recording and reduction system is set out. The column labels are Backsight, Height of instrument, Foresight, Reduced level, and Comments. Sometimes, the Backsight column is labeled "+" and the Foresight column is labeled "–," as these signify the arithmetic operations performed using the items in these columns. The numbers in circles signify the steps for entering the data.

The instrument is then turned and the foresight is observed, with the value entered in the Foresight column at ④. The foresight value is subtracted from the height of the instrument for this setup, which gives the elevation of the point beneath the foresight rod, and this value is recorded on the same line as the foresight at ⑤. The nature of the point is noted in the Comments column (TP is "turning point"). This completes the first setup.

The instrument is then moved to the second setup. The backsight for this setup is the foresight of the previous setup and, therefore, is recorded on the same line as the previous foresight, at ⑥. The backsight value is added to the elevation of the backsight point (at ⑤) and recorded at ⑦. A side shot observation made to a point along the way is recorded in the Foresight column at ⑧, although some versions of the recording format have an additional column to separate foresights from side shots. The value of the side shot is subtracted from the height of the instrument (at ⑦) for the setup to give the elevation value of the point, recorded at ⑨. The point is annotated in the Comments column. A second side shot is recorded at ①⓪, the value is subtracted from the height of the instrument for the setup at ⑦, and the elevation is recorded at ①①, with an annotation in the Comments columns. A third side shot is recorded at ①②, the value is subtracted from the height of the instrument for the setup at ⑦, and the elevation is recorded at ①③, with the annotation in the Comments column. Finally, the foresight for the setup is observed and recorded in the Foresight column at ①④, the value is subtracted from the height of the instrument for the setup at ⑦, and the elevation is recorded at ①⑤, with the annotation in the Comments column. This completes the second setup.

The level is then moved forward for the third and final setup, in this case closing back to the bench mark. The backsight is observed to the previous setup's foresight point and recorded at ①⑥, the value is added to the elevation of the backsight point at ①⑤ to give the height of the instrument's collimation line, and it is recorded at ①⑦. The foresight is observed to the bench mark to close the leveling circuit, recorded at ①⑧, the value is subtracted from the height of instrument for this setup at ①⑦ to give the elevation at the foresight point, and this value is recorded at ①⑨. Note that there is a 0.01 misclosure, but this is the reading precision of the leveling.

Now that this leveling is completed, the leveling needs to be checked. All the backsights are added, with the sum recorded at ②⓪. All the true foresights (colored blue) are added, with the sum recorded at ②①. The smaller number is subtracted from the larger number, and this difference is recorded at ②②. Similarly, the starting and ending elevations are copied to ②③ and ②④. The difference is calculated and recorded at ②⑤. These two differences should be equal; otherwise, there has been an error in the calculation of the elevations of the turning points. The differences are also the misclosure of the leveling, as it closed back to the same point. Note also that the elevations of the side shot points are not checked in any way, as a single observation was made to each point, and the calculation of each elevation is similarly unchecked.

Leveling must be closed back to the starting point or to another point of known elevation. Without this in all leveling work, the entire job is unchecked. As leveling is highly prone to a wide range of errors, not closing leveling and not carefully checking the results is risking many inaccuracies.

The computations that are carried out in the Height of Collimation Method are as follows, in sequence.

1. Elevation + backsight value = height of instrument,
2. Height of instrument − foresight = elevation,
3. Height of instrument − side shot = elevation,
4. Sum of backsight − sum of foresights = elevation difference (measured),
5. Starting elevation − finishing elevation = elevation difference (calculated), and
6. Elevation difference (measured) must equal elevation difference (calculated).

If the leveling closed back on the starting point, the elevation difference is also the misclosure.

An interactive slideshow guides the reader through this method in detail in Bossler and Hazelton (2015, Chapter 4, Section 2). More about this method can be found in Cole (2014), Ghilani (2021), Kavanagh and Mastin (2013), McCormac et al. (2012), Nathanson et al. (2017), Uren and Price (2010), Bhavikatti (2009), Bannister et al. (1998), Anderson and Mikhail (1997), Moffitt and Bossler (1997), and Whyte and Paul (1997).

Rise and Fall Method. This is the standard international method for recording and reducing leveling observations. It has the advantage of better checking and reduction capabilities, as well as not requiring the final elevations to be known to complete checking and adjustment. Some additional calculation is required, but the method allows checking of the calculation of intermediate observations (side shots). The method focuses on elevation differences, which makes checking and adjustment easier.

Stepping through the process of observing and recording a series of setups (flight of levels), the observation process is discussed, and the measurements are entered in the appropriate places in a suitable booking form, as shown in Figure 7-15.

We start with the first setup, observing a backsight to a bench mark, recording the value in the Backsights column at ①, and noting the bench mark identifier in the Comments column. Elevation does not need to be recorded at this time. The level is then used to read three intermediate observations (side shots) at three points of interest. These are recorded in the Intermediates column on consecutive lines at ②, ③, and ④, respectively, and a description of each point noted in the Comments column. The setup is completed with an observation to a change point (CP or turning point) as the foresight. The value measured is

Back	Int	Fore	Rise	Fall	RL	Comments
① 1.957					①⑨ 123.456	BM SR/62/G17
	② 1.327		⑧ 0.630		②⓪ 124.086	C of G
	③ 2.874			⑨ 1.547	②① 122.539	Edge of Road
	④ 3.066			①⓪ 0.192	②② 122.347	Top of Drain
⑥ 2.584		⑤ 1.448	①① 1.618		②③ 123.965	CP
		⑦ 3.091		①② 0.507	②④ 123.458	BM SR/62/G17
①③ 4.541		①④ 4.539	①⑤ 2.248	①⑥ 2.246	123.458	End
4.539			2.246		123.456	Start
①⑦ 0.002	Rise		①⑧ 0.002	Rise	②⑤ 0.002	Rise

Figure 7-15. Booking form for the rise and fall method. Recording and reduction system set out. Column labels are Backsights, Intermediates (side shots), Foresight, Rises, Falls, and Reduced Level (RL). The numbers in circles signify the steps for entering the data.

entered in the Foresight column on the next line at ⑤, and an annotation is added to the Comments column.

The level is then moved forward to the next setup, and the backsight observed to the change point is used as previous foresight. The value is entered in the Backsights column at ⑥, then the foresight is observed, and in this case, it is on the original bench mark. The foresight value is entered in the Foresights columns at ⑦, and the BM identifier is noted in the Comments column. At this point, the leveling work is completed, and the work can be reduced and checked.

The first step is to calculate the elevation differences between each observation, in sequence, going down the page. The difference is calculated between the values on adjacent lines, going down and to the right as needed. So, the difference between the first backsight at ① and the first intermediate at ② is calculated and found to be +0.630, which is a rise [as the second observation is less than the first, see Figure 7-11(a)], and so, is placed in the Rises columns at ⑧. The elevation difference between the first and the second intermediate sights at ② and ③ is calculated and found to be −1.547, a fall [as the second value is larger than the first, see Figure 7-11(b)], which is placed as an unsigned value in the Falls column at ⑨. The elevation difference between the second and the third intermediate sights at ③ and ④ is calculated, is found to be −0.192, and is placed (without sign) at ①⓪. The elevation difference between the third intermediate sight at ④ and the foresight at ⑤ is calculated, is found to be +1.608, and is placed in the Rises columns at ①①. This completes the calculations for the first setup.

For the second setup, the elevation difference between the backsight at ⑥ and the foresight at ⑦ is calculated, is found to be −0.507, and so, is placed as an unsigned value in the Falls column at ①②. This completes the reduction of the observations to elevation differences, and it will be noted that this is done using values from adjacent lines only.

The checks are now done by summing the Backsights, Foresights, Rises, and Falls columns individually and placing the sums at ①③, ①④, ①⑤, and ①⑥, respectively. The smaller value of the Backsights and Foresights sums is placed beneath the larger, and the difference is calculated and placed at ①⑦. The smaller value of the Rises and Falls sums is placed beneath the larger, and the difference is calculated and placed at ①⑧. If the sum of the Rises is greater than the sum of the Falls, then there is an overall rise. If the sum of the

Backsights is greater than the sum of the Foresights, then there is an overall rise. The differences should be the same; otherwise, there has been an error in calculating the rises and falls or in calculating the checks.

At this point, the overall elevation difference over the entire flight of levels has been calculated and checked, so it can now be used for further calculations. This is the case for base leveling, where the overall elevation difference is what is required. However, if the elevations of the observed points are required, these can be calculated as follows.

The elevation of the starting point (the SR/62/G17 bench mark) is placed in the RL column on the same line as the backsight observation to the bench mark, at ①⑨. From this point, the elevations are calculated by applying the elevation difference at each line to the previous line's elevation, again keeping calculations between adjacent lines. So, the elevation of the C of G intermediate point is calculated by adding the elevation difference at ⑧ to the elevation of the SR/62/G17 bench mark at ①⑨, and the result is placed at ②⓪. This process is continued down the sheet. The elevation of the Edge of Road point is calculated by adding the elevation difference at ⑨ to the elevation at ②⓪, and the result is placed at ②①. The elevation of the Top of Drain intermediate point is calculated by adding the elevation difference at ①⓪ to the elevation at ②①, and the result is placed at ②②. The elevation of the CP point is calculated by adding the elevation at ②② to the elevation difference at ①①, and the result is placed at ②③. The elevation determined for the final point (the original bench mark) is calculated by adding the elevation at ②③ to the elevation difference at ①② and placing the elevation at ②④.

The start-and-end elevations are copied to below the final elevation and the difference calculated and placed at ②⑤. This elevation difference must match the elevation differences at ①⑦ and ①⑧; otherwise, there has been an error in the calculation of the elevations. It is critically important that all these checks are completed, so that the reliability of the observations has some degree of support. The thorough checks also give confidence that the reduction process has not introduced errors into the data. Note that the reductions of the intermediate observations are independently checked, although the observations themselves are not checked through this process. Note also the flexibility of this method, which can be used for base leveling, general topographic leveling, and construction surveying.

The computations that are carried out in the Rise and Fall Method can be summarized in sequence as follows (note that the upper and lower lines refer to each of the adjacent lines used for each calculation):

1. Rod observation (upper line) − Rod observation (lower line) = Elevation Difference (separated into Rises and Falls depending on sign),
2. Sum of Backsights − Sum of Foresights = Overall Elevation Difference,
3. Sum of Rises − Sum of Falls = Overall Elevation Difference,
4. Elevation (lower line) = Elevation (upper line) + Rise (lower line) − Fall (lower line),
5. End Elevation − Start Elevation = Overall Elevation Difference,
6. Identical calculation of the three Overall Elevation Differences (serves as a check).

Two interactive slideshows guide the reader through this method in detail in Bossler and Hazelton (2015, Chapter 4, Section 3). To read about the method in more detail, see Uren and Price (2010), Bhavikatti (2009), Bannister et al. (1998), and Whyte and Paul (1997).

Grid Leveling. This recording method is used for topographic surveys. A grid sheet is drawn up at a suitable scale for each setup, and the side shots are recorded on the grid sheet at their appropriate 2D location, as shown in Figure 7-16. The backsight and foresight are recorded and used for the overall reduction of a line of setups prior to reducing the individual sheets. The grid sheets are reduced on the sheet and can also be contoured. Picking a suitable scale allows the contours to be traced directly onto the finished map.

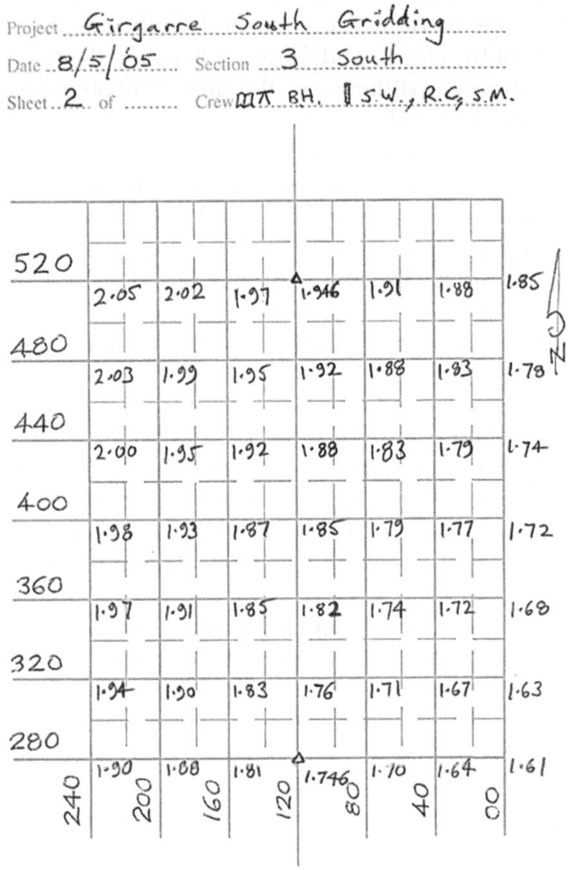

Figure 7-16. Grid-leveling recording sheet. On this sheet, a 40 m grid was used and the location of points is noted by the values in meters along the grid lines. After the level was set up at the center of the area to be measured, a backsight was read to the turning point on the lower edge (1.746 m), and then observations were made to rods carried to the various points (to 0.01 m), before the foresight was measured to the turning point at the upper edge (1.946 m). The sheet is reduced by calculating the height of collimation of the level and subtracting each observation, the reduced level being written on the sheet in a different color and at an angle.

The people holding the leveling rods walk the grid, carefully pacing the appropriate distances, and are checked by the level and the use of optical squares.

See Bossler and Hazelton (2015, Chapter 4, Section 4) for a detailed explanation of this method.

Basic Leveling Processes

Leveling processes can be divided into three categories based on the way that measurements are gathered. This is not the only categorization possible, but it does connect with different methodologies used for vertical measurement, horizontal location, and recording methods used.

Point Leveling. When the elevations of widely separated points are required, point leveling is used. This usually takes the form of base leveling. New or unknown points are connected to points with known elevations by flights of levels that are run twice. There are advantages to using the same turning points for both flights, as this allows fine-grained checking by a comparison of the elevation differences between the two runs. By calculating the mean of the overall elevation differences, the most probable estimate of the overall elevation difference can be calculated.

Point leveling is characterized by a network of interconnected points across a region. It is commonly used for establishing bench marks and control for later work and is commonly done to meet second-order criteria (see subsequent Adjustment Section) or better. Point leveling is discussed in Cole (2014), Ghilani (2021), Kavanagh and Mastin (2013), McCormac et al. (2012), Nathanson et al. (2017), Uren and Price (2010), Bhavikatti (2009), Bannister et al. (1998), Anderson and Mikhail (1997), Moffitt and Bossler (1997), and Whyte and Paul (1997), although it will be described by a number of different names.

Line Leveling. Much leveling is undertaken for support of construction work, and the points whose elevation is determined tend to fall along well-defined lines. Longitudinal sections and cross sections covering a largely linear construction project, such as a road, rail line, pipeline, channel or drainage line, are commonly used to determine the overall shape of topography on such a construction project as well as provide a means for setting out aspects of the construction.

The traditional method of location is termed *station and offset* in the United States. A centerline is established, usually along the centerline of the project, and distances are measured along this. When using feet in the United States, 100 ft is termed a *station*, and distances along the centerline are recorded as the number of stations, plus the distance past that station, in feet. So, a distance that is 2,684.23 ft from the start of the centerline, with the distance running along the centerline (a horizontal distance but running around curves), will be 26 stations plus 84.23 ft, recorded as 26 + 84.23. Distances to either side of the centerline are measured at right angles to the centerline and are termed *offsets*. These distances are usually less than 100 ft, and are recorded as the distance only, but noting whether the offset was to the left or right of the centerline (using "L" or "R"). Left and right are taken while facing the direction that the stationing distance along the centerline increases. In other parts of the world, the distance along the centerline is called *running distance* or *chainage*, and when using metric distances, it is given as a simple distance. So, the location (in feet) to an object in the survey may be 23 + 45.43, 56.34 R. In a metric surveying, it might be 2,598.34, 14.23 R (in meters). Note that this creates a kind of flexible 2D coordinate system, where the centerline follows the project, and the offsets are always measured orthogonal to the centerline.

A survey using line leveling will also require tapes to locate the points measured, usually one along the centerline, and another for measuring offsets. Distances may also be measured using the stadia capabilities of the level's telescope, or the distance measurement capability of a digital level, but the distance may not be along the centerline or the offset line. In the record of the leveling, two or three additional columns for Station and Offset (possibly Left and Right columns) may be added to provide the 2D location for each measured point.

Line leveling is discussed in Cole (2014), Ghilani (2021), Kavanagh and Mastin (2013), McCormac et al. (2012), Nathanson et al. (2017), Uren and Price (2010), Bhavikatti (2009), Bannister et al. (1998), Anderson and Mikhail (1997), Moffitt and Bossler (1997), and Whyte and Paul (1997), although it will be described by a number of different names, often varying with the individual purpose of the survey.

Area Leveling. This is used for topographic surveying, where a region is covered with points, commonly on a regular grid. Various techniques may be used to help with the 2D location of the measured points, such as the grid leveling recording method previously discussed. For area leveling where a large number of points are being measured, the linear recording methods of the Height of Instrument and the Rise and Fall methods are highly inefficient. Using the recording system in a digital level can also be problematic, as the 2D location of measured points needs to be connected to their elevation value, and this requires additional inputs or using a field book to connect record numbers in digital storage with location information.

Area leveling does not have to be done using a regular grid or lattice. Points may be selected at locations that allow the best shape of the terrain to be determined using the minimum number of points. Such collections of points are contoured by developing a triangulated irregular network. However, determining the 2D locations of these points can be difficult, leading to it being more commonly used with total stations.

ERRORS IN LEVELING

In leveling, there are many sources of errors. Some of these are provided in Figure 7-17.

To provide a simple set of criteria for base leveling (i.e., leveling that establishes bench marks), the Federal Geographic Data Committee (FGDC) has developed national standards and specifications for the procedural and equipment requirements to meet the various levels of the standard (FGDC 2004). The US national standards are similar to those around the world, as well as ISO standards for survey work (ISO 17123-2). Survey companies and local or state governments have other standards that may need to be met based on project requirements.

An extensive discussion of error causes, error propagation, precision of equipment, calculation of misclosures, and checking of leveling for errors can be found in Bossler and Hazelton (2015).

As discussed in Chapter 5, independent errors are cumulative and propagate as shown in Equation (5-14).

Errors also propagate through successive measurements, and as a result, the overall error in a flight of levels is proportional to the square root of the number of setups, which is usually proportional to the distance covered in base leveling [Equation (5-24a, b)].

For example, if we can read a metric leveling rod with a standard deviation of 0.001 m for a single observation over, say, 50 m, then the standard deviation (σ) of the elevation difference (ΔEL) obtained from a single setup can be calculated as follows:

$$\sigma_{\Delta EL} = \sqrt{\sigma_{BS}^2 + \sigma_{FS}^2} \tag{7-1}$$

$$\sigma_{\Delta EL} = \sqrt{0.001^2 + 0.001^2} = \sqrt{0.000002} = 0.0014 \text{ m}$$

If each setup has the same amount of error, then 10 setups, covering 1,000 m, then the error for that flight of levels has a standard deviation of

$$\sigma_{\Delta EL} = \sqrt{10(0.0014)^2} = \sqrt{0.00002} = 0.0044 \text{ m}$$

The allowed misclosure for double-run third-order leveling is 0.012 m for 1 km, according to FGCS (2004), which is approximately 2.5 to 3 standard deviations at the above level of reading precision. Therefore, leveling rods need to be read to this precision to allow a reasonable probability of achieving third-order leveling.

Figure 7-17. Some sources of error in leveling. Leibig's Law of the Minimum indicates that the least precise data place an upper limit on data quality.

As discussed in greater detail in Chapter 5, various other errors that occur in leveling are either blunders (and so can be detected by checking everything and closing the leveling), systematic errors (and so can be minimized by following proper practices) or random errors. Random errors have a distribution that has a standard deviation, which can then be included in the aforementioned calculations. Random errors can also be reduced by using better equipment, observing better measurement practices, and repeating measurements.

Many simple practices can be followed to reduce blunders, but relentless and effortful checking of the work is key. Wherever possible, observe important points twice, close leveling, and double-run all leveling that establishes new marks for any later use.

Refraction, a systematic error, is a constant concern in leveling, but the impact of adverse refraction effects can be minimized by avoiding long sights (e.g., <90 m), keeping backsight and foresight lengths balanced, and avoiding having a line of sight closer than 0.5 m (1.5 ft) to the ground at any point.

Blunders and systematic errors caused by the level being out of adjustment can be reduced by keeping sight lengths equal, as well as testing and adjusting the level at least once per day. Undertaking a two-peg test on the first setup of the day's work is a sensible and easy practice.

Errors in the leveling rod can include the scale not being zero at the point where the rod meets the ground (the footplate), the scale not being consistent over joints in the rod, telescopic rods not being fully extended at least to the line of sight, loose footplates and footplates that are not perpendicular to the leveling rod, rod bubbles out of adjustment, and rods bending or moving while being observed, often caused by wind.

Errors can also be introduced by poor choice of ground marks. Marks that are not pointed or that have multiple possible points are poor choices for turning points, as they can allow the rod to be placed at different elevations for consecutive foresight and backsight observations. Marks that are unstable, whether for the time required to move the level to the next setup, or a bench mark that moves significantly over years, introduce significant errors into leveling.

Poor observing practices can also introduce errors. Compensator bias in automatic and digital levels can accumulate over multiple setups if not corrected. Wind and vibration can move compensators off level, shifting the line of sight. Digital levels can have problems reading rods in poor light and dappled light, and through moving branches. Observing leveling runs in loops, rather than double-run sections, can introduce significant errors in later calculations and adjustment, as well as making it much harder to find errors in the original work.

When reducing leveling, there is a significant amount of subtraction involved, and this arithmetic operation is more susceptible to error than addition. Reducing leveling with a calculator can introduce finger errors. Leveling is a highly error-prone activity, so all checks must be completed before observations can be accepted.

A detailed study of the causes and nature of the many errors that can occur in leveling will repay the investment of time many fold, as being able to avoid and minimize errors, as well as catching them quickly when they inevitably occur, is an invaluable skill.

ORTHOMETRIC AND DYNAMIC CORRECTIONS

At the beginning of this chapter, it was stated that leveling can be used to measure elevation differences directly. This is actually not quite true, even when the error sources discussed previously are taken into account. Even the term *elevation* is ambiguous. Usually, it is used to mean *orthometric height*, which is the vertical distance above the adopted datum, that is, the length of the (curved) plumbline. For example, the North American Vertical Datum of 1988 (NAVD 88) is usually given as orthometric heights, specifically *Helmert orthometric heights*. But NAVD 88 can also be given as *dynamic heights*, and, in fact, NGS datasheets for all leveled marks give both NAVD 88 orthometric and dynamic heights. A dynamic height is actually not a height at all (in the geometric sense of the word) but rather a measure of gravity potential energy recast in units of length. A change in dynamic height gives true change in hydraulic head, and they are used in applications where that is important over large areas, such as the Great Lakes. See Chapter 2 for more information on orthometric and dynamic heights and Hofmann-Wellenhoff and Moritz 2006, Chapter 4, for additional details.

Figure 7-18 is a schematic illustrating dynamic and orthometric heights. It shows a hypothetical island with five bench marks (BMs), two at "sea level" (BM 1 and 5, with dynamic and orthometric heights of 0), one at the highest point (BM 3), and two partway between sea level and the high point (BM 2 and 4). BM 1 and 5 are on a single specific level (equipotential) surface that is the geoid, which corresponds to mean sea level. BM 2 and 4 are on the same level surface, so they must have the same dynamic height. However, because the level surfaces are not parallel, the length of the plumbline (which is the orthometric height) is not the same, so they have different orthometric heights. The plumblines are perpendicular to all level surfaces and, therefore, must be curved because the level surfaces are not parallel (as shown).

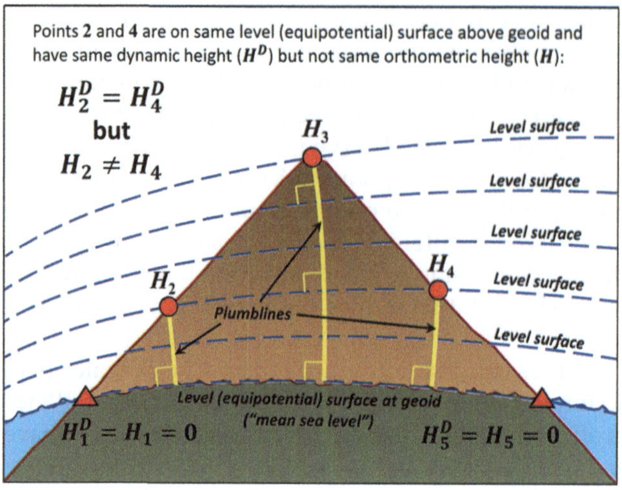

Figure 7-18. Schematic illustrating dynamic and orthometric heights.

Leveling by itself will not yield true differences in orthometric or dynamic heights. The reason is that gravity equipotential surfaces at the instrument and rod locations are not parallel. The plumbline passing through the level instrument is usually not parallel to the one passing through the rod. This is directly analogous to the error caused by not holding a rod perfectly vertical. However, in this case, the rod actually is vertical, but vertical at the location of the rod is slightly inclined with respect to vertical at the instrument. These differences in vertical vary with location, and the net effect is that they cause systematic errors in leveling. Because of this, leveling is a path-dependent process. In other words, leveling will give different results if different leveling routes are taken. This makes it impossible to use leveling to determine unique "elevations" (e.g., orthometric, dynamic, or other "physical" gravity-based heights).

Getting different results depending on the leveling route is obviously undesirable. In many routine leveling applications over small areas, this effect is small and can often be ignored (but not always). For high-accuracy leveling over large areas (such as control surveys), it should be taken into account. This can be done in two ways. One is to convert leveled height differences into geopotential number differences and then convert the final geopotential numbers into the desired type of height. The other is to apply an orthometric or dynamic correction, depending on the type of height desired. The latter approach is covered here.

Both methods for getting orthometric and dynamic heights require knowing the acceleration of gravity at the bench marks along a leveling route. But making gravity measurements as part of leveling is completely impractical in most situations. Fortunately, in the United States, this is not necessary, because NGS has created a model of surface gravity for CONUS and Alaska that is specifically intended for use with NAVD 88. This means that it is possible to get a gravity value anywhere in CONUS or Alaska and to use that value to compute orthometric and dynamic corrections. The NGS utility used for computing NAVD 88 modeled gravity is available on the "NAVD 88 Gravity Computations" webpage (NGS 2022a).

With NAVD 88 surface gravity values in hand, both orthometric and dynamic corrections can be computed. The corrections can be positive or negative, and they are added to the leveled height difference to obtain the corrected value. The orthometric correction from A to B [Equation (7-2)] can be computed as

$$OC_{AB} = \frac{[K(g_A - g_B) - 2\Delta n_{AB}][2H_A + \Delta n_{AB}]}{2(Kg_B + H_A + \Delta n_{AB})} \tag{7-2}$$

where

H_A = Orthometric height at mark A,
Δn_{AB} = Leveled height difference from A to B,
g_A and g_B = Surface gravity at adjacent marks A and B (m/s^2, or ft/s^2 if using feet), and K = 2,358,000 s^2 is a constant factor for computing Helmert orthometric heights (based on a topographic density of 2,670 kg/m^3), and the orthometric height at B is $H_B = H_A + \Delta n_{AB} + OC_{AB}$.

The dynamic correction from A to B is given by

$$DC_{AB} = \left(\frac{g_A + g_B}{2\gamma_{45}} - 1\right)\Delta n_{AB} \tag{7-3}$$

where γ_{45} is the 9.806199 m/s^2 is normal (modeled) gravity on the GRS 80 ellipsoid at 45 degrees latitude (an arbitrary value used by convention).

Because leveling is path-dependent, these corrections should always be applied to adjacent marks along a leveling route. The orthometric correction can be computed from Equation (7-2) using NAVD 88 modeled gravity. Alternatively, NGS has an online tool (NGS 2022b) that computes the leveled height difference between marks (i.e., removes the OC). The tool can be used by manually entering gravity values for two locations. But if the marks of interest are NGS-published leveled bench marks, the permanent identifiers (PID) of the marks (available on NGS datasheets) can be entered instead. The tool then automatically takes the published orthometric height and modeled gravity of the two marks and removes the OC between the marks. For example, using NGS bench marks M 504 (PID = FQ0543) and L 504 (PID = FQ0544), the tool gives a leveled height difference from M 504 to L 504 of 15.358 m. But the orthometric height difference is 15.377 m. This corresponds to $OC = +0.019$ m. This is quite large, considering that the marks are only 1.9 km apart, larger even than the maximum allowable section misclosure for third-order leveling according to FGCS (2004) (i.e., 0.012 m $\times \sqrt{1.9} = 0.017$ m). Using Equation (7-3) with the NAVD 88 modeled gravity, the dynamic correction is $DC = -0.022$ m, slightly larger in magnitude than the OC and opposite in sign. Incidentally, these two marks are separated in a mostly east–west direction. This should lay to rest the common misconception that orthometric corrections apply only in the north–south direction. The reason for this belief is discussed in the next paragraph.

Where gravity values are not available (whether modeled or measured), the actual orthometric correction cannot be computed. Instead, a crude estimate of the nonparallelism of level surfaces can be used, based on the normal ellipsoid. Applying this correction gives normal orthometric heights. However, these corrections are extremely small and apply in a north–south direction only. Normal orthometric heights were used prior to NAVD 88 but are used only by NGS today in small areas where modeled surface gravity is not available, such as for the Puerto Rico Vertical Datum of 2002 (and other small island vertical datums). The normal orthometric correction is considered obsolete. Because of this, as well as its limited applicability and crudeness, this type of correction is not covered here. In addition, normal orthometric heights will not be used at all for the future North American-Pacific Geopotential Datum of 2022 (NAPGD2022).

ADJUSTMENT

The purpose of adjustment is twofold: to provide a consistent set of adjusted measurements, so that all information derived from the measurements is consistent, and to provide a maximum probability estimate of the true values of the measurements. Adjustment does not provide the "true" or "correct" values of the measurements, merely the most probable estimate of them.

Leveling measurements can be considered to come in three basic configurations: double-run leveling; circuit leveling; and networks, as shown in Figure 7-19.

Double-run leveling that uses common points on the two runs can be adjusted by taking the mean of the pairs of elevation differences between common points. As the mean provides the maximum probability estimate of the measured quantity, it is the preferred method. By keeping the common points reasonably close, the method also allows far easier error trapping.

With the ready availability of adjustment software, any network beyond the simplest leveling should be adjusted using least squares (see Chapter 5). Care must be taken to develop good estimates of the errors in each measurement, as these are critical to achieving a high-quality adjustment result. For more details of least-squares adjustment specifically tailored to leveling, including calculating the error parameters, see Bossler and Hazelton (2015), particularly Chapter 5 for errors and error propagation, and Chapter 7, Section 4 for least-squares adjustment. Chapter 5 of this manual also provides equations for constructing a stochastic model for the adjustment.

Adjusting the leveling around a loop or circuit, if there are no other considerations, can be done more crudely using linear interpolation. As stated previously and in Chapter 5, the errors in leveling propagate as the square root of the number of setups, which is not a linear process, so linear interpolation should be used with care. Before considering linear interpolation, ensure that the leveling meets the requirement of the appropriate order of leveling, especially the allowable misclosure. This will provide an indication that blunders and systematic errors have been removed or minimized and that only random errors should remain. The misclosure is then distributed around the loop, and as this misclosure should be small, the corrections made to the measurements should also be small.

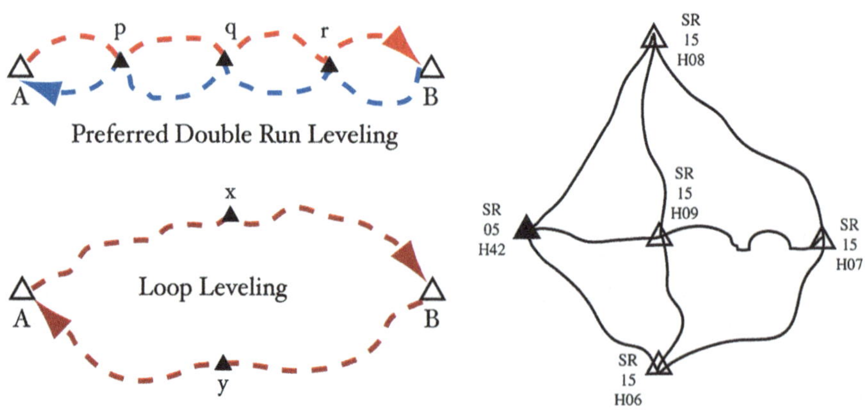

Figure 7-19. Leveling configuration. (Top left) Double-run leveling, which is the preferred method for base leveling. (Lower left) Loop or circuit leveling. (Right) A leveling network.

INTRODUCTION TO TOTAL STATIONS

The modern total station is a surveying instrument that integrates an angle-turning instrument (an electronic theodolite), a distance measuring instrument [electronic distance measurement (EDM)], and a data recording and computation capability. By integrating all these capabilities in a single instrument, it is possible to locate points in 3D rapidly and with quantifiable precision. The computational capabilities of modern total stations also allow a list of the coordinates of the points to be set out to be uploaded to the instrument and for the instrument to calculate the appropriate angles and distance to locate the point.

Robotic capabilities have been added to total stations over the last two decades, allowing robotic total stations to track prisms and operate without direct operator control. This has allowed one-person operation, with the surveyor controlling the instrument by a remote unit on the prism pole. (Although a one-person operation is possible, the authors recommend a two-person survey crew for increasing safety.)

In this discussion of total stations, their general construction and operation is dealt with, together with the types of measurements taken and their associated errors, computational and storage capabilities, some common accessories and additional capabilities, and how to operate one.

DESIGN OF MODERN TOTAL STATIONS

Before total stations, EDM equipment tended to be large and heavy. They were used for control and separately from the angle-turning instrument. As the EDM became smaller, instruments were mounted on the theodolite, albeit as a separate unit. As electronic theodolites were developed, it was a small step to connect the EDM to the theodolite's electronic system and then to integrate the EDM within the theodolite, as shown in Figure 7-20. The first true total station was the Zeiss Reg ELTA 14 (ELectronic TAcheometer) in 1968. Hewlett Packard used the term *Total-Station* to describe their HP 3810 stand-alone EDM unit with an integrated tilt sensor and an HP-35 calculator microprocessor and then their HP 3820 instrument (a true total station) in 1977. The name stuck and became the generic term for these instruments.

Discussion of the use of total stations can be found in all introductory surveying textbooks such as Cole (2014), Ghilani (2021), Kavanagh and Mastin (2013), McCormac et al. (2012), Nathanson et al. (2017), Uren and Price (2010), Bhavikatti (2009), Bannister et al. (1998), Anderson and Mikhail (1997), Moffitt and Bossler (1997), and Whyte and Paul (1997). As use of a total station in part builds on the practices for angle-reading instruments, discussion of theodolites in older texts remains valuable.

Tacheometry is a method of measuring distances optically using a theodolite and a leveling rod, and this allowed 3D locations to be determined quickly. Specialist tacheometers were once built to allow faster work (tache = "fast"; metry = "measurement"). They have been displaced by total stations in recent times and are now obsolete.

Total stations have a horizontal circle in the lower part of the instrument that allows electronic measurement of the angle turned around the instrument's vertical axis. They have a vertical circle in one of the standards supporting the telescope, and this allows the angle of the telescope to be measured electronically. The distance measurement system is built into the telescope and standards and allows distances to be measured along the line of sight. Some type of keyboard and display screen is included, together with internal storage and calculation capabilities. Instruments can also be connected to a data recorder/controller, by cable or by wireless, and aspects of instrument operation managed from the controller.

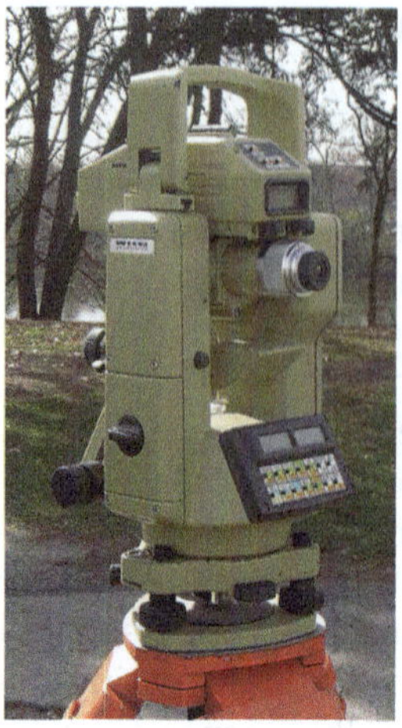

Figure 7-20. Wild T-1602 electronic theodolite and DI-1001 EDM unit, an early combination instrument.

TOTAL STATION EQUIPMENT

Tripods

See the discussion in the section under leveling.

Data Collectors

Total stations generate significant amounts of data. With the availability of large amounts of low-cost data storage, total stations record not only the raw observations but also the derived information, such as coordinates. In addition, status information about the total station, such as settings for temperature, pressure, and atmospheric corrections, whether the curvature and refraction corrections were applied, point description codes, and any other status information, may also be recorded with each observation. Information about each point occupied by the instrument may also be included. This adds up to significant amounts of data stored, often within the instrument. For many modern instruments, even digital photos can be taken and stored.

As the instrument may have a small and awkward keyboard, it can be difficult to manage data within the instrument. One solution is to use an external data collector, which allows a better user interface, larger keys, and the ability to add a calculator and/or alphabetical keypad to the system. An external data recorder is shown in Figure 7-21.

Data collectors, whether internal or external, also allow data to be taken into the field from a CAD-based design and allow points to be set out from wherever the instrument may be located. Once the instrument's position can be calculated, using data measured in the field, the angles and distances required to place the points can be calculated quickly and easily. A range of other calculations can be done in the field, using only the data collected and the total station.

Prisms

Also known as *reflectors*, prisms are glass corner-cube reflectors that return light and IR radiation from the total station along the same path, allowing the distance to be measured. In addition, the prism provides a point of aim for the total station's angle measurements.

Figure 7-21. Topcon 360 degree prism and remote data collector on prism pole with bipod.

Prisms are designed to add a certain amount of length to the measured line, termed the *prism offset*. In some cases, this is designed to be zero, whereas in others, it may be 34.4 mm (to allow the prism to be balanced over the ground point, as in standard Leica prisms), 23.1 mm (for the Leica 360° prism), 17.5 mm for some micro-prisms, and zero for the 360° prism in Figures 7-21 and 7-22. Note that as the prism is as much a part of the distance measurement system as the total station, calibration of the entire system needs to be done to ensure that the calibration is done correctly. The prism offset is also determined by the instrument offset at the total station, as the objective is to get the overall system as close to zero as possible. Total stations can be set to automatically correct for the prism offset, as discussed later.

In many cases, a single prism will be used, mounted on a tripod. Longer distances may need multiple prisms placed on the tripod to ensure a sufficient return signal. For topographic work, a single prism can be mounted on a pole and carried around the

Figure 7-22. Trimble 360 degree prism. Note the use of radio for hands-free communication.

landscape to points of interest. The prism needs to be pointed at the total station, but a 360° prism can also be used to ensure that the prism is always facing the instrument. The possible distance that can be measured is less, as the prism is smaller, but for most topographic surveys, it is sufficient.

Orientation

Total stations can be placed in one of two orientations. In theodolites, the vertical circle was often obvious on one of the standards, and one of the orientations had this to the left of the telescope (as seen from the eyepiece end), and was termed *face left*, *direct*, or *Face I*. The opposite orientation had the circle to the right of the telescope and was termed *face right*, *reverse*, or *Face II*. Figure 7-23 shows instruments in each position. Modern total stations may not have the size distinction but may have almost perfectly symmetrical instruments, including dual keypads.

The simplest way to think about the orientation is that one of them, which can be called *face left*, *direct*, or *I* (FLDI for short in this chapter), will have the vertical angle around 90° with the telescope approximately horizontal, whereas the other orientation is called *face right*, *reverse*, or *II* (FRRII for short), and will have the vertical angle around 270° with the telescope approximately horizontal. In the instrument's display, there will usually be an indication of which orientation the instrument is in (in its particular description, often I or II), and this may also be recorded with each observation.

Measuring using both orientations is important because doing so in the correct sequence (discussed previously) will minimize internal errors in the instrument (e.g., collimation errors), producing a higher quality set of observations.

Figure 7-23. (Left) Leica T2 theodolite in face left, direct or I orientation (FLDI). (Right) Leica T16 theodolite in face right, reverse, or II orientation (FRRII).

TOTAL STATION EXTENSIONS

Total stations have the ability to support significant additional capabilities such as the following.

Reflectorless

Some total stations can measure distances without requiring a reflector, for example the instrument shown in Figure 7-24. Although the longest distances that are measured by most total stations in reflectorless mode are between 100 and 400 m (300 to 1,200 ft), these seem to be sufficient for common work. (Some reflectorless systems can measure up to 3,500 m but have large power demands.) This capability allows a total station to measure a far wider range of objects and far more complex structures far more easily than was possible in the past.

Reflectorless capabilities are limited by the reflectivity of the surface to which observations are made. Although a bright white surface allows longer reflectorless distances to be made, darker surfaces can severely limit the distances possible. For example, one instrument that the author has used can measure 100 m to a bright white surface, but cannot manage 8 m to a dull, matt black surface. Instruments with coaxial EDM and optics are far better for reflectorless work, as offset EDM working at short ranges can miss the point where the telescope's crosshairs are pointed and, therefore, measure the wrong object.

Robotic

High-end total stations may include a robotic capability. A robotic total station has its horizontal and vertical motions controlled by stepping motors, allowing fine movements to point the telescope precisely. Various systems are included in the total station to allow the instrument to track a prism as it moves, to find the prism if it loses contact with it, and to center the telescope on the prism.

The purpose of a robotic total station is to allow the total station to operate largely autonomously, whereas a single person moves the prism around and manages the total

Figure 7-24. Leica TCR 307 reflectorless total station.

station from the prism pole, using equipment as shown in Figures 7-21 and 7-22. The operator (at the prism) manages the total station remotely, recording the observations and adding feature codes to observations. If the total station includes imaging capabilities, the view through the telescope can be relayed to the screen of the remote data collector, which allows finer control by the operator.

As part of construction machine control systems (see Chapter 12), robotic total stations are often used for tracking and controlling the construction equipment. Robotic total stations are also used for structure deformation tracking and landslide monitoring.

Imaging

Some total stations include one or more cameras with a view through the telescope. Some modern instruments, like the Trimble SX-10 and SX-12, make use of imaging and do not even have an eyepiece for looking through the telescope. Two cameras can be used to give different fields of view: a narrow view for finely pointing the telescope; a wide-angle view for locating points. Images can be used to direct the total station for subsequent measurements, as the image records directions in 2D. Images can also be recorded to verify what point was actually measured and can record a great deal of additional information during a survey. Images do increase the storage required.

Targeting

Robotic total stations often have the ability to target the instrument on a prism. The instrument can center itself consistently on the prism, allowing precise measurements. Although the crosshairs may not be exactly centered on the prism, it will be consistently placed on the prism.

Scanning

In the last several years, total stations with laser scanner capabilities have come onto the market. The first appears to have been the Leica "MultiStation," which allowed the operator to select an area on the total station's image touch screen and direct the total station to record a grid of measured 3D points across the selected area. Although the scanner is somewhat slow, it is designed for fairly small areas. The Trimble SX10 and SX12 total stations include a faster scanner, running at up to 26,600 points per second for distances up to 600 m and the Leica Nova MS60 can scan up to 30,000 points per second for distances up to 1,000 m (see Chapter 8 for information on terrestrial laser scanners).

Global Navigation Satellite System Connection

Many total stations allow a GNSS receiver to be connected to the instrument. This is usually done by attaching the antenna to a bracket on the total station's standards. The antenna carries its own batteries, and a Bluetooth connection allows data transfer to the data collector. If the total station is being used for several minutes to measure points, this may be long enough to allow a rapid-static position to be determined by the GNSS receiver. This can tie together the control for the overall survey, especially with 3D location. The total station can allow high-precision azimuths to be carried through the survey, as azimuth is difficult to derive from GNSS locations to a level of precision that matches the total station's capabilities.

Programmable

Higher-end total stations sometimes have the capacity to program the total station to undertake specific operations. These operations may be computational or may be directing a specific set of instrument operations. It is possible to have a robotic total station act as a laser scanner, under programmatic control; programmed total stations can undertake repetitive deformation measurement work to a set of predetermined points and continue this indefinitely without human intervention. A program to have a total station reduce a set of sun observation and produce an azimuth directly has been produced.

SETTING UP THE TOTAL STATION

Setting up a total station has some similarities with setting up a level but includes the added complication of setting up exactly over a point. The setting-up process can be considered in a number of stages. Setting up the reflector or prism on a tripod follows a similar course.

Tripod

Open the tripod and set the length of the legs so that the instrument will be at a comfortable height. Allow for the need to have a clear line of sight to the various targets and objects that need to be measured. If there is any amount of slope, place the tripod with two legs on the downhill side of the ground mark and one leg uphill.

At this time, place the legs in approximately the right position, as the tripod will still need to be shifted to be over the mark. Look for places where the tripod's feet could be either pressed into the ground, or where there may be cracks or other discontinuities in smooth hard surfaces. Try to place the center of the tripod head over the mark and have the tripod approximately level.

Instrument

The instrument is now placed on the tripod and *immediately* secured with the tripod screw. *Never* leave the instrument sitting on the tripod without either keeping a firm grip on it, or having it securely screwed onto the tripod. Adjust the foot screws of the tribrach to about the middle of their run and the instrument to about the middle of the tripod. Align the tribrach so that its triangular shape lines up with the triangular shape of the tripod head (if the tripod has a triangular-shaped head).

If the task is to set just a tribrach, such as for a prism, the tribrach can be placed without an instrument in it, in most cases. The next task is to place the instrument vertically above the ground mark.

One useful characteristic to note is that for the average tripod setup, changing the length of one of the legs will move the tripod head approximately on the surface of a sphere with its center on the ground in the middle of the tripod's feet. This makes leveling and centering the instrument much simpler when using an optical or laser plummet.

Optical Plummet

Looking through the optical plummet, find the ground mark in the field of view. Carefully lift two of the tripod legs and move the tripod so that the ground mark is reasonably close to the optical plummet's crosshairs or other aiming mark (sometimes a small circle). It does not matter if the tripod is disleveled a little in the process. If there is some difficulty in getting close to the mark, the foot screws can be used to move the mark the last part of the way.

Once the optical plummet is correctly sighted onto the ground mark, press the tripod shoes firmly into the ground. Check that the optical plummet is still approximately on the ground mark, and if necessary, adjust with the foot screws. If the optical plummet is within a centimeter or two (a half inch or so) of the ground mark, it is satisfactory at this stage and can be left as is.

Watching the small circular bubble on the tribrach, adjust the length of the tripod legs so as to bring the bubble to the center of the circle on the level vial. This may take some fine adjustment of the legs to get right.

At this point, the optical plummet should show that the instrument is close to being centered over the ground mark, while the tribrach circular bubble is very close to the middle of the circle. If this is not the case, adjust the legs until the bubble is centered. If the ground mark is well away from the aiming mark in the optical plummet, either move the legs a little to bring the ground mark to the aiming point of the optical plummet or adjust the leveling screws if the ground mark is fairly close to the plummet's aiming point.

When the instrument is close to being leveled and over the ground mark, loosen the tripod head screw a little, and looking through the optical plummet, gently slide the tribrach (and instrument) on the tripod head until the optical plummet's aiming mark is exactly over the ground mark. Do not rotate the tribrach, as this will dislevel the instrument and throw everything off. Move the tribrach forward/back, left/right, as seen from the observer looking into the optical plummet.

There follows a cycle of adjusting the foot screws a little to get the bubble exactly centered and adjusting the position of the tribrach to be directly over the ground mark. When the optical plummet shows the ground mark as close to the aiming point as it is possible to get, and if the tribrach is level, the job is done and the tripod screw can be tightened firmly.

Laser Plummet

Not all total stations have laser plummets, but if the instrument has one, switch it on. This should project a small red spot onto the ground beneath the instrument. The brightness of the spot is usually adjustable. Pick up and move two of the tripod's legs until the spot is on or very close to the ground mark. Then adjust the length of the tripod's legs until the circular bubble on the instrument's tribrach is centered. The laser spot should still be close to the ground mark, but if it is not, move the legs a little to bring it onto the mark (or move the laser spot by adjusting the tribrach foot screws) and relevel the instrument by adjusting the length of the tripod legs.

With the instrument approximately level and centered over the mark, fine-tune the leveling with the foot screws, and then loosen the tripod head screw slightly and slide the tribrach (avoiding rotating it) on the tripod head until the laser spot is in the middle of the ground mark. The cycle of adjusting the leveling of the tribrach with the foot screws and moving the tribrach on to the ground mark continues until the tribrach is level and the instrument is centered.

Final Leveling

With the instrument approximately leveled and over the ground mark, the final leveling can be done. The circular bubble on most tribrachs is an approximate device, designed to show when the tribrach is level to within perhaps ± 5 arcminutes or better. Five arcminutes at an instrument height of 1.5 m (5 ft) is 0.002 m (0.09 in.), which means that from the

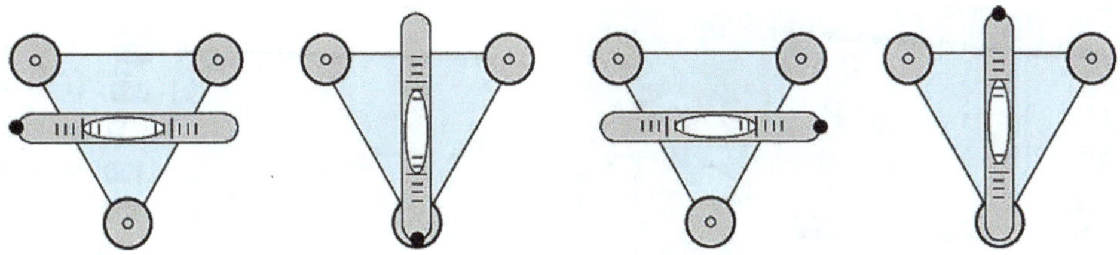

Figure 7-25. Leveling an instrument with a physical bubble tube. The sequence as described usually goes left to right as shown in the diagrams. When the instrument is level, the bubble will remain in the same location in the bubble tube, regardless of the instrument's orientation. This location will not necessarily be the center of the bubble tube. The small black dot depicts the adjustment screw at one end of the bubble tube, so that it can be seen that the bubble remains at the same location within the bubble tube. Checking the bubble location at various orientations is the final step of the leveling process.

average tripod setup height, the optical plummet or laser plummet can get the instrument vertically above the ground mark to the millimeter level quite easily.

Using the plate bubble on a total station, it is quite easy to get the instrument to within ±5 arcseconds of being vertical, possibly even better. This is one or two orders of magnitude better than can be achieved using a circular bubble. As a consequence, the circular bubble is an approximate tool for leveling the instrument, getting the line from the tribrach to the ground mark approximately vertical over that distance, and allowing the placement of instrument and target as close to vertically above the ground mark as is usually required. The total station, on the contrary, needs to project lines of measurement out some thousands of feet or meters from the instrument and, therefore, requires much higher levels of precision in its verticality.

Leveling the instrument with the plate bubble is shown in Figure 7-25 and involves lining up the bubble tube parallel to two of the foot screws, centering the plate bubble with those two foot screws, then rotating the instrument 90° so that the bubble tube is lined up with the third foot screw, and centering it again with that one foot screw. The instrument is then rotated 90° so that it is parallel to the first pair of foot screws, but the instrument is facing in the opposite direction. The bubble is then brought to midway between the bubble positions on the two pointings in this orientation. The same midway point is sought for the instrument orientation over the single foot screw. When the bubble remains in the same position in the bubble tube in all orientations of the instrument, even those not over the foot screws, it is properly leveled. With the instrument properly leveled and positioned vertically above the ground mark, measurement work can begin.

Many modern total stations include an electronic level for fine leveling rather than a physical bubble tool, and it is displayed on the screen during the leveling process as some form of "bubble" image. This may be a pair of orthogonal "bubble tubes" on screen, or a large circular "bubble." Both types are shown in Figure 7-26, but the circular bubble tends to be quicker and easier to use. Adjust the foot screws a fine amount until the bubble on the screen is level, and it is important that the instrument is rotated 180° and the leveling checked. Electronic levels are not perfect, and some difference is common in the bubbles.

Heights

If there is any vertical location work to be done, the height of the instrument's trunnion axis above the ground mark, as well as the height of the center of any prism above its ground mark, needs to be measured. This can be done in several ways. A pocket tape can be used,

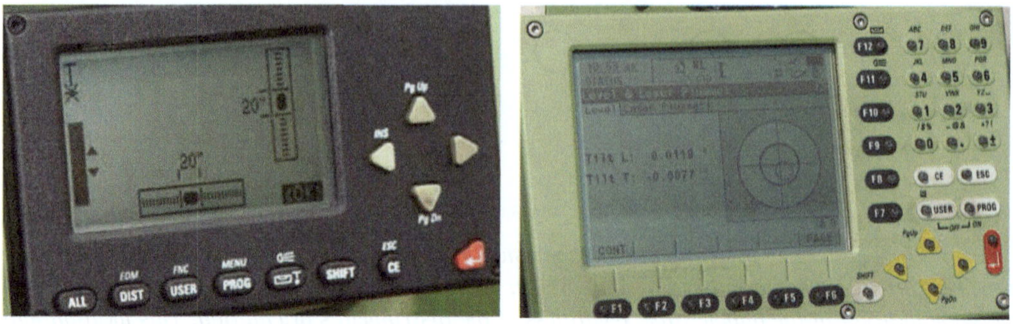

Figure 7-26. (Left) Two linear bubbles on the Leica TCR 307 total station. (Right) Circular bubble and tilt readings on the Leica TPS 1,200 total station.

but this is less precise because the tape has to bend around the tripod head to reach the ground mark. However, there are attachments that allow a tape to be suspended beneath the tripod at a set distance below the prism or instrument trunnion axis. The tape can then be drawn straight down to the ground mark. For some instruments, there are specific tapes that can be attached to one edge of the tribrach and pulled directly to the ground mark. The graduations on the tape are corrected to provide the vertical distance from the mark to the trunnion axis. Measuring poles can also be threaded through suitable tribrachs to give a vertical measurement.

Zeroing Instrument

Many instruments need to be turned through a zero point before they can begin measuring angles. Both the horizontal and vertical circles may need this done. Once done, angles start to register on the instrument's screen.

For topographic surveying, data about the instrument point need to be entered, so that the data collector has the information stored. Usually, there is a screen to allow this information to be entered either on the instrument or on the data collector. The instrument may also need a specific azimuth set to a specific point, such as the last instrument point. Usually, a key sequence does this once the instrument is pointed at the appropriate target.

BASIC TOTAL STATION PROCEDURES

With the advent of GNSS, the need for high-precision angle measurement in geodetic work has decreased markedly. Today, the highest-precision instruments are largely used for high-precision industrial metrology. Despite this decrease in angle measurement, it is still important to be able to measure angles precisely and reliably. The most common means of locating a total station's position, so that it can be used for locating many additional points, is by means of a traverse or similar control network through all the instrument points.

Measuring Horizontal Angles

Although measuring a single angle at a point is a critical part of traversing and measuring control networks, it also occurs in other applications of angle-measuring instruments. Using a total station to measure a single angle is a relatively simple process using the standard Face Left/Swing Right; Face Right/Swing Left method.

A single angle can be measured in either of two ways: measuring the larger angle between the two directions or measuring the smaller angle. Which is chosen will depend on a number of factors, but two of the more important factors are which of the two sights is the clearer one and which is the longer one. The clearer sight should be chosen as the backsight. If there is little difference between the two sights, it is usually better to choose the longer sight. If one of the sights is quite short, especially if it is so short that the size of the target will subtend a significant angle at the instrument, the longer sight is preferable. The objective is to have the best possible sight as the "backsight" (the starting direction for the angle), as this allows the greatest precision for all subsequent measurements made at that point.

For the simplest observation, set the instrument to read 0° 00′ 00″ when pointing at the backsight target, with the instrument in the FLDI position. Check that everything is correct (telescope pointing, angle reading, instrument level), call out the horizontal circle reading to the note keeper or record it in the data collector, and then loosen the horizontal clamp and rotate the instrument clockwise (to the right) until the telescope points at the second target. Gently clamp the instrument and move the crosshairs on to the target. (Note some modern total stations do not require clamping.) Read the angle, check that the telescope is pointing correctly and leveled, and call out the reading to the note keeper (for example, 123° 45′ 50″) or record it in the data collector.

Loosen the instrument's clamps and turn the telescope through 180°, so that the instrument is in the FRRII position. Rotate the instrument counterclockwise (to the left) until the telescope is pointing at the foresight again. Gently clamp the instrument and move the crosshairs onto the target. Read the instrument, check that it is pointing correctly and leveled, and call out the reading to the note keeper (for example, 303° 45′ 58″), or record the value to the data recorder. Unclamp the instrument, rotate it counterclockwise back to the backsight. Gently clamp the instrument and move the crosshairs onto the target. Read the instrument, check that it is pointing correctly and leveled, and call out the reading to the note keeper (for example, 180° 00′ 06″), or record the observation to the data collector.

If recording manually, it is a good practice for the note keeper to call out the measurement as soon as it has been written down, reading it back from the field book, rather than reciting it from memory. The instrument operator should check the read-back value against the observed measurement. If there is a difference, it must be resolved before the instrument is moved to the next direction.

Although the aforementioned discussion suggested setting the instrument to read 0° 00′ 00″ at the first backsight target, such "zero-setting" with a modern total station is technically not required. Today's total stations are directional instruments. In other words, they can read a horizontal direction to the backsight (for example, 13° 12′ 06″) and then read a horizontal direction to the target (for example, 136° 57′ 56″). The total station will then difference the target direction with the backsight direction and compute and store the horizontal angle (in this case, equal to 123° 45′ 50″). Two horizontal directions can be readily differenced to find the horizontal angle, and zero-setting of the instrument is not required.

Reducing Horizontal Angles

At this point, the observations for the set of measurements are complete. However they were recorded, they can be represented in a tabular format, such as in Table 7-1. If they were being recorded manually, they can be recorded directly into this tabular format.

In Table 7-1, the data were entered as they were measured, in the top two lines. The next step is the reduction of the angles. On the third line, the mean value of each direction is calculated, but the 180° difference is omitted. In general, the FLDI degrees value will be adopted, whereas the minutes and seconds will be used to calculate the mean value.

Table 7-1. Recording Horizontal Angle Observations.

At Pt. 6	Backsight (5)	Foresight (7)
L/R	0° 00′ 00″	123° 45′ 50″
R/L	180° 00′ 06″	303° 45′ 58″
Means	0° 00′ 03″	123° 45′ 54″
Abstract	0° 00′ 00″	123° 45′ 51″

So, 0° 00′ 00″ and 180° 00′ 06″ are considered to be 0° 00′ 00″ and 0° 00′ 06″, and the mean is 0° 00′ 03″. Similarly, 123° 45′ 50″ and 303° 45′ 58″ have the 123° used (checking that 123° and 303° differ by 180°), and the 45′ 50″ and 45′ 58″ are used to obtain a mean value of 123° 45′ 54″.

The final step in calculating the angle is to abstract the difference between the two directions. In this case, simply subtract the mean for the first direction from the mean of the second direction. So, 123° 45′ 54″ – 0° 00′ 03″ = 123° 45′ 51″. This is written into the fourth line of the tabulation.

Other patterns for recording are possible, even graphic representations. Data recorded in the data collector may have other ways to represent them. But the essential method is the same.

It is also perfectly sensible to set the initial backsight to be the azimuth of the line to the backsight point, as determined by measurements so far, and, therefore, transfer the azimuth to the foresight line. The same observational approach is used. The azimuth can be used, but bear in mind that the final adjustment of the traverse may alter the azimuths slightly, therefore requiring changes to all the observations. In this case, there may be no real advantage to carrying the azimuths. Weigh the advantages of having approximate azimuths in the field, and if there is no pressing need, set 0° 00′ 00″ as the backsight and correct all the horizontal angle observations afterward.

For more complex networks, many directions can be read at a single point in one large set of directions. For more details of this, see the various surveying textbooks such as Cole (2014), Ghilani (2021), Kavanagh and Mastin (2013), McCormac et al. (2012), Nathanson et al. (2017), Uren and Price (2010), Bhavikatti (2009), Bannister et al. (1998), Anderson and Mikhail (1997), Moffitt and Bossler (1997), and Whyte and Paul (1997). The general procedure is as shown in Figure 7-27.

Measuring Vertical Angles

Traditionally, the measurement of horizontal and vertical angles and distances were separate, simply because the measurement technologies and computing techniques tended to use them separately. However, total stations allow everything to be measured in a single pointing, so this may be the best way to work, especially if the observations are being recorded in the data collector.

Vertical angle observations differ from horizontal observations in two critical ways. First, the zero point on the circle is preset, usually so that it is parallel to the upward vertical, rather than being arbitrary. Second, the circle cannot be rotated to different positions so as to minimize the effect of variations in the markings on the circle.

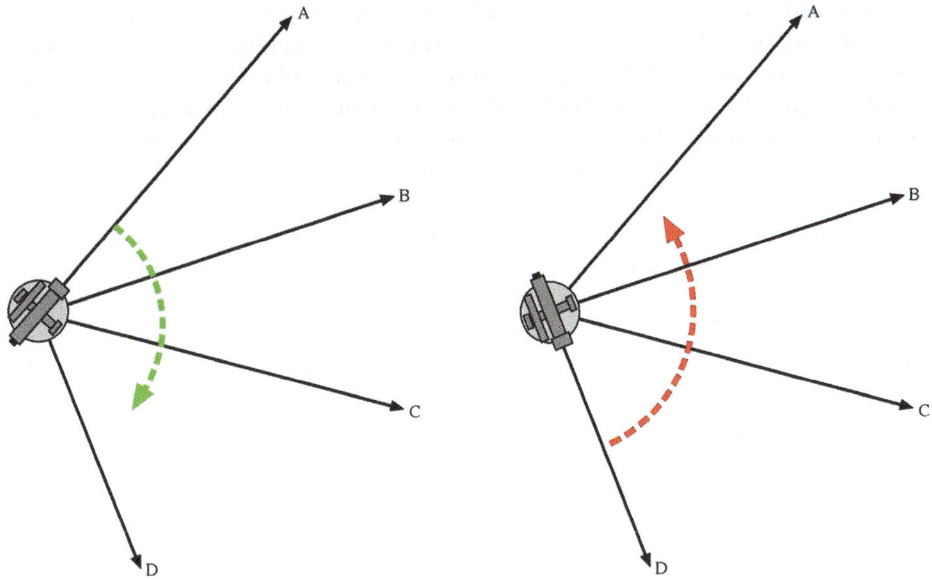

Figure 7-27. Measuring more than two directions at a point. The instrument is set in FLDI, 0° 00′ 00″ set on Point A, and the instrument turned to record the direction to Points B, C, and D. This is a round of angles, but it is not a complete measurement set. After measuring Point D, the instrument is transited and turned counterclockwise to point at Point D. The direction to Point D on FRRII is measured, followed by observations to Points C, B, and A in order. This is one set of observations. Multiple sets may be measured if needed. Note that the angle between Points D and A is not measured.

The zero point for measuring vertical angles has traditionally been the upward vertical (or zenith), largely because of the nature of astronomical observations, the constructions used for spherical trigonometry based on the measured angles, and the advantages of avoiding negative angular quantities and the potential for errors that may be caused. Vertical angles from the upward vertical are commonly referred to as zenith angles. Although some total stations may be able to be configured to read 0° when the telescope is horizontal (producing vertical angles named altitude angles), there is little advantage to this, as the total station will usually take care of any computations that use the vertical angle.

The objective of observing vertical angles is to obtain pairs of observations from the two instrument orientations. There are two common observation methods: one for observations in conjunction with horizontal observations, the other for vertical observations by themselves.

When vertical angles are observed at the same time as horizontal angles, the instrument has its orientation changed between pairs of rounds of angles, and this works well with vertical angles, because when vertical angles are observed on their own, the usual routine is to change instrument position between pairs of observations, with single observations at either end of the sequence, that is,

Direct	Reverse	Reverse	Direct	Direct	Reverse	Reverse	Direct
Left	Right	Right	Left	Left	Right	Right	Left
I	II	II	I	I	II	II	I

This gives four sets of observations. If only one point is being observed, the telescope should be moved away from the point and then brought back by a separate movement. This means that the observer takes their hand off the slow-motion screw, so that the return movement is guided by the eye and not by the memory of the movement away. This allows the point of aim to be shifted to follow the effects of refraction and, therefore, over a number of observations to minimize the variation in refraction.

If multiple points are being observed, they should all be observed together for each set. So if the aforementioned procedure is being used for points A, B, and C, the three points will be observed in some convenient order with the instrument in the FLDI position and then observed again in some convenient order in the FRRII position. The points are all observed again in the FRRII position, and then the instrument is transited and the points observed in the FLDI position. This operation is a combination of the reading process shown in Figure 7-27 and the observing sequence as described.

The effect of atmospheric refraction is not only to make the line of sight curved, but also to add variation to that line of sight. The same effect that makes stars twinkle will cause the target to appear to move around over time. In general, the warmer the atmosphere, the faster the movement. Making several observations over time, as well as spending a little time to watch the movement and trying to find a mean position, will help minimize the effect of this movement. The result will be a mean vertical angle that has minimal effects of refraction variation and, therefore, should be closer to the "theoretical" amount of refraction.

Reducing Vertical Angles

The process of reducing vertical angles takes several steps. Given a set of observations from both instrument positions, pairs of observations are gathered, usually those closest in time. For each pair, the observation made with the instrument in the FRRII position is reduced to the equivalent observation. To do this, the observation is subtracted from 360° and the value written in the "Reduced" column.

From this pair of observations (the FLDI observation and the reduced FRRII observation), compute the mean value. This will be the correct zenith angle, with the effect of vertical circle index error removed.

For multiple pairs of observations, take the mean of all the mean results to get the overall mean vertical angle.

Table 7-2 shows a set of observations and reductions for four sets of vertical angles to a single point. Note that the FLDI observations (denoted as F/L) do not need to be reduced, but this is done in the table as a convenience. Similarly, calculation of the vertical collimation error (the difference between the reduced F/L and F/R observations and shown in parentheses) is not necessary, but it has been done to show how it can be accomplished. A thorough note taker will watch the collimation error as the observations are collected, making sure that it remains consistent throughout the observations. The observations can be recorded in this type of format or can be extracted from a data collector and reduced in this manner.

What can be seen in the vertical angles in Table 7-2 is that there is significant variation between individual observations. This is mainly the result of variability in refraction, which can often be larger than all the other observational random errors. One reason why it is important to make several observations of vertical angles is that observations are spread throughout the range of fluctuations caused by variation in the atmospheric refraction, allowing the mean to smooth out these fluctuations.

Deflection of the Vertical

When a total station is set up and leveled, the vertical that the bubbles use is the local gravity vertical (i.e., the plumbline passing through the vertical axis of the instrument). It

Table 7-2. Recording and Reducing Vertical (Zenith) Angle Observations.

Position	Observation	Reduced	Mean
F/L	87° 23′ 11″	87° 23′ 11″	(38″)
F/R	272° 37′ 27″	87° 22′ 33″	87° 22′ 52″
F/R	272° 37′ 22″	87° 22′ 38″	(39″)
F/L	87° 23′ 17″	87° 23′ 17″	87° 22′ 57″
F/L	87° 23′ 20″	87° 23′ 20″	(38″)
F/R	272° 37′ 18″	87° 22′ 42″	87° 23′ 01″
F/R	272° 37′ 23″	87° 22′ 37″	(39″)
F/L	87° 23′ 16″	87° 23′ 16″	87° 22′ 56″
		Overall mean	87° 22′ 57″
		Collimation error	38″.5

is approximately at right angles to the geoid (approximately because the equipotential "level surface" at the instrument is, in general, not parallel to the geoid surface). However, this line is usually not the same as the normal to the reference ellipsoid, so observations made with the total station as the setup will not be quite the same as if it were set up with its upward axis parallel to the ellipsoidal normal. This difference between the two "verticals" is termed *deflection of the vertical* and it varies from place to place with variations in the direction of gravity.

In almost all cases, the difference in horizontal angle observations will be so small as to be less than the random errors in the observations, and the difference in vertical angle observations will usually be less than the effect of vertical refraction. However, if it is considered necessary to correct one's total station observations from geoidal (regular) to ellipsoidal, the following process can be undertaken.

Deflection of the vertical is presented as the amount of deflection in the meridian (north–south), termed *xi* (ξ, positive to the north) and in the prime vertical (east–west), termed *eta* (η, positive to the east).

Knowing the latitude and longitude of the instrument's location, visit the NGS webpage for the current deflection of the vertical calculation (currently DEFLEC18, but to be updated when the modernized NSRS is released) and determine the two components of the deflection of the vertical at the instrument's location (NGS 2020b).

Within the United States, deflection of the vertical is less than one arcminute in either ξ or η, so a simplified calculation can be applied to calculate the total deflection, d, and its azimuth, θ, at the instrument point.

$$d = \sqrt{\xi^2 + \eta^2} \tag{7-4}$$

$$\theta = \text{ATAN2}(\xi, \eta) \tag{7-5}$$

where ATAN2 is the regular two-argument arctangent ($\tan^{-1}$) function common in spreadsheets and programming languages.

The orientation of the instrument, Az, must be an azimuth on the same basis as the deflection of the vertical, which will be true north. The tilt in the direction that the instrument's telescope is pointing, q, can be calculated using

$$q = \tan^{-1}(\tan(d)\cos(Az - \theta)) \tag{7-6}$$

The transverse tilt of the instrument, p, can be calculated using

$$q = \tan^{-1}(\tan(d)\cos(Az - \theta + 90°)) \tag{7-7}$$

If p and q are calculated in arcseconds, the corrected vertical angle observed in this direction at the particular instrument station can be calculated using

$$z_c = z_{obs} + q + \frac{p^2}{2}\cot(z) \tag{7-8}$$

The observed horizontal reading in this direction can be corrected using

$$\Delta Az = p\cot(z) \tag{7-9}$$

where ΔAz is the correction to the observed horizontal reading (in arcseconds), and z is the observed vertical angle (as a zenith angle).

These corrections will usually need to be applied only when correcting higher-precision astronomical observations for latitude and longitude and when connecting GNSS observations to high-precision total station observations. For the vast majority of survey work, the corrections will be too small to make a difference.

Measuring Distances

Distances measured by total stations are not usually measured using the speed of light directly. Instead, a sinusoidal wave is superimposed on the carrier wave from the EDM unit, and the phase difference between the transmitted and reflected signals is calculated. Using two (or more) different frequencies of superimposed waves allows the distance to be calculated. Distance measurement usually takes only a few seconds and can be updated rapidly to track moving objects.

The sinusoidal wave has to have a carefully defined length when traveling through the air. However, this length is affected by the speed of the overall beam, and this variation in speed is caused by changes in the refractive index of the air. The EDM is usually calibrated at a specific refractive index. Refractive index can be calculated using the temperature and pressure of the air through which the signal is traveling, and in extreme cases, by the humidity of the air (for infrared and visible-light EDM, this is when the temperature is above about 27° C/80 °F and the relative humidity is above 80%). So, the temperature and pressure of the air in which the distance is being measured should be measured. Most total stations allow direct entry of the temperature and pressure (and sometimes relative humidity), and the instrument calculates the atmospheric correction to the distance in parts per million (ppm) and applies this to every measured distance.

Total stations typically compute the atmospheric correction to the distance (Δs) in ppm using the following equation (Ghilani, 2021):

$$\Delta s\ (\text{ppm}) = \frac{273.15}{1013.25} \cdot \frac{N_g P}{t + 273.15} - \frac{11.27e}{t + 273.15} \tag{7-10}$$

where

t = Dry bulb temperature in °C,
P = Pressure in hPa,
e = Partial water vapor pressure in hPa (defined subsequently), and
N_g = Group refractivity in standard air for the electronic distance measurement, calculated as follows:

$$N_g = 287.6155 + \frac{4.88660}{\lambda^2} + \frac{0.06800}{\lambda^4} \tag{7-11}$$

where λ is the wavelength of the light expressed by the EDM in micrometers (μm). Typical values of λ for EDMs are 0.6328 μm for red laser and 0.900 to 0.930 μm for infrared.

The partial water vapor pressure (e) in hPa is found by

$$e = \frac{h}{100} \cdot 10^{[7.5t/(237.3+t)+0.7858]} \tag{7-12}$$

where h is the relative humidity in percent.

Note that pressure data from weather services are the estimated pressure at sea level, not the local pressure, which varies with elevation. As can be seen from the equations, all total stations should have a barometer and thermometer carried with them, and they should be allowed to settle to local conditions during instrument setup. This meteorological data should be checked, entered into the total station, and updated at regular intervals to minimize errors in the observations. They should be taken frequently and logged in the instrument throughout the day as field and weather conditions change.

For most total stations, measuring the distance is as simple as ensuring that the telescope is pointing to the middle of the prism and pressing a button. The distance measured is the slope distance between the EDM unit and the prism. Many total stations offer a single button to measure and record in a single step.

On pressing the button, the total station automatically applies the atmospheric correction and computes the slope distance (s) by the following:

$$s = s_0 \cdot \left[1 + (\Delta s) \cdot 10^{-6}\right] + K \tag{7-13}$$

where s_0 is the uncorrected slope distance, and K is the prism offset.

The slope distance can be reduced to the horizontal distance and the vertical height difference between the total station's trunnion axis and the middle of the prism, and the entire bundle of observations and reduced values is usually recorded as a single entity.

Reducing Distances

With total station measurements, the slope distances are reduced by the machine to horizontal and vertical components, as shown in Figure 7-28. The reduction is multiplying

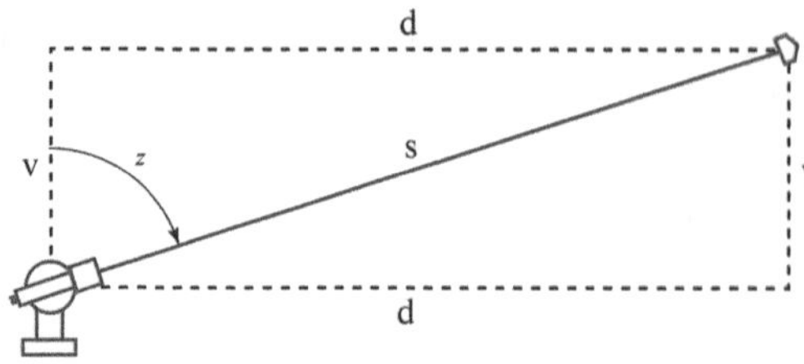

Figure 7-28. The measured slope distance, s, is reduced to its horizontal (d) and vertical (v) components, using the vertical angle (z), measured by the total station.

the slope distance by the sine and cosine of the vertical angle measured by the total station:

$$\text{horizontal distance } (d) = \text{slope distance } (s) \times \sin(\text{vertical angle}, z) \tag{7-14}$$

$$\text{vertical height difference } (v) = \text{slope distance } (s) \times \cos(\text{vertical angle}, z) \tag{7-15}$$

The vertical angle will be automatically adjusted for any collimation error, which should be checked and updated regularly following check and adjustment procedures supplied by the vendor of the instrument. Similarly, the vertical angle may be automatically corrected for refraction if the refraction and curvature correction is switched on.

For the elevation of the measured point, the height of the total station's trunnion axis above the ground point, the elevation of that ground point, and the height of the prism above the target ground mark are combined with the measured vertical height difference. The various components of this computation need to be stored in the instrument. Total stations allow the prism height to be changed between observations, so there is a need for communication between the person with the prism and the instrument operator. If a robotic total station is used, then the operator is at the prism, so a changed prism height can be entered directly into the data collection.

Refraction and Curvature

Refraction causes the angle at which light arrives at the telescope to differ from the straight-line connection between the target and the telescope. The angle varies with the amount of refraction and the distance the light has to travel. It is highly difficult to predict the amount of refraction, and the potential range is large. No easy signs in the environment exist that give a guide to refraction.

The refraction correction formula that is embedded in total stations is based on a simple, standard model that assumes that the line of sight is well clear of the surface. However, the behavior of refraction close to the surface is highly difficult to predict, and no total station attempts to model refraction any better than the standard model. This means that variation in refraction can have a significant difference in elevations measured using the total station. As the refraction changes during the day, the measured elevations will also change. Fortunately, the effect of refraction is relatively small, typically about seven times smaller than that because of the curvature, and it is in the opposite direction of curvature (as discussed subsequently). Nonetheless, the combined error caused by curvature and refraction can be significant and can exceed 0.1 m for long sight distances (usually more than about 1,200 m or 4,000 ft).

Earth curvature is simple and easy to calculate and does not vary. The total station correction for this allows the elevation to be corrected to fit a geoid, using a spherical approximation to the geoid. Any variations in the geoid are usually smaller than the total station's measurement errors, and far smaller than the uncertainty caused by refraction.

Refraction makes objects appear higher than they really are, and this error is, therefore, in the opposite direction of errors because of the curvature. It can, thus, be said that refraction reduces the total effect of Earth curvature, and total stations most typically compute a combined correction for both curvature and refraction (h_{cr}) using the following approximation:

$$h_{cr} = \left(\frac{1-k}{2\cdot R}\right)\cdot(s\cdot\sin z)^2 \tag{7-16}$$

where

k = Mean refraction coefficient;
R = Radius of Earth;
s = Measured slope distance; and
z = Measured vertical angle, as shown in Figure 7-28.

By plugging in typical values $k = 0.14$ and $R = (6.378) \times 10^6$ m, and by recognizing that $s \cdot \sin z$ equals the horizontal distance (d) according to Equation (7-14), then

$$h_{cr} = (6.75)\cdot 10^{-8}\cdot[\mathrm{m}]^{-1}\cdot(d)^2 \tag{7-17}$$

Total stations can be set to compute h_{cr} and automatically apply this correction to the observed vertical height difference, v, by extending Equation (7-15) with the following:

$$v = s\cdot\cos z + h_{cr} \tag{7-18}$$

Although the effects of refraction (and its combination with curvature) can be approximated by the aforementioned equations, the best approach to minimize error is to observe the vertical angle from both ends of the line at approximately the same time (in the same atmospheric conditions). When shooting uphill, v will be greater than actual, and when shooting downhill, v will be less than actual. Averaging the two shots will minimize the errors. This approach, known as taking reciprocal vertical angles, is recommended for precise trigonometric leveling.

Point Codes

Point codes are used to describe each point observed by the total station during a topographic or similar survey. Many schemes are available for point codes, and the purpose is to connect the point descriptor directly to the observations in the data collector, and for post-processing. Early point code schemes were purely numerical, as data collectors and total stations tended to have numerical keys. With the expansion of data collectors and more extensive keypads on total station, it is easier to use alpha-numeric codes.

Point codes are often hierarchical, in that the initial characters refer to general features, whereas later characters offer more detail. For example, 37 may refer to a concrete curb, whereas 37-15 is the initial tangent point of a curve on the curb, 37-16 is a point on the curve on the curb, and 37-17 is the end tangent point on the curb. Such a point sequence allows automated determination of the curve's parameters and plotting of the curb by joining the points. Note that these examples are simplified. The point codes may have many characters to allow sufficient flexibility to cover a wide range of objects and may add additional control codes to add instructions to post-processing software back at the office.

Different code systems are used by different organizations. For example, a box culvert may be designated BCUL, BXC, DCEF, or CULV, together with different additional attribute data, by just five departments of transportation in the United States. It is, therefore, important to ensure that the coding system used matches the one expected by the post-processing software. Most advanced post-processing packages allow the use of extensive look-up tables to translate input codes into the software's preferred system. With the appropriate use of point codes, much of the processing of the point data into lines, polygons, and various features can be undertaken automatically by the post-processing software. Features can be automatically annotated, symbolized, and placed on the correct layer in CAD software, as well as set up for later analysis. This allows significant time savings in post-processing and analysis.

Total stations have different means of uploading and applying point codes systems. Refer to the manual that accompanies the particular total station and use the processing software, as the various methods differ significantly. Similarly, the various point code schemes are often designed for significantly different types of surveys.

Because it adds a level of complex input, point codes may be easier to manage on a separate data collector, as these devices often have higher quality keypad systems. Which is better is a matter of personal preference and may also differ significantly between jobs. The introduction of tablet devices with touch screens, screen-based keyboards, and custom-entry screens can simplify data entry during field measurement.

Electronic Distance Measurement Calibration

Calibrating the distance measurement part of a total station can be done in several ways. The ideal method involves measuring the various components of a calibrated baseline (CBL). In the United States, CBLs are managed by NGS, and details about each CBL can be found on the NGS website (NGS 2022c), along with CBL publications, such as Dracup et al. 2019 and Fronczek 1980. However, for simpler calibration, the primary concern is the zero-offset and prism offset calibration, which should be performed with all changes in the prisms used by a total station. The prisms are as much a part of the overall measurement system as is the total station unit.

For a simple check, set out three points in a straight line, where the distances between the points are integral multiples of 60 m. An example line is shown in Figure 7-29. The reason for the 60 m is that this allows the small cyclic error in the EDM to remain constant for all measurements on this line and, therefore, removes that error as a factor. Note that prisms should be mounted on tripods, as using a prism pole introduces too much error. Ensure that meteorological data are taken and the correction set in the total station.

Set up tripods, place the total station on one of the endpoints, measure the full length of the line (180 m in Figure 7-29) several times, and take the mean. Then, place the total station at the middle point, measure the two partial lines (120 and 60 m in Figure 7-29) several times, and calculate the mean distances. Using the reduced horizontal distances, add the

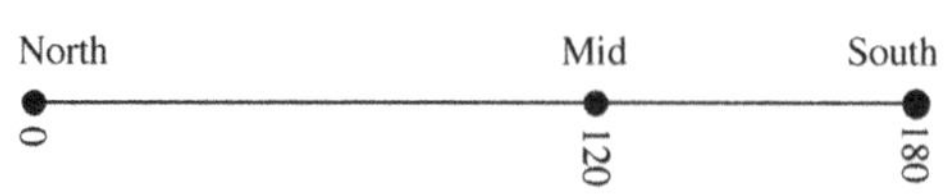

Figure 7-29. Test line for quick and easy determination of combined zero and reflector offset. Distances are in meters. The line segments are integral multiples of 60 m.

two partial measured distances and subtract the overall measured distance. The difference will be the combined zero-offset and prism offset error.

Many total stations allow this correction to be entered in the instrument and applied to all measured distances. This can be used to permit different prisms to be used and to ensure that the correct prism offsets are applied. As setting up a "dual offset" prism the wrong way can introduce about 30-mm error into every line measured, it is worthwhile checking this error frequently.

The definitive work on EDM and its calibration can be found in Rüeger (1996).

TOTAL STATION PROCESSES

Traversing

A common method of tying together total station observations is a traverse. A traverse is a collection of 2D or 3D measured vectors that link together the instrument points. As such, they link together the observations made at each instrument point, providing them with a known point (the instrument point) and a known azimuth for all subsequent observations made from the instrument point.

Traverse observations, which consist of the angle between the traverse lines, and distances and height differences along those lines, as shown in Figure 7-30, should be completed before any additional observations are made at the instrument point. This allows subsequent observations to be connected to the traverse, even if the instrument is disturbed.

Traverses must be closed, so that blunders can be detected. This allows the vector sum (2D or 3D) to be determined. Any misclosure can be assessed for errors, and if appropriate, the observations can be adjusted.

Note that a traverse is the weakest form of providing acceptable control for a survey, in part because it is the minimal method of connecting the points. As the precision requirements for the work increase, the required level of precision may not be able to be attained by the traverse. This limitation is caused by the geometry of the traverse and not by the precision of the individual measurements. In this case, a network must be used to provide sufficient precision.

The means of determining the quality and precision of a given traverse consist of basic checks for blunders, and the use of error propagation tools, usually within a least-squares adjustment. Calculating the misclosure of the traverse, which is the resultant vector when the vectors representing the traverse are summed, is the first basic check, as blunders will often become clear in the misclosure. The misclosure in the measured angles of the traverse, termed the *angular misclosure*, can be quickly obtained in the field as the traverse is

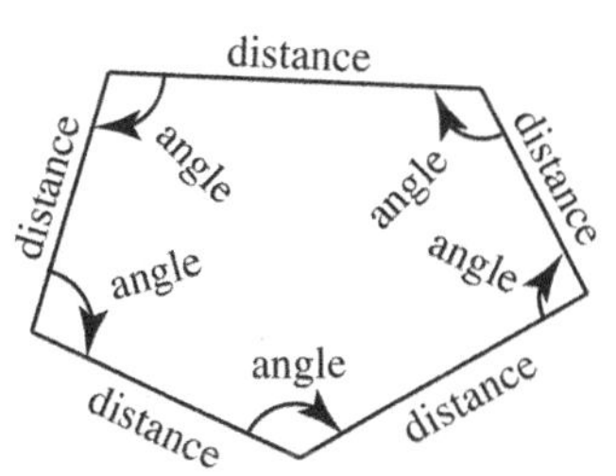

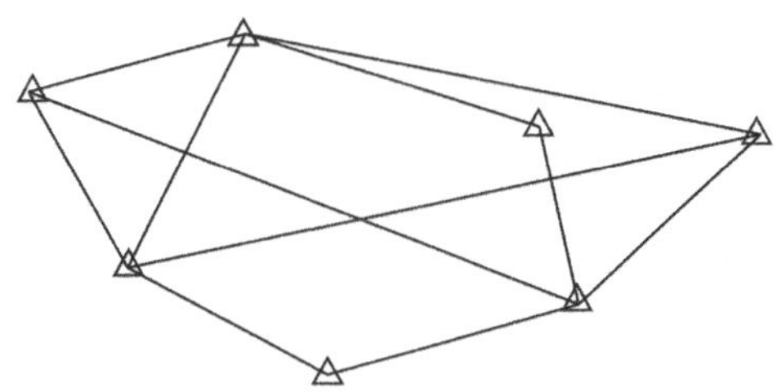

Figure 7-30. (Left) Traverse showing which measurements are made during regular traversing. (Right) Network showing interconnected points. All shown lines have their distances measured, and all angles at the points marked with triangles are measured.

completed, and is a first check. Searching for blunders (and other) errors in a completed body of measurements is termed *error trapping* and is an important part of checking survey work to ensure that it meets job requirements and broader standards.

The easiest way of determining the quality of a traverse is to consider the size of the misclosure as a proportion of the total length (perimeter) of the traverse. Various jurisdictions require specific proportional misclosure levels to be achieved for various types of work. These may be expressed as a proportional ratio, for example, 1 in 10,000, or as a certain number of parts per million (ppm). A traverse that is allowed a 1 in 10,000 misclosure will be allowed a misclosure vector of 1 m in a 10,000 m traverse; this is equivalent to 100 ppm. Getting into this more deeply requires the use of local accuracy determination, which really requires the use of a least-squares adjustment. See Chapter 5 for more on least-squares adjustments and discussion on error propagation.

Much more information about traverses can be found in the usual surveying textbooks, for example, Cole (2014), Ghilani (2021), Kavanagh and Mastin (2013), McCormac et al. (2012), Nathanson et al. (2017), Uren and Price (2010), Bhavikatti (2009), Bannister et al. (1998), Anderson and Mikhail (1997), Moffitt and Bossler (1997), and Whyte and Paul (1997).

Networks

Networks are more complex interconnections of measurements. Although the network can be decomposed into small "traverses," these cannot be adjusted individually, as there are too many vectors in common. A least-squares adjustment (see Chapter 5) is required to obtain consistent results across the entire network.

Networks have the advantage that the precision of the 3D location of the control points can be maintained at very high levels (i.e., highly precise locations). This can be critical for deformation monitoring and high-precision construction projects. However, the costs increase for the control. Cutting corners can increase the overall job costs dramatically (e.g., one particular project in Philadelphia managed to keep its control costs low initially, perhaps saving a few tens of thousands of dollars, but this led to $300 million overruns, lawsuits, and multiyear delays).

Standards of accuracy and precision for networks require greater consideration. The addition of GNSS technology led to geodetic quality increasing to the point where non-GNSS technology could not compete. As older networks, observed with a range of technology and methods, are merged with GNSS observations, paying attention to the precision of measurement of all measurements is of critical importance to the success of the effort. This requires a more advanced understanding of measurement errors and least-squares adjustment, in some cases beyond the common textbooks (e.g., Ghilani 2017, Mikhail and Gracie 1981).

Topographic Surveys

When a large number of points across the landscape are required to be located, this is termed a *topographic survey*. The final product may be a map of some type and total stations are particularly useful for this type of work. The original name of total stations, "electronic tacheometers," was an indication that they were designed for this type of work. By making a single observation with the instrument in one orientation (usually FLDI) and allowing the instrument to make corrections for collimation error, refraction and curvature, calibration, and prism height, the 3D location of points can be determined as quickly as the prism can be moved from point to point. If the description of each point can be included as a point code, it is possible to automate a significant part of the subsequent reduction and plotting work, allowing rapid determination of volumes and areas, and allow a map to be moved to a CAD system for further design work.

Although the precision of the measured points deteriorates because of a single observation and the uncertainty of how steadily the pole can be held, precisions of several centimeters are usually quite acceptable for the needs of these surveys. Higher precision work needs stable prism location and multiple observations, but this is not needed for general topographic work.

Layout Work

When points are required to be placed, such as for engineering construction, a total station is an extremely useful tool. The required coordinates of the point to be placed can be uploaded, and the on-board calculation capability can calculate the angles and distance that are required to be used to set the point.

All placed points need to be checked, and an independent check requires different measurements. One simple approach is to set up the total station at a different point and measure the locations of the placed marks. If close agreement to the required precision is achieved, one can be confident that the placement has been satisfactory. If there is no close agreement, the errors can be corrected in the field before anything serious happens.

REFERENCES

Anderson, J. M., and E. M. Mikhail. 1997. *Surveying: Theory and practice.* 7th ed. Columbus, OH: McGraw Hill.

Bannister, A., S. Raymond, and R. Baker. 1998. *Surveying*. 7th ed. New York: Pearson Prentice Hall.

Bhavikatti, S. S. 2009. *Surveying and leveling, volume II.* 2nd ed. New Delhi: IK International Publishing House.

Bossler, J. D., and N. W. J. Hazelton. 2015. *Leveling and vertical location*. Anchorage, AK: Wollindina Media.

Cole, G. M. 2014. *Surveyor reference manual*. 6th ed. Belmont, CA: Professional Publications.

Dracup, J. F., C. J. Fronczek, R. W. Tomlinson, K. L. Fancher, and C. Geoghegan. 2019. *Establishment of calibration base lines*. NOAA Technical Memorandum NOS-NGS 8. Silver Spring, MD: National Oceanic and Atmospheric Administration, National Geodetic Survey. Originally published 1982, revised 1983, 2014, 2019.

FGCS (Federal Geodetic Control Subcommittee). 2004. *FGCS specification and procedures to incorporate electronic digital/bar-code leveling systems (FGCSVERT)*. Version 4.1. Reston, VA: FGCS.

Fronczek, C. J. 1980. *Use of calibration base lines*. NOAA Technical Memorandum NOS-NGS 10. Rockville, MD: National Oceanic and Atmospheric Administration, National Geodetic Survey.

Ghilani, C. D. 2017. *Adjustment computations: Spatial data analysis*. 6th ed. Hoboken, NJ: Wiley.

Ghilani, C. D. 2022. *Elementary surveying*. 16th ed. New York: Pearson.

Hofmann-Wellenhoff, B., and H. Moritz. 2006. *Physical geodesy*. 2nd ed. New York: Springer.

Kavanagh, B., and T. Mastin. 2013. *Surveying: Principles and applications*. 9th ed. New York: Pearson.

McCormac, J., W. Sarasua, and W. Davis. 2012. *Surveying*. 6th ed. Hoboken, NJ: Wiley.

Mikhail, E. M., and G. G. Gracie. 1981. *Analysis and adjustment of survey measurements*. New York: Van Nostrand Reinhold.

Moffitt, F. H., and J. D. Bossler. 1997. *Surveying*. 10th ed. New York: Pearson.

Nathanson, J. A., M. T. Lanzafama, and P. Kissam. 2017. *Surveying fundamentals and practices*. 7th ed. New York: Pearson.

NGS (National Geodetic Survey). 1982. *Corrections applied by the national geodetic survey to precise leveling observations*. NOAA Technical Memorandum NOS-NGS 34. Silver Spring, MD: National Oceanic and Atmospheric Administration, National Geodetic Survey.

NGS. 2020a. *Geodetic leveling*. NOAA Manual NOS-NGS 3. Silver Spring, MD: National Oceanic and Atmospheric Administration, National Geodetic Survey. Originally published 08/1981, Revised 10/2020.

NGS. 2020b. "Computation of DEFLEC18 deflection of the vertical." *DEFLEC 18*. Accessed February 9, 2022. https://geodesy.noaa.gov/GEOID/DEFLEC18/computation.html.

NGS. 2022a. "NAVD 88 gravity computations." *NGS Geodetic Tool Kit*. Accessed February 9, 2022. https://geodesy.noaa.gov/TOOLS/Navdgrav/navdgrav.shtml.

NGS. 2022b. "LVL_DH." *NGS Geodetic Tool Kit*. Accessed February 9, 2022. https://geodesy.noaa.gov/TOOLS/LVLDH/lvldh.shtml.

NGS. 2022c. "EDMI calibration base line program." *Calibration Basel Line*. Accessed February 9, 2022. https://geodesy.noaa.gov/CBLINES/calibration.shtml.

Rüeger, J. M. 1996. *Electronic distance measurement: An introduction*. 4th ed. Berlin: Springer.

Uren, J., and W. F. Price. 2010. *Surveying for engineers*. 5th ed. Houndmills, UK: Macmillan.

Whalen, C. T. 1979. *Control leveling*. NOAA Technical Rep. NOS 73 – NGS 8. Silver Spring, MD: National Oceanic and Atmospheric Administration, National Geodetic Survey.

Whyte, W., and R. Paul. 1997. *Basic surveying*. 4th ed. London: Routledge.

CHAPTER 8
TERRESTRIAL LASER SCANNING

Michael J. Olsen

INTRODUCTION

Terrestrial laser scanning (TLS) has become a powerful, new surveying technology to support a wide range of engineering applications, including topographic mapping, asset management, deformation monitoring, quantity calculations, safety evaluations, and many more. A TLS operates by rapidly sending out laser pulses across a window of interest (typically a 360 degree panorama), generating a point cloud of the scene (Figure 8-1). The scanner encodes the horizontal and vertical angles relative to an internal reference for each pulse. The range to the object is either calculated via time of flight or differences in phase as the light is reflected back to the scanner. Multiple returns from an individual pulse can occur when the footprint of a pulse intersects multiple objects, resulting in several distinct data points from a single pulse. (However, note that, in general, TLS has limited pulses with multiple returns compared with other platforms because of the viewing geometry of the scene and because weaker returns are filtered from the data). These returns result in *X*, *Y*, *Z* coordinates for each return from the pulse across the objects that are within view and range of the scanner (Figure 8-2). Acquisition rates of lidar systems have increased substantially over the last few decades (Figure 8-3), with systems now capable of gathering more than 1 million points per second.

In addition to the coordinates, an intensity value (return-signal strength) is recorded for each laser return (see Figure 8-1 for an example of a point cloud colored by intensity information). Kashani et al. (2015) provide a review of how intensity information can be utilized for a variety of applications as well as current procedures for radiometric calibration of this intensity information such that it can be used as a measure of an object's reflectivity. Intensity values can provide an indication of the reflectivity of the material; however, intensity values can degrade based on geometric (e.g., range, incidence angle), environmental conditions, surface color (Bolkas and Martinez 2018, Bolkas 2019), and other factors.

Many scanners have also integrated or externally mounted cameras, providing red-green-blue (RGB) color information for the point cloud. Although this information is often used for visualization purposes, it also provides additional attributes to the data. Further, some software enables users to map or extract features from these photographs, which are then linked to 3D coordinates via the point cloud. Figure 8-4 shows an example of a point cloud with photographic color applied for a cultural heritage analysis.

Although TLS is a very exciting technology that has revolutionized surveying, ultimately, it still requires the use of conventional survey (e.g., GNSS, total station) for many surveying

Figure 8-1. Example point cloud of an urban scene, colored by intensity (return signal strength). Green indicates higher intensity, and red indicates lower intensity. Note the shadows (black) that expand from the scan origin (black circle) as well as the decrease in resolution with distance from the scanner.

engineering applications to establish control coordinates and aid the registration. However, TLS reduces the workload using these traditional techniques and can make the field process more efficient and yield significantly more data. Further, the high-resolution capabilities of TLS provide immense value in quality control and help identify problems in the survey control when they occur.

> *TLS is a highly efficient surveying technology; however, rigorous survey control is still required for surveying engineering.*

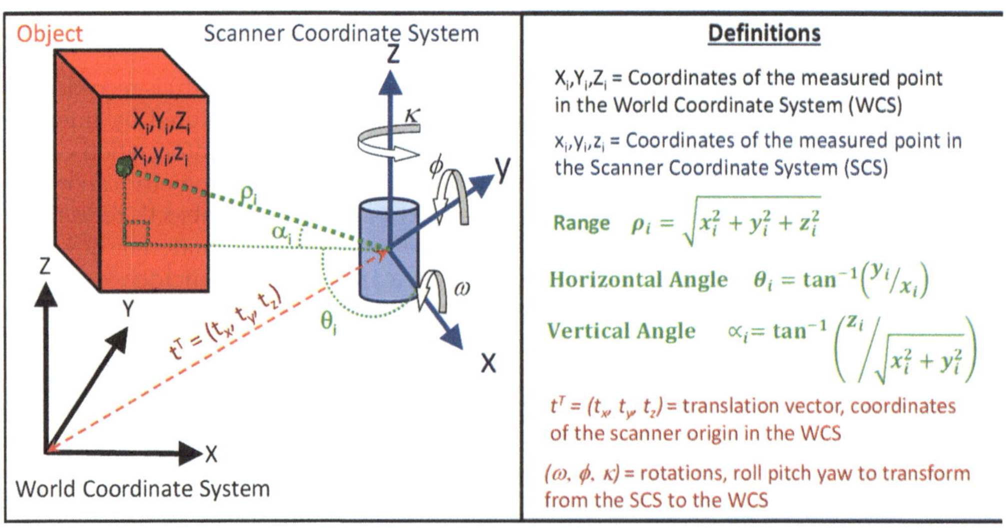

Figure 8-2. Schematic illustrating TLS basic geometry observations and equations as well as parameters for a rigid body transformation.

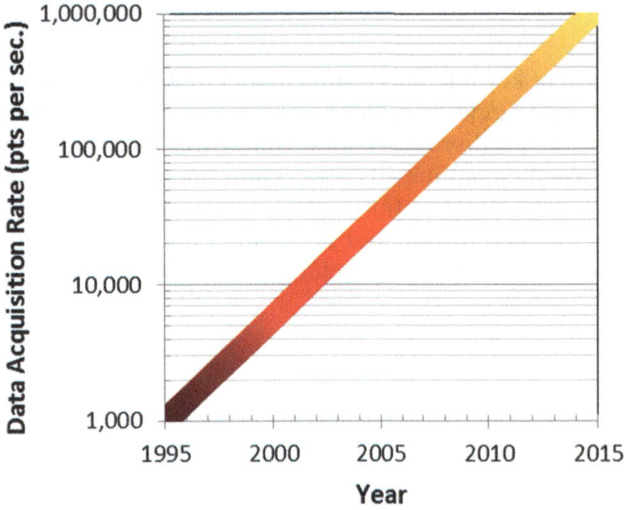

Figure 8-3. Increase in TLS time-of-flight data acquisition rates with time.

Several terms are sometimes used to describe terrestrial laser scanning, such as *3D laser scanning*, *ground-based lidar*, *high definition surveying*, *terrestrial lidar*, *static lidar*, *tripod-based lidar*, *t-lidar*, or *ladar* (laser detection and ranging). For simplicity, we use the *term terrestrial laser scanning* (TLS) throughout this chapter.

A key advantage of TLS over traditional surveying is the density of data obtained, which is orders of magnitude more than traditional surveying. Although methods such as total station surveying can achieve slightly higher accuracies per individual point, TLS still provides highly accurate data that can result in an improved model because of the inherent variability in the terrain or objects of interest that exists between sample points. These features are captured and resolved in TLS but cannot be efficiently captured in traditional surveying.

Figure 8-4. Sample RGB-colored laser scan of the "Hall of the 500" in the Palazzo Vecchio in Florence, Italy, completed by CISA3 (2011) for the Lost Battle of Anghiari by Leonardo Da Vinci.

OVERVIEW

After providing a list of key references, this chapter first describes common applications of TLS applicable to Civil Engineering. Next, we provide an overview of the different types of TLS systems followed by a discussion of the differences in other lidar platforms such as airborne and mobile. We then outline a TLS workflow and explore important steps such as Planning, Field Procedures, Registration and Georeferencing, Processing, and Analysis. Best practices are highlighted throughout the chapter and designated as arrowed bullet points for quick reference. The chapter concludes with a discussion of the future of TLS.

KEY REFERENCES

This section provides key references for those interested in learning more about TLS and applications in civil engineering.

1. *Airborne and Terrestrial Laser Scanning*, edited by George Vosselman and Hans-Gerd Maas. ISBN 978-1-4398-2798-7.
2. *Topographic Laser Ranging and Scanning*, edited by Jie Shan and Charles K. Toth. ISBN 978-1-4200-5142-1.
3. *Non-Destructive Techniques for the Evaluation of Structures and Infrastructure*, Belen Riveiro and Mercedes Solla. ISBN 978-1-138-02810-4.
4. USIBD Manual of Building Documentation, US Institute of Building Documentation. http://www.usibd.org/Standards.
5. ASTM Committee E57 on 3D Imaging Systems, https://www.astm.org/COMMITTEE/E57.htm.

Although not specific to TLS, the following references provide useful information that is relevant to TLS as well as summaries of example applications relevant to civil engineering.

6. *Manual of Airborne Topographic Lidar*, ASPRS, http://www.asprs.org/Press-Releases/Order-the-ASPRS-Airborne-Topographic-Lidar-Manual.html.
7. *Guidelines for the Use of Mobile Lidar in Transportation Applications*. http://www.trb.org/Main/Blurbs/169111.aspx. National Academies of Science. http://learnmobilelidar.com. (Olsen et al. 2013d, O'Banion et al. 2013).

For a comprehensive glossary of terminology commonly used in TLS, the reader is referred to http://learnmobilelidar.com/glossary/.

APPLICATIONS IN CIVIL ENGINEERING

A key phrase associated with any form of lidar technology is "Collect Once, Use Many!" This is certainly true with TLS data. Many costs can be shared by several entities that can utilize the data. Table 8-1 provides a sampling of various applications utilizing TLS, but it is by no means comprehensive. Figure 8-5 presents example applications of TLS in civil engineering. NCHRP Report No. 748 provides a detailed discussion related to applications of mobile laser scanning (MLS) in transportation, many of which are relevant to TLS.

Note that certain TLS systems will perform better for some applications than others, and it is important to evaluate this when selecting a scanner to utilize on a project. Currently, a wide range of scanners that support a broad range of civil engineering applications are available, resulting in a high return on investment.

Table 8-1. Sampling of Applications of TLS in Civil and Construction Engineering.

Application	Examples and description	Sources
Topographic surveying and earthwork quantities	Terrain mapping Earthwork volume calculation	Jaselskis et al. (2003), Duffell and Rudrum (2005), Slattery et al. (2012)
Roadway design	Collecting baseline data and site features through virtual surveying	Kim et al. (2008), Yen et al. (2008a), Mabey (2009)
Machine guidance and control	Automated machine control (e.g., Pavement milling machine)	Singh (2008)
As-built surveys and construction quality control	Road grade and cross slope determination	Olsen et al. (2012e), Su et al. (2006)
	Quality control	Wang et al. (2014a), Maalek et al. (2015)
Site monitoring or progress monitoring	Estimates of percent completion	El-Omari and Moselhi (2008, 2011), Bosché et al. (2009), Turkan et al. (2012), Turkan et al. (2013), Turkan et al. (2014), Bosché et al. (2014b), Bosché et al. (2014a), Maalek et al. (2015)
	Construction	Taneja et al. (2011)
	Earth movements and In-situ infrastructure monitoring	Olsen (2015)

(Continued)

Table 8-1. (*Continued*).

Application	Examples and description	Sources
Transportation	Clearances, inventories, pavement analyses, infrastructure monitoring, accident investigation, and so on	Jaselskis et al. (2003), Hiremagalur et al. (2007), Singh (2008), Yen et al. (2008a), Yen et al. (2008b), Belton and Bae (2010), Olsen et al. (2013d), Olsen et al. (2012e), O'Banion et al. (2013).
	Bridge surveys	Tang and Akinci (2012a, b), Rashidi et al. (2020)
	Sign inventory and reflectivity	Gonzalez-Jorge et al. (2011), Ai and Tsai (2016)
	Sight distance evaluations	Jung et al. (2018)
	Connected/driverless vehicles	Brenner (2009)
	Pavement markings	Che et al. (2019b), Jung et al. (2019)
Pavements	Pavement condition assessment (e.g., roughness, cracking)	Johnson and Johnson (2012)
	Pavement roughness evaluations	Chin and Olsen (2015)
	Assessment of road grade and cross-slope attributes	Kim et al. (2008)
	Flatness defects in concrete	Tang et al. (2011), Puri et al. (2018)
Asset management	Roadside feature inventory (e.g., signs, striping)	Duffell and Rudrum (2005)
	ADA feature inventory	Kingston et al. (2006)
	Geotechnical asset management	Metzger et al. (2013), Cunningham et al. (2015), Oomen et al. (2015)

Hazards and Risk Assessment	Post-disaster reconnaissance	Kayen et al. (2006), Stewart et al. (2009), Kayen et al. (2010), Olsen and Kayen (2012), Olsen et al. (2012c, d), Yim et al. (2014), Kashani et al. (2014a, b), Berman et al. (2020), Wartman et al. (2020)
	Forensic engineering	Collins et al. (2009)
	Natural hazards	Olsen et al. (2012d), Olsen et al. (2013a)
Earth Sciences	General overview	Telling et al. (2017)
	Slope stability, landslide movement, rockfalls	Collins and Stock (2015), Rosser et al. (2007), Carrea et al. (2015), Jaboyedoff et al. (2012), Olsen et al. (2012a), Conner and Olsen (2014), Wang et al. (2013), Abellán et al. (2014), Tonini and Abellán (2014), Olsen et al. (2015), Grant et al. (2018), Dunham et al. (2017), Bolkas et al. (2018), Bunn et al. (2019)
	Ground settlements	Olsen et al. (2016a)
	Landscape evolution	Starek et al. (2010, 2011, 2013)
	Seacliff erosion	Collins and Sitar (2004), Rosser et al. (2005), Collins et al. (2007), Young et al. (2008, 2010), Olsen et al. (2009, 2011, 2013b, 2016b)
	Beach erosion	Paris et al. (2013)
	Streambank erosion	Starek et al. (2013), Barnhart and Crosby (2013)
Facilities	Facility operations	Taneja et al. (2011), Vamsi et al. (2014)
	Powerlines	Jung et al. (2020)
	Pipelines	Son et al. (2015)
	Reverse engineering and inspection	Son et al. (2002)
	Building information modeling	Tang et al. (2010), Bosché et al. (2014a), Bosché et al. (2014b)

(*Continued*)

Table 8-1. (*Continued*).

Application	Examples and description	Sources
Structural analysis	Measurement of structural deformations	Gordon et al. (2003, 2004a, 2004b), Tsakiri et al. (2006), Monserrat and Crosetto (2008), Olsen et al. (2010)
	Bridge structural condition assessment	Turkan et al. (2016), Rashidi et al. (2020)
	Transmission towers	Conde et al. (2015)
	Crack detection or material loss	Laefer et al. (2009), Olsen et al. (2010), Laefer et al. (2010), Laefer et al. (2014)
	Beam stress evaluation	Lee and Park (2008, 2011)
	Dam monitoring	Gonzalez-Aguilera et al. (2008)
	Masonry walls	McInerney et al. (2012), Riveiro et al. (2015b)
	Arch bridge analysis	Riveiro et al. (2011)
	Tunneling	Lindenbergh et al. (2009), Han et al. (2013b), Nuttens et al. (2014), Wang et al. (2014b)
Historical structures	Documentation, preservations, and analysis	CyArk (2021), Olsen et al. (2013a, c)

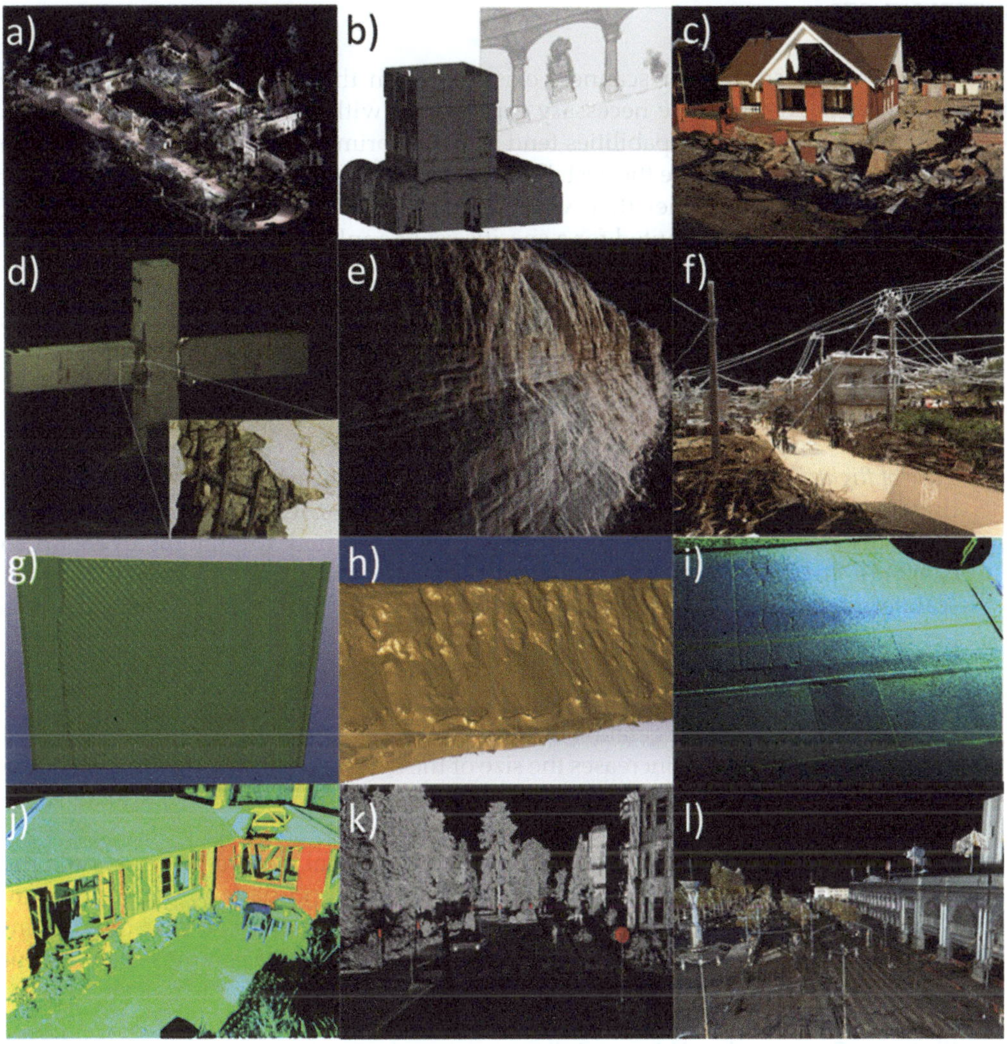

Figure 8-5. Sample of TLS applications: (a) TLS scans of Balboa Park, San Diego, utilized for cultural heritage and wireless signal modeling; (b) 3D solid and wireframe model of a courtyard of the Palazzo Medici, Florence Italy, utilized for renovations (data courtesy of CISA3); (c) tsunami scouring and damage to a house in Dichato, Chile; (d) structural deformations for a lab test; (e) TLS scans utilized for coastal erosion surveys in Torrey Pines State Park, California; (f) powerlines in Bungmati, Nepal; (g) Micron resolution scans of composite materials used in lightweight bridges; (h) 3D model of seacliffs in Beverly Beach State Park, Oregon, highlighting erosion patterns from surface runoff and wave interaction; (i) Pavement cracking and settlement at the Oregon State University Campus; (j) damage to a house in Canterbury, New Zealand, from seismically induced rockfalls; (k) extraction of reflective street signs on the Oregon State University Campus; (l) survey of the street and rail network in front of the San Francisco Ferry Building.
Sources: (d) Olsen et al. (2010), (e) Olsen et al. (2009, 2011, 2016b), (l) data courtesy of Leica Geosystems.

SYSTEM TYPES

A variety of terrestrial laser scanners exist today on the market. System costs vary substantially depending on the necessary capabilities, with most systems ranging from \$40k to \$400k (USD). Range capabilities tend to be the primary driver of cost; however, a variety of features can influence the cost. Because of the wide range of types of scanners, it is important to find the scanner that works efficiently for the intended applications of interest. Each scanner will be suited for a specific set of applications.

Basic categories of scanners include the following:

- *Phase based*: These scanners calculate range based on the phase shift that occurs as the light is transmitted and returned. The advantage to these scanners is that they have high acquisition rates and are, in general, more precise than time-of-flight systems. A disadvantage of these scanners is that they are limited in range (typically <100 m).
- *Time of flight*: Time-of-flight scanners calculate range based on the two-way travel time of the laser pulse. These scanners are suited for longer-range applications (typically <300 m; however, some scanners are suited for ranges even up to 6,000 m). In the past, time-of-flight scanners were slower than phase-based scanners; however, recent scanners can scan at competitive rates (>1,000,000 points per second).
- *Full waveform*: Modern time-of-flight scanners are capable of recording the full waveform as its light is returned to the scanner. This information can be useful to distinguish pulses (e.g., vegetation filtering or classification). However, the additional information significantly increases the size of the datasets, and there is currently very limited software support for full-waveform analysis of terrestrial laser scanners. See Mallet and Bretar (2009) for more information.
- *Flash lidar*: Flash lidar systems transmit a highly divergent beam onto a scene and the reflected light is focused onto a focal plane array in a grid pattern, with each array element measuring the time of flight. This technique produces a structured point cloud simultaneously, similar to a photograph with rows and columns of pixels but with range information. These systems can capture data in a frame quickly (e.g., 30 frames per second).
- *Photon or Geiger mode*—This capability has yet to be implemented on a TLS because the laser pulse footprint (spot size) is generally sufficiently small; however, this implementation may be a possibility for TLS in the future. In this approach, individual light particles are detected to provide a higher-resolution point cloud. It also minimizes the angular uncertainty of where the pulse returned from within the laser pulse footprint.

Data Structure and Scan Patterns

Scanners operate on fixed angular increments because of their mode of acquisition (Figure 8-6). This means that the sample spacing on the targets of interest will vary based on the geometric configuration of those objects relative to the scanner. In addition to the angular sampling, the effective resolution will be influenced by the range because the laser spot size increases with distances (Lichti and Jamtsho 2006). Hence, objects close to the scanner will have dense sample spacing, whereas objects further away will have sparser data (Figure 8-7). Sample spacing (and data quality) also degrades on oblique surfaces, as discussed later.

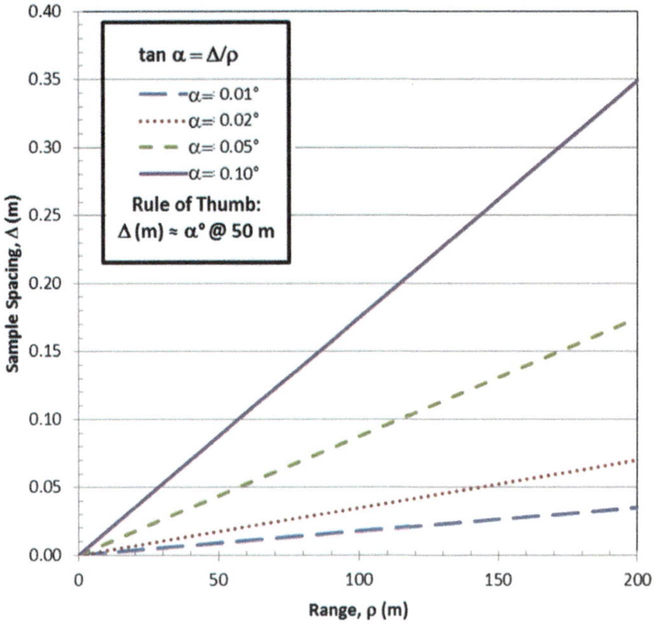

Figure 8-6. Scan sample spacing for various angular resolutions and ranges from the scanner. Note: These sampling intervals are for orthogonal targets only. Oblique targets will result in increased spacing.

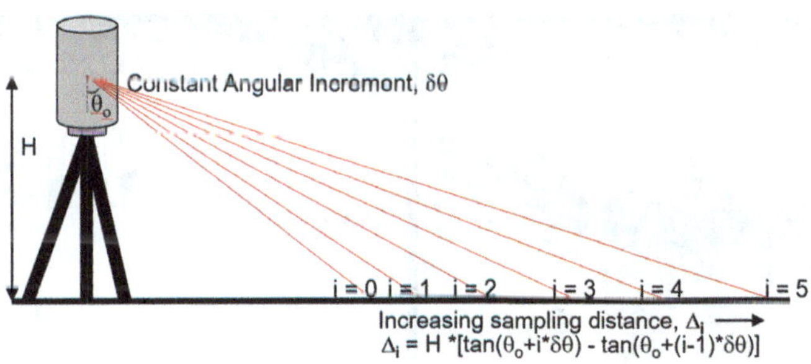

Figure 8-7. Increased spacing and obliquity of scan points with distance from scanner on flat surfaces. Source: Chin and Olsen (2015) © ASCE.

Some systems can provide data in an organized, structured fashion, which can be useful to simplify data processing. For example, Figure 8-8 shows an example of the field of view of a Leica ScanStation II. (Note that some scanners are not capable of scanning directly overhead, which is an important consideration for some applications such as bridge clearances or building interior or façade scanning). Figure 8-9 shows the data that were collected in a spherical coordinate system that has been mapped to a 2D image. Having structured data can help with the rapid generation of surface models, feature extraction, or calculations in the point cloud. However, note that specific file formats for data interoperability such as ASTM E57 are necessary to preserve this structure (Huber 2011 describes the ASTM E57 format). Otherwise, this information is lost when data are transferred as an unstructured point cloud between software packages.

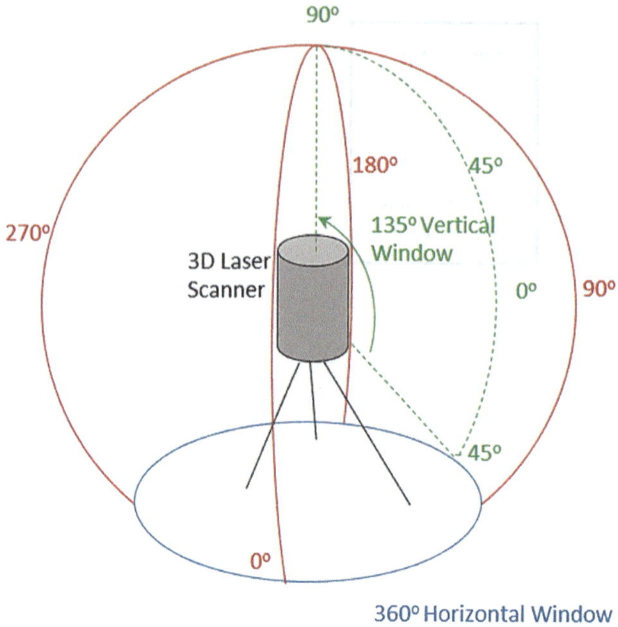

Figure 8-8. Sample field of view of a Leica ScanStation 2, which operates on a spherical coordinate system.
Source: Olsen et al. (2013c), reproduced with permission from Bar Publishing.

Figure 8-9. Structured point cloud of a rectangular courtyard in the Palazzo Vecchio projected to a 2D image colored by: (a) RGB photographs with magenta in locales with no data, (b) intensity values.
Source: Olsen et al. (2013c), reproduced with permission from Bar Publishing.
Note: Black = weaker returns, white = stronger returns.

DATA QUALITY CONSIDERATIONS

This section discusses some common limitations, errors, and artifacts in TLS data.

More noise or errors may occur in lidar data than in some other sources because of the higher resolution of the data. However, noise and errors still exist in all other data sources! The advantage to lidar is that it can be seen and therefore often corrected.

Several aspects can affect the quality of laser scan data and should be considered when working with TLS data. The following is a brief discussion of common artifacts or limitations encountered with TLS data.

- *Line of Sight*—Scanners operate on the line-of-sight principle, meaning "what you see is what you get." Hence, occlusions (shadows or holes) exist in the data behind objects and will not be captured in a scan unless the scanner is repositioned. Multiple scans are required to formulate a complete model (see Registration Strategies for more details).
- *Traffic Obstacles*—Traffic or pedestrians will create some noisy artifacts in the dataset. The impact of these is minimized provided that the objects are moving quickly or far away from the scanner. These can be removed via filtering or manual editing (see Processing Workflows for more details).
- *Multipath Errors*—Highly reflective surfaces such as mirrors or water surfaces can lead to multipath effects where another object is reprojected beyond the surface (Figure 8-10a). The scanner knows only the time of travel and the angle at which the pulse is fired, elongating the path of the laser.
- *Mixed Pixels*—Mixed pixels occur when two objects are close together (<1/2 the laser pulse length) and the laser scanner cannot distinguish between the edge of one object and the other object behind it (Figure 8-10b). In this case, artificial points are created in between the objects.
- *Blooming*—Blooming occurs with reflective objects at long ranges (Figure 8-10c). With blooming, reflective objects become larger in the point cloud than they really are because the laser spot size (footprint) increases with range and the laser pulses capture the edges of a reflective object. However, the lidar unit cannot distinguish where in the pulse footprint the return came from.
- *Saturation*—Saturation is an effect that occurs when a highly reflective target is located close to the scanner (Figure 8-10d). In this case, the range to the object is underestimated because the scanner cannot fully encode the peak as it exceeds its detection threshold. As a result, the reflective object tends to "pop-out" of the surface where it should be.
- *Discoloration of Point Cloud due to Parallax*—Often, the camera is not directly aligned with the scanner. Hence, parallax occurs, and the image does not perfectly match the point cloud, leading to some shifts. Rigorous calibration can minimize these artifacts. Scanners with external cameras may require a calibration or adjustment each time the camera is disconnected from the scanner, whereas scanners with fixed, internal cameras only periodically require calibration. These incorrect colorations are most common on edges of objects such as buildings or in the branches of a tree with the sky background (Figure 8-10e). Further, the motion of trees from wind can also contribute to this incorrect coloration because color images are often captured before or after the scan has been completed rather than simultaneously.
- *Problematic Surfaces*—Several surfaces can be problematic for scanning because of either too low or too high reflectivity of scanned objects. The impact of these will vary heavily depending on scanning geometry and somewhat depending on the scanner wavelength and other system characteristics.
  - *Highly Reflective at Close Ranges*—These objects return too much energy and create saturation problems (Figure 8-10d).
  - *Dark*—Light is absorbed in dark surfaces resulting in a significant decrease in range capabilities.
  - *Wet*—This can be problematic for scanners that operate at the near-infrared range. Intensity values are significantly degraded in wet areas, and in many cases, data may not be obtainable (Figure 8-10f). Similarly, poor results will occur with steam, precipitation, or heat rising from surfaces, which can result in refraction.

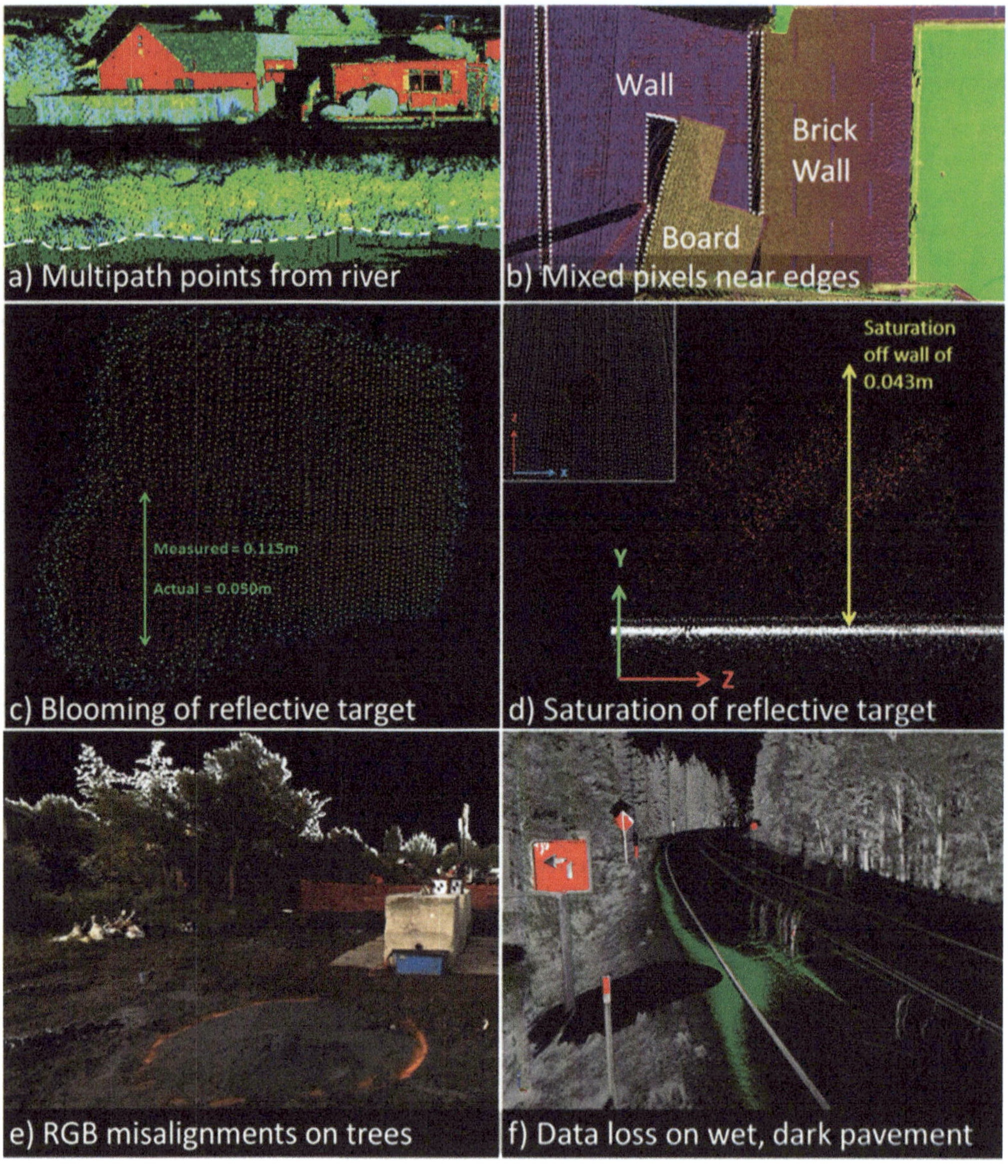

Figure 8-10. Examples of common data artifacts in point clouds: (a) multipath points (dark green) producing an artificial reflection of the cliff below the Avon River (Christchurch, New Zealand), (b) mixed pixels between a board and a wall in a damaged home, (c) blooming of a reflective circular target resulting in it appearing over twice its size, (d) saturation of a reflective circular target scanned at close range resulting in points that are offset from the wall surface, (e) misalignments of RGB color information from poor camera calibration on the tops of trees. However, note that artifacts at the top of trees will still occur because the leaves and stems do not occupy the full pixel so the sky color will dominate those pixels and be mapped to the lidar points because it is not resolved in the pixel but can be detected in the lidar pulse; and (f) data dropout along the Pacific Coast Highway (101) in Oregon after the road was freshly paved and the scan obtained right after rainfall, producing a dark, wet surface. A few (green points) were obtained on the road surface except for at the locations of the reflective striping.
Source: (c) and (d) Modified from Williams et al. (2013).

- *Mirrors*—Mirrors will cause continued reflections and reprojections of the area scanned (Figure 8-10a).

- *Data Quality Influenced by Scanning Geometry*—Scanning surfaces obliquely lead to irregular point spacing (Figure 8-11). They also significantly degrade the accuracy of the data points. Data degrade significantly for incidence angles >45° (Laefer et al. 2009, Olsen et al. 2009). See Schaer et al. (2007) for further discussion on scanning geometry and its influence on accuracy.
- *Edge Loss Effects*—Laser scanning will not precisely capture the edge of objects because the beam can be split (see Tang et al. 2009).
- *Variability in Resolution*—Resolution will vary throughout the point cloud based on range to the scanner as well as incidence angles (Figure 8-11). The significant variability of point density throughout the point cloud of a TLS scan results in many challenges in producing models. Neighborhood search sizes, rasters, or grids use fixed cell sizes. Triangular Irregular Networks (TINs) can be adaptable to data density; however, for complex surfaces, algorithms need regular cell sizes. In some cases, filtering is done on the point cloud for more uniform sampling (see Processing Workflows).
- *False Returns from Incoming Solar Light*—With some units, light from the sun or other sources can be misinterpreted as TLS returns.
- *Spurious Points*—Depending on the wavelength and sensitivity characteristics of the scanner, a point cloud can contain spurious points from moisture, dust, and so on in the air. Many of the scanners most sensitive to these will apply an automatic filter based on the proximity of the points or intensity values of the points to minimize these artifacts (see Processing Workflows).
- *Processing Algorithms in Development*—Although data processing automation has made tremendous advances in terms of efficiency and quality, many artifacts will still need to be cleaned up manually (see Processing Workflows).
- *No Direct Intelligence/Attributes with Data*—In general, individual scan points do not have attributes aside from the XYZ coordinates, intensity, and color information. Fortunately, the data are often easy to interpret visually. However, adding intelligence to the data through classification procedures can simplify processing and interactivity with the data (see Processing Workflows).

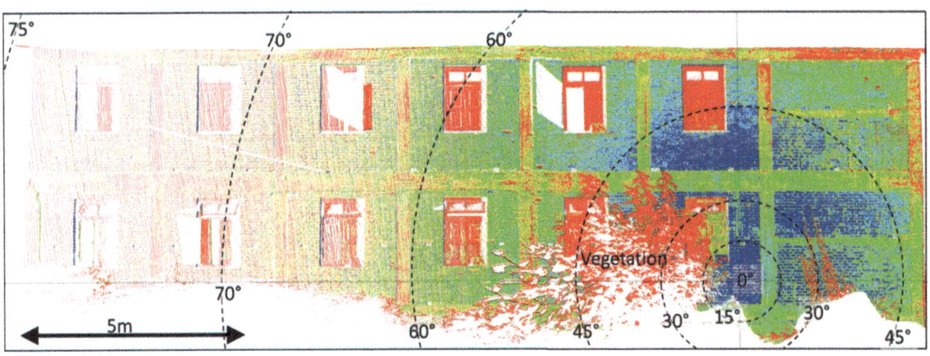

Figure 8-11. Data degradation in resolution accuracy with increasing obliqueness (increasing angle of incidence values, which are plotted as approximate contours).
Note: The scanner was positioned within 3 m of the wall at the 0 degree angle of incidence contour. The point cloud is colored by intensity. Blue = weaker returns, red stronger returns.

DIFFERENCES TO AIRBORNE AND MOBILE LIDAR

Airborne laser scanning (ALS) has become prolific since the 1990s to support a wide variety of applications. Although there are many similarities in the technology, there are some key differences between TLS and airborne lidar. First, airborne lidar is collected kinematically, whereas TLS is collected statically from a tripod. [Note that in some cases, a TLS system can be mounted to a wagon or vehicle (Olsen et al. 2009) to improve efficiency, but without extra components such as an inertial measurement unit (IMU) used for MLS, the scanner must remain stationary during each scan.] Hence, airborne lidar can cover larger areas more efficiently; however, TLS can provide much higher accuracies (particularly local or relative accuracy). In addition, TLS provides improved information on vertical features such as cliffs or building façades that are difficult to capture in airborne lidar without sophisticated flight planning. TLS data are also much of higher resolution (millimeters to centimeters) than airborne lidar datasets (decimeters to meters). However, airborne lidar datasets tend to have relatively more uniform data sampling than TLS, where resolution is high (tens of thousands of points per square meter), close to the scanner, and on orthogonal features and degrades far from the scanner and on oblique surfaces.

MLS is also a kinematic solution similar to airborne lidar. However, MLS comes closer to achieving similar accuracies as TLS. In many cases, TLS data are used to supplement MLS surveys in areas that are not captured from the roadway or where additional detail is needed.

Table 8-2 provides a comparison of these platforms. Chang et al. (2012) provide a detailed comparison of ALS, MLS, and TLS for transportation applications. Young et al. (2010) provide a comparison of ALS and TLS for seacliff erosion evaluations. Each of these techniques provides point clouds that can be utilized for a variety of applications and, in some cases, can be blended together to produce the desired results. However, it is critical to understand the capabilities and limitations of each data source when using them in this context.

Note that although this discussion focuses on airborne, mobile, and terrestrial platforms, a variety of other platforms exist, including backpack and handheld scanners, which may be a more efficient and cost-effective solution for a project.

Because of the many similarities between the lidar platforms, relevant references related to work in airborne or MLS will sometimes be included in this chapter when appropriate. Also note that MLS and ALS are covered in detail in Chapters 9 and 10, respectively.

TLS WORKFLOWS

Figure 8-12 shows common steps in a TLS workflow. Note that the steps will vary project to project. Each of these steps is described in the following sections in more detail.

PLANNING AND PRELIMINARY SITE SURVEYS

Table 8-3 is a checklist of questions to consider when performing planning or preliminary site surveys. Some basic scan planning software tools are available to aid with determining scanner locations. Some guidance is available for planning surveys. For example, Caltrans Manual of Surveys, Chapter 15, provides plans for the spacing of scan positions and targets for highway surveys. Olsen et al. (2009) present a case study that outlines scan spacing, distance from object, and resolution requirements for coastal bluff scans to provide an efficient survey that maintains a prescribed level of accuracy based on the scanner utilized. Similar analyses could be conducted through field efforts or simulations for other types of

Table 8-2. Comparison of Various Lidar Platforms.

Attribute	Airborne	Mobile	Terrestrial
Accuracy	>5-cm RMS (vertical) sufficient for planning, asset management, but not for engineering design	0.05–0.5 m RMS (3D). Varies depending on the system. Some systems are capable of providing data sufficient for engineering design	0.005–0.05 m RMS (3D). In general, capable of providing data sufficient for engineering design and analysis
Footprint	dm (>0.5 m)	mm to cm	mm to cm
Level of detail (resolution)	Lower resolution. Fairly uniform across the site, typically 1–80 points per square meter	High resolution. Highly variable. >1,000 points per square meter close to the scanner, but degrades quickly	Highest resolution (user controlled). Highly variable. >1,000 points per square meter close to the scanner, but degrades quickly
View	Poor coverage on vertical features (e.g., building facades, cliff faces, overhangs)	Limited to objects visible from the roadway	Good coverage of vertical surfaces, good coverage of horizontal surfaces with sufficient setups
Coverage	Very fast. Can cover large areas relatively quickly. Requires good GNSS conditions	Able to collect data at highway speeds. Slower than ALS. May need static TLS to fill in the gaps. Requires good GNSS conditions	Slower. Good for localized sites, buildings (inside and outside), and other locales with no GNSSs. Can be mounted in a stop-and-go fashion to improve coverage
Range	Typically 1,000–2,000 m	Typically <75 m from the vehicle	Typically <100 m; however, some scanners can collect up to 6,000 m
Safety	General flight safety requirements	Travel with traffic with minimal footprint and obstruction	In general safer than traditional survey equipment; however, still requires one to set up on the ground in a stationary position. Also, some systems use noneye safe lasers
Logistics	Requires flight planning, trained pilot	Requires coordination with traffic	Very flexible
Post-processing efficiency	Well-established workflows and software	Processing capabilities of software vary depending on the application	Processing capabilities of software vary depending on the application

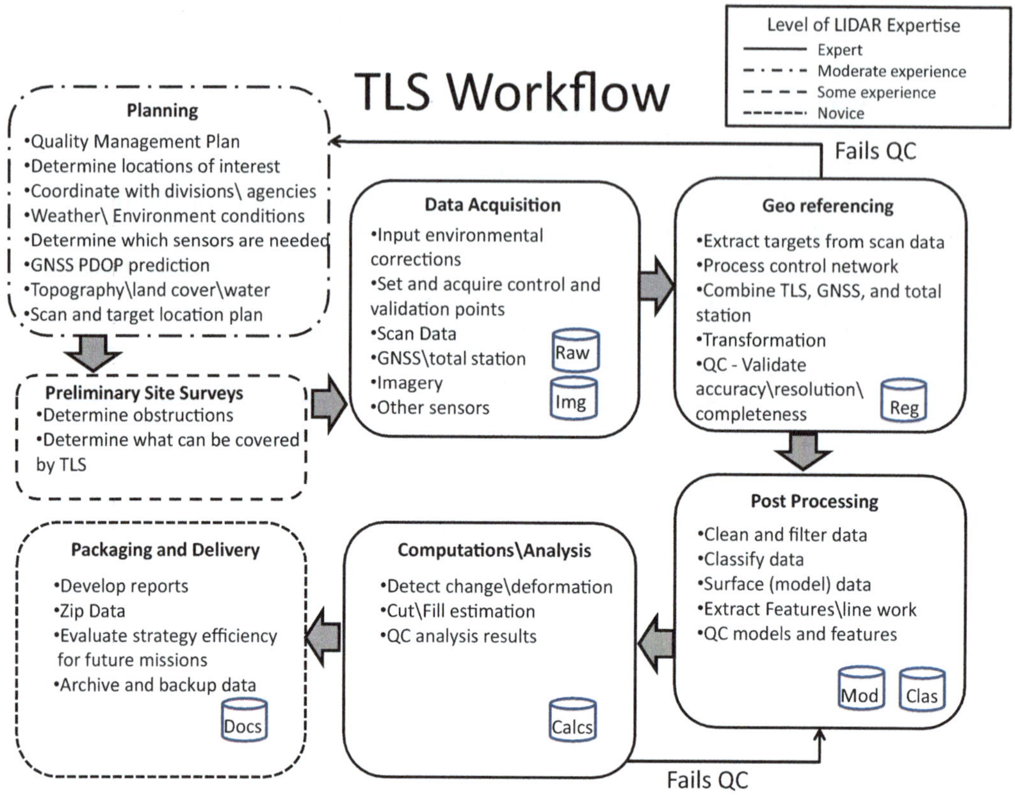

Figure 8-12. Example of steps in a TLS workflow. The canisters represent new datasets that are produced during each stage.
Source: Modified from Olsen et al. (2013d), reproduced with permission from the National Academies Press, Washington, DC.

scenes. O'Banion and Olsen (2019) provide an approach to estimate the amount of scanning necessary based on the desired quality for terrain modeling. Starek et al. (2010) provide guidance for optimizing TLS surveys for dynamic terrain monitoring. Finally, Zhang et al. (2016) provide procedures and considerations for planning TLS surveys for construction.

FIELD PROCEDURES

In the prior section, we discussed important aspects for planning. Although a plan is certainly an important element, it is vital to adapt your plan to the feature(s) of interest and obstacles encountered while scanning. In particular, one should pay careful attention to visibility and perform additional scans as needed to fully capture the objects of interest in case of unexpected objects blocking the view. Sometimes a small shift of the scanner location can substantially increase the field of view, so one should always look around before positioning the instrument. During this quick visual assessment, one should also make sure the scanner can see sufficient targets. If using a GNSS antenna mounted directly above the scanner for positioning, then one should ensure a clear view of the sky.

In addition to the number of location of scans, one also should consider adapting the resolution of the scanner, as necessary. Sometimes, one high-resolution scan can sufficiently

Table 8-3. Questions and Considerations for Planning a Site Survey with TLS.

Category	Questions and considerations
Location	Where are you able to view the object(s) of interest? Where can you physically set up in a stable location? Is an on-site power source available? Where do you have access to set up equipment (e.g., scanner, targets)? Are there onsite vibration and/or other disturbances? Are there major obstructions that need to be worked around? Will there be downtime with reduced traffic (vehicle and pedestrian) at the site? Presence of water at the site. For example, scans on the coast need to work around tide schedules.
Object(s) of interest	What are you scanning? Which objects are the most important? What material is it made of? What is its reflectance? Is the surface wet or dark, which can impede laser returns? How large is it? Are there limits to the visibility of the object?
Project requirements	What are the resolution\accuracy requirements of the project? Where are data gaps intolerable? Where are they acceptable? How does the end user need the data? What end products are required? Will future surveys be needed (e.g., deformation monitoring)? If so, it may be worthwhile to place permanent targets and survey control
Field crew	How many people are available to assist with target setup, survey control, scanning, and so on What training/experience has the field crew had with scanning? What experience do they have with this specific application?
Survey control	Availability of survey control and GNSS visibility? Where will control points be located? How will targets be tied to survey control? Where should validation points be obtained? How many are needed?

cover a scene and will result in easy processing. However, most of the time, multiple scans will be required.

In the author's opinion, for applications such as topographic mapping, it is often more effective to perform more scans at a lower resolution than just a few scans at very high resolution because there will probably be a lot of visibility constraints owing to the presence of objects in the scene. Multiple scans from different vantage points will help fill in those occlusions so that interpolation will not be necessary over large occlusions. They will also improve the uniformity of the sampling.

Note that resolution is set in horizontal and angular increments, so one must consider the range, angle of incidence, and desired sampling on the objects of interest to set the scanner resolution, as discussed previously.

An important consideration during acquisition is to ensure appropriate overlap (scene-dependent, but at least 20%) between scans to provide adequate data for surface matching algorithms as well as validation purposes. This also helps ensure that data are collected below where the scanner is positioned. One should also always be sure to scan beyond the area of interest. This will ensure adequate coverage in the area of interest but is also important for some processing algorithms to avoid boundary effects (e.g., ground filtering, discussed later in the chapter).

During scanning, one can typically monitor the data quality because most systems have an onboard display or can transmit preview data to a laptop. Verify that problem areas (wet or dark surfaces) are providing adequate returns if they are important for the project. Also, one should complete a preliminary registration in the field to ensure that the appropriate area is covered and an adequate amount of targets are scanned, enabling one to correct potential issues in the field.

Leveling/Inclination Sensors

For instrument setup, it is important to discuss some fundamental differences between TLS and common survey devices. Scanners cannot be leveled to the same level of accuracy as a total station or other instruments. However, some scanners are equipped with level compensators. In such a case, the scanner is set up near level (within approximately 20 arcminutes), and the compensator can correct each scan line to be level. The advantage to this approach is that data can be internally corrected in the field from vibrations from passing vehicles. Should the scanner deviate too far from level, the compensators disengage, and the scan must be leveled using alternate techniques.

The second approach is using dual-axis inclination sensors/tilt compensators within the scanner. In this approach, the scanner does not need to be setup level; it can deviate by as much as 10 degrees, depending on the scanner. A variety of techniques are implemented to adjust the scan based on the inclination readings. In some cases, separate cycles of readings are obtained; in other cases, the inclination readings during the scan are averaged and used to correct the scan. For further treatment on the subject, see Lichti (2007) and Silvia and Olsen (2012).

Although level compensators or inclination sensors can provide a correction for an out-of-level setup, a leveled setup will be more stable and reduce potential problems with offsets in weight and vibration of the scanner during operation. One does not need to spend as much time to precisely level the scanner; however, one should still take the effort to approximately level the scanner in a stable configuration. This may be more challenging in a stop-and-go approach where the scanner is on a mobile platform. To summarize:

- Utilize level compensators or inclination sensors for improved quality control.
- Even if an instrument has level compensators or inclination sensors, one should still attempt to level the scanner and ensure that it is in a stable configuration.

The advantage to level compensators or inclination sensors is that field procedures can be implemented much more efficiently because one does not need to precisely level the scanner. For scanners without these sensors or capabilities, scans can be leveled as part of the registration process. However, the absence of these sensors removes a source of redundancy that can help achieve better results or provide data for validation of the registration. Hence, it is strongly recommended to utilize a scanner with these capabilities. It is also important to understand whether scans are constrained to the field leveling (with corrections) or are allowed to adjust in the registration (see Registration Strategies). Sometimes, the registration can provide poorer leveling than the sensors, particularly with poor target geometry or in cloud-to-cloud solutions.

Field Notes

Although field notes are not as intensive as traditional surveys given the automated nature of scanning, it is vital to keep good field notes describing the equipment used, problems encountered, map of scan positions, locations of targets, environmental settings, field crew, and other pertinent information.

Data Backup Strategies

Given the large file sizes in scanning, strict data backup and archiving protocols are important. During acquisition, high-quality media with fast read/write times should be used to minimize data loss. After acquisition, an immediate back up of the scan data to a data storage system must be done and the raw files archived permanently. These raw files can be used if there is a problem or if there is data loss, enabling easy recovery later in the processing. Also, software evolves quickly, and if the data are available in the future, one may be able to register the scans from the raw data with higher accuracy.

Care of Equipment

Not only can the scanners be expensive to replace or repair, but batteries, wiring, and other parts can also be expensive as they are often custom-made for the manufacturers. Further, because scanner technology evolves rapidly, some of these parts may become difficult to find as models are phased out. Each manufacturer will describe specific procedures for care of the instrument; particularly procedures and cloths for removing dust from windows.

REGISTRATION STRATEGIES

TLS is a line-of-sight technique. Hence, multiple scans are often required to generate a complete model of the object(s) of interest. Figure 8-13 shows an example of three scans of seacliffs on a coastline and the occlusions (shadows) that occur. Registration is the process of tying multiple scans together into a single coordinate system. In some literature, particularly within the field of computer science, this may be referred to as an "alignment." In many engineering applications, the scans should also be georeferenced. The georeferencing can either be completed simultaneously with the registration or as a separate registration, depending on the workflow. It also helps link the scan data to other sources such as GIS geodatabases or aerial ortho-photographs. This section discusses common approaches to performing the registration. TLS data can be acquired and registered using a variety of techniques. In addition to this chapter, Lichti and Skaloud (2010) provide a

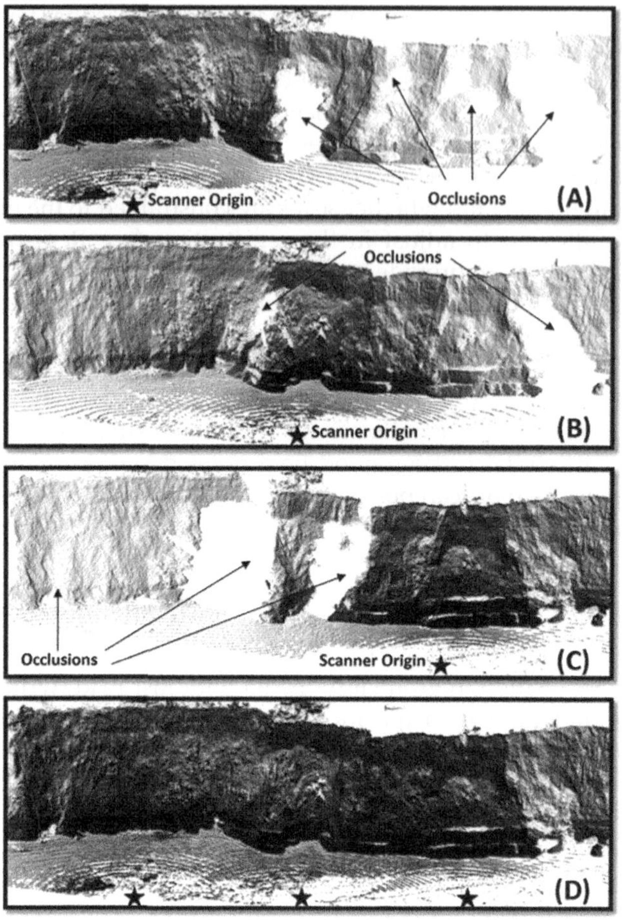

Figure 8-13. Occlusions present in scans A, B, and C. D shows the combined model after registration. Source: Olsen et al. (2011) © ASCE.

comprehensive review of registration and calibration methodologies. This chapter later summarizes the differences between these methodologies.

The following objectives should be completed during the registration and georeferencing process:

- Minimize and optimally distribute error to avoid error propagation,
- Provide consistency within a survey,
- Provide consistency between surveys (e.g., Change Detection—see Analyses section), and
- Enable integration with other data sources.

Calibration Procedures

Scanner calibration for TLS is typically completed annually by manufacturers. However, calibration should always be done when there is concern over instrument quality owing to damage. It is also recommended to perform a calibration before and after a major project to check the stability and reliability of the instrument.

These calibration processes are especially important for systems that utilize multiple lasers. The procedures and appropriate correction models will also vary by scanner model, so they should be carefully developed and applied. For a full treatment on the subject and examples of calibration processes, the reader is referred to work by Lichti and Skaloud (2010). Calibration procedures have evolved from established procedures in surveying (Lichti and Lampard 2008) and photogrammetry and adapted to TLS. Many systematic errors can be treated as perturbations, which can be modeled as adaptive corrections. Common systematic errors of interest (Lichti and Skaloud 2010) that can be accounted for in a calibration procedure include the following:

- *Ranging Errors*: TLS utilize EDMs, which need to be calibrated (Salo et al. 2008). (Also, atmospheric effects such as temperature, pressure, and relative humidity need to be accounted for at each field site.)
- *Internal Axis Offsets* including the vertical offset of the laser axis to the trunnion (horizontal) axis, the nonorthogonality of the collimation (line of sight) axis to the trunnion axis, and the nonorthogonality of the horizontal encoder and the vertical laser axis.
- *Vertical Circle Eccentricities* (elevation angle).
- *Stability and Rotating Mirror Anomalie*s, which vary significantly by scanner model.

Some example work in TLS calibration include Gielsdorf et al. (2004), Kersten et al. (2005), Lichti and Franke (2005), Reshetyuk (2006), Lichti and Licht (2006), Lichti (2007, 2008, 2010a, b), Lichti et al. (2011), Chow et al. 2013, and Glennie and Lichti (2010, 2011).

Transformation Points versus Validation Points

In this section, we define two types of targets to be utilized in scanning: transformation control targets (TCT) and validation control targets (VCT). TCTs are used to compute the transformation(s) for the registration and georeferencing processes. VCTs serve as an independent check on the data quality by surveying key features with another technology.

Although techniques for placing control targets and performing these adjustments are well defined, the density of TCTs and VCTs required is not clearly defined. Typically, the number of additional targets required has been empirically determined, from field trials and survey experience. Manufacturers and public road agencies have developed standards, which have been proven to meet accuracy requirements (e.g., Caltrans Manual of Surveys, Chapter 15). However, the density and pattern of TCTs and VCTs should adapt to different observing criteria (e.g., in a tunnel versus open road) as is recommended in the TRB guidelines, NCHRP Report No. 748 (Olsen et al. 2013d). The number of TCTs and VCTs depends on the accuracy requirements, size, and other aspects of the project.

- TCTs and VCTs should be spread throughout the area of interest.
- In addition to discrete points as VCTs, it is important to use features or cross sections at strategic locations in the dataset to perform a more rigorous analysis and identify potential problems with the systems.
- Although error on many projects is typically evaluated only in 1D (vertical), it is important to evaluate and quantify the 3D errors.

Rigid-Body Coordinate Transformations

For terrestrial laser scanning, a rigid-body coordinate transformation [three translations and three rotations (Figure 8-2)] is applied uniformly to all points belonging to a scan position during the registration. (This differs from kinematic scanning where every point is georeferenced uniquely.) Note that in this example, rotations are performed relative to the scan origin and must be done in the appropriate sequence for obtaining consistent results. Alternate implementations can achieve the same results; however, the key is to be systematic throughout the process. Also note that the transformation process of the scan is the same, independent of the method used for georeferencing to estimate the translations and rotations. The following parameters are of interest in the transformation:

- *Translation* X, Y, Z: Scans are collected in a scanner coordinate system with the scanner origin being located at coordinates 0, 0, 0.
- *Rotation X, Y, Z* (Roll, Pitch, Yaw): Leveling from the instrument is sometimes constrained and other times is solved for when determining the transformation. See the section on leveling for more discussion on this topic.
- *Scale*: Often not included in the transformation directly; however, scaling owing to environmental conditions can be accounted for on many modern scanners by inputting temperature, pressure, and relative humidity readings into the instrument during the acquisition stage.

We now describe the rigid body transformation process (see Figure 8-2 for a schematic and definitions of the terminology).

To register a scan to the project coordinate system, we first apply the rotations about the Scanner's Origin (0, 0, 0). We then define the rotation about the x-axis as roll (ω), the rotation about the y-axis as pitch (ϕ), and the rotation about the z-axis as yaw (κ). Note that these three rotations can also be summarized as a single rotation about a vector (quaternion), which is described subsequently. Scanners with inclination sensors can measure the roll and pitch of the instrument, whereas yaw is related to the scanner's bearing or azimuth relative to an external project coordinate system.

Given the laser scan points, $\boldsymbol{p}$, from a single scan position with coordinates (x_i, y_i, z_i) for the ith point, registration is completed by rotating the points in that scan position around the x, y, and z axes centered at the scan origin:

$$\boldsymbol{p}' = \boldsymbol{R}(w,f,k)\boldsymbol{p}+\boldsymbol{t} = \boldsymbol{R}_z\boldsymbol{R}_y\boldsymbol{R}_x\boldsymbol{p}+\boldsymbol{t} = \boldsymbol{T}(w,f,k,t)\boldsymbol{p} \tag{8-1}$$

where

$\boldsymbol{R}_z$, $\boldsymbol{R}_y$, and $\boldsymbol{R}_x$ = Rotation matrixes around each individual axis,

$\boldsymbol{R}$ = Combined rotation matrix around the scan origin (shown subsequently),

$\boldsymbol{t}^T = (\boldsymbol{t}_x, \boldsymbol{t}_y, \boldsymbol{t}_z)$ = Vector to translate the scan origin (0, 0, 0) to its location in the new coordinate system, and

p' = Matrix of the new point coordinates in the new coordinate system.

Note that each scanner software package and setup will handle these rotations differently.

$$\boldsymbol{R}(w,f,k)=\begin{bmatrix} \cos f\cos k & \sin w\sin f\cos k-\cos w\sin k & \cos w\sin f\cos k+\sin w\sin k \\ \cos f\sin k & \sin w\sin f\sin k+\cos w\cos k & \cos w\sin f\sin k-\sin w\cos k \\ -\sin f & \sin w\cos f & \cos w\cos f \end{bmatrix} \tag{8-2}$$

Note that the rotation and translation matrix can also be combined into a single 4 × 4 transformation matrix, as follows:

$$\boldsymbol{T}(w,f,k,t)=\begin{bmatrix} \cos f\cos k & \sin w\sin f\cos k-\cos w\sin k & \cos w\sin f\cos k+\sin w\sin k & t_x \\ \cos f\sin k & \sin w\sin f\sin k+\cos w\cos k & \cos w\sin f\sin k-\sin w\cos k & t_y \\ -\sin f & \sin w\cos f & \cos w\cos f & t_z \\ 0 & 0 & 0 & 1 \end{bmatrix} \tag{8-3}$$

When utilizing this form, $\boldsymbol{p}$ must be changed into a 4×1 vector by inserting 1 into the last element.

Now, for the inverse. Suppose the 4×4 matrix for the scan transformation is available with the scan. How is six parameters for rotation and translation recovered?

Translation. Translation is straightforward to pull out of the matrix:

$$t_x = R[1,4], t_y = R[2,4], t_z = R[3,4] \tag{8-4}$$

Rotations. The rotations can be found with a little math to simplify some of these elements:

$$w = \tan^{-1}\left(\frac{R[3,2]}{R[3,3]}\right), f = \tan^{-1}\left(-\frac{R[3,1]}{\sqrt{R[1,1]^2 + R[2,1]^2}}\right), k = \tan^{-1}\left(\frac{R[2,1]}{R[1,1]}\right) \tag{8-5}$$

Note that to obtain the correct value, the atan2 function should be used rather than the generic atan function to account for the quadrant of the parameters. For most terrestrial scanning purposes, w and f will typically be very small because the setup will be more or less level (although this will not be the case in some applications such as when the scanner is mounted upside down). However, k will range from $-180°$ to $+180°$.

In many cases, scanners utilize quaternions (rotation, β, about a vector, v, which is the scanner's internal axis) instead of Euler angles for computational efficiency as well as stability to avoid gimbal lock (loss of a degree of freedom when two other degrees of freedom can produce an identical rotation). A rotation vector and angle can be represented as a quaternion:

$$q = \{q_w \quad q_x \quad q_y \quad q_z\} \tag{8-6}$$

where

$$q_w = \cos\left(\beta/2\right), q_x = v_x \sin\left(\beta/2\right), q_y = v_y \sin\left(\beta/2\right), q_z = v_z \sin\left(\beta/2\right) \tag{8-7}$$

A rotation matrix, R, can be generated from the quaternion (following normalization of the quaternion):

$$R = \begin{pmatrix} 1 - 2q_y^2 - 2q_z^2 & 2q_xq_y - 2q_wq_z & 2q_xq_z + 2q_wq_y \\ 2q_xq_y + 2q_wq_z & 1 - 2q_x^2 - 2q_z^2 & 2q_yq_z - 2q_wq_x \\ 2q_xq_z - 2q_wq_y & 2q_yq_z + 2q_wq_x & 1 - 2q_x^2 - 2q_y^2 \end{pmatrix} \tag{8-8}$$

Also, Euler angles (w, f, k) can be extracted from this matrix using the aforementioned formulas if desired.

Different scanners will implement slightly varying procedures for the registration. In some cases, the rotations will be determined based on the scanner coordinate system, and in other cases, they are defined based on the global coordinate system. Hence, it is important to understand how these transformations are being implemented with one's scanner and software.

Often, the scans are first transformed individually into a local coordinate system and then the merged scans are jointly georeferenced into a geodetic coordinate system.

In the final solution, all points from a scan position are georeferenced together as a rigid body. However, each point will have different amounts of overall positioning error because of scanning geometry such as angle of incidence, range, and so on.

Coordinate Systems

Many TLS processing packages do not yet have full coordinate system support as is common in surveying or GIS software; however, in recent years, many major packages have made this support available. The lack of coordinate support can lead to difficulties in effectively applying scale factors to the data to convert the ground measurements obtained with the scanner to grid measurements (see Chapter 3 for more on grid versus ground), particularly over long distances (Olsen and Gillins 2013). It also means that the end user must pay careful attention to this when processing the data because they are left to manage the coordinate system in the registration.

Ensure that the coordinate system utilized is fully understood, and make sure that various packages are handling that coordinate system correctly.

For georeferenced scans, coordinates will be very large. Some scan software packages were programmed using floating point precision and may truncate data. In such a case, an offset is required to translate the coordinates so they can fit well within floating point precision (e.g., coordinates less than 10,000 m for precision at the millimeter level).

Direct Georeferencing

Direct georeferencing is a process of setting up the scanner over a point with a known coordinate, leveling it, and orienting it (either through a digital compass or through backsighting). Some new systems such as multifunction total stations with scanning capabilities can be directly georeferenced following a typical surveying workflow (as per Chapter 7). Alternatively, a GNSS receiver can be mounted atop the scanner instead of setting up over a control point. For example, Mohamed and Wilkinson (2009), Wilkinson et al. (2010), and Wilkinson et al. (2015) provide a methodology to utilize two GNSS receivers for direct georeferencing. Mårtensson et al. (2012) evaluated the measurement uncertainty using network RTK-GNSS positioning for TLS.

See Lichti et al. (2005) for a comprehensive treatise of error sources for directly georeferenced scans. Note that not all scanners have the capability to be set up directly over a point, backsighting capabilities, built-in digital compass, integration with GNSSs, and so on, necessary to achieve these results. In addition, a common limitation is that direct georeferencing does not take advantage of the information available in the overlap between scans; hence, shifts can be observed in the data, whose magnitude depends on the accuracy of the components used to directly georeference the scans. The key advantage to direct georeferencing is that the registration processing is greatly simplified.

Target-Based Registration

Target-based registration is the most common approach to register scans in surveying engineering. In a target-based registration, common target points are linked between scans and the transformation parameters are found through a least-squares adjustment to minimize error. Some software provides the ability to weight individual targets. As an alternative to a least squares adjustment, Han (2010) presents a noniterative approach that yields similar results to a least squares adjustment for the registration of TLS data.

A variety of targets can be utilized. In general, two basic types of targets are common: reflective objects or patterns. For obtaining the best results, either high-resolution scans of the targets need to be completed or they need to be positioned close to the scanner to be extracted from a panoramic scan completed at a sufficient resolution. Scan software will fit shapes to these objects to extract their center points with high accuracy because they perform a least-squares, best fit of the many points obtained across the target. For the best results, subcentimeter resolution should be obtained on each target; however, this varies depending on the size of the target used. Fan et al. (2015) provide a rigorous assessment of target-based georeferencing errors and overall influence on the registration results.

Individual manufacturers will use specific targets for their scanners. Extreme caution should be used when using third-party or custom targets. Although such targets can sometimes be used, one needs to rigorously evaluate their performance with each scanner to avoid problems. In many cases, the manufacturer has optimized their processing workflow to a specific target type. Further, some manufacturers automatically apply corrections (e.g., range or angle of incidence corrections to minimize saturation and blooming problems) to measurements on targets or require a certain reflectivity range to identify a target. Third-party targets may not have the same specifications even if they look similar. This can lead to difficulties extracting the targets, or worse, it can lead to incorrect coordinates for the target center.

For maximum efficiency and quality, use and test the appropriate targets corresponding to your scanner.

Becerik-Gerber et al. (2011) analyzed various target types and layouts and their influence on registration accuracy and efficiency in both indoor and outdoor scenes. They stress the importance of following strict labeling protocols, placing redundant targets, and minimal human interaction with the targets. Their tests showed improved accuracy on targets with time-of-flight scanners compared with phase-based ones. They also note that the distance between the scanner and the targets substantially affects the acquisition time. However, it is important to note that targets need to be placed throughout the scene of interest to avoid overfitting to targets at close range, resulting in poorer registration results far from the scanner.

Reflective Targets. Reflective targets can be discs, cylinders, spheres, or other shapes. Pesci and Teza (2008) provide a rigorous assessment of reflective targets and artifacts for TLS. The advantage to reflective targets is that, given their strong return signal (intensity), they can easily be identified in the point cloud and extracted. Many software packages can do this automatically; however, they can be prone to false positives from other reflective objects in the scene (e.g., headlights on a car). When cylinders or spheres are used, they can be advantageous because they can be captured from all sides without any repositioning. With only partial coverage of the object, a mathematical model can be utilized to extract the center coordinates. Disc targets, however, may need to be rotated to be visible from multiple scan positions. Centering targets over a control point and ensuring stability during scanning can sometimes be a difficult task and can introduce additional error (Becerik-Gerber et al. 2011).

One should be careful when utilizing reflective targets. Reflective targets can be problematic at close range (<20 m but varies based on the scanner model) owing to saturation effects, which will typically cause the scanner to underestimate the range to the target. At far ranges, these targets will create blooming effects where the target appears enlarged in the point cloud; however, in general, blooming effects are symmetric, so they often cancel out when extracting the target center. See the previous section on Data Quality Considerations for more information on scan artifacts. Next, although reflective constants can be applied in some TLS software, most constants are held as proprietary information or not utilized, resulting in introduced error when not using the manufacturer targets because their reflectivity will differ. Finally, extremely reflective objects such as prisms should not be used with TLS. These objects will return too much energy and could potentially damage the scanner.

Pattern Targets. Pattern targets can either be existing, well-defined objects in the scene (e.g., pavement markings) or placed. Some scan manufacturers provide templates of targets that can be printed and placed throughout the scene. (Note that a high-quality laser printer should be used.) These inexpensive targets have a high contrast (e.g., black and white checkerboard) so that they are easy to detect using intensity information. An advantage of this approach is that the targets do not need to be positioned over a control point. Although most often these targets are used as additional tie points between scans, their coordinates can be obtained using a reflectorless total station to link them to control points, if desired. Another advantage is that a high degree of redundancy can be achieved by placing many of these targets throughout the scene. Sometimes these patterns are permanently placed at a site when repeat scanning for monitoring or other purposes is required. Some software can automatically extract targets of a specific pattern within a scan. A disadvantage is that the target most likely will not be orthogonal to all scan positions.

Target Registration Process. Regardless of the types of targets, common targets between scans are matched together, and then a scan is matched to the coordinate system of another scan or the project coordinate system in the registration. In software, matches can be made automatically based on geometric arrangements or by providing a unique name or ID for each target. Once the matches are obtained, the scan transformation parameters are found by a least-squares adjustment. In some software, weightings can be applied to individual target matches. This is very helpful if some targets are extracted with lower confidence than others (e.g., targets farther from scanner that are at a lower resolution).

The following are suggestions of best practices when utilizing targets:

- In placing targets, avoid repetitive patterns and consistent spacing between targets. This will help improve results in geometric-based auto-target matching routines that can be confused when similar target arrangements are found between scans.
- Targets positioned over control points or reference marks ideally should be distributed across the project site. For each scan, targets should be distributed throughout as well so that the solution is not overly fit to close targets where the magnitude of the rotations has less of a significant effect on the error compared with targets further from the scan origin.
- Avoid placing targets close together whenever possible (e.g., front and back of an object). These can be confused as the same target and introduce error into the registration.
- When targets need to be relocated because of limited supply, ensure that all scans requiring targets have been completed before relocating the target.
- Be sure to have a sufficient number of targets per scan. (A total of three targets are needed that can be referenced with other scans or the control survey. However, it is ideal to use many more in case of unexpected problems.) A high level of redundancy is important.

- Make sure that targets are stable throughout all the scans. Make notes in the field notes of any target that is suspected to be disturbed.
- Avoid using duplicate names for targets within a project. This will avoid erroneous matches in post-processing.
- Even when working on an arbitrary coordinate system, it is also helpful to utilize a total station to establish leveling as well as provide a control network to link the scans.
- Verify that all target centers were properly extracted in software and quickly remove false targets. When a target has limited samples on it, distinguish it by appending its name a suffix such as "app" for approximate so it can be given a lower weight or excluded from the registration.
- When manually inserting vertices to extract targets in software, use the intensity information rather than photographic information because the photograph will often be offset slightly owing to parallax.

In the author's experience, improved registration results can be achieved utilizing a substantial amount of pattern targets because of the high level of redundancy and effectiveness of algorithms in extracting the center of the targets. However, one should take care to avoid using duplicate target names in the same scene and perform rigorous outlier analysis before removing targets suspected to be unstable during the scanning.

When an insufficient number of targets are available, one can manually insert vertices at distinct points on objects in the scene to serve as "targets." (Note that edges of intensity contrasts or the tip of a well-defined object (e.g., pole) tend to work well for this process.) However, in general, these targets will be of lower quality and should be weighted as such in the registration process.

Figure 8-14 shows an example of a TLS scan that is registered through the use of targets. It is linked to a network of control points that were measured with GNSS equipment and interconnected with a total station. Reflectorless total station measurements were taken from the control points to the targets.

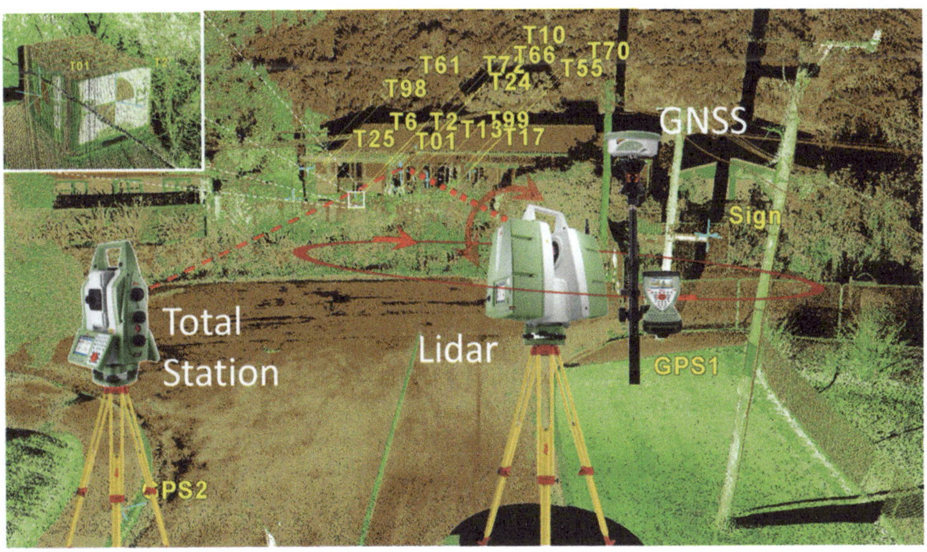

Figure 8-14. Example of the usage of multiple geospatial technologies to georeference a scan. Source: Olsen and Gillins (2015).

Cloud-to-Cloud Surface Matching

Another approach to registration is to register a point cloud to another through "cloud to cloud" surface matching algorithms. In this approach, transformation parameters are estimated for one-point cloud (e.g., one scan position or merged scan positions) based on correspondence with a second-point cloud in areas of overlap. A variety of algorithms are available to accomplish this, most of which are based on the iterative closest point (ICP) algorithm initially developed by Besl and McKay (1992), which was developed for shape registration rather than point clouds. Rusinkiewicz and Levoy (2001) and Grün and Akca (2005) provide a summary of variants to this approach. These algorithms are an active area of research; hence, novel approaches to estimate positioning with improved precision are developed frequently. Many approaches assume that the point cloud has been approximately registered; however, some approaches do not require an a-priori registration of the point cloud to be registered and can estimate an initial pose (Beinat 2006). Table 8-4 provides a representative sampling of several techniques and variants and is by no means comprehensive.

The basic concept of the ICP algorithm is to minimize the squared distances between matching points in one (slave) point cloud to be registered to another that serves as a reference (master) to determine the transformation parameters in a least-squares adjustment. The match points are obtained by finding the closest point in the master scan to each selected point in the slave scan. The process is repeated until the transformations become minimal or a predetermined number of iterations has been reached. Sampled points are often obtained randomly but sometimes are distributed geometrically across the scan scene. The algorithm requires that the initial pose of each scan be reliably estimated either manually or through another algorithm.

Scans are typically matched in a pairwise fashion for cloud-to-cloud surface matching. However, global registration processes can overcome some of the limitations of the typical pair-wise process by linking the various pair-wise registrations together. This process can help avoid error propagation when scans are obtained in closed loops around the objects of interest.

Limitations of Cloud-to-Cloud Surface Matching. A key limitation of point-to-point correspondence is that there is no guarantee that points in the first scan represent the same objects and point in space as the corresponding points in Scan B. To help improve issues with the ICP algorithm falling victim to incorrect local minimums, numerous new variants have been developed (Table 8-4). Several algorithms have functioned on point-to-plane correspondence rather than point to point. New algorithms have developed techniques to extract geometric primitives such as cylinders or spheres from the point cloud for the registration; others have applied computer vision algorithms to extract points of interest from RGB color or intensity information to serve as the match points. Some algorithms tend to perform better on man-made objects, whereas others function better with topography.

The use of cloud-to-cloud surface matching has significant limitations, particularly in surveying engineering where clear data lineage is often required. Cloud-to-cloud surface matching can exhibit instability over large areas without externally surveyed control points or targets as it often caters to local accuracy over network accuracy (Olsen et al. 2011). It can also require substantial computation time for finding appropriate match points through many iterations until convergence is reached. Outliers from vegetation or other noise can also be problematic. Finally, the algorithm is prone to local minimums, requiring significant user intervention to resolve. Nonetheless, cloud-to-cloud surface matching can be used as additional constraints with target-based resection and result in an improved registration.

Table 8-4. Representative Sampling of Cloud-to-Cloud Registration Techniques.

Category	Description	Reference
ICP and variants	Matching shapes	Besl and McKay (1992)
	Introduction of tangent plane for distance minimization	Chen and Medioni (1992), Zhang (1994)
	Review of various approaches in point selection, matching, weighting, outlier rejection, error metrics, and minimization strategies	Rusinkiewicz and Levoy (2001)
	Review of ICP in robotics	Pomerleau et al. (2015)
	Weighting scheme based on point precision and incidence angle	Boström et al. (2008)
Surface matching	3D surfaces are used instead of point clouds	Grün and Akca (2005)
Feature descriptors	Approximate curvature measures derived from eigenvalue decomposition	Bae and Lichti (2008)
	Gaussian curvature	Beinat et al. (2006)
	Key points from range image	Barnea and Filin (2008), Theiler et al. (2014)
	Key points from RGB values	Xie et al. (2013), Kyparissi and Georgopoulos (2013)
	Key points from intensity images	Kang et al. (2007), Han et al. (2013a)
Geometric primitive matching	Planes, spheres, cylinders, and tori are segmented and used as registration objects	Bae and Lichti (2007), Rabbani et al. (2007)
	Reflective features present in the scene	Böhm and Becker (2007)
	Utilizes segmented planar patches	Brenner et al. (2008), Theiler et al. (2012)
Global registration	Considers multiple scans simultaneously	Williams and Bennamoun (2001), Beinat et al. (2006), Theiler et al. (2015), Chaudhury et al. (2015)
Initial pose estimation	Determines the initial pose of arbitrarily oriented point clouds followed by fine-tuning via an ICP variant	Makadia et al. (2006), Mellado et al (2014), Theiler et al. (2012)

It can also be a useful backup when other techniques fail (e.g., insufficient targets). The following are some words of caution:

- Do not rely on cloud-to-cloud surface matching as the primary technique for registration. It can be used as a supplement to other approaches or a backup when a few scans have insufficient targets. However, these cloud-to-cloud techniques can have significant error propagation, which is unsuitable for surveying engineering applications.
- When performing change analysis, avoid using ICP to "register" one epoch to another unless there are significant referencing errors in the dataset. Using ICP to register one epoch to another will result in loss of any overall changes at the site(s) of interest.
- Significant overlap (i.e., >20%) is required for obtaining the best results with cloud-to-cloud surface matching.
- Utilize global registration techniques and match scans with multiple neighbors that have reasonable overlap.

Mixed Approaches

Many newer methodologies have been developed, which combine the techniques described in the previous sections. Some software support least-squares adjustments with target resections and cloud to cloud as additional constraints to compute the transformations.

In Olsen et al. (2009, 2011), a GNSS receiver was used to determine the *X*, *Y*, *Z* position of the scan origin. Level compensators were then utilized to determine the rotation about the *X* and *Y* axes. The rotation about the *Z*-axis was initially estimated by backsighting or usage of a digital compass followed by a global least-squares adjustment of the scans using automated *PointReg* software developed by the authors.

Mohamed and Wilkinson (2009) and Wilkinson et al. (2010) utilize two GNSS receivers mounted to the scanner to perform direct georeferencing to estimate the position and the orientation. In a more recent approach, Wilkinson et al. (2015) utilize features visible in imagery coacquired with the scanner to improve the orientation information.

These mixed approaches can overcome the limitations of the previously discussed approaches and result in more efficient field acquisition and data processing.

Comparison

Although many approaches exist to register point clouds, there are various trade-offs in efficiency in the field and office for each technique as well as accuracy capabilities. Hence, it is important to identify which methodology will best meet project requirements.

Note that the information provided in Table 8-5 is, in general, for time-of-flight systems, and a lot of factors, site conditions, and workflows can significantly influence these. For example, scanning using permanent targets in stable locations (when possible) will have minimal field time in repeat surveys for applications such as deformation monitoring.

Registration Quality Control

This section discusses several measures of scan georeferencing quality, including evaluating error statistics from the registration, qualitative checks, and independent validation techniques. The amount of time spent on these quality control measures should be representative of the project's QA/QC requirements and quantity of data generated. It should consider potential future applications of the data and legal implications. For example, in the case of monitoring applications, one will want to spend significant time to make sure that scan data are consistent so that detected change between surveys are reliable.

Table 8-5. Comparison of Various Point Cloud Registration Techniques and Approaches.

Parameter	Direct	Target resection	Feature-based	Cloud to cloud	Mixed
Field time	Slow	Slow to medium	Medium to fast	Medium to fast	Medium to fast
Processing time	Very fast	Fast	Slow	Slow	Medium to fast
Accuracy	Medium to high	High	Medium	Medium	Medium
Overlap required	None	Minimal	Significant	Significant	Moderate
Benefits	Data can be validated in the field, minimal processing time	Achieves highly accurate results. Registration process is straightforward if a strict naming convention is used and a sufficient number of targets are obtained. Well-defined match points. Follows a more typical survey workflow	Provides flexibility in the field collection, and no targets are needed because it considers objects present in the scene	Provides flexibility to the field collect and no targets are needed because it considers objects present in the scene. Often, a good fallback approach if other techniques fail	Simple and can take advantage of the benefits of the other methods that are combined

(Continued)

Table 8-5. (*Continued*).

Parameter	Direct	Target resection	Feature-based	Cloud to cloud	Mixed
Challenges	Requires careful setup in the field. Some parameters can be difficult to estimate accurately depending on the scanner. Not possible with all scanners	Logistic and efficiency challenges with targets, targets can shift during scanning, difficult to see targets far away, problems with reflective targets (saturation) at close range. Requires additional scanning of targets far away, increased field time in positioning targets. Extra equipment needed. Targets are discrete points and do not utilize the full point cloud. Potential safety concerns in adequately distributing targets. Sensitive to errors in control. Difficult for long-range scans because only large targets can be captured	Requires extraction of common features, significant overlap between scans. Typically limited to well-defined geometric features (e.g., spheres)	Increased processing time, Significant overlap required between scans, estimation of initial pose can be complicated, falls into local minimums and may not converge on the best solution without additional user input	Some methods require high-quality input data (e.g., GNSS coordinates), which cannot be obtained at all sites

Uncertainty Analysis. A current limitation in TLS is that formal uncertainties per point are often not available. Recent research has focused on developing theoretical uncertainty models for TLS. These uncertainties are useful for improved point selection in registration, spatial filtering, and modeling. Bae et al. (2009) developed a closed-form model of positional uncertainty of TLS point clouds. The model is used for surface normal estimation (with associated uncertainty), registration, finding an optimum local neighborhood size, and resampling the point clouds. Hartzell et al. (2015) developed models for per point total propagated uncertainty estimates based on the range and angle of incidence effects. O'Banion et al. (2019) developed intuitive visualization techniques to explore these uncertainty estimates in the point cloud.

Quantitative Error Reporting from Registration Results. Most TLS software will report the errors calculated during the registration for the TCTs. It is important to holistically evaluate these errors and report them at a confidence interval (typically RMS or 95% confidence). It is also important to determine the 3D error and separate the error in the horizontal and vertical directions for many applications. These values can differ significantly from manufacturer specifications, which often report the ranging and angular accuracy components separately and are representative of ideal conditions.

When performing target registration quality control, outliers may be removed from the registration process with caution. Often, these are the result of misnaming or having targets that are very close together. Use of additional targets will result in slight increases in error in the overall registration because of having to fit more constraints but will provide higher confidence in the results and likely a better registration. Hence, only clear outliers should be removed from the registration. A simple strategy is to remove targets with the highest error one by one and see how the registration and error reported for each of these targets and the others change. Although, in general, this approach works well when there are few, clear outliers, it can present difficulties when the erroneous target is not the one with the highest error.

Note that the errors reported from the target matches are not necessarily representative of actual errors in the point cloud. For a target-based registration, a fitting process is performed to determine the center point of the target more accurately. Hence, these error statistics tend to underpredict the actual errors of individual points in the point cloud because some noise can be filtered through the fitting process. Often, targets are located close to the scanner to be captured precisely, which means that they can be modeled at higher accuracy and resolution than the points further away in the dataset.

When performing cloud-to-cloud registrations, RMS fit values between scans are provided. These values, however, can be influenced by obstructions (e.g., vegetation, people, etc.) in the scan scene, so if a substantial amount of these objects are present, they should be screened from the error analyses unless they are the objects of interest. These also can be sensitive to thresholds established for outlier removal (Olsen et al. 2012b).

The cloud-to-cloud technique will consider a broader number of points in the dataset across a larger area compared with the target constraints. This usually helps improve a target registration, especially when targets are constrained in poor geometric configurations or limited in number. However, significant noise in the point cloud can be problematic for the surface matching algorithms and can introduce some erroneous results. It also can be influenced by the lower resolution available with distance from the scanner. Also, many algorithms utilize random sampling, which causes bias in the results toward the dense clusters of points close to the scanner rather than appropriately distributing samples across the scene.

Note that the target errors and the cloud-to-cloud errors are not the same error metric and cannot be directly compared. Cloud-to-cloud RMS errors will typically be higher due to the fact that there is additional uncertainty in the point pair matching process based on the point density of the scan.

Validation Control Points. Given the limitations of registration errors, VCTs can help provide a more realistic estimate of error in the dataset. (See Registration Strategies for a discussion on TCTs versus VCTs.) This section provides some examples of approaches that can be utilized (modified from NCHRP Report No. 748) (Olsen et al. 2013d).

An external dataset (e.g., topographic points acquired with a total station or RTK GNSS, depending on accuracy requirements) is useful for validation of the georeferencing quality (Figures 8-15 and 8-16). However, this requires additional field time, equipment, and personnel. Hence, one needs to strategically plan the location of these validation sample points.

To overcome point density\sampling limitations, Skaloud and Lichti (2006) present a planar-based evaluation approach. In this approach, one acquires sample points, using another survey methodology, on planar features (at the desired interval) visible in the TLS data. 3D network accuracy can be assessed by measuring the offsets of the TLS points from multiple planar surfaces facing different directions. The local accuracy can be determined by evaluating the residuals following a least-squares fit of the TLS points to the plane. A potential limitation of this method would be that in reality, some surfaces may not be planar at the mm scale. Useful surfaces include road surfaces, sidewalks, and so on for vertical evaluation and curbs, buildings or walls for horizontal evaluation.

Ideally, a validation dataset should be an order of magnitude more accurate than the network or local accuracy specification requested. However, this is a challenge for the highest accuracy TLS datasets because there are only limited technologies that will meet this criterion and often are not practical. For example, control for a project will normally be set using a GNSS [long-duration static occupations for the highest accuracy (sub-centimeter), RTK for faster evaluation (a few centimeters)], or total station/digital leveling (sub-centimeter) surveying, depending on the required accuracy of the resultant product (see Chapter 11). Further, although instruments such as a total station provide very high local accuracy, coordinates still must be tied to control for network accuracy evaluations. Leveling can provide high vertical accuracy (sub-millimeter), but it does not provide the ability to assess horizontal accuracy. When performing a VCT survey, the TCTs utilized should be completely independent for evaluating network accuracy. However, if only local accuracy is of interest, then they can be tied to the same base TCTs.

Figure 8-15. Reflectorless total station shots on a cliff surface to evaluate accuracy. Source: Cunningham et al. (2015).

Figure 8-16. GNSS coordinates obtained on points of inflection on a stair well for an accuracy assessment of scans of a coastal cliff.
Source: Olsen (2009).

Visual Verification. Visual checks and verifications should always be performed in addition to the quantitative approaches even when the results are satisfactory. Although large misalignments are, in general, easy to spot when visually inspecting the data, subtle misalignments require extra attention and should be thoroughly checked.

One common approach is to color each scan a different color and examine areas of overlap to ensure a good match (Figure 8-17). When one color is predominant, it indicates a problem in the georeferencing of one of the scans or passes. Rather, the colors should blend together in areas of overlap. Note that there may still be some sections where one color tends to dominate because of increased point density or because that location was only visible from one scan.

Another approach is that a visual examination of cross sections (a few centimeter thick) at various locations and orientations throughout the point cloud can help evaluate the effectiveness of georeferencing. Scans that are misaligned will appear as multiple, disjointed profiles, rather than a single, cohesive profile.

In addition, some software packages can color-code points by deviations between scans or datasets to highlight areas of disagreement.

In monitoring applications, one can compare datasets between epochs on features that are not expected to move. One can evaluate the consistency between multiple surveys to locate georeferencing errors. However, caution must be exercised because some objects assumed to be stable may, in fact, be moving because of landslides or creep. When a long time series of data are available, these differences are easier to resolve.

These visual validations can provide additional insights on potential georeferencing or distortion errors that may not be discovered when focusing solely on the numbers. However, when performing these visual verifications, it is probable that one will find some points from noise that are above the error tolerance. It is important to remember that at 95% confidence, 5% of the dataset will exceed the error tolerance, so overemphasis should

Figure 8-17. Example of visual verification showing (top) misaligned scans and (bottom) properly aligned scans.

not be placed on a few stray points when the majority are satisfactory. For example, in a scan of 10 million points, 500,000 points can be above the error threshold!

Point Density Evaluations. Similar to accuracy, point density should be evaluated throughout the dataset, particularly for objects of interest. Selecting appropriate resolutions for scanning is critical. The USIBD Level of Detail specification provides guidance to assist with some applications. NCHRP Report No. 748 also provides guidance on accuracy and resolution requirements for many transportation applications. On the one hand, insufficient point density will cause many algorithms to fail in subsequent processing as well as make it difficult for one to resolve features in the point cloud. On the other hand, a very high point density can result in datasets being too cumbersome to process efficiently and can confuse some algorithms. Some recent algorithms have been developed to reduce point density efficiently to improve modeling workflows, as discussed later.

Point density at each location can be evaluated by this procedure outlined in NCHRP Report No. 748 (Olsen et al. 2013d):

1. Drawing a polygon (e.g., a square of dimension 1× 1 m) on a planar feature,
2. Selecting all points on the planar surface that are within the polygon extents (excluding points that do not belong to the surface), and
3. Calculating point density as the number of points found in Step 2 divided by the 2D area of the polygon drawn in Step 1.

As with accuracy, point density checks should be conducted at locations throughout the entire dataset. The frequency of evalutions depends on the variability observed in the point density as well as the spatial frequency of the objects of interest. In general, one can use similar intervals as those for accuracy evaluations. The results for each test location can then be statistically evaluated to ensure that 95% of the samples meet the appropriate point density requirements for the features of interest. If different point density requirements are established for various objects (e.g., pavement, signs, cliff), the samples should be categorized and summary statistics for point densities reported for each feature category individually.

When surface or solid models are delivered, point densities can be calculated by the number of points on the model divided by the surface area of the model.

Although it may be tempting to calculate a quick estimate of point density for the dataset by dividing the total number of points in the dataset by the 2D projected area of coverage, such an approach is not recommended because of the following factors: (1) The actual point density will vary substantially across the dataset, as described previously, and (2) such an approach does not account for the 3D nature of the data and also does not account for vertical features.

A continuous, color-coded point density map with summary statistics is a helpful deliverable and can provide a general overview of point density quality throughout the dataset. Note that if a 2D map is provided, point densities will be overestimated in locations with vertical features (e.g., buildings or walls).

Data Completeness. Another consideration for quality control is the frequency of data gaps in the scan survey. Although most of the data may meet resolution requirements, there will be many cases where there will be data gaps (i.e., shadows, occlusions) because of visibility constraints (Figure 8-13). The acceptable quantity and locations of these gaps could be important to consider when there are critical features. A simplified metric can be obtained in 2D by rasterizing the scan data (meeting a sampling threshold per cell) and comparing the area of the cells with data to the total desired coverage area. For more sophisticated procedures on calculating completeness, the reader is referred to Yoo et al. (2010). These procedures, however, require 3D models of the scene to be created, which can require a substantial amount of effort to produce if they are not needed otherwise.

On simple surfaces such as a road surface, building facades, or objects that are orthogonal to the TLS, high values of completeness should be obtainable. However, the completeness quality will degrade with distance from wherever the scanner is positioned because the occlusion from a small object blocking the scanner will enlarge with distance. Further, small data gaps from moving objects (e.g., pedestrians) passing in front of the scanner are less of a problem than larger gaps from static objects (e.g., parked vehicles). To help counteract this, collection should be done during low traffic periods or by spacing scans close together to ensure a high degree of overlap that can help fill in these occlusions.

PROCESSING WORKFLOWS

Once scans are registered or georeferenced, several post-processing techniques are performed. A variety of procedures are necessary to post-process TLS data depending on the application. For these tasks, there are often semiautomated algorithms; however, most require some manual editing to achieve the engineering quality results. Although some software used for ALS post-processing can be used for TLS processing, TLS software often follows different workflows and will vary widely depending on the application and products desired. Also, in general, TLS post-processing is more interactive and graphics-intensive than ALS post-processing, which is performed tile by tile.

Filtering

Following registration, point-cloud data are typically filtered to eliminate unwanted features. These can include "artificial low points," for example where the laser reflects off of a water surface onto another surface and then back to the scanner. It can also include data points from objects passing in the scanner view (e.g., passing cars), unwanted vegetation, or, more generally, anything that is not needed by the end user. Figure 8-18 provides some examples of noise artifacts present in merged TLS data.

Filtering processes vary substantially in the degree of user interaction required and the type and quantity of noise present (Figure 8-19). Some features such as trees and vegetation can be removed through ground-filtering processes; however, many ground-filtering approaches have been developed for ALS and suffer some limitations when being applied to TLS because of the highly variable point density of the TLS dataset.

Some work such as that of Olsen et al. (2013c) has utilized common photo editing software by exploiting the organized scan structure (see System Types discussed previously) and found it to be successful in editing point clouds as well as improving RGB texture mapping.

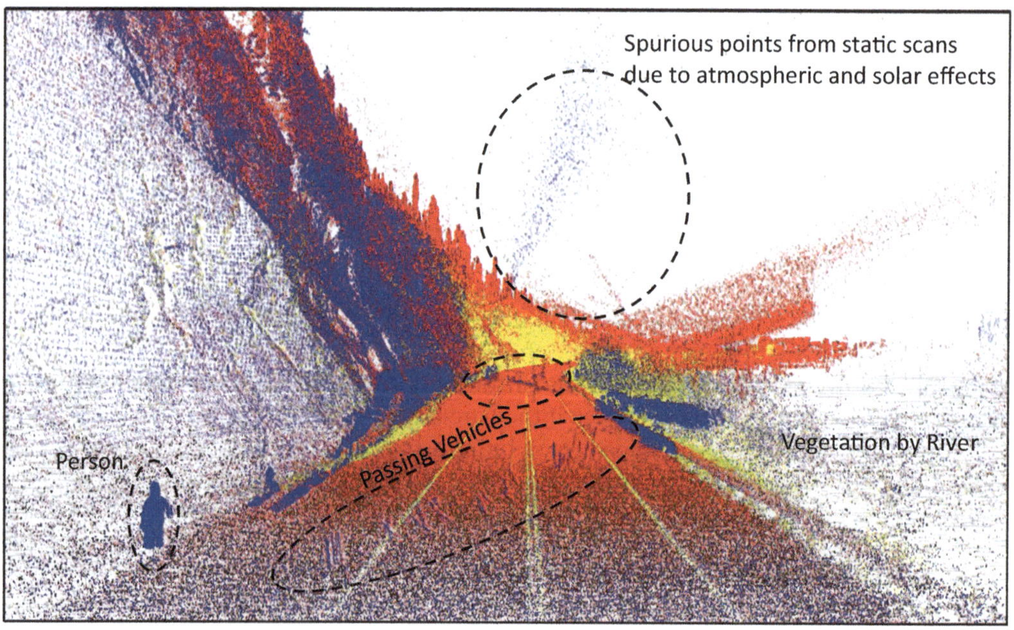

Figure 8-18. Examples of common noise artifacts removed from the scan data. Source: Cunningham et al. (2015).

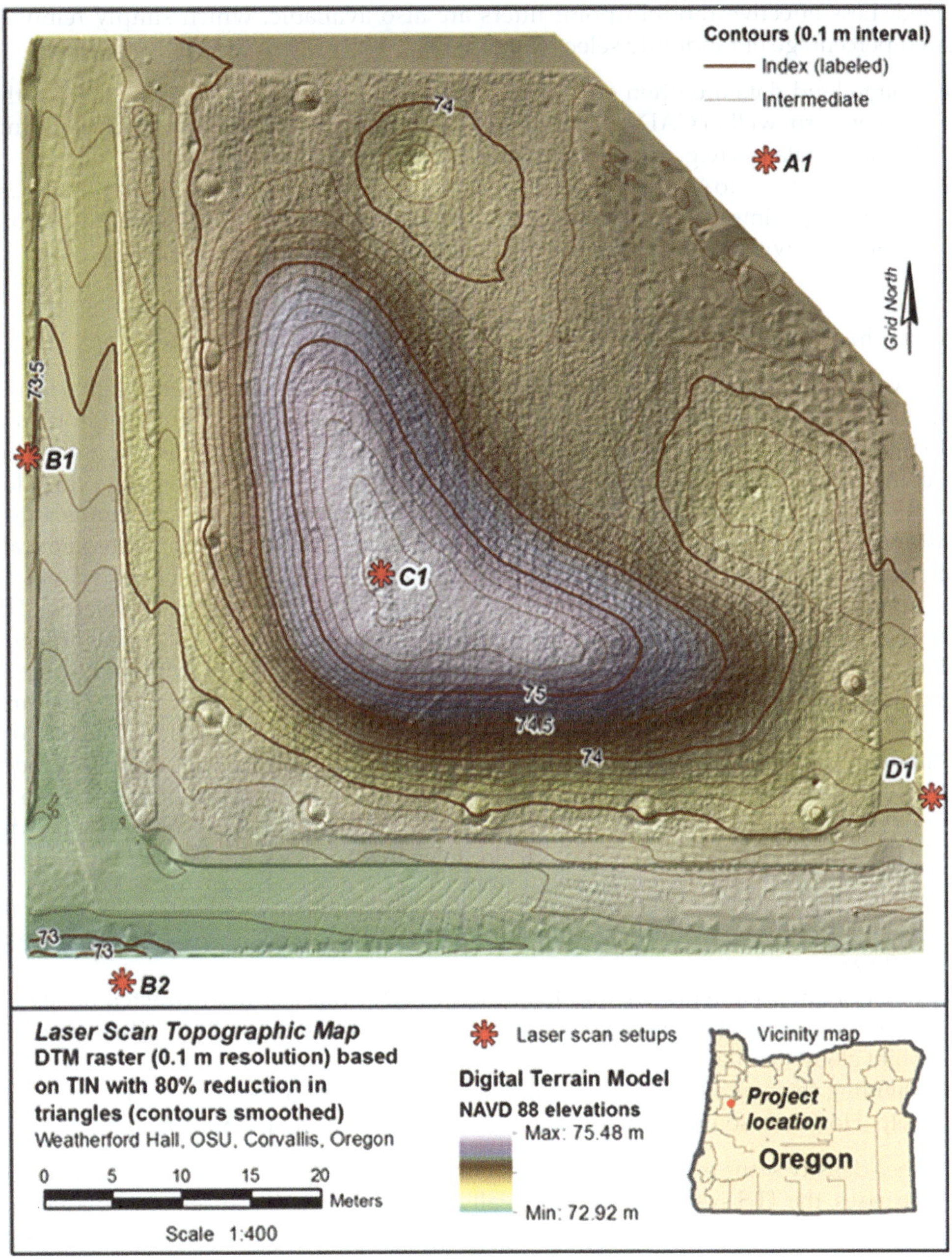

Figure 8-19. Contours from a DTM (no breaklines) of the Weatherford Hall Grassy Knoll site at the Oregon State University Campus.
Source: Courtesy of Michael Dennis.

Density Filters

Some filters also exist to enable the point cloud to be easier to work with and reduce the size of the datasets for engineering analyses. These can include distance threshold filters, which remove points that are closer than a specified distance, and planar filters that remove points in planar areas. Spatial structuring filters such as Octree or K-D tree can also be utilized. Less effective random point filters are also available, which simply remove a desired percentage of randomly selected points.

- Point-cloud data are often subsampled or statistically filtered to create a DTM that will perform well in CAD or other engineering packages, which may not be designed for large datasets (e.g., file size, number of vertexes).
- Software exists to optimize point density by removing points in flat areas and preserving points in areas of distinct change. This can enable one to efficiently perform an analysis with minimal data loss.

Topographic Mapping and Digital Terrain Modeling

A key product from TLS datasets are topographic maps (Figure 8-19) and digital terrain models (DTMs) [see review of capabilities and limitations of ALS and TLS by Pirotti et al. (2013a)]. These maps can be created by a variety of ways. One approach is through virtual surveying where the software operator selects points from the point cloud (e.g., curb flow line) to create breaklines and then selects some topographic points as inputs into a Constrained Delaunay Triangulation. Breaklines will need to be extracted semiautomatically or manually, if desired. TINs will actually be 2.5D datasets, not 3D. Hence, they will not model details on vertical surfaces in the point cloud (e.g., buildings, steep slope faces). This issue is particularly important when analyzing building facades, steep slopes, or slopes with overhanging features that are not properly resolved in 2.5D data. Although some 3D triangulation (meshing) techniques have been developed to improve modeling features, analysis will likely need to be conducted in specialized 3D software for reliable calculations.

Although CAD and GIS software offers basic support for point clouds and TIN models with few to tens of millions of vertices, many engineering analysis and design packages may not support the high-density TINs created by TLS. A few potential solutions provided in NCHRP Report No. 748 (Olsen et al. 2013d) to enable analysis include the following:

- TINs with frequent, planar surfaces can often be significantly optimized to reduce the triangle count with minimal effects on the model accuracy, using readily available software.
- Rather than use the point cloud to create the TINs, one can use extracted breaklines from the full point cloud and then use the breaklines with a subsampled version of point cloud (similar to a photogrammetric process).
- Dividing the overall dataset into individual tiles prior to creating the TIN may also help. However, some software may not be able to work with multiple tiles.

Higher-density scans will be needed to obtain ground points in areas of high vegetation. In some cases, the TLS system may not actually see the ground because of its oblique look angle. For example, downslope sections of slopes adjacent to roadways that are not visible from the roadway will not be captured unless they are visible and captured from another vantage point. For a seamless DTM, interpolation is often necessary to fill in holes or data gaps within the dataset, which can influence the analysis (Olsen et al. 2015).

For general basemap creation, natural terrain mapping will not require as high a resolution as pavement surfaces, particularly because sediment will erode or be deposited across natural terrain surfaces. However, in slope stability and erosion analyses, one will want to capture sufficient detail on the terrain to capture these effects.

Ground Filtering. Ground filtering is a common task for creating DTMs with lidar datasets. Although, in general, automated ground and other surface extraction algorithms work well for airborne lidar data, these can be problematic to implement on TLS datasets that have variable density and contain steep topographic features. For a review of various ground-filtering techniques, the reader is referred to Meng et al. (2010) and Briese (2010). Qin et al. (2021) provide sample ground-filtering datasets to support machine-learning applications.

Some specialized techniques have been developed for TLS to minimize the aforementioned problems. Some filters are geometry based; however, such approaches have difficulty in distinguishing between low vegetation and ground (Fan et al. 2014). Belton and Bae (2010) present an approach that considers elevation variance and implements region growing to generate a ground model. Brodu and Lague (2012) present a supervised classification approach with multiscale dimensionality analysis that considers a variety of geometric indicators. They utilize this approach for ground filtering in a riparian environment with effective results. Users of the approach can share parameter settings for specific scene types. This approach has been integrated into an open-source package, Cloud Compare (Girardeau-Montaut 2011). Che and Olsen (2017) present a fast, efficient approach that exploits the scanline data structure to find density anomalies, enabling the classification of ground candidates. Others have utilized multireturn (Pirotti et al. 2013b), intensity (Pirotti et al. 2013a), sightline Panholzer and Prokop (2013), or RGB (Lau et al. 2015) information for the classification. Finally, TLS data can be supplemented with GNSS observations Guarnieri et al. (2009), Rodríguez-Caballero et al. (2016) to aid in the ground classification.

Ultra-High-Resolution Modeling. Highly detailed surface models may be desired for applications such as change analysis (discussed later). In these cases, the data can be gridded or triangulated for the analysis depending on the orientation of the object. Figure 8-20 shows an example texture mapped surface model for monitoring rock slopes adjacent to a highway in Glitter Gulch, Alaska, using an approach developed by Olsen et al. (2013b). The derived model has a typical triangle edge length of 5 cm. Figure 8-21 shows an example DTM for a small section of a watershed in the HJ Andrews Experimental Forest. The DTM had a final resolution of 2 cm after ground filtering to remove the vegetation and interpolation to fill in the data gaps. Note that despite the dense vegetation present across the site, a high-resolution DTM was obtainable because the scans were typically positioned within 5 m of one another, enabling the laser to sample the ground surface. However, note that interpolation is still necessary to model the ground under larger objects such as tree trunks. This interpolation was completed using a focal statistics median operator.

Figure 8-20. Texture-mapped triangulated surface model using the methodology. Source: Olsen et al. (2013b).

Figure 8-21. (a) Example DTM (2 cm resolution) from the HJ Andrews Forest Watershed 1, (b) photograph, (c) closeup view of the DTM to highlight stream geomorphology, (d) point cloud. For reference between the models, point cloud, and photograph, two rocks are labeled X and Y.

When developing these higher-resolution topographic models, the following should be considered:

1. Many common engineering analysis packages cannot handle these large files; hence, specialized software is often required for analysis. However, a variety of visualization packages are available for viewing these models and performing basic measurements.
2. It is often more effective to do several lower-resolution scans than a few high-resolution scans across the site to create a more uniform point cloud and fill in more of the gaps.
3. In particular, TLS is often done oblique to the ground, so data for topographic mapping should be limited to areas close to the scanner.

Geometric Modeling

Another common task in TLS in civil engineering is to model structures such as buildings and bridges. Currently, algorithms exist to extract well-defined geometric primitives such as planes, spheres, cylinders, and so on from a point-cloud dataset automatically (Huber et al. 2010). Additional semiautomatic algorithms are available for shape fitting once points have been isolated to the object of interest. Figure 8-22 shows an example of objects modeled directly from a point cloud.

Fitting processes can be used to yield a high degree of accuracy because it can remove some sensor noise. Another aspect to consider is that the detail provided by the scanner can show imperfections (e.g., a wall that is assumed to be planar but is bulging in some sections). This information will be lost when generating a CAD/BIM model from the data. Bosché et al. (2015) provides a discussion of Scan to BIM workflows and limitations.

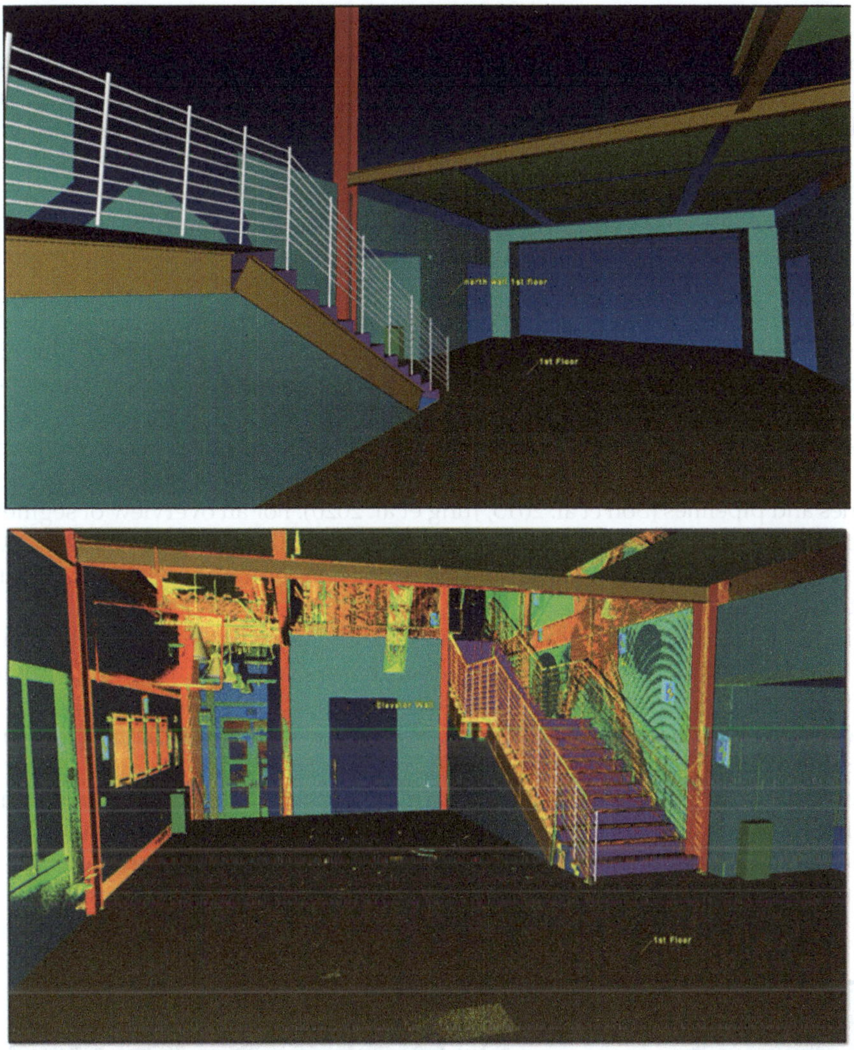

Figure 8-22. Examples of models generated of Kearney Hall Atrium (Oregon State University Campus) with a TLS scan showing various features, including steel shape beams and columns, planes, cones, and cylinders.
Source: Courtesy of Michael Dennis.

Sometimes it is more efficient to draw objects using the point cloud as a visual and snapping reference in CAD software. This latter approach can also help enforce rules (e.g., walls should be perfectly vertical) that may be slightly different in reality. Hence, it is important to identify the intended application(s) and accuracy needs prior to performing the modeling to ensure that the appropriate type of modeling is performed. Be aware of the following when using simplified representations:

- Often features are extracted into CAD files or GIS vector features (such as SHP files). Note that the point cloud will still have dramatically more information on the scene than these simplified versions. Some techniques will classify the point cloud directly, whereas others will extract features as modeled objects.
- Simplified models will lose much of the detail available in the point cloud (e.g., surface imperfections), which may or may not be important depending on the application.

Feature Extraction

The points in a point cloud natively do not have attributes other than *XYZ* coordinates and intensity values. RGB color from coacquired imagery can be mapped to the point cloud through automatic processes. Attributes such as what the point represents (i.e., point classification) are applied later through manual, semiautomatic, and/or automatic processing. For a detailed summary of techniques, see Che et al. (2019a).

Ordinarily, the first step in the process is point-cloud segmentation. Segmentation is the process of dividing the point cloud into distinct sections based on common geometric properties (e.g., planes, cylinders, spheres, etc.) as shown in Figure 8-23. For example, Che and Olsen (2018) developed an efficient approach to utilize normal information to segment smooth surfaces within the point cloud. This reduction and simplification can dramatically reduce file sizes and processing times. (However, note that it will remove details such as deviations from the idealized geometric shape that may be important.) As an example, a planar feature found in a TLS dataset can be simplified from millions of sample points to three or more vertices after segmentation. Other examples include automated extraction of powerlines and pipelines (Son et al. 2015, Jung et al. 2020). For an overview of segmentation processes, the reader is referred to Vosselman and Klein (2010), Wang and Shan (2009) and Che et al. (2019a), which provide an overview of common approaches to point-cloud segmentation.

Classification then builds on the segmentation to identify what the point represents (e.g., ground, vegetation, building, pole, etc.). Classification processes can be implemented on the point cloud directly or solely on the objects extracted from the point cloud. Finally, these extracted objects can be incorporated into CAD, BIM, GIS, or CIM platforms (Tang et al. 2010) for information modeling (Chapter 14).

Many algorithms for feature extraction are still under research and development in various fields such as geomatics, computer science, robotics, and geoinformatics. However, many features can be extracted through semiautomatic processes in commercial software where the user identifies features at specific intervals and the software interpolates between. Note that many of these procedures are specific to a set of objects and often limited to

Figure 8-23. Example segmentation result where each color represents a distinct smooth surface in the point cloud using the technique of Che and Olsen.
Source: Che and Olsen (2018).

relatively small datasets (<10 million points). A few completely automated procedures exist for specific tasks that do not require a moderate degree of user intervention to resolve problems; however, these are often available only in specialized software packages. In many cases, manual editing and extraction of features from the dataset should be performed to achieve quality results, with the following considerations:

- Although automated and semiautomated software packages are available to perform some tasks, often some manual edits are required to improve the results and ensure that they are satisfactory for engineering purposes.
- Advances in automated algorithms occur rapidly, so it is important to stay current with software.

Segmentation and Classification Procedures. Point-cloud segmentation and classification research is rapidly progressing for a wide variety of applications. This section presents an overview of existing work on the segmentation and classification of point clouds. Table 8-6 is organized with respect to common types of approaches, including geometry, color, image, intensity, shape, and structure. However, note that some techniques will use combinations of these approaches. Note that multiple algorithms are often implemented separately to extract the wide range of objects of interest to civil engineers (Vosselman 2013).

Although primarily developed for robotics, the Point Cloud Library (PCL) (Rusu and Cousins 2011) is a recent, powerful, open-source library for segmentation through the extraction of geometric primitives (e.g., planes, cylinders), laser intensity, normal vectors as well as raw RGB color, if available. This library also implements some sophisticated computer vision algorithms.

Examples of Transportation Feature Extractions. Algorithms and software to extract transportation-related objects from point-cloud datasets using many of the aforementioned techniques are evolving rapidly. Belton and Bae (2010) present approaches for the extraction of various road features after extracting ground points in the terrain. Some commonly extracted objects include the following:

- Signs (McQuat 2011, Gonzalez-Jorge et al. 2011, Ai and Tsai 2016)
- Curbs (McQuat 2011, Zhou and Vosselman 2012)
- Automobiles (McQuat 2011)
- Barriers (Pu et al. 2011)
- Building walls (Pu et al. 2011, Rutzinger et al. 2009)
- Trees (Pu et al. 2011, Lehtomäki et al. 2011).
- Streetlights (Yu et al. 2015)
- Tunnel luminaires (Puente et al. 2014)
- Cross sections
- Poles (Pu et al. 2011, Lehtomäki et al. 2011, Rutzinger et al. 2009)
- Culverts (Lin and Hyyppa 2010)
- Roadway surface (Wang et al. 2012, Yang et al. 2013)
- Pavement Markings Guan et al. (2014), Riveiro et al. (2015a), Che et al. (2019b), Jung et al. (2019)

As an example, Pu et al. (2011) describe automated algorithms to recognize features within a point cloud such as traffic signs, trees, building walls, and barriers using characteristics such as size, shape, orientation, and basic topological relationships to classify the point cloud. The authors indicate that, although poles were recognized with an accuracy of 86%, the other categories were not extracted as successfully and would need to be integrated with imagery for improved extraction.

Point Cloud to Finite-Element Models. Conversion of lidar point-cloud data into solid models, necessary for engineering analysis can be a tedious, manual process. However,

Table 8-6. Sampling of Common Point-Cloud Segmentation Approaches.

Type	Advantages	Limitations	Example algorithms	Reference
Geometric	Generic—function with any point cloud, most widely available. Works well in urban environments where well-defined features are prominent	Computationally costly, require finely tuned parameters, limited to small data sets, sensitive to variable point density and data gaps common in TLS data	RANSAC (random sampling and consensus)	Fischler and Bolles (1981), Schnabel et al. (2007)
			Hough Transform	Hough (1962), Ballard and Brown (1982), Vosselman (1999), Maas and Vosselman (1999), Rabbani (2006)
			Region growing (based on proximity, slope, curvature, and surface normal) from a seed location	Ballard and Brown (1982), Rabbani et al. (2007), Pu and Vosselman (2006), Moussa and El-Sheimy (2010)
			Detecting surface discontinuities	Wang and Shan (2009)
			K-means clustering	Chehata et al. (2008)
			Voxelation	Douillard et al. (2011)
			Curvature	Son et al. (2015)
			Multiscale morphological analysis	Brodu and Lague (2012), Rodríguez-Caballero et al. (2016)
			Fuzzy parameters in relative height differentials	Biosca and Lerma (2008)
			Tensor voting of normal vector clusters	Lin and You (2006)
			Support vector machine classification of the DTM	Serna and Marcotegui (2014)

Color	Enables application of computer vision and image processing algorithms, some objects are readably segmentable by color	Color information not always available in point cloud, Obtaining high-quality images with scans can sometimes be time-consuming, Color alone not sufficient to distinguish multiple objects—some features can have more than one color, others can share the same color	Spectral filtering	Lichti (2005)
			Graph theory segmentation and union (also considers normals)	Strom et al. (2010)
			Mean shift smoothing algorithm to cluster sections of images followed by PCA for classification.	Sok and Adams (2010)
			Superpixel clustering (SLIC) followed by normal vector evaluation through SVM. A k-nearest neighbor algorithm is utilized for refinement	Mahmoudabadi et al. (2013)
Intensity	Inherent property of laser scan data, intensity helps distinguish contracts between surfaces that may not be distinguished by geometry alone. Computationally efficient	Intensity values are affected by a variety of factors. Requires radiometric calibration for optimal results	Conditional random field to classify buildings, low vegetation, tree, natural ground, and asphalt	Niemeyer et al. (2012)
			Full waveform analysis of the intensity amplitude, the cross-section per illuminated area, and the backscatter coefficient followed by an SVM classifier	Mallet and Bretar (2009)

(*Continued*)

Table 8-6. (*Continued*).

Type	Advantages	Limitations	Example algorithms	Reference
Data structure	Efficient and enables the exploitation of computer vision and image-processing algorithms	Requires a structure point cloud	Scan line segmentation followed by a surface-growing process between adjacent scan lines	Jiang and Bunke (1994), Hoover et al. (1996), Sithole and Vosselman (2003), Sithole (2005)
			Calculations of gradients from the range image followed by region-growing image segmentation	Gorte (2007)
			Smoothed surface normal and range panorama analysis	Zeibak and Filin (2009)
			Mean-shift algorithm to segment intensity, range, surface normals, and true color channel panoramas	Barnea and Filin (2013)
			Extraction of trees using range, intensity, LUV, and HSV color panoramas	Barnea and Filin (2012)
			Use of computer-vision algorithms to segment based on HDR color, normalized intensity, range, and normal components represented as 2D panoramas	Mahmoudabadi et al. (2016)

recent research has provided some promising solutions. Tang et al. (2010) provide a detailed review of automated reconstruction techniques available to create building information models from point clouds. Hinks et al. (2013) describe an automated process to voxelize scan data of building facades to create solid models that can successfully be implemented in computational modeling. Conde et al. (2015) present an approach to generate finite element models from point clouds for analyzing transmission towers. Riveiro et al. (2015b) developed an automated approach for creating structural models of masonry walls, including detecting individual bricks or rocks through image-processing tools.

Commercial software also enables the creation of geometric primitives (e.g., plane, sphere) and standardized steel shapes (e.g., W14 X40) from point-cloud data, which can then be converted into solid models (e.g., Yim et al. 2014). Bosché and Haas (2008), Bosché (2010), Walsh et al. (2013), Calbaleiro et al. (2014), Barazzetti et al. (2015), and Truong-Hong and Laefer (2015) have developed approaches to automatically extract or segment structural members from the point cloud. Tang et al. (2012a, b) describe a novel approach to automatically extract some information from point clouds such as bridge clearances.

Processing Quality Control

Classification Accuracy is also an important aspect to examine when certifying deliverables. However, at this time, it is difficult to recommend appropriate thresholds for acceptance criteria given the rapidly changing capabilities of algorithms. Nonetheless, ground classifications should be achievable with >90% accuracy. NCHRP Report No. 748 (Olsen et al. 2013d) provides some preliminary guidance for evaluating classification accuracy.

Points in a TLS data set can be classified through semiautomated and in some cases automated processes. Typical classification categories for point clouds include terrain, vegetation (low, medium, high), buildings, power lines, and so on. Note that ground filtering/vegetation removal algorithms tend to perform much better with airborne lidar data than with mobile or static TLS techniques because of the look angle. In addition to geometric accuracy, in the case of a classified point cloud, one will need to evaluate the classification accuracy to determine levels of omission and commission. For example, a point that is classified as ground but is really vegetation can lead to problems in DTM creation, depending on the robustness of subsequent processing algorithms.

The classification accuracy of a point cloud can be determined by manually classifying points in several representative sample sections and comparing them with the classifications provided by the service provider. Statistics can then be developed on how well the points were classified for each category into a confusion matrix (Figure 8-24) showing the quantities and types of incorrect classifications. Ideally, the matrix will be a diagonal matrix with 100% along the diagonal, showing that all data points were classified into the appropriate category.

ANALYSES

TLS enables a significant amount of analyses through both visual and quantitative analyses that support a wide range of engineering analyses. In some cases, analyses can be conducted visually by virtually exploring a scene and examining the spatial relationships between objects. Figure 8-25 provides several example analyses completed with TLS data.

Visualization Analysis

Although point clouds are typically viewed on 2D computer monitor displays, a variety of technology exists that enables a user to explore the point cloud in 3D. Figure 8-26 shows

PREDICTED / ACTUAL	Ground	Vehicle	Vegetation	Building
Ground	90%	1%	9%	0%
Vehicle	3%	87%	8%	2%
Vegetation	7%	12%	75%	3%
Building	0%	0%	6%	95%

Figure 8-24. Example of a confusion matrix comparing predicted versus actual classification accuracies by categories.
Source: Adapted from Olsen et al. (2013d), reproduced with permission from the National Academies Press, Washington, DC.

an example of an immersive Virtual Reality system (GeoMat CAVE) where users can more freely and intuitively explore the point cloud. This system uses the open-source LidarViewer software in the virtual reality user interface (VRUI) developed by Oliver Kreylos, UC-Davis (Kreylos et al. 2008 and Kreylos 2022). In addition, these systems provide tools to extract measurements, classify points, extract geometric primitives, and color points based on differences to surfaces.

Change Analysis

The wealth of data provided in a point cloud enables one to perform detailed change analyses in applications such as slope stability analyses, structural deformation monitoring, geomorphological analyses, or construction progress monitoring and tracking. *CloudCompare* (Girardeau-Montaut 2013) is a powerful, open-source application available for performing change analyses. Three basic types of analyses can be conducted, including cloud to cloud, cloud to model, and model to model. Olsen (2015) presents an approach for in situ change detection (Liscan), which significantly reduces computation time and enables operators to utilize the change information directly in the field for an improved site investigation.

In addition to computing the 3D distance deviations between two surfaces, volumetric analyses can be performed. One can generate a 3D model of a proposed design and estimate quantities of materials needed based on the point cloud of a site. It can also be used to determine quantities of materials placed or removed during construction. Similarly, it can be used in geomorphological analyses to compute volumes of sediment eroded or accreted.

Anil et al. (2013) and Tang et al. (2011) focused on deviation analysis for scans of individual segments of a single building (e.g., walls) to the associated as-is features in the BIM. Tang et al. (2010) utilized scanning to determine flatness defects in concrete. Anil et al. (2013) also discuss challenges in efficiently performing deviation analysis, including data interoperability, speed, and file sizes.

Teza et al. (2007) present a method to analyze landslide movement using a piecewise alignment method where the new dataset is grouped into subareas of 5×5 m and compared with a baseline DTM. The local displacement vector for each subarea is the translation parameters obtained using ICP registration between the new scan and the baseline model. Similar approaches have been utilized for analyzing postearthquake deformations.

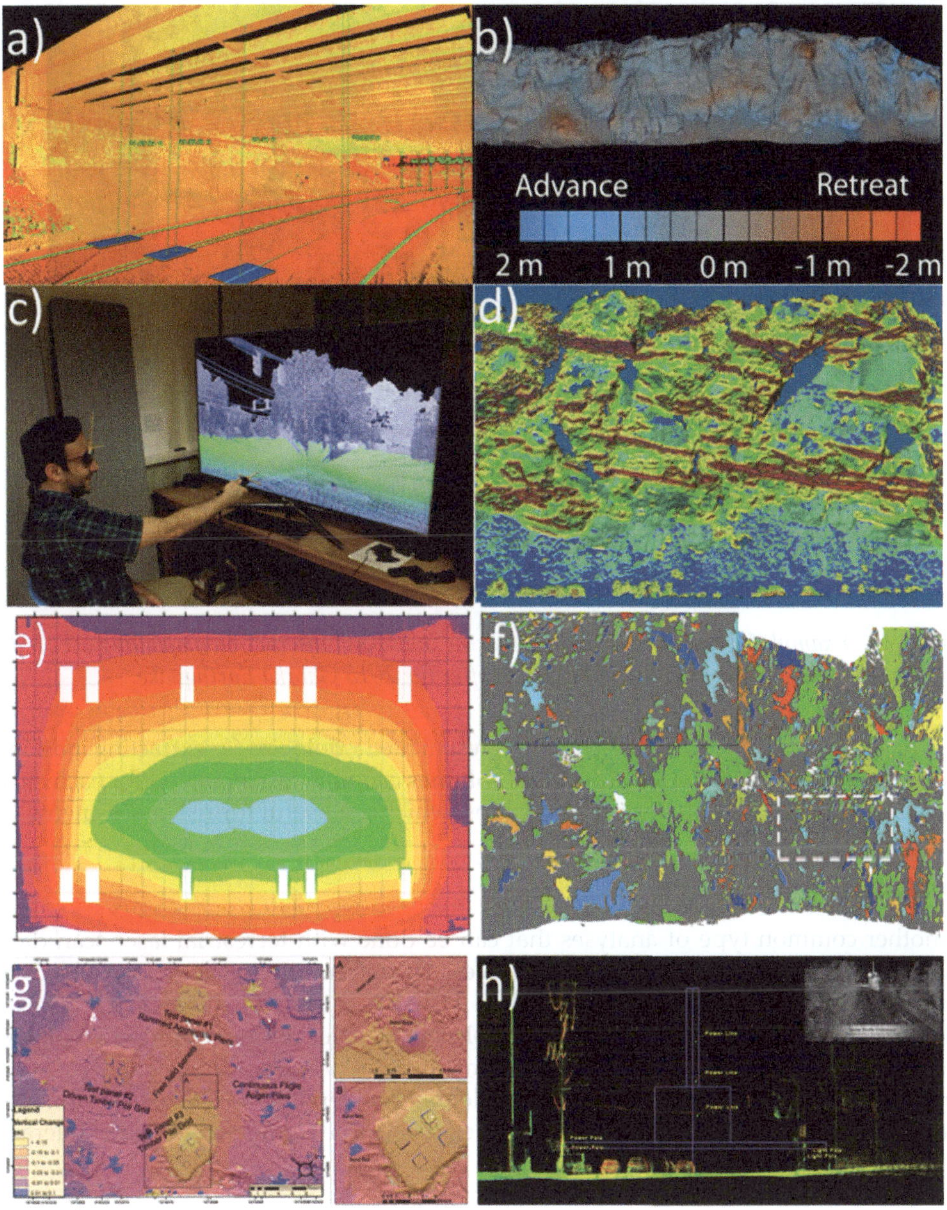

Figure 8-25. Sample applications of TLS: (a) bridge clearances; (b) coastal erosion and landslide movement of the Johnson Creek Landslide; (c) intersection sight visibility analysis; (d) Rockfall Activity Index (RAI) for slopes on the Glenn Highway, Alaska; (e) deformations on a wall of the Gamou Wastewater Plant from tsunami impact in Japan; (f) rockfall clustering for slopes in Glitter Gulch, Alaska, for magnitude frequency risk analysis; (g) measurement (1 cm DTM) of blast-induced liquefaction settlements from the Ground Improvement Trials in Christchurch, New Zealand; (h) clash detection analysis of utilities and vegetation for the transport of the Space Shuttle Endeavour through Los Angeles.
Sources: (a) Courtesy of Oregon DOT, (b) Conner and Olsen (2014), (c) Kashani et al. (2015), (d) Cunningham et al. (2015), Dunham et al. (2017), (e) Chock et al. (2013), (f) Olsen et al. (2015), (g) Olsen et al. (2016a), (h) courtesy of David Evans and Associates (DEA).

Figure 8-26. Example VR system with head tracking and wand interaction displaying a TLS point cloud of a damaged Temple in Bungmati, Nepal, after the 2015 Ghorka Earthquake.

Carrea et al. (2015) and Olsen et al. (2015) present methodologies for extracting individual rockfall events from time series TLS scans. Volumes can then be calculated for each cluster to produce magnitude frequency relationships, which are useful for risk assessments.

Clash Detection

Another common type of analyses that can be done with terrestrial laser scan data is clash detection. Clash detection is the process of determining whether an object will intersect with existing objects. One example is a truck requiring adequate clearance to pass underneath a bridge. Another common example will be determining if a proposed pipe will fit within tight spaces in an industrial plant. In similar cases, these analyses can be conducted digitally in a safe environment to identify problematic objects so that economical solutions can be found and delays avoided. They can also be completed to tighter tolerances than many other techniques.

Reverse Engineering

TLS can also be used to scan in existing objects for custom-part fabrication. Existing objects can be scanned so that new objects can be custom-made to properly connect with those existing objects.

Intensity Analyses

TLS can be used to evaluate the reflectivity of objects; however, this requires that rigorous radiometric calibration processes be completed first (Kashani et al. 2015). Research is in progress to determine if these systems can provide reliable measurements for retro-reflectivity of signs or pavement markings.

Visibility Analyses

TLS can also be used to determine what is visible from a location and objects that will obstruct a view. As one example, architects can use the point cloud to evaluate the available functional space in an area. As another, transportation engineers can utilize the point cloud to evaluate the visibility for sight-distance analyses at locations such as intersections or curves.

Structural Analyses

Although TLS provides a substantial amount of information and leads to more complete surveys than traditional technologies, it is still limited to only capturing information on the surface. For advanced structural analyses, it is often combined with other nondestructive technologies such as acoustic emission, ground-penetrating radar, multispectral imaging, thermography, geophysics, and so on. In such cases, TLS provides the geometric base model where data from these other sources, which tend to be located to specific locales or cross sections, can be linked and integrated together for an analysis. For a variety of examples of integrating TLS with other technologies for structural analysis, the reader is referred to the text by Riveiro and Solla (2016).

BEST PRACTICES

The following is a summary of some best practices for quality lidar surveys in addition to those already presented throughout the chapter. As always, the extent to which these should be implemented will depend on the project requirements.

- Be systematic! Having established workflows and systematic procedures can help move data efficiently.
- Always use a lot of redundancy in the control! See Chapter 11 for more on control.
- Perform reconnaissance and planning, including but not limited to scanner locations considering point density and overlap between scans. Estimate the necessary amount of scanning based on the quality required.
- Ensure that there is adequate overlap between scans so that occlusions can be filled and to help with the registration process and quality control.
- Fully document what was done (field and office) so steps can be retraced.
- Store the native, raw data. Although lots of datasets are typically created throughout the processing stage, having the source data and the basic georeferencing information stored will enable one to quickly re-create the point cloud in the event of data loss further down the processing pipeline. In addition, raw data can be reprocessed in the future if better models are developed.
- Invest in hardware, data storage, and networking capabilities.
- Have strict backup and archive protocols. Working with lidar requires well-developed data management techniques and supporting hardware.
- Do not assume manufacturer specs reflect reality! Test and statistically quantify every error in each project. Error will vary significantly throughout the point cloud and between scene types, so checks should be implemented throughout.
- Utilize independent validation points for high-accuracy projects. Determine where to perform these checks and how frequently to evaluate error throughout the project extents.

- Report and understand your uncertainty properly! If there is 1 cm error RMS in the data set, do not report that 1 mm of change can be detected! Typically, uncertainty results are reported at 95% confidence. However, note that approximately 5% of the data will have a higher amount of uncertainty, which is still a large amount of data points!
- Define the coordinate system, datum\reference frame, and epoch properly (see Chapters 2 and 3).

FUTURE CHANGES

Lidar technology evolves rapidly in both hardware and software capabilities and will likely continue to improve in the years to come. Scanners are rapidly improving in terms of speed, accuracy, range, portability, and many other features. Hence, it is always important to seek out up-to-date information.

Some of the key challenges that are being addressed in research and development include the following:

1. *Better integration of data from multiple sensors.* Newer scanners have improved cameras and are being integrated with IMUs and GNSS for kinematic scanning capabilities.
2. *Availability of individual point uncertainty estimates.* Currently, a single accuracy value is typically reported for a TLS dataset. However, as discussed previously, many factors influence accuracy, and each point will have a unique accuracy.
3. *Improved data interoperability.* Recently, ASTM E57 was developed to help improve data interoperability between software packages. Some TLS point cloud data are also delivered in ASPRS LAS format; however, this format does not preserve the scan structure or the link to the coacquired imagery. The ALS and TLS industries have been separated to some degree; fortunately, developments in MLS and UAS will probably help bridge these gaps.
4. *Enhanced data workflows.* TLS software evolves quickly and regularly adds new features.

The impact on TLS on the surveying engineering community, as well as many others, has been substantial in improving the efficiency, reliability, accuracy, and safety of data capture. To date, relatively few have become adequately versed in this technology. However, barriers to its usage, including the steep learning curve, are eroding. This technology will continue to redefine workflows in engineering (and other fields) and help us transition to state-of-the-art 3D practices. To date, we have only seen the beginning of what this technology has to offer us!

ACKNOWLEDGMENTS

The author acknowledges the support of the National Science Foundation under Grant No. 1351487 in preparing this chapter. Leica Geosystems, David Evans and Associates, and Maptek I-Site provided hardware and/or software utilized to generate figures in this chapter. The author appreciates the assistance of various funding agencies, collaborators, and students in assisting with the acquisition and analysis of data shown in the examples presented throughout this chapter. Any opinions, findings, and conclusions or recommendations expressed in this material are those of the author and do not necessarily reflect those of these supporting entities.

REFERENCES

Abellán, A., M. Jaboyedoff, T. Oppikofer, N. J. Rosser, M. Lim, and M. J. Lato. 2014. "State of science: Terrestrial laser scanner on rock slopes instabilities." *Earth Surf. Processes Landforms* 39: 80–97.

Ai, C., and Y.-J. Tsai. 2016. "An automated sign retroreflectivity condition evaluation methodology using mobile lidar and computer vision." *Transp. Res. Part C Emerging Technol.* 63: 96–113.

Anil, E., P. Tang, B. Akinci, and D. Huber. 2013. "Deviation analysis method for the assessment of the quality of the as-is building information models generated from point cloud data." *Autom. Constr.* 35: 507–516.

Bae, K.-H., and D. D. Lichti. 2007. "On-site self-calibration using planar features for terrestrial laser scanners." *Int. Arch. Photogramm. Remote Sens. Spatial Inf. Sci.* 36 (Part 3/W52): 14–19.

Bae, K.-H., and D. D. Lichti. 2008. "A method for automated registration of unorganised point clouds." *ISPRS J. Photogramm. Remote Sens.* 63 (1): 36–54.

Bae, K.-H., D. Belton, and D. D. Lichti. 2009. "A closed-form expression of the positional uncertainty for 3D point clouds." *IEEE Trans. Pattern Anal. Mach. Intell.* 31 (4): 577–590.

Ballard, D. H., and C. M. Brown. 1982. *Computer vision*. Englewood Cliffs, NJ: Prentice Hall.

Barazzetti, L., F. Banfi, R. Brumana, G. Gusmeroli, M. Previtali, and G. Schiantarelli. 2015. "Cloud-to-BIM-to-FEM: Structural simulation with accurate historic BIM from laser scans." *Simul. Modell. Pract. Theory* 57: 71–87.

Barnea, S., and S. Filin. 2013. "Segmentation of terrestrial laser scanning data using geometry and image information." *ISPRS J. Photogramm. Remote Sens.* 76: 33–48.

Barnea, S., and S. Filin. 2008. "Keypoint based autonomous registration of terrestrial laser point-clouds." *ISPRS J. Photogramm. Remote Sens.* 63 (2–3): 19–35.

Barnea, S., and S. Filin. 2012. "Extraction of objects from terrestrial laser scans by integrating geometry image and intensity data with demonstration on trees." *Remote Sens.* 4 (1): 88–110.

Barnhart, T. B., and B. Crosby. 2013. "Comparing two methods of surface change detection on an evolving thermokarst using high-temporal-frequency terrestrial laser scanning, Selawik River, Alaska." *Remote Sens.* 5: 2813–2837.

Becerik-Gerber, B., F. Jazizadeh, G. Kavulya, and G. Calis. 2011. "Assessment of target types and layouts in 3D laser scanning for registration accuracy." *Autom. Constr.* 20: 649–658.

Beinat, A., F. Crosilla, and F. Sepic. 2006. "Automatic morphological pre-alignment and global hybrid registration of close range images." *Int. Arch. Photogramm. Remote Sens. Spatial Inf. Sci.* 36 (Part 5): 25–27.

Belton, D., and K.-H. Bae. 2010. "Automating post-processing of terrestrial laser scanning point clouds for road feature surveys." *Int. Arch. Photogramm. Remote Sens. Spat. Inf. Sci.* 38: 74–79.

Berman, J. W., J.Wartman, M. J. Olsen, J. Irish, S. Miles, T. Tanner, K. R. Gurley, L. Lowes, A. Bostrom, J. Dafni, M. Grilliot, A. Lyda, and J. Peltier. 2020. "Natural hazards reconnaissance with the NHERI RAPID facility." *Front. Built Environ.* 6:185, section Wind Engineering and Science, Special Issue on Natural Hazards Engineering Research Infrastructure (NHERI) 2016-2020: Mitigating the Impact of Natural Hazards on Civil Infrastructure and Communities. doi: 10.3389/fbuil.2020.573067.

Besl, P. J., and N. D. McKay. 1992. "A method for registration of 3-D shapes." *IEEE Trans. Pattern Anal. Mach. Intell.* 14 (2): 239–256.

Biosca, J. M., and J. L. Lerma. 2008. "Unsupervised robust planar segmentation of terrestrial laser scanner point clouds based on fuzzy clustering methods." *ISPRS J. Photogramm. Remote Sens.* 63: 84–98.

Böhm, J., and S. Becker. 2007. "Automatic marker-free registration of terrestrial laser scans using reflectance features." In *Proc., 8th Conf. on Optical 3D Measurement Techniques*, 338–344. Bellingham, WA: SPIE.

Bolkas, D. 2019. "Terrestrial laser scanner intensity correction for the incidence angle effect on surfaces with different colours and sheens." *Int. J. Remote Sens.* 40 (18): 7169–7189.

Bolkas, D., and A. Martinez. 2018. "Effect of target color and scanning geometry on terrestrial lidar point-cloud noise and plane fitting." *J. Appl. Geod.* 12 (1): 109–127.

Bolkas, D., I. Vazaios, A. Peidou, and N. Vlachopoulos. 2018. "Detection of rock discontinuity traces using terrestrial lidar data and space-frequency transforms." *Geotech. Geol. Eng.* 36 (3): 1745–1765.

Bosché, F. 2010. "Automated recognition of 3D CAD model objects in laser scans and calculation of as-built dimensions for dimensional compliance control in construction." *Adv. Eng. Inf.* 24 (1): 107–118.

Bosché, F., M. Ahmed, Y. Turkan, C. T. Haas, and R. Haas. 2015. "The value of integrating Scan-to-BIM and Scan-vs-BIM techniques for construction monitoring using laser scanning and BIM: The case of cylindrical MEP components." *Autom. Constr.* 49 (B): 201–213.

Bosché, F., A. Guillemet, Y. Turkan, C. T. Haas, and R. G. Haas. 2014b. "Tracking the built status of MEP works: Assessing the value of a Scan-vs-BIM system." *J. Comput. Civ. Eng.* 28 (4): 05014004.

Bosché, F., and C. T. Haas. 2008. "Automated retrieval of 3D CAD model objects in construction range images." *Autom. Constr.* 17 (4): 499–512.

Bosché, F., C. Haas, and B. Akinci. 2009. "Automated recognition of 3D CAD objects in site laser scans for project 3D status and visualization and performance control." *J. Comput. Civ. Eng.* 23 (6): 311–318.

Boström, G., J. G. Gonçalves, and V. Sequeira. 2008. "Controlled 3D data fusion using error bounds." *ISPRS J. Photogramm. Remote Sens.* 63 (1): 55–67.

Brenner, C. 2009. "Extraction of features from mobile laser scanning data for future driver assistance systems." In *Advances in GIScience* 25–42. Berlin: Springer.

Brenner, C., C. Dold, and N. Ripperda. 2008. "Coarse orientation of terrestrial laser scans in urban environments." *ISPRS J. Photogramm. Remote Sens.* 63 (1): 4–18.

Briese, C. 2010. "Extraction of digital terrain models." In *Airborne and terrestrial laser scanning*, edited by G. Vosselman and H.-G. Maas, 83–133. Caithness, UK: Whittles Publishing.

Brodu, N., and D. Lague. 2012. "3D terrestrial lidar data classification of complex natural scenes using a multi-scale dimensionality criterion: Applications in geomorphology." *ISPRS J. Photogramm. Remote Sens.* 68: 121–134.

Bunn, M., B. Leshchinsky, M. J. Olsen, and A. Booth. 2019. "A simplified, object-based framework for efficient landslide inventorying using lidar digital elevation model derivatives." *Remote Sens.* 11 (3): 303.

Calbaleiro, M., B. Riveiro, P. Arias, J. C. Caamaño, and J. A. Vilán. 2014. "Automatic 3D modelling of metal frame connections from lidar data for structural engineering purposes." *ISPRS J. Photogramm. Remote Sens.* 96: 47–56.

Carrea, D., A. Abellán, M. H. Derron, and M. Jaboyedoff. 2015. "Automatic rockfalls volume estimation based on terrestrial laser scanning data." *Eng. Geol. Soc. Territ.* 2: 425–428.

Chaudhury, K. N., Y. Khoo, and A. Singer. 2015. "Global registration of multiple point clouds using semidefinite programming." *SIAM J. Optim.* 25 (1): 468–501.

Chang, J. C., D. J. Findley, M. K. Tsai, C. M. Cunningham. 2012. "Infrastructure Investment Protection with LiDAR." FHWA/NC/2012-15 Report, 81 pp.

Che, E., J. Jung, and M. Olsen. 2019a. "Object recognition, segmentation, and classification of mobile laser scanning point clouds: State of the art review." *Sensors* 19 (4): 810.

Che, E., and M. J. Olsen. 2017. "Fast ground filtering for TLS data via scanline density analysis." *ISPRS J. Photogramm. Remote Sens.* 129: 226–240.

Che, E., and M. J. Olsen. 2018. "Multi-scan segmentation of terrestrial laser scanning data based on normal variation analysis." *ISPRS J. Photogramm. Remote Sens.* 143: 233–248.

Che, E., M. J. Olsen, C. E. Parrish, and J. Jung. 2019b. "Pavement marking retroreflectivity estimation and evaluation using mobile lidar data." *Photogramm. Eng. Remote Sens.* 85 (8): 573–583.

Chehata, N., N. David, and F. Bretar. 2008. "lidar data classification using hierarchical k-means clustering." In Vol. 37 of *Int. Society for Photogrammetry and Remote Sensing Congress Beijing*, 325–330.

Chen, Y., and G. Medioni. 1992. "Object modelling by registration of multiple range images." *Image Vision Comput.* 10 (3): 145–155.

Chin, A., and M. J. Olsen. 2015. "Evaluation of technologies for road profile capture, analysis, and evaluation." *J. Surv. Eng.* 141 (1): 04014011.

Chock, G., L. Carden, I. Robertson, M. J. Olsen, and G. Yu. 2013. "Tohoku tsunami-induced building failure analysis with implications for U.S. tsunami and seismic design codes." *Earthquake Spectra* 29 (S1): 99–126.

Chow, J. C. K., D. D. Lichti, C. Glennie, and P. Hartzell. 2013. "Improvements to and comparison of static terrestrial lidar self-calibration methods." *Sensors* 13 (13): 7224–7249.

CISA3 (Center of Interdisciplinary Science for Art, Architecture and Archaeology). 2011. "Battle of Anghiari." *CISA3 Projects*. Accessed February 8, 2022. http://cisa3.calit2.net/cisa3/projects/anghiari.php.

Collins, B. D., R. Kayen, D. Minasian, and T. Reiss. 2009. *Terrestrial lidar datasets of New Orleans, Louisiana, Levee failures from Hurricane Katrina, August 29, 2005*. Data Series 470. Washington, DC: USGS.

Collins, B., R. Kayen, T. Reiss, and N. Sitar. 2007. *Terrestrial lidar investigation of the December 2003 and January 2007 activations of the Northridge Bluff Landslide, Daly City, California*. Open-File Rep. 2007-1079. Washington, DC: USGS.

Collins, B., and N. Sitar. 2004. "Application of high resolution 3d laser scanning to slope stability studies." In *Proc., 39th Annual Symp. on Engineering Geology and Geotechnical Engineering*, 79–92. Butte, MT, USA.

Collins, B. D., and G. M. Stock. 2015. "lidar based Rock-Fall hazard characterization of cliffs." Accessed February 9, 2022. http://www.nps.gov/yose/learn/nature/upload/Collins-Stock-2012-ASCE.pdf.

Conde, B., A. Villarino, M. Cabaleiro, and D. Gonzalez-Aguilera. 2015. "Geometrical issues on the structural analysis of transmission electricity towers thanks to laser scanning technology and finite element method." *Remote Sens.* 7 (9): 11551–11569.

Conner, J., and M. J. Olsen. 2014. "Automated quantification of distributed landslide movement using circular tree trunks extracted from terrestrial laser scan data." *Comput. Geosci.* 67: 31–39.

Cunningham, K., M. J. Olsen, J. Wartman, and L. Dunham. 2015. *A platform for proactive, risk-based slope asset management, Phase II*. Rep. No. 2013-M-UAF-0042. Seattle : Pacific Northwest Transportation Consortium.

CyArk (2021). "CyArk." Accessed February 8, 2022. https://www.cyark.org/.

Douillard, B., J. Underwood, N. Kuntz, V. Vlaskine, A. Quadros, P. Morton, et al. 2011. "On the segmentation of 3d lidar point clouds." In *Proc., IEEE Int. Conf. on Robotics and Automation*, 2798–2805. Piscataway, NJ: IEEE.

Duffell, C. G., and D. M. Rudrum. 2005. "Remote sensing techniques for highway earthworks assessment." In *GeoFrontiers: Site Characterization and Modeling, Geotechnical Special Publication 138*, edited by P. W. Mayne, E. Rathje, J. DeJong, A. Rechenmacher, J. Yamamuro, S. Sharma, and C. Willson, 1–13. Reston, VA: ASCE.

Dunham, L., J. Wartman, M. J. Olsen, M. S. O'Banion, and K. Cunningham. 2017. "Rockfall Activity Index (RAI): A lidar-derived, morphology-based method for hazard assessment." *Eng. Geol.* 221: 184–192.

El-Omari, S., and O. Moselhi. 2008. "Integrating 3D laser scanning and photogrammetry for progress measurement of construction work." *Autom. Constr.* 18 (1): 1–9.

El-Omari, S., and O. Moselhi. 2011. "Integrating automated data acquisition technologies for progress reporting of construction projects." *Autom. Constr.* 20 (6): 699–705.

Fan, L., W. Powrie, J. A. Smethurst, P. M. Atkinson, and H. Einstein. 2014. "The effect of short ground vegetation on terrestrial laser scans at a local scale." *ISPRS J. Photogramm. Remote Sens.* 95: 42–52.

Fan, L., J. A. Smethurst, P. M. Atkinson, and W. Powrie. 2015. "Error in target-based georeferencing and registration in terrestrial laser scanning." *Comput. Geosci.* 83: 54–64.

Fischler, M. A., and R. C. Bolles. 1981. "Random sample consensus: A paradigm for model fitting with applications to image analysis and automated cartography." *Commun. ACM* 24 (6): 381–395.

Gielsdorf, F., A. Rietdorf, and L. Gruendig. 2004. "A concept for the calibration of terrestrial laser scanners." In *Proc., FIG Working Week*. International Federation of Surveyors, FIG, Copenhagen, Denmark.

Girardeau-Montaut, D. 2013. "Cloud compare: 3D point cloud and mesh processing software, open-source project." Accessed February 9, 2022. http://www.danielgm.net/cc/.

Glennie, C., and D. D. Lichti. 2010. "Static calibration and analysis of the Velodyne HDL-64E S2 for high accuracy mobile scanning." *Remote Sens.* 2 (6): 1610–1624.

Glennie, C., and D. D. Lichti. 2011. "Temporal stability of the Velodyne HDL-64E S2 scanner for high accuracy scanning applications." *Remote Sens.* 3 (3): 539–553.

Gonzalez-Aguilera, D., J. Gomez-Lahoz, and J. Sanchez. 2008. "A new approach for structural monitoring of large dams with a three-dimensional laser scanner." *Sensors* 8 (9): 5866–5883.

Gonzalez-Jorge, H., B. Riveiro, J. Armesto, and P. Arias. 2011. "Geometric evaluation of road signs using radiometric information from laser scanning data." In *Proc., 28th Int. Conf. Int. Symp. on Automation and Robotics in Construction*, 1007–1012. International Association for Automation and Robotics in Construction (IAARC), Seoul, Korea.

Gordon, S., D. D. Lichti, I. Chandler, M. P. Stewart, and J. Franke. 2003. "Precision measurement of structural deformation using terrestrial laser scanners." In *Proc., Optical 3D Measurement Techniques*, 322–329. Bellingham, WA: SPIE.

Gordon, S., D. D. Lichti, J. Franke, and M. P. Stewart. 2004a. "Measurement of structural deformation using terrestrial laser scanners." In *Proc., 1st FIG Int. Symp. on Engineering Surveys for Construction Works and Structural Engineering*. International Federation of Surveyors, FIG, Copenhagen, Denmark.

Gordon, S., D. Lichti, J. Franke, and M. Stewart. 2004b. "Modeling point clouds for precise structural deformation measurement." In *Proc., 20th Int. Society Photogrammetry and Remote Sensing Congress 2004: Geo-Imagery Bridging Continent*.

Gorte, B. 2007. "Planar feature extraction in terrestrial laser scans using gradient based range image segmentation." In *Proc., Int. Society Photogrammetry and Remote Sensing Workshop on Laser Scanning*, 173–177. International Society of Photogrammetry and Remote Sensing, Hannover, Germany.

Grant, A., J. Wartman, C. I. Massey, M. J. Olsen, M. S. O'Banion, and M. Motley. 2018. "The impact of rockfalls on dwellings during the 2011 Christchurch, New Zealand, earthquakes." *Landslides* 15 (1): 31–42.

Grün, A., and D. Akca. 2005. "Least squares 3D surface and curve matching." *ISPRS J. Photogramm. Remote Sens.* 59 (3): 151–174.

Guan, H., J. Li, Y. Yu, C. Wang, M. Chapman, and B. Yang. 2014. "Using mobile laser scanning data for automated extraction of road markings." *ISPRS J. Photogramm. Remote Sens.* 87: 93–107.

Guarnieri, A., A. Vettore, F. Pirotti, M. Menenti, and M. Marani. 2009. "Retrieval of small-relief marsh morphology from terrestrial laser scanner, optimal spatial filtering, and laser return intensity." *Geomorphology* 113 (1–2): 12–20.

Han, J.-Y. 2010. "A non-iterative approach for the quick alignment of multistation unregistered lidar point clouds." *IEEE Geosci. Remote Sens. Lett.* 7 (4): 727–730.

Han, J.-Y., N.-H. Perng, and Y.-T. Lin. 2013a. "Feature conjugation for intensity-coded lidar point clouds." *J. Surv. Eng.* 139 (3): 135–142.

Han, S., H. Cho, S. Kim, J. Jung, and J. Heo. 2013b. "Automated and efficient method for extraction of tunnel cross sections using terrestrial laser scanned data." *J. Comp. Civil Eng.* 27 (3): 274–281.

Hartzell, P. J., P. J. Gadomski, C. L. Glennie, D. C. Finnegan, and J. S. Deems. 2015. "Rigorous error propagation for terrestrial laser scanning with application to snow volume uncertainty." *J. Glaciol.* 61 (230): 1147–1158.

Hinks, T., H. Carr, L. Truong-Hong, and D. F. Laefer. 2013. "Point cloud data conversion into solid models via point-based voxelization." *J. Surv. Eng.* 139 (2): 72–83.

Hiremagalur, J., K. S. Yen, K. Akin, T. Bui, T. A. Lasky, and B. Ravani. 2007. *Creating standards and specifications for the use of laser scanning in Caltrans projects*. Rep. No. CA07-0964. Washington, DC: National Academies of Sciences, Engineering, and Medicine.

Hoover, A., G. Jean-Baptiste, X. Jiang, P. J. Flynn, H. Bunke, D. B. Goldgof, et al. 1996. "An experimental comparison of range image segmentation algorithms." *IEEE Trans. Pattern Anal. Mach. Intell.* 18 (7): 673–689.

Hough, P.V.C. 1962. US Patent 3 066 954. Method and means for recognizing complex patterns.

Huber, D., B. Akinci, P. Tang, A. Adan, B. Okorn, and X. Xiong. 2010. "Using laser scanners for modeling and analysis in architecture, engineering, and construction." *44th Annual Conference on Information Sciences and Systems (CISS), 2010*, pp. 1–6. doi: 10.1109/CISS.2010.5464818.

Huber, D. 2011. "The ASTM E57 file format for 3D imaging data exchange." Accessed February 9, 2022. http://www.ri.cmu.edu/pub_files/2011/1/2011-huber-e57-v3.pdf.

Jaboyedoff, M., T. Oppikofer, A. Abellán, M.-H. Derron, A. Loye, R. Metzger, et al. 2012. "Use of lidar in landslide investigations: A review." *Nat. Hazard.* 61: 5–28.

Jaselskis, E. J., Z. Gao, A. Welch, and D. O'Brien. 2003. "Pilot study on laser scanning technology for transportation projects." In *Proc., 2003 Mid-Continent Transportation Research Symp.*, Ames, IA, USA.

Jiang, X. Y., and H. Bunke. 1994. "Fast segmentation of range images into planar regions by scan line grouping." *Mach. Vision Appl.* 7 (2): 115–122.

Johnson, W., and A. Johnson. 2012. "Operational considerations for terrestrial laser scanner use in highway construction applications." *J. Surv. Eng.* 138 (4): 214–222.

Jung, J., E. Che, M. J. Olsen, and C. Parrish. 2019. "Efficient and robust lane marking extraction from mobile lidar point clouds." *ISPRS J. Photogramm. Remote Sens.* 147: 1–18.

Jung, J., E. Che, M. J. Olsen, and K. C. Shafer. 2020. "Automated and efficient powerline extraction from laser scanning data using a voxel-based subsampling with hierarchical approach." *ISPRS J. Photogramm. Remote Sens.* 163: 343–361.

Jung, J., M. J. Olsen, D. S. Hurwitz, A. G. Kashani, and K. Buker. 2018. "3D virtual intersection sight distance analysis using lidar data." *Transp. Res. Part C Emerging Technol.* 86: 563–579.

Kang, Z., S. Zlatanova, and B. Gorte. 2007. "Automatic registration of terrestrial scanning data based on registered imagery." In *Proc., 2007 FIG Working Week*. International Federation of Surveyors, FIG, Copenhagen, Denmark.

Kashani, A. G., P. S. Crawford, S. K. Biswas, A. J. Graettinger, and D. Grau. 2014a. "Automated Tornado damage assessment and wind speed estimation based on terrestrial laser scanning." *J. Comput. Civ. Eng.* 29 (3): 04014051.

Kashani, A. G., A. Graettinger, and T. Dao. 2014b. "3D data collection and automated damage assessment for near real-time Tornado loss estimation." In *Construction Research Cong.: Construction in a Global Network*, edited by D. Castro-Lacouture, J. Irizarry, and B. Ashuri, 1209–1218. Reston, VA: ASCE.

Kashani, A., M. J. Olsen, C. E. Parrish, and N. Wilson. 2015. "A review of lidar radiometric processing: From ad hoc intensity correction to rigorous radiometric calibration." *Sensors* 15 (11): 28099–28128.

Kayen, R., R. T. Pack, J. Bay, S. Sugimoto, and H. Tanaka. 2006. "Terrestrial- lidar visualization of surface and structural deformations of the 2004 Nigata Ken Chuetsu Japan, Earthquake." *Earthquake Spectra* 22 (S1): 147–162.

Kayen, R., J. P. Stewart, and B. Collins. 2010. "Recent advances in terrestrial lidar applications in geotechnical earthquake engineering." In *Proc., 5th Int. Conf. on Recent Advances in Geotechnical Earthquake Engrg and Soil Dynamics*, San Diego, CA.

Kersten, T. P., H. Sternberg, and K. Mechelke. 2005. "Investigations into the accuracy behaviour of the terrestrial laser scanning system Mensi GS100." In Vol. I of *Proc., Optical 3-D Measurement Techniques VII*, edited by A. Grün and H. Kahmen, 122–131. SPIE, Bellingham, WA.

Kim, J. S., J. C. Lee, I. J. Kang, S. Y. Cha, H. Choi, and T. G. Lee. 2008. "Extraction of geometric information on highway using terrestrial laser scanning technology." *ISPRS Ann. Photogramm. Remote Sens. Spatial Inf. Sci.* 37 (5): 539–544.

Kingston, T., V. Gikas, C. Laflamme, and C. Larouche. 2006. "An integrated mobile mapping system for data acquisition and automated asset extraction." *Int. Arch. Photogramm. Remote Sens. Spat. Inf. Sci.* 36 (5), 5 pp.

Kreylos, O., G. W. Bawden, and L. H. Kellogg. 2008. "Immersive visualization and analysis of lidar data." In Vol. 5358 of *Proc., Int. Symp. on Advances in Visual Computing, Part I. LNCS*, edited by G. Bebis, et al., 846–855. Berlin: Springer.

Kreylos, O. 2022. "Vrui VR Toolkit." Accessed February 7, 2022. https://web.cs.ucdavis.edu/~okreylos/ResDev/Vrui/index.html.

Kyparissi, S., and A. Georgopoulos. 2013. "Point Cloud Registration based on image correspondences." *Int. J. Heritage Digital Era* 2 (3): 471–489.

Laefer, D. F., M. Fitzgerald, E. M. Maloney, D. Coyne, D. Lennon, and S. W. Morrish. 2009. "Lateral image degradation in terrestrial laser scanning." *Struct. Eng. Int.* 19 (2): 184–189.

Laefer, D. F., J. Gannon, and E. Deely. 2010. "Reliability of crack detection methods for baseline condition assessments." *J. Infrastruct. Syst.* 16 (2): 129–137.

Laefer, D. F., L. Tuong-Hong, H. Carr, and M. Singh. 2014. "Crack detection limits in unit based masonry with terrestrial laser scanning." *NDT & E Int.* 62: 66–76.

Lau, C. L., S. Halim, M. Zukeppli, A. Mohd-Azwan, W. L. Tang, and A. K. Chong. 2015. "Terrain extraction by integrating terrestrial laser scanner data and spectral information." *Int. Arch. Photogramm. Remote Sens. Spatial Inf. Sci.* XL-2/W4: 45–51.

Lee, H. M., and H. S. Park. 2008. "Estimation of deformed shape of beam structure using 3D coordinate information from terrestrial laser scanning." *Comput. Aided Civ. Infrastruct. Eng.* 29(1): 29-44.

Lee, H. M., and H. S. Park. 2011. "Gage free stress estimation of a beam like structure based on terrestrial laser scanning." *Comput. Aided Civ. Infrastruct. Eng.* 26(8): 647–658.

Lehtomäki, M., A. Jaakkola, J. Hyyppä, A. Kukko, and H. Kaartinen. 2011. "Performance analysis of a pole and tree trunk detection method for mobile laser scanning data." *Int. Arch. Photogramm. Remote Sens. Spat. Inf. Sci.* 38: 197–202.

Lichti, D. D. 2005. "Spectral filtering and classification of terrestrial laser scanner point clouds." *Photogramm. Rec.* 20 (111): 218–240.

Lichti, D. D. 2007. "Modelling, calibration and analysis of an AM-CW terrestrial laser scanner." *ISPRS J. Photogramm. Remote Sens.* 61 (5): 307–324.

Lichti, D. D. 2008. "A method to test differences between additional parameter sets with a case study in terrestrial laser scanner self-calibration stability analysis." *ISPRS J. Photogramm. Remote Sens.* 63 (2): 169–180.

Lichti, D. D. 2010a. "A review of geometric models and self-calibration methods for terrestrial laser scanners." *Boletim de Ciências Geodésicas* 16 (1): 3–19.

Lichti, D. D. 2010b. "Terrestrial laser scanner self-calibration: Correlation sources and their mitigation." *ISPRS J. Photogramm. Remote Sens.* 65 (1): 93–102.

Lichti, D. D., J. Chow, and H. Lahamy. 2011. "Parameter de-correlation and model-identification in hybrid-style terrestrial laser scanner self-calibration." *ISPRS J. Photogramm. Remote Sens.* 66 (3): 317–326.

Lichti, D. D., and J. Franke. 2005. "Self-calibration of the iQsun 880 laser scanner." In Vol. I of *Proc., Optical 3-D Measurement Techniques VII*, edited by A. Grün and H. Kahmen, 112–121. SPIE, Bellingham, WA.

Lichti, D. D., S. J. Gordon, and T. Tipdecho. 2005. "Error models and propagation in directly georeferenced terrestrial laser scanner networks." *J. Surv. Eng.* 131 (4): 135–142.

Lichti, D. D., and S. Jamtsho. 2006. "Angular resolution of terrestrial laser scanners." *Photogramm. Rec.* 21 (114): 141–160.

Lichti, D. D., and J. Lampard. 2008. "Reflectorless total station self-calibration." *Surv. Rev.* 40 (309): 244–259.

Lichti, D. D., and M. G. Licht. 2006. "Experiences with terrestrial laser scanner modelling and accuracy assessment." *Int. Arch. Photogramm. Remote Sens. Spatial Inf. Sci.* 36 (Part 5): 155–160.

Lichti, D. D., and J. Skaloud. 2010. "Registration and calibration." In *Airborne and terrestrial laser scanning*, edited by G. Vosselman and H.-G. Maas, 83–133. Caithness, UK: Whittles Publishing.

Lin, C.-C., and R.-J. You. 2006. "Planar feature extraction from lidar data based on tensor analysis." In *Proc., 27th Asian Conf. on Remote Sensing*, 1–7. Asian Association on Remote Sensing (AARS), Tokyo, Japan.

Lin, Y., and J. Hyyppa. 2010. "Geometry and intensity based culvert detection in mobile laser scanning point clouds." *J. Appl. Remote Sens.* 4: 043553.

Lindenbergh, R., L. Uchanski, A. Bucksch, and R. Van Gosliga. 2009. "Structural monitoring of tunnels using terrestrial laser scanning." *Rep. Geod.* 2 (87), 231–238.

Maalek, R., D. D. Lichti, and J. Ruwanpura. 2015. "Robust classification and segmentation of planar and linear features for construction site progress monitoring and structural dimension compliance control." *ISPRS Ann. Photogramm. Remote Sens. Spatial Inf. Sci.* II-3/W5: 129–136.

Maas, H. G., and G. Vosselman. 1999. "Two algorithms for extracting building models from raw laser altimetry data." *ISPRS J. Photogramm. Remote Sens.* 54 (2–3): 153–163.

Mabey, S. 2009. *McKim & Creed Deploy TITAN for mobile LIDAR Scanning of Five Major Sections of North Carolina DOT Highways 2009*. Houston, TX, USA: Ambercore.

Mahmoudabadi, H., M. J. Olsen, and S. Todorovic. 2016. "Efficient terrestrial laser scan segmentation exploiting data structure." *J. Photogramm. Remote Sens.* 119: 135–150.

Mahmoudabadi, H., T. Shoaf, and M. J. Olsen. 2013. "Superpixel clustering and planar fit segmentation of 3D lidar point clouds." In *Proc., ComGeo*. San Francisco: IEEE.

Makadia, A., A. Patterson, and K. Daniilidis. 2006. "Fully automatic registration of 3D point clouds." In *Proc., of the 2006 IEEE Computer Society Conf. on Computer Vision and Pattern Recognition*. New York: IEEE, 8 pp.

Mallet, C., and F. Bretar. 2009. "Full-waveform topographic lidar: State-of-the-art." *ISPRS J. Photogramm. Remote Sens.* 64 (1): 1–16.

Mårtensson, S.-G., Y. Reshetyuk, and L. Jivall. 2012. "Measurement uncertainty in network RTK GNSS-based positioning of a terrestrial laser scanner." *J. Appl. Geodesy* 6 (1): 25–32.

McInerney, J., I. Trzcinski, and M. J. DeJong. 2012. "Discrete element modelling of masonry using laser scanning data." In *Proc., 8th Int. Conf. on Structural Analysis of Historical Constructions*, Wroclaw University, Wroclaw, Poland.

McQuat, G. J. 2011. "Feature extraction workflows for urban mobile-terrestrial lidar data." M.S. thesis, Queen's University, Dept. of Geography, http://qspace.library.queensu.ca/bitstream/1974/6530/1/Mcquat_Gregory_J_201105_MSc.pdf.

Mellado, N., N. Mitra, and D. Aiger. 2014. "Super 4PCS fast global pointcloud registration via smart indexing." *Comput. Graphics Forum* 33 (5): 205–215.

Meng, X., N. Currit, and K. Zhao. 2010. "Ground filtering algorithms for airborne lidar data: A review of critical issues." *Remote Sens.* 2 (3): 833–860.

Metzger, A. T., M. J. Olsen, J. Wartman, L. Dunham, A. Stuedlein, and K. Cunningham. 2013. *A platform for proactive risk-based slope asset management—Phase I interim project report*. Rep. No. 2012-M-0002. Seattle : Pacific Northwest Transportation Consortium.

Mohamed, A., and B. Wilkinson. 2009. "Direct georeferencing of stationary lidar." *Remote Sens.* 1 (4): 1321–1337.

Monserrat, O., and M. Crosetto. 2008. "Deformation measurement using terrestrial laser scanning data and least squares 3D surface matching." *ISPRS J. Photogramm. Remote Sens.* 63 (1): 142–154.

Moussa, A., and N. El-Sheimy. 2010. "Automatic classification and 3d modeling of lidar data." In *Proc. Int. Society for Photogrammetry and Remote Sensing Commission III Symp. PCV*. Hannover Germany.

Niemeyer, J., F. Rottensteiner, and U. Soergel. 2012. "Conditional random fields for lidar point cloud classification in complex urban areas." *ISPRS Ann. Photogramm. Remote Sens. Spat. Inf. Sci.* 1 (3): 263–268.

Nuttens, T., C. Stal, H. De Backer, K. Schotte, P. Van Bogaert, and A. De Wulf. 2014. "Methodology for the ovalization monitoring of newly built circular train tunnels based on laser scanning: Liefkenshoek Rail Link (Belgium)." *Autom. Constr.* 43: 1–9.

O'Banion, M. S., M. J. Olsen, and G. V. Roe. 2013. "Learn Mobile Lidar." Accessed February 9, 2022. http://learnmobilelidar.com.

O'Banion, M. S., and M. J. Olsen. 2019. "Efficient planning and acquisition of terrestrial laser scanning-derived digital elevation models: Proof of concept study." *J. Surv. Eng.* 145 (1): 06018003.

O'Banion, M. S., M. J. Olsen, C. E. Parrish, and M. Bailey. 2019. "Interactive visualization of 3D coordinate uncertainties in terrestrial laser-scanning point clouds using OpenGL shader language." *J. Surv. Eng.* 145 (1): 04018012.

Olsen, M. J. 2015. "In-Situ change analysis and monitoring through terrestrial laser scanning." *J. Comput. Civ. Eng.* 29 (2): 04014040.

Olsen, M. J., J. C. Allan, and G. R. Priest. 2012a. "Johnson Creek landslide movement and erosion quantification through 3D laser scanning." In *Geo-Congress: State of the Art and Practice in Geotechnical Engineering*, edited by R. D. Hryciw, A. Athanasopoulos-Zekkos, and N. Yesiller. Reston, VA: ASCE.

Olsen, M. J., S. Butcher, and E. P. Silvia. 2012b. *Real-time change and damage detection of landslides and other earth movements threatening public infrastructure*. OTREC-RR-11-23. OTREC Final Rep. 2011-22 and ODOT Final Rep. RS 500-500. Portland, OR: Transportation Research and Education Center.

Olsen, M. J., Z. Chen, T. C. Hutchinson, and F. Kuester. 2013a. "Optical techniques for multi-scale damage assessment in natural hazard analysis." *Geomatics Nat. Hazards Risk* 4 (1): 49–70.

Olsen, M. J., K. F. Cheung, Y. Yamazaki, S. M. Butcher, M. Garlock, S. C. Yim, et al. 2012c. "Damage assessment of the 2010 Chile earthquake and tsunami using ground-based lidar." *Earthquake Spectra* 28 (S1): 179–197.

Olsen, M. J., and D. T. Gillins. 2013. "Influence of coordinate system selection for coastal deformation analyses." In *Proc., 2nd Joint Int. Symp. on Deformation Monitoring*. Nottingham, UK.

Olsen, M. J., D. T. Gillins, and S. van Ballegooy. 2016a. "Terrestrial laser scanning deformation analyses of blast-induced liquefaction settlements." In *Proc., FIG Working Week 2016*, "Recovery from disaster." International Federation of Surveyors, FIG, Copenhagen, Denmark.

Olsen, M. J., E. Johnstone, N. Driscoll, S. A. Ashford, and F. Kuester. 2009. "Terrestrial laser scanning of extended cliff sections in dynamic environments: A parameter analysis." *J. Surv. Eng.* 135 (4): 161–169.

Olsen, M. J., E. Johnstone, N. Driscoll, F. Kuester, and S. A. Ashford. 2016b. "Fate and transport of seacliff failure sediment in southern California." *J. Coastal Res.* 76 (10076): 185–199.

Olsen, M. J., E. Johnstone, and F. Kuester. 2013b. "Hinged, pseudo-grid triangulation method for long, near linear cliff analysis." *J. Surv. Eng.* 139 (2): 105–109.

Olsen, M. J., E. Johnstone, F. Kuester, S. A. Ashford, and N. Driscoll. 2011. "New automated point-cloud alignment for ground based lidar data of long coastal sections." *J. Surv. Eng.* 137 (1): 14–25.

Olsen, M. J., and R. Kayen. 2012. "Post-earthquake and tsunami 3D laser scanning forensic investigations." In *Forensic Eng. 2012: Gateway to a Safer Tomorrow*. CD-ROM, American Society of Engineers, Reston, VA.

Olsen, M. J., F. Kuester, B. Chang, and T. C. Hutchinson. 2010. "Terrestrial laser scanning based structural damage assessment." *J. Comput. Civ. Eng.* 24 (3): 264–272.

Olsen, M. J., I. Madin, A. Chin, and J. Conner. 2012d. "Natural Hazards Subchapter (10.9)." In *Manual of Airborne Topographic lidar, ASPRS*, edited by M. Renslow, 407–422.

Olsen, M. J., K. Ponto, J. Kimball, F. Kuester, and M. Seracini. 2013c. "2D open-source editing techniques for 3D laser scans." In *Proc., 38th Annual Conf. on Computer Applications and Quantitative Methods in Archaeology*, edited by F. Contreras, M. Farjas, and F. J. Melero, 455–461. Oxford, UK: Arcaheopress.

Olsen, M. J., G. V. Roe, C. Glennie, F. Persi, M. Reedy, D. Hurwitz, et al. 2013d. *Guidelines for the use of mobile lidar in transportation applications*. TRB NCHRP Final Rep. 748. Washington, DC: National Cooperative Highway Research Program.

Olsen, M. J., R. Singh, K. Williams, and A. Chin. 2012e. "Transportation Engineering Subchapter (10.3.2)." *Manual of Airborne Topographic lidar, ASPRS*, edited by M. Renslow, 331–343.

Olsen, M. J., J. Wartman, M. McAlister, H. Mahmoudabadhi, M. S. O'Banion, L. Dunham, et al. 2015. "To fill or not to fill: Sensitivity analysis of the influence of resolution and hole filling on point cloud surface modeling and individual rockfall event detection." *Remote Sens.* 7 (9): 12103–12134.

Olsen, M. J., and D. Gillins. 2015. "How can Geomatics Technologies Benefit Geotechnical Studies?" *6th International Conference on Earthquake Geotechnical Engineering, 6ICEGE*. Christchurch, NZ.

Olsen, M. J., 2009. "Methodology for assessing coastal change using terrestrial laser scanning." *PhD Dissertation, Structural Engineering*. University of California at San Diego, La Jolla, CA.

Oomen, et al. 2015. "Sustainable geotechnical asset management." Accessed August 4, 2016. http://www.mtri.org/geoasset/.

Panholzer, H., and A. Prokop. 2013. "Wedge-filtering of geomorphologic terrestrial laser scan data." *Sensors* 13: 2579–2594.

Paris, P., M. J. Starek, E. Hardin, O. Kurum, M. Overton, and H. Mitasova. 2013. "Lines in the sand: Geomorphic and geospatial characterization and interpretation of sandy shorelines and beaches." *Geogr. Compass* 7 (5): 315–343.

Pesci, A., and G. Teza. 2008. "Terrestrial laser scanner and retro-reflective targets: an experiment for anomalous effects investigation." *Int. J. Remote Sens.* 29 (19): 5749–5765.

Pirotti, F., A. Guarnieri, and A. Vettore. 2013a. "State of the art of ground and aerial laser scanning technologies for high-resolution topography of the earth surface." *Eur. J. Remote Sensing* 46: 66–78.

Pirotti, F., A. Guarnieri, and A. Vettore. 2013b. "Ground filtering and vegetation mapping using multi-return terrestrial laser scanning." *ISPRS J. Photogramm. Remote Sensing* 76: 56–63.

Pomerleau, F., F. Colas, and R. Siegwart. 2015. "A review of point cloud registration algorithms for mobile robotics." *Found. Trends Rob.* 4 (1): 1–104.

Pu, S., M. Rutzinger, G. Vosselman, and S. Oude Elberink. 2011. "Recognizing basic structures from mobile laser scanning data for road inventory studies." *ISPRS J. Photogramm. Remote Sens.* 66: S28–S39.

Pu, S., and G. Vosselman. 2006. "Automatic extraction of building features from terrestrial laser scanning." *ISPRS J. Photogramm. Remote Sens.* 36 (5): 25–27.

Puente, I., H. Gonzalez-Jorge, J. Martinez-Sanchez, and P. Arias. 2014. "Automatic detection of road tunnel luminaires using a mobile lidar system." *Measurement* 47: 569–575.

Puri, N., E. Valero, Y. Turkan, and F. Bosché. 2018. "Assessment of Compliance of dimensional tolerances in concrete slabs using TLS data and the 2D continuous wavelet transform." *Autom. Constr.* 94: 62–72.

Qin, N., W. Tan, L. Ma, D. Zhang, and J. Li. 2021. "OpenGF: An ultra-large-scale ground filtering dataset built upon open ALS point clouds around the world." In *Proc., IEEE/CVF Conf. on Computer Vision and Pattern Recognition*, 1082–1091.

Rabbani, T. 2006. "Automatic reconstruction of industrial installations using point clouds and images." Ph.D. thesis. *Publications on Geodesy 62, Netherlands Geodetic Commission*. Accessed February 9, 2022. https://repository.tudelft.nl/islandora/object/uuid%3A83466092-f531-41e2-8ef3-5985b608aeac.

Rabbani, T., S. Dijkman, F. van den Heuvel, and G. Vosselman. 2007. "An integrated approach for modelling and global registration of point clouds." *ISPRS J. Photogramm. Remote Sens.* 61 (6): 355–370.

Rashidi, M., M. Mohammadi, S. Sadeghlou Kivi, M. M. Abdolvand, L. Truong-Hong, and B. Samali. 2020. "A decade of modern bridge monitoring using terrestrial laser scanning: Review and future directions." *Remote Sens.* 12: 3796.

Reshetyuk, Y. 2006. "Calibration of terrestrial laser scanners for the purposes of geodetic engineering." In *Proc., 3rd IAG/12th FIG Symp.* International Federation of Surveyors, FIG, Copenhagen, Denmark.

Riveiro, B., H. González-Jorge, J. Martínez-Sánchez, L. Díaz-Vilariño, and P. Arias. 2015a. "Automatic detection of zebra crossings from mobile lidar data." *Opt. Laser Technol.* 70: 63–70.

Riveiro, B., P. B. Lourenço, D. V. Oliveira, H. Gonzalez-Jorge, and P. Arias. 2015b. "Automatic morphologic analysis of quasi-periodic masonry walls from lidar." *Comput. Aided Civ. Infrastruct. Eng.* 31 (4): 305–319.

Riveiro, B., P. Morer, P. Arias, and I. de Areaga. 2011. "Terrestrial laser scanning and limit analysis of masonry arch bridges." *Constr. Build. Mater.* 25 (4): 1726–1735.

Riveiro, B., and M. Solla. 2016. *Non-destructive techniques for the evaluation of structures and infrastructure.* Boca Raton, FL: CRC Press.

Rodríguez-Caballero, E., A. Afana, S. Chamizo, A. Sole-Benet, and Y. Canton. 2016. "A new adaptive method to filter terrestrial laser scanner point clouds using morphological filters and spectral information to conserve surface micro-topography." *ISPRS J. Photogramm. Remote Sens.* 117: 141–148.

Rosser, N. J., S. A. Dunning, M. Lim, and D. N. Petley. 2007. "Terrestrial laser scanning for quantitative rock fall hazard assessment." *Landslides* 10: 409–420.

Rosser, N. J., D. N. Petlety, M. Lim, S. A. Dunning, and R. J. Allison. 2005. "Terrestrial laser scanning for monitoring the process of hard rock coastal cliff erosion." *Q. J. Eng. Geol. Hydrol.* 38: 363–375.

Rusinkiewicz, S., and M. Levoy. 2001. "Efficient variant of the ICP algorithm." In *Proc., 3-D Digital Imaging and Modelling*, 145–152. New York: IEEE.

Rusu, R. B., and S. Cousins. 2011. "3D is here: Point cloud library." In *Proc., IEEE Int. Conf. on Robotics and Automation*, 1–4. New York: IEEE.

Rutzinger, M., S. O. Elberink, S. Pu, and G. Vosselman. 2009. "Automatic extraction of vertical walls from mobile and airborne laser scanning data." *ISPRS Ann. Photogramm. Remote Sens. Spatial Inf. Sci.* 38: W8.

Salo, P., O. Jokinen, and A. Kukko. 2008. "On the calibration of the distance measuring component of a terrestrial laser scanner." *Int. Arch. Photogramm. Remote Sens. Spatial Inf. Sci.* 37 (Part B5): 1067–1071.

Schaer, P., J. Skaloud, S. Landtwig, and K. Legat. 2007. "Accuracy estimation for laser point cloud including scanning geometry." *Int. Arch. Photogramm. Remote Sens. Spatial Inf. Sci.* 36 (Part 5/C55): 279–353.

Schnabel, R., R. Whal, and R. Klein. 2007. "Efficient RANSAC for point cloud shape detection." *Comput. Graphics Forum* 26 (2): 214–226.

Serna, A., and B. Marcotegui. 2014. "Detection, segmentation and classification of 3d urban objects using mathematical morphology and supervised learning." *ISPRS J. Photogramm. Remote Sens* 93: 243–255.

Silvia, E. P., and M. J. Olsen. 2012. "To level or not to level: Laser scan inclination sensor evaluation." *J. Surv. Eng.* 138 (3): 117–125.

Singh, R. 2008. *Engineering automation, key concepts for a 25 year time Horizon.* V3, ODOT Rep. Salem, OR: Oregon Department of Transportation.

Sithole, G. 2005. "Segmentation and classification of airborne laser scanner data." Ph.D. thesis, *Publications on Geodesy 59, Netherlands Geodetic Commission.* Accessed March 17, 2009. http://www. ncg.knaw.nl/Publicaties/Geodesy/pdf/59Sithole.pdf.

Sithole, G., and G. Vosselman. 2003. "Automatic structure detection in a point cloud of an urban landscape." *Proc., 2nd GRSS/ISPRS Joint Workshop on Remote Sensing and Data Fusion over Urban Areas, Urban*, 67–71.

Skaloud, J., and D. Lichti. 2006. "Rigorous approach to bore-sight self-calibration in airborne laser scanning." *ISPRS J. Photogramm. Remote Sens.* 61 (1): 47–59.

Slattery, K. T., D. K. Slattery, and J. P. Peterson. 2012. "Road construction earthwork volume calculation using three-dimensional laser scanning." *J. Surv. Eng.* 138 (2): 96–99.

Sok, C., and M. D. Adams. 2010. "Visually aided feature extraction from 3D range data." In *Proc., IEEE Int. Conf. on Robotics and Automation*, 2273–2279. New York, New York, USA: IEEE.

Son, H., C. Kim, and C. Kim. 2015. "Fully automated as-built 3D pipeline extraction method from laser scanned data based on curvature computation." *J. Comput. Civ. Eng.* 29 (4): B4014003.

Son, S., H. Park, and K. H. Lee. 2002. "Automated laser scanning system for reverse engineering and inspection." *Int. J. Mach. Tools Manuf* 42 (8): 889–897.

Starek, M. J., H. Mitasova, and R. S. Harmon. 2010. "Optimization of terrestrial laser scanning survey design for dynamic terrain monitoring." In *AGU Fall Meeting. American Geophysical Union*. Washington DC, USA.

Starek, M. J., H. Mitasova, and R. Harmon. 2011. "Modeling landscape evolution at multiple scales using airborne, ground, and laboratory laser scanning." *Geosphere* 7 (6): 1340–1356.

Starek, M. J., H. Mitasova, K. W. Wegmann, and N. Lyons. 2013. "Space–time cube representation of stream bank evolution mapped by terrestrial laser scanning." *IEEE Geosci. Remote Sens. Lett.* 10 (6): 1369–1373.

Stewart, J. P., J. Hu, R. E. Kayen, A. J. Lembo, Jr., B. D. Collins, C. A. Davis, and T. D. O'Rourke. 2009. "Use of airborne and terrestrial LIDAR to detect ground displacement hazards to water systems." *J. Surv. Eng.* 135 (3): 113–124. ASCE.

Strom, J., A. Richardson, and E. Olson. 2010. "Graph-based segmentation of colored 3D laser point clouds." In *Proc., IEEE/RSJ Int. Conf. on Intelligent Robots and Systems*. New York, New York, USA: IEEE.

Su, Y. Y., Y. M. A. Hashash, and L. Y. Liu. 2006. "Integration of construction as-built data via laser scanning with geotechnical monitoring of urban excavation." *J. Constr. Eng. Manag.* 132: 1234–1241.

Taneja, S., et al. 2011. "Sensing and field data capture for construction and facility operations." *J. Constr. Eng. Manage.* 137 (10): 870–881.

Tang, P., and B. Akinci. 2012a. "Automatic execution of workflows on laser-scanned data for extracting bridge surveying goals." *Adv. Eng. Inf.* 26 (4): 889–903.

Tang, P., and B. Akinci. 2012b. "Formalization of workflows for extracting bridge surveying goals from laser-scanned data." *Autom. Constr.* 22 (3): 306–319.

Tang, P., B. Akinci, and D. Huber. 2009. "Quantification of edge loss of laser scanned data at spatial discontinuities." *Autom. Constr.* 18 (8): 1070–1083.

Tang, P., B. Akinci, and D. Huber. 2011. "Characterization of laser scanners and algorithms for detecting flatness defects on concrete surfaces." *J. Comput. Civ. Eng.* 25 (1): 31–42.

Tang, P., D. Huber, B. Akinci, R. Lipman, and A. Lytle. 2010. "Automatic reconstruction of as-built building information models from laser-scanned point clouds: A review of related techniques." *Autom. Constr.* 19 (7): 829–843.

Telling, J., A. Lyda, P. Hartzell, and C. Glennie. 2017. "Review of Earth science research using terrestrial laser scanning." *Earth Sci. Rev.* 169: 35–68.

Teza, G., A. Galgaro, N. Zaltron, and R. Genevois. 2007. "Terrestrial laser scanner to detect landslide displacement fields: A new approach." *Int. J. Remote Sensing* 28 (16): 3425–3446.

Theiler, P. W., J. D. Wegner, and K. Schindler. 2012. "Automatic registration of terrestrial laser scanner point clouds using natural planar surfaces." In *ISPRS Annals, 22nd ISPRS Cong*. International Society of Photogrammetry and Remote Sensing, Hannover, Germany.

Theiler, P. W., J. D. Wegner, and K. Schindler. 2014. "Keypoint-based 4-points congruent sets–Automated marker-less registration of laser scans." *ISPRS J. Photogramm. Remote Sens.* 96: 149–163.

Theiler, P. W., J. D. Wegner, and K. Schindler. 2015. "Globally consistent registration of terrestrial laser scans via graph optimization." *ISPRS J. Photogramm. Remote Sens.* 109: 126–138.

Tonini, M., and A. Abellán. 2014. "Rockfall detection from lidar point clouds: A clustering approach using R." *J. Spat. Inf. Sci.* 8: 95–110.

Truong-Hong, L., and D. F. Laefer. 2015. "A semi-automatic member detection for metal bridges." *IABSE Symp. Rep.* 104 (21): 214–215

Tsakiri, M., D. Lichti, and N. Pfeifer. 2006. "Terrestrial laser scanning for deformation monitoring." In *Proc., 3rd IAG/12th FIG Symp.* International Federation of Surveyors, FIG, Copenhagen, Denmark.

Turkan, Y., F. Bosché, C. T. Haas, and R. G. Haas. 2012. "Automated progress tracking using 4D schedule and 3D sensing technologies." *Autom. Constr.* 22: 414–421.

Turkan, Y., F. Bosché, C. T. Haas, and R. G. Haas. 2013. "Toward automated earned value tracking using 3d imaging tools." *J. Constr. Eng. Manage.* 139 (4): 423–433.

Turkan, Y., F. Bosché, C. T. Haas, and R. Haas. 2014. "Tracking of secondary and temporary objects in structural concrete work." *Constr Innovation Inf Process Manage.* 14 (2): 145–167.

Turkan, Y., S. Laflamme, and L. Tan. 2016. "Terrestrial laser scanning-based bridge structural condition assessment." In *Trans Project Reports, Paper 199*. Accessed February 9, 2022. http://lib.dr.iastate.edu/intrans_reports/199.

Vamsi, S. K., Y. Turkan, and T. Ptur. 2014. "Toward automated spatial change analysis of MEP components using 3D point clouds and as-designed BIM models." In Vol. 2 of *Proc., 2nd Int. Conf. on 3D Vision*, 145–152. New York: IEEE.

Vosselman, G., and Klein, R. 2010. "Visualizing and Structuring of Point Clouds." *In Airborne and terrestrial laser scanning*, edited by G. Vosselman and H.-G. Maas, 45–81. Caithness, UK: Whittles Publishing. Boca Raton, FL: CRC Press.

Vosselman, G. 2013. "Point cloud segmentation for urban scene classification." *Int. Arch. Photogramm. Remote Sens. Spatial Inf. Sci.* XL-7/W2: 257–262.

Vosselman, G., B. G. H. Gorte, G. Sithole, and T. Rabbani. 2004. "Recognising structure in laser scanner point clouds." *Int. Arch. Photogramm. Remote Sens. Spatial Inf. Sci.* 46 (8): 33–38.

Vosselman, G. 1999. "Building reconstruction using planar faces in very high density height data." *International Archives of Photogrammetry and Remote Sensing*, 32 (Part3/2W5): 87–92.

Walsh, S. B., D. J. Borello, B. Guldur, and J. F. Hajjar. 2013. "Data processing of point clouds for object detection for structural engineering applications." *Comput.-Aided Civ. Infrastruct. Eng.* 28 (7): 495–508.

Wang, G.-Q., J. Joyce, D. Phillips, R. Shrestha, and W. Carter. 2013. "Delineating and defining the boundaries of an active landslide in the rainforest of Puerto Rico using a combination of airborne and terrestrial lidar data." *Landslides* 10 (4): 503–513.

Wang, H., Z. Cai, H. Luo, C. Wang, P. Li, W. Yang, et al. 2012. "Automatic road extraction from mobile laser scanning data." In *Proc., Computer Vision in Remote Sensing, 2012 Int. Conf.*, 136–139. New York, New York, USA: IEEE.

Wang, J., and J. Shan. 2009. "Segmentation of lidar point clouds for building extraction. In Proc., *American Society for Photogramm, Remote Sens, Annual Conf.*, 9–13. American Society for Photogramm, Remote Sens, Bethesda, MD, USA.

Wang, J., W. Sun, W. Shou, X. Wang, C. Wu, H.-Y. Chong, et al. 2014a. "Integrating BIM and lidar for real-time construction quality control." *J. Intell. Rob. Syst.* 79 (3): 417–432.

Wang, W., W. Zhao, L. Huang, V. Vimarlund, and Z. Wang. 2014b. "Applications of terrestrial laser scanning for tunnels: A review." *J. Traffic Transp. Eng.* 1 (5), 325–337.

Wartman, J., J. W. Berman, A. Bostrom, S. Miles, M. J. Olsen, K. R. Gurley, J. Irish, L. Lowes, T. Tanner, J. Dafni, M. Grilliot, A. Lyda, and J. Peltier. 2020. Research Needs, Challenges, and Strategic Approaches for Natural Hazards and Disaster Reconnaissance, Frontiers in Built Environment, 6:182, section Wind Engineering and Science, Special Issue on Natural Hazards Engineering Research Infrastructure (NHERI) 2016-2020: Mitigating the Impact of Natural Hazards on Civil Infrastructure and Communities. https://doi.org/10.3389/fbuil.2020.573068

Wilkinson, B., A. Mohamed, and B. Dewitt. 2015. "Dual-antenna terrestrial laser scanner georeferencing using auxiliary photogrammetric observations." *Remote Sens.* 7 (9): 11621–11638.

Wilkinson, B., A. H. Mohamed, B. Dewitt, and G. H. Seedahmed. 2010. "A novel approach to terrestrial lidar georeferencing." *Photogramm. Eng. Remote Sens.* 76 (6): 683–690.

Williams, J., and M. Bennamoun. 2001. "Simultaneous registration of multiple corresponding point sets." *Comput. Vision Image Understanding* 81 (1): 117–142.

Williams, K., M. J. Olsen, G. V. Roe, and C. Glennie. 2013. "Synthesis of transportation applications of mobile lidar." *Remote Sens.* 5: 4652–4692.

Xie, J., Y. Hsu, R.S. Feris, and M.T. Sun. 2013. "Fine registration of 3D point clouds with ICP using an RGB-D camera." In *Proc., Int. Symp. on Circuits and Systems*. New York: IEEE.

Yang, B., L. Fang, and J. Li. 2013. "Semi-automated extraction and delineation of 3D roads of street scene from mobile laser scanning point clouds." *ISPRS J. Photogramm. Remote Sens.* 79: 80–93.

Yen, K. S., K. Akin, and B. Ravani. 2008a. *Accelerated project delivery: Case studies and field use of 3D terrestrial laser scanning in Caltrans projects*. Research Rep. No. UCD-ARR-08-06-30-0. Advanced Highway Maintenance and Construction Technology Research Center, Davis, CA.

Yen, K. S., T. Swanston, B. Ravani, and T. A. Lasky. 2008b. *Enhancement and field evaluation of an advanced clearance measurement system for highway structures, including generation of clearance diagrams*. No. CA08-0937.

Yim, S. C., M. J. Olsen, K. F. Cheung, and M. Azadbakht. 2014. "Tsunami modeling, fluid load simulation and validation using geospatial field data." *J. Struct. Eng.* 140 (8): A4014012.

Yoo, H. J., F. Goulette, J. Senpauroca, and G. Lepere. 2010. "Analysis and improvement of laser terrestrial mobile mapping systems configurations." *Int. Arch. Photogramm. Remote Sens. Spat. Inf. Sci.* 38(5): 633–638.

Young, A. P., and S. A. Ashford. 2008. "Instability investigation of cantilevered seacliffs." *Earth Surface Processes and Landforms* 33 (11): 1661–1677.

Young, A. P., M. J. Olsen, N. Driscoll, R. Gutierrez, R. T. Guza, R. E. Flick, et al. 2010. "Comparison of airborne and terrestrial lidar estimates of seacliff erosion in Southern California." *Photogramm. Eng. Remote Sens.* 76 (4): 421–427.

Yu, Y., J. Li, H. Guan, C. Wang, and J. Yu. 2015. "Semiautomated extraction of street light poles from mobile lidar point clouds." *IEEE Trans. Geosci. Remote Sens.* 53 (3): 1374–1386.

Zeibak, R., and S. Filin. 2009. "Object extraction from terrestrial laser scanning data." FIG Working Week 2009, TS 8E.

Zhang, C., V. S. Kalasapudi, and P. Tang. 2016. "Rapid data quality oriented laser scan planning for dynamic construction environments." *Adv. Eng. Inf.* 30 (2): 218–232.

Zhang, Z. 1994. "Iterative point matching for registration of free-form curves and surfaces." *Int. J. Comput. Vision* 13 (2): 119–152.

Zhou, L., and G. Vosselman. 2012. "Mapping curbstones in airborne and mobile laser scanning data." *Int. J. Appl. Earth Obs. Geoinf.* 18: 293–304.

CHAPTER 9
MOBILE TERRESTRIAL LASER SCANNING AND MAPPING

Michael J. Olsen, Jaehoon Jung, Erzhuo Che, Chris Parrish

INTRODUCTION

Mobile mapping systems (MMSs) have become an important surveying tool over the last decade, supporting a wide range of road and rail corridor surveys and related applications (Graham 2010; Olsen et al. 2012, 2013a; Williams et al. 2013). A variety of systems and configurations exist to support different applications, including imaging systems, pavement profiling or condition assessment systems, and mobile terrestrial laser scanning (MTLS). MTLS is often referred to as mobile lidar (ML) and mobile laser scanning (MLS). MTLS systems (Figure 9-1) are the focus of this chapter, because they are the type of system that is most relevant to the surveying engineer; however, it should be noted that other system types can also be useful to surveying engineering. MTLS systems collect accurate, detailed, three-dimensional (3D) point cloud data (Figure 9-2) more efficiently and safely (Yen et al. 2011) for road or rail projects than many other approaches available to the survey engineer. The key value is able to capture a scene holistically, which minimizes the need for multiple data collection efforts and provides context for the measurements.

One of the first mobile mapping systems, GPSVan, was created by the Center for Mapping at Ohio State University in the early 1990s. The GPSVan system consisted of a GPS receiver, gyroscope sensor, distance measurement instrument (DMI), two CCD cameras, and a voice recorder (R. Burtch, unpublished lecture notes, https://pdfcoffee.com/mobile-mapping-pdf-free.html., 2006). The first MTLS unit was constructed in 2003 by mounting a helicopter-based lidar system to a vehicle by turning it on its side (Glennie 2009). Since that time, a variety of imaging and lidar-based systems have been developed to support a wide range of applications. Yen et al. (2010) and Puente et al. (2013) compare many mobile scan systems currently available. For the purposes of this chapter, these systems can be classified loosely into higher-accuracy survey-grade systems and lower accuracy mapping/asset management grade systems.

KEY REFERENCES

The following references, developed by several transportation authorities and research institutions, have been identified as the most helpful resources for understanding mobile lidar technology. A continually updated reference list and a detailed glossary of terminology can be found at the Learn Mobile Lidar Website (O'Banion et al. 2013).

Figure 9-1. Example MTLS system acquiring data (Leica Pegasus: Two).

- Transportation Research Board NCHRP Report No. 748: *Guidelines for the Use of Mobile Lidar in Transportation Applications* (Olsen et al. 2013a). NCHRP Report No. 748 contains detailed, performance-based guidelines. In addition to the report, these guidelines have been developed into an interactive website (O'Banion et al. 2013). The website contains a variety of helpful resources, including e-learning modules, listing of key references, lists of key geospatial standards, a literature database, glossary of terminology, current research projects, and more. Much of the text and examples in this chapter draws on materials from NCHRP Report No. 748, because it is the most comprehensive source on the subject.
- Caltrans: *Survey Manual, Chapter 15* (2018): Detailed specifications for the acquisition of lidar data. Note that this specification has been adapted by other DOTs such as Florida DOT (2015) into its guidelines and specifications.
- Missouri DOT: *Light detection and ranging (lidar) technology evaluation.* This study evaluated the capabilities and limitations of airborne, static, and mobile lidar technology (Vincent and Ecker 2010).
- Queensland Australia Department of Transportation Mainroads (2014a, b): *Mobile Laser Scanning Technical Guideline* and *Best Practice for mobile lidar survey requirements (Discussion Paper).* The first document focuses on achieving more value with MTLS through consistent terminology and understanding of MTLS. The second document provides examples of mobile lidar survey requirements for procurement purposes by addressing requirements for positioning accuracy, point density, multiple passes, control and validation points, and deliverables and documentation. It includes helpful checklists with information needed to support MTLS.
- UC DAVIS AHMCT: This research group, in collaboration with Caltrans, has produced several important documents evaluating the capabilities of mobile lidar including *Using Mobile Laser Scanning to Produce Digital Terrain Models of Pavement Surfaces* (Yen et al. 2010), *LiDAR for Data Efficiency* (Yen et al. 2011), Mobile Terrestrial Laser Scanning Workflow Development, Technical Support and Evaluation (Yen et al. 2014), and *Expanding Mobile Terrestrial Laser Scanning Capability and Capacity throughout Caltrans* (Yen et al. 2018).
- *Airborne and Terrestrial Laser Scanning*: Vossleman and Maas (2010).
- Mobile Lidar Review Papers: Several informative review papers are available covering the state of the art in MTLS. These include the following:

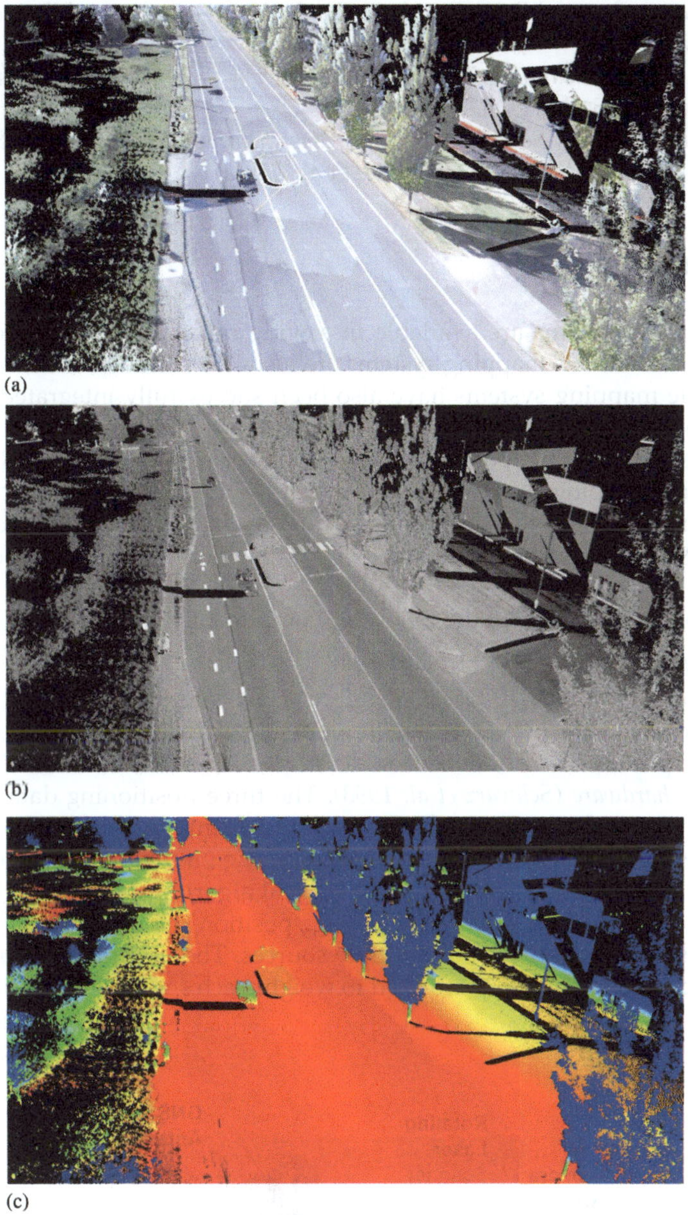

Figure 9-2. Example point cloud acquired from a mobile lidar system, colored by (a) RGB photographs, (b) intensity or return signal strength, (c) elevation.

- *Review of Mobile Mapping and Surveying Technologies* (Puente et al. 2013).
- *Synthesis of Transportation Applications of Mobile LiDAR* (Williams et al. 2013).
- *Applications of Terrestrial Laser Scanning for Tunnels: A Review* (Wang et al. 2014).
- *A Review of LiDAR Radiometric Processing: From Ad Hoc Intensity Correction to Rigorous Radiometric Calibration* (Kashani et al. 2015).
- *Use of Mobile LiDAR in Road Information Inventory: A Review* (Guan et al. 2016).
- *Automated Extraction of Road Features Using LiDAR Data: A Review* (Gargoum and El-Basyouny 2017).
- *LiDAR Point Clouds to 3-D Urban Models: A Review* (Wang et al. 2018).

- *Mobile Laser Scanned Point-Clouds for Road Object Detection and Extraction: A Review* (Ma et al. 2018a).
- *Review of Laser Scanning Technologies and Their Applications for Road and Railway Infrastructure* (Soilán et al. 2019).
- *Object Recognition, Segmentation, and Classification of Mobile Laser Scanning Point Clouds: A State of the Art Review* (Che et al. 2019a).

SYSTEM COMPONENTS

Mobile mapping systems are available in a wide range of configurations. Common platforms on which mobile mapping systems are installed include trucks, cars, SUVs, and vans, but mobile mapping systems have also been successfully integrated on airplanes, boats, ATVs, unmanned aircraft systems (UAS), robots, and human-carried platforms (e.g., backpacks). The principles of operation are usually the same, regardless of the platform on which the mobile mapping system is installed, but the applications of the data can differ. Indoor-based mobile mapping technologies relying on simultaneous localization and mapping (SLAM) algorithms rather than GNSS for positioning are discussed at the end of the chapter.

Most MTLS systems consist of the following components (Figure 9-3):

1. *Mobile platform* (e.g., truck, SUV, ATV).
2. *Rigid platform or frame*: This connects the various components so that MTLS can be transferred from vehicle to vehicle without requiring a full calibration. It also provides housing to protect the various sensors.
3. *Positioning hardware (Schwarz et al. 1993)*: The three positioning data sources given subsequently are blended, typically through a Kalman filter process, to produce a high-accuracy trajectory, taking advantage of the strengths of each component. A Kalman filter (Kalman 1960) is a mathematical algorithm that estimates values at a future time step for unknown variables (e.g., position and orientation of MTLS) using time series measurements from multiple sources. The uncertainty estimates of each data source at each time step are used to weight each source.

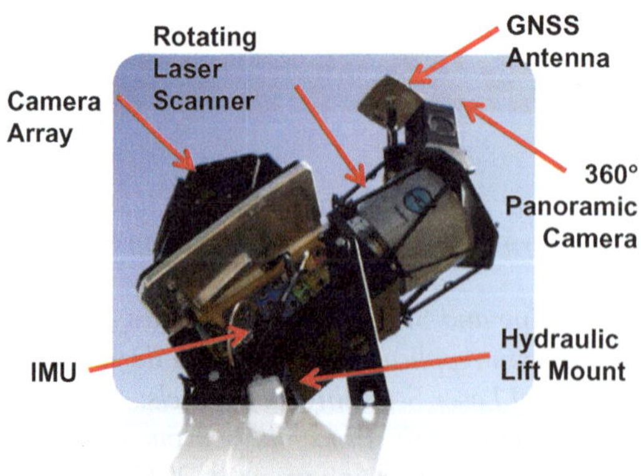

Figure 9-3. Example components of a mobile lidar unit (Topcon IPS2).
Source: Singh et al. (2012), reproduced with permission from the Oregon Dept. of Transportation.

(a) *GNSS*: Most systems have one GNSS sensor onboard for obtaining georeferenced coordinates of the system, typically at 1 Hz. See Chapter 6 for further discussion on GNSS technology.

(b) *Inertial measurement unit (IMU)*: The IMU is the primary component of the inertial navigation system, and it contains orthogonal triads of gyroscopes and accelerometers that measure 3D angular rate and inertial acceleration (specific force), respectively. These measurements are integrated and added to the initial position and orientation to obtain position and orientation of the platform or sensor along a trajectory. IMU measurement rates are high: typically, 100 to 2,000 Hz. The integration of GNSS and IMU data via a Kalman filter enables both position and orientation (roll, pitch, yaw) of the sensor to be accurately determined along the trajectory and also overcomes the problem of inertial sensor drift, which can occur in inertial navigation in the absence of GNSS aiding.

(c) *Distance measurement indicator*: A DMI, or precise odometer, provides an estimate of the distance traveled (e.g., measures tire rotations). It can be used as an additional input to the Kalman filter, along with the GNSS and IMU data, and/or utilized as the triggering device to determine when to capture photographs.

4. *3D laser scanner(s)*: One or more 3D laser scanners operate in a profile mode to capture the scene as the vehicle progresses forward. The geometric orientation (i.e., incidence angle, distance to target) of the scanners relative to the surface of interest (e.g., horizontal ground surface versus vertical building facades) significantly influences the overall data quality because the incidence angle at which the laser strikes the surface causes variation in accuracy. These also affect the intensity readings (Yousif et al. 2010, Kashani et al. 2015). The native data collected by these systems are angles, ranges, and timestamps. Additional discussion of scanning technology and the resulting data can be found in Chapter 8.
5. *Photographic/video camera(s)*: These systems can collect stand-alone imagery or video logs that can be a product by themselves. However, they are also calibrated so that they can map RGB color information to the point cloud. In some software, one can perform digitization and annotations on the photographs, which are then linked back to the point cloud to provide geometric information.
6. *Computer and data storage*: A modified version of a desktop computer is often integrated into the vehicle with the scanner to stream and store the data from the various sensors.
7. *Power source*: A dedicated power source/battery is often included to ensure smooth operation of the mobile lidar unit.
8. *Additional sensors*: Other mobile sensors may be added to MTLS to maximize the efficiency of the data collection effort.

These components can be configured in a variety of ways with the sensors at different orientations for specific applications. Yoo et al. (2010) evaluate several different mobile configurations.

APPLICATIONS

An important aspect of mobile mapping is its ability to efficiently collect data to support a wide range of applications from one platform in a single acquisition (Figures 9-4 and 9-5). It also provides significant safety benefits, particularly for surveying along high-volume, high-speed corridors. NCHRP No. 748 (Olsen et al. 2013a), Williams et al. (2013), and Che et al. 2019 provide a detailed summary of applications for which mobile lidar has been utilized, as well as specific considerations for these applications. The reader is referred to these documents for further detail.

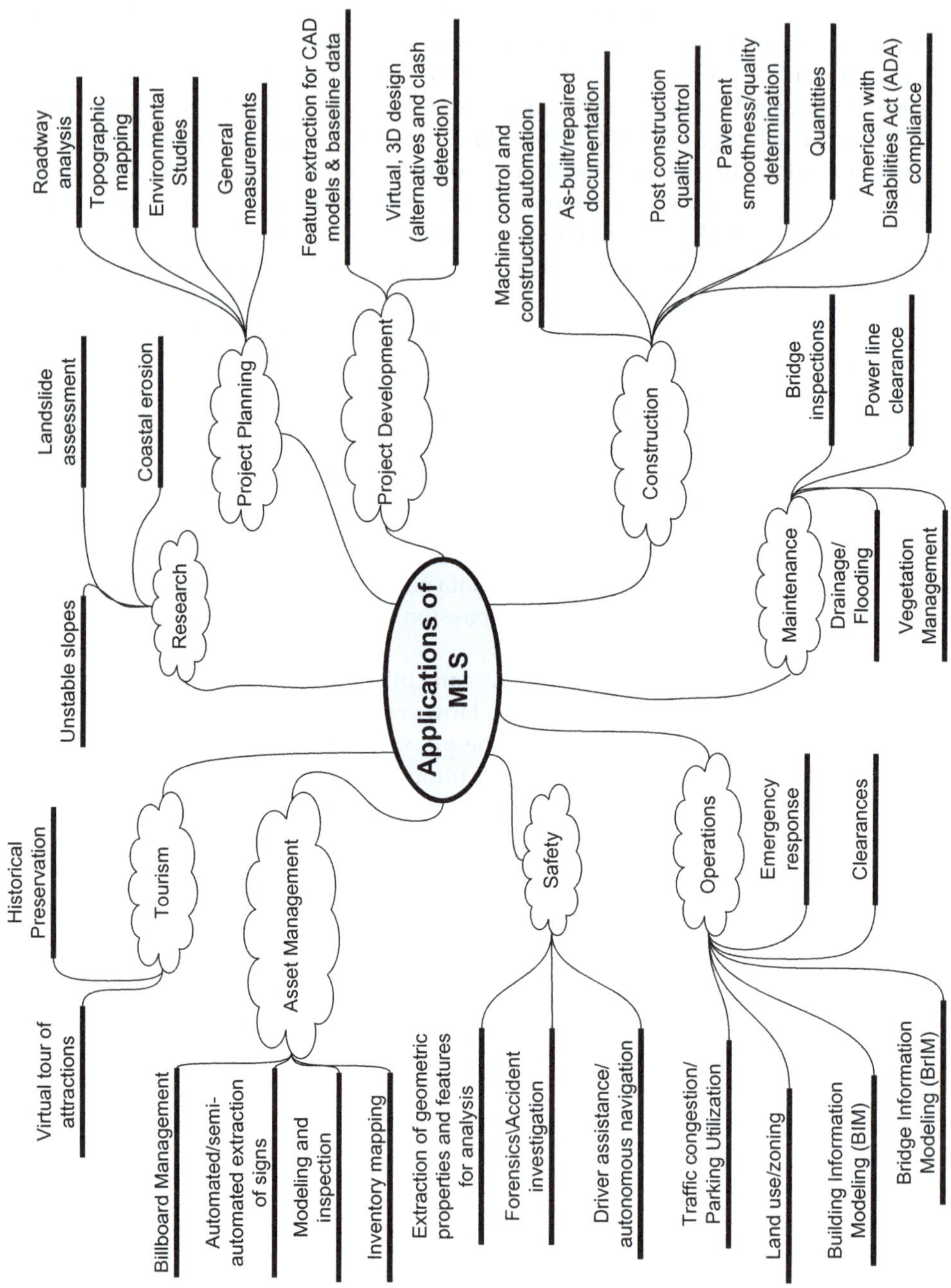

Figure 9-4. Categorization of mobile lidar applications in transportation.
Source: Modified from Olsen et al. (2013a), reproduced with permission from the National Academies Press, Washington, DC.

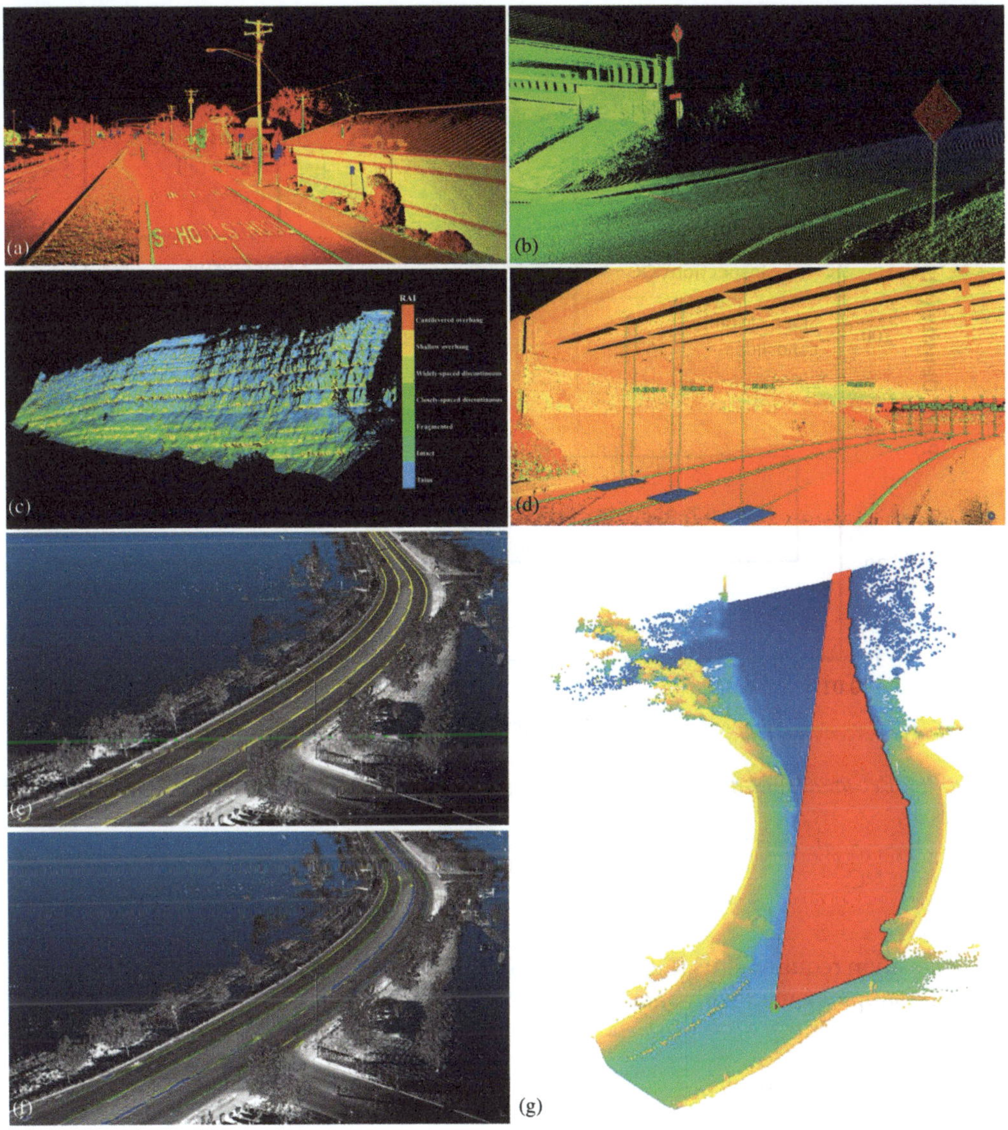

Figure 9-5. Sampling of applications of mobile lidar data: (a) example point cloud showing a variety of assets that can be extracted from the mobile lidar data, including pavement markings, signs, poles, curbs, road surface, and so on; (b) example signs extracted from mobile lidar data based on intensity; (c) rock slope hazard analysis with the Rockfall Activity Index; (d) bridge clearance evaluation; (e) and (f) pavement marking extraction and retro reflectivity evaluation; (g) sight distance analysis (red zone) using mobile lidar data.
Source: (c) Dunham et al. (2017), (d) courtesy of Oregon DOT, (e) and (f) Che et al. (2019), Jung et al. (2019), (g) Jung et al. (2018).

When utilizing mobile lidar data for a specific application, it is important to carefully consider the appropriate accuracy and point density requirements to support that application. These considerations are equally applicable to new data collection and to the evaluation of existing datasets for a new purpose. NCHRP Report No. 748 provides significant guidance in formulating these requirements as well as recommendations for

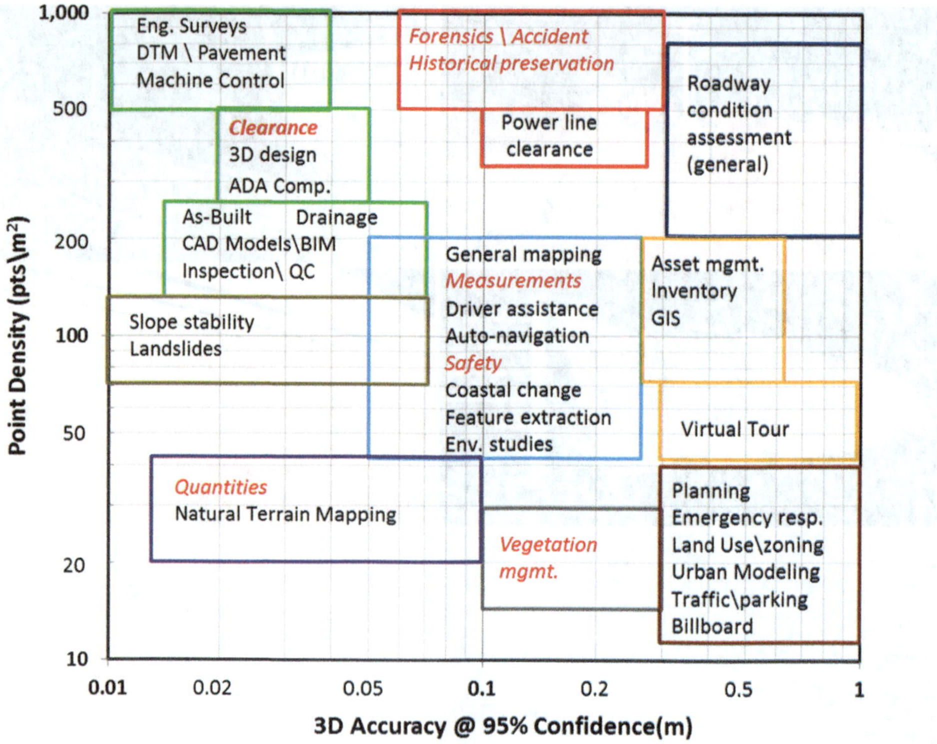

Figure 9-6. Recommended accuracy and point density requirements for a variety of applications of mobile lidar.
Source: Olsen et al. (2013a), reproduced with permission from the National Academies Press, Washington, DC.

data acquisition requirements (Figure 9-6). Note that for some applications, MTLS data may need to be supplemented with data from static laser scanners to provide sufficient coverage.

PROJECT WORKFLOW

Figure 9-7 provides an overview of a typical mobile lidar workflow, modified from NCHRP Report No. 748. Note that the workflow may vary somewhat depending on the specific needs of the application. This section describes processes and important considerations during the various stages of the workflow, including planning, preliminary site surveys, data acquisition, georeferencing, post-processing, computations and analysis, and packaging and delivery.

Planning

Although mobile mapping provides substantial efficiency benefits in data collection, many applications require detailed planning, as with any surveying project. The following are important considerations for planning a mobile lidar survey. The reader is also referred to the planning sections of Chapter 8 on Terrestrial Laser Scanning, Chapter 6 on GNSS Surveying, and Chapter 11 on Survey Control for additional details. Saylam (2009) also provides additional considerations.

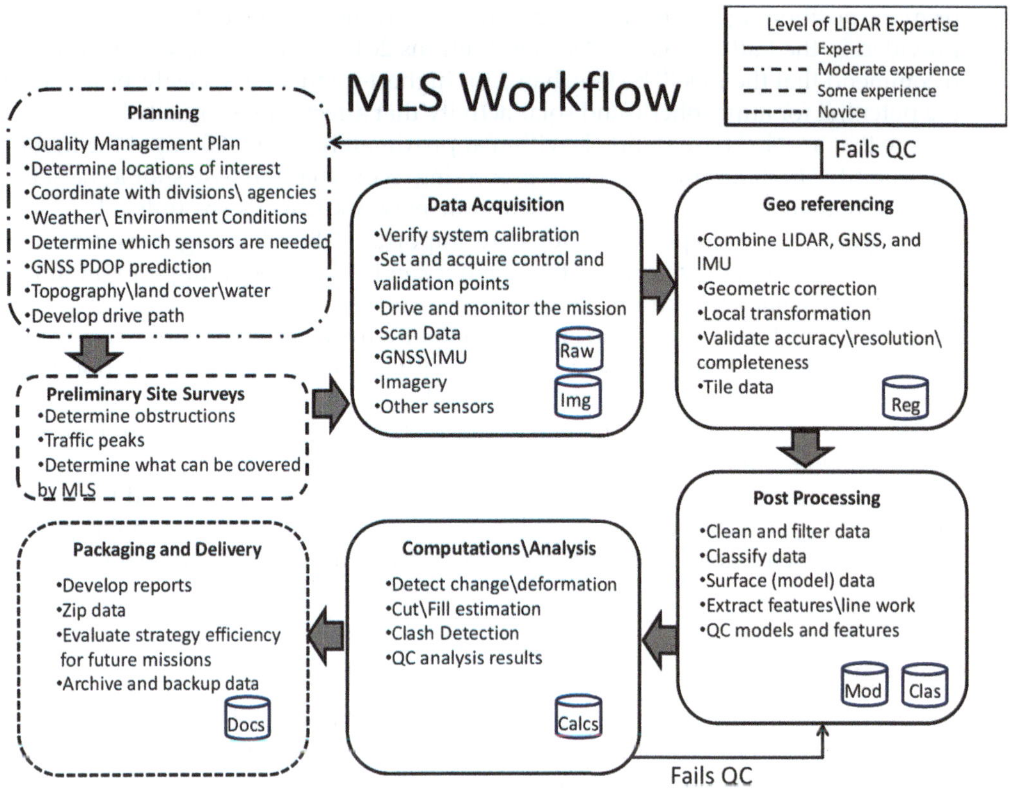

Figure 9-7. Overview of common steps in an MTLS processing workflow.
Source: Modified from Olsen et al. (2013a), reproduced with permission from the National Academies Press, Washington, DC.

1. Evaluate line of sight/view of objects of interest: Can the objects of interest be adequately seen and captured? Are terrestrial laser scans required to supplement the survey? Note that mobile lidar units will not capture data on the downslope side very well, given the oblique angle.
2. Monitor weather conditions: Although the systems are designed to be rugged against moisture, wet surfaces will affect the data quality by resulting in lower-intensity values and more data dropouts, particularly in areas of pooled water. Other effects such as fog and dust can result in ranging issues (see discussion in Chapter 8).
3. Traffic patterns: Ideally, scans should be collected during low-vehicle/trucking traffic volumes to minimize data blockage. Similarly, in urban areas, it is advisable to avoid peak times of pedestrian traffic. Increasing the height of the scanner can help improve the view, but bridge and other clearances need to be carefully considered. A second vehicle can be employed to follow behind MTLS to prevent other vehicles from getting too close behind MTLS and blocking the view.
4. Determine GNSS availability: The detailed protocols in Chapter 6 should be adapted to evaluate GNSS visibility at locations along the drive path that have significant overhead obstructions. For long projects where data will be collected over the span of several hours, these checks should consider the timing of the acquisition and when the vehicle will be at each location. Sometimes, line-of-sight obstructions to the satellite (e.g., buildings and trees) and related multipath may be unavoidable, resulting in degradation of the GNSS data. During these times, the IMU will be the

primary positioning sensor. Recommendations on the lengths of these outages are provided in the *Caltrans Survey Manual* (Caltrans 2018). Further, these specifications recommend that the PDOP be less than 5 for highest-quality data. Lastly, be aware of the potential for atmospheric and solar activity that affects GNSS quality.

5. Ensure that a GNSS base station is within approximately 16 km (10 mi) of the mobile lidar unit for high-accuracy baseline processing. Note that the base should be set to log at a frequency of 1 Hz or higher. Real-time networks can also be utilized.
6. If coacquired photograph/video logs are important, then lighting considerations should be considered. Otherwise, the scanner can be operated at night to minimize noise and data gaps from traffic.
7. Determine which lane to drive in and, for major highways, whether on- or off-ramps also need to be captured. These can significantly increase the data collection time. Determine if multiple passes are to be acquired.
8. Identify potential locations, frequency, and distribution of transformation control targets and validation control targets.
9. Identify strategic locations that have a clear view of the sky to start and end the survey safely. For long projects, pullouts should be considered as intermediate stopping points.
10. Include other sensors on the mobile platform to maximize ROI and value (Olsen et al. 2013b).

Preliminary Site Surveys

Whenever possible, a drive-through of the site should be completed to identify potential issues. Should this not be possible or economical, then, as a minimum, a virtual site visit through mobile (street) and airborne imagery should be completed.

Data Acquisition

Acquisition should be completed by both a driver and an operator for safety reasons. Prior to acquisition, many mobile lidar systems have a calibration process [e.g., driving a test path (Figure 9-8), etc.] to be completed prior to driving the path. Other systems can complete this calibration while the mobile lidar unit is stationary. For rigorous calibration, Heinz et al. (2020) outline procedures to establish an MTLS calibration site.

Because the GNSS data are often processed as post-processed kinematic observations, it is recommended to start and stop the survey/drive path at a safe location with a clear view of the sky such that the forward and reverse processing of the trajectory can begin and end with a high-quality position.

During the drive, the operator monitors the various sensors such as the GNSS positioning quality to ensure that significant GNSS outages did not occur. Most systems also provide a preview of the point cloud as it is collected; however, the actual processing and full point cloud reconstruction is completed in the office.

The driver also should maintain a constant speed to the extent possible within the clear constraints of safety and traffic regulations to ensure consistent spacing between scanlines. Cahalane et al. (2010) evaluate the influence of speed on the point density of MTLS point clouds.

Georeferencing

The key to mobile mapping is to precisely determine the trajectory of the vehicle and the associated sensors. All other data (laser scan point cloud, imagery, etc.) is then calculated based on the trajectory as its origin. Typically, this is done using the IMU to establish the

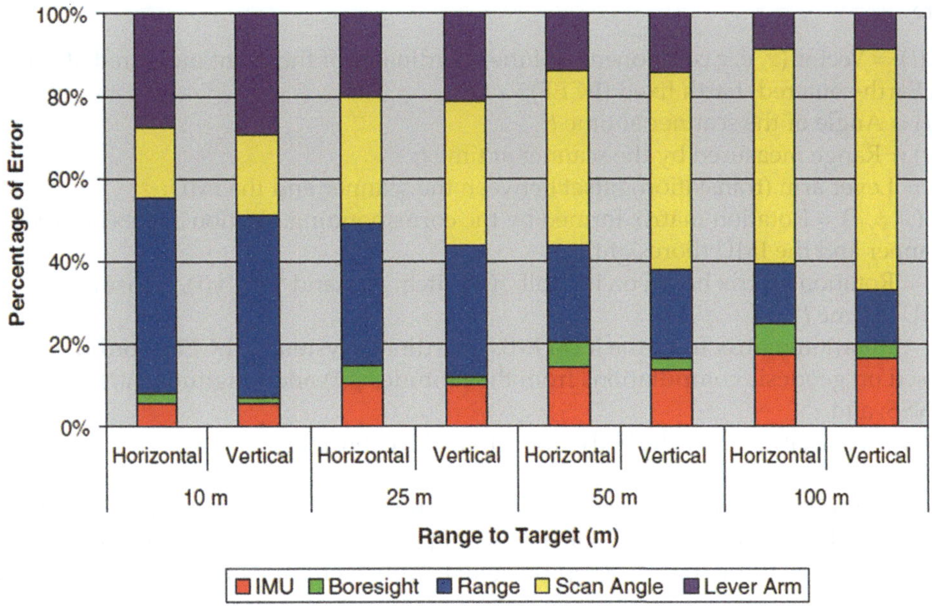

Figure 9-8. Contribution of various error sources to the overall error in MTLS. Assumes ideal scene conditions and orthogonal observations to the target.
Source: (Glennie 2007b), reproduced with permission from Walter de Gruyter GmbH.

reference frame for the other sensors (e.g., scanner, camera). The reader is referred to Glennie (2007a, b) and NCHRP 748 for a detailed discussion of georeferencing for mobile mapping.

When mobile mapping systems are manufactured, they undergo rigorous calibration procedures. Offsets (translations and rotation) between each of the sensors are determined by both measuring the physical offsets and then performing test drives at a calibration course with known targets that can be used to determine the boresight [misalignment angle between the scanner and the IMU (Rieger et al. 2010)]. The calibration should be repeated every time the components are adjusted or removed, before major projects, annually, or when the system has been subjected to excessive movement or vibration (e.g., rough roads).

During a mobile mapping mission, the IMU and GNSS work in concert to provide the positioning and orientation information. The GNSS positioning is not subject to drift, as is unaided inertial navigation, but over short time periods, its accuracy is not as good, and it measures at relatively low rates (e.g., 1 Hz). The IMU has complementary characteristics: Its accuracy is good over short time periods, but is subject to drift, and it measures at high rates (e.g., 200 Hz or higher). Therefore, the integration of the GNSS and IMU data via Kalman filtering enables the best characteristics of both systems to be leveraged and also enables optimal estimates of both position and orientation (pitch, roll, and yaw) at high rates. Combined with the post-processed trajectory, the angles and ranges measured by the laser scanner are then used to compute georeferenced point clouds. The computation of the object coordinates [Earth-centered, Earth-fixed (ECEF)] is completed through the following laser geolocation equation (Baltsavias 1999, Glennie 2007b):

$$x_p^e(t) = x_b^e(t) + R_l^e\left\{\varphi(t), \lambda(t)\right\} R_b^l\left\{r(t), p(t), y(t)\right\} R_s^b(\omega, \phi, \kappa)\left\{x_b^s + \rho(t)\begin{pmatrix} 0 \\ \sin\theta(t) \\ \cos\theta(t)\end{pmatrix}\right\} \tag{9-1}$$

where

$x_p^e(t)$ = Vector (x, y, z components) of the coordinates of the point measured at time t, in Earth-centered, Earth fixed (ECEF);
$\theta(t)$ = Angle of the scanner at time t;
$\rho(t)$ = Range measured by the scanner at time t;
x_b^s = Lever arm (translation) offset between the scanner and the IMU;
$R_s^b(\omega,\phi,\kappa)$ = Rotation matrix formed by the corresponding rotation angles between the scanner and the IMU (boresight);
R_b^l = Rotation matrix based on the roll, $r(t)$, pitch, $p(t)$, and yaw, $y(t)$, measured by the IMU at time t'
R_l^e = Rotation matrix from the local IMU coordinate system to ECEF coordinates based on geodesic computations from the latitude, $\varphi(t)$ and longitude, $\lambda(t)$ of the GNSS; and
$x_l^e(t)$ = Translation from the body frame to ECEF coordinates.

Note that a key point of the georeferencing process is that each point is individually georeferenced; as such, each point will have a unique georeferencing solution and associated accuracy value as error sources are propagated. However, the points are linked together through the trajectory.

For use in engineering applications, the point cloud is often converted from ECEF coordinates to a mapping projection (e.g., state plane coordinate systems) with an associated vertical datum through the geodetic techniques described in other chapters in this manual (Chapters 2 to 4 and Chapter 11).

Geometric Corrections. The quality of the direct georeferencing solution (i.e., the blended navigation solution from the post-processed GNSS-aided IMU data) can vary considerably, as a function of a number of variables, which include the following: integer ambiguity resolution throughout the trajectory (which, in turn, is a function of satellite geometry, ionospheric and tropospheric delay, and multipath), baseline length (distance from base station to the mobile platform's GNSS antenna), IMU alignment, boresight calibration quality, and system processing settings (Mostafa 2002). It is also important to note that direct georeferencing of mobile lidar is usually more difficult than for airborne lidar, because there are often more objects close to the ground (e.g., trees and buildings), which can obstruct GNSS satellite signals, leading to issues such as multipath, loss of lock, and low signal-to-noise ratios. To deal with suboptimal direct georeferencing solutions, a common practice in mobile lidar acquisition is to adjust the data set to control targets placed on or near the roadway that are surveyed using GNSS or other techniques and perform a "geometric correction." Note that an important recommendation of NCHRP 748 is to "limit these adjustments to a rigid body translation and avoid ad hoc transformations" to evade unintentional skewing of the data in an attempt to fit the control points better at the expense of the overall data set. The exception is if these observations are used to correct the trajectory itself and the point cloud is recomputed.

> Note that erroneous control coordinates can result in significant problems when using targets to adjust the data. Control should never be assumed to be perfect, but rather should be checked by using the mobile lidar data.

The current standard practice implemented for surveying engineering to reduce positioning errors in the MTLS direct georeferencing is to place a series of tightly spaced transformation control targets (TCTs) prior to data acquisition (e.g., Toth et al. 2008). These TCTs are often painted markings with a pattern (e.g., chevron, checkerboard) such that they are readily identifiable in the point cloud. The coordinates of the TCTs are often determined through RTK GNSS for efficiency, but for projects with high-accuracy requirements, a total station or level may be used for improved accuracy, particularly in the vertical, which is often the most critical in transportation applications. A corresponding vertical, horizontal, or 3D coordinate is extracted from the point cloud to be utilized in an adjustment process to compute the transformation to minimize error. Several models have been proposed, such as a 12-parameter affine transformation (Wu et al. 2013), a polynomial least-squares collocation (Mao et al. 2015), and a sliding least-squares collocation (Zhang et al. 2014). In some approaches, well-defined geometric objects present in the scene such as linear features, planes, or cylinders are used in place of specific "point" TCTs (Haala et al. 2008, Barber et al. 2008, Hauser et al. 2016). These objects are used in a manner similar to compute the transformation.

Many guidelines and specifications recommend supplementing TCTs with validation control targets (VCTs) that are surveyed as well but withheld from the transformation to validate the process (see Chapter 8 for more on TCTs versus VCTs). Unfortunately, guidance for spacing these TCTs and VCTs is sparse and often dictated by experience, manufacturer recommendations, or field trials. Caltrans (2018) recommends a spacing of approximately 450 m for TCTs and 150 m for VCTs for engineering applications. Spacing for vertical TCTs as tight as 50 m has been suggested (Soininen 2012). NCHRP Report No. 748 (Olsen et al. 2013a) recommends varying the density of TCTs and VCTs to different conditions (e.g., poor GNSS conditions in a tunnel or forest canopy versus open road) that affect the georeferencing quality along the mobile lidar trajectory. Consistent spacing of TCTs and VCTs can lead to either inefficiencies in the surveying process or inadequate control spacing in poor conditions. Certainly, the optimal placement of these targets is a topic requiring further exploration. Insufficient targets lead to poor data quality, whereas too many targets minimize the safety and efficiency benefits of mobile lidar.

In some cases, the positioning accuracy of the data can be improved by supplementing TCTs after the acquisition by surveying distinct features in the scene (e.g., corners of pavement markings or other objects). However, placement of TCTs prior to acquisition is preferable to ensure that points are spatially distributed appropriately and are well defined. In many cases, if a mobile lidar system is available locally, it may be more practical and economically feasible to recollect the mobile lidar data again using a higher-quality system for the section of interest with the marks in place if higher accuracy is needed. Utah DOT (Searle and Sridharan 2014) recently explored the feasibility of this technique to enable data collected primarily for asset management grade data to be utilized in design processes to extract more value from the data. Suitable contours for design were achieved in the case studies with well-defined TCTs that were placed or present before the acquisition but not surveyed until much later; however, the suitability of the data was unclear for other case studies without preplaced TCTs and relying solely on objects present in the scene.

> If considering implementing a similar approach to utilize available mobile lidar data collected for other purposes for design, one should first carefully verify the accuracy of the data, particularly in the vertical, to ensure that it does indeed meet the requisite specifications needed for design as well as ensure that the data are up to date to represent current site conditions.

Given the many shortcomings and challenges with TCTs previously described, the multipass approach (MPA) (Nolan et al. 2015, 2017) serves as a promising alternative to collect data suitable for surveying engineering. For the MPA, multiple passes (typically 4 to 8) with MTLS are completed on the section of interest. A "control polyline" is extracted from the point cloud (e.g., centerline of the road) for each pass. The differences in elevation between the passes along this control polyline are compared to identify problematic sections of data. (A similar approach can be used along the sides of buildings or walls for constraining the horizontal direction, if desired.) The sections in each pass with poor GNSS data are removed. The control polylines from each acceptable pass are then averaged over a small window (typically 1 m) and used as a corrector in the Kalman filter to regenerate the trajectory and subsequently the point cloud. The principal idea behind the multipass approach is that by performing repeat passes, errors such as multipath from GNSS can be minimized because the constellation will vary for each of the different passes. The studies performed by Nolan et al. (2015, 2017) show that it can significantly reduce the number of, and increase the spacing of, TCTs necessary as well as identify problems in the control network. The approach can also be utilized in conjunction with TCTs, if desired, for obtaining very high-accuracy results.

> It is recommended to use the both the MPA with some level of TCTs and VCTs, particularly in poor GNSS areas. The TCTs provide redundancy and the VCTs provide an independent check.

As another alternative, Løvås (2017) proposes and demonstrates the utilization of SLAM (developed through the robotics community), as a mechanism to improve the accuracy of the trajectory (and subsequently the point cloud) in areas with poor georeferencing quality. More details on SLAM can be found in the Indoor Mobile Mapping Technology section of this chapter.

Error Sources in Georeferencing. Given that MTLS integrates data from multiple sensors, and a wide range of systems are available that utilize diverse components, it is important to understand the contribution of each error source to the overall error budget. A system calibration (Glennie and Lichti 2010) should be done before undertaking major projects if the system has been modified, or periodically (e.g., annually), to ensure the generation of high-quality data. Typically, the measurements are weighted in the Kalman filter to process the trajectory based on these errors. More specifically, the measurement uncertainty will be a factor in the computation of the Kalman gain at each epoch and, therefore, how much the measurements are weighted relative to the prediction in determining the current state estimate. A more rigorous discussion on this topic is provided in Glennie (2007b), Habib et al. (2009), and NCHRP Report No. 748 (Olsen et al. 2013a). This section summarizes the information contained in those references. Figure 9-8 from Glennie (2009) shows the relative contribution of each error source for different ranges. Note that errors in the sensor and the calibration ultimately propagate to the point cloud and derivative products.

- *IMU attitude errors*: The IMU is the primary source of the roll, pitch, and yaw information. Several systems are available with attitude accuracy specifications (68% confidence) typically between 0.003 degrees and 0.03 degrees in roll and pitch and 0.004 degrees to 0.1 degrees in heading. The IMU also provides positioning information to supplement the GNSS observations; however, between GPS epochs, the associated relative positioning errors with the IMU are much smaller than the GNSS positioning errors and, thus, receive a higher weight within the Kalman filter, so they are often ignored. Nevertheless, these positioning errors translate into significant drift during long periods of GNSS outages (e.g., tunnels, dense urban environments with tall buildings, forest canopy).
- *Boresight errors*: The boresight angles (rotations) and lever arms (translations) result from the misalignment between the IMU and the laser scanner coordinate frame axes and the fact that their origins are typically offset by distances ranging from several centimeters to up to a meter or more in each axis. A variety of boresight calibration techniques are available. Overlapping strips of lidar data (typically from opposing

directions) are used to compute this correction through a least-squares adjustment of extracted tie points. Example approaches include Skaloud and Lichti (2006), Morin (2002), Talaya et al. (2004), and Habib et al. (2011). Boresight angle errors (68% confidence) can typically be resolved to 0.001 degree in roll and pitch and 0.004 degree in yaw (Morin 2002 or Skaloud and Lichti 2006).

- *Lever arm errors:* The lever arm is the translation vector representing the offset between the optical reference point for the laser scanner and the IMU observation reference point. In general, these offsets can be determined to within a few millimeters in each component (X, Y, Z) through either a calibration process or a combination of physical measurements and engineering specifications. It is also possible to use the Kalman filter to estimate the lever arms and boresight angles or to refine the calibrated values.
- *Laser scanner errors*: Laser scanner error sources include the ranging (distance measurement) and scan angle measurement and are affected by beam width and divergence. These errors in terms of positioning usually increase with distance from the scanner. Ranging error can be affected by the material as well as the geometry of the object relative to the scanner. Highly reflective surfaces close to the scanner cause saturation problems. Oblique surfaces result in increased error in range measurements as the laser bleeds across the surface. A more detailed discussion of these error sources can be found in Chapter 8 and in Morin (2002) and Lichti and Gordon (2004). Ranging error typically varies between 0.005 and 0.025 m (68% confidence) and angular resolution varies between 0.001 degree and 0.09 degree depending on the quality of the system. Angular uncertainty from the beam width varies between 0.004 degree and 0.04 degree.
- *GNSS positioning errors*: The reader is referred to Chapters 6 and 11 for more details regarding sources of error and error budget in GNSS positioning. Commonly errors are estimated at 1 cm + 1 ppm horizontal and 2 cm + 1 ppm vertical (Raquet (1998), Bruton (2000)) for relative differential GPS kinematic positioning with short (<18.5 miles, 30 km) baselines. However, in locations with obstructions, multipath can significantly increase the errors. Although highly scene dependent, multipath error for high-quality antennas typically is around 2 cm (68% confidence) (Lau and Cross 2007, Schön and Dilßner 2007).

In addition to errors related to the sensor, the data ultimately incorporate errors associated with the datum utilized and the control survey. These sources are summarized in Nolan et al. (2017).

- *Horizontal datum error*: The base station used to post-process GNSS data on the mobile lidar will either be placed atop a reference mark or be processed against nearby CORSs or RTN base stations using techniques described in Chapters 6 and 11. If only one reference station is utilized, this error source will primarily affect the network accuracy rather than the relative accuracy. However, when multiple reference station setups are required for longer projects, this error source can affect the relative positioning accuracy of the mobile lidar data as well.
- *Vertical datum or geoid model error:* In most mobile lidar applications, orthometric heights, rather than ellipsoidal heights, are of interest. Network errors in geoid models are poorly understood but are often assumed to be in the range of 0.05 to 0.1 m; however, they can be substantially larger in rugged terrain. The relative accuracy of the current hybrid geoid model in the conterminous United States, Puerto Rico, and the US Virgin Islands (GEOID18) has been estimated to be about 0.02 m (68% confidence) but can be highly variable depending on the location and distance from a constrained bench mark in the model (Ahlgren et al. 2020).

- *Local survey errors*: A combination of different tools is utilized to perform the control surveys to support mobile lidar acquisition, including determining the coordinates of TCTs and VCTs. Although total stations and levels have a higher relative accuracy than mobile lidar, errors are still introduced into the surveying procedures. These errors can be difficult to track down in the data when using a TCT georeferencing procedure but can be readily identified using an approach such as the multipass approach (Nolan et al. 2015, 2017).
- *Map projection errors*: The user is referred to Chapter 3 for a discussion of map projection errors/distortion. Ultimately, mobile lidar systems provide a point cloud wherein the coordinates for the points were derived using the angles and ranges measured from the ECEF coordinates at the scanner [Equation (9-1)]. Resulting coordinates for the points do not exactly match grid measurements within a coordinate system. However, the capabilities of lidar processing software to handle coordinate systems and map projections have increased substantially over the last few years.

Note that in the case of repeat surveys, some of these errors will systematically cancel between surveys when performing change analysis. However, random errors such as the scanner errors and GNSS errors cannot be eliminated through canceling.

Quality Control. Quality control should be planned and implemented in each stage of the mobile lidar workflow. Before the survey, the system calibration needs to be verified. During data acquisition, the data collection mission needs to be monitored, including the status of each sensor, precision of GNSS, data storage, and so on. Then, in the stage of georeferencing and post-processing, more measures are taken for quality control. First of all, the coverage of the mobile lidar point clouds needs to be verified to make sure that data collection covers the area of interest. Sometimes, TLS is used to fill in gaps and supplement the data. Then, the point density is further measured in the point cloud to ensure that the data resolution is sufficient for the project. It is worth noting that because the point density is impacted by many factors, including the angular resolution of the scanner, driving speed of the mobile lidar platform, range, and incidence angle, the point density needs to be measured at a number of spots to systematically cover the entire scene. Furthermore, the accuracy of the mobile lidar point clouds should be analyzed rigorously. The accuracy of mobile lidar can be categorized into local accuracy and network accuracy as defined by FGDC (1998). When the mobile lidar system surveyed the site with multiple passes, the repeatability of the point clouds can be assessed to evaluate the local accuracy. Regarding the network accuracy, the point clouds should be compared against the independent check points of higher accuracy. Some blunders can be detected by conducting a visual assessment. More details about quality control can be found in NCHRP Report No. 748 (Olsen et al. 2013a) and ASPRS Accuracy Standards for Digital Geospatial Data (ASPRS 2015) (also, see Chapter 5 for more details on these standards).

Post-Processing

Following georeferencing, post-processing procedures can vary somewhat depending on the application. Chapter 8 of this manual details many post-processing procedures for lidar data in general. The following specific post-processing procedures are commonly implemented for mobile lidar:

1. *Tiling*: Mobile lidar data can span a very long section of highway, resulting in large volumes of data. To improve processing efficiency, data are often divided into smaller tiles based on a distance along the trajectory.

2. *Filtering*: Noise in the dataset can occur from passing vehicles, pedestrians, spurious points in the sky, reflections off wet surfaces, and many other sources. These can be removed from the dataset through automated or semiautomated procedures.
3. *Classification*: The mobile lidar data do not natively have attributes associated with them. Classification processes can help improve the interpretation of the scene for subsequent analyses and processing. ASPRS LAS file format point classifications (ASPRS 2019) for airborne lidar "Ground," "Low Vegetation," and so on are commonly utilized for mobile lidar as well. Nevertheless, the mobile lidar point cloud often needs to be classified into more detailed categories to support other applications (Figure 9-9). For a detailed discussion of segmentation, classification, and object recognition of mobile lidar data, see Che et al. (2019a) and Ma et al. (2018a).
4. *Digitization:* In many cases, end users are not yet equipped to utilize the full range of information available in the point cloud. A common product is to digitize CAD or GIS line work or point features (e.g., top of curb, flow line, signs) from the point cloud to serve as a base map in engineering plans or as data layers for asset management purposes. As in the process of collecting data for surveying, there can be some ambiguity to the actual physical reference locations for extracting these features as well as the naming convention used. Open Laser Scanning Forum (Open LSEF) provides a standard definition and nomenclature for extracting transportation assets from point clouds to help minimize confusion and take advantage of the high-accuracy data provided by mobile lidar.
5. *Modeling*: Additional modeling work may include the generation of TINs or creating solid objects from the point clouds. These procedures are discussed in more detail in Chapter 8.

Table 9-1 summarizes references on different processing strategies, including road surface extraction, road marking extraction, pole-like object extraction, asset condition assessment, general segmentation and classification, and sign detection

NCHRP Report No. 748 (Olsen et al. 2013a) also provides detailed considerations for processing mobile lidar data for different applications.

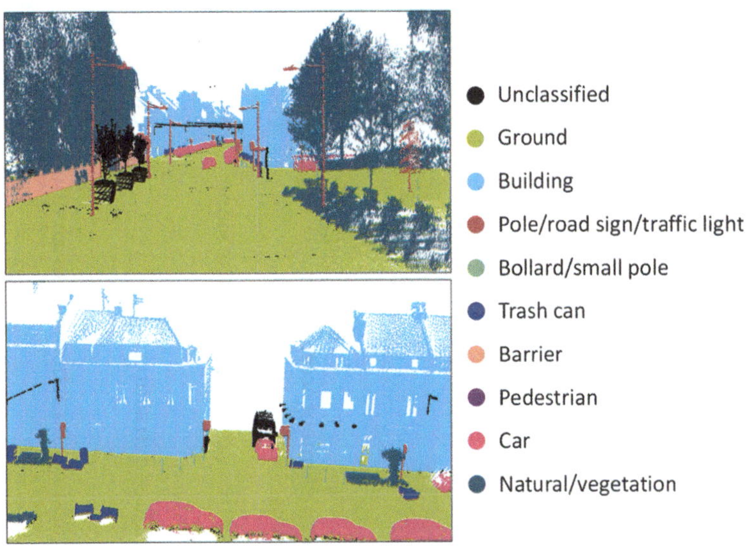

Figure 9-9. Example of classified mobile lidar point clouds.
Source: Che et al. (2019a).

Table 9-1. Summary of References for Different Processing Applications.

Applications	References
General segmentation and classification	Kingston et al. (2006), Pu et al. (2011), Aijazi et al. (2013), Gonzalez-Jorge et al. (2013), Bremer et al. (2013), Serna and Marcotegui (2014), Lehtomäki et al. (2015), Luo et al. (2015), Yang et al. (2015), Weinmann et al. (2015), Hackel et al. (2016), Landrieu et al. (2017), Li et al. (2017), Yang et al. (2017), Luo et al. (2018), Sun et al. (2018), Xiang et al. (2018), Ma et al. (2018a), Gargoum and Basyouny (2019), Che and Olsen (2019), Che et al. (2019a, 2021).
Asset management/condition assessment	Chang et al. (2006), Grafe (2008), McQuat (2011), Burns and Jones (2012), Teo and Yu (2015), Guan et al. (2016a), Ai and Tsai (2016a), Soilán et al. (2019)
Topography and drainage (cross slope, culverts)	McQuat (2011), Vaaja et al. (2011), Lin and Hyyppä (2010), Tsai et al. (2013), Shams et al. (2018)
Road surface extraction	Chen et al. (2009), Wang et al. (2012), Holgado-Barco et al. (2014), Wang et al. (2015), Wu et al. (2016), Zai et al. (2018), Ma et al. (2019b), Che et al. (2021)
Road marking extraction and evaluation	Guan et al. (2014, 2015), Yao and Hu (2014), Guo et al. (2015), Riveiro et al. (2015), Zhang et al. (2016), Kumar et al. (2014), Yu et al. (2015), Soilán et al. (2017), Yan et al. (2016), Chen et al. (2009), Yang et al. (2017b), Jaakkola et al. (2008), Smadja et al. (2010), Yang et al. (2012), Jung et al. 2019, Che et al. 2019b, Wen et al. (2019), Rastiveis et al. (2020)).
Pole-like object extraction	Lehtomäki et al. (2012), Li and Elberink (2013), Cabo et al. (2014), Yu et al. (2015), Teo and Chiu (2015), Rodríguez-Cuenca et al. (2015), Guan et al. (2016b), Wang et al. (2017), Li et al. (2018);

Walls/facades	Becker and Haala (2007), Rutzinger et al. (2009), Jochem et al. (2011), Aldosari et al. (2020).
Powerlines	Jung et al. (2020)
Clearances	Vasquez et al. (2012), Gargoum et al. (2018b).
Sign detection and evaluation	Yang and Dong (2013), Riveiro et al. (2016), Soilán et al. (2016), Tan et al. (2016), Zhou and Deng (2014), Li et al. (2016), Pu et al. (2011), Wen et al. (2016), Vu et al. (2013), Wu et al. (2015), Yang et al. (2015), Sairam et al. (2016), Ai and Tsai (2014, 2016a), Yang et al. (2017a)
Sight distance analysis	Castro et al. (2017), Jung et al. (2018), Gargoum et al. (2018a), Gónzalez-Gómez and Castro (2018), Iglesias et al. (2019), Ma et al. (2018b), Ma et al. (2018b, 2019a), Gargoum and El-Basyouny (2020), Gargoum and Karsten (2021)
Bicycle lanes	Ma et al. (2020b)
Pedestrian infrastructure	Ai and Tsai (2016b), Ai and Hou (2019), Hou and Ai (2020)
Autonomous vehicles	Brenner (2009), Barry (2011)
Cultural heritage/digital preservation	Kerstem et al.(2009)
Rockfalls	Lato et al. (2009), Metzger et al. (2013)

Computations/Analysis

After the desired products are generated in the processing stage, they can be utilized for detailed analysis specific to each application. These products are utilized in a wide range of engineering software, including general-purpose software such as CAD or GIS to more specific software to perform analyses such as hydrographic modeling, slope stability, forensic analysis, and structural analysis. Although the data products are smaller and easier to manage, these analyses are often performed by specialists rather than the surveying engineer. However, it is important that the surveying engineer has a basic understanding of these processes and how the data are utilized to support these analyses so that the appropriate data are utilized and the data are not misused. Additional information on and examples of computation and analyses can be found in Chapter 8.

Packaging/Delivery

The reader is referred to the NCHRP Report No. 748 (Olsen et al. 2013a) for additional discussion of deliverables as well as important information to include in the final reporting. In addition to derived products and analysis desired, the user should request the point cloud, trajectory, control survey report, error analyses, and other important information. Given the many files produced in the MTLS workflow, an organized file structure and naming convention is critical. A checklist is an important part of the QA/QC process to verify the completeness and correctness of organization of the data and products (e.g., Figure 9-10).

INDOOR MOBILE MAPPING TECHNOLOGY

Recently, there has been a growing interest in building indoor mapping for efficient and effective management and operation of existing buildings. However, indoor mapping with conventional static scanning systems is becoming more challenging as modern buildings are getting larger and more complicated, resulting in difficult, time-consuming workflows to register multiple scans collected from different positions (Jung et al. 2015). Alternatively, a mobile mapping system is receiving great attention because of its simple and efficient mapping process. For most mobile mapping systems, manual registration using cumbersome targets are not necessary; the user can simply operate the mobile mapping system simply by walking around the buildings, and the resulting 2D or 3D measurements of the surrounding environments are automatically registered along the trajectory of the system. Although the resolution and details captured by the mobile mapping system can be lower than a static mapping system, recent devices are capable of providing sufficient accuracy for many building analysis purposes (Otero et al. 2020). Chapter 14 of this manual discusses Building Information Models and the process to convert a point cloud into a geometric model.

Indoor mobile mapping systems have one important difference from outdoor systems. Most outdoor mobile mapping systems utilize GNSS signals to track the trajectory of a mobile system as it travels. However, GNSS signals cannot penetrate the buildings and, thus, cannot be used for localizing the mobile system in a building indoor environment. SLAM was originally developed for the navigation of autonomous robots using artificial intelligence and has been extensively researched over the last two decades (Taheri and Xia 2021). The task of a robot tracking its trajectory in known environments is called *localization*, where a prior given map is an essential requirement for successful localization. However, an advanced mobile robot is capable of capturing and exploring unknown environments using sensor data. SLAM refers to a process where a robot builds a map of the environment

Better by DESIGN MDOT Design Division

Michigan Department of Transportation
Survey QA/QC Checklist

Revision Date: 3/15/2021

Refer to Chapter 10 in the Survey Standards of Practice for the Standard Naming Conventions

Item #	QA/QC Review by Consultant	MDOT Survey Project Manager:______ Consultant Firm:______ MDOT Job Number:______ Design Project Manager (if sub):______ Consultant Project Manager:______ Control Section:______
		PWISE - MAIN PROJECT FOLDER
1		**XXXXXX_MDOT_0222_20YY-MM-DD.pdf** MDOT Form 222 (5/01) Survey Notes Receipt and Transmittal filled out with pertinent information and signed (e-signature is acceptable). Once all files are uploaded to ProjectWise, change state to 'Next' and send the auto-generated e-mail to the MDOT Survey PM.
2		**XXXXXX_Align_ROW_Portfolio_20YY-MM-DD.pdf** Bookmarked Adobe PDF of of Surveyor's Alignment & ROW Report and pertinent alignment/ROW documents **(See end of this document for items to be included in the portfolio):**
3		**XXXXXX_Survey_Portfolio_20YY-MM-DD.pdf** Bookmarked Adobe PDF of Surveyor's Project Report and pertinent administrative, control, mapping, LiDAR, structure, hydraulic and miscellaneous project documents **(See end of this document for items to be included in the portfolio):**
		PWISE - CONTROL FOLDER
7		**XXXXXX_MDOT_Monument_*monument number*.doc** MDOT Monument Establishment Data Sheets of all Primary Control Points established and or used as part of this project. This form is available at: http://mdotwiki.state.mi.us/design/index.php/Chapter_10_-_Deliverables
8		**XXXXXX_Images_Control.zip** Digital photos relating to site control. The name of each file should relate to the control point being photographed. All files are to be saved in one zip file.
9		**Total Station RAW Data** RAW data pertaining to Primary and Intermediate Control
10		**GPS RAW Data** RAW data pertaining to Primary and Intermediate Control
11		**GPS RINEX Data** GPS RAW data pertaining to Primary and Intermediate Control converted to RINEX format
12		**MSRN RINEX Data** The RINEX files for the MSRN stations used by OPUS
13		**Vertical RAW Data** Electronic raw data files for all vertical control established.
		PWISE - MTL (MOBIL TERRESTRIAL LIDAR) FOLDER
24		**Point Cloud Files** Provided on electronic media of an appropriate size to contain all project information without separating on more than one media. Two complete (2) copies have been provided to MDOT Lansing Survey Support Unit. A third (3rd) copy has been provided for use by the MDOT Region Surveyor or Design Engineer. (See Appendix F)
25		**XXXXXX_MTL_Mission_Plan.tpm** and **.kml / .kmz** MTL Planning files - Trajectory plan with planned target(control and validation) files in TopoDot Mission planning format, and kml/kmz format
26		**XXXXXX_MTL_Survey_3D_*20YY-MM-DD*.dgn** 3D MicroStation DGN file containing all mapping, meeting MDOT Cadd requirements
27		**XXXXXX_MTL_Terrain_3D_*20YY-MM-DD*.dgn** 3D MicroStation DGN triangle file containing the terrain surface triangles
28		LiDAR data has been tiled and saved as colorized .POD (Point Tools/MicroStation point cloud file) files using photo images
29		**XXXXXX_POD_Tile_Index_*20YY-MM-DD*.dgn** DGN index file showing tile layout and naming of POD files
30		LiDAR data has been tiled and submitted in .LAS/.LAZ (version 1.2 or later) format with RGB values and intensity values (File formats XYZIRGB or XYZRGBI) colorized from photo images
31		**XXXXXX_LAS_Tile_Index_*20YY-MM-DD*.dgn** DGN index file showing tile layout and naming of .LAS files
32		Photo mosaic/Images along route that support the LiDAR .POD/.LAS point cloud
33		Original unedited raw LiDAR dataset in manufacturer's file format
34		**XXXXXX_CntrltoMTL_Diff_3D_*20YY-MM-DD*.dgn** DGN file containing mapping locations of all Control Targets, Validation Targets, GNSS Base Stations, Primary Control points, Intermediate Control points and Benchmarks used or established for the project. The points comply with MDOT features and coding. The DGN also contains values showing the XYZ differences to the corresponding point in the final LiDAR Point Cloud. File is named
35		**XXXXXX_DTM_Compare.dgn** and **.kmz** 3D MicroStation DGN and KMZ file providing the graphical representation of the resulting differences between the edited project DTM surface and the targets and validation points
		PWISE - INTERIM MTL FOLDER
36		**Processing Trajectories, Scan Line Processing and Point Cloud Registration (Workflow of tasks done prior to data extraction/mapping tasks)**
37		**XXXXXX_Targets.txt** Coordinate list of the Target points used for the Mobile mapping and the adjustment reports establishing these points.
38		**XXXXXX ValidationPts.txt** Coordinate list of the Validation points used for the Mobile mapping and the adjustment reports establishing these points.
39		**XXXXXX_PtCloud_Compare.dgn** and **.kmz** 3D MicroStation DGN and a KMZ file providing the graphical representation of the resulting differences between the project Point Cloud dataset and the targets and validation points
40		**XXXXXX_MTL_Interim_Report_*20YY-MM-DD*.pdf** Narrative report describing the actions taken to process and refine the trajectories and the results obtained. Including, but not limited to, the following items:
41		Smooth Best Estimated Trajectory (SBET) resulting accuracies report (precision) for Trajectories
42		Describe and show any areas of the project where the data, as collected, **exceeded the maximum elapsed time or distance traveled of uncorrected IMU drift** due to GNSS signal loss or obstruction.
43		Total Number of all Targets/Points/Validations
44		Number of Control Targets
45		Number of Object Points (i.e. Pick Points)

Figure 9-10. Example of checklist of mobile lidar project deliverable files.
Source: Select portions extracted from the Michigan Department of Transportation (MDOT) Survey QA/QC Checklist, reproduced with permission from MDOT.

without any prior knowledge and at the same time localizing its position within this built map (Thrun 2002). During the travel of the system, SLAM enables the recording of a 2D or 3D map data. This map continually improves and is refined as the user's path intersects with a previous portion of the path, enabling adjustment of the trajectory. As a result of these benefits, SLAM has been widely adopted for indoor scanning and mapping.

Sensors for Indoor Mapping

Most mobile mapping systems are equipped with internal sensors (i.e., odometers, compasses, or IMU) for tracking the trajectory. Because these internal sensors are not

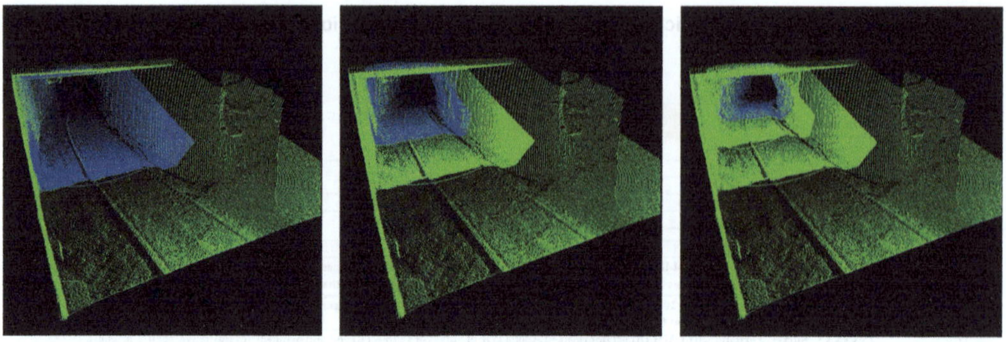

Figure 9-11. Example of scans collected with the stop-and-go mobile scanning approach of Nüchter et al. (2004).

capable of capturing the environments while their self-localization quickly degrades as the system travels, a sensor fusion technique is commonly used to integrate different types of external sensors into SLAM problems. In general, the state of the art of SLAM sensor techniques can be categorized into vision-based SLAM and lidar-based SLAM (Taheri and Xia 2021). Vision-based SLAM uses a stream of visual observations, which is an attractive choice owing to its low cost (Davison et al. 2007). Depending on the camera types, vision-based SLAM is further categorized into monocular, stereo, and RGB-D SLAM. Utilizing one camera concerning the SLAM problem is called *monocular SLAM*. Utilizing multiple cameras is called *stereo vision SLAM* and is capable of reconstructing the 3D maps of environments, but its performance and applicability are sensitive to variations in light and shadow of the environment (Pinto et al. 2012). In addition, 3D maps obtained by visual SLAM are often sparse and nonuniformly distributed point clouds, which makes it more challenging to identify and model indoor structures with the point-cloud data alone. Alternatively, an RGB-D camera, such as the Microsoft Kinect sensor, has been used widely in SLAM, because it can provide colorized and relatively dense-environment maps (Suarez and Murphy 2012). The main limitations of the Kinect sensor are its low depth resolution and large measurement error, resulting in relatively low mapping quality for large areas (Khoshelham and Elberink 2012, Suarez and Murphy 2012).

Lidar-based SLAM is a comparatively more sophisticated approach that has been adopted in various intelligent robots as well as commercial mobile mapping systems (Taheri and Xia 2021). To obtain dense 3D data, 2D lidar sensors are extensively used in lidar-based SLAM. One way to acquire the 3D data with 2D lidar is by stop-and-go scanning (Figure 9-11): A scanner mounted on a pan-tilt unit scans part of the environment, moves to the next position, and scans again. Subsequently, using a scan matching method, consecutive scans taken at different positions can be automatically aligned to generate a consistent map. Another approach is through continuous scanning with multiple 2D lidar units (Figure 9-12): One is mounted horizontally for navigation, and one or more others are mounted to collect vertical scan lines of the environment along the trajectory (Jung et al. 2015).

SLAM Algorithms

Because of sensor imperfections and noises, a mobile mapping system's self-localization quickly degrades as it travels, which can later lead to significant errors in the resulting scan data (Nüchter et al. 2007). To address this problem, widely used solutions are based on two approaches, filter-based SLAM and optimization-based SLAM (Taheri and Xia 2021).

Figure 9-12. Example of data collected with a continuous scanning approach.

Most filter-based SLAM solutions are based on a probabilistic approach to model various sources of uncertainties (Thrun 2002). The Kalman filter (KF) method is one of the most well-known filter-based approaches that assume the Gaussian noise for the robot motion and mapping. Indeed, even many newly developed SLAM solutions still depend on this basic algorithm. Prior to operation, users need to determine the types of features that should be detected by sensors (Arras 2003, Mahon et al. 2008). The KF can be employed for any sensor readings, provided that recognizable features exist in the environment. If the system motion is near linear, the KF can produce a reliable outcome. However, if the system motion is nonlinear, a large error in localization and mapping is incurred (Welch and Bishop 1995, Ryu 2012). To address the problem of nonlinearity, the extended Kalman filter (EKF) and the unscented Kalman filter (UKF) can be used. The EKF is the nonlinear version of KF that uses the Taylor series expansions to linearize the state estimate. On the contrary, the UKF is a deterministic sampling approach, depending on a minimal set of carefully chosen samples, called *sigma points*, to approximate the state estimate. Compared with the EKF that approximates the state mean and covariance to the first-order accuracy (Taylor series expansion), the UKF can achieve the third-order accuracy for Gaussian inputs and the second order for non-Gaussian inputs (Wan and Van Der Merwe 2000). However, conventional EKF-based methods are not suitable for the estimation of maps with high dimensions and are also very slow in environments with many complex features (Thrun et al. 2004). Alternatively in 2016, an adaptive Kalman filter (AKF) was proposed by Jiang et al. (2016), which is capable of adjusting the parameters of KF in real-time processing and, thus, improving the filtering efficiency and the accuracy of localization and mapping results. However, it is still an issue that KF-based approaches assume a Gaussian noise hypothesis that renders it difficult to estimate a large environment with many landmarks because of the problems involved in the data association (Taheri and Xia 2021).

Another approach is known as the *particle filter* (PF) method. The PF approximates the state estimate using a set of particles generated from Monte Carlo simulation to address the accumulated nonlinearity errors. The PF does not consider the system to be linear, nor does it need the restrictions of the Gaussian noise assumption concerning the observation. PF can excellently handle nonlinearity and non-Gaussian models, the computation of PF is simple, and it can estimate any likelihood distributions. However, PF is prone to loss of the correct particles. Further, the accuracy and efficiency of the PF highly depends on the number of particles (Weingarten 2006, Aulinas et al. 2008); a larger number of particles lead to better approximation, but it increases computational costs. An alternative to reduce PF's computational cost is the Factored Solution to SLAM (FastSLAM), which factors the SLAM problem into a system motion and a map component. FastSLAM uses PF to estimate the system motion and a few low-dimensional EKF to estimate the map component (landmark positions). In other words, FastSLAM uses a hybrid technology that is a combination of the EKF and PF. Because each particle defines a different system motion hypothesis, no covariance between the features needs to be computed such that the algorithm can be more efficient (Montemerlo et al. 2002, Montemerlo et al. 2003, Neuhaus 2007, Aulinas et al. 2008); $O(M \log N)$ is the computational complexity of FastSLAM, where M is the number of landmarks and N is the number of particles. On the contrary, the computational complexity of standard EKF-based SLAM is quadratic $O(N^2)$. Further, using PF for estimating the system motion enables less memory consumption in comparison with KF-based methods.

One important limitation of the described filter-based techniques is that the collected data are discarded after use (Thrun and Montemerlo 2006). To produce an accurate, consistent map, the data should be revisited for optimization. Loop closure is an important compound of mapping that enables the robot to recognize a visited place and, therefore, optimize its estimated trajectory to reduce its accumulated location and mapping errors (Taheri and Xia 2021). The graph-based SLAM is an effective solution for this problem. It uses a graph concept to represent data with nodes and edges: Nodes represent system poses, and edges correspond to a spatial constraint between nodes. The process of graph-based SLAM can be decomposed into two steps, front end and back end: the front end is to collect constraints from the sensor measurements, detect a loop, and add them to the graph as edges; and the back end is responsible for optimization to calculate a maximum likelihood map and its corresponding set of poses given the constraints (Latif and Neira 2013). The commonly used process for the front end is a scan-to-scan matching method, where its solution is the rigid-body transformation that corresponds to the motion of the robot between two locations. The popular iterative closest point (ICP) algorithm finds the closest point pairs in two different scans and iteratively minimizes their relative transform (Lu and Milios 1997, Surmann et al. 2003). However, the ICP algorithm is slow (Li et al. 2010) and its motion estimation can be poor (Chung and Huang 2007). Many follow-up studies have been conducted to improve the performance of ICP by increasing the convergence speed and matching rates. The correlative scan matching (Olson 2009) and Gauss–Newton-based scan matching (Kohlbrecher et al. 2011) are some of the effective approaches to overcome the limitations of the conventional ICP methods. Once such a graph is constructed, the next step is to compute the best configuration of the nodes with respect to the given constraints. Because the constraints are, in general, nonlinear, they can be linearized and solved using the least-squares method. Because the graph-based SLAM naturally forms a huge sparse graph, many current optimization methods take advantage of sparse linear solvers to speed up the process. Figure 9-13 compares indoor mobile lidar data without and with the graph optimization.

Loop closures can greatly reduce the accumulated mapping errors by adding redundant constraints to the graph optimization. However, after traveling long distances in a large environment, the robot may fail to recognize a visited location. Data association is a process that the SLAM associates a currently observed landmark with the previously observed landmarks (Taheri and Xia 2021). The incorrect data association can occur when a robot fails to differentiate two similar-looking features from different places within the same environment. Because the least-squares optimization is weak for detecting outliers, a single incorrect data association can corrupt the entire mapping result. Although SLAM techniques have greatly evolved over the last two decades, data association is still a challenging problem undergoing refinement.

(a)

(b)

Figure 9-13. Indoor mobile lidar data: (a) without, (b) with graph optimization.

Indoor Mobile Mapping Systems

Over the last few decades, there have been significant advances in SLAM-based indoor mobile mapping systems. Broadly, these can be categorized into a handheld, backpack, and trolley scanners (Otero et al. 2020). Figure 9-14 shows some examples of commercial mobile mapping systems (GeoSlam 2021, Viametris 2021, NavVis 2021). Many of these systems can also be utilized in exterior environments.

Handheld scanners are the most maneuverable because they are carried in the arms and, thus, designed not to be too heavy for the surveyor. Handheld scanners can be categorized further into three different configurations: one is equipped with 2D lidar and a small engine that rotates the 2D lidar to collect 3D point clouds of the surrounding environment. Another

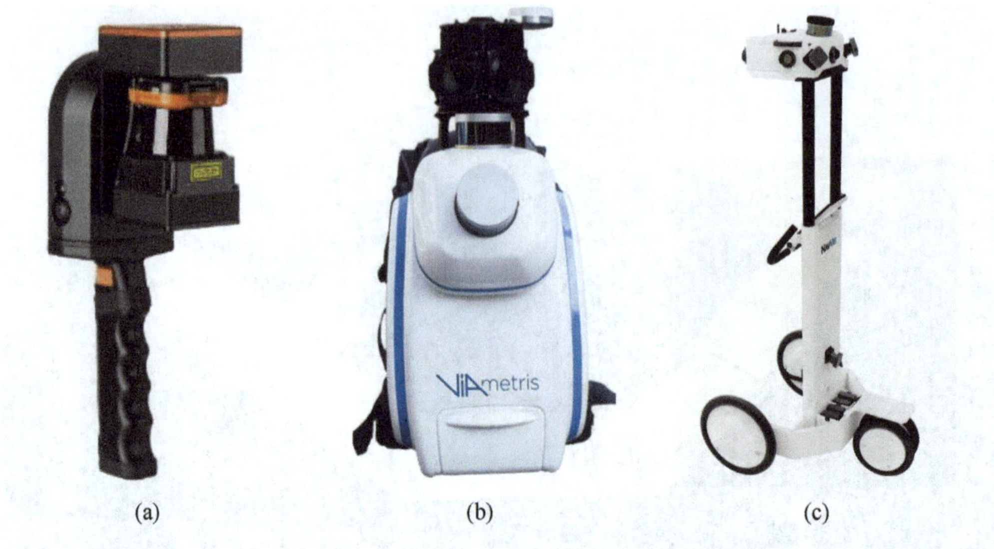

Figure 9-14. Examples of commercial mobile mapping systems: (a) handheld (GeoSLAM ZEB Go), (b) backpack (Viametris bMS3D), (c) trolley (NavVis M6) scanners.
Source: Courtesy of (a) GeoSLAM, (b) Viametris, (c) NavVis, reproduced with permission.

configuration is a tablet-type scanner that uses RGB-D sensors. These are portable and easy to carry, but the measurement range is very short (usually 0.5 to 4 m) and accuracy is very low. Finally, smartphones and tablets have a lot of potential for indoor mapping because they are equipped by default with RGB and IMU sensors that can be used for SLAM approaches. However, the accuracy and resolution of the mapping results can vary depending on the different types of smartphones. Because of Apple's inclusion of a lidar sensor in its 2020 mobile phone (iPhone), 3D mapping with a smartphone has been receiving great attention for its suitability for engineering applications. Although the quality of 3D maps obtained by the iPhone cannot be considered a replacement to a high-end lidar system, its low cost and easy setup will increase the number of users and possible

Figure 9-15. 3D data of a total station collected by an iPhone lidar sensor.

applications of lidar. Figure 9-15 provides an example of 3D data collected by iPhone. Compared with other types of scanners, handheld devices allow more flexibility in the operation because of their small size and weight. However, because of the limit on the low weight and battery, its operation time is relatively short and commonly provides results with lower precision.

Although backpack scanners are typically heavier than handheld devices, they are still highly portable. Unlike the handheld systems, most backpack systems use a lidar sensor rather than an RGB-D camera because the backpack configuration is more comfortable for carrying a heavier sensor and the short measurement range of RGB-D is a limiting factor. Backpacks equipped with 2D lidar mainly adopt a continuous scanning approach, collecting 2D scanlines within a plane along to the trajectory. Because 2D lidar does not provide enough geometric information with overlaps required for SLAM approaches, this configuration usually requires an IMU sensor to track the trajectory. On the contrary, some other backpack scanners carry a hybrid lidar sensor that can fire multiple laser pulses in different directions. Although the density of its scan data is still lower than other typical 3D lidar sensors, the scan data can be used for registration procedures for tracking the trajectory using SLAM algorithms.

Finally, trolley scanners are the heaviest mobile indoor mapping system that cannot be carried by the user's body. Because the weight is no longer a limiting factor, the design is very flexible in the choice of the lidar sensor (regardless of its weight), enabling high-precision mapping. Like the backpack configuration, trolley scanners equipped with 2D lidar collect the 3D data using the continuous scanning approach, but typically they carry multiple lidar sensors deployed with different orientations. Often, one is employed to obtain a horizontal section that can be used for SLAM to track the trajectory of the system, and the others are included to obtain vertical or slant sections to capture the 3D environment. Another configuration is a trolley scanner equipped with 3D terrestrial lidar, which enables the most accurate and dense point clouds among all the previously noted indoor mapping systems. However, its high cost is a potentially limiting factor. Further, most trolley scanners are hard to operate in a complex environment that includes stairs or a lot of furniture.

LOOKING FORWARD

Kinematic surveying and mapping systems will continue to expand in application and utility in the future. A wide range of technological advances are expected in the future. Sensors are rapidly becoming smaller, lightweight, and less expensive [e.g., MEMS inertial sensors, <$1,000 GNSS receivers capable of carrier-phase based positioning (RTK or PPK), OEM components, and sensors such as the Microsoft Kinect]. The combination of these sensors is likely to drive costs for mobile lidar and mobile mapping down significantly in the near future, along with expanding vehicle-based lidar for autonomous vehicles.

These technologies provide significant safety and efficiency benefits over traditional surveying approaches. However, it is critical to note that "these systems still require fundamental surveying practices to provide control to ensure data of sufficient accuracy for engineering purposes," as described in the other chapters in the manual. Hence, they do not replace, but rather augment, the surveying engineer's tools, technologies, and practices.

Mobile mapping technology is critical to provide the base map to geospatially link a variety of other sensor information together so that the information can have added context. NCHRP Synthesis 446 (Olsen et al. 2013b) discusses the use of mobile lidar and other geospatial technologies for transportation applications. Mobile lidar, potentially in conjunction with other sensors such as UAS-based lidar (Chapter 10), provides the base map to which many of these other data sources can be linked and analyzed in context. This geospatial base map becomes increasingly important with recent technological advances affecting the role of civil engineers, such as smart cities, structural health monitoring, location-based services, and connected (autonomous) vehicles.

ACKNOWLEDGMENTS

This material is based on work supported by the National Science Foundation under Grant No. CMMI-1351487. Oregon DOT provided funding support for some of the examples provided in this chapter through SPR799 and SPR809. In particular, Lloyd Bledstoe, Dan Wright, and Rhonda Dodge assisted with the data collection and processing, respectively, for some of the data utilized in these examples. Any opinions, findings, and conclusions or recommendations expressed in this material are those of the author(s) and do not necessarily reflect the views of the National Science Foundation or others included in the acknowledgments.

REFERENCES

Ahlgren, K., G. Scott, D. Zilkoski, B. Shaw, and N. Paudel. 2020. *GEOID18*. NOAA Tech. Rep. No. NOS NGS 72. Silver Spring, MD: National Oceanic and Atmospheric Administration, National Geodetic Survey.

Ai, C., and Y. J. Tsai. 2016a. "An automated sign retroreflectivity condition evaluation methodology using mobile lidar and computer vision." *Transp. Res. Part C Emerg. Technol.* 63: 96–113.

Ai, C., and Y.-C. J. Tsai. 2014. "Critical assessment of an enhanced traffic sign detection method using mobile LiDAR and INS technologies." *J. Transp. Eng.* 141 (5): 04014096.

Ai, C., and Y.-J. Tsai. 2016a. "An automated sign retroreflectivity condition evaluation methodology using mobile LIDAR and computer vision." *Transp. Res. Part C Emerging Technol.* 63: 96–113.

Ai, C., and Y. J. Tsai. 2016b. "An automated sidewalk assessment method for the Americans with disabilities act compliance using 3-D mobile LiDAR." In *Proc., 95th Transportation Research Board Annual Meeting*. Washington, DC: Transportation Research Board.

Ai, C., and Q. Hou. 2019. *Improving pedestrian infrastructure inventory in Massachusetts using mobile LiDAR*. Final Report 53. Rep. No. 19-007. Washington, DC: Federal Highway Administration, US Department of Transportation.

Aijazi, A. K., P. Checchin, and L. Trassoudaine. 2013. "Segmentation based classification of 3D urban point clouds: A super-voxel based approach with evaluation." *Remote Sens.* 5 (4): 1624–1650. doi:10.3390/rs5041624

Aldosari, M., A. Al-Rawabdeh, D. Bullock, and A. Habib. 2020. "A mobile LiDAR for monitoring mechanically stabilized earth walls with textured precast concrete panels." *Remote Sens.* 12 (2): 306.

Arras, K. O. 2003. "Feature-based robot navigation in known and unknown environments." Ph.D. thesis, Swiss Federal Institute of Technology Lausanne, School of Architecture, Civil and Environmental Engineering.

ASPRS (American Society for Photogrammetry and Remote Sensing). 2015. "ASPRS positional accuracy standards for geospatial data." *Photogramm. Eng. Remote Sens.* 81 (3): 1–26.

ASPRS. 2019. "LAS specification version 1.4-R15." Standards Committee. Accessed February 2, 2022. http://www.asprs.org/wp-content/uploads/2019/07/LAS_1_4_r15.pdf.

Aulinas, J., Y. R. Petillot, J. Salvi, and X. Lladó. 2008. "The SLAM problem: A survey." In *Proc., 11th Int. Conf. of the Catalan Association for Artificial Intelligence*, 363–371. https://dl.acm.org/doi/10.5555/1566899.1566949.

Baltsavias, E. P. 1999. "Airborne laser scanning: basic relations and formulas." *ISPRS J. Photogramm. Remote Sens.* 54: 199–214.

Barber, D., J. Mills, and S. Smith-Voysey. 2008. "Geometric validation of a ground-based mobile laser scanning system." *ISPRS J. Photogramm. Remote Sens.* 63 (1): 128–141.

Barry, K. 2011. "BMW tests an autonomous vehicle." Accessed September 17, 2013. http://www.wired.com/autopia/2011/08/bmw-tests-an-autonomous-vehicle/.

Becker, S., and N. Haala. 2007. "Grammar supported facade reconstruction from mobile LIDAR mapping." *Int. Arch. Photogramm. Remote Sens. Spat. Inf. Sci.* 38: 229–234.

Bremer, M., V. Wichmann, and M. Rutzinger. 2013. "Eigenvalue and graph-based object extraction from mobile laser scanning point clouds." *ISPRS Ann. Photogramm. Remote Sens. Spatial Inf. Sci.* II-5 (W2): 55–60.

Brenner, C. 2009. "Extraction of features from mobile laser scanning data for future driver assistance systems." In *Advances in GIScience. Lecture Notes in Geoinformation and Cartography*, edited by M. Sester, L. Bernard, and V. Paelke, 25–42. Berlin: Springer. https://doi.org/10.1007/978-3-642-00318-9_2.

Bruton, A. M. 2000. "Improving the accuracy and resolution of SINS/DGPS airborne gravimetry." UCGE Rep. No. 20145. Ph.D. thesis, University of Calgary, Dept. of Civil Engineering.

Burns, S., and W. S. Jones. 2012. "Capturing the road of tomorrow—UDOT's transportation asset and highway safety data program." In *Proc., Geospatial Transportation Mapping Association Webinar*.

Cabo, C., C. Ordoñez, S. García-Cortés, and J. Martínez. 2014. "An algorithm for automatic detection of pole-like street furniture objects from mobile laser scanner point clouds." *ISPRS J. Photogramm. Remote Sens.* 87: 47–56.

Cahalane, C., T. McCarthy, and C. P Mc Elhinney. 2010. "Mobile mapping system performance–an initial investigation into the effect of vehicle speed on laser scan lines." In *RSPSoc No. 2005*. Nottingham, UK: Remote Sensing and Photogrammetry Society.

Caltrans. 2018. "Terrestrial laser scanning specifications." Chap. 15 in *Surveys manual*. Sacramento, CA: Caltrans.

Castro, M., C. De Santos-Berbel, and L. Iglesias. 2017. "A comprehensive methodology for the analysis of highway sight distance." In *Proc., CitA BIM Gathering Conference on Transport Infrastructure and Systems, Rome*, 193–199. Boca Raton, FL: CRC Press.

Chang, J. R., K. T. Chang, and D. H. Chen. 2006. "Application of 3D laser scanning on measuring pavement roughness." *J. Test. Eval.* 34 (2): 83–91.

Che, E., J. Jung, and M. Olsen. 2019a. "Object recognition, segmentation, and classification of mobile laser scanning point clouds: State of the art review." *Sensors* 19 (4): 810.

Che, E., and M. Olsen. 2019. "An efficient framework for mobile Lidar trajectory reconstruction and Mo-Norvana segmentation." *Remote Sens.* 11 (7): 836.

Che, E., M. J. Olsen, and J. Jung. 2021. "Efficient segment-based ground filtering and adaptive road detection from mobile Lidar data." *Int. J. Remote Sens.* 42 (10): 3633–3659.

Che, E., M. J. Olsen, C. E. Parrish, and J. Jung. 2019b. "Pavement marking retroreflectivity estimation and evaluation using mobile lidar data." *Photogramm. Eng. Remote Sens.* 85 (8): 573–583.

Chen, X., B. Kohlmeyer, M. Stroila, N. Alwar, R. Wang, and J. Bach. 2009. "Next generation map making: geo-referenced ground-level LIDAR point clouds for automatic retro-reflective road feature extraction." In *Proc., 17th ACM SIGSPATIAL Int. Conf. on Advances in Geographic Information Systems*, November 4–6, 488–491. Seattle, WA: ACM.

Chung, S. Y., and H. P. Huang. 2007. "Relative-absolute information for simultaneous localization and mapping." In *Proc., 2007 IEEE Int. Conf. on Robotics and Biomimetics*, 1641–1646. New York: IEEE.

Davison, A. J., I. D. Reid, N. D. Molton, and O. Stasse. 2007. "MonoSLAM: Real-time single camera SLAM." *IEEE Trans. Pattern Anal. Mach. Intell.* 29 (6): 1052–1067.

Dunham, L., J. Wartman, M. J. Olsen, M. O'Banion, and K. Cunningham. 2017. "Rockfall activity index (RAI): A lidar-derived, morphology-based method for hazard assessment." *Eng. Geol.* 221: 184–192.

FGDC (Federal Geographic Data Committee). 1998. "Geospatial positioning accuracy standards part 3: National standard for spatial data accuracy." Accessed February 2, 2022. http://www.fgdc.gov/standards/projects/FGDC-standards-projects/accuracy/part3/chapter3.

Gargoum, S., and K. El-Basyouny. 2017. "Automated extraction of road features using LiDAR data: A review of LiDAR applications in transportation." In *Proc., 4th Int. Conf. on Transportation Information and Safety*, 563–574. New York: IEEE.

Gargoum, S., and K. El-Basyouny. 2019. "A literature synthesis of LiDAR applications in transportation: feature extraction and geometric assessments of highways." *GISci. Remote Sens.* 58 (8): 1–30.

Gargoum, S. A., and K. El-Basyouny. 2020. "Analyzing the ability of crash-prone highways to handle stochastically modelled driver demand for stopping sight distance." *Accid. Anal. Prev.* 136: 105395.

Gargoum, S. A., K. El-Basyouny, and J. Sabbagh. 2018a. "Assessing stopping and passing sight distance on highways using mobile LiDAR data." *J. Comput. Civ. Eng.* 32 (4): 04018025.

Gargoum, S., L. Karsten, and K. El-Basyouny. 2018b. "Network level clearance assessment using LiDAR to improve the reliability and efficiency of issuing over-height permits on highways." *Transp. Res. Rec.* 2672 (3): 45–56.

Gargoum, S. A., and L. Karsten. 2021. "Virtual assessment of sight distance limitations using LiDAR technology: Automated obstruction detection and classification." *Autom. Constr.* 125: 103579.

GeoSlam. 2021. "The smart way to map and understand spaces." Accessed April 7, 2021. https://geoslam.com/.

Glennie, C. 2007a. "Reign of point clouds. A kinematic terrestrial LIDAR scanning system." *Inside GNSS* 2 (7): 22–31.

Glennie, C. 2007b. "Rigorous 3D error analysis of kinematic scanning LIDAR systems." *J. Appl. Geod.* 1 (3): 147–157.

Glennie, C. 2009. "Kinematic terrestrial light-detection and ranging system for scanning." *Transp. Res. Rec.* 2105 (1): 135–141.

Glennie, C., and D. D. Lichti. 2010. "Static calibration and analysis of the Velodyne HDL-64E S2 for high accuracy mobile scanning." *Remote Sens.* 2 (6): 1610–1624.

Gónzalez-Gómez, K., and M. Castro. 2018. "Suitability testing of LiDAR processing software aimed at 3-D sight distance estimations." *Transp. Res. Procedia* 33: 163–170.

Gonzalez-Jorge, H., I. Puente, B. Riveiero, J. Martinez-Sanchez, and P. Arias. 2013. "Automatic segmentation of road overpasses and detection of mortar efflorescence using mobile LiDAR data." *Optics Laser Technol.* 54 (30): 353–361.

Grafe, G. 2008. "Kinematic 3D laser scanning for road or railway construction surveys." In *Proc., 1st Int. Conf. on Machine Control and Guidance*. ETH, Zurich.

Graham, L. 2010. "Mobile mapping systems overview." *Photogramm. Eng. Remote Sens.* 76 (3): 222–228.

Guan, H., J. Li, Y. Yu, C. Wang, M. Chapman, and B. Yang. 2014. "Using mobile laser scanning data for automated extraction of road markings." *ISPRS J. Photogramm. Remote Sens.* 87: 93–107.

Guan, H., J. Li, Y. Yu, M. Chapman, and C. Wang. 2015. "Automated road information extraction from mobile laser scanning data." *IEEE Trans. Intell. Transp. Syst.* 16: 194–205.

Guan, H., J. Li, S. Cao, and Y. Yu. 2016a. "Use of mobile LiDAR in road information inventory: A review." *Int. J. Image Data Fusion* 7: 219–242.

Guan, H., Y. Yu, J. Li, and P. Liu. 2016b. "Pole-like road object detection in mobile LiDAR data via supervoxel and bag-of-contextual-visual-words representation." *IEEE Geosci. Remote Sens. Lett.* 13 (4): 520–524.

Guo, J., M.-J. Tsai, and J.-Y. Han. 2015. "Automatic reconstruction of road surface features by using terrestrial mobile lidar." *Autom. Constr.* 58: 165–175.

Haala, N., M. Peter, J. Kremer, and G. Hunter. 2008. "Mobile lidar mapping for 3D point cloud collection in urban areas-a performance test." *Int. Arch. Photogramm. Remote Sens. Spatial Inf. Sci.* 37 (B5): 1119–1127.

Habib, A., K. I. Bang, A. P. Kersting, and D. C. Lee. 2009. "Error budget of LiDAR systems and quality control of the derived data." *Photogramm. Eng. Remote Sens.* 75 (9): 1093–1108.

Habib, A. F., A. P. Kersting, A. Shaker, and W.-Y. Yan. 2011. "Geometric calibration and radiometric correction of LiDAR data and their impact on the quality of derived products." *Sensors* 11 (9): 9069–9097.

Hackel, T., J. D. Wegner, and K. Schindler. 2016. "Fast semantic segmentation of 3D point clouds with strongly varying density." *ISPRS Ann. Photogramm. Remote Sens. Spatial Inf. Sci.* 3 (3): 177–184.

Hauser, D., C. Glennie, and B. Brooks. 2016. "Calibration and accuracy analysis of a low-cost mapping-grade mobile laser scanning system." *J. Surv. Eng.* https://doi.org/10.1061/(ASCE)SU.1943-5428.0000178, 04016011.

Heinz, E., C. Holst, H. Kuhlmann, and L. Klingbeil. 2020. "Design and evaluation of a permanently installed plane-based calibration field for mobile laser scanning systems." *Remote Sens.* 12 (3): 555.

Holgado-Barco, A., D. Gonzalez-Aguiler, P. Arias-Sanchez, and J. Martinez-Sanchez, 2014. "An automated approach to vertical road characterisation using mobile LiDAR systems: Longitudinal profiles and cross-sections." *ISPRS J. Photogramm. Remote Sensing* 96: 28–37.

Hou, Q., and C. Ai. 2020. "A network-level sidewalk inventory method using mobile LiDAR and deep learning." *Transp. Res. Part C Emerging Technol.* 119: 1027772.

Iglesias, L., C. De Santos-Berbel, V. Pascual, and M. Castro. 2019. "Using small unmanned aerial vehicle in 3D modeling of highways with tree-covered roadsides to estimate sight distance." *Remote Sens.* 11 (22): 2625.

Jaakkola, A., J. Hyyppä, H. Hyyppä, and A. Kukko. 2008. "Retrieval algorithms for road surface modelling using laser-based mobile mapping." *Sensors* 8: 5238–5249.

Jiang, X., T. Li, and Y. Yu. 2016. "A novel SLAM algorithm with Adaptive Kalman filter." In *Proc., Int. Conf. on Advanced Robotics and Mechatronics*, 107–111. Piscataway, NY: IEEE.

Jochem, A., B. Höfle, and M. Rutzinger. 2011. "Extraction of vertical walls from mobile laser scanning data for solar potential assessment." *Remote Sens.* 3 (4): 650–667.

Jung, J., E. Che, M. J. Olsen, and C. Parrish. 2019. "Efficient and robust lane marking extraction from mobile lidar point clouds." *ISPRS J. Photogramm. Remote Sens.* 147: 1–18.

Jung, J., E. Che, M. J. Olsen, and K. C. Shafer. 2020. "Automated and efficient powerline extraction from laser scanning data using a voxel-based subsampling with hierarchical approach." *ISPRS J. Photogramm. Remote Sens.* 163: 343–361.

Jung, J., M. J. Olsen, D. S. Hurwitz, A. G. Kashani, and K. Buker. 2018. "3D virtual intersection sight distance analysis using LiDAR data." *Transp. Res. Part C Emerging Technol.* 86: 563–579.

Jung, J., S. Yoon, S. Ju, and J. Heo. 2015. "Development of kinematic 3D laser scanning system for indoor mapping and as-built BIM using constrained SLAM." *Sensors* 15 (10): 26430–26456.

Kalman, R. E. 1960. "A new approach to linear filtering and prediction problems." *Trans. ASME J. Basic Eng.* 82 (D): 35–45.

Kashani, A., M. Olsen, C. Parrish, and N. Wilson. 2015. "A review of LiDAR radiometric processing: From ad hoc intensity correction to rigorous radiometric calibration." *Sensors* 15 (11): 28099–28128.

Kerstem, T. P., G. Buyuksalih, I. Baz, and K. Jacobsen. 2009. "Documentation of Istanbul historic peninsula by kinematic terrestrial laser scanning." *Photogramm. Rec.* 24 (126): 122–138.

Khoshelham, K., and S. O. Elberink. 2012. "Accuracy and resolution of Kinect depth data for indoor mapping applications." *Sensors* 12 (2): 1437–1454.

Kingston, T., V. Gikas, C. Laflamme, and C. Larouche. 2006. "An integrated mobile mapping system for data acquisition and automated extraction." *Int. Arch. Photogramm. Remote Sens. Spatial Inf. Sci.* 36: 5–C55.

Kohlbrecher, S., O. Von Stryk, J. Meyer, and U. Klingauf. 2011. "A flexible and scalable SLAM system with full 3D motion estimation." In *Proc., Int. Symp. on Safety, Security, and Rescue Robotics*, 155–160. Piscataway, NY: IEEE.

Kumar, P., C. P. McElhinney, P. Lewis, and T. McCarthy. 2014. "Automated road markings extraction from mobile laser scanning data." *Int. J. Appl. Earth Observ. Geoinform.* 32: 125–137.

Landrieu, L., H. Raguet, B. Vallet, C. Mallet, and M. Weinmann. 2017. "A structured regularization framework for spatially smoothing semantic labelings of 3D point clouds." *ISPRS J. Photogramm. Remote Sens.* 132: 102–118.

Latif, Y., and J. Neira. 2013. "Go straight, turn right: Pose graph reduction through trajectory segmentation using line segments." In *Proc., European Conf. on Mobile Robots*, 144–149. Piscataway, NY: IEEE.

Lato, M., J. Hutchinson, M. Miederichs, D. Ball, and R. Harrap. 2009. "Engineering monitoring of rockfall hazards along transportation corridors: Using mobile terrestrial LIDAR." *Nat. Hazards Earth Syst. Sci.* 9 (3): 935–946.

Lau, L., and P. Cross. 2007. "Investigations into phase multipath mitigation techniques for high precision positioning in difficult environments." *J. Nav.* 60 (3): 457–482.

Lehtomäki, M., A. Jaakkola, J. Hyyppä, A. Kukko, and H. Kaartinen. 2012. "Performance analysis of a pole and tree trunk detection method for mobile laser scanning data." *Int. Arch. Photogramm. Remote Sens. Spat. Inf. Sci.* 38 (5/W12): 197–202.

Lehtomäki, M., A. Jaakkola, J. Hyyppä, J. Lampinen, H. Kaartinen, A. Kukko, et al. 2015. "Object classification and recognition from mobile laser scanning point clouds in a road environment." *IEEE Trans. Geosci. Remote Sens.* 54 (2): 1226–1239.

Li, F., S. O. Elberink, and G. Vosselman. 2017. "Semantic labelling of road furniture in mobile laser scanning data." *Int. Arch. Photogramm., Remote Sens. Spatial Inf. Sci.* 42.

Li, D., and S. O. Elberink. 2013. "Optimizing detection of road furniture (pole-like objects) in mobile laser scanner data." *ISPRS Ann. Photogramm. Remote Sens. Spat. Inf. Sci.* 1: 163–168.

Li, Y., J. Fan, Y. Huang, and Z. Chen. 2016. "LIDAR-incorporated traffic sign detection from video log images of mobile mapping system." *Int. Arch. Photogramm, Remote Sens. Spatial Inf. Sci.* 41.

Li, F., S. Oude Elberink, and G. Vosselman. 2018. "Pole-like road furniture detection and decomposition in mobile laser scanning data based on spatial relations." *Remote Sens.* 10 (4): 531–559.

Li, X., W. Cui, and S. Jia. 2010. "Range scan matching and particle filter based mobile robot SLAM." In *Proc., IEEE Int. Conf. on Robotics and Biomimetics*, 779–784. New York: IEEE.

Lichti, D. D., and S. J. Gordon. 2004. "Error propagation in directly georeferenced terrestrial laser scanner point clouds for cultural heritage recording." In *Proc., FIG Working Week, Int. Federation of Surveyors, FIG*, Copenhagen, Denmark.

Lin, Y., and J. Hyyppä. 2010. "Geometry and intensity based culvert detection in mobile laser scanning point 484 clouds." *J. Appl. Remote Sens.* 4 (1): 043553.

Løvås, M. 2017. "Increasing the accuracy of positioning in mobile mapping systems: A method supported by simultaneous localization and mapping," M.S. thesis. Norwegian University of Science and Technology, Dept. of Civil and Environmental Engineering. http://hdl.handle.net/11250/2459948.

Lu, F., and E. Milios. 1997. "Robot pose estimation in unknown environments by matching 2d range scans." *J. Intell. Rob. Syst.* 18 (3): 249–275.

Luo, H., C. Wang, C. Wen, Z. Cai, Z. Chen, H. Wang, et al. 2015. "Patch-based semantic labeling of road scene using colorized mobile LiDAR point clouds." *IEEE Trans. Intell. Transp. Syst.* 17 (5): 1286–1297.

Luo, H., C. Wang, C. Wen, Z. Chen, D. Zai, Y. Yu, et al. 2018. "Semantic labeling of mobile LiDAR point clouds via active learning and higher order MRF." *IEEE Trans. Geosci. Remote Sens.* 56 (7): 3631–3644.

Ma, L., Y. Li, J. Li, C. Wang, R. Wang, and M. Chapman. 2018a. "Mobile laser scanned point-clouds for road object detection and extraction: A review." *Remote Sens.* 10 (10): 1531.

Ma, L., Y. Li, J. Li, Y. Yu, J. Marcato Jr., W. N. Goncalves, et al. 2020a. "Capsule-based networks for road marking extraction and classification from mobile LiDAR point clouds." *IEEE Trans. Intell. Transp. Syst.* 22 (4): 1981–1995.

Ma, Y., Y. Zheng, J. Cheng, and S. Easa. 2018b. "Real-time visualization method for estimating 3D highway sight distance using LiDAR data." *Transp. Eng.* 145 (4): 04019006.

Ma, Y., Y. Zheng, J. Cheng, and S. Easa. 2019a. "Analysis of dynamic available passing sight distance near right-turn horizontal curves during overtaking using LiDAR data." *Can. J. Civ. Eng.* 47 (9): 749–758.

Ma, Y., Y. Zheng, S. Easa, and J. Cheng. 2020b. "Semi-automated framework for generating cycling lane centerlines on roads with roadside barriers from noisy MLS data." *ISPRS J. Photogramm. Remote Sens.* 167: 396–417.

Ma, Y., Y. Zheng, S. Easa, H. Mingyu, and J. Cheng. 2019b. "Automated method for detection of missing road point regions in mobile laser scanning data." *ISPRS Int. J. Geo-Inf.* 8 (12): 525–554.

Mahon, I., S. B. Williams, O. Pizarro, and M. Johnson-Roberson. 2008. "Efficient view-based SLAM using visual loop closures." *IEEE Trans. Rob.* 24 (5): 1002–1014.

Mao, Q., L. Zhang, G. Li, and Q. Hu. 2015. "A least squares collocation method for accuracy improvement of mobile LiDAR systems." *Remote Sens.* 7 (6): 7402–7424.

Metzger, A. T., M. J. Olsen, J. Wartman, L. Dunham, A. Stuedlein, and K. Cunningham. 2013. "A platform for proactive risk-based slope asset management—Phase I." Interim Project Report. Accessed February 2, 2022. http://depts.washington.edu/pactrans/wp-content/uploads/2012/12/PacTrans-2-739439-Metzger-Andrew-Multi-Project.pdf.

Mcquat, G. J. 2011. "Feature extraction workflows for urban mobile-terrestrial LiDAR data." M.S. thesis, Queen's University, Dept. of Geography. http://qspace.library.queensu.ca/bitstream/1974/6530/1/Mcquat_Gregory_J_201105_MSc.pdf.

Montemerlo, M., S. Thrun, D. Koller, and B. Wegbreit. 2002. "FastSLAM: A factored solution to the simultaneous localization and mapping problem." In *Proc., National Conf. on Artificial Intelligence*, 593–598. Palo Alto, CA: Association for the Advancement of Artificial Intelligence.

Montemerlo, M., S. Thrun, D. Koller, and B. Wegbreit. 2003. "FastSLAM 2.0: An improved particle filtering algorithm for simultaneous localization and mapping that provably converges." In *Proc., 18th Int. Joint Conf. on Artificial Intelligence, JCAI'03*, 1151–1156. San Francisco: Morgan Kaufmann.

Morin, K. W. 2002. "Calibration of airborne laser scanners." UCGE Rep. No. 20179." M.S. thesis, University of Calgary, Dept. of Civil Engineering.

Mostafa, M. M. 2002. "Quality control of direct georeferencing data." In *Proc., ISPRS Commission I Mid-Term Symp. in conjunction with Pecora 15/Land Satellite Information IV Conf.* Denver, CO: ISPRS.

NavVis. 2021. "Scalable reality capture." Accessed April 13, 2021. http://www.navvis.com/M6.

Neuhaus, F. 2007. "A full 2D/3D GraphSLAM system for globally consistent mapping based on manifolds." Master's thesis, Universitat Joblenz-Landau, School of Computer Science.

Nolan, J., R. Eckels, M. Evers, R. Singh, and M. J. Olsen. 2015, "Multi-pass approach for mobile terrestrial laser scanning. "*ISPRS Ann. Photogramm. Remote Sens. Spatial Inf. Sci.*, II-3/W5: 105–112. https://doi.org/10.5194/isprsannals-II-3-W5-105-2015.

Nolan, J., R. Eckels, M. J. Olsen, K. S. Yen, T. A. Lasky, and B. Ravani. 2017. "Analysis of the multipass approach for collection and processing of mobile laser scan data." *J. Surv. Eng.* 143 (3): 04017004-1.

Nüchter, A., K. Lingemann, J. Hertzberg, and H. Surmann. 2007. "6D SLAM—3D mapping outdoor environments." *J. Field Rob.* 24 (8–9): 699–722.

Nüchter, A., H. Surmann, K. Lingemann, J. Hertzberg, and S. Thrun. 2004. "6D SLAM with an application in autonomous mine mapping." Accessed April 7, 2021. http://kos.informatik.uni-osnabrueck.de/download/mine/.

O'Banion, M. S., M. J. Olsen, and G. V. Roe. 2013. "Learn mobile lidar." Accessed February 2, 2022. http://learnmobilelidar.com.

Olsen, M. J., G. V. Roe, C. Glennie, F. Persi, M. Reedy, D. Hurwitz, et al. 2013a. *Guidelines for the use of mobile Lidar in transportation applications.* TRB NCHRP Final Rep. 748. http://www.trb.org/Publications/Blurbs/169111.aspx.

Olsen, M. J., G. V. Roe, and J. Raugust. 2013b. *Use of advanced geospatial data tools, technologies, and Information in DOT projects.* NCHRP Synthesis 446, Topic 43-09. Washington, DC: National Academies of Sciences, Engineering, and Medicine.

Olsen, M. J., R. Singh, K. Williams, and A. Chin. 2012. "Transportation applications subchapter." *ASPRS LIDAR Manual.*

Olson, E. B. 2009. "Real-time correlative scan matching." In *Proc., Int. Conf. on Robotics and Automation*, 4387–4393. Piscataway, NY: IEEE.

Otero, R., S. Lagüela, I. Garrido, and P. Arias. 2020. "Mobile indoor mapping technologies: A review." *Autom. Constr.* 120: 103399. doi:10.1016/j.autcon.2020.103399.

Pinto, A. M. G., A. P. Moreira, and P. G. Costa. 2012. "Indoor localization system based on artificial landmarks and monocular vision." *Telkomnika* 10 (4): 609–620.

Pu, S., M. Rutzinger, G. Vosselman, and S. Oude Elberink. 2011. "Recognizing basic structures from mobile laser scanning data for road inventory studies." *J. Photogramm. Remote Sens.* 66: S28–S39.

Puente, I., H. González-Jorge, J. Martínez-Sánchez, and P. Arias. 2013. "Review of mobile mapping and surveying technologies." *Measurement* 46: 2127–2145.

Raquet, J. F. 1998. "Development of a method for kinematic GPS-carrier phase ambiguity resolution using multiple reference receivers." UCGE Rep. No. 20116. Ph.D. thesis, University of Calgary, Dept. of Civil Engineering.

Rastiveis, H., A. Shams, W. A. Sarsua, and J. Li. 2020. "Automated extraction of lane markings from mobile LiDAR point clouds based on fuzzy inference." *ISPRS J. Photogramm. Remote Sens.* 160: 149–166.

Rieger, P., N. Studnicka, M. Pfennigbauer, and G. Zach. 2010. "Boresight alignment method for mobile laser scanning systems." *J. Appl. Geod.* 4 (1): 13–21.

Riveiro, B., L. Díaz-Vilariño, B. Conde-Carnero, M. Soilán, and P. Arias. 2016. "Automatic segmentation and shape-based classification of retro-reflective traffic signs from mobile LiDAR data." *IEEE J. Sel. Top. Appl. Earth Obs. Remote Sens.* 9: 295–303.

Riveiro, B., H. González-Jorge, J. Martínez-Sánchez, L. Díaz-Vilariño, and P. Arias. 2015. "Automatic detection of zebra crossings from mobile LiDAR data." *Optics Laser Tech.* 70: 63–70.

Rodríguez-Cuenca, B., S. García-Cortés, C. Ordóñez, and M. Alonso. 2015. "Automatic detection and classification of pole-like objects in urban point cloud data using an anomaly detection algorithm." *Remote Sens.* 7 (10): 12680–12703.

Rutzinger, M., S. O. Elberink, S. Pu, and G. Vosselman. 2009. "Automatic extraction of vertical walls from mobile and airborne laser scanning data." *Int. Arch. Photogramm. Remote Sens. Spatial Inf. Sci.* 38 (W8): 361–372.

Ryu, K. 2012. "Autonomous robotic strategies for urban search and rescue." Ph.D. thesis, Virginia Polytechnic Institute and State University, Dept. of Mechanical Engineering.

Sairam, N., S. Nagarajan, and S. Ornitz. 2016. "Development of mobile mapping system for 3D road asset inventory." *Sensors* 16: 367.

Searle, J. L., and R. Sridharan. 2014. *Evaluate use of supplied mapping grade lidar for design.* UDOT University Report F-I-80-4(148)148.

Saylam, K. 2009. "Quality assurance of LIDAR Systems–Mission planning." In *Proc., ASPRS Annual Meeting, Baltimore.*

Schön, S., and F. Dilßner. 2007. "Challenges for GNSS-based high precision positioning – some geodetic aspects." In *Proc., 4th Workshop on Positioning, Navigation, and Communication.*

Schwarz, K., M. Chapman, M. Cannon, and P. Gong. 1993. "An integrated INS/GPS approach to the georeferencing of remotely sensed data." *ISPRS J. Photogramm. Remote Sens.* 59 (11): 1667–1674.

Serna, A., and B. Marcotegui. 2014. "Detection, segmentation and classification of 3D urban objects using mathematical morphology and supervised learning." *ISPRS J. Photogramm. Remote Sens.* 93: 243–255.

Shams, A., W. A. Sarasua, A. Famili, W. J. Davis, J. H. Ogle, L. Cassule, et al. 2018. "Highway cross-slope measurement using mobile lidar." *Transp. Res. Rec.* 2672 (39): 88–97.

Singh, R., M. J. Olsen, and G. V. Roe. 2012. "Mobile LiDAR: Theory to practice." In *Proc., ASCE Workshop Course Notes, SPAR 2012.*

Skaloud, J., and D. Lichti. 2006. "Rigorous approach to bore-sight self-calibration in airborne laser scanning." *ISPRS J. Photogramm. Remote Sens.* 61: 47–59.

Smadja, L., J. Ninot, and T. Gavrilovic. 2010. "Road extraction and environment interpretation from LiDAR sensors." In *IAPRS, Vol. XXXVIII, Part 3A*, edited by N. Paparoditis, M. Pierrot-Deseilligny, C. Mallet, O. Tournaire, 281–286. Saint-Mandé, France: International Society for Photogrammetry and Remote Sensing.

Soilán, M., R. Riveiro, J. Martínez-Sánchez, and P. Arias. 2016. "Traffic sign detection in MLS acquired point clouds for geometric and image-based semantic inventory." *ISPRS J. Photogramm. Remote Sens.* 114: 92–101.

Soilán, M., B. Riveiro, J. Martínez-Sánchez, and P. Arias. 2017. "Segmentation and classification of road markings using MLS data." *SPRS J. Photogramm. Remote Sens.* 123: 94–103.

Soilán, M., A. Sánchez-Rodríguez, P. del Río-Barral, C. Perez-Collazo, P. Arias, and B. Riveiro. 2019. "Review of laser scanning technologies and their applications for road and railway infrastructure monitoring." *Infrastructures* 4 (4): 58–87.

Soininen, A. 2012. "Mobile accuracy and control: How much control do we need?" Accessed January 12, 2017. https://www.terrasolid.com/download/presentations/2012/mobile_accuracy_and_control.pdf.

Suarez, J., and R. R. Murphy. 2012. "Using the Kinect for search and rescue robotics." In *Proc., Int. Symp. on Safety, Security, and Rescue Robotics*, 273–274. Washington, DC: IEEE.

Sun, Z., Y. Xu, L. Hoegner, and U. Stilla. 2018. "Classification of MLS point clouds in urban scenes using detrended geometric features from supervoxel-based local contexts." *ISPRS Ann. Photogramm. Remote Sens. Spatial Inf. Sci.* 4 (2): 271–278.

Surmann, H., A. Nüchter, and J. Hertzberg. 2003. "An autonomous mobile robot with a 3D laser range finder for 3D exploration and digitalization of indoor environments." *Rob. Auton. Syst.* 45 (3): 181–198.

Taheri, H., and Z. C. Xia. 2021. "SLAM; definition and evolution." *Eng. Appl. Artif. Intell.* 97: 104032.

Talaya, J., R. Alamus, E. Bosch, A. Serra, W. Kornus, and A. Baron. 2004. "Integration of a terrestrial laser scanner with GPS/IMU orientation sensors." In *Proc., 20th ISPRS Congress, Int. Society of Photogrammetry and Remote Sensing*. Hannover, Germany.

Tan, M., B. Wang, Z. Wu, J. Wang, and G. Pan. 2016. "Weakly supervised metric learning for traffic sign recognition in a LIDAR-equipped vehicle." *IEEE Trans. Intell. Transp. Sys.* 17 (5): 1415–1427.

Teo, T.-A., and C.-M. Chiu. 2015. "Pole-like road object detection from mobile lidar system using a coarse-to-fine approach." *IEEE J. Sel. Top. Appl. Earth Obs. Remote Sens.* 8: 4805–4818.

Teo, T. A., and H. L., Yu. 2015. "Empirical radiometric normalization of road points from terrestrial mobile LiDAR system." *Remote Sens.* 7 (5): 6336–6357.

Thrun, S. 2002. "Probabilistic robotics." *Commun. ACM* 45 (3): 52–57. doi: 10.1145/504729.504754.

Thrun, S., W. Burgard, and D. Fox. 2006. *Probabilistic robotics*. Boston: MIT Press, 647.

Thrun, S., D. Koller, Z. Ghahramani, H. Durrant-Whyte, and A. Y. Ng. 2004. "Simultaneous mapping and localization with sparse extended information filters: Theory and initial results." In *Algorithmic Foundations of Robotics V*, 363–380. Berlin: Springer.

Thrun, S., and M. Montemerlo. 2006. "The graph SLAM algorithm with applications to large-scale mapping of urban structures." *Int. J. Rob. Res.* 25 (5–6): 403–429.

Tsai, Y. J., C. Ai, Z. Wang, and E. Pitts. 2013. "Mobile cross-slope measurement method using Lidar technology." *Transp. Res. Rec.* 2 (2367): 53–59.

Toth, C., E. Paska, and D. A. Grejner-Brzezinska. 2008. "Using road pavement markings as ground control to support the QA/QC of LiDAR data." *Int. Arch. Photogramm. Remote Sens. Spatial Inf. Sci.* 37 (B1): 189–195.

Vaaja, M., J. Hyyppä, A. Kukko, H. Kaartinen, H. Hyyppä, and P. Alho. 2011. "Mapping topography changes and elevation accuracies using a mobile laser scanner." *Remote Sens.* 3 (3): 587–600. doi:10.3390/rs3030587

Vasquez, R. 2012. "Endeavour moves through the cloud." *Lidarnews*. November 2, 2012.

Viametris. 2021. "Mobile mapping solutions for all measurements professionals." Accessed April 7, 2021. https://viametris.com/.

Vincent, R., and M. Ecker. 2010. *Light detection and ranging (LIDAR) technology evaluation*. Missouri Dept. of Transportation. https://rosap.ntl.bts.gov/view/dot/18379.

Vosselman, G., and H. G. Maas. 2010. *Airborne and terrestrial laser scanning*. Scotland, UK: Whittles. Print.

Vu, A., Q. Yang, J. A. Farrell, and M. Barth. 2013. "Traffic sign detection, state estimation, and identification using onboard sensors." In *Proc., 16th Int. IEEE Conf. on Intelligent Transportation Systems-(ITSC)*, October 6–9, 875–880. The Hague: Netherlands.

Wan, E. A., and R. van der Merwe. 2000. "The unscented Kalman filter for nonlinear estimation." In *Proc., Adaptive Systems for Signal Processing, Communications, and Control Symp.*, 153–158. Washington, DC: IEEE.

Wang, H., Z. Cai, H. Luo, C. Wang, P. Li, W. Yang, et al. 2012. "Automatic road extraction from mobile laser scanning data." In *Proc., Int. Conf. on Computer Vision in Remote Sensing*, 136–139. Washington, DC: IEEE.

Wang, J., R. Lindenbergh, and M. Menenti. 2017. "SigVox—A 3D feature matching algorithm for automatic street object recognition in mobile laser scanning point clouds." *ISPRS J. Photogramm. Remote Sens.* 128: 111–129.

Wang, H., H. Luo, C. Wen, J. Cheng, P. Li, Y. Chen, et al. 2015. "Road boundaries detection based on local normal saliency from mobile laser scanning data." *IEEE Geoscience and Remote Sensing Lett.* 12 (10): 2085–2089.

Wang, R., J. Peethambaran, and D. Chen. 2018. "LiDAR point clouds to 3-D urban models: A review." *IEEE J. Sel. Top. Appl. Earth Obs. Remote Sens* 11: 606–627.

Wang, W., W. Zhao, L. Huang, V. Vimarlund, and Z. Wang. 2014. "Applications of terrestrial laser scanning for tunnels: A review." *J. Traffic Transp. Eng.* 1 (5): 325–337.

Weingarten, J. 2006. "Feature-based 3D SLAM." Ph.D. thesis, Swiss Federal Institute of Technology, School of Engineering, Lausanne. doi:10.5075/epfl-thesis-3601.

Weinmann, M., B. Jutzi, S. Hinz, and C. Mallet. 2015. "Semantic point cloud interpretation based on optimal neighborhoods, relevant features and efficient classifiers." *ISPRS J. Photogramm. Remote Sens.* 105: 286–304.

Welch, G., and G. Bishop. 1995. *An introduction to the Kalman filter*. UNC-CH Computer Science Tech. Rep. 95041. Chapel Hill, NC: University of North Carolina.

Wen, C., J. Li, H. Luo, Y. Yu, Z. Cai, H. Wang, et al. 2016. "Spatial-related traffic sign inspection for inventory purposes using mobile laser scanning data." *IEEE Trans. Intell. Transp. Syst.* 17: 27–37.

Wen, C., X. Sun, J. Li, C. Wang, Y. Guo, and A. Habib. 2019. "A deep learning framework for road marking extraction, classification and completion from mobile laser scanning point clouds." *ISPRS J. Photogramm. Remote Sens.* 147: 178–192.

Williams, K., M. J. Olsen, G. V. Roe, and C. Glennie. 2013. "Synthesis of transportation applications of mobile LIDAR. *Remote Sens.* 5 (9): 4652–4692. https://doi.org/10.3390/rs5094652.

Wu, C. T., C. Y. Hasiao, and C. S. Chen. 2013. "Improvement of LiDAR data accuracy using 12-parameter affine transformation." *J. Chin. Soil Water Conserv.* 44 (4): 293–301.

Wu, B., B. Yu, C. Huang, Q. Wu, and J. Wu. 2016. "Automated extraction of ground surface along urban roads from mobile laser scanning point clouds." *Remote Sens. Lett.* 7 (2): 170–179.

Wu, S., C. Wen, H. Luo, Y. Chen, C. Wang, and J. Li. 2015. "Using mobile LiDAR point clouds for traffic sign detection and sign visibility estimation." In *Proc., IEEE International Geoscience and Remote Sensing Symposium (IGARSS)*, July 26–31, 565–568. Milan, Italy.

Xiang, B., J. Yao, X. Lu, L. Li, R. Xie, and J. Li. 2018. "Segmentation-based classification for 3D point clouds in the road environment." *Int. J. Remote Sens.* 39 (19): 6182–6212.

Yan, L., H. Liu, J. Tan, Z. Li, H. Xie, and C. Chen. 2016. "Scan line based road marking extraction from mobile LiDAR point clouds." *Sensors* 16: 903.

Yang, B., and Z. Dong. 2013. "A shape-based segmentation method for mobile laser scanning point clouds." *ISPRS J. Photogramm. Remote Sens.* 81: 19–30.

Yang, B., Z. Dong, Y. Liu, F. Liang, and Y. Wang. 2017a. "Computing multiple aggregation levels and contextual features for road facilities recognition using mobile laser scanning data." *ISPRS J. Photogramm. Remote Sens.* 126: 180–194.

Yang, B., Y. Liu, Z. Dong, F. Liang, B. Li, and X. Peng. 2017b. "3D local feature BKD to extract road information from mobile laser scanning point clouds." *ISPRS J. Photogramm. Remote Sens.* 130: 329–343.

Yang, B., Z. Dong, G. Zhao, and W. Dai. 2015. "Hierarchical extraction of urban objects from mobile laser scanning data." *ISPRS J. Photogramm. Remote Sens.* 99: 45–57.

Yang, B. S., L. N. Fang, Q. Q. Li, and J. Li. 2012. "Automated extraction of road markings from mobile lidar point clouds." *Photogramm. Eng. Remote Sens.* 78 (4): 331–338.

Yao, Y., and Q. Hu. 2014. "Automatic extraction method study of road marking lines based on projection of point clouds. In *Proc., 22nd Int. Conf. on Geoinformatics (GeoInformatics)*, July 25–27, 1–4. IEEE, Kaohsiung, Taiwan.

Yen, K. S., K. Akin, A. Lofton, B. Ravani, and T. A. Lasky. 2010. *Using mobile laser scanning to produce digital terrain models of pavement surfaces*. Davis, CA: Advanced Highway Maintenance and Construction Technology Research Center.

Yen, K. S., B. Ravani, and T. A. Lasky. 2011. *LIDAR for data efficiency*. Davis, CA: Advanced Highway Maintenance and Construction Technology Research Center.

Yen, K. S., B. Ravani, and T. A. Lasky. 2014. *Mobile terrestrial laser scanning workflow development, technical support and evaluation*. AHMCT Research Report: UCD-ARR-14-06-14-01. Davis, CA: Advanced Highway Maintenance and Construction Technology Research Center.

Yen, K. S., S. Donecker, B. Ravani, and T. A. Lasky. 2018. *Expanding mobile terrestrial laser scanning capability and capacity throughout Caltrans*. AHMCT Research Report: UCD-ARR-18-02-14-01. Davis, CA: Advanced Highway Maintenance and Construction Technology Research Center.

Yoo, H. J., F. Goulette, J. Senpauroca, and G. Lepere. 2010. "Analysis and improvement of laser terrestrial mobile mapping systems configurations." *Int. Arch. Photogramm. Remote Sen. Spat. Inf. Sci.* 38 (5): 633–638.

Yousif, H., J. Li, M. Chapman, and Y. Shu. 2010. "Accuracy enhancement of terrestrial mobile LiDAR data using theory of assimilation." *Int. Arch. Photogramm. Remote Sens. Spat. Inf. Sci.* 38 (5): 639–645.

Yu, Y., J. Li, H. Guan, F. Jia, and C. Wang. 2015. "Learning hierarchical features for automated extraction of road markings from 3-D mobile LiDAR point clouds." *IEEE J. Sel. Top. Appl. Earth Obs. Remote Sens.* 8: 709–726.

Zai, D., J. Li, Y. Guo, M. Cheng, Y. Lin, H. Luo, et al. 2018. "3-D road boundary extraction from mobile laser scanning data via supervoxels and graph cuts." *IEEE Trans. Intell. Transp. Syst.* 19: 802–813.

Zhang, H., J. Li, M. Cheng, and C. Wang. 2016. "Rapid inspection of pavement markings using mobile lidar point clouds." *Int. Arch. Photogramm. Remote Sens. Spatial Inform. Sci.* XLI-BI, 717–723: 41.

Zhang, L., Q. Mao, Q. Li, and P. Zhang. 2014. "An accuracy-improvement method for GPS/INS kinematic levelling for use in linear engineering surveying projects." *Measurement* 54: 22–30.

Zhou, L., and Z. Deng. 2014. "LIDAR and vision-based real-time traffic sign detection and recognition algorithm for intelligent vehicle." In *Proc., IEEE 17th Int. Conf. on Intelligent Transportation Systems (ITSC)*, October 8–11, 578–583. Qingdao, China.

CHAPTER 10
AERIAL SURVEYING TECHNOLOGY

Michael J. Starek, Benjamin E. Wilkinson

AERIAL PHOTOGRAMMETRY

INTRODUCTION

Photogrammetry has been defined as "the art, science and technology of obtaining reliable information about physical objects and the environment through the process of recording, measuring, and interpreting photographic images and patterns of EM radiant energy and other phenomena" (Slama et al. 1980). Although many products can be derived from it, the development of *topographic maps* is the primary objective of aerial photogrammetry. Topographic maps illustrate both planimetric location of ground features and a depiction of terrain elevations. Photogrammetry is used extensively by federal agencies such as the United States Geological Survey, state agencies such as departments of transportation, and private surveying and engineering companies to develop these maps at various scales and for a variety of purposes (Wolf et al. 2014, Mikhail et al. 2001). The advent of the digital age has brought two more essential photogrammetric products: *digital elevation models* (DEMs) and *orthophotos*. DEMs are typically digital arrays where each cell conveys horizontal (established by the row and column location of the cell) and vertical (established by the digital number associated with the cell) information about a scene. DEMs enable the generation of orthophotos, which are digital photos corrected for the effects of camera tilt and terrain variations such that accurate horizontal measurements can be made on them. DEMs and orthophotos also have a wide range of applications and are used extensively within geographic information systems (GISs). They are also often coupled with surveyed or designed features (i.e., linework and symbols) to serve as base maps for engineering projects.

The development of the principles of photogrammetry transpired long before the invention of the camera. For example, the principles of optical projection were described by Aristotle, Leonardo da Vinci graphically portrayed them in 1492, and a mechanical device for creating perspective drawings (including stereoscopic views) was constructed by German artist Albrecht Durer in 1525. A year following the announcement of the Daguerreotype photographic process in 1839, the use of photography for topographic mapping was demonstrated by French Geodesist Fracois Arago. Although kites and balloons were used for aerial photography in the late 1800s by French Army Corps of Engineers Colonel Aimé Laussedat, the "father of photogrammetry," most photogrammetric work was based on terrestrial photos until the invention of the airplane in 1903. During World War I, aerial photos were widely used for reconnaissance, and World War II spurred advancements in both photogrammetric hardware and processing techniques because of the need for rapid production of topographic information over large areas (Walker and Alspaugh 2013, Wolf et al. 2014).

Advancements have continued over past decades, with important developments made regularly and introduced into modern photogrammetric collection and processing workflows. Furthermore, several advancements originating from the computer vision community related to speeding up and automating many procedures have been integrated into photogrammetric processes (Granshaw and Fraser 2015). Photogrammetry is a rich and complex discipline. In this chapter, only the fundamental concepts are described, and the reader is encouraged to consult the references listed for additional details. Particularly, the *Manual of Photogrammetry* (McGlone 2013) published by the American Society for Photogrammetry and Remote Sensing is a most comprehensive source covering in detail all the essential topics of photogrammetry.

CAMERAS

Cameras consist of three basic components: the lens or lens assemblage used to focus nonconverging light rays reflected from object points; the aperture, an opening to allow light rays to enter the interior of the camera body; and media used to record where the focused light rays strike the focal plane of the camera. Primitive cameras did not have lenses but relied on having a very small aperture, hence the term *pinhole camera*, to prevent highly divergent light rays from object points being recorded and, thus, resulting in sharp images. However, pinhole cameras allow only a small amount of light to enter, leading to long exposure times for recording the image. The introduction of lenses allowed for a larger aperture and, therefore, more light entering the enclosure and faster exposure times. Lenses are characterized by their *focal length*, which is defined as the distance from the lens to the focal plane for which the image of an object point at infinite distance from the camera is brought into focus. The aperture is characterized by its *f-stop*, the ratio of the focal length to the diameter of the opening. The larger the *f*-stop, the smaller the aperture. Smaller apertures lead to a larger *depth of field*, range in distances from object to camera without decreased sharpness, but also require the aperture to be open longer for sufficient exposure, which leads to blurring if the camera or object is moving relative to each other. Aerial photography is not significantly affected by the depth of field, however, because object distance variations are small relative to the mean object distance. Film, consisting of a supporting material coated with light-sensitive emulsions, was once widely used as a recording media for aerial mapping photography, with the exposed and processed film photographs often scanned to produce digital images. However, several large-format digital mapping cameras are available, which use arrays of light-sensitive solid-state sensors, typically *charge-coupled devices* or *complementary metal-oxide semiconductors*, in place of film. An important parameter of a camera system is the angle measured at the lens of the camera covering the extent of the imaged scene, or the *angular field of view* (AFOV). A function of the focal length and the dimensions, or *format*, of the effective focal plane, AFOV, α, can be calculated by $\alpha = 2\tan^{-1}(d/2f)$, where d is the distance measured across the format (of the film or digital sensor array). Most digital cameras have nonsquare formats, and, thus, the AFOV is specified for both the along-track (in the direction of flight) and the across-track (perpendicular to the direction of flight) dimensions. Note that d and f must be in the same units in the equation given. Thus, for digital cameras, one must know about either the physical dimensions of the sensor array or physical distance between the sensor-array elements, or the *pitch* of pixels, for conversion. Usually, the pitch is only a few micrometers. Figure 10-1 shows a large format digital mapping camera, Leica DMC III, which has a 391 mp (26,112 × 15,000) CMOS sensor, 92 mm focal length, 3.9 μm pitch, and a nominal 57 degree across-track AFOV.

Figure 10-1. Leica DMC III airborne digital camera.
Source: Courtesy of Leica Geosystems, reproduced with permission.

Cameras can capture panchromatic or color photography. Panchromatic photography results from using recording media sensitive to all visible light usually leading to black-and-white photographs. Color photography results from recording separate images associated with typically red, green, and blue light, which are combined to create a single-color image. When using film, colors are captured separately via a stack of specific color-sensitive emulsions and filters. In digital cameras, one approach is to use a color filter array (CFA), with the most widely used being the Bayer filter. CFAs allow only certain colors (usually red, green, and blue) to be transmitted to associated sensors in the sensor array, with colors alternated along rows and columns of the pixels. To produce a color image, CFA *demosaicing* is applied to interpolate the raw data, resulting in red, green, and blue channels for each pixel in the image. Avoiding the degradation introduced by demosaicing, large-format digital mapping cameras often simultaneously collect high-resolution panchromatic images and low-resolution single-color images using separate CCD or CMOS arrays that can be combined to produce a single multiple-channel image. The principles described here can be extended to generating images with more than three channels, and furthermore, light outside of the visible spectrum can be collected. Often, mapping cameras are capable of collecting near-infrared (NIR) imagery, which is particularly beneficial when mapping vegetation because of its high reflectivity of NIR light. *Multispectral cameras* collect imagery with channels for specific wavelengths to facilitate the identification of spectrally discriminable materials. *Hyperspectral cameras* collect many channels (often hundreds) of adjacent spectral bands to capture a large portion of the spectrum of emitted or reflected electromagnetic (EM) energy. These types of sensors allow for the identification of specific objects and materials based on their "spectral signature." The field of *remote sensing* focuses on the detection and classification of specific objects. It is a discipline so closely linked to photogrammetry that professional societies traditionally combine the two (e.g., the American Society for Photogrammetry and Remote Sensing and the International Society for Photogrammetry and Remote Sensing) (Green 2008). However, the distinction should be made: photogrammetry focuses on the geometric aspects of aerial imaging. In other words, it is more concerned with exactly where things are than exactly what they are.

Camera Distortion and Calibration

For simplicity, many of the basic mathematical concepts of photogrammetry assume pinhole camera geometry. That is, that recorded light rays travel in a straight line from object points to the focal plane. However, theory dictates that when using lenses, distortion (deviation from the pinhole model) always exists in images, and in practice, other distortions occur because of imperfections in camera manufacturing. The effects of this distortion can be modeled and eliminated from the data via calibration allowing for precision measurements. Further, cameras produced specifically for photogrammetry are *metric cameras*, cameras designed so that distortion parameters are minimal and stable over time. Accompanying calibration reports contain the numerical values of calibration parameters for a given camera. Table 10-1 lists the essential camera calibration parameters. Although additional distortion parameters can be used to model effects theoretically inherent in digital camera photography, these may rarely be significant (Fraser 2013).

The *calibrated focal length* is the distance from the rear nodal point of the lens to the focal plane of the camera. Note that this definition is different from that given for the property of a lens; thus, the term *principal distance* more appropriately describes the parameter. The *principal point coordinates* designate the location where a line from the rear nodal point of the lens intersects the focal plane perpendicularly. The principal point is usually very close to the center of the photograph, and when accounting for lens distortion, the principal point often serves as the origin for image coordinate measurements for convenience. *Symmetric radial lens distortion coefficients* describe lens distortion occurring radially from the principal point. This distortion can warp images away from and/or toward the principal point depending on the radial distance from it, and it is always present when using lenses. *Decentering distortion coefficients* describe the distortion caused by irregularities and misalignment of the lens components (Wolf et al. 2014). The correction for the effect of these parameters, $(\partial x, \partial y)$, is shown in Equation (10-1), where (x, y) are the distorted image coordinates, $(\bar{x}, \bar{y}) = (x - x_0, y - y_0)$, and $r^2 = \bar{x}^2 + \bar{y}^2$.

$$\partial x = -x_0 + \bar{x}(k_1 r^2 + k_2 r^4 + k_3 r^6) + (1 + p_3 r^2)\left[p_1(2\bar{x} + r^2) + 2p_2\bar{x}\bar{y}\right] \tag{10-1}$$

$$\partial y = -y_0 + \bar{y}(k_1 r^2 + k_2 r^4 + k_3 r^6) + (1 + p_3 r^2)\left[p_2(2\bar{y} + r^2) + 2p_1\bar{x}\bar{y}\right]$$

Lens distortion parameters for a particular camera can be determined either by using laboratory instruments designed to directly measure the various parameters or through field methods. For nonmetric cameras, it is recommended to perform the in situ method of *analytical self-calibration*. In analytical self-calibration, the distortion parameters are included

Table 10-1. Camera Calibration Parameters.

Parameter	Symbology
Calibrated focal length (principal distance)	f
Principal point coordinates	x_0, y_0
Symmetric radial lens distortion coefficients	k_1, k_2, k_3
Decentering distortion coefficients	p_1, p_2, p_3

as unknowns in analytical aerotriangulation, or bundle adjustment, which is described in a subsequent section. Regardless of the method employed, it is critical to compensate for camera distortion because the resulting errors in image measurements propagate to the photogrammetric products and can significantly reduce their quality.

BASIC PRINCIPLES OF AERIAL PHOTOGRAMMETRY

Aerial photographs can be classified by their orientation with respect to a mapping datum. *Vertical photos* are those that are pointed nominally downward, and *oblique photos* are those purposely pointed away from vertical. Vertical photography is typically used in topographic mapping, with oblique photography typically being used for specialized applications such as generating 3D textured models of urban areas. In practice, perfectly vertical photography is not possible, although with stable flight and specialized camera mounts, deviations are usually less than 1 degree. For planning purposes and lower-accuracy applications, perfectly vertical photography can be assumed for simplicity, enabling an assortment of calculations.

The *scale* of a photograph is the ratio of a distance measured in the image (ab) to the distance measured in object space (AB). For vertical photography over flat, level terrain, the scale is given in Equation (10-2), where H' is the height of the camera above ground.

$$s = ab/AB = f/H' \tag{10-2}$$

Note that the units of ab and f are the same, and, similarly, the units of AB and H' are the same. Photo scale allows the calculation of ground distances based on measurements in the vertical photo. The dimension of the area on the ground captured in vertical aerial photography is the *ground coverage*, G, and can be computed for vertical photography over flat, level terrain by $G = d/s = dH'/f = 2H'\tan(\alpha/2)$, where d is the format dimension of the camera. For nonsquare formats, this can be calculated for both the height and the width components of the image footprint projected to the ground. As the tilt of the camera deviates from vertical and/or the terrain deviates from flat and level, the scale and ground coverage are not uniform. For a specific point with elevation h above datum, the corresponding scale in the image is $s = f/(H-h)$, where H is the camera height above datum. Similarly, the average photo scale is $s_{avg} = f/(H-h_{avg})$, where h_{avg} is the average height above datum for all locations imaged in the photo.

In digital imagery, the object space distance covered by a single pixel is referred to as the *ground sample distance* (GSD). For example, if a 100 m-long soccer field is imaged in a vertical digital aerial photo and is measured in the photograph as 500 pixels from end to end, then the corresponding nominal GSD is $100\text{m}/500\text{pixels} = 0.2\text{m}$. Although it is tempting to quantify the resolution of a camera in terms of its GSD, the resolving power of a camera is the result of all its constituent parts (along with environmental effects) and can be better conveyed using the *modulation transfer function*, a measure of the ability of the camera to effectively capture the spatial frequencies of objects, with the upper limit governed by the Nyquist rate determined by the spacing of the photosensitive elements in the sensor array (Honkavaara et al. 2006).

When objects have different elevations, their locations in vertical aerial photos are shifted relative to one another. This effect can be readily observed in the appearance of buildings or other tall structures in vertical aerial photos: The top of buildings appears to be displaced radially outward from the center of the photo relative to the base. This phenomenon is termed *relief displacement* and always exists if terrain undulations or vertical features exist

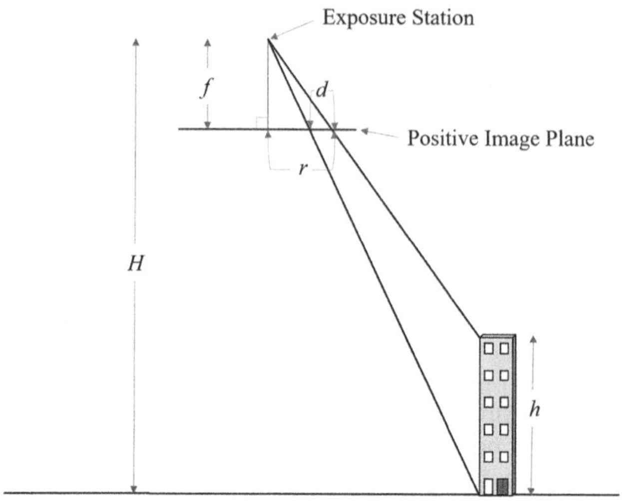

Figure 10-2. Relief displacement of a vertical photograph of a building.

in the photography. Relief displacement prevents accurate planimetric measurements to be made on raw aerial imagery, necessitating the production of orthophotos. However, relief displacement can be exploited to obtain information about the scene being photographed, particularly the height of features. If we consider relief displacement, d, to be the photo distance from the base of a feature at datum to the top of the feature, then the height of the feature above datum, h, is given by Equation (10-3), where r is the radial distance measured in the photo from the principal point to the top of the feature (Figure 10-2).

$$h = dH/r \tag{10-3}$$

Note that in Equation (10-3), datum can be artificially located at the base of the feature. Then, one can replace d with d', the distance from the base of the feature to the top of the feature in the photo, and H with the flying height above the base of the feature, H', and the resulting h will be the height of the feature relative to its base.

The main method humans and many animals use for perceiving depth is stereoscopic vision. When eyes are directed at an object, the lines-of-site, or *optical axes*, of the eyes intersect. The angle between the optical axes at the object is the *parallactic angle*. The smaller the parallactic angle, the farther away the object. The eyes and associated neurological mechanisms can fuse similar images automatically, and the parallactic angle, θ, formed when the image of an object is fused (along with other cues) generates a perception of the distance of the object from the perceiver. The concepts from stereoscopic vision can be applied to *stereo photography*, photos taken with overlap in the imaged scenes. Two photos taken from different vantages with overlap in coverage are called *stereopairs*. The image coordinates of a feature in two vertical aerial photos taken from separate positions will vary in the direction of the offset between *exposure stations*, locations where the photos were taken. The direction is aligned with the direction of flight, often nominally parallel with the x-axis in the photo coordinate system. This discrepancy in the image location of conjugate features is termed *parallax* and is a function of the height of the feature, with taller features resulting in more parallax. Elementary calculations can be made using the principles of parallax, under the assumption of vertical, aerial stereo photography. If (x_1, y_1) and (x_2, y_2) are the image coordinates of some feature in Stereo photos 1 and 2, respectively, where x is parallel with the direction of flight, B is the *airbase*, which is the distance between the exposure stations of two vertical aerial photos, and $p = x_1 - x_2$ is the parallax between two image locations of the feature, then the 3D position of the feature relative to the nadir projection of the principal point of Photo 1 at datum is shown in Equation (10-4). Figure 10-3 illustrates the stereoscopic parallax.

$$X = Bx_1/p$$

$$Y = By_1/p \tag{10-4}$$

$$Z = H - Bf/p$$

From Equation (10-4), the 3D position of multiple features can be calculated, forming a model of the scene in the stereo overlap area and allowing for further dimensional calculations such as distance, area, and volume. The difference in height between two objects in a stereo overlap area can be calculated using the difference in the parallax between the two features. Equation (10-5) gives the equation for height difference by difference in the parallax, where h_A and h_C are the heights of objects A and C, respectively; and $\Delta p = p_A - p_B$ is the parallax associated with object A minus the parallax associated with object C. Note the convenience of Equation (10-5) when h_C is on the ground and assumed to be zero, allowing the determination of feature A height above ground. A comparison of Equation (10-5) with Equation (10-4) shows that using differences in the parallax also eliminates the need for a knowledge of B and f, which can be advantageous.

$$h_A - h_C = \Delta p(H - h_C)/p_A \tag{10-5}$$

Several mechanical/optical devices have been developed and used historically for precise measurement of parallax in stereopairs, enabling 3D mapping of features sufficient for many applications. However, errors in photo coordinate and parallax measurements, parallax direction inconsistencies relative to photo coordinate systems, flying height estimates, tilt, camera distortion, and so on make these methods insufficient for high-precision photogrammetric mapping. For the best results, *analytical photogrammetry* should be employed.

ANALYTICAL PHOTOGRAMMETRY

Analytical photogrammetry can be considered a suite of methods to calculate the 3D coordinates of points from photography using rigorous mathematical models and usually the solution of large redundant systems of equations using *least squares*. Least-squares estimation is a rich and complex subject, and several texts such as Ghilani (2010) and Mikhail and Ackermann (1976) serve as excellent references on the subject (in addition, see Chapter 5 for more on least squares). Instead of attempting to cover least squares here, the focus is on general methods and models.

Although others exist, the most popular mathematical model used for analytical photogrammetry is the *collinearity condition*, which reflects the property that a light ray from an object traces a straight line from the object, through lens, and to the image (at both the positive and the negative image plane locations), as shown in Figure 10-4. This condition can be assumed to hold if distortions, such as lens distortion and atmospheric refraction, have been appropriately handled. Note that only the positive image plane is shown in

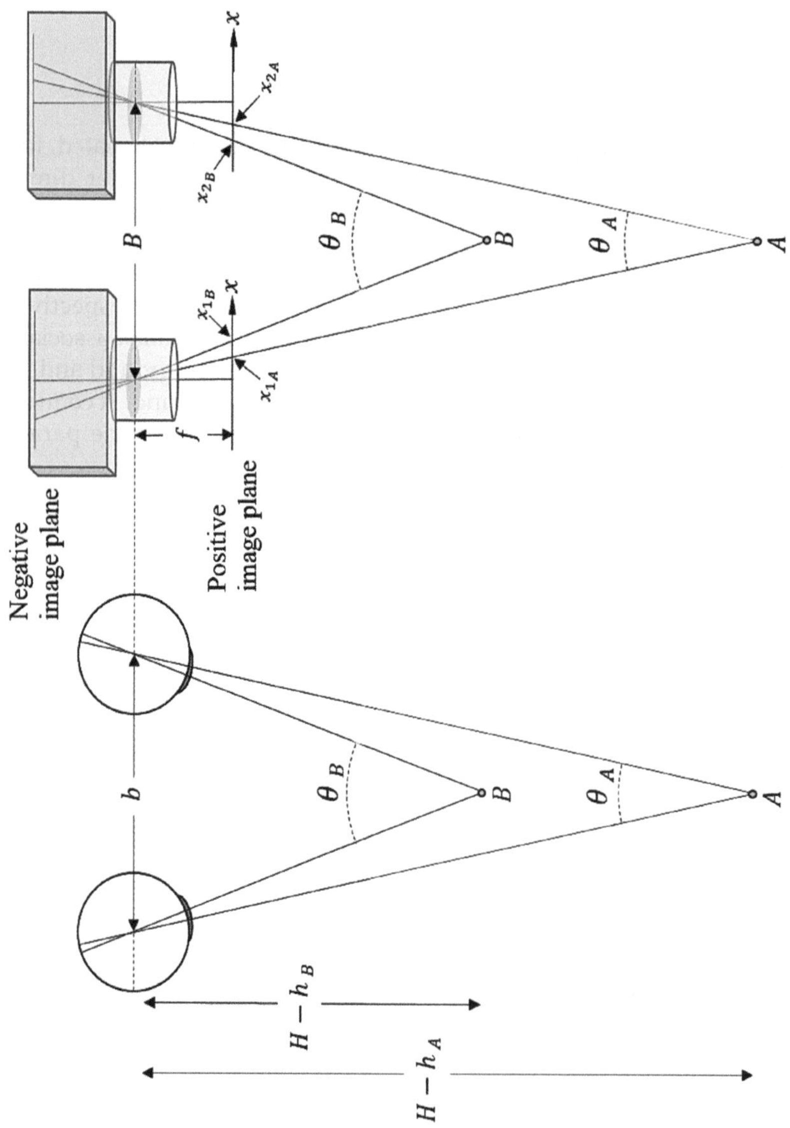

Figure 10-3. (Left) Parallax measurements for eyes, (Right) stereo photography.
Note: $\theta_A < \theta_B$ *and* $(x_{1A} - x_{2A}) = p_A < (x_{1B} - x_{2B}) = p_B$. *Note that b, the distance between eyes, is often referred to as the eye base.*

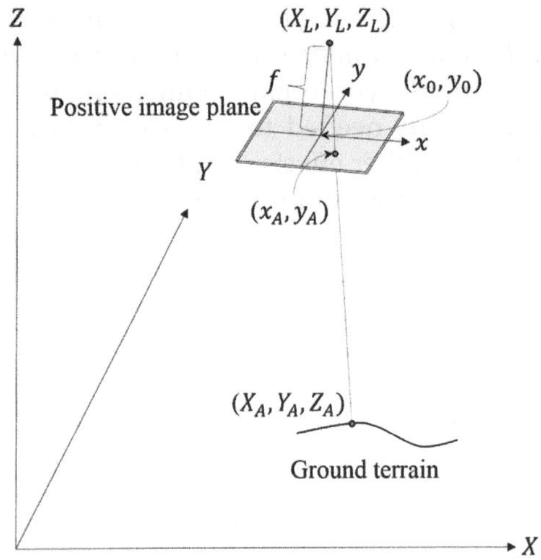

Figure 10-4. Collinearity condition.

Figure 10-4. Although the negative image plane is where the digital array or film is physically located, illustrating the positive image plane, located at a distance *f* below the lens in the camera coordinate system, is advantageous in graphical depictions of collinearity and other relationships. The mathematical relationships are the same for either representation. Equation (10-6) shows the collinearity equations,

$$x_A = -f\frac{m_{11}(X_A - X_L) + m_{12}(Y_A - Y_L) + m_{13}(Z_A - Z_L)}{m_{31}(X_A - X_L) + m_{32}(Y_A - Y_L) + m_{33}(Z_A - Z_L)} \tag{10-6}$$

$$y_A = -f\frac{m_{21}(X_A - X_L) + m_{22}(Y_A - Y_L) + m_{23}(Z_A - Z_L)}{m_{31}(X_A - X_L) + m_{32}(Y_A - Y_L) + m_{33}(Z_A - Z_L)}$$

where

$$M = \begin{bmatrix} m_{11} & m_{12} & m_{13} \\ m_{21} & m_{22} & m_{23} \\ m_{31} & m_{32} & m_{33} \end{bmatrix}$$

is the 3D rotation matrix from the object space or mapping coordinate system to the camera (x, y, f) coordinate system. M is conventionally defined by sequential rotation angles ω (about the X-axis), ϕ (about the once-rotated Y-axis), and κ (about the twice rotated Z-axis) in photogrammetry. The coordinates (x_A, y_A) are the image coordinates of point A, (X_L, Y_L, Z_L) are the object space coordinates of the camera lens, and (X_A, Y_A, Z_A) are the object space or mapping coordinate system coordinates of point A. Note that (x_A, y_A) are assumed to have been *refined*, corrected for camera distortion via the model in Equation (10 1).

An *analytical stereomodel* is a numerical model of the overlap area of a stereopair of photographs that enables the calculation of 3D point positions. Stereomodels can be used for a number of different purposes, including point positioning (control densification), contour map generation, DEM generation, and orthophoto production. Forming a stereomodel requires three steps, all analytical photogrammetric procedures: *analytical interior orientation*, *analytical relative orientation*, and *analytical absolute orientation*. Analytical interior orientation enables photo coordinate refinement by establishing the image coordinate system and coefficients needed to apply camera distortion corrections via Equation (10-1) and to correct for other effects manifesting in image coordinate deviation from the collinearity condition such as atmospheric refraction. Relative orientation is the establishment of the translation and angular orientation offsets between the exposure stations of the stereo pair, using *pass point* observations, the image coordinates (in both images) of object-space features found in the overlap area of the stereopair. As the name implies, this step establishes only the relative orientation between photos, and because no control is used in the step, there is an ambiguity in the stereomodel scale. This ambiguity requires constraining an unknown component of the relative orientation parameters, typically the airbase, B. The remaining unknown parameters, three rotations and two translations, are solved via least squares using more than five tie points, each contributing a pair of collinearity condition observation equations for each photo. The result of relative orientation is a stereomodel in an arbitrary coordinate system. Transformation from the arbitrary coordinate system to a mapping coordinate system is achieved in absolute orientation. At least three 3D ground control points (GCPs) must be located in the stereomodel to solve for seven 3D conformal coordinate transformation parameters using least squares. Once the transformation parameters are found, all points in the stereomodel (including additional points not used in orientation) can be transformed to the mapping coordinate system, allowing accurate 3D measurements, feature digitization, and map compilation.

For achieving the highest overall accuracy, a *bundle adjustment* is used to simultaneously perform relative and absolute orientation for a block of photos using a single least-squares adjustment. Bundle adjustment (also known as bundle block adjustment) is extremely flexible. Pass points between multiple photos along flight lines, *tie points* (conjugate points between photos across adjacent flight lines) and ground control points, and points with presurveyed (X_A, Y_A, Z_A) coordinates serve as observations to solve for all camera stations' *exterior orientation parameters* (EOPs), comprising the absolute 3D positions (X_L, Y_L, Z_L) and 3D rotation angles (ω, ϕ, κ) of the camera at the times of exposure, and 3D ground coordinates of all pass points. In appropriate circumstances, *analytical self-calibration* can be used to additionally estimate the interior orientation parameters by including calibration parameters as unknowns in the bundle adjustment. Analytical photogrammetric products have a range of accuracies depending on the methods and hardware used. Under typical conditions, ground point coordinate accuracies are 1/15,000 of the flying height for the horizontal components and 1/10,000 for the vertical components. Under special circumstances, a much higher accuracy is achievable.

Stereoplotters

Stereoplotters are devices used for 3D photogrammetric mapping. Historically, stereoplotters consisted of large optical–mechanical devices allowing operators to view and compile maps in three dimensions by physically simulating the geometry of the camera stations at the times of exposure using projectors and other optical devices. Nowadays, photogrammetrists use *softcopy stereoplotters* for analytical photogrammetric operations and 3D map compilation. Softcopy stereoplotters, such as those shown in Figure 10-5, consist of specialized software and 3D-vision-capable computers. Multiple devices are available for 3D vision in computers. One of the most common uses shutter glasses worn by the operator. The computer screen alternates displaying two digital images from a stereopair and LCD shutter glasses worn by the operator alternate opacity between the left and the

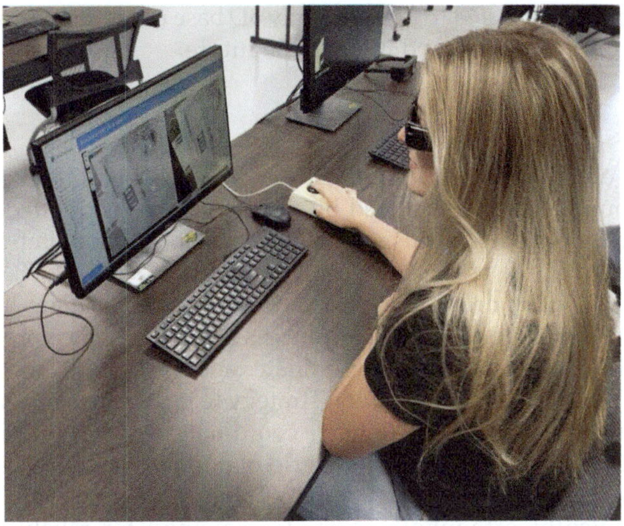

Figure 10-5. Softcopy stereoplotter.

right eyepieces. When the "right" image is displayed on screen, the right eyepiece is transparent and the left eye piece is opaque, and when the "left" image is displayed on screen, the left eyepiece is transparent and the right eyepiece is opaque. The shutter and screen synchronously alternate between left and right up 120 times per second, giving the operator a 3D visual perception of the stereomodel. Rows of pixels in each image must be aligned with one other for this process to work; thus, *epipolar resampling* is applied to the stereopair images based on the relative orientation of the photos. The operator can navigate in the stereomodel horizontally and vertically with a floating mark allowing 3D drawing of lines, symbols, and other features.

Digital Elevation Models

As previously mentioned, collinearity equations can be used to solve for the object space coordinates of conjugate image points in an oriented stereopair of photographs. Doing this for many points over a large area results in a 3D *point cloud*, which can be converted and stored as a DEM. There are two types of DEMs: digital surface models (DSMs), which encompass all features in the images; and digital terrain models (DTMs), which include only the ground terrain. DEMs may be stored in a number of ways, with the most common being stored as a regular grid of elevations or as a triangulated irregular network (TIN). Storing as a regular grid has the advantage that the DEM can be represented as a digital image with each pixel assigned an elevation, and image processing algorithms can easily be implemented on them. Storing as a TIN has the advantage that areas with insignificant relief variations can be stored more efficiently because only a few nodes are needed to effectively represent them.

DTMs are frequently coupled with surveyed or designed features (i.e., linework and symbols) to form 2.5D base maps. The horizontal position of the features is drawn on the surface of the DTM, but the feature itself is not modeled in 3D. For example, a pole may be depicted as a flat symbol placed at its horizontal position on the surface of the DTM. Many departments of transportation have been developing these types of 2.5D base maps for engineering projects over the last three decades. They are used extensively for road design, earthwork calculations, construction staking, machine control, and more. Some pilot studies have also been conducted to develop fully 3D base maps for design, and these types of base maps will most likely be used more in the future.

Orthophotos

An orthophoto can be formed using a DEM and an oriented photo covering the same area. First, the ground area to be covered by the orthophoto is divided into a horizontal normal grid of groundels, square cells with some constant distance between their centers. Then, for each groundel, the associated ground coordinates (X_A, Y_A, Z_A) provided by the DEM, and the EOPs of the oriented photo can be used to calculate image coordinates (x_A, y_A) using Equation (10-6). The digital numbers (brightness or colors) in the image can be queried using these calculated image coordinates and stored in the respective groundels. This step involves some sort of interpolation strategy (e.g., the nearest neighbor, bilinear or bicubic interpolation), as the calculated image coordinates will not coincide perfectly with actual image pixel coordinates. The resulting grid of populated groundels is an image with accurate horizontally spaced pixels that does not have the effects of relief displacement or image tilt. Consequently, orthophotos provide a method for reliable spatial analysis, enabling accurate measurements of positions, distances, areas, and angles of imaged features.

PROJECT PLANNING

As with all operations in surveying and mapping, planning prior to collection and processing are critical to ensure quality geospatial products that address clients' needs. In project planning for airborne photogrammetry, the first step is to determine exactly what is required by the client. What information do they want, how do they want it delivered, and what accuracy is acceptable? Once these requirements are determined, the flight plan and ground survey plan are designed. The flight plan for an aerial photogrammetric mission consists of the lines along which the aircraft flies and the exposure locations where the camera is triggered. This must take into consideration the intended scale and the *endlap* and *sidelap* of the photos. Endlap is the percentage of overlap in successive photos along the flight strip, and sidelap is the percentage of overlap of adjacent strips of photos. At least 50% endlap is required for stereo coverage. Because terrain variations and aircraft tilt can affect the amount of overlap, planned endlap is typically 60% for manned flights to ensure stereo coverage. A larger nominal endlap (70% to 90%) is used for smaller, more dynamic aircraft, such as unmanned aircraft systems (UASs), because their flight paths are less stable and more influenced by wind gusts, and because a larger endlap is often required in processing software typically used for UAS imagery (more details on UASs are given later). Gaps in stereo coverage can be devastating because they prevent orthophoto production in the missed area requiring reflight or compromised products. Because endlap dictates the along-strip stereo coverage, lower planned percentages are typically established for sidelap with a common nominal value of around 30% to ensure overlap of strips sufficient for observing tie points. As with endlap, planned sidelap is also typically higher in small UASs. Note that to maintain planned endlap and sidelap, the flying height above datum may need to be adjusted across the project area in locations with large relief variations.

The size of the features that need to be identified in the photogrammetric products is an important consideration in project planning. The ability to capture the features of a certain size is dictated by the recording media of the camera, the flying height, and the focal length. Calculating the nominal GSD required for capturing the intended features is a useful procedure, keeping in mind the Nyquist rate. The target GSD informs the selected camera, focal length, and flying height. The accuracy of photogrammetric products is greatly influenced by the base-to-height ratio of the imagery, B/H'. Larger base-to-height ratios lead to more accurate products, with the conventional planned value being 0.6. One should keep this in mind when planning: when a constant endlap percentage is held, as the focal length increases, the base-to-height ratio decreases, thus negatively affecting the products' vertical accuracy.

Establishing ground control points can contribute up to 50% of the total project cost and should be carefully planned prior to image acquisition. GCPs should be widely distributed throughout the project area, and the points should be marked on the ground with artificial targets easily identifiable in the imagery prior to flights or be located on easily identifiable existing features such as road paint stripes. Targets usually have symmetric shapes where the center is located at the intersection of two linear features, with examples including crosses, chevrons, or "T" patterns. The points may be surveyed before or after the photography as long as the targets do not move between phases. The accuracy of ground control points is a limiting factor of the accuracy of the final products; thus, the customers' requirements should be considered when planning the number and distribution of GCPs and the accuracy of the ground survey. It is also common to establish additional *checkpoints*, surveyed ground points not used in analytical aerotriangulation but for measuring the accuracy of the final product. Several mapping standards are used in the United States to check the accuracy of photogrammetrically derived maps and, therefore, dictate the amount and accuracy of the control points used. These include, but are not limited to, the *ASPRS Positional Accuracy Standards for Digital Geospatial Data* (ASPRS 2015) and the *Federal Geographic Data Committee National Standard for Spatial Data Accuracy* (Congalton and Green 2008). Another mapping standard that is sometimes used, but should be considered a "legacy" standard, is the *US National Map Accuracy Standards* (NMASs). The *ASPRS Positional Accuracy Standards for Digital Geospatial Data* includes accuracy standards for digital photogrammetric products, light detection and ranging or lidar, and digital mapping data in general. Practitioners should consult these standards prior to planning and ensure that products meet or exceed the desired level of accuracy. (See Chapter 5 for more on ASPRS (2015) and Chapter 11 for establishing survey control.)

In addition to GCPs, airborne control (i.e., direct georeferencing) is often used. This typically consists of fixing a combined Global Navigation Satellite System (GNSS) antenna and inertial measurement unit (IMU) to the aircraft and camera, respectively. GNSS/IMU units directly measure the EOPs of the camera during times of exposure. Precise location and angular attitude (*leverarm* and *boresight angles*, respectively) of the GNSS/IMU relative to the camera must be known to implement airborne control (see Chapter 9 for more details on kinematic surveying). Although airborne control can be used to fully control analytical photogrammetric operations, GCPs are still included often to strengthen the solution and ensure product fidelity.

AERIAL LASER SCANNING

INTRODUCTION

Aerial laser scanning (ALS), commonly referred to as airborne lidar, is a method to rapidly collect topographic data over large areas at submeter 3D spatial resolution. Over the last few decades, ALS has evolved from a developmental technology into a well-established mapping solution. By the 1980s, NASA had been experimenting with airborne laser altimetry systems for such applications as mapping terrain, measuring sea surface height, and monitoring changes in the Greenland ice sheets (Krabill et al. 1984). Variations on the basic lidar design, such as large-footprint full-waveform systems, have appeared, but it is the small-footprint form that has become the most widely used for topographic mapping because of the higher-resolution spatial sampling that it offers (Slatton et al. 2007). During the mid-to-late 1990s, airborne lidar systems became available from commercial manufacturers that were compact, lightweight, and power efficient enough to be operated from a light single-engine aircraft. The progression of lidar also relied on improvements in differential GNSS technology and inertial measurement unit (IMU) technology for precise kinematic positioning and orientation of the airborne platform. Finally, an increase in computing power and data storage helped pave the way for the rise in lidar as a tool for wide area, dense topographic mapping. Today, lidar system technology continues to evolve at a rapid pace, but the fundamental operating principles and data products remain consistent. The purpose of this section is to provide an overview on ALS technology including basic principles, data processing, and survey considerations for end-user applications.

FUNDAMENTAL PRINCIPLES

ALS is a light detection and ranging method that pulses a laser to measure the range between an airborne platform and the Earth's surface many thousands of times per second. Light travels in a vacuum at a speed of 299,792,485 m/s (Vosselman and Maas 2010). By accurately timing the round trip time of travel (T_r) of the light pulse from the laser to the reflecting surface (Figure 10-6), it is possible to determine the range (ρ) from the laser to the target as follows:

$$\rho = \frac{c}{n}\frac{T_r}{2} \tag{10-7}$$

where c is the velocity of light in a vacuum, and n is a correction factor equal to the refractive index of the atmospheric propagation medium, which depends on air pressure, temperature, and humidity. For standard air, $n = 1.00025$ is commonly used. Because it is active, unlike aerial photography, lidar does not depend on ambient light, which makes it operable

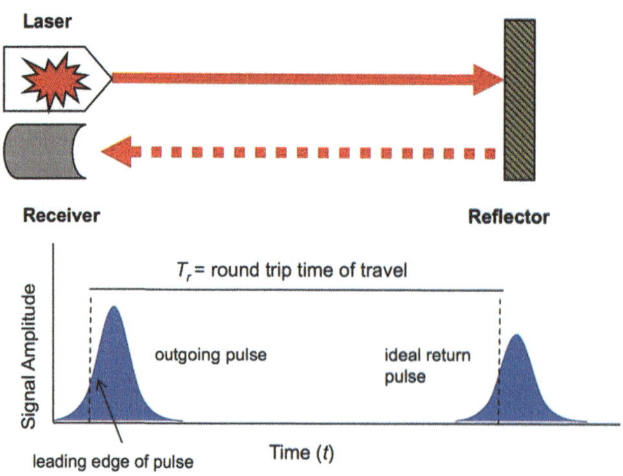

Figure 10-6. Lidar measures the round trip time of travel of an emitted and reflected laser pulse to derive the range to the target. The bottom figure shows a Gaussian-shaped pulse reflected from a surface oriented perpendicular to the incident laser beam. Reduction in return signal amplitude stems from atmospheric attenuation, backscattering, and other factors.

during day or night. Using an oscillating mirror, rotating mirror, or other scanning mechanisms, the laser pulses can be made to sweep through an angle, tracing out a line or other pattern on the reflecting surface. With the scan line oriented perpendicular to the direction of flight, a single-axis oscillating scan mirror produces a sawtooth pattern of points within a strip (swath) centered directly along the flight path [Figure 10-7(a)]. Sensor position and orientation information is combined with the scan angle and measured ranges to compute the x,y,z coordinates of the illuminated spots on the ground surface.

Multiple echo detection lidar systems can detect more than one pulse echo that may occur as the emitted laser pulse interacts with the underlying surface. Early commercial lidar systems had the ability to record only two discrete returns (echoes) per an emitted laser pulse: the first return pulse echo and the last return echo. Here, the last return corresponds to the last detectable signal above the sensor noise threshold from a series of returns. Modern "discrete-return" lidar systems can record multiple returns per an emitted laser pulse, including first and last. This multireturn capability is particularly useful for surveying topography under forest canopy because the last returns have a higher probability of reflecting from the true ground surface. It is important to mention that, although a lidar sensor might be able to record multiple returns per an emitted laser pulse, this does not mean that it always will do so. For example, over exposed, flat terrain such as a parking lot or beach, only one return per an emitted laser pulse will likely be recorded regardless of the multireturn detection capability of the lidar system.

Determination of Position and Orientation

A GNSS receiver onboard the aircraft is used to determine aircraft position and subsequently the laser head positioning, which is rigidly attached (Figure 10-7). GNSS update rates are usually at 1 to –10 Hz epochs. The measurements are referenced to the phase center of the GNSS antenna. Precisely surveyed lever arm offsets between the GNSS antenna reference point and the laser head reference point are used to translate the positional origin to the sensor head. The aircraft receiver measurements must be differentially corrected, typically using carrier-phase measurements for better accuracy, to remove inherent errors caused by GNSS signal propagation delay through the atmosphere, satellite clock bias, and ephemeris inaccuracies among others. By operating at least one GNSS base station on the ground within the survey area, relative positioning techniques can be applied between the aircraft's GNSS receiver and the base station to improve the aircraft trajectory (see Chapter 6 for more details on relative positioning). In relative positioning theory, the shorter the distance between the aircraft and base station, the higher the accuracy in the resulting GNSS baseline. Often times, continuous GNSS base stations within the survey vicinity are used for post-processing (Slatton et al. 2007).

As explained in Chapter 6, signal propagation delay through the ionosphere (the upper atmosphere) can be greatly reduced by employing geodetic quality, dual-frequency GNSS receivers. The L1 and L2 carrier phases of the observations can be combined within a double-difference solution using kinematic GNSS post-processing software to differentially correct the onboard receiver measurements. One such program is the Kinematic and Rapid Static (KARS) program developed by the National Geodetic Survey (Mader 1992).

The largest remaining errors in the GNSS observables are usually caused by the larger proportion of path length between the airplane and the GNSS satellite that traverses through the troposphere as the satellite elevation angle decreases (Slatton et al. 2007). These effects are difficult to remove by differential corrections because of the significant differences between troposphere ambience at the base station and aircraft. The work by Shan et al. (2007) suggests that calibration of the tropospheric effects is possible by employing multiple base stations with the aircraft flying at a range of altitudes.

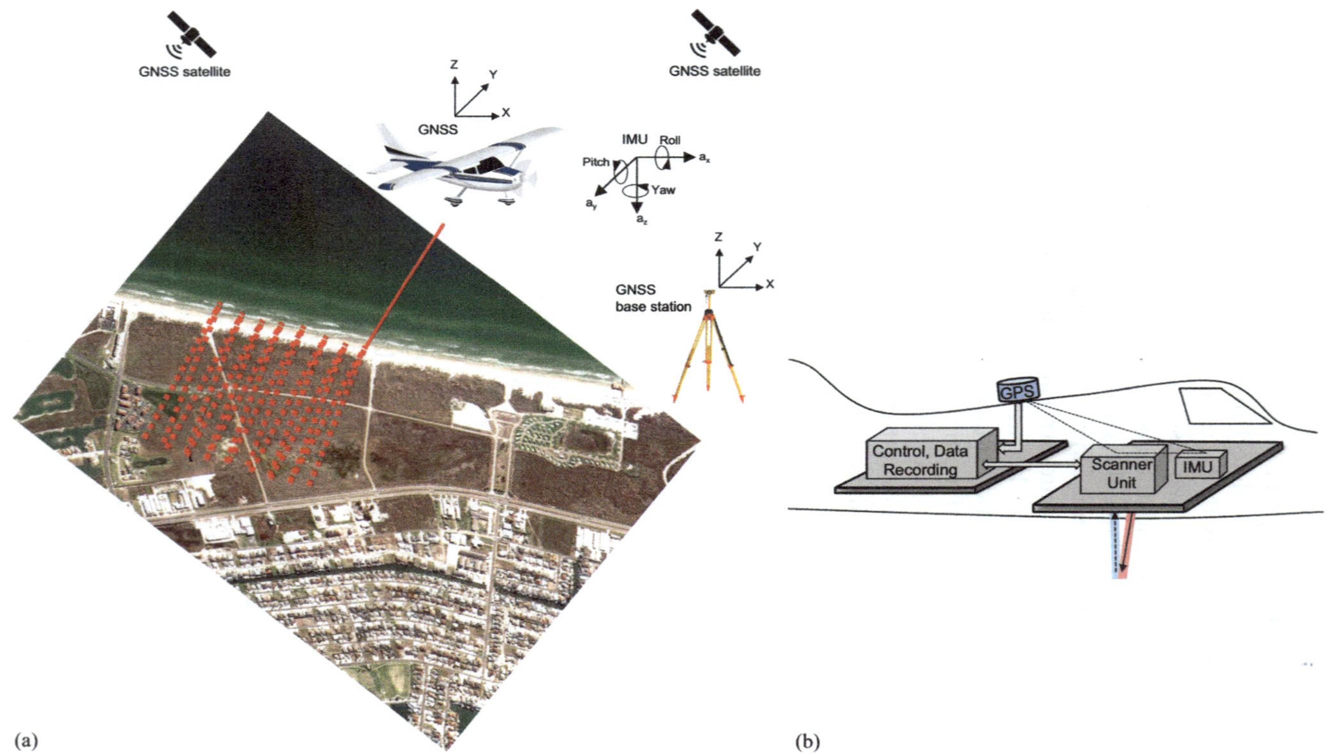

Figure 10-7. (a) Illustration of a light aircraft collecting topographic lidar over a beach. An onboard oscillating mirror distributes infrared laser pulses generating a saw-tooth pattern. The airborne GNSS and IMU data are integrated in the Kalman filter to produce a blended navigation solution comprising the sensor position and orientation (typically specified using the roll, pitch, yaw Tait-Bryan angles) along the trajectory, (b) onboard components of an aerial laser scanner.

Sensor orientation (roll, pitch, and yaw) observations are collected by an inertial measurement unit (IMU) onboard the aircraft, typically mounted in the lidar unit, to sense changes in motion. A typical IMU consists of a set of triaxial, solid-state accelerometers to measure the aircraft's accelerations (a_{xs}, a_{ys}, a_{zs}) plus the gravitational force, and a set of triaxial mounted gyroscopes to measure angular velocity. The acceleration measurements and angular velocity measurements can be integrated to obtain position and orientation measurements. Relative to the GNSS, the IMU provides very high update rates (e.g., 200 Hz) but suffers from long-term drift because of random walk. As such, an aircraft trajectory obtained solely from the IMU will contain significant errors that will continually grow if not corrected. To the contrary, the GNSS provides much lower sampling rates but more stability in the long term. The GNSS can be used to aid the inertial solution by updating the IMU position estimates within a Kalman filter to derive a "best estimate" of aircraft position and orientation at the time of each laser shot. With this approach, trajectory reconstruction can be performed to better than 5 to 10 cm positional accuracy.

The Kalman filter is a recursive estimation algorithm that minimizes the variance to provide a linear estimate of the state of a system. For cases where the true error model is Gaussian, the Kalman filter is an optimal estimator in a minimum mean square error (MMSE) sense. Excellent discussions on recursive state estimation with Kalman filters and GNSS-aided inertial navigation are provided in Gelb (1974) and Rogers (2003), respectively.

PROPERTIES OF AERIAL LASER SCANNERS

Aerial laser scanners consist of the following basic components (Figure 10-7b):

- *Lidar scanner assembly* includes the laser unit, optical scanner, transmit and collecting optics, and receiver electronics. The laser system is typically mounted over a hole in the aircraft's fuselage and continuously sends out pulses steered by an optical scanner. Scanning mechanisms include oscillating mirror, rotating polygon, and Palmer scan among others. A small fraction of the light from each pulse is directed to a photodetector in the receiver unit to start the clock that measures the round-trip travel time of the light to and from the reflecting surface. After passing through collection optics and spatial and spectral filters, the returning pulses are focused onto another photodetector, usually an avalanche photodiode or a photomultiplier tube (Carter et al. 2007). Modern scanner assemblies provide a roll compensation to compensate for aircraft roll. This feature helps avoid gaps in sampling coverage between adjacent swaths caused by roll and can enable more precise planning of flight line overlap for contiguous coverage.
- *Airborne GNSS* measures aircraft position at a continuous sample rate (e.g., 1 Hz). The standard is a dual-frequency receiver with the antenna mounted at an exposed position on top of the aircraft to provide an undisturbed view to the GNSS satellite constellation.
- *IMU* records acceleration data and rotation rates at a continuous sampling rate (e.g., 200 Hz). Acceleration data are used to support the interpolation of platform position with the GNSS, whereas rotation rates are used to determine platform orientation. The IMU is either fixed directly to the lidar sensor head or close to it on a stable platform.
- *Control and data recording unit* is responsible for time synchronization and control of the whole system. It stores ranging and positioning data gathered by the lidar scanner, IMU, and GNSS. Modern lidar systems can operate at >300 kHz pulse rates creating massive amounts of data, producing tens of gigabytes per an hour or greater (Vosselman and Maas 2010).

Other supporting components may include an operator laptop or tablet for communicating with the control and data recording unit to set mission parameters and monitor survey progress and a flight management system that provides a means for the pilot to display preplanned flight lines and support mission flight operations.

Topographic and Bathymetric Aerial Laser Scanners

ALS systems can be loosely categorized into two main classes: topographic lidar systems and bathymetric lidar systems. Topographic lidar systems are targeted for landcover mapping and typically operate in the near-infrared portion of the EM spectrum between approximately 800 and 1,550 nm. Spectral bandwidths range between 0.1 and 0.5 nm (Vosselman and Maas 2010). Near-infrared wavelengths are strongly reflected by green vegetation because of the cellular structure of healthy leaves and are also heavily absorbed by water. As such, these lidar systems are not able to penetrate water.

In contrast, bathymetric lidar systems are designed for mapping below the water surface to acquire digital water depth (bathymetry) data. These systems pulse a laser operating in the green wavelength range of the EM spectrum where visible light is less attenuated by water enabling penetration. A common operating wavelength of commercial bathymetric lidar systems is 532 nm because this wavelength can be generated by frequency doubling 1,064 nm near-infrared light output from a commonly used lasing material called neodymium yttrium aluminum garnet (Nd:YAG).

Conventional topo-bathymetric lidar systems employ two lidar sensors: one operating in the near-infrared (e.g., 1,064 nm) part of the spectrum and another operating in the blue-green (e.g., 532 nm) part of the spectrum for bathymetric mapping (Figure 10-8). The near-infrared laser pulse is mostly absorbed by the water surface (does not penetrate), but some of the emitted energy is backscattered to the sensor allowing measurement of the distance and location from the survey platform to the water surface. This lidar sensor can also be used as a devoted channel for topographic mapping. The green laser pulse is refracted as

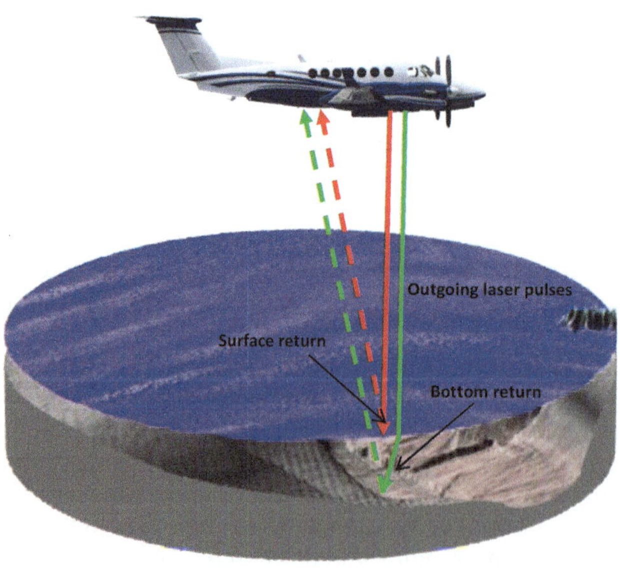

Figure 10-8. Basics of a conventional bathymetric lidar system. Near-infrared laser pulse is emitted to detect the water surface. Green laser pulse is emitted to map below the water. Refraction must be accounted for in computing the distance and location to green laser footprint on the seabed.

it propagates from air into the water column and can reflect from the seabed dependent on pulse attenuation and scattering through the water column. Georeferenced location of the illuminated spot on the seabed can be computed by means of the sensor position and orientation, scan angle, distance to water surface, time-of-flight of the green laser pulse echo from the seabed, and refraction angle of the green laser pulse obtained via Snell's law (Wozencraft and Millar 2005). The output is a 3D point cloud analogous to topographic lidar data. Over water only, points may stem from the water surface, water column, seabed, and objects on or below the water surface. Water depth is calculated by taking the vertical difference of the sea surface location from the seabed location to get depth below the water surface. For coastal zone surveys, tidal adjustments are often applied to reference the water depths relative to a local tidal datum. Single-frequency bathymetric lidar systems using full-waveform digitization and other modalities also exist.

Penetration depth is a function of water clarity and will vary depending on the properties of the lidar system (e.g., pulse energy), water conditions at the time of the survey, and seabed reflectivity. Increased water turbidity reduces penetration. Compared with topographic scanning, conventional laser bathymetry requires a higher pulse energy to compensate for the effects of refraction, absorption, and scattering as the green laser pulse propagates through the water column. A higher laser pulse energy results in longer pulse widths and lower pulse repetition frequency (in general). Higher-energy pulses provide more signal-to-noise ratio and are better for deeper water mapping. In clear water conditions with a high-energy green laser, mapping can be conducted in water depths of up to about 50 m (Wozencraft and Millar 2005). Higher-energy laser pulses are not conducive for shallow water mapping because longer pulse lengths have reduced range resolution. In shallow water, the longer pulse width can exceed the distance between the water surface and the bottom. This can result in convolution (mixing) of the return pulse signal as it interacts with the water surface and bottom, making it difficult for the sensor to distinctly separate the two surfaces. For mapping shallow water, bathymetric lidar systems with shorter pulse widths are preferred. Bathymetric lidar systems can also work in topographic mode for scanning the land surface as green laser pulses will reflect from the landcover. To be efficient, they need a high pulse rate.

Compared with traditional hydrographic surveying, aerial bathymetric laser scanning is highly efficient in shallow water. The methodology offers significant cost savings both in productivity and in operational expense when dense measurements are needed (Guenther et al. 2000). For data acquisition in shallow relief bathymetry and complex areas, the survey speed may be up to 20 times quicker than traditional acoustic soundings (sonar) obtained from ships (Vosselman and Maas 2010). The dependence of the methodology on water clarity is presently a limiting factor that precludes its use in water with high turbidity (e.g., Mississippi River inlet) and in deeper water further offshore. Other factors that impact bathymetric lidar performance include sun glint that can increase ambient noise, rough seas with high waves that decrease measurement accuracy, and white water from breaking waves that can impede the laser pulse causing a data gap in the surf zone.

Ranging Modalities

Aerial laser scanners can be further characterized based on the type of echo detection and ranging method employed by the lidar system. Commercial lidar systems designed for topographic mapping have traditionally employed discrete-return ranging. This method employs time-of-flight receiver hardware that records only the stop signal at a certain rise time of the return pulse echo after it exceeds a signal threshold (Figure 10-9). It provides no information about the echo's shape such as width. The instrumentation typically includes features such as different channels for strong and weak signals and constant fraction

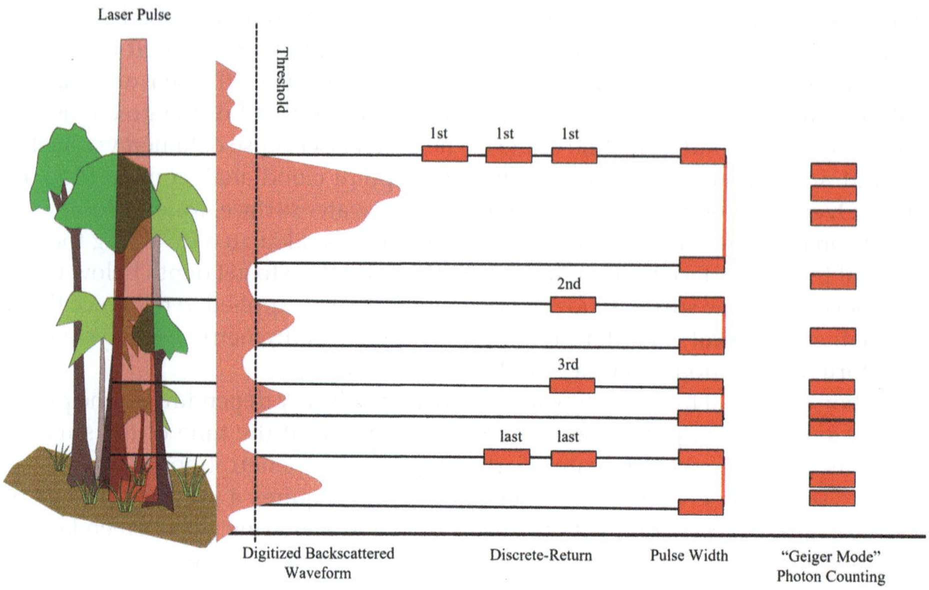

Figure 10-9. Lidar ranging modalities: full waveform digitization, discrete-return ranging, photon counting. Discrete-return example shows point locations for single return, two returns (first and last), and four returns computed using a leading-edge amplitude threshold to detect a discrete echo; common approach.
Source: Adapted from David Harding, NASA.

discriminators in the range measuring circuitry to determine the timing location (Shrestha et al. 2007). As previously mentioned, discrete-return lidar systems typically allow the recording of more than one return echo per an emitted laser pulse (up to five returns is common for commercial systems at the time of this writing). The ability of the sensor to distinguish successive reflections from a single transmitted pulse is governed primarily by the laser pulse duration but is also affected by pulse shape, detector design, and the receiver circuitry. Topographic lidar systems usually employ longer-duration laser pulses (e.g., 10 ns duration $\approx$ 3 m pulse width) to obtain high signal-to-noise reflections. At present, most commercial discrete-return lidar systems have limitations in their range measurement resolution (e.g., some have range resolutions upward of 2 m or greater), which produces gaps in the measurements along the sensor line of sight. These gaps are often referred to as "dead zones," which means that the sensor receiver is not capable of detecting the next consecutive range return within a certain time interval after the previous measurement is taken (Ussyshkin and Theriault 2011).

In contrast to discrete-return ranging, full waveform echo digitization records the entire shape of the backscattered laser pulse using a digitizer operating at very high temporal resolution (e.g., gigahertz). This provides a complete intensity profile (waveform) of the pulse echo (Figure 10-9). This shape information provides a detailed picture of the pulse interaction with the landcover along the transmit path of the laser pulse. The waveform shape can be analyzed to extract additional value from a lidar survey beyond the 3D point cloud data. For example, the waveform shape can be analyzed to better estimate vertical structure of forest canopy and improve biomass estimation. Besides this full profile information, waveform digitization also gives the user more flexibility on range determination and point cloud generation by enabling software post-processing for discrete echo detection. Such techniques involve waveform shape fitting algorithms that are used to determine the number of distinct waveform peaks (echoes) and timing point locations for range computation (Parrish et al. 2011). This flexibility, for example, can be used to potentially improve the detection of more returns from below canopy. In most cases, full-waveform-processed point clouds will provide a denser, more richly detailed point cloud because of the ability to extract more returns. However, such detail may not be needed for many engineering and surveying applications where the main need is bare-earth terrain or a digital surface model only. This is why discrete-return ranging systems have been a commercial standard in topographic lidar mapping for some years. Lidar systems may also employ both or a combination of the two approaches. For example, some commercial systems perform echo digitization and then employ online waveform processing, which is a signal processing technique to detect echoes and determine ranges and amplitudes for each echo. Alternatively, certain lidar systems may record the entire waveform but perform echo detection solely through software implementation during data post-processing.

Traditional lidar systems are referred to as "linear mode" systems. Both discrete-return ranging and full waveform digitization, as previously described, fall under this category. Linear-mode lidar sends individual, high-powered laser pulses to illuminate the surface and measure a range to a target. This concept relates to the linear response of the receiver performance and its dependence on high signal-to-noise ratio pulse echoes for range computation. However, an alternate approach for lidar mapping uses a detector that is sensitive to even one received photon and is appropriately termed a single photon-counting lidar (SPL) system. SPL systems range based on a single photon detection event and, therefore, do not provide a measure of return signal strength (echo intensity) common to linear mode systems. Presently, there are two commonly employed hardware approaches to detect single-photon events, Geiger-mode avalanche photodiode (GmAPD) (Johnson et al. 2003) and microchannel-plate photomultiplier tube (MC-PMT) (Eisaman et al. 2011). These two approaches have a high gain to ensure single-photon sensitivity, and SPL systems based on both detector types are being developed. The biggest difference between these two detectors are their respective dead times. The dead time of a PMT is much shorter than that of GmAPD, and so systems with MC-PMT are able to record multistop events. However, the short dead time increases the data volume, which in return, limits the size of detector arrays for PMT systems (Wang 2017). The major advantage of these new SPL systems for aerial surveying is that they can collect data at a higher speed than conventional lidar with equivalent point density. This can potentially enable more efficiency and lower costs for large-scale geographic coverage compared with linear mode lidar (Corbley 2015). A major disadvantage with SPL systems is that such systems are also very sensitive to returns from background solar noise and system dark current, which causes a high noise rate in the collected data (Wang 2017). Point clouds generated from SPL systems can be quite noisy and require additional post-processing and advanced techniques to clean the data. As the noise handling algorithms and detection hardware improves, photon-counting lidar technology will likely find growing appeal to the surveying and mapping profession. Figure 10-9 shows the different lidar ranging modalities: discrete-return, full waveform, and SPL.

Another emergent lidar technology is flash lidar. Flash lidar cameras appear similar to 2D digital cameras and use no mechanical scanning components. These sensors use a 3D focal plane array to sample incoming photon streams from a laser "flash" and generate 3D depth with location and reflective intensity images. Such systems are good for real-time target tracking and lidar-aided navigation but will likely find more appeal in surveying and mapping as detector resolutions increase.

Finally, the discussion thus far has described time-of-flight-based lidar systems. An alternate ranging approach is employed by some lidar systems called phase-based ranging; however, such methods are usually limited to shorter scan distances and are, at present, rarely employed for ALS.

Beam Divergence

The laser beam continually widens as it propagates away from the scanner. Beam divergence controls this spreading. Small-footprint lidar systems employ relatively narrow laser beam divergence and are the most widely used for topographic mapping because of the higher sampling resolution. Modern systems can provide a switchable divergence, for example, from 0.3 to 0.8 mrad. Which beam divergence is best for a given mapping application depends on the structure and spatial distribution of the landcover. Having the capability of more than one divergence provides a valuable flexibility to optimize the data collection for bare-surface detection in a variety of circumstances and/or for characterizing canopy structure (Shrestha et al. 2007, Slatton et al. 2007). Large-footprint lidar systems employ a wider beam divergence to create larger pulse diameters (several to tens of meters). One application of large-footprint waveform lidar systems is mapping of vegetation structure and estimating forest attributes (e.g., biomass).

Laser Scanner Characteristics

Aerial laser scanners with standard scanning mechanisms, like an oscillating mirror, share fundamental relationships. The swath width sw of a scanner is given by (Vosselman and Maas 2010)

$$sw = 2h\tan\left(\frac{\theta}{2}\right) \tag{10-8}$$

where θ is the full scan angle, and h is the height above ground. Most airborne scanners allow angles ranging between about 5 degrees and 75 degrees.

Laser beam divergence governs the size of the illuminated pulse footprint on the reflecting surface for a given propagation distance. Assuming the spot shape to be a circle, the following relationship describes the cross-sectional diameter d of the laser footprint on the reflecting surface:

$$d = 2h\tan\left(\frac{\gamma}{2}\right) \tag{10-9}$$

where γ (full angle) is the beam divergence, and h is the height above the ground. Typical beam divergences are between 0.1 and 1 mrad (Vosselman and Maas 2010). So, for example, a 0.2 mrad divergence will have a footprint of 0.12 m at a 600 m flying height above ground.

Pulse repetition frequency (PRF) is a very important property of a laser scanner because it governs the point density on the sampled terrain. Laser pulse repetition frequencies of current airborne lidar systems can exceed 500 kHz. Assuming an oscillating mirror scanner, the area F covered over a certain flight time interval t at a cruising speed v and swath width sw is given by

$$F = (v)(t)(sw) \tag{10-10}$$

The number of laser pulses P generated during the time interval t is $P = \text{PRF}\ t$. The mean point density, P_m, can then be calculated by means of P and the area covered during t as follows:

$$P_m = \frac{(\text{PRF})(t)}{F} = \frac{\text{PRF}}{(v)(sw)} = \frac{\text{PRF}}{2vh\tan\left(\frac{\theta}{2}\right)} \tag{10-11}$$

Hence, the highest point density is reached with a high PRF, low cruising speed v, low survey height h, and small scan angle θ. For example, at a survey height of 600 m and a scan angle of 40 degrees, the swath width sw is approximately 437 m. Based on the P_m equation, a PRF of 167 kHz at a cruising speed of 65 m/s at the given swath width will result in a mean point density of about 6 points/m^2.

Equation (10-11) represents an idealized mathematical description of a homogeneous point distribution over the area F on a perfectly flat terrain. In reality, the point distribution depends on several factors, including surface topography and scan pattern (Figure 10-10). The scan pattern from an oscillating mirror on flat terrain will vary in the along-track and cross-track directions. Furthermore, the maximum achievable PRF for a given lidar system is usually not a fixed value because it depends on the survey height h. This is necessary to balance the reduction in pulse energy with increasing PRF and to reduce potential ambiguities that can arise from a high PRF when flying high above ground. A high PRF at high flying heights can result in multiple pulses outgoing and incoming, thereby creating potential ambiguities in echo detection (Vosselman and Maas 2010). Multiple-time-around processing is a technique employed by some system manufacturers to address this issue in high PRF lidar systems.

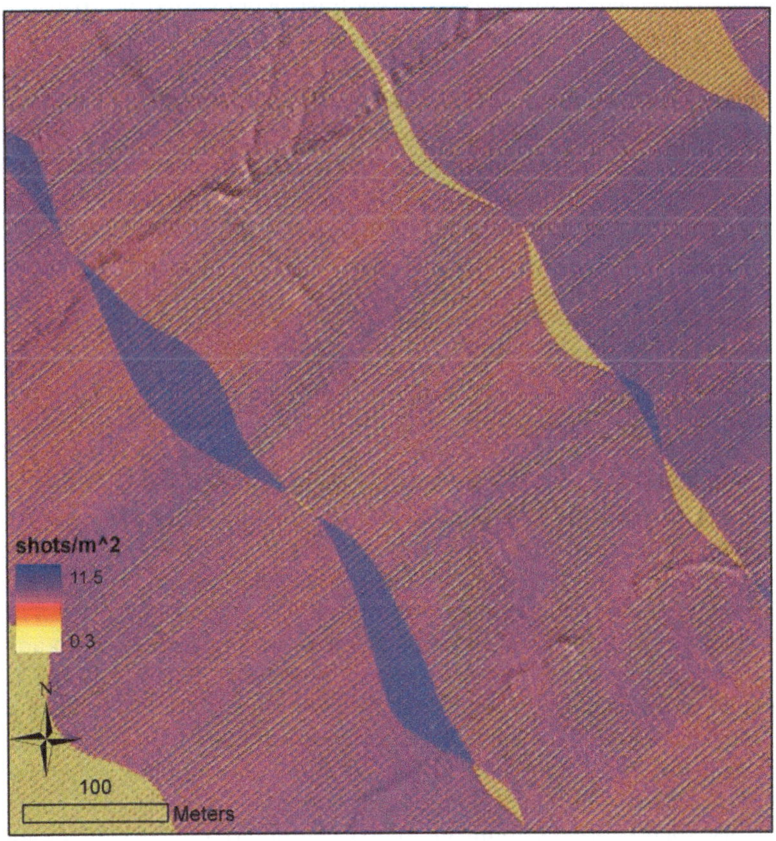

Figure 10-10. Raster color coded by point density (the number of points in square meter) showing the variability in the scan pattern and point spacing. The variability in point spacing stems from the scanning mechanism, overlapping flight lines, and effects of the terrain among other factors. Source: UNAVCO.

OPERATIONAL ASPECTS OF AIRBORNE LASER SCANNING

Project Planning and Execution

ALS survey design is dictated by project specifications. These include area of coverage, desired point density, and accuracy requirements. Flight planning must, therefore, account for lidar system performance (PRF, scan angle, beam divergence), flight parameters (velocity and altitude), and characteristics of the underlying topography and landcover. This process is usually supported by manufacturer-specific software that allows input of lidar scan settings, flying height, cruising speed, and flight line overlap to estimate survey characteristics such as swath width and average point density. Some software will allow input of a digital terrain model of the project area to more realistically estimate variability in swath coverage and sampling density. This is particularly useful in designing a flight plan over mountainous terrain to avoid gaps in coverage.

The objective in flight planning is to design as efficient a flight path as possible (typically parallel flight lines) to cover the entire project area while meeting point density requirements. As always, economic factors must also be taken into consideration when designing an ALS survey. For example, the full scan angle is an important parameter choice. Over flat terrain, larger scan angles may be the best choice because they produce larger swath widths for a set flying height, thereby reducing the number of flight lines required to collect the data. This reduces flight time and thereby survey costs. In contrast, over forested or urban terrain, smaller scan angles may be more appropriate to reduce shadowing of the laser pulse by the land cover. Similarly, flying height can increase swath width and reduce flight time at the expense of reduced point density. To avoid gaps in data coverage and improve the mean point spacing, adjacent flight lines are typically designed to ensure swath overlap with a minimum of 20% (Vosselman and Maas 2010). For densely vegetated areas, swath overlap as high as 50% or greater may be used to obtain as many points as possible from the underlying ground surface. Figure 10-10 shows an example of the effects of swath overlap on point density.

Project planning must also take into consideration the location of GNSS base stations for differential correction of the aircraft trajectory. For high-accuracy ALS data collection (referred to as "geodetic grade"), it is standard to utilize two or more base stations with the airplane at a maximum distance of 20 to 25 km from a base station during the survey. In many cases, such stringent requirements are not practical, but a general rule of thumb is to have a base station within 40 km of the airplane (Slatton et al. 2007). Similarly, bearing in mind IMU drift characteristics, the maximum length of a flight line will typically be 30 to 40 km dependent on the selected aircraft velocity (Vosselman and Maas 2010). Survey areas are, therefore, usually broken up into different survey blocks. Flight cross lines are also typically designed at regular distances during a survey, at least one per a flight block. During data processing, the cross lines are used for checking the accuracy of merged scan lines and performing calibration adjustments.

Once the flight plan is designed, the objective of field operations is to adhere to this mission plan as precisely as possible for collecting the data. The ideal time for a survey mission depends on project needs (e.g., leaf on/leaf off conditions, free of snow, low tide for coastal mapping) as well favorable weather conditions such as reduced wind and no storms. Rain and fog can absorb and heavily scatter the outgoing pulse. Furthermore, factors in regard to the collection of GNSS base station data for differential correction of the aircraft trajectory, such as good satellite geometry dilution of precision (Chapter 6), should also be taken into consideration. Once in the field, collecting quality lidar observations requires the collaboration of an experienced onboard lidar operator and pilot to ensure that data are collected along the previously setup flight plan. The aircraft must be equipped with a real-time navigation display to enable the pilot and operator to follow the flight plan (i.e., to stay online and at the planned altitude) very accurately. In areas with large changes in the terrain surface height, the flight plan must include preplanned climbs and descents of the aircraft to maintain uniformity in the ground point spacing. In particular, rapid climbs and descents must be avoided to maintain minimal pitch angles and minimal roll turns employed to help maintain matching altitudes on overlapping flight lines.

Finally, it should be mentioned that ALS can be performed without an onboard operator by using software that links the lidar system with the aircraft navigation system to control laser firing and scan patterns. At the time of this writing, UASs are progressing ALS toward fully autonomous flight and data collections for some applications.

Calibration

Even the best ALS instruments require careful and frequent calibration. It is presently not feasible to design ALS system electronics and mechanical components that are completely immune to the high vibration, variable pressure and temperature, impacts of rough landings, varying orientation, and varying electrical supply typically encountered in a light aircraft operating at low altitudes. Also, there are natural aging and wearing processes of components, such as the laser pump diodes and optical scanner that would cause variations over time even if the system was mounted on an optical bench in a clean stable environment. The only way to know the performance of a system during any given survey is to design the data collection to reveal calibration errors or changes in performance.

Manufacturers provide calibration parameters for the units they build, including values for the offsets among such components as the IMU, the scanner, and an exterior point to which the GNSS antenna is to be referenced. These mechanical dimensions typically are stable to the millimeter level or less, and there is, in general, no need for operators to measure these parameters between regularly scheduled factory cleaning and boresight alignment servicing. However, the manufacturers also typically provide estimates of such parameters as the zero point and scale of the scanner and biases (alignment offsets) in the roll, pitch, and yaw gyroscopes that can and do change significantly with time. Accurately knowing these parameters on a flight-by-flight basis is important because erroneous values can result in artifacts in the delivered observations and derived products that users may interpret as "signal" in their applications. For example, scanner scale error produces a well-known "smile/frown" topographic artifact along a scan line (Shrestha et al. 2007).

Quality data must be collected with an ALS system that is calibrated regularly, preferably on a flight-by-flight basis or before field deployment. Kinematic GNSS is often used to survey ground lines on hard surfaces such as roads and runways, together with orthogonal ALS flight lines, to adjust calibration parameters on a flight-by-flight basis or before field deployment. Depending on the particular project goals and area to be imaged, procedures in the field are adapted to achieve the required accuracy in the most cost-efficient manner. For example, the greater the flying height, the more critical the calibration parameters such as scanner scale and orientation biases become. If other constraints require operating at a higher altitude, greater care in determining the calibration parameters must be exercised. For more details on ALS calibration procedures, the reader is referred to the discussions presented in Skaloud and Lichti (2006), Shrestha et al. (2007), and Vosselman and Maas (2010).

Quality Control and Accuracy Reporting

For quality control, ground truth elevation and position data are collected within the project area (e.g., using RTK GNSS) for checking the accuracy of ALS-derived elevation products. Flat surfaces such as roads and parking lots in the survey area can serve as a control for height accuracy, whereas readily identifiable and distinct linear features in the data such as roof planes and building edges can serve as a control for planimetric position accuracy. These data can also be used in the calibration procedures described previously to adjust any observed vertical and horizontal biases. Elevation data may also be collected under forest canopy, in vegetated areas, and in other terrain types (e.g., steep slopes) to measure the effects of varying landcover on lidar-derived bare-earth elevation products.

Following ASPRS guidelines, vertical accuracy should be reported at the 95% confidence level for data tested by an independent source of higher accuracy (ASPRS 2015). Checkpoints should be well distributed throughout the data set and distributed to reflect the geographic area of interest. ASPRS recommends collecting a minimum of 20 checkpoints (30 preferred) in each of the major land cover categories representative of the area. Representative landcover categories include exposed terrain, tall weeds and dense vegetation, forested areas fully covered by trees, and urban areas with dense human-made structures. Assuming we have a set of checkpoints, the RMSE can be computed according to Equation (5-34) in Chapter 5. Equation (5-37) represents ASPRS lidar elevation accuracy reported at 95% confidence level based on the computed RMSE value. It is important to mention, however, that the aforementioned assumes normally distributed lidar elevation error, which commonly occurs in exposed terrain. In areas of nonstandard error distributions such as vegetated landcover, the ASPRS standard uses the 95th percentile to report vertical error (ASPRS 2015). Similar approaches can be applied to compute and report horizontal accuracy under the ASPRS standard. Furthermore, the reader is recommended to review the USGS Lidar Base Specification publication (Heidemann 2014) for additional information on assessing and reporting lidar horizontal and vertical positional accuracy.

Data Processing

After a lidar survey mission, typically three datasets are present (Vosselman and Maas 2010):

- GNSS base station data;
- Navigation data obtained from onboard GNSS and IMU; and
- Laser ranges, amplitude data, scanner parameters (e.g., scan angle), and echo counts.

Processing of these data consists of two main stages. The first stage consists of post-processing the base GNSS and airborne GNSS/IMU data. Ground GNSS observations are used to differentially correct the airborne GNSS observations. These corrected GNSS observations are then integrated with IMU measurements via a Kalman filter to obtain a blended navigation solution. This provides an accurate determination of platform position and orientation (POSE) and subsequently the laser reference point at the instant each laser pulse leaves the aircraft. Refer to the previous section for a more detailed discussion on this process.

The second stage of processing merges post-processed navigation data and ranging measurements. Precise time tagging is critical to accurately integrate these data. For example, each lidar system component might sample at the following rates:

- GNSS at 10 Hz,
- IMU at 200 Hz,
- Laser pulse repetition frequency at 300 kHz, and
- Scanning mirror oscillation rate at 30 Hz.

These measurements must be time-synchronized. Time tagging is performed by a control and data recording unit through GNSS time synchronization [Figure 10-7(b)]. With the help of time tags, POSE information is combined with the measured scan angle and range for each detected pulse echo to compute 3D coordinates of the sample point on the reflecting surface (Baltsavias 1999, Wehr and Lohr 1999). This process is referred to as geolocation of the laser footprint (or pulse echo). The computed x, y, z coordinates are initially georeferenced relative to the reference system of the GNSS base station coordinates, for example, NAD 83 (2011). In general, the coordinates of the base station are determined or transformed into the reference system desired by the end user to meet project requirements. During this phase of processing, different types of correction data (calibration data, mounting parameters, atmospheric correction) are also taken into account. Calibration parameters might also be adjusted if they are determined no longer optimal (Vosselman and Maas 2010, Slatton et al. 2007).

The first stage of data processing is typically performed with GNSS-aided inertial navigation processing software developed by the maker of the position and orientation system used onboard the aircraft. The second stage of processing is typically performed with proprietary software developed by the manufacturer of the lidar system.

Data Products

Point Cloud. The raw output generated from a processed ALS survey is a densely sampled set of x, y, z measurements of the underlying reflecting surface called a 3D point cloud (Figure 10-11). Point cloud data provide a three-dimensional representation of the ground and landcover that also reflects the scan geometry of the survey. Airborne lidar point cloud data are sometimes referred to as "2.5D" data, because lidar samples from an aerial nadir (top-down) perspective typically resolves one *z-value* (height or elevation) per a given *x,y* coordinate on an object, as opposed to true 3D data that might have several *z-values* for a single *x,y* position (e.g., oblique viewing 3D terrestrial laser scanning, according to Chapter 8). However, there are caveats, because airborne lidar can sample vertically along walls on the sides of buildings or other objects such as telephone poles, particularly on the edge of a scan line. Therefore, for practical purposes, ALS is considered 3D data.

Lidar point cloud data are commonly stored and shared in the binary LAS format or compressed LAS format (LAZ), which has become a standard format for lidar data in recent years (ASPRS 2019). Other data formats exist as well, such as the ASTM E2807 *Standard Specification for 3D Imaging Data Exchange* known as the E57 format (Huber 2011). Point

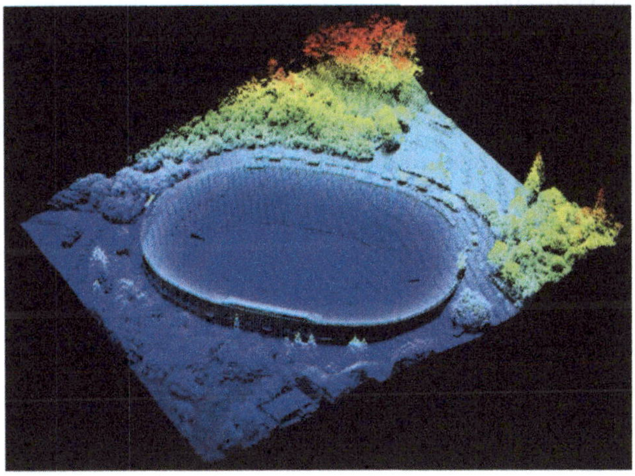

Figure 10-11. Perspective view of 3D point cloud data colored by elevation showing the University of California at the Berkeley football stadium. Point density is >10 points/m². Lidar survey conducted by the National Science Foundation (NSF)–supported National Center For Airborne Laser Mapping (NCALM).

cloud data are typically processed to derive DSMs or bare-earth DTMs. The latter is often a desired final mapping product.

Filtering and Terrain Modeling. After reducing and editing the observations, the next major step in ALS data processing is point cloud filtering. The most common goal of filtering ALS observations is to identify and remove points associated with vegetation and buildings and extract "bare earth ground" points, which are subsequently interpolated to form DTMs (Figure 10-12). Many different filtering algorithms exist to remove nonground points from lidar data, several of which are reviewed in Sithole and Vosselman (2004) and Meng et al. (2010). In general, these filters operate iteratively to refine bare-surface estimates based on different criteria such as the point's vertical offset and slope. Commercial software is available, which can implement ground point filtering. The capabilities and quality of those filters vary widely, and care must be taken to employ the appropriate algorithm and parameter settings depending on the types of features to be removed or preserved for a given application. Filter tuning is an iterative process of adjusting parameter values to assess a filter's performance in terms of the number of segmented ground points and classification accuracy. Classification accuracy is often measured in a qualitative sense based on a visual inspection of the filtered point cloud, or a shaded-relief DEM derived from the ground point set, focusing on identifiable areas of ground and nonground features. Quantitative accuracy assessment can be performed by manually segmenting a portion of the cloud into ground and nonground points and then measuring how well the classified points match, or by other means such as assessing DEM error generated from a filtered point cloud relative to ground control.

No single filter (even an adaptive one) is optimal for all terrain features. Misclassification error will vary with the terrain, data characteristics (e.g., point density), and method implemented. Vegetation that is both short and dense can be particularly difficult to remove completely because it greatly reduces the number of ALS returns from the ground. In some cases, the laser pulse may have never reached the ground surface because of vegetation occlusion. In such cases, a filter may still return a set of ground points, but technically all these points will be misclassified as they represent "near-ground points." Finally, it is important to have the filtered data inspected by qualified personnel who have knowledge of the project area and can identify problems prior to using the observations for project applications (Slatton et al. 2007).

After filtering is applied, the ground point elevations can be interpolated into a regular grid of elevation measurements to create a bare-earth DTM for representation of the continuous topography. For multireturn point cloud data, the last (and single) return data are typically utilized to minimize the probability of land cover biasing the surface elevations. Alternatively, the first (and single) return data can be interpolated into a regular grid of elevation measurements to create a DSM of the landcover and exposed terrain. Numerous spatial interpolation methods exist, including deterministic methods like inverse distance weighted averaging (IDW) or the spatial statistical approach of Kriging (Cressie 1991). The method that best suits an application depends on the complexity of the terrain and characteristics of the ALS point cloud. When point density is high, differences between interpolation methods can be minimal over noncomplex terrain. In some cases, the DTM or DSM can be created by a simple binning process such as taking the mean value of all point elevations falling within the grid cell. The desired spatial resolution at which to interpolate the DTM or DSM (e.g., 1 m) should be based on the average input point density distributed across the area of interest. As a rule of thumb, the spatial resolution of the elevation model (post spacing) should not exceed the average point spacing. The accuracy of lidar-derived DEM products should be compared with that of elevation control data collected in the area, such as with that of RTK GNSS. Figure 10-13 shows an example of a DSM, DEM, and contour map derived from ALS point cloud data.

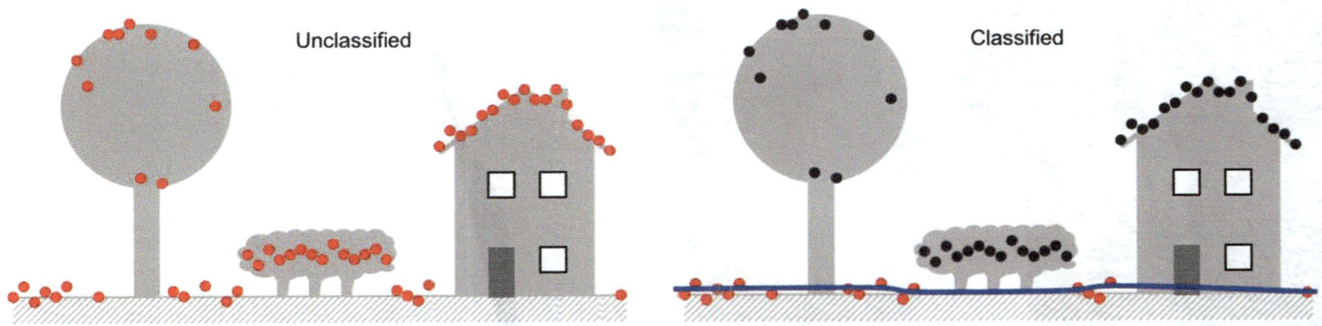

Figure 10-12. *Example of (left) nonfiltered point cloud and (right) filtered point cloud segmented into ground and nonground points. Image adapted from Claus Brenner.*

Figure 10-13. (Top) ALS-derived digital surface model, (middle) bare-earth digital elevation model, and (bottom) contour map of Tellulah, Georgia.
Source: Courtesy of Ramesh Shrestha, NCALM, reproduced with permission.

Intensity. In addition to recording ranges and generating point clouds, airborne lidar systems usually record the return signal strength of the pulse echo (e.g., 8 bit or 12 bit values). As discussed in Chapter 8, these amplitude measurements are commonly referred to as lidar "intensity" values. The strength of a pulse echo characterizes to some degree the reflectance of the illuminated spot on the ground, but the return signal depends on several factors, including variations in path length, surface roughness and orientation, incidence angle, beam divergence, object spectral properties (directional reflectance), object optical density, attenuation of the signal through the atmosphere, and ALS system characteristics. Lidar intensity is, therefore, a relative measure of the reflectivity of the objects at the specific wavelength of the laser (Vosselman and Maas 2010, Starek et al. 2007). Recorded intensity data are often provided as complementary information to the 3D point cloud data and can be stored in the LAS format. Although intensity calibration that accounts for the terrain's directional spectral properties remains an open area of lidar research, normalization of intensities based on path length is straightforward to implement. It is commonplace for ALS system manufactures to provide intensity normalization as part of its standard data reduction software workflow. Some ALS system manufacturers provide a calibrated relative reflectance value. For example, RIEGL's V-Line instruments provide a relative reflectance reading for each detected target as an additional attribute. The relative reflectance provided is a ratio of the actual amplitude of that target to the amplitude of a white flat target at the same range, orientated orthonormal to the beam axis, and with a size in excess of the laser footprint (Pfennigbauer and Ullrich 2010). A normalized intensity image resembles a black and white photograph if shown in gray scale. Intensity data may serve to help identify objects that are difficult to recognize or segment by elevation and 3D structure alone (Figure 10-14).

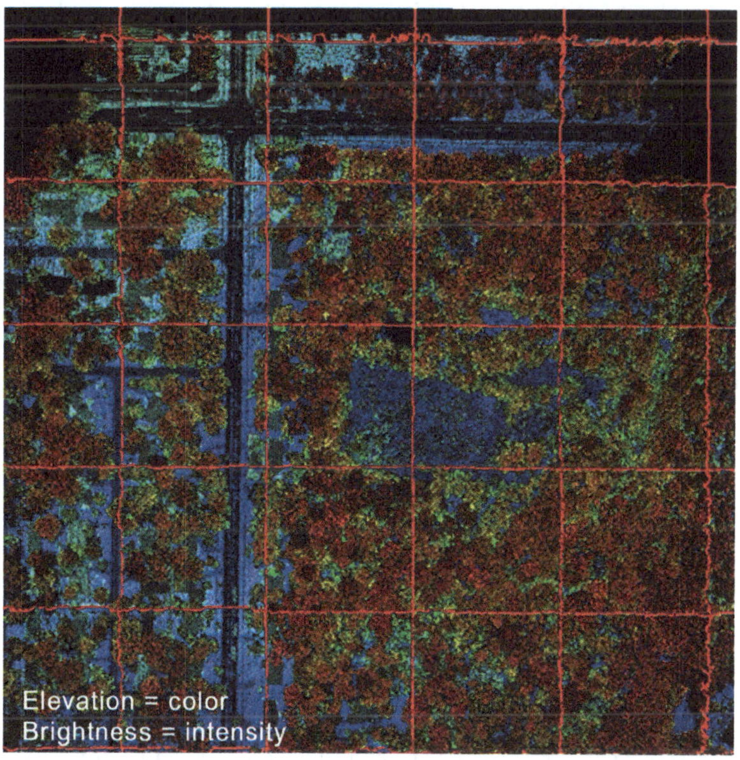

Figure 10-14. Nadir view of ALS point cloud colored by elevation showing an urban and forested region. Brightness is determined by the point's intensity value. Area is approximately 500 m × 600 m.

Because the terrain and landcover consist of objects that range from minimally reflective to highly reflective in the near-IR operating wavelengths of topographic lidar, the dynamic range of the return signal can exceed upward of three orders of magnitude. Most photodetectors in a lidar system are not capable of a linear response over so large a range and try and employ automatic gain control techniques to reduce range walk as a function of signal strength. Range walk is when bright returns produce a reduction in the perceived range to the object causing artifacts in the point cloud data. Example of a common range walk artifact includes white paint stripes that appear to float above road surfaces because of their high reflectivity. However, sometimes more subtle artifacts can be observed, such as vertical offsets at the intersection of wet and dry sand on beaches (Shrestha et al. 2007). System manufacturers will provide intensity lookup tables with each ALS unit to eliminate range walk or implement high and low power receiver channels in the lidar unit to compensate, among other techniques.

AERIAL LASER SCANNING PERFORMANCE

Error Budget

Various error sources propagate into the final point coordinates derived from an ALS survey. The main sources of error that contribute to the final accuracy are the following:

- Erroneous calibration of the GNSS, IMU, and scanner assembly;
- Inaccuracies in estimation of the aircraft trajectory and sensor orientation; in particular, GNSS-induced positional error can be one of the largest contributors;
- Ranging uncertainty/bias, scanner jitter, and angular measurement bias;
- Surface complexity of the target; it is important to note that over sloping terrain, an error in position results in an error in elevation, and as such, vertical errors tend to grow as the slope grows; similarly, when examining lidar-derived DEMs, vertical errors will, in general, be greater over vegetated surfaces than exposed surfaces;
- Multipathing of the laser pulse as it propagates, leading to an increased time delay and longer range than should be recorded;
- Coordinate transformation and geoid model; and
- Growth of illuminated laser footprint as range increases because of beam divergence.

This list is not exhaustive, but it is clear that many sources can contribute to the final accuracy of the computed coordinates. For obtaining quality data, great care must be taken in regard to position and orientation determination, data processing, calibration, and survey planning. With rigorous calibration and survey methodology using a high-precision ALS, accuracy values in the range of 0.05 to 0.10 m vertical (z) and 0.15 to 0.30 m horizontal (x, y) over flat terrain can be regularly achieved (Slatton et al. 2007). In standard practice, ALS positional accuracies (at 68% confidence) typically range between 0.1 and 0.2 m in height and 0.2 and 1.0 m in position for flying heights $<$2,000 m above ground (Vosselman and Maas 2010).

Advantages and Limitations of Aerial Laser Scanning

ALS is a powerful method for rapid collection of topographic data over a large spatial extent. However, like any technology, ALS has some limitations. Because it is an active technology using emitted laser light, it is weather dependent. ALS systems operating in the NIR cannot penetrate dense fog or clouds and, therefore, should not be operated during rainfall, which will scatter and absorb the pulse. Dense vegetation, such as short, thick scrub brush and deep multilayered forest canopy, can completely occlude the laser pulse to the underlying surface. To try and overcome this will require high swath overlap and low flying heights to sample as densely as possible. This may not be feasible due to project constraints, and it may prove futile for some types of landcover. Over smooth water, total reflection of the laser pulse away from the scanner can occur at scan angles away from nadir. This often results in data dropouts over water surfaces. Similar effects can sometimes be seen on wet road surfaces or other smooth surfaces. Furthermore, lidar intensity and derived products are relative measures that provide only limited spectral information about the surface reflectance properties. To compensate, many commercial data providers perform ALS surveys with onboard digital mapping cameras to provide textured point clouds with multispectral information content (e.g., RGB). This imagery can also be orthorectified using a DSM derived from the ALS data providing an additional data product for clients. Some manufacturers of ALS systems provide a digital imaging payload to go along with the lidar unit. Another consideration for ALS is economics. For cost-effectiveness, ALS is better suited for mapping at regional geographic scales (hundreds of square kilometers and upward), as opposed to smaller areas, given the cumulative expense for field operations, equipment, expertise, and processing time.

In spite of these limitations, ALS is a proven technology that has become a standard method in the surveying community for efficient and high-quality production of elevation data and DEMs. ALS has some advantages relative to other topographic mapping methods, including (Vosselman and Maas 2010) the following:

- High measurement density and high accuracy providing decimeter point spacing (>10 points/m^2 is commonplace for high pulse rate systems at low flying heights) with achievable vertical accuracies <0.1 m.
- Fast data acquisition enabling large areal extent mapping at dense resolution during day and night operation, weather permitting.
- Active ranging approach that takes advantage of direct georeferencing and established calibration procedures, thereby reducing the need for extensive ground truth data or control target layout.
- Recording of multiple returns (echoes) per transmitted laser pulse and full-waveform digitization capabilities enable ALS to penetrate canopy, even dense canopy at high sampling resolution, allowing for the measurement of underlying structures and bare-earth surface reconstruction. This multireturn capability is a very important and powerful difference between lidar and photogrammetry. The latter is a passive optical method that is inherently a single-return method based on corresponding pixel overlap. As such, it is limited in terms of canopy penetration and can be prone to errors over canopy stemming from a disparity in the parallax owing to movement such as branches waving in the wind.
- Complementary surface model products can be generated from the multireturn information: digital surface models and bare-earth digital terrain models.

UNMANNED AIRCRAFT SYSTEMS

INTRODUCTION

UASs represent a new frontier in aerial photogrammetry and remote sensing. Improvements in the availability and accuracy of miniaturized positioning and orientation sensors (GNSSs and IMUs, respectively), autonomous navigation, image post-processing software and algorithms, and in the quality of consumer grade digital cameras and miniaturized sensors have resulted in an explosion in the civilian use of UASs for mapping applications. The advantages of UASs for photogrammetry and remote sensing compared with traditional piloted aircraft and satellites include the following: ability to fly at much lower altitudes enabling hyperspatial image resolutions (<1 cm); flexible and repeatable data acquisition to achieve high temporal resolution, target ideal weather conditions and events, or for rapid response needs; data acquisition efficiency and relative inexpensiveness for mapping at localized geographic scales (Turner et al. 2012, Woodget et al. 2015). Through exploitation of high image overlap, automated structure-from-motion (SfM) photogrammetric techniques can be applied to post-process the imagery and derive accurate 2D and 3D geospatial data products about the observed terrain. Spectral data can also be acquired from multispectral sensors onboard the UAS for performing traditional remote sensing tasks such as mapping vegetation health. Small lidar sensors and IMUs can also be carried onboard to enable lidar mapping with UASs. With these capabilities, UASs are being utilized for a variety of surveying and remote sensing tasks in support of various industries, including precision agriculture, topographic surveying, coastal mapping, environmental monitoring, oil and gas development, civil engineering and construction work, transportation engineering, inspection surveying, and many others (Colomina and Molina 2014, Gillins et al. 2018, Starek et al. 2019).

PLATFORMS AND SENSORS

The majority of current aerial mapping performed by the commercial sector is conducted with small UAS platforms that can be equipped with lightweight, consumer-grade cameras providing a cost-effective and efficient means for conducting localized aerial surveys. Currently, the Federal Aviation Administration (FAA) of the United States classifies small UASs as weighing less than 25 kg (55 lbs) including payload. Often these platforms weigh only a few kgs with payload. UASs can be broadly classified into two types: rotary craft and fixed-wing (Figure 10-15). Rotary craft typically provide more flexibility in sensor payload integration, enable more stabilized flight in windy conditions, provide vertical take-off and landing (VTOL) capability, and are better suited for inspection surveying (e.g., hovering), videography, and flying lower or slower to collect higher-resolution imagery with reduced motion blur. In contrast, fixed-wing UASs are usually more efficient at mapping larger areas because of more effective lift generation for a given payload, thereby maximizing battery endurance (Colomina and Molina 2014). The majority of current UAS activity is centered on small rotary platforms because of their widespread availability, ease of use, and lower cost compared with commercial fixed-wing platforms designed for mapping. There are also hybrid platforms integrating rotary and fixed-wing design that provide the advantages of both, such as VTOL with fixed-wing flight capability (Figure 10-15). With continued technological progression, the speed and efficiency of these craft will improve as will battery endurance, and performance differences in platform types will become more blurred (Starek et al. 2019).

Main system components of a UAS include the GNSS receiver for position, IMU for orientation, radio link for communication, onboard processor, and sensor/camera for data acquisition. Autonomous flight is performed using an autopilot that is fed by the onboard GNSS, IMU, and other sensors. UASs often include other navigation aiding sensors such as magnetometers and pivot tubes, as well as sense-and-avoid sensors such as optical flow cameras or lidar, coupled with built-in artificial intelligence for avoidance response. UAS imaging sensors can range from simple RGB visual cameras to near-infrared (NIR) or thermal imagers to sophisticated multispectral, hyperspectral sensors, or lidar sensors. For many common mapping applications, high-resolution digital images from a three-band RGB camera will suffice. Consumer-grade RGB cameras can also be easily modified using filters to capture near-infrared wavelengths, considerably reducing the cost required to

Figure 10-15. Examples of different small UAS commercial platforms from the 2015 to 2020 era: (a) DJI Phantom 4 Pro quad-copter rotary UAS equipped with a 20 MP RGB camera, (b) fixed-wing commercial grade mapping system called the SenseFly eBee that comes with a suite of consumer grade digital cameras modified for visible and near-infrared wavelength capture, (c) WingtraOne hybrid UAS that has a vertical take-off and landing capability that transitions to a fixed wing during flight.
Source: Starek et al. (2019), reproduced with permission from Taylor & Francis Group.

capture such data (Greenwood 2015). Commercial UASs include easy-to-use onboard autopilots for controlling image acquisition, which is important for image stitching and photogrammetric applications. The onboard radio is also essential because it dictates the strength and length of the command and control link between the operator's ground control station (GCS or remote controller; typically, a laptop, tablet, or smart phone with mission control software for automated flight) and the UAS, which has implications for both data collection and safety (Starek et al. 2019).

AERIAL MAPPING WITH UNMANNED AIRCRAFT SYSTEMS

UAS-collected imagery is processed to derive 2D and 3D geospatial data products. Two-dimensional image mosaics (aerial maps) are the most commonly created products from imagery collected by a UAS. The simplest way to create a mosaic from aerial imagery is by using photostitching software, which combines a series of overlapping aerial photographs into a single image called an "image mosaic." However, this approach does not account for perspective distortion and camera lens distortion, thereby resulting in nonuniform scale and inaccurate distance measurements for geospatial applications. An orthomosaic is a mosaicked series of overlapping aerial images that have been geometrically corrected (orthorectified) using an elevation model of the terrain to give them a uniform scale. Orthorectification removes distortion from the images by adjusting for topographic relief, lens distortion, and camera tilt. Orthomosaics are a fundamental GIS data product used in a variety of planar mapping applications. Another commonly created product from UAS imagery is a DSM. A DSM provides a digital representation of the three-dimensional landcover and exposed topography. It is created from 3D point measurements (called a point cloud) derived from the overlapping UAS imagery using specialized photogrammetric processing. The DSM is a raster data type that can be easily ingested into a GIS for further analysis such as to measure object height or volume of a topographic feature. Furthermore, the generated DSM can be used by photogrammetric software to orthorectify the UAS imagery (Starek et al. 2019). Details on flight planning, survey design, and photogrammetric processing are described subsequently.

Mission Planning

Planning a mapping mission with a small UAS entails a number of considerations. A first-order decision to be made is whether the flight will be conducted under autonomous control with mission planning and control software using GNSS waypoints or conducted manually. The decision of whether to use manual or autonomous control hinges on your application. As a general rule, autonomous control is more useful when one is trying to fly in a systematic pattern to acquire imagery for aerial mapping and photogrammetry purposes. In contrast, manual control is usually used for events that require reacting to information in real time such as inspection surveying (e.g., wind turbine, bridges) or video monitoring and streaming of events in real time (e.g., search and rescue operations). Both types of missions can be flown in either manner, or indeed, in a hybrid of manual and automatic control (Greenwood 2015, Gillins et al. 2016).

Prior to conducting any UAS flight operation, it is important to analyze the area to be mapped before liftoff. The area should be visually inspected by walking or driving the premises or otherwise evaluated before the mission starts so as to identify obstacles such as power lines, water bodies, large trees, sensitive areas, or other potential pitfalls. It is good practice to use existing satellite imagery to inspect the area and plot out a flight before takeoff. Most mission planning software enables satellite imagery overlays by linking to web repositories such as Google Earth. Pilots should know how to competently fly their UASs, even if they plan to use them primarily for autonomous mapping missions. This means understanding the launch and recovery procedure, flight control software, trouble shooting, and safety procedures.

Mission planning should also ensure that the flights adhere to regulations such as those from FAA. For example, airspace classes vary and may be restricted near airports, military zones, and sporting stadiums. Mission planning must be aware of the airspace class. In addition, pilots need to be properly certified by FAA, and UASs should remain within the visual line of sight of pilots and ground observers unless the pilots have sufficient experience and special regulatory permission for beyond line-of-sight operations (Starek et al. 2019). See more discussion on FAA regulations at the end of the chapter.

Flight Design

Designing a flight plan for autonomous mapping is done using mission planning and flight control software. These are often proprietary software provided by UAS manufacturers; however, there are open-source solutions. Mission Planner is a current open-source software package widely used by the hobbyist community that works with various compatible autopilots. Also, numerous "flight apps" are designed to work on smart phones and tablets for programming flight missions. The functionality for most of these software, regardless of modality, is usually similar. UAS aerial mapping missions are usually flown in a specific pattern of parallel lines, commonly described as transects, which are connected by a series of geopositional waypoints (Starek et al. 2019). These flight patterns are a method of ensuring that the UAS captures an adequate quantity of images at the desired amount of image overlap. Overlap is essential for high-fidelity photogrammetric processing to derive mapping products and is described in more detail subsequently.

To design a flight plan, the pilots first connect their ground control station (laptop, tablet, smartphone, etc.) to the UAS's flight controller via either a ground control radio link, Wi-Fi, Bluetooth, direct link with a USB cable, or other method. The flight planning process then usually proceeds in the following manner. The pilot defines an area to be mapped with a polygon, specifies the camera model (or inputs information about the camera such as sensor size and focal length), sets the desired operational altitude (which sets the GSD), and then defines overlap to control how the camera will be triggered to take photographs. Once these factors are entered, the mission planning software generates a series of transects with waypoints and displays the estimated GSD, required number of photographs, and other useful information (Figure 10-16). The pilot can then change the distance between each photo, the amount that photos will overlap, the altitude of operation, and other parameters such as the start and home waypoint. Different wind conditions (to assess impact on flight endurance), flight routes, and other factors can be simulated with some mission planning software before flight, making it relatively easy to plan (Greenwood 2015). Satellite and aerial imagery, as well as other ancillary data, like a DSM, can typically be integrated to help the planning process. One should have a working internet connection to plan flights in the field if requiring baseline imagery. Otherwise, the mission can be planned in the office and saved to a computer for access in the field.

Once the mission plan is completed to satisfaction, it is uploaded to the UAS controller and the system is ready for launch. The operator can then launch the drone using the manufacturer's prescribed launch procedure, such as hand launching of a small fixed wing [Figure 10-14(a)]. During flight, the mission control software will display flight diagnostics on the computer or tablet screen including parameters such as altitude, wind speed, GNSS status, battery status, and ground signal status. Most of these software monitor image acquisition during flight to ensure that proper overlap and coverage is maintained.

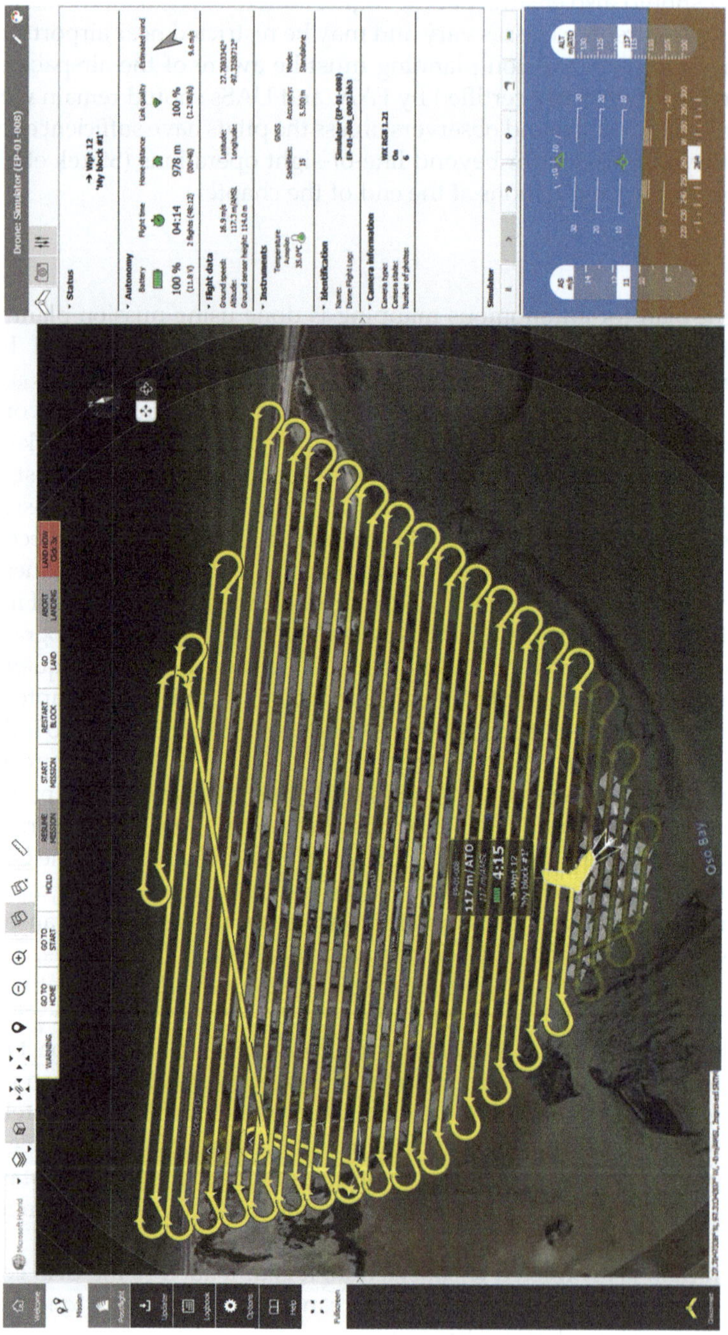

Figure 10-16. Example of a UAS flight design with parallel transects created in proprietary mission planning software called eMotion 3 by senseFly. The software shows image capture and status of the UAS during flight. The area of coverage is the island campus of Texas A&M University-Corpus Christi.

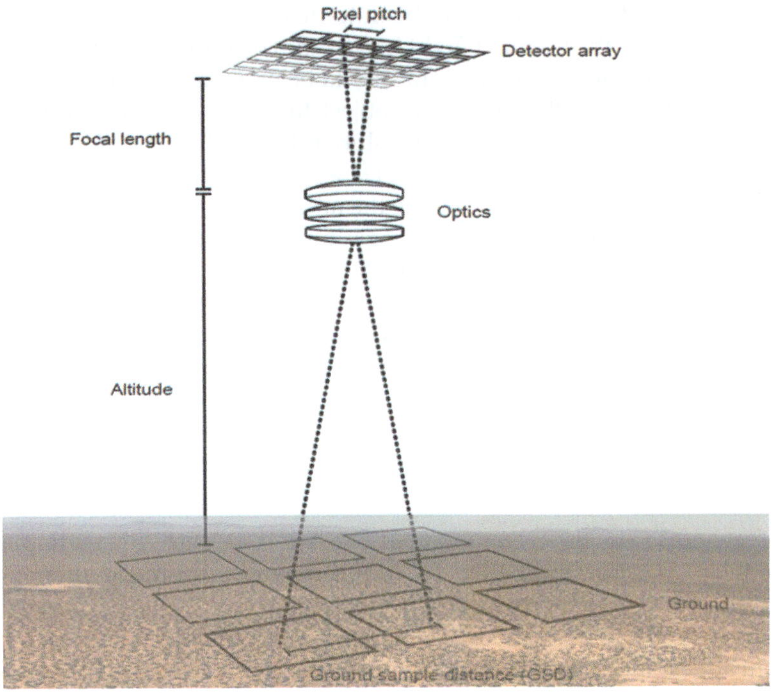

Figure 10-17. Relationship between ground sample distance and flying height for a given camera's pixel pitch and focal length.
Source: Starek et al. (2019), reproduced with permission from Taylor & Francis Group.

Provided that the link is maintained between the ground control station and craft, the software will typically enable one to change flight parameters in-flight, such as adjusting flight line orientation for a certain wind direction (Starek et al. 2019).

Imagery collected during flight is typically stored onboard the platform using a SD memory card located in the camera or via other data storage means. There are services that aim to provide real-time wireless transfer of collected imagery to a cloud architecture during flight for downstream rapid processing using a cellular network or other wireless connection modes. Similar approaches can be implemented for wireless transfer of image data from the platform to the ground control station. However, the feasibility of these approaches will depend on bandwidth, available wireless connectivity, size and amount of imagery, and other factors. Presently, the most common operating mode for image acquisition is onboard storage and download of imagery after landing (Starek et al. 2019).

Ground Sample Distance and Overlap. For conducting aerial photogrammetry missions with small UASs, the two main parameters that govern flight design are GSD and image overlap. GSD is the projected pixel area on the ground and is a function of the camera focal length, physical size of the individual sensor elements inside the camera (called pixel pitch), and flight height above ground level (Figure 10-17). GSD can be estimated as follows:

$$\mathrm{GSD} = \frac{(\text{pixel pitch})(h)}{f} \tag{10-12}$$

where f is the focal length of the respective camera, and h is the flight height above ground level. During flight design, mission planning software will allow one to specify the specific camera model and focal length or it will be auto-detected by the software. The software then allows one to adjust the nominal flight height above ground level to achieve a desired GSD for a given camera model. The camera should be maintained as much as possible at a constant flight height above the ground or object to ensure a consistent GSD. This will ensure the same level of spatial detail or resolution for the images. However, changes in relief will alter the GSD for a set flight height. Dependent on project GSD requirements, the flight altitude may need to be adjusted during the mission to compensate for large variations in terrain elevation. Certain UASs can be programmed to use elevation models to follow the terrain for image acquisition (Starek et al. 2019).

Acquiring sufficient overlap is crucial for small UAS mapping applications because of the type of photogrammetry employed to process the imagery, called SfM. There are two types of overlap to be concerned with, frontal overlap with respect to the flight direction (or endlap) and side overlap between flight lines (or sidelap). The ideal image acquisition plan depends on the type of terrain or object to be reconstructed. The following guidelines are stipulated by the widely used drone photogrammetry software Pix4Dmapper (Pix4D 2018a):

- For the general case, the recommended overlap is at least 75% endlap and at least 60% sidelap. It is recommended to take the images with a regular grid pattern (Figure 10-18).
- For areas covered by trees and dense vegetation, these scenes will often have a very different appearance between overlapping images because of their complex geometry. This makes it difficult to extract common characteristic points (keypoints) between the images. Recommendations to address this include increasing the overlap between images to at least 85% endlap and at least 70% sidelap. In addition, one can increase the flying height to reduce perspective distortion. This can improve the appearance of dense vegetation, making it easier to detect visual similarities between overlapping images in such areas.
- For cases where the terrain is flat with homogeneous visual content such as agricultural fields, image texture is reduced, making it difficult to extract keypoints between the images. To achieve good results, it is recommended to use the same grid image acquisition plan as for the general case but to increase the overlap between images to at least 85% endlap and at least 70% sidelap. In most cases, flying higher can also improve the results.
- For corridor mapping of roads, railways, or rivers, and so on, it is recommended to fly a dual track flight path with a forward and return track. Overlap settings should be at least 85% endlap and at least 60% sidelap. Nadir images or oblique images may be used with an angle between 0 degree and 45 degree pointing in both tracks to the center of the corridor. For flat terrain, it is recommended to use nadir images. If a dual track image acquisition plan is not possible, a single-track image acquisition plan can be used provided the overlap is set high enough. At least 85% endlap is recommended.
- For 3D building modeling, a specific image acquisition plan is necessary. It is recommended to first fly around the building with a 45 degree camera angle and then to fly around it a second and third time, increasing the flight height and decreasing the camera angle with each round. It is recommended to take one image every 5 degrees to 10 degrees to ensure enough overlap, depending on the size of the object and distance to it.
- For mapping of urban areas, it is recommended to fly a grid pattern with perpendicular flight lines (referred to as a double grid). This helps improve the reconstruction of building edges and facades from all directions. Recommended overlap is the same as for the general case.

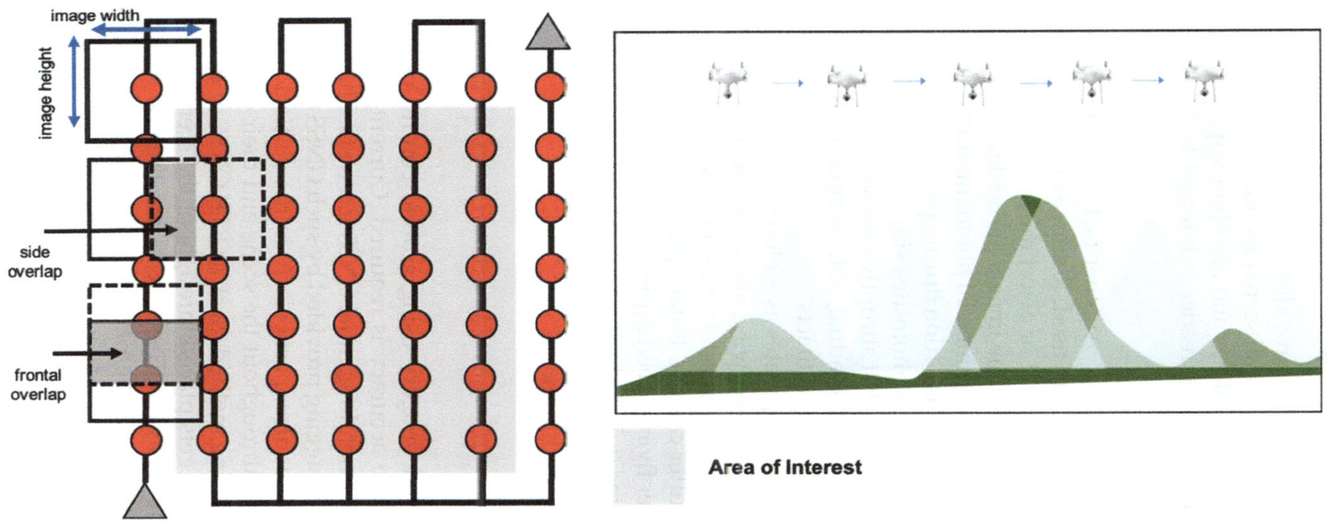

Figure 10-18. Grid image acquisition plan showing the meaning of frontal overlap (endlap) and side overlap (sidelap). Source: Starek et al. (2019), reproduced with permission from Taylor & Francis Group.

- For multiple flights, the following considerations should be taken for image acquisition. Make sure that each individual flight plan captures images with enough overlap and that there is sufficient overlap between two flight plans. The different plans should be acquired as much as possible under the same ambient conditions (sun direction, weather conditions, no new buildings, etc.).

The aforementioned guidelines are provided only as example suggestions for certain cases and are by no means inclusive of all scenarios. Image acquisition plans will vary by project requirements and UAS capabilities and are often impacted by weather. For details on flight design for UAS mapping, the reader is referred to Pix4D (2018a) and Carrivick et al. (2016).

Rolling and Global Shutter Cameras. Another factor to consider when designing UAS mapping missions is the type of camera used. Rolling shutter cameras read the image line by line or in groups. Camera movement or object movement during this reading time can create artifacts such as motion blur in the image, particularly for UASs moving at a rapid velocity relative to the terrain. Some commercial photogrammetry software provide settings to attempt to correct rolling shutter distortion during processing. In contrast, global shutter cameras control incoming light to all photosites simultaneously. A global shutter camera can work using a hard shutter mode that abruptly exposes and then obstructs all photosites at once. An alternate technique is a gradual, soft electronic shutter. Hard global shutter cameras fully address rolling shutter artifacts.

Recognizing whether a camera operates as a global or rolling shutter can impact flight design for mapping applications. Fixed-wing platforms are limited in their ability to reduce velocity, and so, higher flying heights relative to the terrain may be necessary to reduce motion blur for a rolling shutter camera. Rotary platforms provide more control on flight velocity, allowing for a slower flying speed during image acquisition to reduce motion blur. If necessary, a rotary platform equipped with a rolling shutter camera can be programmed to come to a stop, hover, and acquire the image at each desired photo site with the trade-off that this will greatly increase flight time.

Ground Control

Georeferencing of acquired imagery using ground control points (GCPs) is usually necessary when high absolute accuracy is required. Currently, most small UASs come equipped with single-channel, nondifferential GNSS receivers. The raw positional accuracy stemming from the image geotagging provided by such GNSS receivers is typically meter-scale but can be substantially worse, particularly in the vertical (Jóźków and Toth 2014). Control targets can be placed throughout the scene and their coordinates acquired using traditional surveying engineering methods such as RTK GNSS or from other sources. GCPs can consist of aerial control targets placed throughout the scene or static features that are easily identifiable within the imagery. These GCPs can then be identified in the imagery during data processing to improve the photogrammetric solution and more accurately georeference the derived mapping products (Figure 10-19). For ease of identification, it is recommended that the GCP target size be 5 to 10 times the size of the imagery GSD (Pix4D 2018b).

When establishing a ground control network for a small UAS photogrammetric survey, the GCPs should be distributed as evenly as possible throughout the landscape to minimize the error in scale and orientation. At a minimum, three GCPs are required for them to be taken into account for a 3D reconstruction. Pix4D (2018b) recommends a minimum number of five GCPs, each of which is identifiable in five or more images. As a general guideline,

Figure 10-19. (Left, middle) different types of aerial GCP targets placed in a scene. (Right) Identifiable features in the scene that can be used as GCPs (yellow arrows).
Source: Courtesy of FHWA, reproduced with permission.

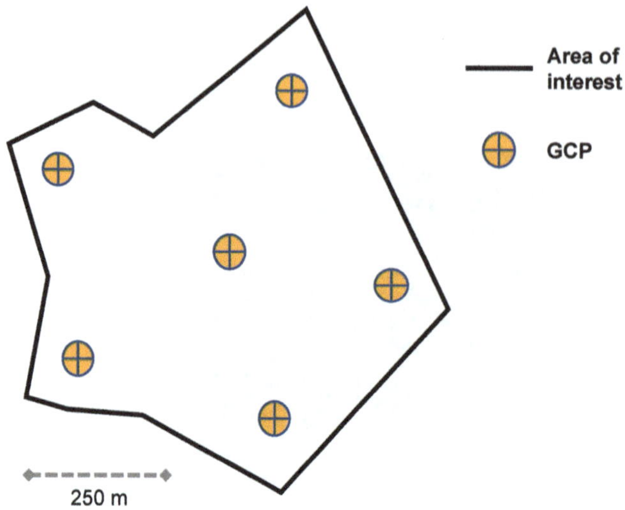

Figure 10-20. Example geometry for a ground control network.

they state that 5 to 10 GCPs are usually enough, even for larger projects conducive to a small UAS survey. However, the optimal number and distribution will depend heavily on the size of the area to be mapped and the complexity of the terrain as well as feasibility in their placement (McMahon et al. 2021). In areas where the topography of the area is complex (large changes in elevation), more GCPs will usually lead to better (more accurate) reconstruction. Furthermore, GCPs should not be placed exactly at the edges of the survey area, as they will be visible only in a few images. For corridor mapping, GCPs should be staggered on either side of the corridor feature to be mapped. Figure 10-20 shows an example geometry for a ground control network.

UAS photogrammetric measurements are only as reliable as the control network for which they are based during data processing. Measurements moving away from the control network, in general, cannot be trusted. In essence, your area of high accuracy is constrained by the boundary of your control. Furthermore, it is best practice to include additional GCPs to serve as independent check points for evaluating reconstruction accuracy. Similar to target GCPs, these should be distributed as evenly as possible throughout the survey area to assess error. Following ASPRS guidelines, a minimum of 20 checkpoints should be utilized for statistical evaluation of RMSE (ASPRS 2015). In practice, the number of checkpoints needed will depend on accuracy validation requirements, terrain complexity, and feasibility in their placement. Finally, more and more UASs are being equipped with differential-enabled GNSS receivers (RTK and/or PPK capable) that greatly improve the accuracy of the onboard geotagging of the imagery. These can reduce and, in some cases, eliminate the need for ground control dependent on accuracy requirements (Jóźków and Toth 2014). This has the advantage of potentially eliminating the workload required in establishing a control network, although it is still recommended to include some GCPs as independent checkpoints for accuracy evaluation.

STRUCTURE-FROM-MOTION PHOTOGRAMMETRY

Most small UAS photogrammetric surveys are processed using a technique called SfM/Multi-View Stereo (MVS) photogrammetry (or SfM for short). Traditional photogrammetry requires metric cameras precisely calibrated. However, metric cameras are expensive and not conducive for the widespread use of UASs for mapping applications. SfM exploits information from multiple overlapping images to extract 3D object information and negate the need for precise camera calibration. SfM derives three-dimensional structure from two-dimensional image sequences through movement of the camera, thereby providing different perspective views of the scene. By using the UAS as the moving platform, SfM can be implemented with an onboard camera by acquiring images with sufficient overlap as previously explained. At the time of this writing, UAS-SfM is becoming a widely used and powerful tool for surveying engineering. (Note that UAS-lidar is also becoming a powerful tool in surveying engineering and provides data similar to MLS discussed in Chapter 9, but it requires more expensive sensors and UAS capable of lifting heavier payloads; UAS-SfM is currently more popular, differs from traditional photogrammetry, and is discussed in more detail subsequently.)

Several commercial SfM photogrammetry software options are available to process UAS imagery. Two commonly used commercial SfM software at the time of this writing are Pix4Dmapper and Agisoft Metashape. Although each SfM software package has its own interface, workflows, and internal algorithms, most SfM software packages follow similar processing steps. The typical SfM image processing workflow is as follows (Westoby et al. 2012, Slocum and Parrish 2017, Javadnejad et al. 2021):

- Stage 1 (Keypoint Extraction and Matching): Image sequences are input into the software and a keypoint detection algorithm, such as the scale invariant feature transform (SIFT), is used to automatically extract features and find keypoint correspondences between overlapping images using a keypoint descriptor. Keypoints are points of interest (e.g., features of high-contrast, interesting texture) on the UAS images that can be easily recognized by the SfM software's automated keypoint extraction and correspondence algorithms. SIFT is a well-known computer vision algorithm that allows for feature detection regardless of scale, camera rotations, camera perspectives, and changes in illumination (Lowe 2004).
- Stage 2 (Bundle Adjustment and Sparse Point Cloud Creation): A least-squares iterative bundle adjustment (BA) is performed to minimize the errors in the keypoint correspondences by automatically solving for camera interior orientation (IO) parameters and performing aerial triangulation (AT) to resolve the relative orientation parameters (position and angular orientation) of the camera at the time of image acquisition. Based on this reconstruction, the matching points are verified, and their 3D coordinates simultaneously calculated to generate a sparse point cloud. Without any other spatial information, the coordinate system is arbitrary in translation and rotation and has inaccurate scale.
  - Sub Step (Point Cloud Georeferencing). To further constrain the problem and develop a georectified point cloud, GCPs and/or camera geolocations provided from an onboard GNSS receiver are introduced to constrain the solution. This process resolves the exterior orientation (EO) parameters of the camera and the coordinates of the sparse point cloud with respect to a geodetic coordinate system. The solution can rely solely on the direct georeferencing provided by the onboard GNSS receiver, or GCPs can be applied to optimize the solution. Orientation information provided by an onboard IMU can also be used by the SfM software to weigh the solution and potentially aid it, but in general, this information is not necessary as orientation is solved directly via BA in Stage 2.
- Stage 3 (3D Point Cloud Densification): Finally, the interior and exterior orientations for each image are used as input into an MVS algorithm, which attempts to densify the point cloud by projecting every pixel at the full image scale or projecting pixels at a reduced image scale. Image scale refers to the size of the image at which additional 3D points are generated. For example, ½ image scale densification uses images that are half the size of the original image size.

Figure 10-21. Densified 3D point cloud textured by the RGB pixel values generated from a small UAS-SfM survey of a section of interstate highway in Texas. Point density is >1,000 points/m².
Source: Courtesy of FHWA, reproduced with permission.

The base data product output from the UAS-SfM image processing workflow is a densified set of *X–Y–Z* coordinates of the imaged scene. This set, called a point cloud, is typically colorized by the RGB pixel values of the digital camera. UAS-SfM point clouds can have high point density (easily exceeding 1,000 points/m^2) because of the high camera resolutions and typical low altitudes at which data are collected (Figure 10-21). The 3D point cloud can then be used to generate a digital surface model of the terrain, which can then be used to orthorectify the images and produce an orthomosaic image or a 3D textured mesh. Figure 10-22 summarizes the UAS-SfM processing workflow implemented within the commercial software Pix4Dmapper.

In addition to RGB cameras, consumer-grade cameras can be easily adapted with filters to enable the acquisition of near-infrared and other parts of the spectrum. Furthermore, advancements in sensor miniaturization enable thermal and multispectral sensors to be equipped and flown on small UASs. These sensors can be processed using SfM techniques to derive spectral information, such as reflectance maps, for fusion with the 3D structure information (Javadnejad et al. 2019).

Unmanned Aircraft Systems-Structure-from-Motion Accuracy

Deriving accurate, reliable 3D reconstructions of a scene from UAS-SfM presents interesting challenges. The accuracy of UAS-SfM is adversely affected by a variety of factors, including the following: poor image overlap, inadequate modeling of lens distortion, poor GCP distribution, inaccurate GCP or camera positions, poor image resolution, blurry imagery, noisy imagery, varying sun shadows, moving objects in the scene, user error in manually selecting image coordinates of GCPs, a low number of images, or a low number of GCPs (Smith and Vericat 2015, Slocum and Parrish 2017). Because of the large number of variables involved, understanding UAS-SfM measurement uncertainty in real-world surveying scenarios is an active area of research.

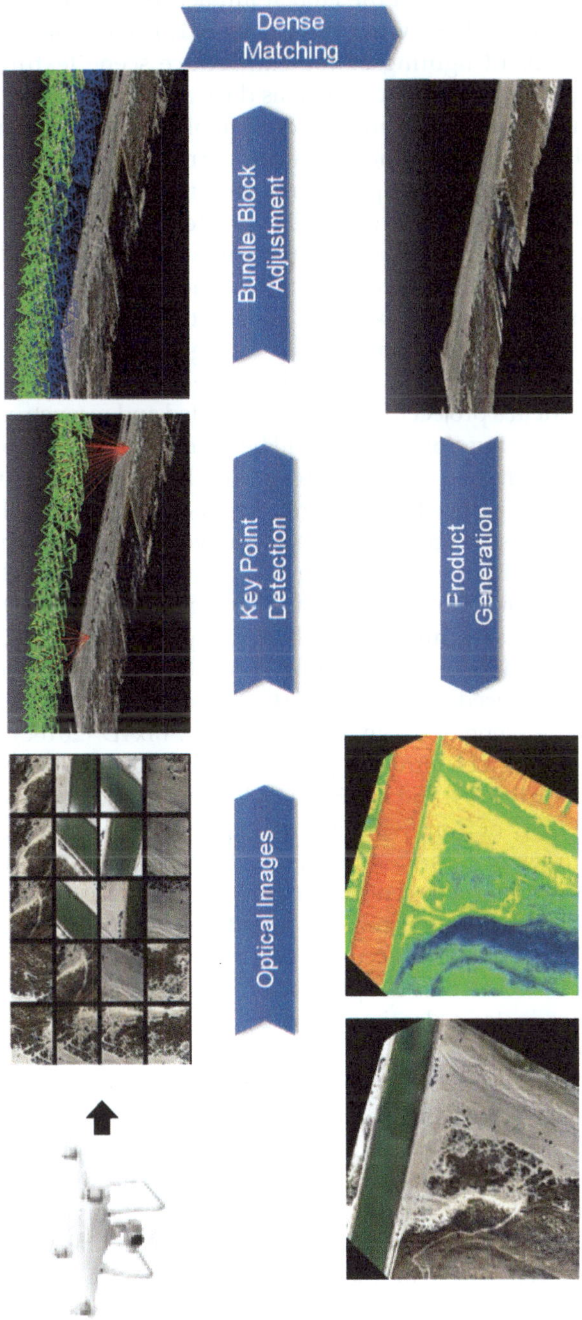

Figure 10-22. SFM workflow to process overlapping UAS image sequences into a densified 3D point cloud, DSM, and orthomosaic. Source: Starek et al. (2019), reproduced with permission from Taylor & Francis.

Scene texture also influences SfM reconstruction accuracy. The quality and uniqueness of detected keypoints in an image and on an object is subject to scene texture as represented by the image. Image texture describes the spatial arrangement of color or intensities in an image or selected region of an image. The lack of texture within a scene, as represented by an image, has been shown to have one of the largest impacts on the accuracy of UAS-SfM point clouds (Javadnejad et al. 2021, Schwind and Starek 2017). Suboptimal environmental conditions such as low ambient lighting or fog can reduce scene texture significantly. Similarly, certain landscapes or built surfaces such as dark asphalt pavement may be more homogeneous in appearance, lacking distinct or identifiable texture compared with other types, and can potentially result in reduced feature correspondence (i.e., matching of distinct keypoints shared between overlapping images) in certain areas. Understanding this relationship is complex and presently not well characterized.

Relative accuracy (i.e., position or distance measurements made within the point cloud itself) will depend on the quality of the reconstruction of the UAS-SfM project, which itself depends highly on the amount of overlap between images, the visual content of the images, and many other parameters as explained previously. As a general guideline, Pix4D (2018c) states that one can expect an error of one to three times the GSD of the original images for the relative position of a point in a project that is correctly scaled and reconstructed. However, this error is not global and can vary. Some points will be more accurate than others based on the number of images on which they are visible (overlap), how close they are to a GCP (if used), and what is the GSD in this area. For sub-centimeter GSD projects, it is harder to reach the ideal error conditions because problems with perspective, vibrations of the camera, blur effect, and so on, are magnified. Furthermore, using GCPs may improve the relative accuracy, especially in areas with lower overlap or with difficult image content (Pix4D 2018c).

Network accuracy increases significantly when using GCPs or RTK/PPK GNSS–corrected image geolocations. When using ground control, the accuracy depends on the accuracy of the survey method used to measure the GCP coordinates, the number of GCPs, and their distribution in the scene. As a general guideline, Pix4D (2018c) states that the expected accuracy of a correctly reconstructed project is 1to 2 times the GSD of the original images horizontally (x,y-coordinates) and 1 to 3 times the GSD vertically (z-coordinate). For example, a project with a GSD of 5 cm will have an expected horizontal accuracy of 5 to 10 cm and a vertical accuracy of 5 to 15 cm (Pix4D 2018c). This error can, in theory, be reached at any scale; however, for sub-centimeter GSD projects, it is harder to obtain the ideal conditions because of the same problems mentioned for relative accuracy. It is important to emphasize that the network accuracy of the results cannot exceed the accuracy of the GCP coordinates or GNSS method employed for image geolocation. The accuracy of UAS-SfM derived data products is also impacted by the variety of other factors listed previously, including original image quality and visual content. Similarly, terrain features such as sharp edges, reflective surfaces, moving objects (e.g., vegetation blowing in the wind), and low texture areas, such as roads, may also be locally less accurate.

In spite of the error complexity, numerous studies have shown the potential of UAS-SfM implemented with consumer-grade digital cameras to generate high-fidelity, accurate topographic survey products at a low cost for a variety of applications (Westoby et al. 2012, Starek et al. 2014, McMahon et al. 2021). Point-cloud accuracies comparable to terrestrial laser scanning (survey-grade like data) are achievable in certain cases with proper use of ground control and a well-designed image acquisition plan as evidenced by some of the prior studies.

REGULATIONS

In the United States, commercial use of UASs for aerial surveying or other applications is regulated by FAA's small UAS rule (called Part 107), which became effective on August

29, 2016. Current highlights of the operational rules at the time of this writing (2021) are summarized as follows:

> Aspiring small UAS commercial pilots must be 16 years old, pass an initial aeronautical knowledge test at an FAA-approved knowledge testing center, and be vetted by the Transportation Security Administration (TSA). Their aircraft must weigh less than 55 pounds and be registered with the FAA. Operation is limited to Class G airspace, within visual line-of-sight, under 400 feet, daylight hours, and at or below 100 miles per hour. The small UAS must yield right of way to manned aircraft and cannot fly from a moving vehicle. Part 107 includes the option to apply for a certificate of waiver, which allows for a small UAS operation to deviate from certain operating rules if the FAA finds that the proposed operation can be performed safely.

Any UAS flight conducted for commercial use must ensure that it adheres to appropriate FAA regulations and that proper safety precautions are taken. A more detailed discussion of UAS regulations and flight safety is beyond the scope of this discussion. However, the reader is referred to the "Know Before You Fly" website (AUVSI and AMA 2020) for up-to-date information on UAS regulations and flight responsibilities.

Finally, it is important to mention that at the time of this writing, UAS technology and regulations are still emerging. The technology, regulations, and capabilities for surveying will rapidly evolve over the coming years. However, the sensor types, data products, and general methods for data acquisition and processing, as discussed here, will likely remain similar.

REFERENCES

ASPRS (American Society for Photogrammetry and Remote Sensing). 2015. "ASPRS positional accuracy standards for geospatial data." *Photogramm. Eng. Remote Sens.* 81 (3): 1–26.

ASPRS. 2019. "LAS specification 1.4-R15." Accessed February 4, 2022. http://www.asprs.org/wp-content/uploads/2019/07/LAS_1_4_r15.pdf.

AUVSI (Association for Unmanned Vehicle Systems International) and AMA (Academy of Model Aeronautics). 2020. "Know before you fly." http://knowbeforeyoufly.org/ (accessed 4 February 2022).

Baltsavias, E. P. 1999. "Airborne laser scanning: Basic relations and formulas." *ISPRS J. Photogramm. Remote Sens.* 54 (2-3): 199–214.

Carrivick, J. L., M. W. Smith, and D. J. Quincey. 2016. *Structure from motion in the geosciences.* Chichester, UK: Wiley.

Carter, W. E., R. L. Shrestha, and K. C. Slatton. 2007. "Geodetic laser scanning." *Phys. Today* 60 (12): 41.

Colomina, I., and P. Molina. 2014. "Unmanned aerial systems for photogrammetry and remote sensing: A review." *ISPRS J. Photogramm. Remote Sens.* 92: 79–97.

Congalton, R., and K. Green. 2008. *Assessing the accuracy of remotely sensed data: Principles and practices.* Boca Raton, FL: CRC Press.

Corbley, K. P. 2015. "Geiger-mode LiDAR delivers fast, wide-area mapping." Accessed February 4, 2022. https://eijournal.com/print/articles/geiger-mode-lidar-delivers-fast-wide-area-mapping.

Cressie, N. A. 1991. *Statistics for spatial data.* New York: Wiley.

Eisaman, M. D., J. Fan, A. Migdall, and S. V. Polyakov. 2011. "Invited review article: Single-photon sources and detectors." *Rev. Sci. Instrum.* 82: 071101.

Fraser, C. 2013. "Automatic camera calibration in close range photogrammetry." *Photogramm. Eng. Remote Sens.* 79 (4): 381–388.

Gelb, A. 1974. *Applied optimal estimation*. Cambridge, MA: MIT Press.

Ghilani, C. 2010. *Adjustment computations: Spatial data analysis*. New York: Wiley.

Gillins, D. T., C. Parrish, M. N. Gillins, and C. Simpson. 2018. *Eyes in the sky: Bridge inspections with unmanned aerial vehicles*. Final Rep. No. SPR 787. Salem, OR: Oregon Dept. of Transportation.

Gillins, M. N., D.T. Gillins, and C. Parrish. 2016. "Cost-effective bridge safety inspections using unmanned aircraft systems (UAS)." In *Geotechnical and Structural Engineering Cong.*, edited by C. Y. Chandran and M. I. Hoit. Phoenix, AZ: ASCE.

Granshaw, S., and C. S. Fraser. 2015. "Editorial: Computer vision and photogrammetry: Interaction or introspection?" *Photogramm. Rec.* 30 (149): 3–7.

Green, K. 2008. "ASPRS: The imaging and geospatial information society." *Photogramm. Eng. Remote Sens.* 74 (5): 561–565.

Greenwood, F. 2015. *How to make maps with drones*, 355–347. Appears in: Drones and aerial observation: New technologies for property rights, human rights, and global development. Tech. Rep. Washington, DC: New America.

Guenther, G. C., A. G. Cunningham, P. E. LaRocque, and D. J. Reid. 2000. *Meeting the accuracy challenge in airborne lidar bathymetry*. Dresden, Germany: EARSel.

Heidemann, H. K. 2014. "Lidar base specification." In *Techniques and methods*, Book 11, Chapter B4. Reston, VA: USGS.

Honkavaara, E., J. Jaakkola, L. Markelin, and S. Becker. 2006. "Evaluation of resolving power and MTF of DMC." *Int. Arch. Photogramm. Remote Sens. Spatial Inf. Sci.* 36 (B1): 6.

Huber, D. 2011. "The ASTM E57 file format for 3D imaging data exchange." In *Three-Dimensional Imaging, Interaction, and Measurement*, Vol. 7864, edited by J. A. Beraldin, G. S. Cheok, M. B. McCarthy, U. Neushaefer-Rube, I. E. McDowall, M. Dolinksy, and A. M. Baskurt; 78640A. Bellingham, WA: International Society for Optics and Photonics.

Javadnejad, F., D. T. Gillins, C. Parrish, and R. K. Slocum. 2019. "A photogrammetric approach to fusing natural colour and thermal infrared UAS imagery in 3D point cloud generation." *Int. J. Remote Sens.* 41 (21): 1–27.

Javadnejad, F., R. K. Slocum, D. T. Gillins, M. J. Olsen, and C. E. Parrish. 2021. "Dense point cloud quality factor as proxy for accuracy assessment of image-based 3D reconstruction." *J. Surv. Eng.* 147 (1): 04020021.

Johnson, S., P. Gatt, and T. Nichols. 2003. "Analysis of Geiger-mode APD laser radars." In *Proc. of SPIE 5086, Laser Radar Technology and Applications VIII*, edited by G. W. Kamerman, 359–368. Bellingham, WA: International Society of Optics and Photonics.

Jóźków, G., and C. Toth. 2014. "Georeferencing experiments with UAS imagery." *ISPRS Ann. Photogramm. Remote Sens. Spatial Inf. Sci.* 2 (1): 25.

Krabill, W. B., J. G. Collins, L. E. Link, R. N. Swift, and M. L. Butler. 1984. "Airborne laser topographic mapping results." *Photogramm. Eng. Remote Sens.* 50 (6): 685.

Lowe, D. G. 2004. "Distinctive image features from scale-invariant keypoints." *Int. J. Comput. Vis.* 60: 91–110.

Mader, G. L. 1992. "Rapid static and kinematic global positioning system solutions using the ambiguity function technique." *J. Geophys. Res.* 97 (B3): 3271–3283.

McGlone, J. C., ed. 2013. *Manual of photogrammetry*. 6th ed. Bethesda, MD: American Society for Photogrammetry and Remote Sensing.

McMahon, C., O. E. Mora, and M. J. Starek. 2021. "Evaluating the performance of sUAS photogrammetry with PPK positioning for infrastructure mapping." *Drones* 5 (2): 50.

Meng, X., N. Currit, and K. Zhao. 2010. "Ground filtering algorithms for airborne LiDAR data: A review of critical issues." *Remote Sens.* 2 (3): 833–860.

Mikhail, E., and F. Ackermann. 1976. *Observations and least squares*. New York: IEP.

Mikhail, E., J. Bethel, and J. C. McGlone. 2001. *Introduction to modern photogrammetry*. New York: Wiley.

Parrish, C. E., I. Jeong, R. D. Nowak, and R. B. Smith. 2011. "Empirical comparison of full-waveform LiDAR algorithms." *Photogramm. Eng. Remote Sens.* 77 (8): 825–838.

Pfennigbauer, M., and A. Ullrich. 2010. "Improving quality of laser scanning data acquisition through calibrated amplitude and pulse deviation measurement." In *SPIE Defense, Security, and Sensing*. Bellingham, WA: International Society for Optics and Photonics.

Pix4D. 2018a. "Designing the image acquisition plan." *Article 202557459*. Accessed May 2018. https://support.pix4d.com/hc/en-us/articles/202557459.

Pix4D. 2018b. "Getting GCPs on the field or through other sources." *Article 202557489*. Accessed May 2018. https://support.pix4d.com/hc/en-us/articles/202557489.

Pix4D. 2018c. "Accuracy of Pix4Dmapper outputs." *Article 202557489*. Accessed May 2018. https://support.pix4d.com/hc/en-us/articles/202558889-Accuracy-of-Pix4Dmapper-Outputs.

Rogers, R. M. 2003. *Applied mathematics in integrated navigation systems*. 2nd ed. Reston, VA: American Institute of Aeronautics and Astronautics.

Schwind, M., and M. J. Starek. 2017. "How to produce high-quality 3d point clouds: Structure-from-motion photogrammetry." *GIM Int. Worldwide Mag. Geomatics* 31 (10): 36–39.

Shan, S., M. Bevis, E. Kendrick, G. Mader, D. Raleigh, K. Hudnut, et al. 2007. "Kinematic GNSS solutions for aircraft trajectories: Identifying and minimizing systematic height errors associated with atmospheric propagation delays." *Geophys. Res. Lett.* 43: L23S07.

Shrestha, R., W. Carter, C. Slatton, and W. Dietrich. 2007. "Research-quality' Airborne laser swath mapping: The defining factors." *National Center for Airborne Laser Mapping (NCALM)*. Accessed February 2022. http://ncalm.berkeley.edu/reports/NCALM_WhitePaper_v1.2.pdf.

Sithole, G., and G. Vosselman. 2004. "Experimental comparison of filter algorithms for bare-Earth extraction from airborne laser scanning point clouds." *ISPRS J. Photogramm. Remote Sens.* 59 (1-2): 85–101.

Skaloud, J., and D. Lichti. 2006. "Rigorous approach to bore-sight self-calibration in airborne laser scanning." *ISPRS J. Photogramm. Remote Sens.* 61 (1): 47–59.

Slama, C., C. Theurer, and S. Henrickson. eds. 1980. "*Manual of photogrammetry*." 4th ed. Bethesda, MD: American Society for Photogrammetry.

Slatton, K. C., W. E. Carter, R. L. Shrestha, and W. Dietrich. 2007. "Airborne laser swath mapping: Achieving the resolution and accuracy required for geosurficial research." *Geophys. Res. Lett.* 34 (23): 1–5.

Slocum, R. K., and C. Parrish. 2017. "Simulated imagery rendering workflow for UAS-based photogrammetric 3D reconstruction accuracy assessments." *Remote Sens.* 9 (4): 396.

Smith, M. W., and D. Vericat. 2015. "From experimental plots to experimental landscapes: Topography, erosion and deposition in sub-humid badlands from structure-from-motion photogrammetry." *Earth Surf. Processes Landforms* 40 (12): 1656–1671.

Starek, M. J., T. Davis, D. Prouty, and J. Berryhill. 2014. "Small-scale UAS for geoinformatics applications on an island campus." In *Proc. of 2014 Ubiquitous Positioning Indoor Navigation and Location Based Service*, edited by A. Wieser, 120–127. Danvers, MA: IEEE.

Starek, M. J., M. Gingras, and G. Jeffress. 2019. "Application of unmanned aircraft systems for coastal mapping and resiliency." In Chap. 7 of *Sustainable Development Goals Connectivity Dilemma: Land and Geospatial Information for Urban and Rural Resilience*, edited by A. Rajabifard, 109–125. Boca Raton, FL: CRC Press.

Starek, M. J., R. K. Vemula, K. C. Slatton, R. L. Shrestha, and W. E. Carter. 2007. "Shoreline based feature extraction and optimal feature selection for segmenting airborne LiDAR intensity images." In *Vol. 4 of Proc., IEEE Int. Conf. on Image Processing*, IV–369. Danvers, MA: IEEE.

Turner, D., A. Lucieer, and C. Watson. 2012. "An automated technique for generating georectified mosaics from ultra-high resolution unmanned aerial vehicle (UAV) imagery, based on structure from motion (SfM) point clouds." *Remote Sens.* 4 (5): 1392–1141.

Ussyshkin, V., and L. Theriault. 2011. "Airborne lidar: Advances in discrete return technology for 3D vegetation mapping." *Remote Sens.* 3 (3): 416–434.

Vosselman, G., and H.-G. Maas. eds. 2010. *Airborne and terrestrial laser scanning*. Boca Raton, FL: CRC Press.

Walker, S., and D. Alspaugh. 2013. "A brief history of photogrammetry." In Chap. 1 of *Manual of photogrammetry*, 6th ed., edited by J. Chris McGlone, 4–10. Bethesda, MD: American Society for Photogrammetry and Remote Sensing.

Wang, X. 2017. "Filtering of single photon-counting lidar data." Ph.D. thesis, University of Houston, Dept. of Civil and Environmental Engineering.

Wehr, A., and U. Lohr. 1999. "Airborne laser scanning—An introduction and overview." *J. Photogramm. Remote Sens.* 54 (2-3): 68–82.

Westoby, M. J., J. Brasington, N. F. Glasser, M. J. Hambrey, and J. M. Reynolds. 2012. "'Structure-from-Motion' photogrammetry: A low-cost, effective tool for geoscience applications." *Geomorphology* 179: 300–314.

Wolf, P., B. Dewitt, and B. Wilkinson. 2014. *Elements of photogrammetry (with applications in GIS)*. 4th ed. New York: McGraw-Hill.

Woodget, A. S., P. E. Carbonneau, F. Visser, and I. P. Maddock. 2015. "Quantifying submerged fluvial topography using hyperspatial resolution UAS imagery and structure from motion photogrammetry." *Earth Surf. Processes Landforms* 40 (1): 47–64.

Wozencraft, J., and D. Millar. 2005. "Airborne LiDAR and integrated technologies for coastal mapping and nautical charting." *Mar. Technol. Soc. J.* 39 (3): 27–35.

CHAPTER 11
SURVEY CONTROL

Daniel T. Gillins

INTRODUCTION

Every surveying engineering project requires control. A properly controlled project ensures that the work is tied together and connected to a common reference system or datum. In addition, it is common practice to set or *densify* control in an engineering project area for future operations. The purpose of this chapter is to discuss why it is important to correctly identify the necessary types of control at the beginning of a survey project; provide recommendations on how to find available local and geodetic control; and mention available tools for establishing new, denser control for a project. This chapter also discusses the Online Positioning User Service (OPUS) of the National Geodetic Survey (NGS). OPUS is a freely available tool and can be used to readily reference surveys to the National Oceanic and Atmospheric Administration (NOAA) continuously operating reference stations (CORS) Network, the backbone control of the United States' National Spatial Reference System (NSRS). OPUS is available at NGS (2021a).

The identification, appropriate use, and setting of control is critical for surveying engineering. Careful planning should take place at the beginning of a project to identify the control to serve as the foundation of all survey operations. Such operations should be based on and tied to the survey control. This way, the engineering work will mesh together, and coordinates will be properly related in the same system (ASCE 1985). Construction plans and right-of-way plats will be based on a common bearing base and will provide consistent data for the lengths and directions of lines. Because control helps tie everything together, it is best practice to set stable, long-lasting monuments. This way, current and even future operations can readily take place.

Before proceeding with the stated material in this chapter, we pause a moment to underscore that its contents are specific to the control from which the survey measurements are referenced. The control allows determination of the coordinates (e.g., latitude, longitude, elevation) and/or basis of bearing for all measured marks in the survey project. The control can be thought of as providing the fundamental *base layer*, which ties the survey together and on which other layers of spatial information can be constructed and meshed together. Property corners and rights-of-way markers are typically referenced to the same control to provide a consistent set of coordinates for the boundary lines in the same system for a project. This allows engineering design, construction, as-built documentation, and more. However, it is important to note that boundary surveying is based on legal principles that ensure property rights are not harmed and bona fide rights are protected. It involves following in the footsteps of original surveys and finding evidence of controlling boundary corner monuments. Such material is outside the scope of this manual and falls under the expertise of a licensed professional land surveyor.

HORIZONTAL, VERTICAL, OR BOTH

Traditionally, surveying engineering control has been separated into two categories: control in the horizontal (i.e., latitude and longitude), and control in the vertical (i.e., elevation or orthometric height). Historic horizontal control was most often established using triangulation (optical) methods, where intervisibility was critical between marks. Hence, horizontal control or *triangulation* marks were often set on the tops of buildings, hilltops, and mountains, providing long-sight distances. On the contrary, vertical control, referred to herein as *bench marks*, was more commonly established by differential leveling, where ease of access was paramount to leap-frog the rods along leveling lines. Thus, bench marks were most frequently set along conveniently accessed road and corridor alignments. These differing requirements typically resulted in two sets of control for a given survey project.

Because of the two categories of control, it was common practice for surveying engineers to search for existing horizontal and vertical controls separately. In fact, surveys were usually done in two pieces, where a horizontal-only survey tied to the horizontal control might be done using a transit to determine latitudes and longitudes. Then, elevations were determined by differential leveling from the vertical control. Survey networks were adjusted separately to determine horizontal coordinates and heights.

Today, Global Navigation Satellite Systems (GNSSs) have changed the way control is established (Chapter 6). First, GNSS outputs coordinates in three dimensions (i.e., latitude, longitude, and ellipsoid height). By using a geoid model, it is also possible to subtract geoid heights into ellipsoid heights to estimate orthometric heights. Second, GNSS does not require intervisibility between marks nor necessarily ease of access. Rather, it requires sufficient overhead visibility for making satellite observations. With the advent of GNSSs, control marks no longer need to be separated into horizontal and vertical categories; instead, they can function as control for both.

Nonetheless, it is important to note that differential leveling and even trigonometric leveling is more accurate than GNSSs for transferring orthometric heights over short distances (i.e., <10 to 20 km) (Chapter 7). As a result, GNSS is more commonly used today to establish horizontal control. Actually, it would be better to say GNSS is used to establish geometric control because it also outputs ellipsoid heights; hence, the term *geometric control* is used in this chapter to indicate control in latitude, longitude, and ellipsoid height. Leveling is commonly done to improve the accuracy of the vertical or orthometric height differences in a surveying engineering project. Moreover, in the United States, existing vertical datums (e.g., NAVD 88) are primarily realized by differential leveling between networks of bench marks. The published orthometric heights for these bench marks represent the physical manifestation of this type of vertical datum. Therefore, the most-accurate method of connecting a survey to the existing vertical datum is to find and tie in the bench marks and then use them as vertical control. Accordingly, separate geometric and vertical control still commonly exist today unless the control mark has been both measured with GNSSs and leveled.

As discussed in Chapter 2, it is worthwhile to pause here to point out that NGS is in the middle of modernizing the vertical datums in the United States and replacing them with NAPGD2022 for orthometric heights and IGLD2020 for dynamic heights. Both NAPGD2022 and IGLD2020 will be accessed using GNSSs and a purely gravimetric geoid model, GEOID2022. In these new datums, both geometric and vertical control can be established by observing marks in convenient locations with GNSSs and using GEOID2022. The development of NAPGD2022 and IGLD2020 is still ongoing, and some of the content on vertical control in this chapter will need to be updated once they are rolled out. Chapters 2 and 7 provide more information on dynamic heights, how they are computed and related to orthometric heights, and why they are used near the Great Lakes for measuring the hydraulic head between locations for water management and power generation. The remainder of this chapter focuses on orthometric heights, which can then be converted to dynamic heights, as discussed in Chapter 2.

Confusion can be avoided if the engineering manager directs the surveying engineer immediately to the appropriate geometric and/or vertical control at the onset of the project. Often, local or geodetic control may be identified for use. In addition to pointing out control, the engineering manager must be familiar with the requirements and must be capable of specifying to the surveying engineer the desired coordinate system for the survey. Possible error can be reduced if the same coordinate system is used from the beginning, avoiding the need for making a transformation or calibration of coordinates. All subsequent engineering designs, plans, and plats can then be done in a commonly understood system.

LOCAL CONTROL

Local control is a generic term and can be any natural or artificial monument on which a survey may reference. Like any control, it must be stable; ideally, it should be long-lasting. It does not necessarily need to be a disk or rod with markings, although such items have great utility and are commonly used in practice for clearly marking survey control. As an example of local control, the right of way of a railroad alignment can be controlled by the center line of its tracks. A deformation survey might be referenced from the beginning point of a large structural wall or substructure. Or, an engineering manager may request that a survey be referenced from some natural, permanent, or highly stable feature. It is also common that when a survey is done to retrace previous work, the monuments from the past project are found and tied in to function as local control.

As the name implies, local control may have coordinates in a local or legacy coordinate system. In some cases, local control may have coordinates in a planar coordinate system or heights in a vertical datum defined by a local city, county, or another municipality. Hence, when using coordinates for local control, it is important for the surveying engineer to understand their sources and limitations.

GEODETIC CONTROL

Although local control coordinates may be required to meet project requirements, it is better practice to use or establish coordinates for control in a well-defined, geodetic coordinate system such as the most-recent realization of the NSRS. By using geodetic control, survey data can be readily georeferenced, meshed with data from other sources in the same common coordinate system, and can be imported into data management software (e.g., geographic information system) (Chapter 14).

The NSRS, as defined by NGS, provides a consistent coordinate system throughout the United States and its territories, and all surveys done on behalf of the federal civilian government must be done in this system. The NSRS enables the determination of geodetic latitude, longitude, ellipsoid and orthometric height, geopotential, acceleration of gravity, and deflection of the vertical at any point in the United States and its territories (Snay and Soler 2008). Many states and territories have adopted the NSRS as its official coordinate system, and geodetic control in this system has great utility and widespread use.

NGS provides two kinds of geodetic control for tying surveys to the NSRS: (1) active control as part of the NOAA CORS Network, and (2) passive control typically consisting of disks or rods. NGS publishes official coordinates and uncertainties for active and passive control via a datasheet. Another popular method for tying surveys to the NSRS is to use (3) the active control and network correctors provided by a real-time network (RTN) aligned with the NSRS. Coordinates for controlling RTN base stations are provided by RTN managers, although some managers have successfully sent GNSS survey data to NGS for publication of coordinates on datasheets. Owing to their widespread popularity, all three of these kinds of geodetic control are discussed in more detail in the following; however, it is worth noting that other governments and organizations also maintain databases of control that may be needed depending on project requirements.

Active Control: National Oceanic and Atmospheric Administration (NOAA) Continuously Operating Reference Stations (CORS) Network

Since the early stages of the development of the GPS, NGS recognized its potential for enhancing the NSRS. In 1986, NGS introduced the Cooperative International GPS Network, consisting of dual-frequency GPS receivers that continuously recorded signals from GPS satellites at a few sites in the United States (Chin et al. 1987). This network was augmented and densified globally, and it eventually evolved into the current International GNSS Service (IGS) network that is under the auspices of the International Association of Geodesy (Snay and Soler 2008). IGS collects, archives, and distributes GNSS data from a global network of GNSS ground tracking stations. These data are used to help define the International Terrestrial Reference System.

In 1994, NGS began building a denser network of GNSS ground tracking stations or CORSs that cover the entire United States, known today as the NOAA CORS Network. (Note that in this manual, only a station in the NOAA CORS Network is referred to as a CORS.) Recognizing the social and economic benefits for precise positioning and navigation, many federal, state, and local partners wished to participate in this endeavor by installing dual-frequency GNSS equipment on stable monuments and uploading data to NGS. Today, the NOAA CORS Network consists of approximately 2,630 stations (roughly 1,800 of which are active today) that are contributed by more than 150 government, academic, commercial, and private organizations. NGS provides an online, interactive map of the network, and a screen capture is shown in Figure 11-1 for the conterminous United States. Stations are also available elsewhere, including in Alaska, Hawaii, and the Pacific. Approximately 50 of the stations in the NOAA CORS Network are also a part of the IGS network.

NGS maintains a webpage for every station in the network (NGS 2021b). It is possible to search for a given CORS from the online map or by querying its identifier, name, or city location. The webpage for each CORS provides coordinates, velocities, a station log with metadata, photographs, time-series plots, and tools for viewing data availability and downloading raw GNSS data.

NGS processes and aligns the stations in the NOAA CORS Network with the latest IGS realization of the International Terrestrial Reference Frame (ITRF). Most recently, in January 2017, IGS released its realization of ITRF2014 (known as IGS14). Afterward, NGS reprocessed all its holdings of daily GPS data files collected at every station in the NOAA CORS Network and IGS Network from 1996 to 2017 (a total of 3,050 stations) in this reference system. The processing was done in three steps. First, NGS processed the data at the IGS stations to solve for Earth Orientation Parameters, hourly tropospheric delay parameters, and weekly IGS station positions. Second, the stations in the NOAA CORS Network were processed and tied to the IGS stations by holding fixed the estimated orbits, tropospheric delay parameters, Earth Orientation Parameters, and IGS station coordinates. Finally, third,

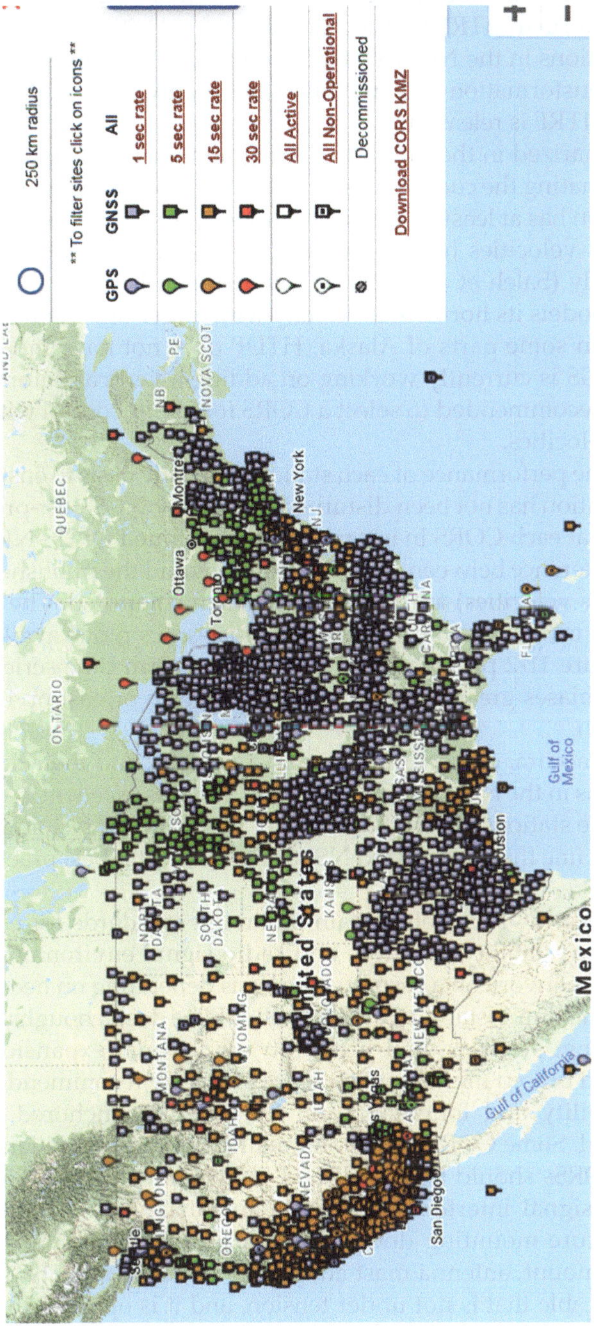

Figure 11-1. Map of the NOAA CORS Network within the conterminous United States.
Source: NGS (2021b). Courtesy of NOAA, reproduced with permission.

the coordinates and velocities were estimated at all the stations in ITRF2014 at epoch 2010.00. This final step required numerous iterations to remove discontinuities in the time series because of antenna changes, geophysical processes (e.g., earthquakes), and disturbances caused by snow accumulation. The final solution was completed in 2019, and it is known as Multi-Year CORS Solution 2 (MYCS2) (Saleh et al. 2021). NGS then used its software HTDP to transform the ITRF2014 coordinates and velocities to NAD 83 coordinates and velocities at all stations in the NOAA CORS Network (see Chapter 4 for more details on such a coordinate transformation and on HTDP). It is highly likely that similar work will be done when the next ITRF is released.

The workflow summarized in the preceding paragraph is meant to illustrate that great care is taken when estimating the coordinates and velocities of stations in the NOAA CORS Network. When a station has at least 3 years of continuous observations for processing, the horizontal and vertical velocities have uncertainties less than 0.7 and 1.0 mm/yr (68% confidence), respectively (Saleh et al. 2021). When a station is "new" and has less than 3 years of data, NGS models its horizontal velocity with larger uncertainties using HTDP. Unfortunately, except in some parts of Alaska, HTDP does not have models for vertical velocities (although NGS is currently working on adding a vertical deformation model). For these reasons, it is recommended to select a CORS for use as control with computed as opposed to modeled velocities.

NGS also monitors the performance of each station on a daily basis to ensure data quality and to verify that the station has not been disturbed. Every day, NGS post-processes the 24 h GNSS data files logged at each CORS in internal software named OPUS-NET (Gillins et al. 2019b). The residual difference between the daily solutions and the published solution (i.e., official coordinates plus velocities) at each station is then computed. The most-recent 90 days of these residuals can be viewed in *short-term time-series* plots, available from each station's webpage. Figure 11-2 provides an example short-term time-series plot. Stations with average absolute biases greater than 2.0 cm in the north or east or 4.0 cm in up are flagged for investigation.

Owing to their high accuracy in coordinates and velocities, and their daily monitoring for performance, stations in the NOAA CORS Network serve as excellent geometric control. It is often said that these stations are the backbone of the NSRS. In fact, in the near future, NGS will use them to define the modernized NSRS. Therefore, it is best practice to reference GNSS surveys to CORSs.

NGS requires that each CORS meets certain minimum standards for equipment, data upload, monumentation, stability, location, and radio signal environment (Soler 2011). CORSs are added to the network on a case-by-case basis depending on needs. Monuments should be designed to maximize stability, and they should be deep enough not to be subject to frost heave, shrinking, swelling, soil instability, or thermal expansion/contraction. Shallow- or deep-drilled braced monuments (Figure 11-3a) are recommended as they offer a high degree of stability and longevity, and they can be anchored in bedrock or unconsolidated material. Some CORSs are monumented on concrete pillars or rooftops of masonry buildings. CORSs should be established in open areas with minimal overhead obstructions or radio signal interferences. GNSS antennas should be attached to the monument using a secure mounting device (e.g., SCIGN mount [Figure 11-3b], CORS adjustable monument mount, antenna mast adapter). The GNSS antenna must be at least dual frequency with a cable that is not under tension, and it is optional to cover it with a radome for environmental protection, although a radome will interfere with antenna phase center variation modeling. Receivers should be at least dual frequency, track satellites down to the horizon (0 degree), sample at a frequency of at least 30 s, and have uninterrupted power supply with at least 30 min of backup power. Tracked data, preferably in 24 h blocks, must be transferred to NGS and made available for NGS to freely distribute. NGS archives raw GNSS data in Receiver Independent Exchange Format.

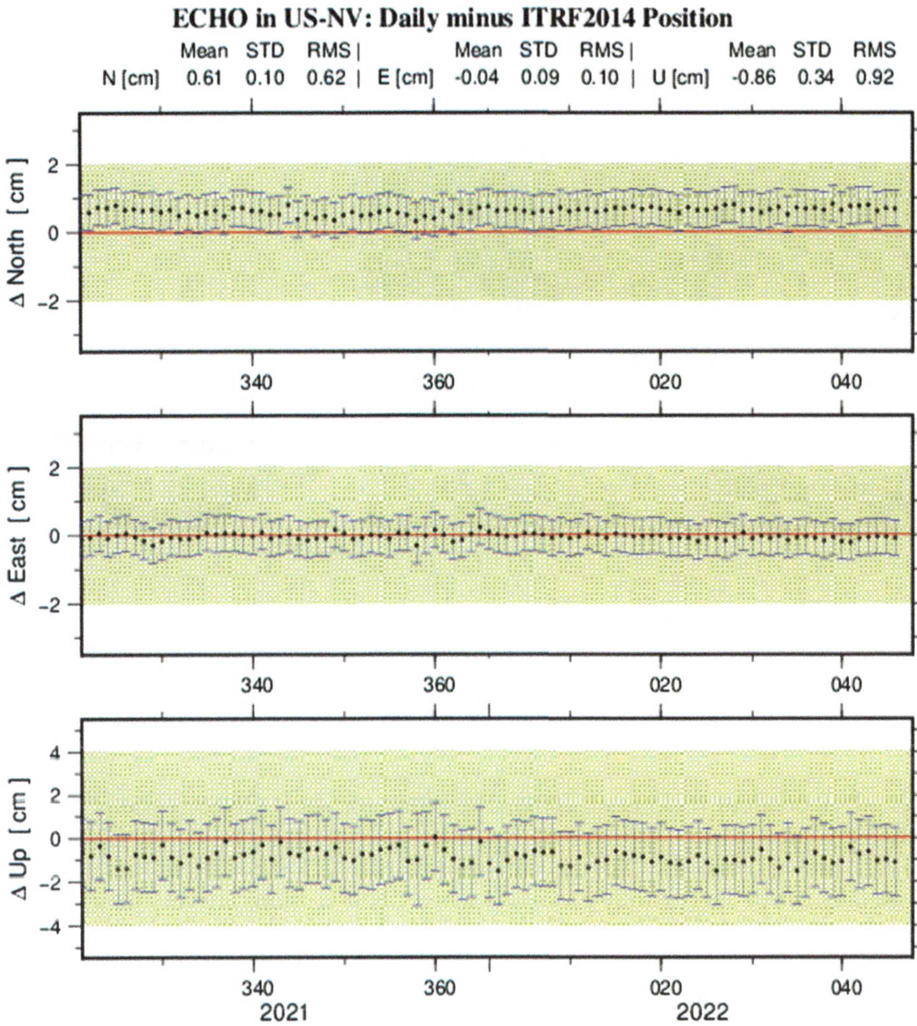

Figure 11-2. Short-term time-series plot for monitoring station "ECHO." The red line shows the published solution for the station, and the black dots and error bars are the residual daily solution differences. The top label provides summary statistics for the most-recent 90 days [mean, standard deviation (STD), and root-mean-square (RMS) error of the residuals in all three components]. Source: Courtesy of NOAA, reproduced with permission.

The surveying engineer should evaluate a CORS before deciding to use it as control. The minimum standards aim to ensure that all stations in the NOAA CORS Network provide high-quality data. Nevertheless, because of the nature of working with data contributed by hundreds of different partners, and the many years of evolution in GNSS technology and understanding since the creation of the network, some stations provide higher-quality data than others. NGS does provide tools for conducting such an evaluation. The following steps are recommended by the author:

1. Choose stations that are within 100 to 300 km of a project site. It is well known that errors in GNSS baseline processing are directly correlated with baseline length (Chapter 6). Tropospheric delay is reduced when baselines are shorter and atmospheric conditions at each end of the line are similar. In addition, mutual satellite visibility at each end of the line for differencing diminishes as baselines grow longer. That said, errors in GNSS processing are more occupation time-dependent than baseline

(a) (b)

Figure 11-3. (a) Station "ECHO," a CORS in Lincoln County, Nevada; the antenna is mounted on a deep-drilled braced monument, (b) A SCIGN mount device for attaching the antenna of the CORS to the monument.
Source: Courtesy of (a) National Geodetic Survey, (b) NOAA, reproduced with permission.

length-dependent (Eckl et al. 2001). Therefore, for short GNSS sessions (i.e., <2 h), choose CORSs within approximately 100 km as control; for moderate GNSS sessions (i.e., 2 to 8 h), choose CORSs within approximately 300 km. Note that even longer baselines can be successfully processed when GNSS sessions are very long in duration (e.g., up to 2,000 km for 24 h sessions).

2. Determine if GNSS data are available at a given CORS during the time of your survey. Of course, if data are unavailable, then the station simply cannot be used as control. NGS provides a tool known as User Friendly CORS (UFCORS) for entering a date and time range to view available data at a given station (NGS 2021c). This tool can also be used to download the raw GNSS data for processing and adding a station to the survey network.
3. As discussed previously and when possible, choose a CORS with computed velocities rather than modeled velocities from HTDP. NGS provides tables of official coordinates with "comp[uted]" versus "htdp" coordinates and velocities on the website for CORS.
4. Review the aforementioned short-term time-series plot for the station, ideally at the time of the project. Stations with large spikes, data gaps, bias from the published "red" line, or large standard deviations should be avoided. A good rule-of-thumb is for the RMS in the short-term time-series plot (Figure 11-2) to be less than 1.0 cm in the north and east and 2.0 cm in the up direction in a local geodetic horizon frame at the station.
5. Examine the formal uncertainties for the official coordinates of the CORS. Standard deviations in the north, east, and up are provided on the station's datasheet, accessible from the webpage for the CORS (more on datasheets are discussed in the following under Passive Control). Stations with unusually large standard deviations (>3 cm) should be avoided. Note that standard deviations are not available for CORSs with modeled velocities.

Active Control: Real-Time Network

The NOAA CORS Network provides raw GNSS data that can be included in a survey network for post-processing, but it does not support real-time GNSS positioning. Real-time kinematic (RTK) GNSS surveying is popular in practice because it is highly efficient and quite accurate. RTK solutions at centimeter-level accuracies can be derived in minutes or less. For example, Allahyari et al. (2018) showed that a 5 min GPS-only RTK solution from a full network corrector of an RTN in South Carolina yielded coordinates with horizontal RMS error equal to 1.2 cm, and vertical (ellipsoid height) RMS error equal to 2.5 cm.

Owing to its popularity, numerous local, regional, and national private and public RTNs have been developed in the United States. The number of networks continues to increase, and RTN base stations (continuous GNSS stations) are installed to increase density and improve network coverage. Some RTNs require a subscription, and others are freely available. Many more RTN base stations than CORSs exist in the United States. Often, RTN base stations are built to the same minimum standards as a CORS and with similar equipment; however, the standards vary by network and some bases are less stable.

RTN managers are responsible for determining the coordinates of the base stations. In some cases, RTN base stations also function as stations in the NOAA CORS Network. In that case, the RTN manager will typically adopt the coordinates of CORSs, as published by NGS. Otherwise, RTN managers are encouraged to derive coordinates for the base stations by aligning their networks to the NSRS.

Henning et al. (2013) provide guidelines for RTNs and recommendations for deriving coordinates for the base stations in the NSRS. These guidelines recommend the following: (1) ensure that at least three base stations or at least 10% of the total number of base stations in the RTN (whichever number is bigger) are also part of the NOAA CORS Network; (2) submit 10 24 h daily data files at each base station to NGS for post-processing using the online tool, OPUS-S (described in greater detail in the following); and (3) find and use the mean coordinates of the 10 solutions.

RTN managers may have used other methods for deriving the coordinates of the base stations. For example, Armstrong (2011) uploaded 5 days of 24 h data files from the Oregon Department of Transportation's RTN to NGS software OPUS-Projects (also described in the following). Then, data were added from several CORSs in Oregon to the network. Finally, OPUS-Projects was used to post-process each daily session and adjust the network. During the adjustment, the published coordinates for the CORSs in the network were held fixed to yield coordinates tied to the NSRS at all other base stations. After completion, vendor software (e.g., Leica Spider) was used to monitor the performance of the RTN bases and check their alignment with the NSRS. On roughly an annual basis, the Oregon Department of Transportation repeats this workflow to decide if the base station coordinates need to be updated.

RTNs broadcast the coordinates of the base stations to end-users. Using a wireless communication method, a surveying engineer can configure a GNSS rover to connect to the server for the RTN, receive network corrector messages, and process baselines in real time using the coordinates of the base stations as reference control. In this manner, RTN base stations are frequently used as control; however, it is important for the surveying engineer to understand the source and error estimates for these coordinates. Any misalignment in the coordinates of the RTN base station with the NSRS will be propagated to the rover. RTN base station coordinates are also sometimes not aligned with local control in a given engineering project. To overcome misalignment issues with local control, some surveying engineers use calibration procedures as discussed in Chapter 4. The surveying engineer should contact the RTN manager for details and information specific to the network.

Raw GNSS data are also usually logged and stored at each RTN base station. RTN managers will typically make these data available as a service for use in post-processing static GNSS campaigns, similar to how data are made available for stations in the NOAA CORS Network by NGS.

Passive Control

Active control, such as CORSs and RTN base stations, provides continuous measurement and monitoring; therefore, the position of such stations is well studied and serves as excellent geometric control. New GNSS surveys should reference these stations. Passive marks are only episodically measured, and their published coordinates may be based on as little as one control survey.

Figure 11-4 shows some common types of passive marks according to a stability classifier of "A" (most reliable and expected to hold well) to "D" (questionable or unknown reliability), as defined by NGS. Stability refers to the ability to maintain a fixed position. Stability A marks are stamped disks set in rock outcrops or large structures with either deep

Figure 11-4. Some examples of passive marks: (a) stability A mark in rock outcrop; (b) stability B rod driven to refusal in sleeve 3 m deep; (c) stability C mark in concrete monument; and (d) stability D mark in concrete sidewalk.
Source: Courtesy of NOAA, reproduced with permission.

foundations or foundations on bedrock. Stability B marks are metal rods (i.e., stainless steel, aluminum, copper, galvanized steel, etc.) driven to refusal and in a sleeve at least 3 m deep or disks set in tunnels, bridge abutments, massive structures or retaining walls, or in an unspecified boulder. The sleeve has markings around the top of its rim. Stability C marks are stamped disks set in the top of a concrete monument, retaining walls, the footings of small/medium structures, in a reinforced slab or mat foundation, or a rod in a sleeve less than 3 m deep. Stability D marks are objects driven in the ground, or disks in a concrete mass, pavement, prefabricated post, and so on.

Regardless of the stability, passive marks are more prone to disturbance and destruction than active control. Because bench marks were often set along linear routes, they are frequently destroyed when a roadway is widened. Some marks (especially Stability D marks) may have been poorly set and are subject to frost heave, soil consolidation, soil contraction/expansion, and so on. Vandalism, snow plows, and other construction and land development activities can disturb or destroy marks.

Because passive marks are not continuously monitored and are infrequently checked, the surveying engineer should exercise caution when using passive marks as control. The engineer should never simply assume that the passive marks have not moved or that their published heights are without error. (Technically, this recommendation even extends to active control, but the assumption is less risky.) One way to check is to find and tie in to multiple passive marks or, better yet, a combination of passive marks and CORS. Then, compare the newly measured coordinate differences with the published coordinate differences. If the residuals are small and on a par with the accuracy of the published coordinates, then there is evidence that the marks have not moved. More discussion on this topic is given in the last section of the chapter.

Despite the aforementioned weaknesses, passive marks do have great utility as geodetic control. Some surveys simply must use passive marks as control to meet project requirements. Other benefits are listed as follows:

- More than one million passive marks with coordinates are present in the national database managed by NGS (known as the NGS integrated database), whereas there are only thousands to tens of thousands of continuous GNSS stations. Even more marks are published by other federal, state, and local governments and other municipalities. Such a large number of marks can be used as check control and are more readily identified (and even used) by lay persons.
- Passive marks have existed for centuries (rather than continuous GNSSs which began in earnest in the 1990s), and much of the US infrastructure was built from these marks. Record of survey plats, maps, and engineering plans more frequently reference these marks.
- Instruments or targets can be set up over passive marks, serving as a good basis for controlling a survey project (especially a non-GNSS survey project). The marks also provide starting control when returning to a survey project.
- As discussed previously, the current vertical datums for the United States (e.g., NAVD 88) were realized by differential leveling between the passive marks. Thus, these bench marks are the physical manifestation of the vertical datums. As noted previously, this point will no longer be relevant when the vertical datums are modernized.

NGS provides tools for researching passive marks in the office and recovering marks in the field. If a passive mark is not in the NGS Integrated Database, then the surveying engineer should contact the responsible agency to gather the required notes and data. Because of their widespread and national use, only tools pertaining to passive marks in the NGS Integrated Database are discussed in the following, but similar tools may exist for other passive control databases.

One useful tool for researching passive marks in the field is the NGS Data Explorer, which can be accessed at NGS (2021d). The tool provides an interactive map that can be zoomed into a project area (Figure 11-5). It is then possible to toggle layers and requests views of certain mark types (i.e., horizontal or vertical control or both). The user can click on a desired point, and a popup window appears where it is possible to view the mark's datasheet.

The beginning portion of an example datasheet is given in Figure 11-6. The fields in the sheet are populated at the time the link is accessed from the Data Explorer. The datasheet presents a great deal of information, including but not limited to a unique and official point identifier (PID) for the mark, its name or designation, its official latitude, longitude, and orthometric height, a description of how these coordinates were derived, and the estimated standard deviations in the north, east, and up along with the correlation in the north and east. It also gives the network and local accuracies, hyperlinks to view photos of the mark, a detailed description of the mark and reference ties, and any outdated or superseded coordinates near the bottom of the sheet. Note that datasheets are available for both CORSs and passive marks.

The datasheet provides descriptors next to the coordinates and heights. For example, note the word "ADJUSTED" next to the latitude/longitude and ellipsoid height in Figure 11-6 and "GPS OBS" next to the orthometric height. It is important to view these descriptors when deciding if the mark is suitable for potential use as geometric and/or vertical control. At the top of every datasheet, a hyperlink is provided to a document that describes the file format and descriptors in detail (i.e., note the link to the dsdata.pdf in Figure 11-6).

For geometric control, NGS recommends using marks with the "ADJUSTED" descriptor next to the latitude and longitude (position) and ellipsoid height. It indicates that the coordinates are from a least-squares adjustment of a survey network. In addition, NGS recommends using geodetic coordinates in the most-recent realization of the NSRS. Positions scaled from a topographic map ("SCALED" descriptor) or derived from a handheld GPS receiver or autonomous GPS observation ("HD_HELD1, HD_HELD2") are not sufficiently accurate for use as geodetic control.

For vertical control, NGS recommends the use of marks with the "ADJUSTED" descriptor next to the orthometric height. It indicates that the orthometric height is from a least-squares adjustment of differential leveling data. In addition, marks with the "GPS OBS" descriptor are recommended if at the top of the datasheet there is a statement, "This is a Height Modernization Survey Station." This statement is meant to denote that the GPS-derived orthometric height meets precise GPS surveying guidelines in Zilkoski et al. (2008), described later. Numerous other descriptors for orthometric heights are in the NGS Integrated Database; surveying engineers should be cautious and aware of their meaning when considering them as vertical control. Published orthometric heights that are only precise to the nearest meter or decimeter (0 or 1 decimal place) are not recommended for use as vertical control.

In addition to providing data via the Data Explorer, NGS also provides shapefiles of geodetic control that can be viewed and queried in GIS. Online tools are also available for searching for datasheets by PID, county, database load date, quadrangle, coordinates, and by survey project identifiers. New beta tools have been recently released for viewing a datasheet in a more modern interface as well as viewing bench marks by leveling project. All these tools can be used to find and evaluate control while in the office and prior to a project.

Finally, NGS has recently released a mobile-friendly tool aimed to help surveying engineers find marks and submit recovery notes and photos while in the field. The Survey Mark Recovery tool can be accessed at NGS (2020). When entering a given PID in the online form, the fields are prepopulated (Figure 11-7a). Then, the engineer may note changes or

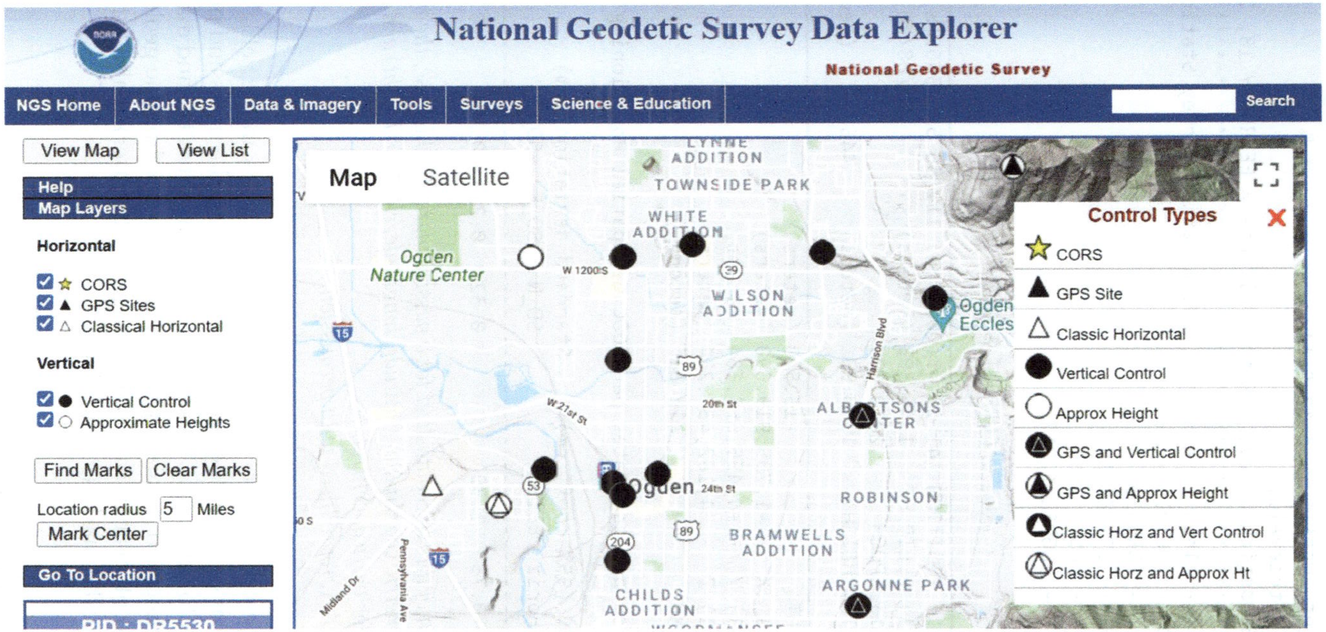

Figure 11-5. Screen capture of the NGS Data Explorer. The circles represent vertical control, and the triangles represent horizontal or geometric control.
Source: Courtesy of NOAA, reproduced with permission.

The NGS Data Sheet

See file dsdata.pdf for more information about the datasheet.

```
PROGRAM = datasheet95, VERSION = 8.12.5.14
Starting Datasheet Retrieval...
1        National Geodetic Survey,   Retrieval Date = FEBRUARY 18, 2022
 DR5530 ***********************************************************************
 DR5530  HT_MOD       -  This is a Height Modernization Survey Station.
 DR5530  DESIGNATION -  WC 29
 DR5530  PID          -  DR5530
 DR5530  STATE/COUNTY-  UT/WEBER
 DR5530  COUNTRY      -  US
 DR5530  USGS QUAD    -  OGDEN (2017)
 DR5530
 DR5530                          *CURRENT SURVEY CONTROL
 DR5530  ___________________________________________________________________
 DR5530* NAD 83(2011) POSITION- 41 13 43.00840(N) 111 56 53.43569(W)   ADJUSTED
 DR5530* NAD 83(2011) ELLIP HT-  1344.393 (meters)        (07/08/20)   ADJUSTED
 DR5530* NAD 83(2011) EPOCH   -  2010.00
 DR5530* NAVD 88 ORTHO HEIGHT -  1360.93  (meters)     4465.0   (feet) GPS OBS
 DR5530  ___________________________________________________________________
 DR5530  GEOID HEIGHT     -         -16.520 (meters)                   GEOID18
 DR5530  NAD 83(2011) X  -  -1,795,919.690 (meters)                    COMP
 DR5530  NAD 83(2011) Y  -  -4,456,658.684 (meters)                    COMP
 DR5530  NAD 83(2011) Z  -   4,182,437.224 (meters)                    COMP
 DR5530  LAPLACE CORR     -          11.81  (seconds)                  DEFLEC18
 DR5530
 DR5530  Network accuracy estimates per FGDC Geospatial Positioning Accuracy
 DR5530  Standards:
 DR5530         FGDC (95% conf, cm)     Standard deviation (cm)     CorrNE
 DR5530            Horiz  Ellip             SD_N   SD_E   SD_h      (unitless)
 DR5530  -------------------------------------------------------------------
 DR5530  NETWORK    0.73   1.39             0.32   0.27   0.71      -0.04313161
```

Figure 11-6. Top portion of an NGS datasheet for a passive mark. In this case, the mark has both a high accuracy set of geometric coordinates (latitude, longitude, ellipsoid height) and an orthometric height that can be used for vertical control.
Source: Courtesy of NOAA, reproduced with permission.

make corrections and enter recovery notes on finding the mark in the field. They can also attach photos of the mark and hit the submit button to send the notes and photos to NGS for review and loading to a database. In addition, there is a button to "find marks near me." This feature opens a map and displays nearby marks, centered on the geolocation of the mobile device (or the center can be manually changed by the user). It also lists the distance and direction to the nearby marks (Figure 11-7b).

PROJECT PLANNING AND CONTROL

The location and number of control stations are important considerations when conducting a survey. Ideally, control should be distributed evenly across a project to ensure that the survey network is well braced and to prevent warping. In addition, it is important

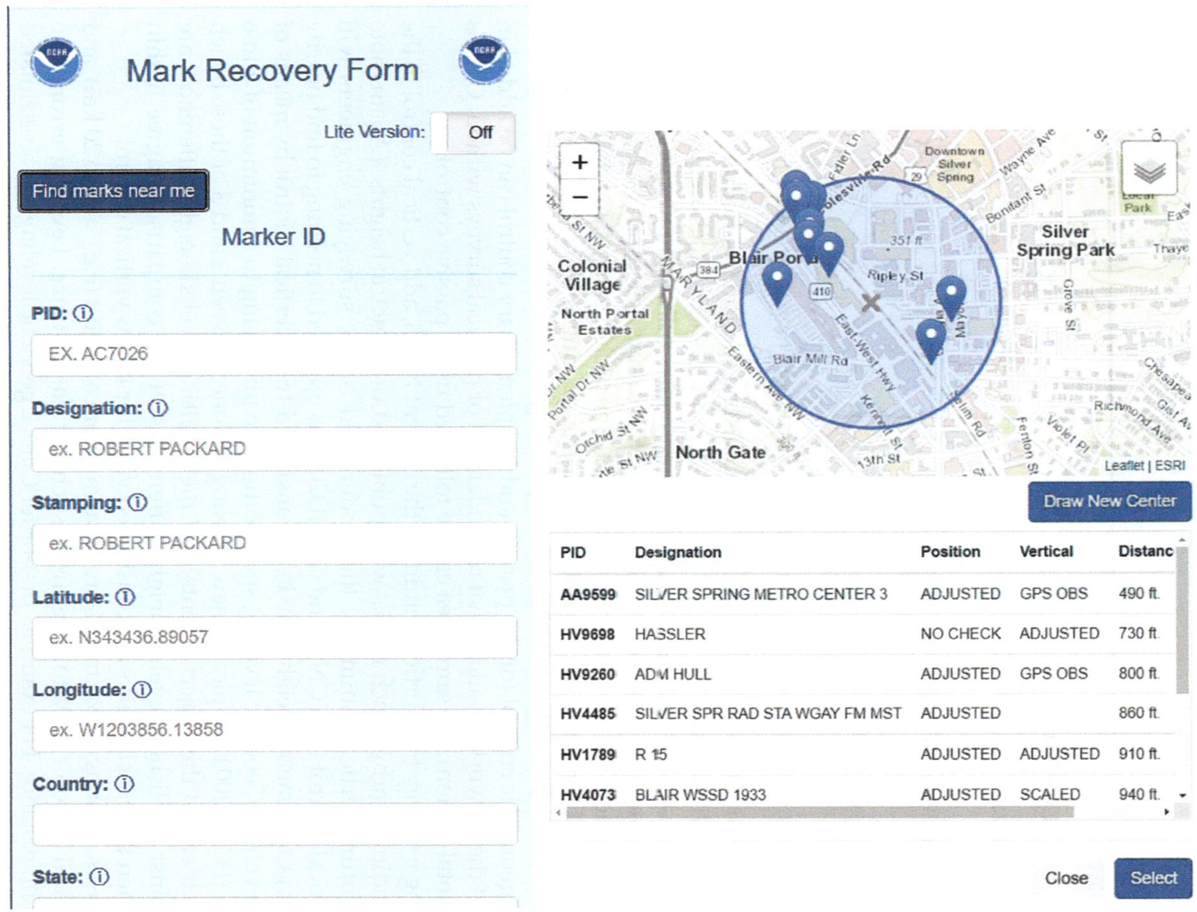

(a) (b)

Figure 11-7. Survey Mark Recovery Tool provides (a) a mobile-friendly form for entering mark recovery information and attaching photos for submission to NGS, (b) interactive map for finding marks near the user.
Source: Courtesy of NOAA, reproduced with permission.

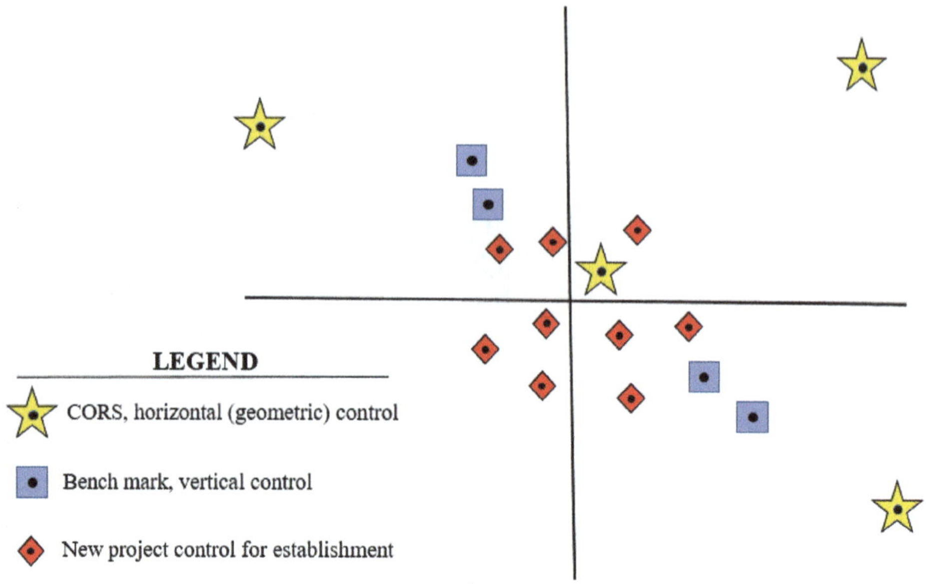

Figure 11-8. Example survey diagram for geometric, vertical, and new control (not to scale.)

to use more than one control mark to evaluate the published coordinates. Redundant control helps identify whether a mark has been disturbed or its published values are in error.

Except for route surveys, a common approach and handy rule of thumb is to divide the project into four quadrants (van Sickle 2015). For GNSS surveys, select one CORS in at least three of the four quadrants, preferably near or outside the project extents. In addition, when possible, select a CORS close to the center of the engineering project to strongly tie the survey to the NSRS (see Figure 11-8 for a simple diagram). These CORSs can serve as the primary geometric control for the project, and supplementary control can be added in the interior of the network for additional redundancy. For classical surveys without GNSSs (e.g., only a total station), the same idea can be followed using passive control.

When tying a survey to a vertical datum in the current NSRS, add vertical control to the project by finding nearby bench marks with published orthometric heights. (Again, note that in the future vertical datum for the modernized NSRS, surveying engineers will establish vertical control with GNSS and GEOID2022 as opposed to needing to find nearby bench marks.) Orthometric heights are best transferred from vertical control by means of differential leveling. Geodetic leveling specifications require tying to a minimum of 4 to 6 bench marks (FGCS 2004). Typically, new leveling is done between 2 and 3 of these bench marks on each end of the project extents, and the resulting leveled height differences are checked against published leveled height differences. If the comparisons agree within tolerances, then the bench marks are considered valid for use as vertical control.

In some cases, existing bench marks are far away from a project (e.g., >10 to 20 km), and it is impractical or too costly to level from such a great distance. Leveling errors also propagate with distance. For example, first-order Class 2 geodetic leveling has an estimated error of 1.8 cm (95% confidence) after 20 km. As discussed later, Zilkoski et al. (2008) provide guidance for using GNSSs and a geoid model to tie in vertical control as opposed to leveling; these guidelines can also be followed to derive orthometric heights on new marks with estimated accuracies of less than 2.0 cm (95% confidence).

For route surveys or long linear projects, control is recommended at the beginning and end of the project and on both sides. In addition, control should be established at appropriate intervals (e.g., 5 to 10 km) along the length of the project.

SETTING NEW CONTROL

After identifying the existing control for a project, it is typical to determine next if additional, new marks should be set near the engineering project. The ensuing survey then establishes coordinates on these new marks, thereby densifying available control in the vicinity of the project for use. When necessary, supplementary control, targets, ground control points, and check points are then further added and tied to the project control to support ensuing construction, photogrammetry, or laser scanning surveys (Chapters 8 through 10 and Chapter 12).

Even a highly accurate, careful survey will prove inadequate if its control is poorly monumented (ASCE 1985). It is best practice to set durable monuments that not only last throughout the life of the engineering project but also last a long time should it be required to do more work or resurvey the project in the future. In addition, surveying engineers should set physical monuments that meet the standards of their clients.

Adequate monumentation varies widely and is based on project requirements. During a topographic survey or a small construction project, an adequate monument may simply be a spike driven into a concrete or gravel road surface. Of course, this spike may not hold its position for a long time, but it may be sufficient for the project's needs. Consider another survey for a major engineering project that will take years to complete. In this case, the control monuments must be highly stable and longer lasting. Monumentation depends on the precision of the survey being carried out and the duration of the project for which the control is needed.

Smith (2010) presents guidelines for setting new or replacement bench marks. Following these guidelines help preserve orthometric heights on existing bench marks that will soon be disturbed or destroyed. Smith (2010) also provides step-by-step guidance on how to install highly stable monuments such as disks in bedrock or a large structure, rods driven in the ground to refusal, or disks in concrete monuments. These types of monuments are stable in both the horizontal and the vertical.

Sound bedrock is the most desirable setting for a mark because it is the most stable setting and underground activity is highly unlikely (Floyd 1978). A surveying engineer can drill a small hole (2.5 cm in diameter by 8 cm deep) into a fairly level spot on an outcrop, and then place a disk (Figure 11-9) in the hole and in the cement mix. Highly stable monuments can also be achieved by setting a disk in massive structures with deep foundations that exceed the depth of frost penetration.

Rods can be driven in the ground when bedrock and large structures are not available (Smith 2010). Using a gasoline-powered reciprocating hammer, a stainless steel rod (17 mm

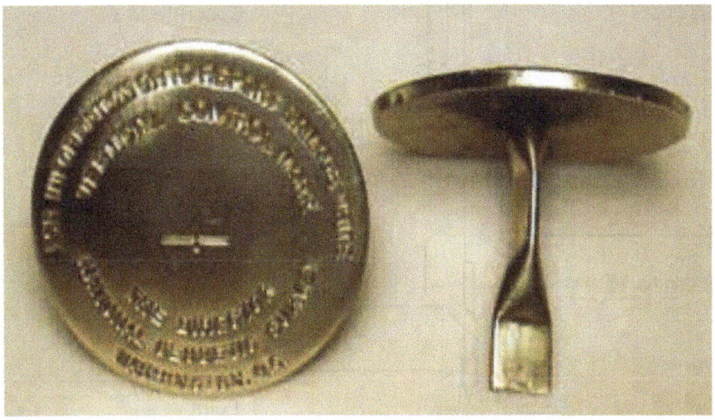

Figure 11-9. Brass disk.
Source: Smith (2010), reproduced with permission from NOAA.

in diameter) can be driven in the ground until refusal or a reduced driving rate has been achieved (e.g., 250 kg of resistance). The upper meter of the rod is encased in a 2.5 cm grease-filled and finned plastic sleeve that is held horizontally stable by back-filled, washed sand (Figure 11-10). The effects of up and down ground movement because of freeze and thaw or wet and dry cycles are buffered from the anchored rod by the grease-filled sleeve, promoting vertical stability. The top of the rod serves as the survey control datum point, and a casement is added around the rod near the ground surface to protect it from damage by impact. The casements are typically 13 to 15 cm in diameter, and they are labeled for identification of the mark.

Disks can be set in a concrete monument as an alternative to driving rods, although they are sometimes less stable. The concrete is poured into a hole dug in the ground that is cylindrical or squared and slightly "bell-shaped" at the bottom to minimize pullout or overturning (Figure 11-11). The monument must also extend well below the frost line and is typically around 1.3 m deep and 0.3 to 0.5 m in diameter. Subsurface rock may prevent these depths, whereas soft soils may require even deeper monuments. In general, the hole is dug with an auger or posthole digger, widened at the bottom and then backfilled with concrete mix. A cardboard form or rolled felt paper is set at the top of the hole and extending 0.4 m deep to support the upper portion of the monument until the cement cures. It is also important to finish the top so that it has a smooth surface. The survey disk is lightly pressed into the top of the monument until the concrete slightly overlaps its edges.

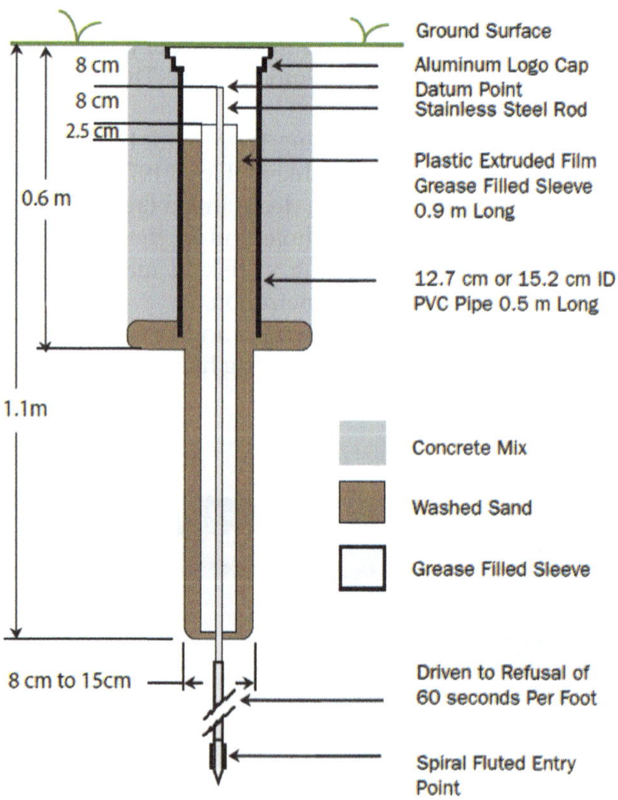

Figure 11-10. Driven rod in grease-filled sleeve.
Source: Smith (2010), reproduced with permission from NOAA.

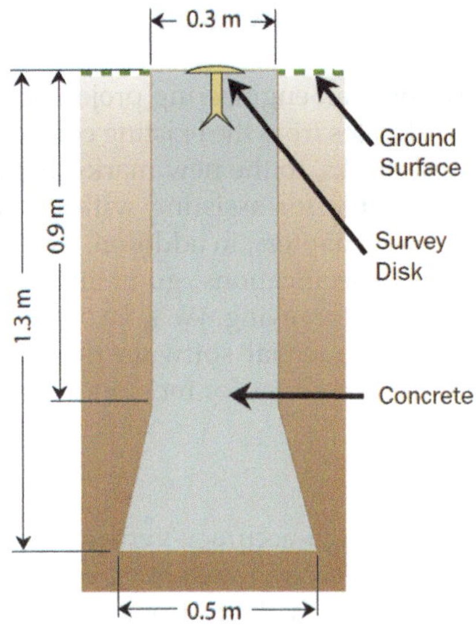

Figure 11-11. Disk in concrete monument.
Source: Smith (2010), reproduced with permission from NOAA.

Unfortunately, monuments are often destroyed during construction. Contractors should be required to protect and preserve all control monuments, especially when other work may follow for which they will be needed. Often, there is a general apathy toward the preservation of monuments once their need for the control is past. Engineers need to be on guard against the loss of survey control; otherwise, unnecessary surveying costs may incur to reestablish it for ongoing or future work. Some local jurisdictions have gone so far as to require developers to post bonds for survey control monuments to ensure preservation during construction. It is good practice to set control outside the construction zone so that the monuments are not easily disturbed or destroyed.

Whenever possible, set control in stable locations and outside an area of possible terrain movement. The most effective precaution for assuring the stability of a mark is to choose a good location for setting it (Floyd 1978). Avoid low, potentially wet areas, slopes, and earthen-fill. Monuments can be subject to movement because of ground subsidence, excessive groundwater withdrawal, removal of oil, gas, sulfur, or other minerals; frost heave; successive wetting and drying of expansive soils; underground excavations; soil consolidation; crustal motion; and earthquakes, landslides, and other earth movements (Floyd 1978, ASCE 1985). Over very short periods of time, these disturbances may be difficult to detect. The cumulative effect of such disturbances over longer time periods can be extensive enough to cause serious errors in engineering surveys and in the design and construction activities based on such surveys (ASCE 1985).

For future GNSS surveying, it is also critical to choose locations for new control with favorable, unobstructed views of the sky for making satellite observations.

After setting new control, the marks should be thoroughly described and referenced (Chapter 13). The description and notes for the mark should be recorded in a control mark sheet (Figure 6-9). It is also best practice to tape distances to a minimum of three distinguished features in differing quadrants or directions from the mark, all within 30 m. These reference ties should also be recorded in the control mark sheet, and they can be used in the future to find the mark (especially marks beneath the ground surface) or even reestablish lost or disturbed marks assuming that the taped distances are sufficiently accurate (i.e., measured to the nearest millimeters to stable, distinguished features).

COMMON TOOLS FOR CONTROL SURVEYING

Once the new control marks near an engineering project set and cure, a control survey can be executed to transfer coordinates from the existing control, such as CORSs, RTN base stations, and/or passive control marks, to the new marks. Numerous tools, technologies, and software programs are available for assisting with a control survey. The manual summarizes many of them in other chapters; in addition, the following section provides a brief overview of some national specifications, guidelines, and freely available online software specific to geodetic control surveying. We give only a summary in the following, and we recognize many other commercial software programs are available. We also encourage the reader to access the references for more information.

Geodetic Leveling Specifications

As mentioned in Chapter 7, for high-accuracy vertical control surveying, the Federal Geodetic Control Subcommittee (FGCS) released specifications and procedures for conducting precise differential leveling with electronic (digital barcode) leveling systems (FGCS 2004). The specifications state requirements for the maximum and average bench mark spacing, distance from control points, and the minimum number of vertical control ties according to the desired leveling order (first, second, or third and Class I or II). The specifications also provide requirements for instruments and rods and for checking the instruments' collimation. FGCS (2004) provides field procedures according to order and class. For example, it states the requirements for running a leveling section forward and backward between two bench marks, the maximum sight lengths, the minimum ground clearance, the maximum section and loop misclosure, and the maximum difference in intermediate forward and backward sight lengths. Finally, it also provides procedures for evaluating the leveling work in the office, including checking the sum of all corrected section misclosures and loop misclosures. By following the specifications, it is possible to measure highly accurate height differences between marks in a survey network. For example, for first-order Class II or second-order Class II work, orthometric height differences will have estimated errors approximately equal to 1.0 mm or 2.1 mm per square root of the running distance in kilometers (68% confidence), respectively (Zilkoski et al. 1992).

GNSS Control Surveying Guidelines: NOS NGS-58 and 59

NGS provided GNSS surveying guidelines for establishing high-accuracy ellipsoid heights and orthometric heights, known as NOS NGS-58 (Zilkoski et al. 1997) and NOS NGS-59 (Zilkoski et al. 2008). (NGS is also currently working on new specifications to replace these guidelines in the near future.) By following the chosen standard in the guidelines, it is possible to measure ellipsoid and orthometric heights to less than 2 cm or 5 cm (95% confidence). Horizontal accuracies (latitude and longitude) will be even smaller, usually one-half of the vertical accuracies.

NOS NGS-58 presents procedures for conducting, processing, and adjusting a static GNSS survey campaign. The procedures first recommend dividing the network into levels of hierarchy: (1) ties to existing geometric control in the NSRS (e.g., CORS), (2) establishment of primary control, and (3) additional densification of optional secondary and local control. NOS NGS-59 also recommends (4) tying into existing vertical control at usually a maximum 20 km spacing for connecting the GNSS survey to the vertical datum.

The guidelines provide recommendations for maximum baseline length, observation session duration, and scheduling and require a number of repeat sessions depending on control type. The guidelines also provide expected tolerances for the precision of repeat measurements and for baseline processing statistics. Finally, they discuss how to perform and evaluate the results of least-squares adjustments of the survey networks.

Surveys that meet the guidelines in NOS-NGS-58/59 can be submitted to NGS for review and publication in the NGS Integrated Database following a process known as Bluebooking (NGS 2021e). This will, in turn, enable updating or establishing new coordinates on datasheets. Such a workflow is known as "height modernization surveying" as it allows updating of heights on bench marks using GNSSs.

Online Positioning User Service-Static

In 2001, NGS introduced the OPUS, which has become widely popular throughout the United States for surveying, engineering, GIS, and mapping (Snay and Soler 2008). Numerous recent research papers are available on OPUS (Wang and Soler 2013, Smith et al. 2014, Soler and Wang 2016, Wang et al. 2017, Gillins and Eddy 2017, Jamieson and Gillins 2018, Gillins et al. 2019b). Several articles have also been compiled into an ASCE book (Soler 2011) with in-depth details on OPUS and its development.

The first version of this service is known today as OPUS-Static (OPUS-S). Users can upload a raw static data file from a dual-frequency GPS receiver for a single observational session on a mark through a free, online portal at NGS (2021a). OPUS-S will post-process the data and send back an email with coordinates referenced to the NSRS.

OPUS-S computes the coordinates by finding the unweighted mean of three single-baseline GPS solutions computed by double-differenced, carrier phase measurements from the observed mark to three stations in the NOAA CORS Network or IGS Network. OPUS-S uses NGS baseline processing software named the Program for Adjustment of GPS Ephemerides (PAGES).

In addition to providing coordinates referenced to the NSRS, OPUS-S provides coordinates referenced to the latest ITRF at the epoch of the survey, error estimates, baseline processing statistics, northings and eastings in the State Plane Coordinate System and in UTMs, the distance to nearby published marks, and more. Error estimates are given in terms of "peak-to-peak" values, which are the range of the three single-baseline solutions in each component (i.e., range in the north, east, and up of the three individual baseline solutions to three reference stations).

Soler et al. (2006) developed an empirical equation for estimating the standard deviation (σ) of the coordinates from an OPUS-S solution in a given component:

$$\sigma(\text{cm}) = \sqrt{\left(\frac{\text{RMS}_\varphi}{1.5}\right)^2 \times \left(\frac{k}{\sqrt{T}}\right)^2 + \left(\frac{p}{1.7}\right)^2} \tag{11-1}$$

$$\begin{cases} \text{east}: & k = 1; & p = p_e \\ \text{north}: & k = 1; & p = p_n \\ \text{up}: & k = 3.7; & p = p_u \end{cases}$$

where

RMS_φ = Overall RMS (cm) for the double-differenced carrier-phase observables for the three single-baseline solutions as given in the OPUS-S solution,
T = Session duration in hours, and
p_e, p_n, and p_u = Peak-to-peak errors (cm) and in the east, north, and up, respectively.

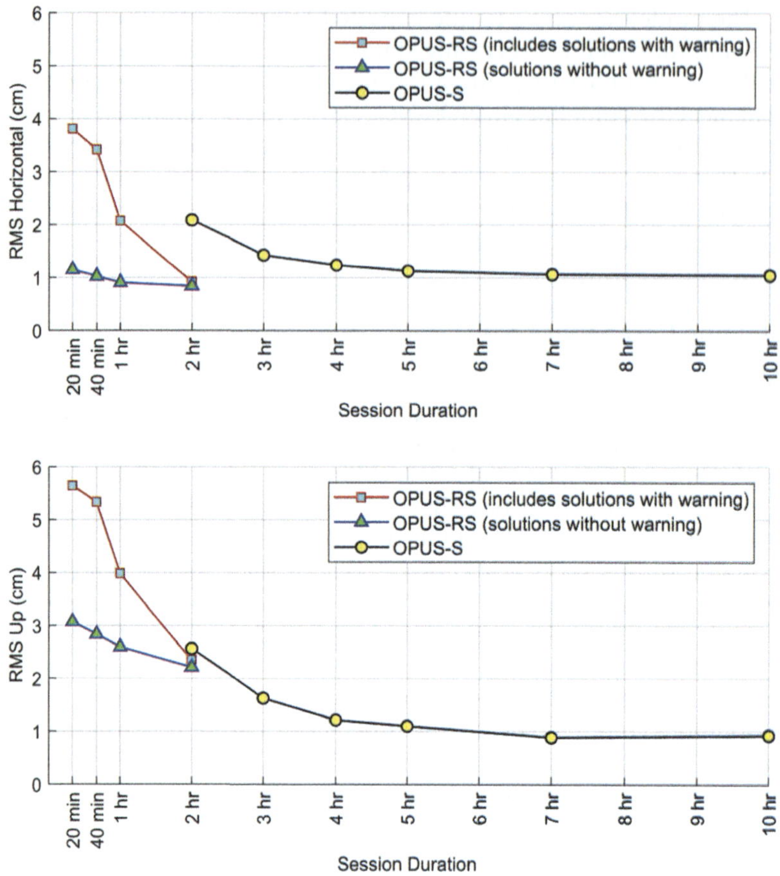

Figure 11-12. Accuracy of OPUS-S and OPUS-RS versus session duration.
Source: Gillins et al. (2019b) © ASCE.

More recently, Gillins et al. (2019b) evaluated the accuracy of OPUS-S using static GPS data collected on 20 bench marks in Oregon. Figure 11-12 presents the resulting error estimates in terms of RMS according to session duration. As should be expected, the accuracy improves as the session duration increases. The curves for OPUS-S become fairly asymptotic and improve marginally after $T = 4$ h; therefore, $T = 4$ h is an optimal session duration for OPUS-S.

Currently, the minimum required session duration for OPUS-S is 2 h. In addition, OPUS-S is capable of only processing GPS observables. NGS is currently working on a new version of PAGES (known as M-PAGES) that can process all viable GNSS signals as well as produce reliable results for shorter session durations. When finished, this new baseline processor will be implemented in all versions of OPUS, and it will improve its accuracy versus session duration.

Online Positioning User Service-Rapid Static

OPUS-S currently cannot produce reliable coordinates for sessions shorter than 2 h in duration; in 2007, NGS released OPUS-Rapid Static (OPUS-RS) for surveying engineers who wish to observe a mark for only 15 min. to 2 h. OPUS-RS is accessed at the same website as OPUS-S, and it functions similarly in that a user can upload raw GPS data and receive a solution via email containing geodetic coordinates for the observed mark tied to the NSRS.

OPUS-RS does have several key differences with OPUS-S. For instance, it uses a different baseline processor named RSGPS developed by the Satellite Positioning and Inertial Navigation group at Ohio State University. OPUS-RS uses session baseline processing of double-differenced carrier-phase GPS measurements from the observed mark to three to nine CORS and/or IGS stations that must meet stringent geometric conditions (Schwarz 2008). The geometry of the control is important because RSGPS spatially interpolates the tropospheric delay at the observed mark by plane fitting to the reference stations. As a result, OPUS-RS may work poorly near the coast or in areas where extrapolation is required because of a lack of control stations.

NGS provides an interactive map for viewing the estimated accuracy and availability of OPUS-RS (Figure 11-13). The user can pan and zoom the map, or they can enter latitude and longitude to view estimates of the accuracy of OPUS-RS in the horizontal or in terms of ellipsoid height for $T = 15$ m or 1 h.

Figure 11-12 also shows error estimates in terms of RMS according to session duration for a regional study of OPUS-RS in Oregon. OPUS-RS provides warning messages when it suspects a poor solution. As shown in Chapter 13 of Soler (2011), OPUS-RS is capable of providing many different types of error and warning messages as it has numerous quality control measures. Some of the most common warning messages occur when the scatter of the single baseline solutions is greater than 5 cm horizontally or 10 cm in ellipsoid height, the software struggles to fix integers, the solution quality indicator for OPUS-RS is less than 1.0, or the normalized RMS from the final adjustment exceeds 1.5 (Schwarz et al. 2009). Surveying engineers should be wary of solutions with warning messages, as can be seen in the curves in Figure 11-12. In other words, solutions from OPUS-RS with warning messages should be used with caution, as errors may be larger than estimated.

Online Positioning User Service-Projects

OPUS-S and OPUS-RS allow processing only a single GNSS session (currently GPS-only) on a mark. Released in 2012, OPUS-Projects is meant for managing, post-processing, and adjusting static GNSS campaigns involving multiple observations on numerous survey marks. It is considered the most rigorous and accurate version of OPUS, and it is free, web-based software designed to manage large sets of static GNSS data files, perform session baseline processing of simultaneous occupations on multiple marks, and conduct least-squares network adjustments of one or more sessions (Hensel et al. 2021). OPUS-Projects is accessed at NGS (2021f).

Similar to OPUS-S, OPUS-Projects currently uses PAGES as its baseline processor. Hence, it also currently requires static data files that are at least 2 h in duration, and it processes only GPS signals. Users may choose from a list of output reference frames (e.g., NAD 83 or ITRF2014) and geoid models, select various tropospheric delay models, and change elevation cutoff angles for session processing. The user can readily add data from stations in the NOAA CORS Network or IGS Network for tying to the NSRS, and they can create custom survey network and session processing designs with customizable weights on constraints.

Surveying engineers can use OPUS-Projects for processing and adjusting a GNSS survey network tied to the NSRS, thereby establishing geodetic control. The latest version of OPUS-Projects also prepares a set of complex files following specifications in the FGCS Bluebook (NGS 2021). The software allows users to submit these files to NGS for review and approval. If the engineer follows NGS guidelines (Zilkoski et al. 1997, 2008), then NGS

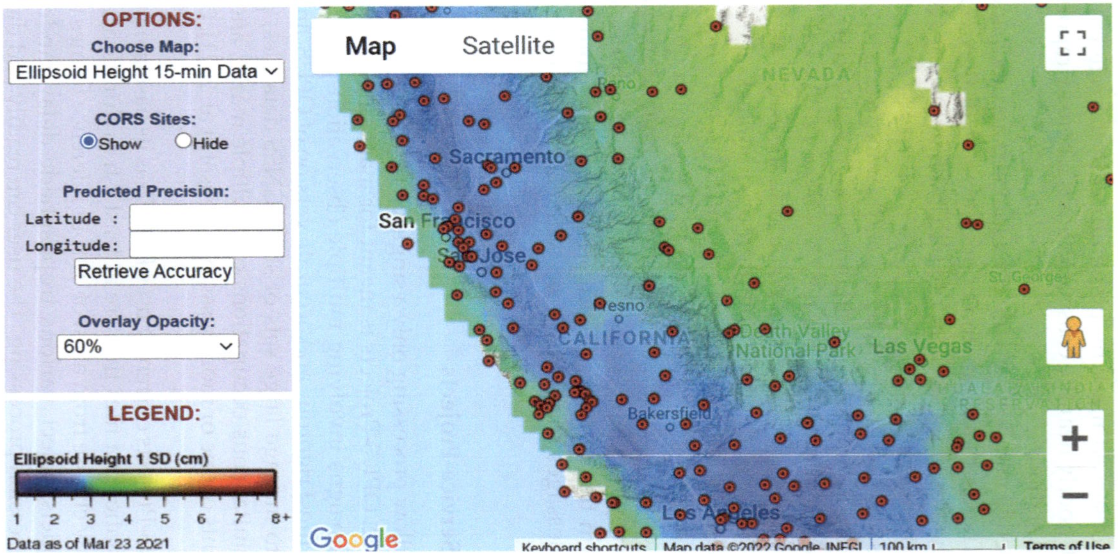

Figure 11-13. Interactive map of the accuracy and availability of OPUS-RS. The red dots are CORSs, and blank areas in the map are where OPUS-RS is unavailable.
Source: NGS (2021g). Courtesy of NOAA, reproduced with permission.

will accept these files and upload the survey into the NGS Integrated Database, thereby updating coordinates on datasheets. As a result, OPUS-Projects is becoming a popular tool for processing and adjusting GNSS survey campaigns according to NGS guidelines to establish geodetic control at airports and from height modernization surveys (Gillins and Eddy 2017, Weaver et al. 2018).

Hensel et al. (2021) provides a user guide with step-by-step instructions for using OPUS-Projects and submitting surveys to NGS. This guide recommends using a hub-and-spoke design for each GNSS session in a project (Figure 11-14). The manual recommends the following:

- Designating a single station within roughly 100 km of the observed marks in the session as the hub. The hub should be a CORS, temporary CORS, or other types of active GNSS stations so that it collects 24 h of data per day.
- If the observed marks in the session are spread across a larger area (e.g., >200 km), then designate more stations as hubs.
- Add data to each session from nearby stations in the NOAA CORS Network or IGS Network to serve as control. Use multiple stations to help reference the network to the NSRS and check for possible bias in the coordinates of the control stations. In addition, ensure that at least one long baseline (e.g., 300 to 1,000 km) exists in each session to help PAGES solve for the wet component of its tropospheric delay modeling parameters. This may require adding data from a distant CORS to the session.
- Process baselines such that every observed mark in the session is directly connected to a hub. Designating a CORS or IGS station as the hub is ideal because its 24 h data file helps ensure that there is sufficient mutual satellite visibility between the hub and the other CORSs that may be hundreds of km away for double differencing. Observations as short as 2 h in duration are adequate for the shorter baselines (i.e., <100 km) between the hub and the passive marks in the session.
- Constrain the hub(s) during session baseline processing.

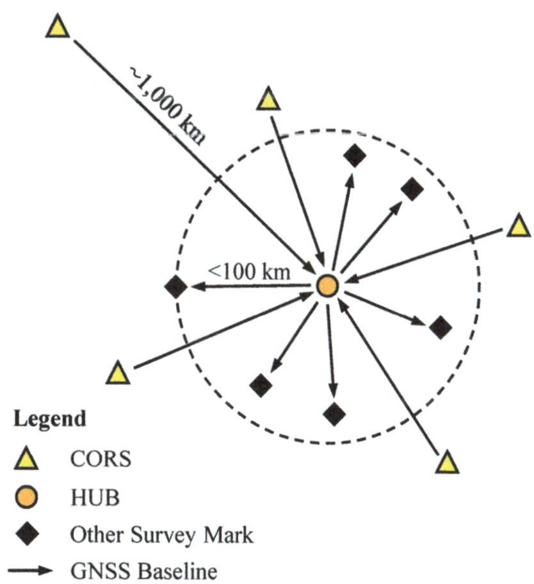

Figure 11-14. Recommended hub network design for session baseline processing in OPUS-Projects.
Source: Gillins et al. (2019b) © ASCE.

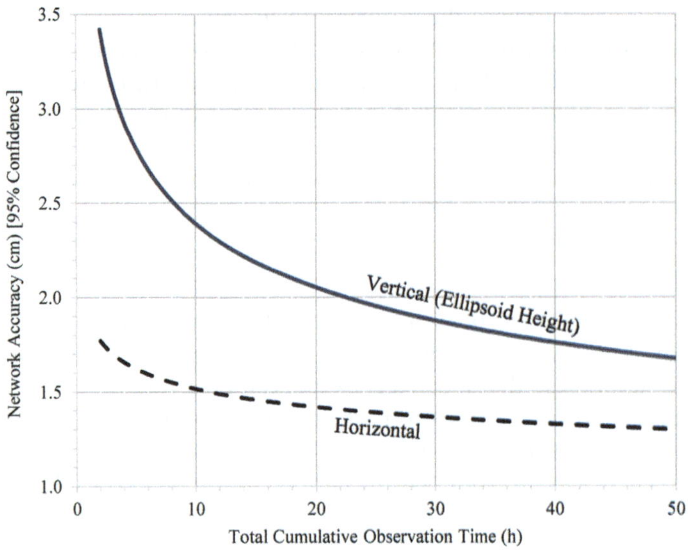

Figure 11-15. Design curves for OPUS-Projects; horizontal and vertical network accuracies versus total cumulative GNSS observation time.

After processing every session in a given survey project, OPUS-Projects provides tools for running horizontal (geometric) and vertical least-squares adjustments of the survey network. The user can select which stations to hold as network constraints. See Hensel et al. (2021) for more details. (See Chapter 5 for more on least-squares adjustments.)

Figure 11-15 provides engineering "design curves" for how long to observe marks with GNSSs during a campaign to achieve a certain horizontal or vertical accuracy at 95% confidence when using OPUS-Projects. This curve is based on the results of an empirical study of 24 GNSS surveys distributed across the United States that were processed and adjusted in OPUS-Projects (Park et al. 2018). The study found that accuracy is most strongly correlated with the total time spent observing the mark with GNSSs. According to the curves, when observing a mark for a cumulative total of 4 h (which can be accomplished in two independent, 2 h sessions), the final ellipsoid height should be expected to be less than 3.0 cm and the horizontal position less than 1.6 cm at 95% confidence.

Based on the curves in Figure 11-15, Table 11-1 provides the recommended number and duration of GNSS static sessions for achieving a certain vertical standard (ellipsoid height)

Table 11-1. Recommended GNSS Sessions to Achieve Certain Vertical (Ellipsoid Height) Accuracy Standards at 95% Confidence in OPUS-Projects.

Vertical standard (cm)	Total time on mark (h)	Recommended number and session duration
3.0	4	(2) 2 h
2.5	8	(2) 4 h
2.0	22	(3) 8 h

at 95% confidence. Such a Table can be useful for planning a GNSS survey campaign to achieve a desired level of accuracy. Since the horizontal errors are smaller (roughly 1.3 to 2.0 times smaller) than the vertical errors, the requirements to achieve vertical standards with GNSS control the survey design. Table 11-1 recommends repeat sessions on the marks to check the precision of each session solution. Because multipathing changes throughout the day, the repeat sessions should be done at a different time of day (e.g., separated by >2 to 3 h) to provide independent satellite observations for checking (Zilkoski et al. 1997). If the repeat sessions are done at the same time of day, there is a likelihood that the session solutions for a given mark will be overly precise and potentially bias the same amount because of similar satellite conditions. Table 11-1 shows that long static sessions are required, especially to achieve the 2.0 cm vertical standard.

The newest version of OPUS-Projects (version 5.0) also allows users to upload previously processed GNSS vectors to the survey network for least-squares adjustment (and submission to NGS if desired). Users can upload GNSS vectors resulting from RTK surveys or from baseline processing in third-party software. The GNSS vectors are uploaded in a standard file format known as the GNSS Vector Exchange (GVX) format (Gillins and Sellars 2021).

OPUS-Projects provides a lot of flexibility for working with uploaded GNSS vectors, but Figure 11-16 presents a simple diagram of the recommended workflow for adding RTK vectors from an RTN to a control network (based on Weaver et al. 2018, Gillins et al. 2019a). According to the diagram, first, observe marks using RTK techniques with an RTN. Note that for the procedure to work, configure the rover to store data as vectors from physical base stations in the RTN to the observed marks. Download the resulting RTK survey solutions to the computer, and they will appear on a map as vectors from RTN base station(s) to the observed marks (Figure 11-16a). Next, convert the RTK solutions into the GVX file format, and upload the file(s) to OPUS-Projects. Afterward, upload static GNSS data files logged at any of the base stations that are not in the NOAA CORS Network or IGS Network to OPUS-Projects. Then, add data from stations in the NOAA CORS Network or IGS Network to the survey network for use as control and then process all static sessions (Figure 11-16b). Finally, use tools in OPUS-Projects to adjust the survey network consisting of both uploaded RTK vectors and session baseline processing solutions (Figure 11-16c). The approach results in an interconnected survey network that is tied to the NSRS.

The accuracy of an RTK solution improves with observation duration; however, the improvement is subtle. Allahyari et al. (2018) found from testing RTNs in Oregon and South Carolina that there was little improvement in accuracy after 180–300 s (3–5 min.). There is much greater benefit in taking repeat, independent RTK observations on marks (Smith et al. 2014). Based on tests in Weaver et al. (2018) and Gillins et al. (2019a), Table 11-2 presents the recommended number of repeat RTK observations on marks when using the full network corrector from an RTN to ensure that ellipsoid heights are less than a given vertical standard at 95% confidence. For this table, "GNSS" indicates using GPS plus additional signals from other viable satellite systems (e.g., GLONASS, Galileo, BeiDou). Like in static GNSS surveys, surveying engineers should check the precision of the repeat observations and remove outliers, and they must also make the repeat observations at different times of the day (e.g., separated by >2 to −3 h).

Other Online Tools for Global Navigation Satellite System Processing

This chapter has provided a brief summary of the many variants of OPUS. Of course, other online software applications are available for processing static GNSS data. Studies have shown that similar accuracies in the global reference frame (e.g., ITRF2014) can be derived when using these applications versus using OPUS-S (Silver 2013, Jamieson and Gillins 2018, El Shouny and Miky 2019, Gökdaş and Özlüdemir 2020).

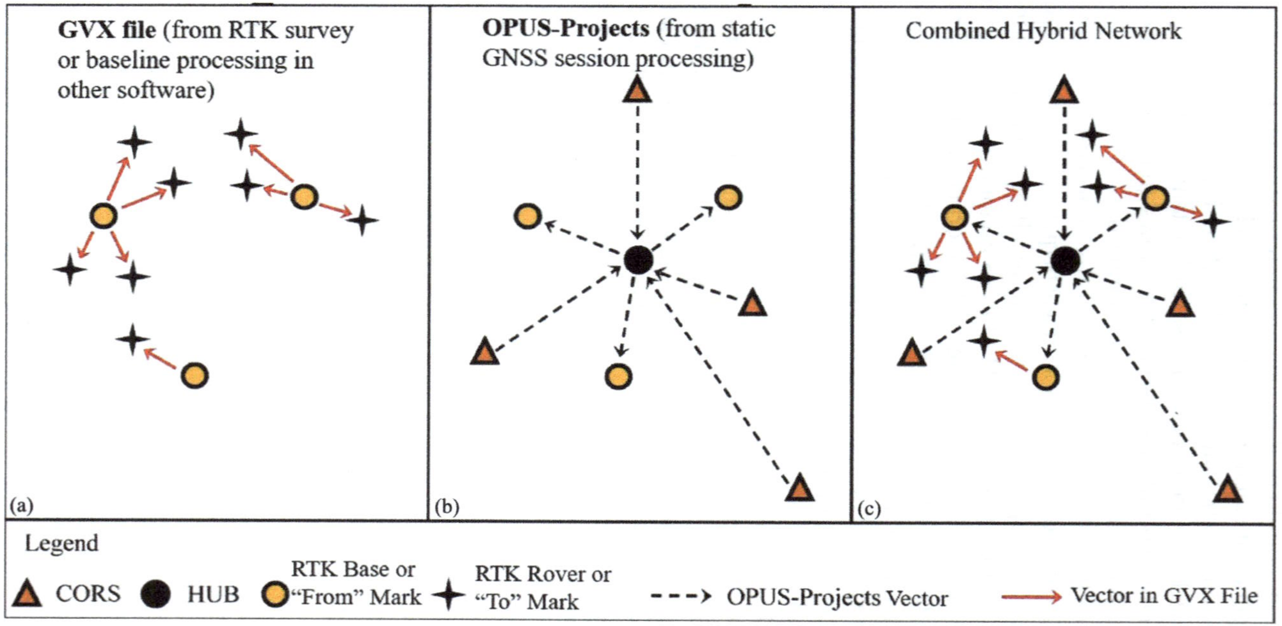

Figure 11-16. (a) Uploading RTK GNSS vectors to OPUS-Projects, (b) processing static GNSS data (if any), (c) combining results for a least-squares survey network adjustment.
Source: Adapted from Weaver et al. (2018) © ASCE.

Table 11-2. Recommended Number and Duration of Network RTK Observations on Marks to Meet Certain Vertical (Ellipsoid Height) Accuracy Standards at 95% Confidence in OPUS-Projects.

	Number of repeat network RTK observations			
Vertical standard (cm)	GPS-only	GPS + more (GNSS)	Observation duration (min)	Minimum epochs used
3.0	3	3	5	300
2.5	4	3	5	300
2.0	6	5	5	300

Some of the available applications include AUSPOS, developed by Geoscience Australia; CSRS-PPP developed by Natural Resources Canada; GAPS, developed by the University of New Brunswick; and Trimble-RTX, commercial software developed by Trimble. All these online applications are capable of processing a static GNSS session on a mark and providing coordinates in the global reference frame. Similar to OPUS-S, AUSPOS uses relative positioning techniques (see Chapter 6) and IGS or other active GNSS stations as the reference stations.

Some of the applications (CSRS-PPP, GAPS, and Trimble-RTX) use precise point positioning (PPP) techniques (see Chapter 6). PPP techniques are based on absolute point positioning because they only process the satellite observations from a single receiver as opposed to relative positioning techniques which require at least two receivers. Because only a single receiver is required, a baseline solution or vector is not generated. Thus, the resulting coordinates from a PPP application are relative to the satellites rather than relative to a reference station. It can be said that PPP opens a new category for geodetic control: control from the satellites rather than an active reference station or passive mark.

ADJUSTMENTS AND EVALUATING CONTROL

Control is never perfect, and a surveying engineer should never assume that published coordinates are without error. Errors in control can occur owing to a variety of reasons, including previous survey blunders, poor modeling, use of lower-quality equipment, or mark movement or disturbances.

It is, therefore, best practice to tie surveys to multiple, redundant control marks like in the diagram in Figure 11-8. Extra control can help identify if any of the marks have coordinates or heights that are in error. A common approach is to difference the newly measured coordinates between the control marks in the network and then compare these differences with published coordinate differences. If the results are similar or have small residuals, then the surveying engineer has greater confidence in the control.

Errors in control can also be found by running minimally and fully constrained least-squares adjustments of the survey network. A minimally constrained adjustment should be done first by holding only the minimum number of necessary constraints fixed or as if errorless. As examples, in a GNSS survey, the coordinates (latitude, longitude, ellipsoid height) of one reliable station in the network are held fixed as control; for a leveling network, the orthometric height of one bench mark is held fixed. By only using minimal constraints, the adjustment is not influenced by any possible error in the control. Rather, a minimally constrained adjustment enables the determination of whether the measurements themselves in the survey network fit together and close internally. If a minimally constrained adjustment fails, it is highly likely that there are errors in the measurements that need to be dealt with first by rejecting the bad measurement and removing it from the network or by making additional measurements.

After a successful minimally constrained adjustment, it is then possible to check the control by running fully constrained adjustments. Fully constrained adjustments are doable only when there is more than the minimum number of necessary constraints or redundant or extra control is in the survey network. A fully constrained network requires the selection of control marks with reliable coordinates and then appropriately weighting the estimated errors of these coordinates. After running a fully constrained adjustment, an *F*-test (statistical hypothesis test) can be run to check if the variance of unit weight differs with the variance of unit weight from the minimally constrained network adjustment (see Chapter 5 for more details).

A good approach for evaluating control is to attempt numerous iterations of fully constrained network adjustments using different combinations of control marks. By following a systematic approach, a control mark with poor coordinates (or poor error estimates) can be isolated and eliminated as a network constraint. Gillins and Eddy (2017) proposed a simple approach that is modified and summarized as follows:

1. Constrain the published coordinates and/or heights of all control marks in the network and perform a fully constrained least-squares adjustment of the survey.
2. For every constraint, compute a *constraint ratio* as defined in Equation (11-2). Individual constraint ratios can be computed for the north, east, ellipsoid heights, or orthometric heights, depending on the constrained component

$$\mathrm{CR}_X = \frac{\left|X_{\text{adjusted}} - X_{\text{published}}\right|}{\sigma_{X,\text{published}}} \tag{11-2}$$

   where the numerator is the absolute residual difference of the adjusted value (i.e., north, east, or height) with the published and constrained value, and σ_X is the standard deviation assigned to the constrained value in the adjustment.
3. Identify all valid constraints, defined as when their $\mathrm{CR} < 3$. A value of 3 is chosen because 99.7% of the data should be within three standard deviations from the mean; it is a good threshold for detecting outliers.
4. Remove any constraints with a $\mathrm{CR} > 3$ and then rerun the fully constrained adjustment of the survey network. For thoroughness, constraints should be removed one at a time.
5. Repeat Steps 1 to 4 until all used constraints have $\mathrm{CR} < 3$. This procedure should result in all valid control being identified for use.

REFERENCES

Allahyari, M., M. Olsen, D. T. Gillins, and M. L. Dennis. 2018. "A tale of two RTNs: Rigorous evaluation of GNSS survey observations in real-time networks." *J. Surv. Eng.* 144 (2): 05018001.

Armstrong, M. 2011. *Guidelines for positioning the Oregon real time network with the NGS national spatial reference system*. Salem, OR: Oregon Dept. of Transportation, Geometronics Unit.

ASCE. 1985. *Engineering surveying manual*, MOP 64. Reston, VA: ASCE.

Chin, M. M., D. R. Crump, and K. A. Berstis. 1987. "The status of the NGS GPS orbit tracking network (abstract)." *EOS Trans. Am. Geophys. Union* 68 (44): 1239.

Eckl, M. C., R. Snay, T. Soler, M. W. Cline, and G. L. Mader. 2001. "Accuracy of GPS-derived relative positions as a function of interstation distance and observing-session duration." *J. Geod.* 75 (12): 633–640.

El Shouny, A., and Y. Miky. 2019. "Accuracy of relative and precise point positioning online GPS processing services." *J. Appl. Geod.* 13 (3): 215–227.

FGCS (Federal Geodetic Subcommittee). 2004. *FGCS standards and procedures to incorporate electronic digital/bar-code leveling systems*. Version 4.1. Reston, VA: US Geological Survey.

Floyd, R. P. 1978. *Geodetic bench marks*. NOAA Manual NOS NGS 1. Silver Spring, MD: National Oceanic and Atmospheric Administration.

Gillins, D. T., and M. J. Eddy. 2017. "Comparison of GPS height modernization surveys using *OPUS-projects* and following NGS-58 guidelines." *J. Surv. Eng.* 143 (1): 05016007.

Gillins, D. T., J. Heck, G. Scott, K. Jordan, and R. Hippenstiel. 2019a. "Accuracy of GNSS observations from three real-time networks in Maryland, USA." In *Proc., FIG Working Week 2019*. Hanoi, Vietnam: https://www.fig.net.

Gillins, D. T., D. Kerr, and B. Weaver. 2019b. "Evaluation of the online positioning user service for processing static GPS surveys: *OPUS-Projects, OPUS-S, OPUS-Net*, and *OPUS-RS*." *J. Surv. Eng.* 145 (3): 05019002.

Gillins, D. T., and I. Sellars. 2021. *GNSS vector exchange file format (GVX): Narrative description of GVX data types and elements*. Technical Notes, version 1.0. Silver Spring, MD: National Oceanic and Atmospheric Administration, National Geodetic Survey.

Gökdaş, O., and M. T. Özlüdemir. 2020. "Velocity estimation performance of GNSS online services (APPS and AUSPOS)." *Surv. Rev.* 53 (4): 1–9.

Henning, W., D. Martin, G. Schrock, G. Thompson, and R. Snay. 2013. *Guidelines for real time GNSS networks*. Version 2.2. Silver Spring, MD: National Oceanic and Atmospheric Administration, National Geodetic Survey.

Hensel, P., M. Schenewerk, J. Prusky, D. Gillins, D. Zenk, K., et al. 2021. *OPUS projects: Online position user service baseline processing and adjustment software*. User guide. Silver Spring, MD: National Oceanic and Atmospheric Administration, National Geodetic Survey.

Jamieson, M., and D. T. Gillins. 2018. "Comparative analysis of online static GNSS post-processing services." *J. Surv. Eng.* 144(4): 05018002.

NGS (National Geodetic Survey). 2020. *Survey mark recovery*. Accessed January 31, 2022. https://geodesy.noaa.gov/surveys/mark-recovery/index.shtml.

NGS. 2021a. *OPUS: Online positioning user service*. Accessed January 31, 2022. https://geodesy.noaa.gov/OPUS/

NGS. 2021b. *CORS map*. Accessed January 31, 2022. https://geodesy.noaa.gov/CORS_Map/.

NGS. 2021c. *User friendly CORS (UFCORS)*. Accessed January 31, 2022. https://www.ngs.noaa.gov/UFCORS/.

NGS. 2021d. *National geodetic survey data explorer*. Accessed January 31, 2022. https://geodesy.noaa.gov/NGSDataExplorer/.

NGS. 2021e. *Input formats and specifications of the National Geodetic Survey data base (The NGS "Bluebook")*. Silver Spring, MD: National Oceanic and Atmospheric Administration, NGS.

NGS. 2021f. *OPUS-Projects*, Accessed January 31, 2022. https://geodesy.noaa.gov/OPUS-Projects/OpusProjects.shtml.

NGS. 2021g. *OPUS-RS Precision map*. Accessed June 7, 2021. https://geodesy.noaa.gov/OPUSI/plots/gmap/OPUSRS_sigmap.shtml.

Park, J., S. Kim, A. Shahbazi, D. Gillins, and M. Dennis. 2018. *Evaluation of static GPS surveying campaigns processed in OPUS-projects.* Final Tech. Rep. FY17 NA293P. Burke, VA: National Geodetic Survey.

Saleh, J., S. Yoon, K. Choi, L. Sun, R. Snay, P. McFarland, et al. 2021. "1996–2017 GPS position time series, velocities and quality measures for the CORS Network." *J. Appl. Geod.* 15 (2): 105–115.

Schwarz, C. R. 2008. "Heuristic weighting and data conditioning in the National Geodetic Survey Rapid Static GPS software." *J. Surv. Eng.* 134 (3): 76–82.

Schwarz, C. R., R. A. Snay, and T. Soler. 2009. "Accuracy assessment of the national geodetic survey's *OPUS-RS* utility." *GPS Solutions* 13 (2): 119–132.

Silver, M. 2013. "A comparison of free GPS online post-processing services." Accessed June 14, 2021. https://www.gpsworld.com/a-comparison-of-free-gps-online-post-processing-services/.

Smith, C. L. 2010. *Bench mark reset procedures.* Silver Spring, MD: National Geodetic Survey, National Oceanic and Atmospheric Administration.

Smith, D., K. Choi, D. Prouty, K. Jordan, and W. Henning. 2014. "Analysis of the TXDOT RTN and *OPUS-RS* from the geoid slope validation survey of 2011: Case study for Texas." *J. Surv. Eng.* 140 (4): 05014003.

Snay, R. A., and T. Soler. 2008. "Continuously Operating Reference Station (CORS): History, applications, and future enhancements." *J. Surv. Eng.* 134 (4): 95–104.

Soler, T. 2011. *CORS and OPUS for engineers: Tools for mapping applications.* Reston, VA: ASCE.

Soler, T., P. Michalak, N. D. Weston, R. A. Snay, and R. H. Foote. 2006. "Accuracy of OPUS solutions for 1- to 4-h observing sessions." *GPS Solutions* 10 (1): 45–55.

Soler, T., and G. Wang. 2016. "Interpreting *OPUS-static* results accurately." *J. Surv. Eng.* 142 (4): 05016003.

van Sickle, J. 2015. *GPS for land surveyors.* Boca Raton, FL: CRC Press.

Wang, G., and T. Soler. 2013. "Using OPUS for measuring vertical displacements in Houston, Texas." *J. Surv. Eng.* 139 (3): 126–134.

Wang, G., M. Turco, T. Soler, T. J. Kearns, and J. Welch. 2017. "Comparisons of *OPUS* and PPP solutions for subsidence monitoring in the greater Houston area." *J. Surv. Eng.* 143 (4): 05017005.

Weaver, B., D. T. Gillins, and M. Dennis. 2018. "Hybrid survey networks: Combining real-time and static GNSS observations for optimizing height modernization." *J. Surv. Eng.* 144 (1): 05017006.

Zilkoski, D. B., J. H. Richards, and G. M. Young. 1992. "Results of the general adjustment of the North American vertical datum of 1988." *Surv. Land Inf. Syst.* 52 (3): 133–149.

Zilkoski, D. B., E. E. Carlson, and C. L. Smith. 2008. *Guidelines for establishing GPS-derived orthometric heights.* NOAA Technical Memorandum NOS NGS-59. Silver Spring, MD: National Oceanic and Atmospheric Administration, National Geodetic Survey.

Zilkoski, D. B., J. D. D'Onofrio, and S. J. Frakes. 1997. *Guidelines for establishing GPS-derived ellipsoid heights (standards: 2 cm and 5 cm).* NOAA Technical Memorandum NOS NGS-58, Version 4.3. Silver Spring, MD: National Oceanic and Atmospheric Administration, National Geodetic Survey.

CHAPTER 12
CONSTRUCTION SURVEYS

Marlee A. Walton

INTRODUCTION

A construction survey includes processes to establish locations and evaluate and estimate quantities for construction activities. It is the process of drawing the design plans on the actual construction site at the designated location. Construction surveying is carried out to layout infrastructure such as buildings, roadways, bridges, tunnels, and various utilities.

The purpose of construction surveys is to control construction activities. The number and extent of surveys conducted is governed by the time available, the standard of construction desired, and the availability of personnel and materials. The quality and efficiency of the construction is directly proportional to the number and extent of surveys and other planning activities.

Construction surveys cover a wide range of tasks. These tasks are performed before construction begins, during the construction process, and after the construction has been completed.

Before Construction

The preconstruction phase can include a location survey of the project area, sometimes called *reconnaissance, preliminary survey*, or *location survey*. This survey provides the basis for selecting acceptable sites or routes and furnishes information for use on subsequent surveys.

For linear projects such as roadways and utilities, stationing is established as the location is being surveyed. In addition, profile and cross-sectional information may be obtained to allow rough calculations of the earthwork involved.

For location surveys, on both linear and site projects, topographic data are typically collected. These usually include the collection of planimetric features and terrain elevation data. Planimetric features include human-made or physical objects in, or adjacent to, the construction site. The features may include such items as existing vegetation, fencing, roads, buildings, utilities, waterways, and railroads. The terrain elevation data were historically collected in a grid-like manner throughout the project area (see Chapter 7). Supplemental elevation points were also picked up to further define the terrain conditions. Over the last three decades, elevation points have been commonly measured along breaklines and at other strategic positions to develop terrain elevation models. Other remote sensing technologies such as lidar and photogrammetry are also popular for efficiently collecting terrain elevation data at a higher resolution (Chapters 8 through 10).

Once the terrain data are collected, the terrain surface is shown with a digital elevation model (DEM). The DEM data can be used to determine earthwork quantities and to facilitate machine guidance and control or stakeless construction. These data can also be used to determine surface water runoff, flood risk, and other urban drainage–related issues.

Sometimes, it is necessary to conduct several location surveys to investigate more than one suitable route. After the alternatives are surveyed and evaluated, the best site is proposed for construction.

Irrespective of whether a location survey is conducted, a preconstruction activity, which is always needed, is the establishment of a network of accurate control points and bench marks that will be used and referenced throughout the construction phase of a project. This control scheme will be used to control the horizontal positions and the elevations of the infrastructure. The horizontal and vertical control network is a key component to laying out engineering and construction works. It is critical that control points for monitoring surveys be placed outside the area of influence or area that is subject to movement from construction activities to accurately monitor any movement. It is also critical to check locations where the new construction meets the existing features for horizontal and vertical accuracy. Doing this before construction begins can identify problems with the design and allow them to be corrected with minimal delays to the project; see Chapter 11 for more details on finding existing control and densifying control for an engineering project.

The final preconstruction activity is typically construction staking, also referred to as a *site layout survey*. Staking is completed by interpreting design plans and marking the locations of the proposed infrastructure. This is performed to ensure that a project is built according to the engineering design plans. The staked reference points guide the construction of the proposed improvements on the property and help ensure that the construction project is completed on schedule, on budget, and as intended.

During Construction

Fieldwork during construction includes tying into existing horizontal and vertical control points and augmenting and replacing control points as necessary to control the construction project. Random checks should be performed on the horizontal and vertical control network points to compare the resultant coordinates with the original coordinates.

Ongoing replacement of construction layout stakes is almost always essential as a project progresses toward completion. Periodic checks of the existing staked-out locations should also be completed to ensure compliance with the design dimensions and to ascertain quantities.

Even as the project is progressing, as-built information should be collected for the purpose of documenting the final installation of the contract items. This not only ascertains quantities but also provides documentation to corroborate the payments made to the contractor.

Sometimes a construction survey will include monitoring existing structures. A contractor may need to provide assurances that activities such as blasting, excavation, and pile driving are not causing disturbances to adjacent buildings. In these cases, specialized reflecting targets are attached to the existing structures. These targets are surveyed on a monthly, weekly, or even daily basis, and their location is compared with an original baseline measurement to determine whether there has been any settlement or movement.

After Construction

After construction has finished, construction surveys can be used to assess the final positions and adequacy of the work, comparing what has actually been built with the way it was planned. This type of construction survey is sometimes referred to as an *as-built* or *as-constructed* survey.

This survey can be used to compute final construction payments. Whenever possible, field measurements for pay quantities should be made concurrently with construction operations to provide better accuracy and reliability. However, some items can be checked only after construction is complete.

HORIZONTAL AND VERTICAL CONTROL

Chapter 11 discusses finding existing horizontal (or geometric) and vertical control and establishing new control. Prior to establishing horizontal and vertical control, the design plans should be obtained and marked up with the approximate location of existing control marks. The marked-up information might include station descriptions, coordinate information, and the accuracy characteristics of the control mark.

Next, the locations where construction control marks will be required can be evaluated. Proper planning of control marks should take into account the number and physical location of project points (i.e., accessibility, spacing, visibility, and geometry), the standard errors one expects to achieve from field measurements, and the number and types of observations to be measured (instrument types and repeat measurements). Such planning can save considerable field time and assure that the required control marks will be established at their optimum locations.

Sometimes the contract specifications call for following certain national surveying specifications (see FGCS 2004 or NOS-NGS 58 and 59, as discussed in Chapter 11) and reporting accuracies at 95% confidence [e.g., see Equations (5-33) and (5-37) in Chapter 5].

If the contract does not call for certain accuracy specifications or specify how many new project control marks are required and the respective distribution within the project area, then local standards and regulations could form the basis for the control network design.

When setting control marks, the material needs to be substantial enough to remain in place and undisturbed for the duration of the construction project. In addition, every effort should be made to protect and preserve these monuments; see Chapter 11 for some example control monuments and more discussion on preserving control.

During construction, the network of control marks is used to guide the construction activities. Measurements are made to these control marks to determine the location of the planned structure or improvements, horizontal and vertical positioning, dimensions, and configuration and to control the elevation of new structures. The most fundamental aspect of a successful survey and all subsequent activities is accurate, reliable, and precise survey control.

Horizontal Control

The distance between horizontal control marks will be controlled by the project specifications and accuracy needs. In general, for construction projects, the spacing between horizontal control marks should be limited to no more than one-quarter mile. Chapter 11 discusses dividing the project into four quadrants and ensuring that there are horizontal control marks in at least three of the four quadrants as well as in the middle.

Supplemental control may also be established from the control network to aid in either the data gathering process or the construction staking. With proper planning, much of the supplemental control may have been included in the control survey during the preconstruction phase. In general, supplementary control marks will be within a distance of approximately 150 m (500 ft) from a control mark.

Vertical Control

Vertical control is an important part of all construction projects. The most critical aspect of the vertical control is that the same datum be used from preliminary surveys through the final construction.

A relatively dense network of vertical control or bench marks will be established for most projects prior to construction. If there has been a delay between the time preconstruction bench marks were established and the construction phase, the condition of the bench marks should be field-checked. Each bench mark should be checked for possible disturbance or settlement and whether it is in satisfactory condition.

As discussed in Chapter 11, bench marks may be established differentially with levels or total stations (see Chapter 7) or with Global Navigation Satellite Systems (GNSSs) (Chapters 6 and 11). The equipment used mainly depends on the project circumstances. When the equipment is selected, accuracy needs are key, and bench mark accuracy should be consistent with the type of construction for which they will be used.

In addition to accuracy, the required density of bench marks will depend on terrain, vegetation, and the type of construction. The bench marks should be of sufficient density to adequately complete the construction activity. In general, a minimum of four bench marks are recommended, although a large project should have more. The bench marks are best located at the four corners of the project area. Linear projects require bench marks at the beginning, end, and supplemented intervals along the line.

Chapter 11 provides details on how to use tools provided by the National Geodetic Survey (NGS) to find existing bench marks with published orthometric heights. From these existing bench marks, temporary bench marks (TBMs) may be established by leveling or GNSSs. TBMs are less permanent and may be required for a limited period or for a specific survey operation such as slope staking. Such monuments are typically not perpetuated after construction.

A minimum of four existing bench marks should be utilized when establishing TBMs to verify that the bench marks have valid orthometric heights and have not moved or been disturbed (see Chapter 11). A sufficient number of TBMs should be set to control a construction project. A minimum spacing of 240 m (800 ft) is typical, although spacing may vary depending on project requirements. In addition, it is common to keep TBMs just outside the construction limits of the project (e.g., no more than 60 m or 200 ft beyond the outer limits is typical, but again, it depends on project requirements). All TBMs should be located properly and comprised of sufficient material such that their integrity will not be compromised throughout the life of the project.

It is important to set stable marks to serve as TBMs. Because these marks must serve as temporary vertical control, they should be set deep enough to not be subject to frost heave or other processes that can cause vertical motion. Although TBMs are meant to serve as temporary control, they must remain stable for the life of the construction project.

TBMs are frequently marked or *guarded* by a flat guard stake that remains partially above the ground and may be driven at a slope, so its top is over the TBM. These guard stakes are flat stakes, and they are typically written on with symbols, numbers, and abbreviations to identify the name assigned to the TBM. Sometimes, other types of markers such as a nail driven into a tree, chiseled marks in concrete, or a pipe driven into the ground may also be used as a reference location.

Bench marks in the project area can be a source of confusion and possible error through misidentification. It is important that each bench mark be uniquely identifiable by name, number, and location and marked with the appropriate identification code. During periods of use, a flagged or painted lath can aid in the speedy location of the bench mark.

CONSTRUCTION SURVEY TASK SEQUENCE

Several tasks need to be completed during construction surveying, as discussed in Dewberry Companies (2002), and Kavanaugh and Slattery (2015). In general, the tasks include the following steps:

1. Compile all data and material, including plans, maps, and control information.
2. Carefully study and check all the compiled data and material to become familiar with the project and to detect any possible errors or omissions.
3. Field-inspect the construction site and discuss proposed staking with the construction contractor, project engineer, or designated representative.
4. Determine in detail the staking required, the established procedures for reporting, and the format of the deliverables required.
5. Make calculations, field notes, and survey drawings for the layout and control of the work.
6. Set and preserve survey monuments and control marks and make checks for accuracy and precision.
7. Set stakes to define the work and complete field checks for the stakes set.
8. Check locations for horizontal and vertical accuracies of the design plans with the existing conditions.
9. Replace and augment control marks, bench marks, and stakes as necessary.
10. Process data collection files and review for errors and conformance to standardized procedures and policies.
11. Measure quantities to ensure that the project has actually been completed in terms of dimensional measurements and collaborate contractor payments.
12. Develop a control mark list that tabulates all control for a project.
13. Submit deliverables.

CONSTRUCTION SURVEY EQUIPMENT

Traditionally, construction survey equipment consisted of the total station and level (see Chapter 7). The introduction of new technology has revolutionized the way construction surveys are done today.

The data collector, also called data controller or electronic field book, computes coordinates from distances and angles and vice versa; performs inverses, intersections, and resections; and determines stationing, offset positions, elevations, and cut and fills. Design plan data can be uploaded into the data collector and readied for the field. In the field, the data collector can record the construction staking information, including coordinates, station and offsets, stake elevations, design grade, and cut/fill information. After the field data are collected, the data can be downloaded to a computer for reference, storage, data processing, and record-keeping purposes.

Although the data collector is referred to as an electronic field book, it is common to still have a field notebook to record sketches and field notes.

Chapter 6 provides details on GNSS surveying. GNSSs can be used for many tasks in construction surveying, such as setting control, measuring, staking, mapping, and DEM development. Real-time kinematic (RTK) GNSS techniques are typically more efficient than other instruments because roving receiver-equipped surveying engineers can move quickly to establish precise locations for control and layout staking points. Unlike traditional surveying equipment, RTK measurements can be taken by a single person holding the GNSS surveying instrument for a handful of seconds to a few minutes at the different points in the area, depending on accuracy requirements. GNSS measurements do not require a line of sight between the receivers, and the units work under most weather conditions. In addition, the recent increase in the coverage of real-time network base stations has lessened the need for as many on-site, temporary ground base stations. However, GNSS measurements do have the disadvantage of requiring a clear view to the sky. When GNSS signals are blocked by tree canopies or other obstructions, alternative surveying methods must be used.

As detailed in Chapter 10, unmanned aircraft systems (UASs) are increasingly complementing other sources of survey data and information on construction projects. UASs can be used in construction for topographic mapping surveys and in the development of the DEM. There are several key advantages tousing UASs to gather data. UASs allow surveys to be carried out quickly, deliver highly accurate data, and provide a highly cost-effective solution. UASs are also quick to mobilize and can survey areas that are difficult to reach.

In addition to the survey-related information, UASs can collect other data that may be beneficial for project management. For example, UASs have been used to obtain construction progress photography at regular intervals throughout a project, to monitor job sites to ensure safe practices, to inspect bridges and other structures, and to show clients an aerial overview of completed projects.

Terrestrial (Chapter 8), mobile (Chapter 9), and airborne lidar (Chapter 10) are also gaining popularity as efficient survey methods for completing topographic surveys, developing a DEM, and monitoring the progress of ongoing construction. Lidar provides high-resolution three-dimensional models by collecting and processing millions of data points in real time. For many challenges such as scanning between trees, lidar has proven to be a valuable technology because it offers a fast, precise, and direct method of three-dimensional mapping while producing accurate data.

Regardless of the instruments selected for a project, all construction survey equipment should meet the following criteria:

- Suitable for the purpose intended,
- Proper integration of all components,
- Capable of achieving the specified tolerances,
- Maintained to meet manufacturer's specifications, and
- Kept in proper adjustment and good operating condition throughout the duration of the project.

Choosing the correct survey equipment and method should be based on the accuracy needs and site conditions of the project.

FIELD NOTES

Field notes are not an accessory to the survey; they are an integral part of the survey. Survey field notes are important because they contain the complete graphic, tabular, and written survey records. Field notes depict each step of the work done in the field, provide a way to check the accuracy of the fieldwork, and offer a means to perpetuate a survey even when stakes and monuments are obliterated or lost; see Chapter 13 for more details on what should be documented in field notes and how should these notes be kept clear.

The use of data collectors does not eliminate the need for field notes and sketches. A field book still needs to be maintained to supplement the digital data.

In the data collector, data are often collected and downloaded in the format of point number, northing, easting, elevation, and code. The point numbers and codes are assigned by the surveying engineer. The codes are used to identify specific features such as curb, sign, tree, control mark, or centerline. A system should be developed to have consistency in the use of point numbers and codes. The data collector also stores a raw data file, which includes data such as GNSS vectors, horizontal angles, vertical angles, and distance measurements. It is important for the surveying engineer to keep the raw observational data and not just the point data. The raw observational data provide a digital record of what occurred in the field, which corrections were applied, and when were the data collected. In addition, the data can be adjusted or reprocessed in the future when better models become available.

Both the field notes and data collector files should be stored and accessible for a proper duration after the construction project is complete.

CONSTRUCTION STAKING AND LAYOUT

Construction staking, also known as a *site layout survey*, is performed to ensure that a project is built according to the engineering design plans (Crume 2014). The staked reference points guide the construction of proposed improvements on the property. Accurate construction staking is a critical step in ensuring the success of a construction project.

The primary functions of construction stakes are to place the earthwork and infrastructure components in the correct location, guide equipment operators, and eliminate unnecessary work. They can also indicate the width of clearing required by showing the limits of the project.

The location of earthwork and infrastructure components is set based on the location of control marks. The location is then marked by setting stakes opposite all important points and at sufficient intervals to define the alignment of the horizontal and vertical planes. The stakes should be located at some convenient or whole-number horizontal distance from the actual construction surface (e.g., 2 or 3 m).

In the construction of linear projects such as roadways or utility corridors, a survey line can be established parallel to and at a convenient horizontal offset distance from the infrastructure. Stakes can be set at intervals to provide convenient measurements but far enough away to avoid displacement.

After a surveying engineer has completed the layout and marking of stakes, it becomes the contractor's responsibility to utilize these stakes to control construction. In general, this operation is accomplished using grade-alignment wires or strings, batter boards, laser devices, or other suitable means.

After staking has been completed, a thorough check of all measurements should be made to make certain that minimal errors exist. Some errors will always occur during construction staking because of weather, equipment, terrain, and human fallibility. The surveying engineer needs to know and apply the acceptable tolerances within those errors. For example, stakes that are set for controlling earthwork activities may need to be only within a 0.3 m horizontal distance from the plan location, whereas a building might need to have accuracies within 3 mm. Each type of stakeout must be analyzed in light of its purpose, how it will be used, and whether the points need to be perpetuated. If the tolerable error is exceeded, the source of error must be found and the measurement corrected.

Construction Staking Equipment

The use of a three-dimensional coordinate system for the design makes staking with total stations or GNSS instruments a viable and cost-effective option for slope staking, particularly in rough terrain.

If a total station and data collector are used for staking, a radial stakeout or a method where many or all offset points are established from one instrument setup will be used. The first requirement for radial slope staking is that a minimum of two control marks be available for the slope staking. A second requirement is that the instrument be within approximately 300 m (1,000 ft) of the furthest slope stake to be set. The control marks will serve as the instrument location and the back azimuth or both will be used in resection to determine the coordinates of the instrument. This radial method can be risky if not followed with a series of checks. Whenever possible, angles and distances, or coordinates, to other known control marks should be checked from the key points of the stakeout. In addition, when using the radial stakeout method, the further away one is from the total station, the more error will accumulate. Care should be taken to meet the required accuracies of a project. One approach to check work is to radially stake out all points from one setup of the instrument, move the instrument to another location, and then remeasure the points. If the new measurements do not agree with the first set of measurements, then the points should be measured again.

If there is an unobstructed view to the sky, RTK GNSS may be used to set stakes depending on accuracy requirements. When using RTK GNSS, it is common practice to measure from a minimum of two control marks and then check if the results agree. Then, the stakes based on the predetermined or calculated coordinates of the stakes' locations are set. After the stakes are set, the surveying engineer should return to the control marks to check the positional accuracy.

Construction Stakes

Construction stakes and finished grade stakes are often set during the survey, and they differ in shape and size so that it is easier to differentiate between the two types (Figure 12-1).

Figure 12-1. Construction stakes and markings.

Colored marking keels are used to mark the stakes. A uniform system of marking should be used, and the information shown on stakes needs to be concise and legible. The stakes are marked so that the construction will conform to the planned line and grade, and the information on the stakes will be properly interpreted by the construction crews.

The construction stakes most typically used are discussed subsequently. It should be noted, however, that advanced technologies have the potential to alter the construction stakeout process radically. For example, machine guidance and control allow construction activities to be completed with minimal to no stakes (more on machine guidance later).

Reference Stakes

Many stakes marking the location of infrastructure components are easily and frequently disturbed, accidentally removed, or covered up during construction. They must be replaced, often more than once, before construction is completed. As an aid in relocating a mark that may become covered or as a means of replacing marks that may be destroyed, measurements are made to nearby permanent or semipermanent objects. This process is known as *referencing* or *witnessing a point*. In many surveys, permanent objects may not be available as witnesses. In such cases, reference stakes are driven to mark locations.

Reference stakes are, preferably, less than 30 m (100 ft) from the permanent point. Every time a point is referenced, a sketch is drawn in the field book.

Slope Stakes

Slope stakes are set to control the construction of earthwork slopes. They are set to mark the intersection of a constructed slope and the original ground, often called the *catch point*. Stakes are not set at the catch point location. Instead, slope stakes are usually set at a constant offset distance from the catch point.

Linear engineering projects such as roadways, railroads, ditches, and canals often use slope stakes to define the limits of excavation and grading. Slope stakes and references are typically set on both sides of the centerline at 15 m (50 ft) stations, terrain breaks, and changes in the typical cross section.

Grade Stakes

Grade stakes are set for vertical elevations when completing the site grading. Grade stakes may consist of rough grade staking to map the general location of improvements at a site or more precise site layout surveys for the actual construction processes.

For projects requiring significant earthwork movement, rough grade stakes are typically used. After a rough grade stake is set, the elevation of the point is determined. The difference in elevation between the stake top and the construction grade is calculated and translates into a *cut* (design elevation lower than the stake) or *fill* (design elevation higher than the stake) The cut or fill is marked on the stake (with a C or F) to guide the construction personnel with the respective reference to the final grade elevation. The information is also noted in the engineer's field book for future use in replacing destroyed stakes.

After the earthwork is roughed in, finish grade stakes are set. These are used when the grade is within a short distance of the final elevation. These stakes indicate the actual finished elevation of the final grade to which the completed facility is to be constructed. This type of stake normally requires a guard stake to protect it and indicate its location. On large projects, it may be impractical to use guard stakes with each stake.

Site Layout Stakes

Site layout staking commences once the site has received inspection and approval. Then, the contractor begins the various stages of construction such as underground utilities, retaining walls, buildings, site lighting, and parking lot or street paving. The site layout staking phase typically begins with those features that are underground such as sanitary sewer lines, storm drain lines, water lines, electrical lines, and many more. Once all underground utilities are installed, the above-ground features are staked for construction. The building corners are staked along with any interior grid lines throughout the building, as well as on-site items such as fire hydrants, curb and gutter, walls, catch basins, and area drains.

Structure Stakes

Stakes set to control the location and elevation of structures are necessary to provide the contractor with accessible and understandable reference points and working lines to line up with, or measure from, without the use of sophisticated surveying equipment. Structure stakes also ensure that the contractor has the information needed to construct a structure to the lines and grades shown in the plans.

In general, all major working lines for abutments, piers, and columns should be referenced with two intersecting lines of stakes. At least, the two stakes nearest to the component should be in line with, and at a set distance from, the component. Outer stakes may be set for line only.

Right-of-Way Markers and Property Boundary Monuments

According to Chapter 1, boundary and right-of-way surveying should only be conducted by a licensed professional land surveyor. Right-of-way markers are monuments placed at boundary lines of public infrastructure projects. The locations of these and property boundary monuments of record, irrespective of whether they are shown on the construction plans, should be referenced to the project control by a licensed land surveyor and checked prior to the start of clearing, excavation, or grading activities.

If, for any reason, the disturbed or destroyed right-of-way markers or property boundary monuments cannot be reset in their original horizontal location, a witness monument or monuments may have to be set to replace the old one. The witness monument should clearly show its relative position to the record marker or monument and to the right-of-way or property boundary line that it is identifying. Final boundary or witness monuments should be documented by a licensed land surveyor on a record of survey map or corner record as required by state law.

EARTHWORK COMPUTATIONS

Earthwork computations involve the calculation of earthwork volumes, the determination of final grades, the balancing of cuts and fills, and the planning of the most economical material. The exactness with which earthwork computations are made depends on the extent and accuracy of field measurements, which, in turn, are controlled by the time available and the type of construction involved. To plan a schedule, the quantity of earthwork and the soil and haul conditions must be known so that the most efficient type and quantity of earthmoving equipment can be chosen and the appropriate time allocated.

For economic reasons, engineers and designers try to optimally balance cut-and-fill volumes and minimize earth movement requirements. Precise balancing is not feasible because of geometric and esthetic design considerations and because of the difficulty in predicting shrinkage and swell. Shrinkage may occur when soil is excavated and then placed and compacted. Swell may occur when placing blasted rock, as solid rock will significantly expand when shattered. Clay can also expand when excavated and placed.

Surveying engineers assist in the process of determining how to optimally balance cut-and-fill volumes for a construction project by taking topographic measurements. For linear routes (roads, sewers, pipelines, etc.), the surveying engineer will at least measure a series of elevations along lines at right angles to the route (i.e., cross sections) at a specific stationing interval. Contour lines can then be interpolated between these elevations and used for estimating earthwork quantities and cuts and fills along the route for the engineering design. The areas of each cross section along a route can be used in equations to estimate volumes.

For other small construction projects, spot elevations are measured at strategic locations to generate contour maps of the site. For greater resolution and accuracy, surveying engineers will use remote sensing techniques such as terrestrial or mobile laser scanning (Chapters 8 and 9) or photogrammetry (Chapter 10) to generate point clouds and digital terrain models of a construction site. UASs are also popular for creating digital elevation models for stockpiles and mining sites (Chapter 10). Earthwork volumes and cuts and fills can then be estimated at design elevations for a given construction project from these high-resolution models. For example, see Figure 12-2.

Some typical types of surveys that involve earthwork include measuring and monitoring settlement wells (i.e., monitoring and observing soil compaction), measuring overexcavation for construction, surveying stockpiles for estimating their volumes, topographic surveying to enable earthwork computations for engineering design, and monitoring construction and earthmoving progress.

AS-BUILT SURVEYS

As-built or as-constructed survey measurements should be collected for all constructed facilities and improvements to confirm dimensions, lines, grades, locations, and materials that may have deviated for the design plan. This provides assurance that all construction is built in accordance with the design standards. This survey can also be used to confirm that the constructed project adheres to accuracy requirements. In some instances, the as-built survey is a requirement for the contractor to obtain final payment.

Oftentimes, this as-built information is overlaid with the original plans. Then, the appropriate information is shown in a specific manner to differentiate it from the original information so that any discrepancies can easily be noted.

MACHINE GUIDANCE AND CONTROL

Traditional survey layout activities can be very time-consuming for the surveying engineer. For example, in highway construction, where cuts and fills can be large, engineers often have to restake the project continuously because the grade and alignment stakes are knocked out or buried during the cut-and-fill process.

The machine guidance and control system involves the integration of positioning tools with construction machinery. The system includes both hardware and software solutions that determine the machine's current position on Earth and then compare that position with the desired design surface. The difference is displayed inside the machine cab on a monitor, allowing the operator to easily and efficiently determine what is required to create

(a)

(b) (c)

Figure 12-2. UAS photogrammetric mapping of a stockpile for earthwork volume estimation. The spatial resolution of the DEM and orthophoto is 7 cm: (a) oblique view of project site, (b) orthophoto for project site, (c) cut (red) and fill (blue) areas.
Source: Courtesy of Land Design Consultants, Inc., reproduced with permission.

the intended design on the ground. Sometimes, the machine is fully automated and makes the required adjustments without operator intervention.

Simple machine guidance techniques have been around for a long time. For example, laser beams, rotating in a fixed plane, at a known vertical offset to finished grade, have long been used to guide earthmoving equipment.

The difference between machine guidance and machine control is the level of machine automation. More sophisticated systems can send signals to machine receptors that automatically open or close the hydraulic valves needed to direct the machine to the proper alignment and grade. The system is connected to and controls the hydraulics while the operator simply drives the equipment and manages the automation.

Before using the machine guidance and control system, a three-dimensional and georeferenced data file portraying all the original features and the existing ground must be created in a DEM. Using the same coordinate system, the proposed design surface is added to the machine guidance and control system.

Sophisticated machine guidance and control systems typically utilize either robotic total station techniques (local positioning system or LPS) or RTK GNSS techniques. The RTK GNSS techniques have been used where a more moderate level of accuracy is needed, although now a laser sensor instrument can be used to improve the vertical accuracy. Total station techniques are typically used where a higher level of accuracy is required.

With total stations, the robotic total station tracks one of the construction machines as it moves by keeping a lock on a machine-mounted reflecting prism. The coordinates of the machine are continuously updated, and the machine's coordinates are compared with the DEM coordinates to determine cut or fill and left or right directions needed. These directions are sent via radio or optical communications to and from the machine control computer.

The drawback of using robotic total stations is that the total station instrument can only direct one machine at a time, and the instrument requires an unimpeded line of sight. Therefore, RTK GNSS guidance and control is often the technique used. With the RTK GNSS system, antennas and receivers are mounted on the construction equipment. These receivers are rovers that obtain a correction from GNSS base stations (see Chapter 6 on RTK GNSS surveying). From this positional information, readings are transmitted to the machine control computer to determine how much up or down and right and left movement is needed.

The cab-mounted monitor display receives and processes the information from both the GNSS receivers and the machine sensors to calculate the machine's current position. The display also contains the design information and can compare the current and desired positions to calculate a horizontal and vertical difference. The display processes all this information at extreme speeds to display the needed adjustments on the screen.

The machine guidance references the position of the cutting edges or pavement molds using GNSS sensors, robotic total stations, lasers, or a combination of these. Initially, it calculates the finishing grade for that location, and then it automatically adjusts the cutting edges or pavement molds for small differences in elevation. For large differences in elevation, it displays cuts or fills on the monitor for the operator. Usually, the operator manually grades the surface until it is close to the designs and then switches to an automatic mode for finish grading (FHWA 2013). Many different types of machines can be equipped with the technology (Figure 12-3), including bulldozers, graders, rollers, milling and paving machines, scrapers, excavators, and trenching machines for a variety of earthwork.

Stakeless grade construction is most often controlled by only GNSS sensors on the excavation equipment, whereas stakeless pavement construction is controlled by a combination of GNSS, laser, and/or total station sensors guiding the paving equipment. Stakeless paving can require multiple total stations operating together to send vertical positions to the paving equipment. This requires more intermediate control points and a more robust control survey than using the staked paving hub method but often has been found to be more efficient and less costly.

Recent advances in machine guidance and control have resulted in layout techniques that significantly improve the efficiency and safety of the construction layout process. For increases in efficiency, machine guidance and control reduce the continual and tedious staking work, which also reduces the chances for costly errors and mistakes. To enhance

Figure 12-3. Some examples of equipment with machine control and guidance systems: (a) grader with reflectors (prisms), (b) roller with a GNSS sensor, (c) paver with prisms, (d) excavator with GNSS sensors.

safety, the machine guidance and control reduces the time operations are working near the construction equipment, providing a significant improvement in personnel safety. Machine guidance can increase productivity by up to 50% on some operations and cut survey costs by as much as 75% (FHWA 2013). The benefits include reduced construction costs on account of improved machine productivity and reduction in fuel, increased potential for digital "as-built" documents, reduced time for staking, fewer errors requiring rework, increased levels of construction accuracy with design, fewer field personnel requirements, reduced needs for string lines, and more.

Caltrans (2018) states that the following types of projects are best candidates for implementing machine guidance and control. Not every project is suitable for this technology.

- Projects with large amounts of earthwork (>3,800 m^3) or paving,
- New route alignments,
- Sites with open views of GNSSs for satellite-based positioning or enough line of sight for control via a robotic total stations, and
- Projects with a 3D design model based on an accurate DEM.

The machine guidance and control system software also has the capability to record the contractor's progress and measurements to allow for an automated progress and as-built measurements.

The surveying engineer has an important role in quality assurance and quality control for the project before the construction begins, during the construction process, and in the as-built stages. The surveying engineer will assist with establishing the geodetic control for the machine guidance system for a given construction project, setting up the control tracking system (e.g., the total stations for location tracking), and checking that the machine is indeed following the 3D design model by making survey spot and check measurements (see NCHRP Project 10-77 for a review and greater details on the use of machine guidance in the transportation industry) (NCHRP 2017).

REFERENCES

Caltrans (California Department of Transportation). 2018. "Implementation of automated machine guidance (AMG)." *Guidelines for surveys implementation of automated machine guidance*. Accessed September 17, 2021. https://ca.gov/-/media/dot-media/programs/right-of-way/documents/ls-manual/surveys-amg-implementation-guidelines-a11y.pdf.

Crume, J. 2014. *Construction staking: Step by step guide*. Scotts Valley, CA: Createspace Independent Pub.

Dewberry Companies. 2002. *Land development handbook planning, engineering, and surveying*. 2nd ed. New York: McGraw-Hill.

FHWA (Federal Highway Administration). 2013. *Automated machine guidance with use of 3D models, executive summary*. TechBrief FHWA-HIF-13-054. Accessed September 17, 2021. https://www.fhwa.gov/construction/pubs/hif13054.pdf.

Kavanaugh, B. F., and D. K. Slattery. 2015. *Surveying with construction applications*. 8th ed. London: Pearson.

NCHRP (National Cooperative Highway Research Program). 2017. *Use of automated machine guidance within the transportation industry*. Project No. 10-77. Washington, DC: AASHTO, Transportation Research Board.

CHAPTER 13
SURVEY RECORDS

Andrew C. Kellie

INTRODUCTION

Prior to construction, maps prepared from engineering surveys provide some of the information necessary for the design process. During construction, engineering surveys provide onsite positioning needed to construct the work planned. Following construction, engineering surveys show conformance with design, document the as-built location, and demonstrate compliance with various governmental regulations. Consequently, engineering surveying incidental to a major construction project may extend over a period of years. The survey records generated will include significant field measurement data. The records also will include maps drawn for various purposes and, in all likelihood, extensive calculations. In addition, they will show how, why, where, when, and by whom the work involved was performed. Survey records for a major project are extensive, and they are the result of a significant investment of money and expertise. Because of this, careful management of all survey records involved in a project is required if the work is to proceed correctly and profitably.

The conduct of surveying engineering in the field, the resulting data reduction, and the plotting and preparation of maps, profiles, and sections involve not only physical work and intellectual effort, but the exercise of professional judgment as well. If survey records are accessible and complete, it is not necessary to duplicate this work if the project is extended or revised.

Surveying engineering often must meet specific contract provisions or conform to the requirements listed in an agency or industrial work order or authorization. Records of the survey can demonstrate that procedures and outcomes required were incorporated into the work delivered. Approval by or acceptance of survey work by the owner or the owner's representative—again, as documented in the survey records—is one way of showing completion of the work as ordered.

Where statutes or regulations govern the practice of surveying, the records of the survey can be used to show conformance with these. In fact, even where surveying engineering is not regulated, survey records may be used to show that work done did not include areas of practice subject to regulation.

Survey records are useful to the surveying engineer in preventing or avoiding rework. When survey control—control marks, reference marks, and bench marks—is documented, it can be located and reused. Instrument calibration results, which are part of the survey records, enable the engineer to correct for systematic error. If procedures and equipment are stated in the survey records, expected random errors can be evaluated and anticipated tolerances determined. The object of properly managing survey records in all these areas is to reduce project costs.

Finally, when problems occur on a project, it is necessary to find what went wrong and why and to devise methods to remedy the difficulty. Survey records do three things in this regard. First, the records provide a review of the work that should help detect problems before they occur. Second, with survey records available, there is little need to spend time determining what was done, when, how, and by whom. Third, approvals obtained for work and included in the survey records address issues of responsibility.

Survey records are also utilized extensively by facility owners after construction and during the operational and maintenance phase of the facility. The records are invaluable many years later when preparing for a future project on or near the site.

TYPICAL SURVEY RECORDS

By its nature, surveying engineering incorporates a host of survey records in written, graphical, and mathematical form. For the purposes of this chapter, survey records will be considered to include (1) company or agency surveying and graphic standards, (2) the construction documents or work order describing the project itself and the surveying engineering services to be provided, (3) numeric data from both field surveys and calculation, and (4) graphic data in the form of both field notes and maps. These are considered separately in the following.

Company Standards as Survey Records

Most companies and agencies have developed their own standards for surveying engineering, graphic products, and general business practices (KTC 2006; NJDOT 2015; ODOT 2015; USACE 2007, 2012), or they have adopted national standards such as those discussed in Chapter 11. These standards are typically published in manual form. Because agency and company work is done in accordance with these standards, they are included by reference in the survey records. Any departure from the standards must be carefully documented, as it involves a change from established policy.

Survey manuals like the ones cited previously provide company or agency standards for monumentation, data acquisition, and staking and setting out. Departures from these standards impact other work. For example, company or agency standards may require the use of physical control monuments for staking infrastructure improvements, particularly in seismically active areas. If such control is not available on a project, the actual control marks employed, the reason for doing so, and the results obtained must be documented in the survey records. Similarly, company standards may require at least two redundant and independent measurements for control established using GNSS positioning (see Chapter 6). If observations do not include this minimum, the survey records must show the reason for departing from established procedures.

Drafting standards employed by the company or required by an agency govern the preparation and distribution of graphic records of the survey. These standards serve to ensure a consistent product and standard content and to document ownership. Any departure from the standards should be shown in the survey records.

Construction Documents as Survey Records

Survey work is typically done under contract (see Chapter 15), and the contract itself must be included in the survey records. The contract (agreement) includes by reference all project documents, that is, project maps prepared by the surveying engineer, design drawings prepared by the engineer or architect, specifications, general and special conditions, and addenda. These documents, and change orders subsequent to the construction agreement, must be included in the survey records. Contract provisions govern the extent of work to be done, methods to be employed, and the outcomes required.

Survey work is also performed by force account. This is typical where surveying is to support agency functions or industrial activities. In such cases, a copy of the work order or authorization and project designation should be included in the survey records.

During the course of the project, construction staking (see Chapter 12) will be based on the design drawings for the project. These are subject to change. During the course of a project, several complete sets of design drawings, each incorporating numerous changes, may be issued. It is important not only that the correct drawings be used for the survey work required but also that the survey record shows the date, revision number, and page number of all drawings used for survey work. Typically, this is shown in the field notes, electronic data files, and dependent calculations.

Numeric Survey Records

Numeric data in survey records provide the basis for survey measurement and describe the outcome of the measurement work itself. Among other things, numeric data describe the equipment used and calibration results; the precision of the distances, directions, and positions obtained; the reference frame and vertical datum employed; and the results of onsite positioning generated during the project.

Data Source. All control networks (see Chapter 11) are based on a reference frame and vertical datum. These define the basic mathematical and geometric relationship between the survey measurements made and the surface of the Earth. Survey records must list the reference frame and vertical datum employed, the reference epoch (e.g., 2010.00), and the dates of observation for the control network.

In most cases, the horizontal and vertical control employed in an engineering survey is that established by the firm involved as project control. Control marks have known precision that is already incorporated into the survey records for the project.

When project control is based on the work of another company or agency, verification of control quality is normally required prior to use. Control schemes are based on professional judgment and experience. Between individuals, both these differ. As a result, supplemental control may be required even where an existing network has been verified and is of acceptable quality. The results of verification and densification must be part of the survey records; see Chapter 11 for more on finding existing control and densifying control.

Survey records must also show information about any assumed local or projected coordinate system (PCS) used in work (see Chapter 3), especially if coordinates are provided as part of the work product. Enough information must be given to unambiguously retrace the survey. If a PCS is used, its relationship with a defined geodetic reference frame (datum) is required, including the specific realization and epoch date, as necessary [e.g., North American Datum of 1983 (2011) epoch 2010.00]. If a custom PCS is used, its complete defining parameters should be provided, which usually consist of projection type, geodetic origin, grid origin, and scale (although this can vary for different projections). Whether an assumed system or PCS is used, the linear unit must be given. If the unit is the foot, the type of foot must be stated (international or US survey).

Use of the State Plane Coordinate System (SPCS) as a basis for surveying is common practice in the United States (see Chapter 3). The same information as any other PCS must be provided. However, for the SPCS, it is not necessary to give the projection parameters, because they are defined in the public record. Instead, the zone must be clearly identified, including the SPCS version, plus (optionally) its numeric code. There have been two versions so far, SPCS 27 and SPCS 83, and a third one (SPCS2022) will be implemented in the near future. An example of a fully identified zone is "SPCS 83 Oregon North Zone (3601)." (Note that the numeric code alone does not distinguish between SPCS 27 and 83.) It is fairly common practice to also provide grid-to-ground information, although how and whether this is done depends on the organization performing the work and project requirements. This is typically provided as a combined factor, which may also include the point-scale factor and/or height ("elevation") factor. The combined factor may be provided as a single representative value for a small project, as a different value for each sheet in a plan set, as point labels on a plat, or for every point in a list of coordinates. Convergence angles may be provided in a similar manner, to give information on the difference between grid and geodetic north for the project. Although including combined factors and convergence angles may not be required, a scale factor *must* be provided if the SPCS has been "scaled to ground." This must consist of both the scale factor itself *and* the coordinates about which the scaling was performed. In situations where a project falls in more than one SPCS zone, it is common practice to choose one zone that best represents the project (e.g., the zone that includes a majority of the project area). The advantage of using a single zone is that coordinates are consistent over the entire project. But for large projects, using a single zone may not be feasible. In such cases, all the zones used must be documented, and the records must clearly indicate which parts of the project are in which zone.

The following is a recommended example of how to document the reference frame, vertical datum, PCS, bearings, and distances for a survey record. It also provides an example statement for documenting how the orthometric heights were derived. This example is modified from NGS (2021).

Basis of Bearings and Coordinates

- *Latitude, longitude, and ellipsoid heights*: The 2011 realization of the North American Datum of 1983 [NAD 83 (2011)], Epoch 2010.00.
- *Orthometric heights*: North American Vertical Datum of 1988 (NAVD 88).
- *Linear unit*: International foot, ift (1 ft = 0.3048 m).
- *Projected coordinate system*: State Plane Coordinate System of 1983 (SPCS 83), Oregon North Zone (3601).
- All distances and bearings shown hereon are projected (grid) values based on the projection definition. The grid bearings shown hereon (or implied by grid coordinates) do not equal geodetic bearings because of meridian convergence.
- Orthometric heights (elevations) were determined using GNSS equipment with NGS hybrid geoid model GEOID18 combined with differential and trigonometric leveling.

Raw Field Data. The extent to which raw field data—actual satellite positioning data, total station measurements, and scanner output—are retained as part of the survey records may be a matter of company policy or may be defined by a contract. Being in electronic form, the data require relatively little physical space to store, and there are certain elements of the raw data that are very useful in tracking mistakes or problems. For example, the prism offset correction used in electronic distance measurements (see Chapter 7) can be applied directly to measured distances in the field or to the results of field measurements during data reduction. It can also be (and has been) accidentally applied in both places in the same project. The result is a series of measurements that are extremely precise but totally inaccurate, and a cause that is potentially difficult to find without access to the raw data files. Likewise, cycle slips in satellite positioning data may cause positional errors that propagate into the control network. An examination of the raw data file provides a way to either identify or correct the problem. The alternative to having access to the raw data is potentially lengthy remeasurement or reobservation at a cost in both time and effort. Another advantage to storing the raw field data is that it can be reprocessed or readjusted in the future when better models or new information comes available. For all these reasons, it is highly recommended to always retain the raw field data for a survey project.

Adjusted Data. Data adjustment is an attempt to rationally distribute corrections among field observations (see Chapter 5). This can be done with software employing different adjustment rules and weighting schemes. The network will likely be constrained to existing control. All these things—adjustment rule, weighting method, software employed, and stations constrained during adjustment—must be included in the survey records.

Adjustment results are typically tabulated in a station list and included in the survey records. The station list must show the date, because a control network is a dated entity. The format and content of the station list is normally specified by either company or agency policy.

Company policy sometimes requires the rotation, translation, or projection of the adjusted coordinates for use in layout (see Chapter 4). Where this is the case, the techniques used must be documented.

The following is a simple example statement that can be included in a survey record for documenting adjustment results. In the record, it can precede a list of coordinates with error estimates at 95% confidence.

> The survey was conducted using post-processed and real-time GNSS, leveling, and total station equipment and methods. A partial list of point coordinates is given below (additional coordinates are available upon request). Accuracy estimates are at the 95% confidence level and are based on an appropriately constrained and weighted least-squares adjustment of redundant observations.

Project Computations. Every project requires at least some computation based on the design drawings and the control system. The computations involve everything from basic coordinate geometry for staking to complex three-dimensional models needed for machine control (see Chapter 12). It is common for both firms and agencies to require all computations incidental to a project to be included in the survey records. The required format varies, but all computations must be dated and clearly identified by the project. Drawing number, revision, and revision date should also be included if appropriate. Company standards often require that this work be on company computation sheets to show ownership of the work.

Two classes of project computation merit additional attention. The first are those computations that involve critical points in alignment, grade, or position. Such computations control much subsequent work, so the survey records must clearly show how and by whom the computations were done.

The second are those computations that control high-cost operations. Models computed for machine control or positioning information developed for piling are examples. Mistakes in either of the aforementioned computations are potentially expensive. Survey records of computations should include not only the calculations themselves but also the checks, reviews, approvals, sketches, and maps that support the solution. Software used in producing the solution must also be noted.

Graphic Survey Records

The graphical record incidental to a project includes the project maps, design drawings, and surveying field notes kept by hand during the work, and both the manual sketches and maps prepared to facilitate surveying engineering on the project site. Interestingly, parts of the graphical record—the design drawings, for example—are generated by the engineer or architect. They become part of the survey record because other work depends on them. During the course of the project, maps may be prepared by the surveying engineer to show work performed to date or prepared as a basis for, or to show the results of, earthwork computations.

Field Notes. Despite the availability of electronic data collection, manually kept field notes are often an effective supplement to the electronic record. For example, ties to survey stations, the proximity of a station to adjacent buildings, fences, or power lines, or the relative location of objects and survey stations on the project site can be efficiently shown on a sketch. Alternatively, the same things can be coded, stored, and downloaded from the data collector.

Written field notes must be linked to electronic field data by reference in the field notes, the data collector file, or preferably both. All computations done in the field—even the trivial ones—should be recorded in the field book and signed by the person making them. This provides a means of tracking a mistake—or, conversely, eliminating the existence of a mistake—in field calculations.

General field note format is well known, and sample field notes for typical surveying operations can be found in surveying textbooks and company and agency survey manuals (Ghilani and Wolf 2012; KTC 2006; NJDOT 2015; ODOT 2015; USACE 2007, 2012). Despite some differences in format and content, basic requirements are obvious. The notes identify the persons doing the work, the equipment used, the work and its location, and the date. Knowing these things fixes responsibility, links measurement errors and corrections to specific equipment, and describes where and when the work was performed. An example set of field notes for differential leveling is shown in Figure 7-15, and for grid leveling, it is shown in Figure 7-16.

Field notes for engineering surveys involving staking or layout must include a reference to the design drawings on which the work is based. At a minimum, this should include the sheet numbers, date, and revision number.

Sketches in the field notes should be drawn with a straight edge and employ normal drafting conventions as to line type, lettering placement, symbols, and organization. The sketch is not drawn to scale because parts of the sketch will need to be exaggerated. This is done to clearly show relationships, measurements, or specific details. Orientation is important and shown with a north arrow. Many firms and agencies require consistent orientation with the north direction toward the top of the field notes. It is very important that the field notes be complete.

Field notes are signed by the person keeping the notes. Some firms also require the construction superintendent to sign the notes to show that the work has been reviewed and discussed. Other firms require that the project engineer sign the field notes following completion of the work to document review and approval. To guard against loss of the physical notes, it is common to scan field notes and save them as digital survey records.

For static GNSS surveys, some firms or agencies require the field surveyor to fill out an observation form or log that documents the observed mark, start/stop time, height of the GNSS antenna, meteorological data, and any comments (see Figure 6-10 for an example).

Maps and Drawings. Engineering surveys proceed from the maps and design drawings as referenced in the construction agreement or work order. It is frequently necessary for the surveying engineer to prepare additional maps to document work during construction (see Chapter 12). For example, it might be useful to prepare a map showing supplemental control marks established on the project site or to show the relationship of column lines on a structure to existing project control. Such maps assist the engineer in planning work to be done, in showing the status of work in progress, or in documenting survey work already completed. They also provide a means of showing the results of calculations necessary to project layout.

In addition to the maps just described, topographic mapping beyond that shown on the project drawings is required on most excavation and embankment projects. For example, to check or control grading, it may be necessary to map the location of the "daylight" line separating areas of excavation and embankment. Further, if required by contract and as discussed in Chapter 12, the engineer may need to periodically determine volumes of excavation and embankment or to compute material quantities in stockpile. Computations done as part of this work can be shown graphically on the topographic maps involved. Both the computations and the maps must be included in the survey records.

Survey records for topographic surveys incidental to volume computation require more than simple archiving of the field data. Payment is based on the quantities estimated, and volume computation can be very sensitive to the computational routines used. For example, if gridding of the field data is used, the gridding method and grid size employed need to be described in the survey records. The surface generated from the grid controls the accuracy of the computed volume, and the grid size determines the precision. The use of multiple volume computation methods is assumed, and the results of each method should be documented in the survey records.

Many of the maps produced during a project and included in the survey records are work maps intended as internal documents for use by the engineering and surveying staff. Should problems arise during or after the project, it is quite possible that even such work maps will come under scrutiny. For this reason, conformance to company graphic standards and accepted standards of professional practice for technical drawings is expected.

PROJECT MONUMENTATION

It is recognized that the control network on which construction layout depends is normally based on physical monuments such as CORSs, RTN base stations, and passive marks or bench marks (see Chapter 11). The importance of physical control marks used on almost all engineering projects—and particularly those projects in seismically active areas—requires careful documentation in the survey records. This includes not only coordinate position but also a detailed description of the physical location of the station itself.

Control marks are expensive to create, but they are central to the correct layout of a project. It is critical that the location of survey monuments, ties, and bench marks set in the ground be perpetuated for future use (see Chapter 11). Simply put, if a control mark cannot be located by the surveying engineer, the time and money invested in mark establishment are wasted. Carefully kept survey records should help minimize costs of finding and using existing control marks.

Control Diagram

Control marks are documented and included in the survey records in several ways. Relative mark location typically is shown on a control diagram prepared in graphical format to company or agency standards. This shows in overview the entire control network used for the project, together with existing adjacent control.

The control diagram must distinguish between pre-existing control marks and new control marks established specifically for the project itself. Marks may be designated by their name or code, and reference text provided to the control record or description for the mark with reference ties. Where marks included in the network were established by others, the name of the agency or firm responsible needs to be included in the survey records.

In addition to relative location, the control diagram shows network geometry. This is done by showing the lines in the network actually observed. Network geometry is a significant control on error distribution and on the strength of the network. It is also important in determining how the network is extended, densified, or used in later work.

The control diagram is a dated work product. As the network is densified or extended, the control diagram must be updated and included as part of the survey records. Depending on company or agency policy, a revision number and date may show the most recent work, or a color or symbol scheme may be used for different iterations of the network. The point is that the survey records should show not only the most recent control network but also the control network actually used as specific sections of the work were done.

The exact information shown on the control diagram is a matter of company policy. For example, error ellipses (see Chapter 5) may be shown on the diagram to assist in network planning, or they may be omitted to prevent clutter. Mark positions may be shown in tabular format on the control diagram, or they may be described by reference. Callouts or supplements may be used to show details, or separate control diagrams may be prepared for each section of the project.

Mark Descriptions

Mark descriptions are fundamental to station recovery and use in the field. Descriptions may use a number of formats—graphic, photographic, narrative—or any combination thereof. This is because failure to adequately describe control marks can result in significant financial loss.

Graphic descriptions document mark location directly in, or from, the field notes. One page of the notes can be used to show a general location. The facing page can show specific location and measured ties. The graphic description is drawn using standard company or agency note keeping format, and the mark description is considered complete when the field notes are submitted and accepted. Scanning the field notes into the project file adds the mark description to the survey records; see Figure 6-9 in Chapter 6 for an example survey record for a control mark in a project, with a basic mark and to-reach description, and reference ties.

Digital photographs are also used to describe mark location. Photographic documentation should include at least two photos. One should show the on-site location, with a note accompanying the photo giving the direction of the camera to the mark. The second photo should show the mark itself. Both photos should be annotated with the date and included within the survey records.

As shown in Figure 6-9, narrative descriptions are common with control established for public agencies. The narrative format is specified by the agency involved. Format normally includes both general and specific descriptions of the mark. The general description gives mark location relative to a prominent intersection or structure and includes the state, county, city, section of the city, and street. The specific description includes distance and direction ties from the mark to a series of reference objects. The material of which the mark is made is stated. Additional information, such as the person and company or agency establishing the mark, date established, and suspected stability, can be included as specified by contract or by company or agency standards. The narrative description is included electronically in the survey records together with the field notes and photos for the mark. As an example, NGS provides detailed descriptions and links to photos of control marks electronically via datasheets (see Chapter 11).

REPORT OF SURVEY

A formal report of survey is required by some agencies as part of any contract for surveying services (NJDOT 2015). Some survey firms prepare a report as part of their standard practice (Robillard et al. 2002). Generating the survey records described in this chapter and including them in the report of survey is actually an ongoing process incidental to the work involved. It is not an afterthought completed following project closeout. Proper and complete survey records provide reliable and up-to-date information for the project. The purpose of the report of survey, then, is to make this information accessible and organized.

Except for agency or firm requirements, there is no standard format for a report of survey beyond that expected for any technical report. The information in the report includes the names, addresses, and contact information for the client, the owner, and the project; the project location and duration; the purpose of the survey work done; the survey records pertinent to the project; and attachments or appendixes supporting specific parts of the report.

Names, addresses, and contact information identify the parties involved with the project. To whom the surveying engineer is responsible for work done, and how to contact that person, must be clearly stated. For example, if survey work is done under subcontract from a general contractor, the surveying engineer is obligated to the general contractor and not to the project owner. In addition, any number of reasons might make it necessary to immediately contact the client or the owner of the project. If this information is readily available in the survey report, both anxiety and delay may be avoided.

The project name and the dates involved are included because the same client may employ the surveying engineer on several different assignments. In addition, if the work involves different sections in a phased project, it is important to know to which part of the project the survey report refers.

Including the purpose of the survey work in the report requires that the person preparing the survey report examine the scope of the work as included in the contract documents. Stating the objectives would, at the beginning of the work, remove any question of exactly what was to be done, how, by whom, and when. Inclusion of contract documents (see Chapter 15)—electronically or by reference—would link directly the survey objectives and the contract documents. In the case of volume computations, such linking would hopefully prevent conflicts of interest as well.

The bulk of the survey report constitutes the survey records themselves. As described previously, this includes the raw and adjusted data for both control and staking, computations incidental to staking and volume computation, field notes and graphics, models prepared for machine control or grading, mark descriptions, and other survey work as required.

The report also must include copies of any reports, maps, or computations filed as part of the work. For example, a municipal storm water management plan may require an as-built survey to verify detention basin capacity. Alternatively, contract provisions may require the surveying engineer to provide the as-built location of utilities relocated during road construction to a specific agency. In both these cases, some form of report, supported by maps and computations, would probably be submitted by the surveying engineer to the regulatory agency. All this material must be included in the survey report.

SUMMARY

Survey records help the surveying engineer track the significant volume of information generated by work in surveying engineering. The primary beneficiary is anyone who has to follow in the footsteps of the work accomplished by the surveying engineer for the multitude of reasons described throughout this chapter. Organized survey records facilitate finding, referencing, and using positioning information.

The survey records for a specific project include company surveying and graphics standards, the contract documents, numeric data (raw observations, adjusted positional data, calculations, and three-dimensional models), and graphic data (field notes, sketches, and maps). Descriptions for the physical monuments that form the basis for an engineering survey must also be part of the survey records.

Organization of survey records for a project can be done by linking individual survey records in the project file. Alternatively, a report of survey can be used to formally organize the survey records. Developing the report of survey is done as the survey work itself is actually in progress. The objective is not only to organize the survey records but also to provide rapid access to both survey data and project documents.

Survey records are not an end in themselves. They simply help get the project done. As the project size increases, the management of information—including survey records—becomes very important in tracking, reporting, and documenting the surveying engineering work accomplished. Survey records—accessible, accurate, organized, and complete—are fundamental to successful and profitable surveying engineering work.

REFERENCES

Ghilani, C. D., and P. R. Wolf. 2012. *Elementary surveying: An introduction to geomatics.* 13th ed. Upper Saddle River, NJ: Prentice Hall.

KTC (Kentucky Transportation Cabinet). 2006. *KTC highway design manual.* Frankfort, KY: KTC.

NGS (National Geodetic Survey). 2021. *Blueprint for the modernized NSRS, part 3: Working in the modernized NSRS.* NOAA Tech. Rep. NOS NGS 67. Silver Spring, MD: National Geodetic Survey, National Oceanic and Atmospheric Administration.

NJDOT (New Jersey Department of Transportation). 2015. *NJDOT survey manual.* Trenton, NJ: NJDOT.

ODOT (Oregon Department of Transportation). 2015. *Survey policy and procedure manual.* Salem, OR: Geometronics Unit, ODOT.

Robillard, W. G., D. A. Wilson, and C. M. Brown. 2002. *Evidence and procedures for boundary location.* 4th ed. New York: Wiley.

USACE (US Army Corps of Engineers). 2007. *Control and topographic surveying.* Manual No. 1110-1-1005. Washington, DC: USACE.

USACE. 2012. *Survey markers and monumentation.* Manual No. 1110-1-1002. Washington, DC: USACE.

CHAPTER 14
INFORMATION SYSTEMS IN CIVIL ENGINEERING

Yelda Turkan, Dimitrios Bolkas, Jaehoon Jung, Matthew S. O'banion, Michael Bunn

INTRODUCTION

This chapter introduces geographic information system (GIS) and building information modeling (BIM) techniques and summarizes the different applications for which they are used in civil engineering.

Geographic Information Systems

A GIS is a software platform that facilitates the visualization, analysis, management, and sharing of digital geospatial information. Compatible information often takes the form of discrete points, lines, and polygons (i.e., vector data) or pixel- and grid-based formats (i.e., raster data) such as a digital elevation model (DEM). Fundamental to all GIS software solutions is support for an ever-growing list of geographic coordinate systems, map projections, and height datums. Modern GIS solutions provide the ability to both manage and analyze survey-grade and mapping-grade data from a variety of terrestrial, airborne, and satellite sources. The ability to integrate these data in a flexible GIS environment can facilitate rigorous analyses, enhanced decision-making, and improved communication with stakeholders.

Maps are often consulted when finding locations and seeking directions—locating places of interest and inferring their spatial relationships. These relationships are probably the most useful application of maps in the civil engineering discipline. Modern civil engineering projects are often complex operations involving multiple moving parts and an incredible number of details. Even simple projects may require work from multiple disciplines, including environmental impact studies, surface water runoff management plans, subgrade inspection, and transportation planning, all in addition to the work of the surveying engineer. Being able to view many, or all, components of a project can help managers make informed decisions and help instruct workers about the tasks that they need to perform. The emerging field of Geodesign brings all these components together by combining geography with design to provide designers with robust tools enabling them to evaluate design alternatives. Ultimately, it helps designers, engineers, city planners, and other stakeholders to make better-informed decisions, which is most helpful when dealing with issues such as population growth, climate change, and increasing demands for resources (Esri 2021).

Building Information Modeling

A BIM is a digital representation of the physical and functional characteristics of a facility. Typically, BIMs can provide a sharable, stable, and organized structure of information requirements and knowledge of the domain context. Complex, object-oriented data structures containing architectural details as well as relevant information from fields such as structural, foundations, electrical, sanitary, heat ventilation and air conditioning (HVAC), and mechanical are included in BIMs. BIMs are principally different from three-dimensional (3D) computer-aided design and drafting (CADD or CAD) because they enable simultaneous and automatic updates of changes in building elements across all views (Azhar et al. 2015). An important aspect of a BIM is the concept of parametric objects. These objects (1) consist of geometric definitions and associated data and rules; (2) have rules that determine when a particular change violates object feasibility regarding size, manufacturability, among others; (3) are defined at different levels of aggregation; and (4) have the ability to link to or receive, broadcast, or export sets of attributes to other applications and models (Eastman et al. 2011).

A building consists of thousands of components that spatially depend on one another (Yarmohammadi and Ashuri 2015). Coordinating the layout of the components of different building systems is an essential activity conducted prior to construction (Hu et al. 2019). Through this coordination, building components are relocated and rerouted to fit within the structural, architectural, and engineering constraints of the building envelope (Janis and Tao 2005). This prevents spatial collisions or clashes between the components during the actual construction phase and, thus, ensuring expected performance regarding comfort and safety (United BIM 2021; Sierra Aparicio 2020). BIM has become a major, must-have tool for design, construction, and facility management in the architecture, engineering, and construction (AEC) domain (Hong et al. 2015) as well as land administration practices such as ownership, taxation, and management (Kalogianni et al. 2020).

However, design coordination is a challenging task as the complexity of building services is continually increasing because of the increased requirements by building users. Modern building projects now include much more than the traditional mechanical, electrical, and plumbing systems such as fire protection, building automation and control systems, process piping, and telephone/datacom that require different specialty contractors for installation (United BIM 2021). In general, the coordination starts with larger and less movable items such as HVAC dry and wet and then evaluates plumbing, fire, electrical, and control systems in order (Korman et al. 2008). In a conventional approach, 2D drawings of different building systems are sequentially overlaid on a glass light table [Figure 14-1(a)] and compared with one another to identify possible conflicts (Korman et al. 2010), which is slow, expensive, prone to errors, and also highly fragmented, whereas the coordination takes place on an as-needed basis and is often not performed sequentially. CAD introduced a 2D model in a digital form, but basically it just mimicked a manual process; there is no automated method to identify the clashes/conflicts, which is possible only with a 3D simulation/model (United BIM 2021).

Extensive efforts have been made to improve coordination efficiency by utilizing BIM solutions. In the construction industry, the BIM is one of the most innovative and integrative technologies that modify the digital process of representing buildings (Muñez et al. 2020). BIM can improve the design process by providing a virtual 3D model of a building, which can be developed before the actual construction begins. Further, parametric BIM modeling allows an efficient revision of design. If any input is updated, it automatically updates all the relevant models such that users can easily modify any part of a 3D model, eliminating the need to redraw the design again (United BIM 2021). BIM-based coordination process enables participants to combine 3D virtual models of different building systems into a

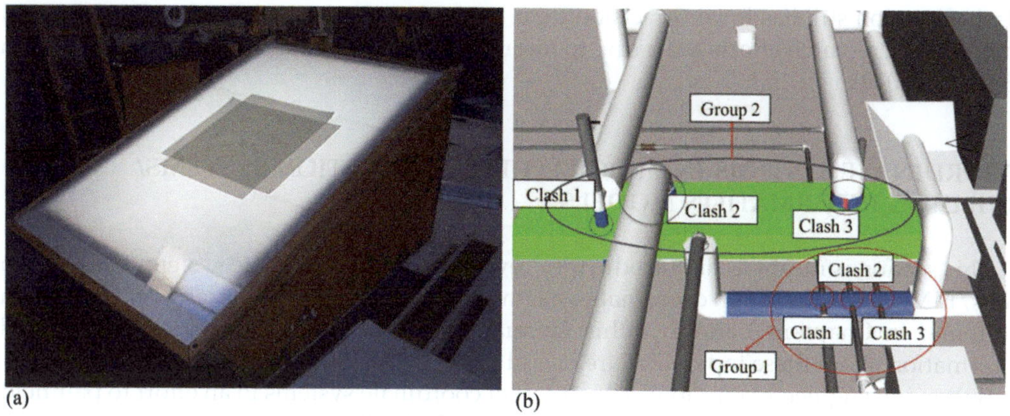

Figure 14-1. (a) Traditional coordination using light tables; and (b) BIM-based coordination. Source: Saul (2021), Hu et al. (2019), reproduced with permission from Elsevier.

single integrated model. The integrated model enables automated detection of system conflicts (Figure 14-1b), which can significantly improve the coordination workflow. In addition, technological advancements have made several new practices possible, such as parallel, remote, and cloud computing–based coordination approaches, which enables a consistent communication channel between different teams, contractors, fabricators, consultants, and engineers (Yarmohammadi and Ashuri 2015, United BIM 2021).

Over the last 15 years, BIM-based coordination has advanced to enable clash-free design, improved work quality, faster design time, increased collaboration, and greater productivity, which has ultimately led to overall lower costs (Bwail 2018). Figure 14-2 shows some critical milestones of the BIM-based coordination. In 2007, the first BIM-based clash detection tool was introduced, and 2 years later, a BIM collaboration format that enables the exchange of comments and information within multiple project trades was developed. In 2018, a cloud-based automated clash detection and visualization tool was presented. However, automated

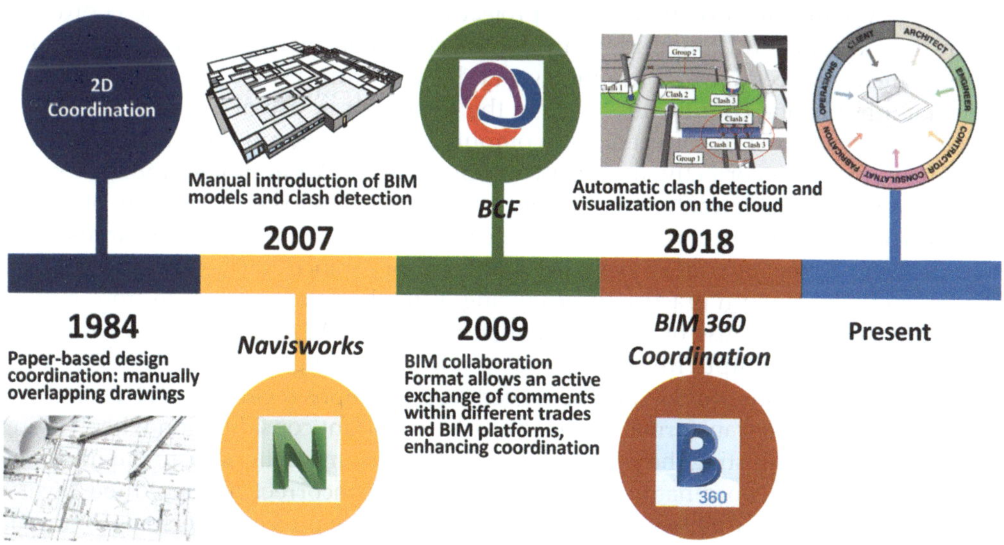

Figure 14-2. Evolution of BIM-based coordination.
Source: Modified from Hu et al. (2019) and Aparicio 2020, reproduced with permission from Elsevier.

clash detection still has lower precision than manual methods (Mehrbod et al. 2019), and many recent research efforts are trying to improve the quality of automated clash detection (Hu et al. 2019).

COORDINATE SYSTEMS IN GEOGRAPHIC INFORMATION SYSTEMS/ BUILDING INFORMATION MODELS

This section is meant as an introductory review on how coordinate systems are used in GISs/BIMs. More details on coordinate systems and coordinate system transformations can be found in Chapters 3 and 4 of this manual. Chapters 3 and 4 provide fundamental information about these topics that are useful to the surveying engineering practice.

Surveying engineers commonly use different coordinate systems in an effort to pair field measurements with coordinate system projections that will minimize linear distortion once their measurements are georeferenced. State plane coordinate systems, scale factors, and, more recently, localized or project-specific coordinate systems are all popular in practice, and a surveying engineer will choose the best coordinate system for the job. GISs use the same coordinate systems as a surveying engineer might, but GIS projects tend to focus on larger spatial areas and may incorporate numerous data sets with different coordinate systems on a single project. For this reason, GIS software tends to have resources aimed at helping users stay organized and work efficiently. These resources are centered around on-the-fly (OTF) projections, where data sets from multiple coordinate systems are displayed in a single coordinate system so that they can simultaneously be viewed and analyzed. OTF projections are simply a coordinate system transformation that is performed as a layer is added to the map, and they are implemented automatically by most GIS software. The same GIS software commonly includes transparency about which coordinate system is being used for display and which coordinate system is being used by each layer. When layers are added to the display in a new coordinate system, users are typically asked to confirm that the coordinate system is different and correct.

Despite the ability of GIS software to perform OTF projections, there is still often a need to perform permanent coordinate transformations. A common GIS application for coordinate system transformations is to project a raster from a geographic coordinate system (i.e., latitude and longitude) into a projected coordinate system (northing and easting in meters or feet) when computing ground slope. Correct slope calculation relies on vertical and horizontal coordinates being in the same unit or use of a conversion factor. Conversion factors exist to correct slopes computed using GISs, but they add complexity and, when applied incorrectly, may lead to false values. An alternate application for coordinate system transformations on rasters is to prepare files to be shared with others. In much of North America, rasters displayed in geographic coordinates will appear distorted and unrealistic. To the unfamiliar viewer, this altered appearance may be misinterpreted as being correct. The modern practice of producing orthophotos and structure-from-motion topography is a major cause of this problem (see Chapter 10), as geographic coordinates are the most common output from nonsurvey grade positioning systems.

The basic information provided in the preceding paragraphs regarding OTF projections and coordinate system transformations leads to an important consideration when editing data in GISs. When adjusting the position of an item on the map, the adjustment may look correct in the map's coordinate system, but it may be incorrect in the edited layer's coordinate system. This problem is most pronounced for the projected-versus-geographic coordinate difference, but it may also appear when using different datums such as NAD 1983 versus NAD 1927. Despite its age, numerous data sets with NAD 1927 as a datum still exist. The possibility of editing problems like this can be investigated by toggling the map to display the layer's coordinate system and looking for misplacement. If a problem does exist, actual locations may need to be field-verified.

A final topic involving coordinate system transformations is georeferencing. Georeferencing is often performed when spatial data need to be placed somewhere on Earth's surface. This task is incredibly common, especially when working with paper maps. What if, for example, a GIS user wants to scan an old paper map to digitize a historical route? They will need to perform georeferencing. In a GIS, georeferencing can be as simple as choosing two common points between the layer being georeferenced and a map with known locations. If the layer is two-dimensional, the GIS will scale, translate, rotate, and warp—if called for but not flip the layer. Additional points may be useful in cases when the locations of common points are less certain. When more than two common points are identified, most GIS software will perform a least-squares fit to the points (see Chapter 5). Most georeferencing fits tend to involve only rotation and translation. If the image was initially made in a projected coordinate system, only scaling, rotation, and translation are necessary for georeferencing. If the georeferenced image was initially made in a geographic coordinate system, then the raster may need to be warped. In these scenarios, it may often be challenging to obtain good georeferencing.

Countless other coordinate system considerations exist. We refer the reader to Chapters 3 and 4 for more detailed information about coordinate systems.

GEOGRAPHIC INFORMATION SYSTEMS/BUILDING INFORMATION MODELING TECHNOLOGIES

This section starts with an overview of the computing hardware and software necessary for GIS and BIM technologies. Next, modeling of 3D point clouds into GISs/BIMs is discussed. Finally, immersive visualization technologies such as virtual reality (VR), augmented reality (AR), and mixed reality (MR) and their implementation together with GISs/BIMs is reviewed.

Computing Hardware and Software

Similar to most commercially available software, GIS/BIM platforms usually provide recommended computer specifications. Such documented specifications are a good place to start when configuring a GIS/BIM workstation; however, data processing and storage efficiencies can be gained by upgrading certain components.

When considering a central processing unit (CPU), the quantity of cores should be given primary consideration. Modern GIS/BIM platforms are becoming increasingly multithreaded. To take advantage of the available parallel processing architecture, a multicore processor is needed. The level of parallel processing varies among different GIS/BIM platforms as well as among individual processing tools in a given platform. In general, a CPU with around 10 cores will achieve high parallel processing performance when available. As for the clock speed of the CPU, preference should be given to the number of cores. It is recommended that first the number of desired cores are constrained, followed by choosing the highest clock speed based on the available budget. Another strategy for gaining parallel processing performance is to utilize a cloud computing solution/platform. Cloud computing solutions provide on-demand access to high-performance computational resources, which are often well maintained. When considering a cloud computing platform, it is important to consider the quantity of data that will need to be exchanged with and/or stored on the cloud for all anticipated workflows. The transferring of large amounts of data back and forth with cloud storage can generate a workflow bottleneck in some cases.

A dedicated graphics processing unit (GPU) card is preferred as opposed to an integrated GPU built into the motherboard (common in some laptops). As for choosing a dedicated GPU, a professional-grade workstation GPU (e.g., Nvidia Quadro series) is recommended as the longevity of the card, and the amount of onboard memory is often increased over that for gaming-grade cards. If anticipating that GIS/BIM workflows will involve a fair amount of three-dimensional (3D) or big data visualization and high-resolution rendering, a GPU at the mid to upper end of the offered workstation range should be considered. In addition, although cloud computing solutions can support graphics processing, interactive 3D visualization performance will be less than that achieved with a GPU-equipped local machine. Internet speed will restrict the ability to visualize 3D data in cloud computing applications, and cloud computing performs best with fast internet connections.

Now that most of the popular GIS/BIM platforms are based on 64-bit architecture, large quantities of random-access memory (RAM) are supported. Investing in large quantities of RAM can dramatically improve performance when working with large data sets. Prior to allocating resources for larger quantities of RAM, confirm that your computer operating system and chosen GIS software are 64 bit. If 32-bit software is being used, no performance increases will be observed above 4 GB of RAM.

When considering storage hardware, preference should be given to solid-state drives (SSDs). SSDs will provide improved data access performance over conventional disk drives. SSDs do come at a higher cost, but the performance increase is significant and data loss is less likely. Data storage safety can be improved with a robust data backup plan that duplicates stored data on additional local drives and/or syncs to cloud-based storage accessed via an internet connection. For the most part, it is recommended that GIS and BIM workflows be performed with data that are located on the local machine running the software. The use of cloud storage or a virtual network drive is discouraged when conducting GIS/BIM operations using software running on a local machine. This practice can result in extended processing times, corrupted data files, or erroneous results. The use of cloud storage is appropriate when using a cloud computing solution for conducting GIS/BIM operations.

Many open-source and closed-source/licensed GIS, CAD, and BIM software platforms/ecosystems are available today. Open source refers to software in which the source code is open to the public for review and, in some cases, open for revision/updating by a community of programmers. Closed-source software is proprietary, supported, and is updated and maintained by the company that owns it. Some proprietary GIS, CAD, and BIM platforms deliver ready-to-use data, imagery, computing services, APIs, machine learning (ML), artificial intelligence (AI), web capabilities, and the ability to integrate with other systems. It is common for proprietary GIS, CAD, and BIM platforms to also support open-source file formats, processing routines, applications, and other plugins/addons.

From Point Clouds to Models

Point cloud technologies from terrestrial laser scanning (TLS) and small unmanned aircraft systems (UAS) are often used to capture accurate geometric information (Volk et al. 2014, Bolkas et al. 2020). From the point clouds, the user can model, usually by hand, the main 3D objects. As discussed in Chapter 8, this is often done with the use of standard primitives such as boxes, spheres, and cylinders. These are manually fitted to the point cloud, whereas some automated tools also exist (Berger et al. 2017).

Figure 14-3 shows an example of UAS photogrammetric and TLS point clouds. The utilization of multiplatform point clouds necessitates data acquisition in a common reference frame and point cloud merging. Figure 14-4 shows a merging approach developed in Bolkas et al. (2020) for 3D building modeling. Finally, Figure 14-5 shows the resulting 3D building models, which were manually modeled from the merged point cloud.

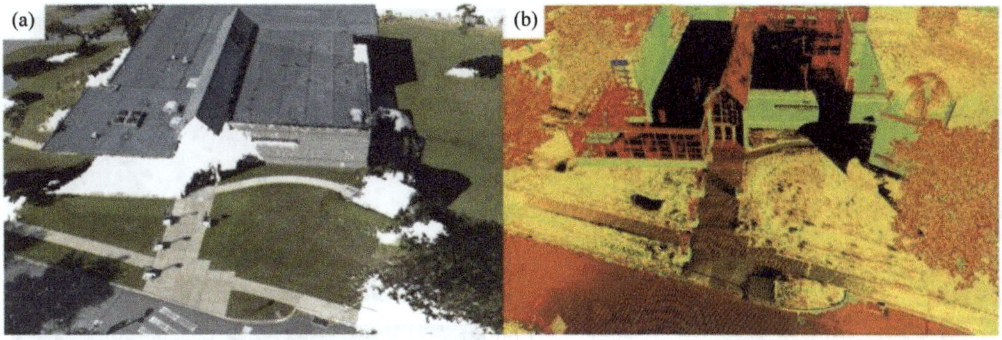

Figure 14-3. Point cloud data sets at Penn State Wilkes-Barre created by (a) UAS, (b) TLS.
Source: Bolkas et al. (2020), reproduced with permission from ISPRS.

Immersive Visualization Technologies

The advent of cost-effective immersive technologies such as head-mounted displays (HMDs) and AR glasses have allowed widespread dissemination and use for visualization in GIS and BIM workflows. First, it is useful to distinguish the different immersive visualization methods, namely, these are often categorized as (1) VR, (2) AR, and (3) MR (Table 14-1). Many of the available VR, AR, and MR solutions support stereoscopic visualization combined with point of view/head tracking. This combination enables a user to experience an enhanced sense of depth and interact with data from a realistic perspective (Brando et al. 2017).

For VR, a complete virtual environment is created, which can be imaginary or based on the physical world. Equipment such as an HMD, interface controllers, or a large projector-based cave automatic virtual environment (CAVE) system is often used to generate a high

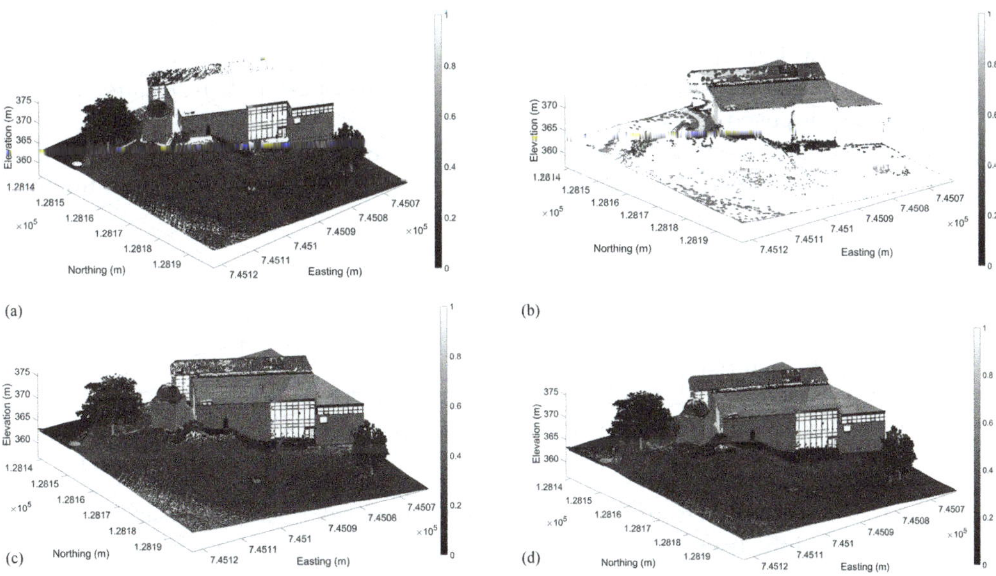

Figure 14-4. Example of TLS, UAS, and merged point clouds: (a) TLS point cloud, (b) UAS point cloud, (c) merged point cloud using 20-cm grid spacing, (d) merged point cloud using 10-cm grid spacing. The color bar shows normalized grayscale color.
Source: Bolkas et al. (2020), reproduced with permission from Taylor & Francis.

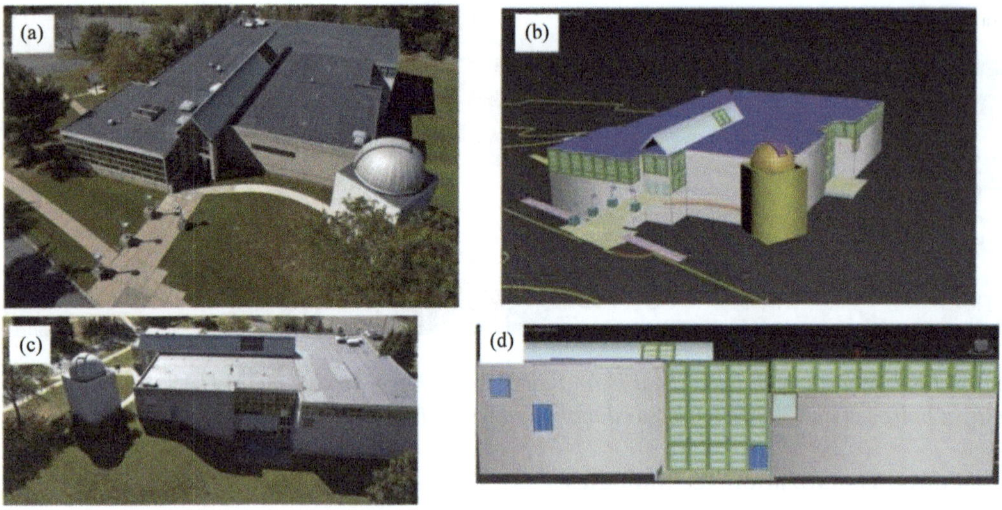

Figure 14-5. Real and modeled buildings: (a) front view picture, (b) front view modeled, (c) side view picture, (d) side view modeled.
Source: Bolkas et al. (2020), reproduced with permission from Taylor & Francis.

level of spatial immersion. The user "virtually exists" in the virtual environment, being able to navigate and interact with virtual objects. Navigation and interactions with the virtual environment often take place using controllers or specially designed gloves. The full immersion inhibits any contact of the user with the physical world. In AR, the user is not fully immersed in the virtual environment; instead, virtual objects are overlaid on top of the physical world, and the user can view them and interact with them using a device. The hardware devices that are often used with AR are smart glasses, smartphones, and tablets. Another interesting form of virtual technology is MR. In MR, virtual objects not only are overlaid on the physical world, but they also interact with the physical world, and the user can interact with both physical and virtual objects. The devices used in MR are typically the same as the ones used in AR.

In the context of GISs and BIMs, all three immersive methods can be utilized based on the task at hand. For instance, VR and AR can be used in the office environment to

Table 14-1. Immersive Methods.

Immersive method	Characteristics	Hardware
Virtual reality (VR)	Fully immersed in the virtual environment. The user interacts with virtual objects only.	(1) HMDs and controllers or gloves (2) CAVE
Augmented reality (AR)	Virtual objects are rendered on the real world and the user can view them and interact with them	AR glasses and headsets, smartphones, tables
Mixed reality (MR)	Virtual objects are rendered and interact with the real world, the user can interact with both the virtual and the physical world	AR glasses and headsets, smartphones, tables

Figure 14-6. Example of using augmented reality in building construction.
Source: Courtesy of Trimble Inc., reproduced with permission.

communicate aspects of a BIM with stakeholders and enhance the decision-making process. In addition, AR and MR can be used to overlay virtual building elements (or other virtual objects) in an actual physical location to assist with project planning, construction, and stakeholder communication. With MR, the users can also visualize and assess the interaction between virtual objects and physical world.

Figure 14-6 shows an example where AR is being used to overlay the proposed building on top of the physical site assisting in field procedures. Figure 14-7 shows another example, where an entire physical site has been re-created in immersive VR. In this case, the user is completely immersed in the virtual environment.

KEY DATA MODELS

Key Data Types for Geographic Information Systems

GISs have the ability to integrate many types of data, including point, vector, tabular, big data, unstructured data, voxels, real-time, IoT, among others. For BIM practitioners, raster, point, and vector are the major data types, but as GIS/BIM integration grows, GIS-supported data becomes more important. Figure 14-8 shows the key data types for GISs. Raster data appear as a grid comprised of cells. Cells, also referred to as pixels, are positioned based on their row and column within the grid and the grid's cell size. Cell sizes are defined in the unit of a coordinate system, such as feet, meters, or degrees, and can be sized as squares or rectangles, although rectangle-shaped cells are highly uncommon. The entire grid is located in space using the coordinate of one grid corner. Vector data appear as a vertex or a set of vertexes connected by lines. In some cases, the vertices and lines may be arranged to create a closed shape or polygon. The coordinates of each vertex determine the spatial positioning of vector data.

Rasters store information in what are called values. Single-band rasters store one value per grid cell, whereas multiband rasters store multiple values per grid cell. Multiband rasters can be easily visualized as a stack of single-band rasters, and in this way, all grid cells require a value for each band. If one cell has three values, all pixels must have three values. This requirement also means that rasters must be rectangular. The number of rows in each column must stay the same for all columns. Not all spatial information is well represented by a rectangle, and a workaround is to assign "No Data" values to pixel values that are not used. No Data values are implemented by assigning a common value

Figure 14-7. Building in virtual reality: (a) zoomed view of the front side in virtual reality, (b) real picture of (a), (c) side view of the building, (d) real picture of (c), (e) front view of the building, (f) real picture of (e).
Source: Bolkas et al. (2020), reproduced with permission from Taylor & Francis.

to relevant pixels that does not correspond to other data values in the same raster. A common example may be something like -9999. Common GIS software such as QGIS or ArcGIS will create data with No Data stored "under the hood" and will make No Data cells invisible.

Functions of raster data are varied and include thematic purposes such as basemaps for maps or numerical purposes such as terrain analysis and using AI and ML for automated feature extraction. Commonly used thematic raster data include aerial or satellite imagery or informational basemaps. Informational basemaps, like mosaicked United States Geological Survey topographic maps, are built from vector data but displayed as rasters to reduce memory requirements and computational demand. A limitation of thematic rasters is that each individual raster represents a single spatial scale—new rasters must be prepared for each scale that may be displayed. For static maps, this limitation is of minimal significance, but as interactive maps become more prevalent, the limitation becomes more noticeable. The workaround is to produce map tiles, which are assemblages of rasters having different scales and sizes, that can be cycled in and out of view as needed. Map tiles have the added benefit of only needing to display tiles in the map view, improving the speed at which a basemap can be displayed. Common numerical raster analyses include

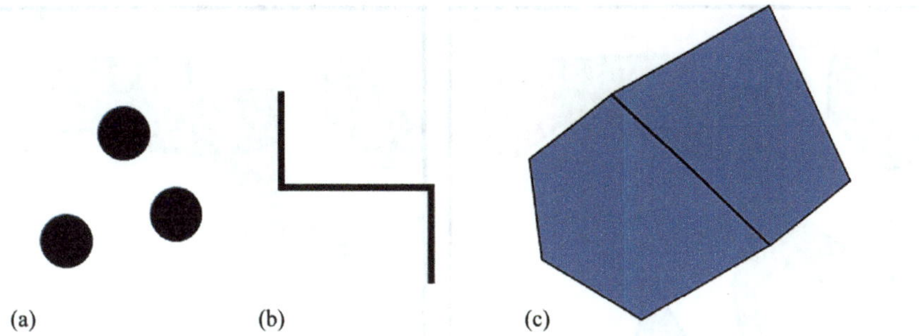

1	1	1	4	4
1	1	1	4	4
1	2	2	3	4
2	2	3	3	4
2	2	3	3	4

(d)

170	238	165	230	52	157
50	170	75	165	90	52
175	240	163	237	56	156
50	175	73	163	88	56
172	243	168	237	165	225
54	172	78	168	79	165

(e)

Figure 14-8. Main data types used in GISs: (a) points, (b) lines, (c) polygons, (d) grid cells, (e) pixels.

calculation of ground slope from DEMs or hydrologic purposes that involve calculation of flow accumulation and flow direction.

Vector and point data store information in a table, referred to as an attribute table. In GISs, these are managed in a database. Each attribute table consists of columns corresponding to stored values and rows corresponding to each shape, or feature, in the data layer. Types of shapes include points, lines, and polygons. Points consist of only one vertex, whereas lines and polygons may have many vertices. Polygons differ from lines in that they are closed regions. No single attribute table can relate to different shapes; for example, if the layer is a point layer, then each row must correspond to a point. In some rare cases, multipart features may exist, and a row may correspond to multiple shapes. Unlike rasters, any feature may lack values in any attribute column.

Locations represented by vector data include contour lines, bench mark locations, land parcels, and countless other items or places. Attributes can be few, like an elevation for each contour line, or numerous, like plat numbers, hyperlinks to external documents, address information, and so on for land parcels. Vector data also benefit from scalability, with features appearing the same regardless of the scale used. The primary limitation for vector data is their increase in computational demand alongside increases in shape vertices and attribute information.

Raster and vector data have clear differences in their applications. Raster data provide uniform coverage and are useful for displaying continuous information. Vector data allow for the incorporation of irregularly spaced data and can contain many more pieces of information than rasters (Figure 14-9).

Key Data Types for Building Information Modeling

In the AEC domain, numerous BIM software applications from different vendors exist, which save the results in their formats (Majcher 2019, Zhu et al. 2018). Because almost all construction firms use different software (Grani 2016), there have been increasing demands for a standardized file format that is compatible with different software platforms to ensure interoperability. Industry foundation classes (IFC) are developed by buildingSMART to

Figure 14-9. Comparison of land-use and land-cover data represented in: (a) polygon, (b) 30 m resolution raster formats.
Source: Price et al. (2006), reproduced with permission from USGS.

meet this need. IFC is an open data schema that became an ISO standard (ISO 16739) (Tobiáš 2015). IFC is vendor-neutral; it is not controlled by a single or a group of suppliers and usable across a wide range of hardware devices, software platforms, and interfaces for many different use cases (Grani 2016). IFC is designed to provide all information about buildings, such as the physical components of buildings, manufactured products, mechanical/electrical systems, as well as more abstract structural analysis models, energy analysis models, cost breakdowns, work schedules, and so on (buildingSMART 2021). The key advantage of the IFC is that it improves collaboration between the various agencies and contractors involved in a construction project by allowing the efficient exchange of information through a standard format. This results in higher quality as well as a reduction in errors, costs, and time for the operation and maintenance of buildings (Muñoz et al. 2020).

There have been many releases of the IFC since the first version was introduced in 1997. After releasing IFC 1.5.1, there were 2×, 2 × 2, and 2 × 3 (the × is short for "extension"), and then IFC4 was released in 2013. This release could have been named IFC 2 × 4, but instead, they simplified the naming of the version to IFC 4, and then the following major version will be IFC 5 (Grani 2016). Even though IFC 2 × 3 is the previous release, it is still the most widespread version (Tobiáš 2015). However, IFC 2 × 3 contains some limitations, which are being addressed by buildingSMART in the following releases. IFC 4 supports geometries and parametric information and includes a simple XML format. The latest

```
ISO-10303-21;
HEADER;
FILE_DESCRIPTION(('ViewDefinition [CoordinationView]'),'2;1');
FILE_NAME('../results_0608/22_Annex_01_F1_3dmodel.ifc','2021-06-11T21:14:39',('GT'),('Georgia Tech'),
FILE_SCHEMA(('IFC2X3'));
ENDSEC;
DATA;
#1=IFCPERSON($,$,'GT',$,$,$,$,$);
#2=IFCORGANIZATION($,'Georgia Tech',$,$,$);
#3=IFCPERSONANDORGANIZATION(#1,#2,$);
#4=IFCAPPLICATION(#2,'0.6','IfcOpenShell','');
#5=IFCOWNERHISTORY(#3,#4,$,.ADDED.,$,#3,#4,1623446079.2526205);
#6=IFCDIRECTION((1.,0.,0.));
#7=IFCDIRECTION((0.,0.,1.));
#8=IFCCARTESIANPOINT((0.,0.,0.));
#9=IFCAXIS2PLACEMENT3D(#8,#7,#6);
#10=IFCDIRECTION((0.,1.,0.));
#11=IFCGEOMETRICREPRESENTATIONCONTEXT($,'Model',3,1.E-05,#9,#10);
#12=IFCDIMENSIONALEXPONENTS(0,0,0,0,0,0,0);
#13=IFCSIUNIT(*,.LENGTHUNIT.,$,.METRE.);
#14=IFCSIUNIT(*,.AREAUNIT.,$,.SQUARE_METRE.);
#15=IFCSIUNIT(*,.VOLUMEUNIT.,$,.CUBIC_METRE.);
#16=IFCSIUNIT(*,.PLANEANGLEUNIT.,$,.RADIAN.);
#17=IFCMEASUREWITHUNIT(IFCPLANEANGLEMEASURE(0.017453292519943295),#16);
#18=IFCCONVERSIONBASEDUNIT(#12,.PLANEANGLEUNIT.,'DEGREE',#17);
#19=IFCUNITASSIGNMENT((#13,#14,#15,#18));
#20=IFCPROJECT('084t0VoleHwwYsc92MtdzY',#5,'../results_0608/22_Annex_01_F1_3dmodel.ifc',$,$,$,$,(#11)
```

Figure 14-10. Plain text contents of an IFC-SPF file.

version is IFC 4.3, which has extended its domain to cover the description of infrastructure constructions such as Railways, Roads, Ports, and Waterways (buildingSMART 2021).

There are several different types of IFC format, but three widely adopted formats are IFC-SPF, IFC-XML, and IFC-ZIP. IFC-SPF is the most widespread format, having the advantage of readable text, where each line typically consists of a single object record (Figure 14-10). IFC-SPF is defined by ISO 10303-21 ("STEP-File") and has a file extension ".ifc." IFC-XML is an XML format defined by ISO 10303-28 ("STEP-XML"), having a file extension ".ifcXML." IFC-XML file format is also human-readable, but the file is represented as an XML document, which renders it easier to query the model. IFC-XML can also be operated with XML tools. Because both the IFC SPF and the IFC-XML files are text files containing a lot of data, there is a need for compression. IFC-ZIP is a ZIP-compressed format consisting of an embedded IFC-SFC file or IFC-XML file and has a file extension ".ifcZIP" (Grani 2016, buildingSMART 2021).

The IFC uses multiple methods to represent 3D objects, including boundary representation (B-Rep), constructive solid geometry (CSG), and sweep volume (Tobiáš 2015). B-rep represents a solid object using planar faces that are located at the boundary of the solid body and completely enclose it. B-rep is usually used for complex objects such as doors (IfcDoor) and windows (IfcWindow). In CSG, a new object is created by one or many Boolean operations (difference, union, and intersection) of simpler objects such as spheres, cones, pyramids, or cylinders. Finally, in sweep volume, a solid object is created by moving a 2D profile along a path that defines the route over which the surface is extruded (Zhu et al. 2018, Das 2019). CSG and sweep volume representations are generated by calculating parameters, whereas B-Rep representations are explicit models (Zhu et al. 2020). Figure 14-11 represents examples of B-Rep, CSG, and sweep volume.

BIM models in IFC are categorized into six groups according to the details they contain by Levels of Development (LODs), from LOD 100 to LOD 500 (Table 14-2). Unlike the Levels of Details in CityGML that defines the degree of abstraction of real-world objects, the LOD framework in IFC is designed to specify the state of development of a BIM element from conceptual (LOD 100) to as-built (LOD 500), such that they are helpful for communication and coordination (Donkers 2013). Table 14-2 includes more details of LOD in IFC.

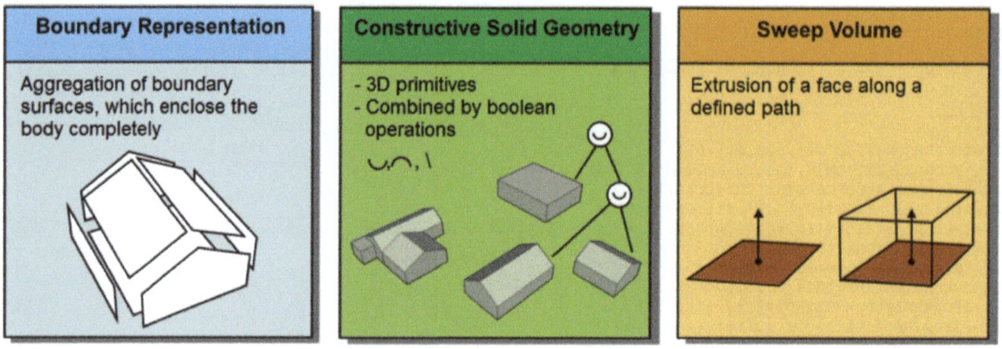

Figure 14-11. Representation of 3D geometry in IFC: (Left) boundary representation; (middle) constructive solid geometry; and (right) sweep volume.
Source: Donkers (2013), reproduced with permission from Technische Universität Münche.

Table 14-2. Levels of development in IFC.

LOD	Description	Content
100	Concept or predesign	Conceptual model including an estimate of areas, volumes, orientation, and cost. Building elements are represented as generic placeholders, such as symbols, that lack physical information
200	Developed or schematic design	A design development model with approximate quantities, size, shape, location, and orientation. It also has nongeometric information related to them
300	Production or design development	Modeled elements are accurate and coordinated, suitable for cost estimation and regulatory compliance check. The information can be used during the construction stage.
350	Construction documentation	It contains the same information as LOD 300, but also includes interfaces with other building components (supports and connections) such that building systems can interact with one another.
400	Installation or construction stage	The details and information contained in the element can be used by suppliers to manufacture the components that are represented.
500	As-built	It represents the existing installation. The model and associated data are suitable for the maintenance and operations of the building.

Source: Adapted from Das (2019).

DATABASE

Databases in Geographic Information Systems

A database is a collection of data with respect to a particular subject. In its simplest form, a database can just be data-recorded in a table format such as an ASCII text or a spreadsheet [Figure 14-12(a)]. Each row represents a record, and each column represents a variable or property about the data set. A database is a collection of one or more tables, following a specific structure and rules, which improves efficiency, accessibility and security, data sharing, querying, and functionality (Longley et al. 2005).

A database in a GIS, often called a geographic database, is nothing more than a simple database but with spatial information about the data (Longley et al. 2005). For instance, consider a database that contains household income information for each state. A geographic database will also include the layout of those states, making it possible to create a map where each state is shaded with different colors representing the different median income categories.

Consider the simplest database form, a table. Figure 14-12 shows an example of preparing the median household income data set for use in a GIS system. Figure 14-12a shows the data sets as downloaded from Census.gov. In Figure 14-12b, the median income and state information has been extracted. In Figure 14-12c, the spreadsheet is converted to a GIS table format. Note that still the table in Figure 14-12c does not contain any geographic information. This is achieved by joining the table of Figure 14-12c with the states' geographic features in Figure 14-13. The join operation is discussed in the following sections. Figure 14-14 shows the final maps that can be created in GISs using geographic data sets and databases.

Accessing and storing information in the database is done through a database management system (DBMS). The DBMS will contain all the necessary functions and capabilities to (Longley et al. 2005):

- Represent real-world objects into specific data models (e.g., points, lines, and polygons).
- Use a query language to query information in the datasets.

(a)

	A	B	C	D
1		Total	Median income	Mean income
2	Label			
3	Alabama			
4	Households			
5	Estimate	1,897,576	51,734	71,478
6	Families			
7	Estimate	1,237,883	66,171	85,630
8	Married-couple families			
9	Estimate	892,050	81,393	101,086
10	Nonfamily households			
11	Estimate	659,693	29,458	42,510
12	Alaska			
13	Households			
14	Estimate	252,199	75,463	98,504
15	Families			
16	Estimate	163,134	91,971	115,639
17	Married-couple families			
18	Estimate	124,608	106,046	130,245
19	Nonfamily households			
20	Estimate	89,065	47,508	61,602
21	Arizona			
22	Households			
23	Estimate	2,670,441	62,055	84,438
24	Families			
25	Estimate	1,740,704	74,468	97,314
26	Married-couple families			
27	Estimate	1,272,331	87,708	112,220
28	Nonfamily households			
29	Estimate	929,737	40,198	55,773

(b)

	A	B
1	State	Median income (dollars)
2	Alabama	51,734
3	Alaska	75,463
4	Arizona	62,055
5	Arkansas	48,952
6	California	80,440
7	Colorado	77,127
8	Connecticut	78,833
9	Delaware	70,176
10	District of Columbia	92,266
11	Florida	59,227
12	Georgia	61,980
13	Hawaii	83,102
14	Idaho	60,999
15	Illinois	69,187
16	Indiana	57,603
17	Iowa	61,691
18	Kansas	62,087
19	Kentucky	52,295
20	Louisiana	51,073
21	Maine	58,924
22	Maryland	86,738

(c)

MedianHouseholdIncome

OBJECTID *	State	Median income (dollars)
1	Alabama	51,734
2	Alaska	75,463
3	Arizona	62,055
4	Arkansas	48,952
5	California	80,440
6	Colorado	77,127
7	Connecticut	78,833
8	Delaware	70,176
9	District of Columbia	92,266
10	Florida	59,227
11	Georgia	61,980
12	Hawaii	83,102
13	Idaho	60,999
14	Illinois	69,187
15	Indiana	57,603
16	Iowa	61,691
17	Kansas	62,087
18	Kentucky	52,295
19	Louisiana	51,073
20	Maine	58,924
21	Maryland	86,738
22	Massachusetts	85,843
23	Michigan	59,584
24	Minnesota	74,593
25	Mississippi	45,792
26	Missouri	57,409
27	Montana	57,153
28	Nebraska	63,229
29	Nevada	63,276
30	New Hampshire	77,933

Figure 14-12. Data preparation for GISs. Median household income by state: (a) data downloaded from Census.gov in spreadsheet format, (b) keeping only the state and median income, (c) spreadsheet converted to a GIS table.

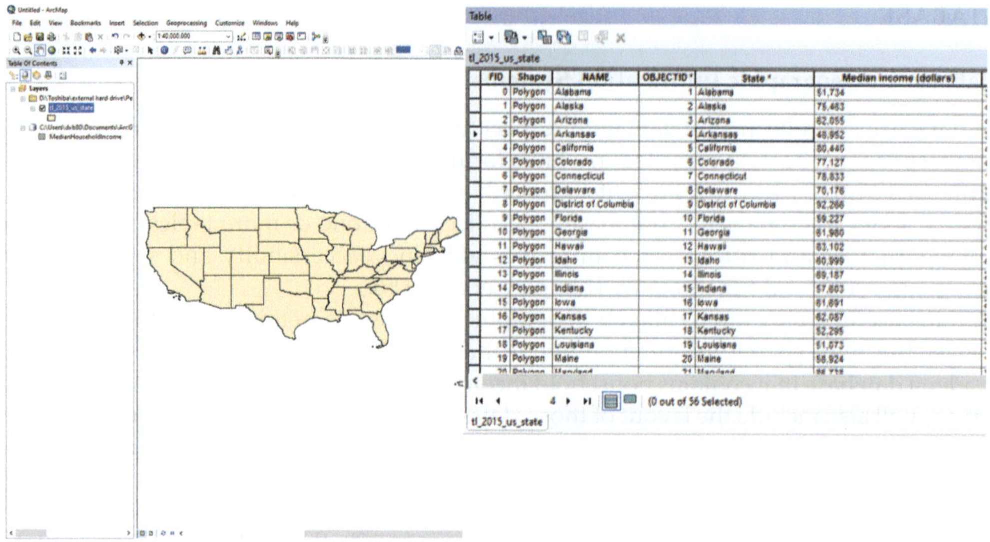

Figure 14-13. Table joined to state geographic features.

- Include indexing to speed up searching.
- Provide controlled access to data and restricting access to all or parts of the database.
- Furnish updating, backup and recovery functions.
- Utilize database administration tools.
- Employ tools for creating, editing, and maintaining databases.
- Give application programming interfaces.

The main three categories of DBMS in GISs are (1) relational, (2) object-oriented, and (3) object relational. Relational DBMSs are based on a set of tables with those tables being able

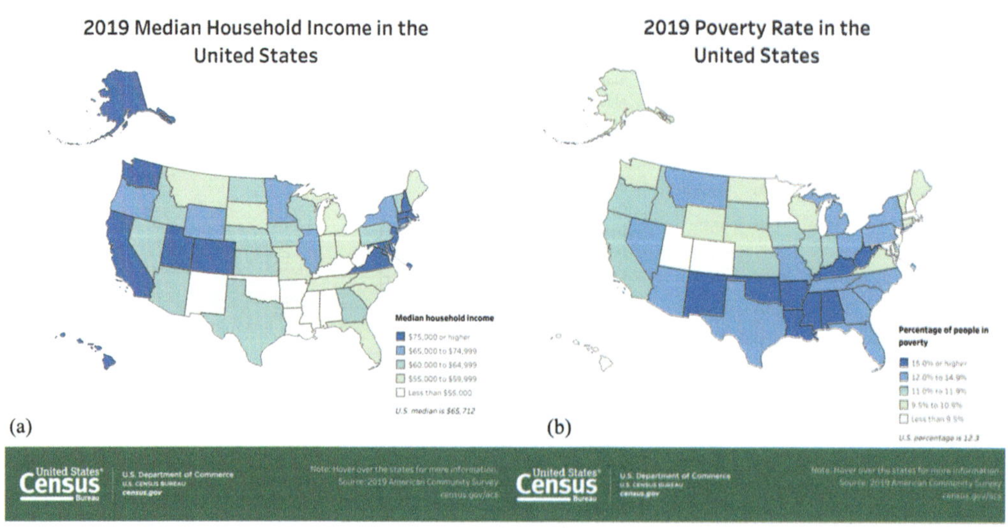

Figure 14-14. Maps generated from geographic databases: (a) 2019 median household income by state, (b) 2019 Poverty rate by state.
Source: Maps retrieved from USCB (2020b). Courtesy of US Census Bureau, reproduced with permission.

to connect to one another using attributes that appear in two or more tables using a foreign key attribute that references a primary key of another table. Object-oriented DBMSs have the ability to store complete objects in the database, and relationships between objects are represented by references via object identifiers and indexing techniques. Relational DBMSs have extended their capabilities to include many of the capabilities found in the object-oriented databases, creating a new, hybrid, category of DBMS, namely, the object-relational DBMS. Object-relational DBMSs can handle both the data attribute data that describe an object and the behavior of the object (Longley et al. 2005). For instance, Esri geodatabases use an object-relational approach that supports both GIS data objects and behaviors such as feature classes, raster data sets, topologies, networks, and other functions (Esri 2020).

ID Fields

In GIS tables, columns are fields or attributes related to the data set (e.g., name, deed number, state, city, address), whereas rows are records (or observations). The first row in a GIS table contains the names/titles of each attribute, which should be short and self-explanatory. Some GIS software allow the user to use an alias and then provide a more comprehensive description for each field. However, each table must have a field that will serve as a unique identifier (ID) in the table; in other words, this ID field will be able to uniquely identify each row in the table. This ID field is often called the "primary key" and must have two properties: (1) each value should be unique, and (2) no nulls exist.

Often the primary key is associated with geographical shapes or entities such as states, counties, parcels and so on. The Federal Information Processing Standard (FIPS) code, which used two digits to identify US States, followed by three digits to identify counties within a state (FCC undated; NBS 1987). In 2008, the FIPS codes were withdrawn by the National Institute of Standards and Technology (NIST) because "they are obsolete or have not been updated to adopt current voluntary industry standards, federal specifications, federal data standards, or current good practices for information security" (Turner 2008). The FIPS was replaced by the American National Standards Institute (ANSI), changing the FIPS codes to FIP Series, which here we call ANSI codes for simplicity (USCB 2020a). The ANSI codes are used to uniquely identify each state, counties and county subdivisions, metropolitan statistical areas (MSAs), congressional districts, voting districts, cities, school districts, and places (e.g., Indian reservations, airports, post offices) (Figure 14-15). For instance, Harvey's Lake borough in Pennsylvania is represented with the following ANSI code: 4207933000, where 42 identifies the state Pennsylvania, 079 the county Luzerne County, and 33,000 the Harveys Lake borough subdivision. Similar types of geocodes are the postal Zone Improvement Plan (ZIP) codes and the Property Index Number (PIN) number to represent land properties. The PIN is a 14-digit number that can be broken down into two numbers for township, two for section, three for block, three for parcel, and four numbers for unit.

Note that when creating a table or data set in a GIS software, a unique ID is automatically created to ensure that each row is uniquely identified for internal processes and operations of the software. To enhance the integrity and usability of their tables and databases, practitioners should not rely on such autogenerated ID fields but use industry standard IDs and geocodes for geographical areas and property parcels.

Joining and Relating Tables

Joins are a method of combining multiple tabular data sets based on common field values. Most GIS software have built-in functions to perform joins between two tables. Types of join vary among software and programming languages but typically consist of

Table

tl_2015_42_cousub

FID	Shape	STATEFP	COUNTYFP	COUSUBFP	COUSUBNS	GEOID	NAME	NAMELSAD
1591	Polygon	42	019	32688	01214994	4201932688	Harmony	Harmony borough
1636	Polygon	42	115	32720	01217086	4211532720	Harmony	Harmony township
2358	Polygon	42	053	32704	01216453	4205332704	Harmony	Harmony township
2241	Polygon	42	027	32792	01216127	4202732792	Harris	Harris township
1406	Polygon	42	043	32800	01215162	4204332800	Harrisburg	Harrisburg city
1665	Polygon	42	003	32832	01215809	4200332832	Harrison	Harrison township
2052	Polygon	42	009	32840	01215894	4200932840	Harrison	Harrison township
2504	Polygon	42	105	32848	01216973	4210532848	Harrison	Harrison township
2574	Polygon	42	019	32896	01214995	4201932896	Harrisville	Harrisville borough
1195	Polygon	42	119	32936	01215631	4211932936	Hartleton	Hartleton borough
1198	Polygon	42	119	32952	01217131	4211932952	Hartley	Hartley township
2018	Polygon	42	079	33000	01215391	4207933000	Harveys Lake	Harveys Lake borough
56	Polygon	42	021	33080	01215024	4202133080	Hastings	Hastings borough
832	Polygon	42	091	33088	01215470	4209133088	Hatboro	Hatboro borough
324	Polygon	42	091	33120	01216855	4209133120	Hatfield	Hatfield township
1599	Polygon	42	091	33112	01215471	4209133112	Hatfield	Hatfield borough
371	Polygon	42	045	33144	01216383	4204533144	Haverford	Haverford township

Figure 14-15. Example of ANSI codes in the US county subdivision data set.

two types: inside and outside. Regardless of the join type, tables can be joined only if they have a field, called the join field, with at least one value in common with a join field value from the other table. Inside joins combine only the rows from each table that share values of the join field. Outside joins combine tables regardless of whether a join field is shared. Columns in the joined table corresponding to values not shared between the individual tables are filled left empty. Outside joins can retain rows corresponding to both tables and only one of the tables. Basic examples of inside and outside joins are illustrated in Figure 14-16.

In Figure 14-16, Tables A and B are combined into a single table. In a GIS, this task is complicated because Table A, and even Table B, may be connected to spatial data. For example, the rows in Table A might have corresponding lines representing the highway location. GIS software overcomes this complication by appending one table to the other. In the case of our example, Table B may be appended to Table A, and the line features of the respective layer will have the attributes from both tables. If, for example, the highway line of Highway 26 was now queried in the GIS, a user will be able to see that it was designated as a US route.

Table A

Highway	Length
26	46
30	125
120	25
136	85

Table B

Route	Designation
26	US Route
120	State Route
34	State Route

Inside Join

Highway	Length	Route	Designation
26	46	26	US Route
120	25	120	State Route

Outside Join

Highway	Length	Route	Designation
26	46	26	US Route
30	125		
120	25	120	State Route
136	85		
		34	State Route

Figure 14-16. Inside and outside joins. Join fields for Table A and Table B are "Highway" and "Route," respectively.

What is necessary to connect two spatial data sets and their attributes? An alternate approach to combining tables with a join is to connect the tables with what is called a relate. Relates use a common field, similar to joins, but temporarily append both tables with attributes from the other table. Another alternative to joins available in GIS software is called *spatial join*. Spatial joins are similar to traditional joins but use a common location instead of a common attribute. The primary forms of spatial joins are intersections, unions, and merges. These spatial joins are discussed in a later section of this chapter.

Query Languages

Querying is a very important component of a database and, of course, an important and useful operation in GISs. Queries are used to select a subset of data from the database. Although several languages exist for querying, the most common language used in relational databases such as the databases often used in GISs is the structured query language (SQL). With the SQL, a user can "ask" a question by typing a certain expression and retrieve a subset of the data sets based on the entered expression. For instance, a user can select parcels with a specific land use or zoning attribute (e.g., commercial, residential, park). In Figure 14-17, parcels with "state" ownership have been selected for the state of Washington.

Fields and Data Types

Each geospatial data set in a GIS is associated with its respective table that is used to store important information (attributes), and table fields are the columns that give structure

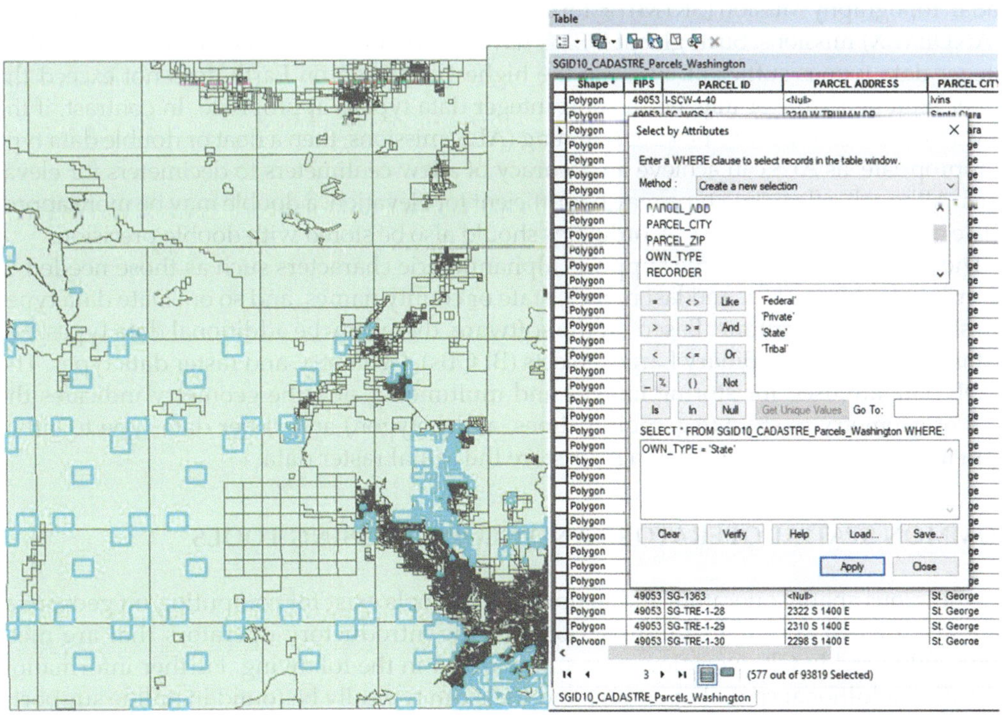

Figure 14-17. Example showing how to use SQL to select all parcels with "state" ownership in the state of Washington.

Table 14-3. Integer and Real Number Data Types.

Data type	Range	Precision
Short integer	−32,768 to 32,767	1–5
Long integer	−2,147,483,648 to 2,147,483,647	6–10
Float (single precision)	approximately -3.4×10^{38} to 1.2×10^{38}	1–6
Double (double precision)	approximately -2.2×10^{308} to 1.8×10^{308}	7+

to the table based on the attributes that we want to store for a geospatial data set. Utilizing the information existing in the fields, the GIS allows the user to query and conduct any kind of analysis. When fields are added in a GIS table, it is important to select the appropriate data type that best serves the properties of the geospatial data set and the purpose for which this data set was created. Therefore, the number of fields and data type in the fields in a GIS table is directly associated with the geospatial data set and the application at hand. Although there may be differences between GIS software, the main field data types are (1) integer numbers, (2) real numbers, (3) string (text), and (4) dates. Table 14-3 provides the main numeric data types and their ranges, and Figure 14-18 shows an example of the available fields in Esri ArcGIS when creating a new polygon data set.

It is important for the user to identify the proper data type for each field. For instance, consider that we need to extract and store elevations from a DEM that covers the entire Earth, and it is derived from satellite means; for example, consider the DEMs derived from the Shuttle Radar Topography Mission (SRTM) or TerraSAR-X add-on for Digital Elevation Measurement (TANDEM-X) missions. Such global DEMs have an accuracy at the few meters level and an integer data type is sufficient. Because the highest elevation on Earth does not exceed the upper limit of 32,767 for integers, a short integer data type is appropriate. In contrast, if the elevations come from airborne laser scanning (ALS) missions, then a float or double data type is appropriate, as ALS can achieve an accuracy of a few centimeters to decimeters for elevations. Although a float data type may be sufficient for elevation, a double may be more appropriate as Easting and Northing coordinates should also be stored with double precision.

The string or text data type supports alphanumeric characters such as those needed to describe an address, first and last names, state or county names, and so on. Date data types can store date and/or time. Based on the software, there may be additional data types. For instance, Esri also uses Binary Large Objects (BLOBs), Geometry, and raster data types. The BLOB data are used for storing images and multimedia, and the geometry indicates the type of geometry to store (e.g., points, lines, and polygon) and raster data type to either store in the field the link to an image or store the actual raster data.

COMMON SPATIAL OPERATORS AND GEOPROCESSING TOOLS

Numerous spatial operators and geoprocessing tools exist for computing on geospatial data and the list continues to grow. Some of the introductory operators that are most commonly used in GIS workflows are summarized in the following. Further information regarding additional spatial operators and tools can typically be found in online support/help resources for a given GIS platform. An investigation of these online resources is highly encouraged as they often provide valuable insight into the architecture of a given operator/tool and its capabilities.

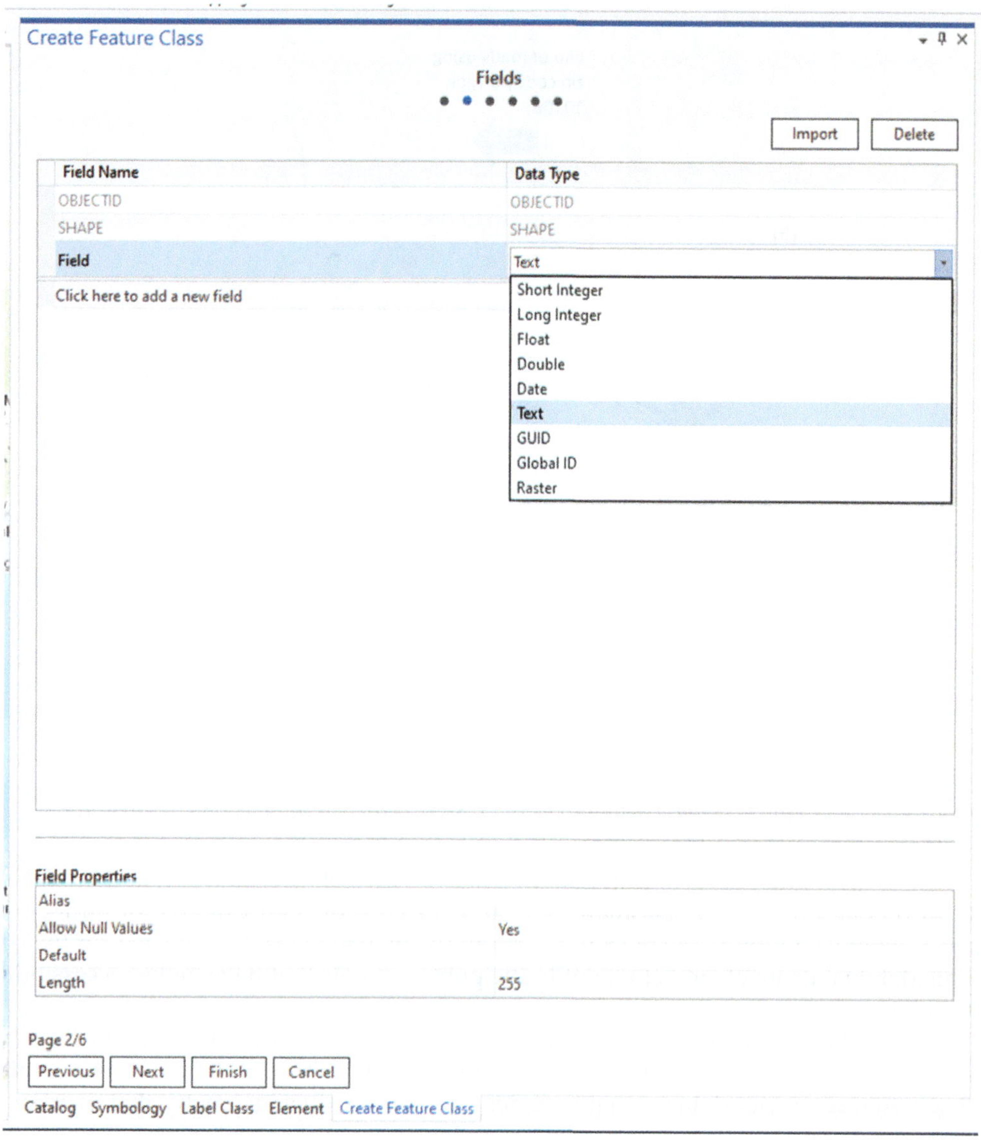

Figure 14-18. Example of available data types in Esri ArcGIS Pro.

Vector Operators

Vector operators provide the ability to manipulate vector data (i.e., points, lines, and polygons) and incorporate vector elements into geospatial analyses. The most common spatial operators associated with vector data are as follows: buffering, clipping, intersection, union, and dissolve (Figure 14-19).

Buffering is used to generate a polygon with a specified offset that surrounds a given vector-based feature. For example, a 30 m buffer applied to a point will result in a circle with a radius of 30 m. This buffer can then be used to efficiently identify if other features are located within the buffered region or include/exclude this region for a subsequent geospatial analysis. Clipping of vector data is used to manipulate the extent of input data.

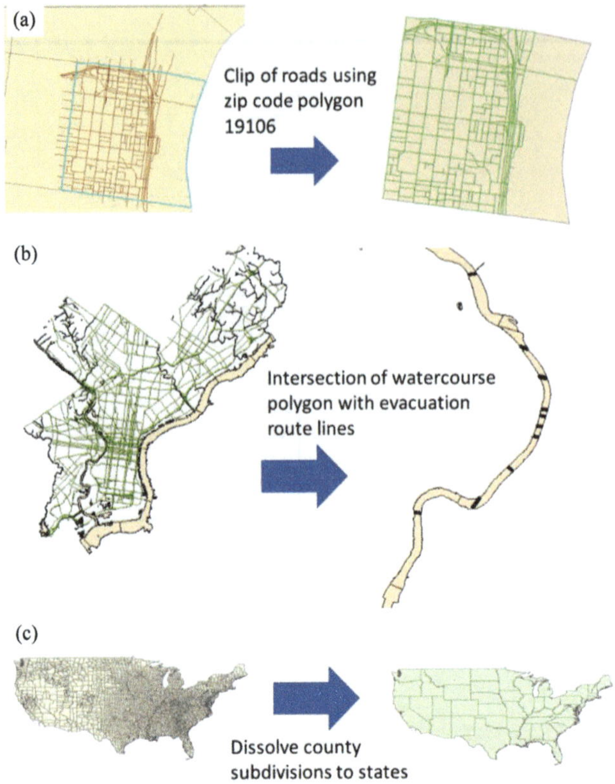

Figure 14-19. Example of geoprocessing tools in GISs: (a) clip, (b) intersection, (c) dissolve.

Clipping is most often used to trim data to match an area of interest (AOI) and/or isolate data in preparation for spatial analysis. Figure 14-20 shows an example of buffer tool analysis where a buffer of 150 m (500 ft) around the bike network was created for the city of Philadelphia, Pennsylvania, to identify the public schools that have higher accessibility to the bike network.

When working with multiple vector data sets, the intersection, union, and dissolve operators can be used to manipulate and merge chosen features. The intersect operator will generate an output data set that includes only the features/geometry that overlap in the original input data. If there is a need to combine data into a single vector data set, the union or dissolve operators can be utilized. The difference between union and dissolve is that union will preserve aspects of the original input data, such as individual feature boundaries, and dissolve will completely merge all the geometry into one vector feature. For example, if there is a property subdivision plan and each property boundary is represented by individual polygon features, the union operator facilitates the merging of the polygon features into a single data set where each property boundary will remain visible. Alternatively, if the dissolve operator is utilized, the output data set would include one large polygon feature that represents the collective coverage of all the individual parcels.

Topology

Topology plays an integral role in the management of geospatial data integrity in all modern GIS platforms. Most of the critical topological rules and constraints are automatically

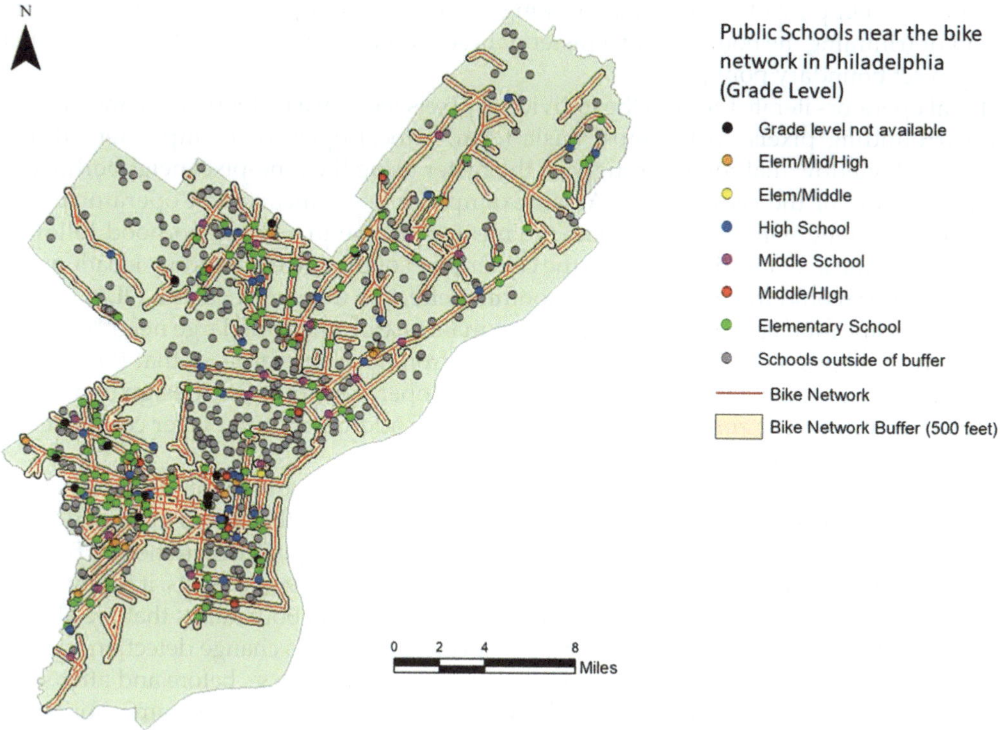

Figure 14-20. Example of using the buffer tool for analysis in GISs. A buffer of 150 m (500 ft) around the bike network was created for the city of Philadelphia, Pennsylvania, to identify the public schools that have higher accessibility to the bike network.

accounted for by the various data structures and file types (e.g., shapefiles and geodatabases) supported by available GIS solutions. With that said, a proficient GIS user must be able to designate and/or refine topology when necessary. An example of using defined topology to inform a geospatial analysis is the calculation of a least-cost path through an urban transportation network. In addition to correct network geometry, topological rules related to connectivity and travel direction must be accounted for to ensure a valuable result.

Raster Operators

Raster data sets can be manipulated and computed on in a variety of ways. The most common spatial operators used with raster data are as follows: resampling, masking/clipping, focal and zonal operators/statistics, and raster calculator/raster math.

Resampling is used to change the pixel size/spatial resolution of a raster data set. Resampling uses various interpolation approaches (further discussed in the next section) to either increase or decrease the pixel size of a given raster. Resampling can be useful for decreasing the overall data storage size or aligning raster pixels with another data set; however, thought should be given to the original data set and how it is affected by the resampling. For example, when using resampling to increase pixel size (decrease the spatial resolution) of a raster, pixel detail is inherently lost. It is also worth mentioning that when resampling is used to decrease the pixel size of a raster such as a DEM, it may appear that the spatial resolution has increased; however, the level of detail provided by the pixels is still limited by the original raster data set.

Similar to clipping of vector data, masking or raster clipping is useful for adjusting the extent or trimming the boundary of a raster data set based on an additional data layer such as a project boundary polygon.

Focal operators iterate through the individual pixels/cells of a raster data set and analyze the surrounding pixels. In the most basic form, a focal operator is implemented as a 3×3-pixel window that roves throughout the raster using the nine-pixel neighborhood to alter the value of the center pixel. As an example, a 3×3 mean focal operator uses a rectangular roving window to calculate the mean of a nine-pixel neighborhood followed by assigning the average pixel value to the center pixel of the current window location. The focal operator would therefore have a smoothing effect on a raster such as a DEM or land surface classification layer. Focal operators can make use of small and large neighborhoods with varied shapes. Zonal operators are different from focal operators in that they do not rely on a roving analysis window. Zonal raster operators perform designated raster calculations on specific zones that are specified by an additional input layer comprised of raster or vector-based data.

Raster calculator operations allow pixel-by-pixel computations and comparisons using two or more raster data set layers. Raster calculator tools often require the user to manually key in the mathematical operation, which can include arithmetic, equality statements, Boolean operators, and basic logic such as an If statement. To avoid ambiguous results, it is best to use raster that have the same spatial resolution and individual pixel boundaries that are aligned or snapped together. A basic use case for a raster calculator tool is change detection analysis using two DEMs of the same location collected at different times (e.g., before and after some terrain altering event). Assuming that the DEMs are referenced to the same horizontal coordinate system and vertical datum, a raster calculator operation can be performed that subtracts one raster from the other. The resulting output raster will have pixel values that represent the positive or negative height difference throughout the analyzed extent.

Automated Feature Identification in Imagery

Automated feature identification in imagery using AI and ML approaches is an emergent GIS capability that is rapidly evolving with continued breakthroughs in the field of deep learning. Leveraging automated image classification tools can greatly improve productivity in situations where workflows require the analysis of numerous digital images and/or when there is a large quantity of features that must be identified/classified in imagery. In general, most GIS-related image classification tools are categorized as either supervised or unsupervised approaches. Supervised classification of imagery requires input from the user in the form of training data that are representative of the locations and data values for the desired classes. A supervised classification routine then utilizes the training data to attempt a classification of the remaining pixels in the imagery. Unsupervised classification does not require training data and relies on the automatic clustering of imagery pixels based on a variety of approaches rooted in machine learning and statistics. In conventional implementations of unsupervised image classification, it is then up to the user to assign the identity of clusters/classes. As AI/ML capabilities continue to develop along with the continued growth of massive databases of imagery classification training data, AI/ML-based GIS solutions are quickly approaching the ability to provide robust, fully automated image classification results. Given the quickly evolving nature of AI/ML capabilities, it is recommended that users investigate the AI/ML image classification/segmentation capabilities of a given GIS platform and take the time to understand how to properly use the specific tools. A pitfall of implementing an AI/ML GIS solution is blindly trusting the results without having a good understanding of what they represent and what are their inherent limitations.

Interpolation Techniques

Interpolation is the process of finding/predicting values based on other known values in the vicinity. Interpolation is often performed in one, two, or three dimensions. In surveying engineering applications, spatial interpolation is often sought, where there are points with known easting and northing (x, y) coordinates indicating the location in space, and the third dimension represents the known values. Often the third dimension represents topography, but it can also represent temperature, slope, pressure, gravity, or any other physical variable. In addition, they can also represent a statistical variable within a GIS such as population density, crime, income, unemployment, land value, and so on. The known value data can be scattered (random) or regularly spaced (e.g., data collected in a grid), and interpolated points can also be sought in scattered or regularly spaced locations. Figure 14-21a shows an example of 1D linear interpolation, where points have known x, y values and interpolated points are sought in certain locations. In 1D interpolations the x-axis usually represents time. Figure 14-21b shows a spatial interpolation example, where there are scattered data points with known x, y, and z coordinates. Using the x- and y-coordinate values of the data points, interpolation is applied to derive interpolated z-values in new locations. In this example, the new locations are on a grid.

The interpolation process is often implemented to derive a continuous surface from scattered data sets. In surveying engineering, spatial interpolation and estimation of a continuous surface from scattered elevation data is often desired. There are several interpolation methods to derive a continuous surface, and this section covers the most common ones, namely (1) natural neighbor, (2) bilinear, (3) quadratic, (4) cubic, (5) spline, (6) inverse distance weighted (IDW), and (7) kriging. It is important to note that methods (1) through (6) are deterministic methods, as the interpolated values are derived from explicitly defined mathematical formulas. Instead, kriging is a geostatistical method, as it is based on a statistical model that depends on the autocorrelation of the input points. Such geostatistical methods also have the ability to provide measures of uncertainty for the interpolated values. It is important to understand that different interpolation methods can produce different results. The spatial distribution of the scattered points (used as input in the interpolation) is an important factor, affecting the outcome of the interpolation. Some discussion about differences in accuracy based on the interpolation methods can be found in Caruso and Quarta (1998), Hagan and West (2006), Guo et al. (2010), Schwendel et al. (2012), Rishikeshan et al. (2014), Polat et al. (2015), and Bolkas et al. (2016). For instance,

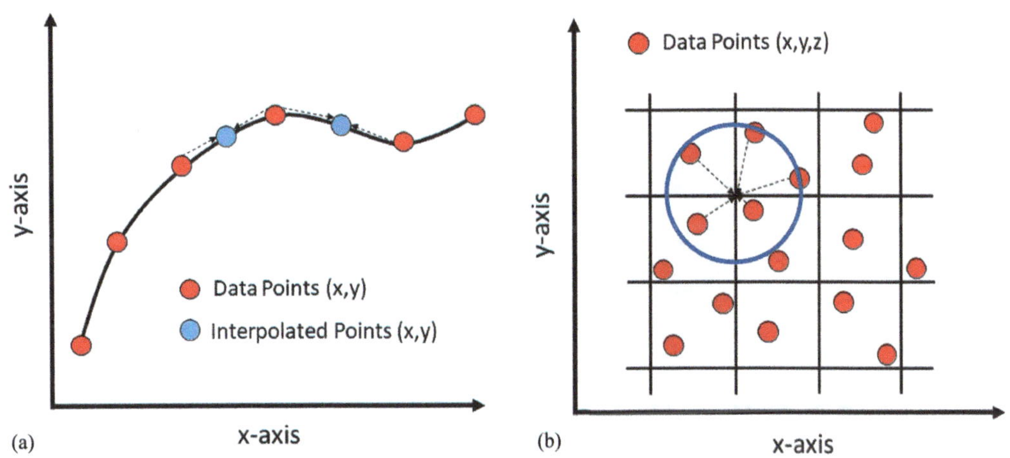

Figure 14-21. Interpolation examples: (a) 1D interpolation, (b) 2D interpolation.

Rishikeshan et al. (2014) assessed the different results of DEM interpolation in a GIS using different interpolation methods (Figure 14-22), and it can be seen that based on the interpolation method, different results can be obtained.

Topographic Operations

From topographic data representing elevations such as DEMs, we can extract several important topographic operations. This subsection discusses the hillshade, slope, aspect,

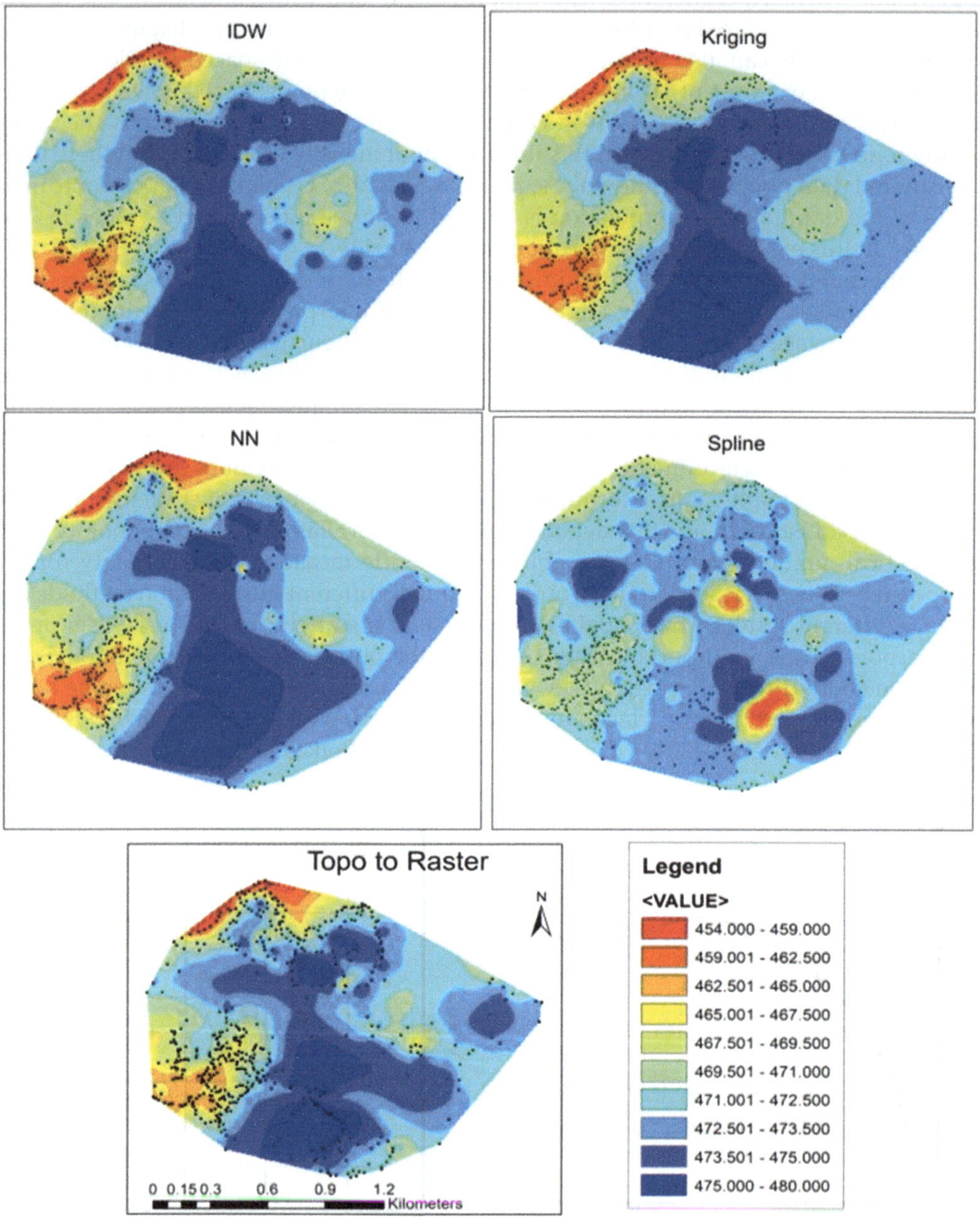

Figure 14-22. Different interpolation's results and their ellipsoid heights in meters. Source: Rishikeshan et al. (2014), reproduced with permission from IEEE.

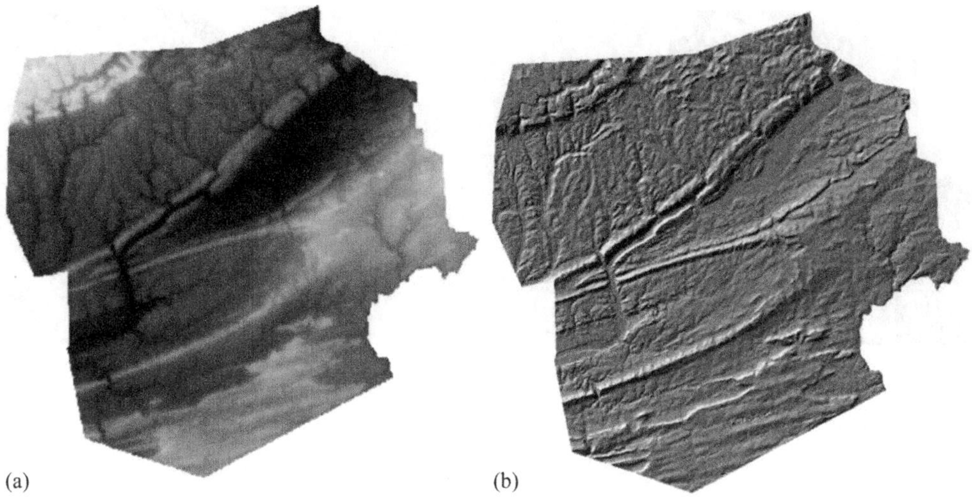

Figure 14-23. Hillshade example in Luzerne County, Pennsylvania: (a) DEM data set, (b) Hillshade.

and curvature operations. A hillshade is used to illuminate a surface, giving it a 3D appearance. Hillshades are very useful to improve the depiction and understanding of terrain undulations in a raster data set. Two main parameters are needed to create a hillshade from a DEM, namely (1) altitude and (2) azimuth angles. The altitude is the height of the light source (i.e., the sun) above the surface's horizon and can range from 0 to 90 degrees. The azimuth is the angle of the sun relative to the true north. For example, the default values are usually set to 45 degree altitude and 315 degree azimuth. Figure 14-23a shows an example of a DEM data set and the corresponding hillshade in Figure 14-23b.

Another important topographic information that is often needed in engineering projects is the slope of the terrain. The slope is the gradient of maximum change in elevations. The slope at a point is calculated using its neighboring values from the following expression (Burrough and McDonell 1998):

$$\text{slope (radians)} = \tan^{-1}\left(\sqrt{\left(\frac{dz}{dx}\right)^2 + \left(\frac{dz}{dy}\right)^2}\right) \tag{14-1}$$

where x and y are the easting and northing of the point, and z is the elevation.

A parameter that often accompanies slope data is the aspect. The aspect represents the slope direction. The output raster will contain the slope direction in degrees from 0 degree being due north and up to 360 degrees, measured in a clockwise direction. Another parameter, the curvature, is the second derivative of topography, that is, the slope of the slope, meaning that it depicts the rate of change of slope values on a surface.

Figure 14-24 shows an example of the slope and aspect operators. The data are derived from airborne lidar in Pennsylvania, and the spatial extent is about 3.2 km by 3.2 km (2 mi by 2 mi). From the point cloud, the last returns were kept to extract points representing the ground surface. Figure 14-24a shows the ground surface, as interpolated using a natural neighbor interpolation. Figure 14-24b shows the corresponding terrain slope and Figure 14-24c shows the aspect.

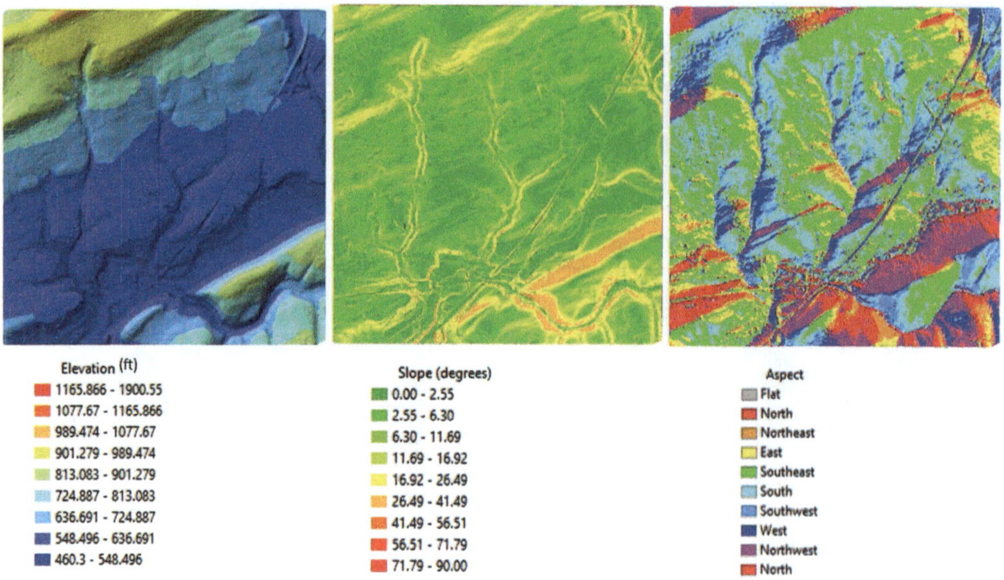

Figure 14-24. Elevation, slope, and aspect in a lidar data set in Luzerne County, Pennsylvania.

EXAMPLE GEOGRAPHIC INFORMATION SYSTEMS APPLICATIONS AND ANALYSES

The mapping capabilities of modern GIS software have always been relevant to civil engineering, although up until recently, engineers have typically thought of CAD software such as AutoCAD and MicroStation for such applications. CAD software is especially useful for plan preparation, particularly when the presentation of measurements is of critical importance, but it lacks the focus on databases and spatial analysis capabilities of GIS software.

As examples for common practices in civil engineering, GIS software can be incorporated into geohazard studies (e.g., landslides, subsidence, seismic liquefaction) at an early stage, typically before any site visits have been performed. An important first step involves a search for any existing data relevant to the hazard. In nearly all cases, prehazard topography for the area of interest is available. Good sources of United States topographic data in GIS-ready formats include the United States Geological Survey National Map, National Science Foundation–supported Open Topography, and many state-run websites such as the Alaska Elevation Portal (DGGS undated). Once topography has been obtained, practitioners can search for additional data. For landslide-related studies, landslide inventory features and geologic maps provided by state geological surveys such as the Washington State Department of Natural Resources or Oregon's Department of Geology and Mineral industries are readily available. These additional layers can be viewed along with topography to help interpretations underlying causes or implications of the geohazard that may not have been initially apparent. Although not always the case, these interpretations sometimes show that the geohazard may extend beyond its obvious extent (field surveys should attempt to capture the full extent of a hazard—this helps engineers design proper mitigation/repairs.)

Hydrologic studies are another civil engineering topic that benefits greatly from GISs. Several civil engineering disciplines form their designs on what water does and what it does not do. Highway engineers may find themselves concerned with whether surface runoff will flow from their roadway and down drainage ditches or how large a culvert should be to allow a creek to pass through, even in the most severe of storms. Alternatively, geotechnical engineers may be interested in improving the drainage of water from a soil slope or armoring slopes subjected to erosion. These applications rely on the compilation of multiple data sets. In the case of culvert sizing, an engineer may have historic measurements at several locations along the stream (point data) and historical stream channel locations (line data) and may be interested in the extent of the contributing watershed. In addition, raster data for topography and spatial estimates of predicted rainfall may be needed.

As part of a state-wide geotechnical urban levee evaluation, a private engineering firm used a commercial GIS solution to store, manage, and analyze all surface and subsurface investigation data along designated urban levee alignments. Armed with georeferenced geologic maps, surveyed subsurface soil borings, and associated soil testing results (e.g., density, permeability, strength) the firm was able to not only organize and manage all acquired data but also perform customized geospatial analysis routines. Analysis results such as a soil layer interpolation raster helped in the identification of potential problem zones susceptible to seepage and/or slope stability issues. Given the nature of GIS platforms available today, these results were rapidly transformed into easy-to-interpret maps that were used for subsequent project planning and discussion of mitigation techniques.

EXAMPLE BUILDING INFORMATION MODELING APPLICATIONS

Building Information Modeling for Infrastructure Projects

Heavy civil contractors as well as several DOTs started adopting BIMs in their projects given the evidence and documented benefits of using BIMs in the building construction industry. A Federal Highway Administration (FHWA) initiative on BIM for Infrastructure provides information on the benefits of using BIMs as well as best practices. Many state DOTs recognize the benefits associated with the bigger picture of BIMs as a data-centric approach both for project delivery and for asset management practices. For example, Iowa DOT incorporated a BIM for the I-80/I-380 Systems Interchange project in Eastern Iowa. Figure 14-25 shows the detailed BIM that they developed for Ramp B using OpenBridge modeler.

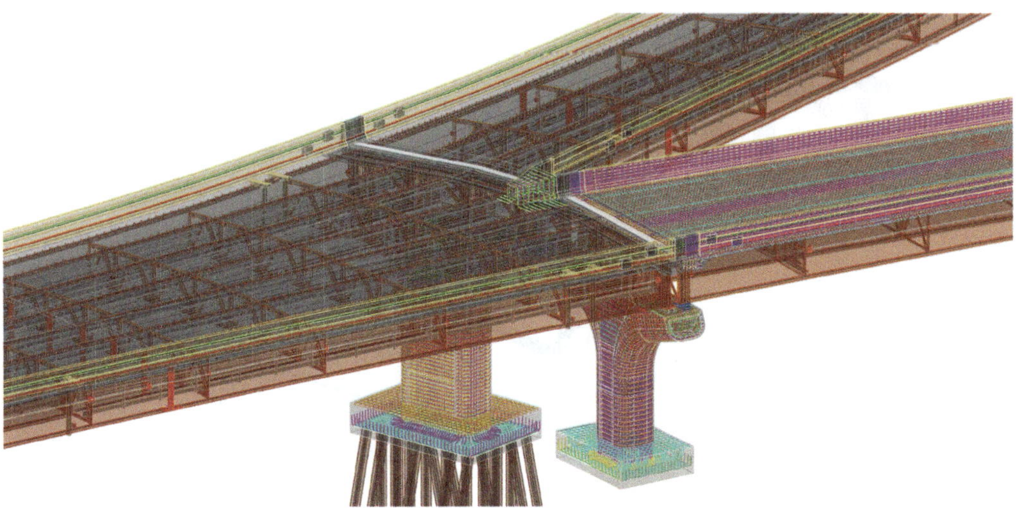

Figure 14-25. Detailed BIM for Ramp B developed using OpenBridge Modeler. Source: FHWA (2020), reproduced with permission from FHWA.

Iowa DOT stated several benefits of using BIMs in their projects, and one of the major benefits of using BIMs is the ability to bring multiple disciplines together during the design process enabling Iowa DOT to coordinate and address any conflicts during design (FHWA 2020).

Guo et al. (2014) explored civil information modeling (CiM), another term used when BIMs are applied to civil infrastructure projects, practices at the Department of Transportations (DOT)s. They conducted interviews of perspective DOTs and reviewed the driving forces and challenges as well as the benefits of using BIMs throughout the project development and the operation and maintenance stages. Civil integrated management (CIM), an approach to assist DOTs to transition to data-based project and program delivery by providing appropriate stakeholder access to an agency-wide data pool, was explored in Guo et al. (2017). The BIM is one of the major components of the CIM approach, which addresses the collection, organization, and management of data and information throughout transportation projects and programs.

BIMs have been implemented in numerous bridge construction projects across the nation, and the benefits of using BIMs for bridge design and construction have been well documented. Important to note here that bridge information modeling (BrIM) is another term when BIMs are used for bridge projects. For example, Shim et al. (2011) implemented BIMs for digital mock-up, parametric modeling and 4D and 5D simulations from planning to construction phases of bridge projects. Fanning et al. (2015) compared bridge construction projects (with and without BIMs) to better understand the impact of BIMs on the project performance and concluded that BIM implementation may lead to 5% to 9% cost savings from reducing Request for Information (RFI) and change orders on bridge construction projects. Also, BIM implementation can improve the efficiency in design and construction of complex bridge projects substantially (Liu et al. 2014). The benefits of using BIMs in bridge design and construction have been well documented and implemented in the industry. However, BIM implementation in existing bridges is rare mainly because of the challenges associated with converting 2D as-built drawings to 3D models (Volk et al. 2014).

Several studies looked into using BIMs during operations and maintenance phase of bridges given the better visualization and interoperability capabilities that they offer (Azhar et al. 2015, Marzouk and Hisham 2012). Al-Shalabi et al. (2015) proposed a 3D BIM-based bridge inspection framework that implements a BIM, mobile devices, and cloud computing (Figure 14-26). Building on Al-Shalabi et al. (2015) study, Xu and Turkan (2019) proposed a

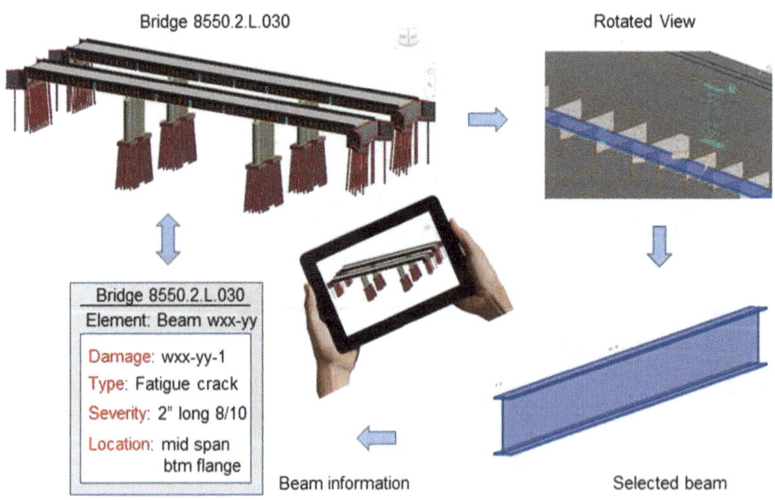

Figure 14-26. BIM-based bridge inspection framework.
Source: Al-Shalabi et al. (2015), reproduced with permission.

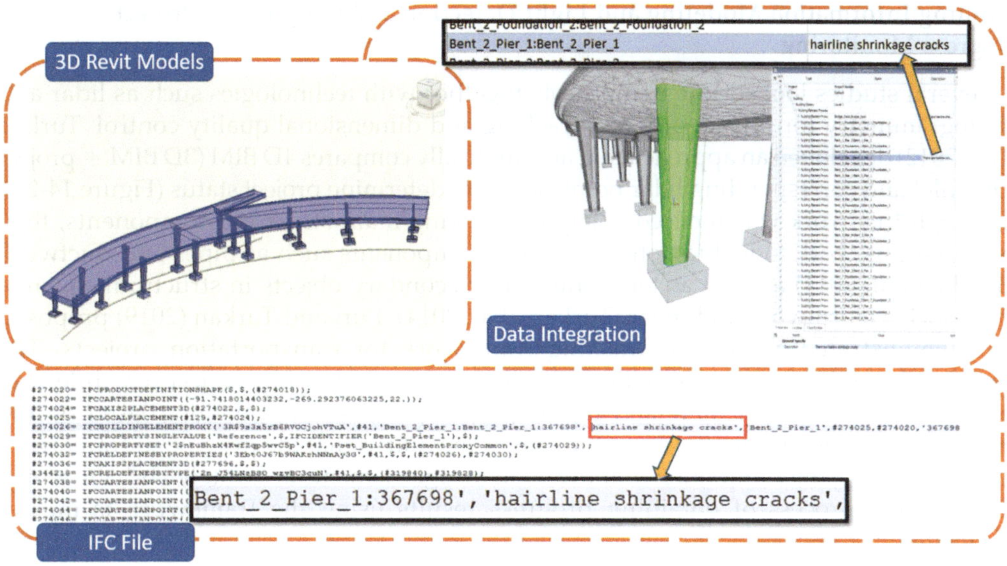

Figure 14-27. Procedure for integrating bridge inspection data.
Source: Xu and Turkan (2019), reproduced with permission from Emerald Publishing Ltd.

novel systematic bridge inspection framework that implements both BIM and Unmanned Aircraft System (UAS) technologies. In this framework, UASs were used to rapidly collect digital images and to detect cracks and other defects automatically with the help of computer vision algorithms (Figure 14-27). The resulting defect information such as type and severity of the damage, were then assigned to individual bridge elements in BIM. This enables all bridge information, including its 2D as-built drawings and 3D models, to be stored in a central, object-oriented database, that is, a BIM that can be accessed both from the home office and in the field (Figure 14-28).

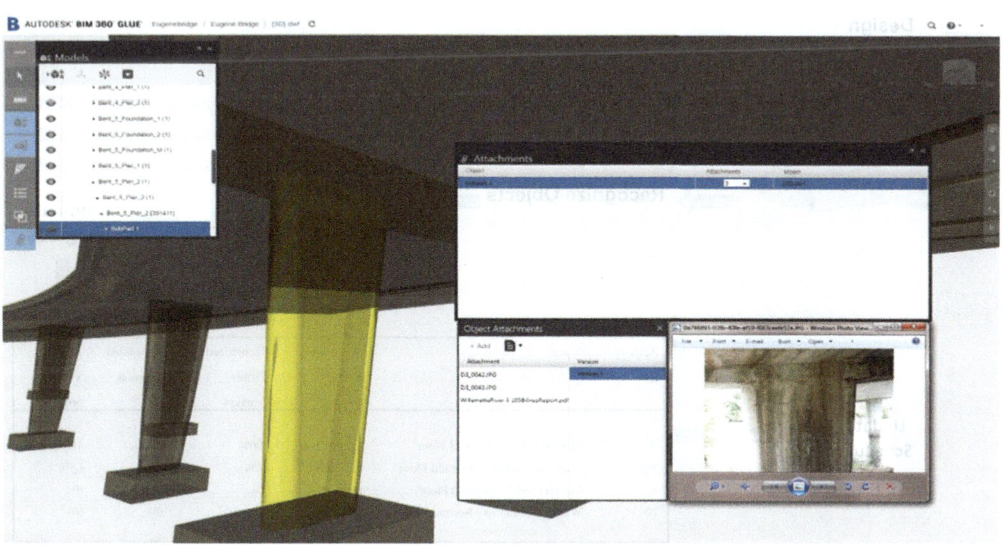

Figure 14-28. Bridge information management in Autodesk BIM 360 Glue.
Source: Xu and Turkan (2019), reproduced with permission from Emerald Publishing Ltd.

Building Information Modeling and Light Detection and Ranging for Project Progress Monitoring

Several studies looked into using BIMs together with technologies such as lidar and photogrammetry for project progress tracking and dimensional quality control. Turkan et al. (2012) developed an approach that automatically compares 4D BIM (3D BIM + project schedule) and corresponding lidar point clouds to determine project status (Figure 14-29). Although this work was focused solely on tracking main structural components, they extended this work to include other building components such as pipes and ductwork (Bosché et al. 2014) as well as temporary and secondary objects in structural concrete work such as formwork and rebar (Turkan et al. 2014). Puri and Turkan (2019) proposed and demonstrated a progress tracking framework for transportation projects. The framework combines the information included in 3D design BIM, project schedule, and mobile lidar data to calculate and report project progress information. Based on the results they obtained from the case study, they concluded that the developed framework enables tracking the completion of individual bridge elements accurately and efficiently, and that the process of obtaining progress results using the framework is mostly automated.

Scan-to-Building Information Modeling: Converting Point Clouds into Building Information Models

In another line of work, a process called scan-to-BIM, converting lidar point clouds into as-built BIMs has been explored by a number of researchers as well as software vendors. Scan-to-BIM enables the development of BIMs for existing buildings that can be used for a variety of purposes, including restoration of heritage buildings (Brumana et al. 2020), renovation projects, and operations and maintenance. However, current solutions provide manual or semiautomated solutions to fit 3D solid objects into lidar point clouds. Son et al.

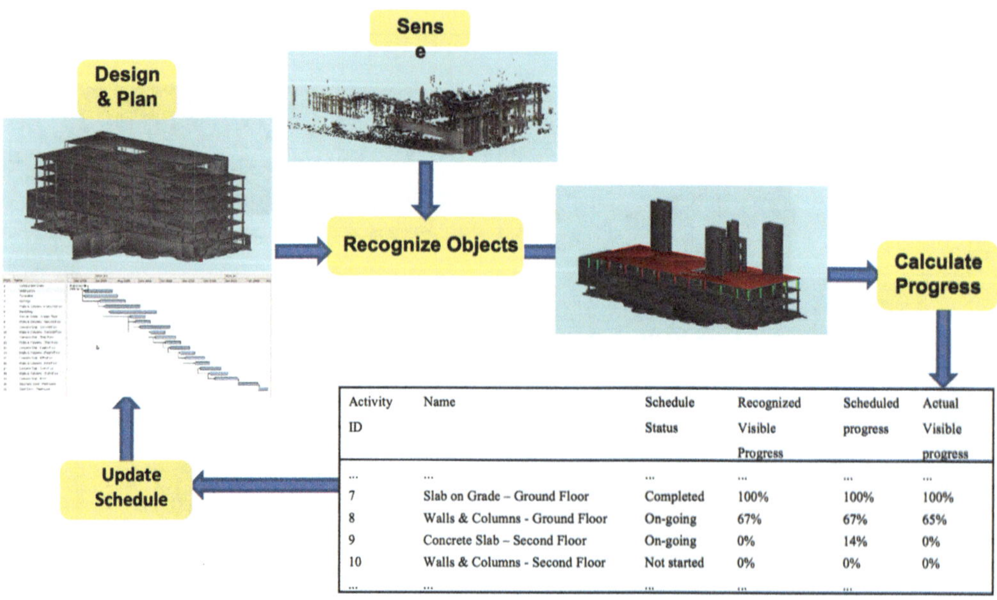

Figure 14-29. Automated progress tracking using 4D BIM and lidar.
Source: Turkan et al. (2012), reproduced with permission from Elsevier.

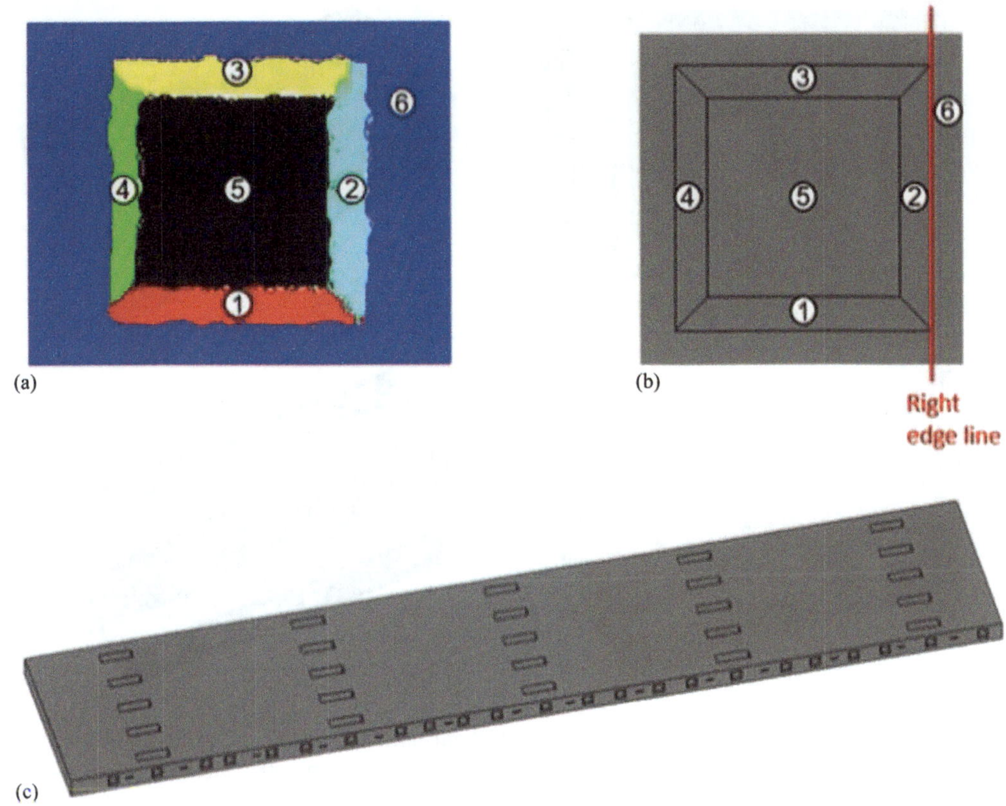

Figure 14-30. As-is BIM reconstruction of the precast panel: (a) matching scan points to different as-designed surfaces, (b) extraction of edge lines as the intersection lines of as-is planes, (c) reconstructed as-is BIM model of the precast panel.
Source: Wang et al. (2019), reproduced with permission.

(2015) provided a comprehensive overview of techniques to retrieve MEP components from lidar point clouds, while identifying challenges associated with it. Wang et al. (2019) developed an application-oriented framework and demonstrated it on two illustrative examples. Figure 14-30 shows the steps required to convert lidar point clouds (scans) into as-is BIM. Recent advances in laser scanning technologies and indoor navigation have enabled acquisition of distant and dense 3D information at very high speed, facilitating fast and accurate conversion of building structures into 3D as-built BIMs. Jung et al. (2015, 2016, 2017, 2018) proposed a framework that incorporates an indoor mobile mapping technology for efficient as-is BIM generation: (1) collecting data; (2) preprocessing data to filter and register the point-cloud data in a common coordinate system; (3) geometric modeling to represent the 3D building components with a simplified representative 3D shape, and finally (4) creating an as-built BIM to transform the low-level geometric model into a semantically rich BIM by attributing an object category, material properties, and topological relationships between objects. Figure 14-31 shows an example of the Scan-to-BIM process for indoor environments. Several research groups are currently exploring the feasibility of using deep learning algorithms to accelerate and fully automate the scan-to-BIM process to enable large amounts of lidar point clouds to be converted into BIMs, thus facilitating a wider adoption of BIMs in industry.

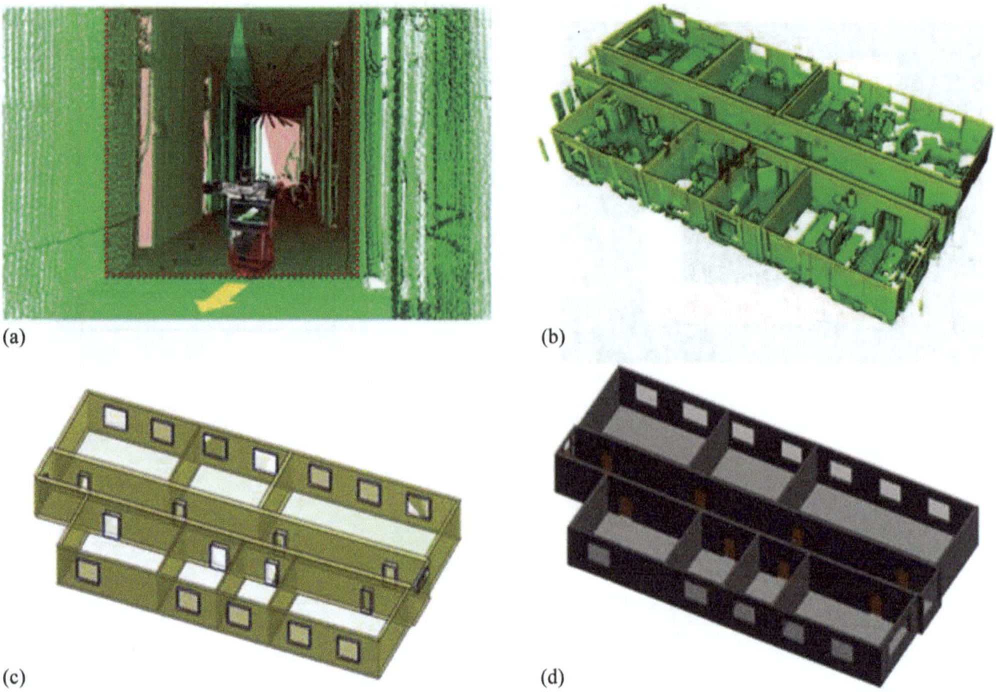

Figure 14-31. Process of Scan-to-BIM: (a) point-cloud data collection using an indoor mobile mapping system with a laser scanner, (b) data preprocessing, (c) geometric modeling, (d) as-built BIM creation.

REFERENCES

Al-Shalabi, F. A., Y. Turkan, and S. Laflamme. 2015. "BrIM implementation for documentation of bridge condition for inspection." In *Proc., Canadian Society for Civil Engineering 5th Int./11th Construction Specialty Conf.* Vancouver: University of British Columbia.

Azhar, S., M. Khalfan, and T. Maqsood. 2015. "Building information modelling (BIM): Now and beyond." *Constr. Econ. Build.* 12 (4): 15–28.

Berger, M., A. Tagliasacchi, L. M. Seversky, P. Alliez, G. Guennebaud, J. A. Levine, et al. 2017. "A survey of surface reconstruction from point clouds." *Comput. Graphics Forum* 36 (1): 301–329.

Bolkas, D., J. Chiampi, J. Chapman, and V. F. Pavill. 2020. "Creating a virtual reality environment with a fusion of sUAS and TLS point-clouds." *Int. J. Image Data Fusion* 11 (2): 136–161.

Bolkas, D., G. Fotopoulos, A. Braun, and I. N. Tziavos. 2016. "Assessing digital elevation model uncertainty using GPS survey data." *J. Surv. Eng.* 142 (3): 04016001.

Bosché, F., A. Guillemet, Y. Turkan, C. T. Haas, and R. G. Haas. 2014. "Assessing the value of a Scan-vs-BIM framework for tracking the built status of MEP works." *J. Comput. Civ. Eng.* 28 (4): 05014004.

Brando, G., D. Rapone, E. Spacone, M. S. O'Banion, M. J. Olsen, A. R. Barbosa, et al. 2017. "Damage reconnaissance of unreinforced masonry bearing wall buildings after the 2015 Gorkha, Nepal, earthquake." *Earthquake Spectra* 33 (S1): 243–273.

Brumana, R., D. Oreni, L. Barazzetti, B. Cuca, M. Previtali, and F. Banfi. 2020. "Survey and scan to BIM model for the knowledge of built heritage and the management of conservation activities." In *Digital transformation of the design, construction and management processes of the built environment*, edited by B. Daniotti, M. Gianinetto, and S. Della Torre, 391–400. Cham, Switzerland: Springer.

buildingSMART. 2021. Accessed April 13, 2021. https://www.buildingsmart.org/.

Burrough, P. A., and R. A. McDonell. 1998. *Principles of geographical information system*. Oxford, UK: Oxford University Press.

Bwail, K. 2018. "How has BIM changed MEP coordination?" Accessed April 13, 2021. https://medium.com/@xscad/how-has-bim-changed-mep-coordination-d63bd88ec8cc.

Caruso, C., and F. Quarta. 1998. "Interpolation methods comparison." *Comput. Math. Appl.* 35 (12): 109–126.

Das, D. 2019. "GeoBIM: Data formats & models for converging 3D GIS & BIM." Accessed April 13, 2021. https://avineonlab.com/geobim-data-formats-models-for-converging-3d-gis-bim/?lang=en.

DGGS (Alaska Division of Geological and Geophysical Surveys). Undated. "Division of Geological and Geophyscial Surveys Elevation portal." Accessed February 7, 2022. https://elevation.alaska.gov/.

Donkers, S. 2013. "Automatic generation of CityGML LoD3 building models from IFC models." Ph.D. thesis, Delft University of Technology, Dept. of Civil and Environmental Engineering.

Eastman, C., P. Teicholz, R. Sacks, and K. Liston. 2011. *BIM handbook: A guide to building information modeling for owners, managers, designers, engineers and contractors*. Chichester, UK: Wiley.

Esri. 2020. "The geodatabase is object relational." Accessed February 7, 2022. https://desktop.arcgis.com/en/arcmap/latest/manage-data/gdb-architecture/the-geodatabase-is-object-relational.htm.

Esri. 2021. "The geodatabase." Accessed February 7, 2022. https://www.esri.com/en-us/arcgis/geodesign.

FCC (Federal Communications Commission). Undated. "Federal Information Processing System (FIPS) Codes for States and Counties." Accessed February 7, 2022. https://transition.fcc.gov/oet/info/maps/census/fips/fips.txt.

Fanning, B., C. M. Clevenger, M. E. Ozbek, and H. Mahmoud. 2015. "Implementing BIM on infrastructure: Comparison of two bridge construction projects." *Pract. Period. Struct. Des. Constr.* 20 (4): 04014044.

FHWA (Federal Highway Administration). 2020. "Tech brief Hybrid Approach to BIM in Iowa." Accessed February 7, 2022. https://www.fhwa.dot.gov/construction/bim/pdfs/fhwa_hif_20_084.pdf.

Grani, H. K. 2016. "What is IFC and what do you need to know about it?" Accessed April 13, 2021. https://blog.areo.io/what-is-ifc/.

Guo, F., C. T. Jahren, Y. Turkan, and H. D. Jeong. 2017. "Civil integrated management: An emerging paradigm for civil infrastructure project delivery and management." *J. Manage. Eng.* 33 (2): 04016044.

Guo, F., Y. Turkan, C. T. Jahren, and H. D. Jeong. 2014. "Civil Information Modeling adoption by Iowa and Missouri DOT." In *15th Int. Conf. in Computing in Civil Engineering and Built Environment*, edited by R. I. Issa and I. Flood, 463–471. Reston, VA: ASCE.

Guo, Q., W. Li, H. Yu, and O. Alvarez. 2010. "Effects of topographic variability and lidar sampling density on several DEM interpolation methods." *Photogramm. Eng. Remote Sens.* 76 (6): 701–712.

Hagan, P. S., and G. West. 2006. "Interpolation methods for curve construction." *Appl. Math. Finance* 13 (2): 89–129.

Hong, S., J. Jung, S. Kim, H. Cho, J. Lee, and J. Heo. 2015. "Semi-automated approach to indoor mapping for 3D as-built building information modeling." *Comput. Environ. Urban Syst.* 51: 34–46.

Hu, Y., D. Castro-Lacouture, and C. M. Eastman. 2019. "Holistic clash detection improvement using a component dependent network in BIM projects." *Autom. Constr.* 105: 102832.

Janis, R. R., and W. K. Tao. 2005. *Mechanical and electrical systems in buildings*. Hoboken, NJ: Pearson/Prentice Hall.

Jung, J., S. Hong, S. Yoon, J. Kim, and J. Heo. 2016. "Automated 3D wireframe modeling of indoor structures from point clouds using constrained least-squares adjustment for as-built BIM." *J. Comput. Civ. Eng.* 30 (4): 04015074.

Jung, J., C. Stachniss, S. Ju, and J. Heo. 2018. "Automated 3D volumetric reconstruction of multiple-room building interiors for as-built BIM." *Adv. Eng. Inf.* 38: 811–825.

Jung, J., C. Stachniss, and C. Kim. 2017. "Automatic room segmentation of 3D laser data using morphological processing." *ISPRS Int. J. Geo-Inf.* 6 (7): 206.

Jung, J., S. Yoon, S. Ju, and J. Heo. 2015. "Development of kinematic 3D laser scanning system for indoor mapping and as-built BIM using constrained SLAM." *Sensors* 15 (10): 26430–26456.

Kalogianni, E., P. van Oosteom, E. Dimopoulou, and C. Lemmen. 2020. "3D land administration: A review and a future vision in the context of the spatial development lifecycle." *ISPRS Int. J. Geo-Inf.* 9 (2): 107.

Korman, T. M., L. Simonian, and E. Speidel. 2008. "Using building information modeling to improve the mechanical, electrical, and plumbing coordination process for buildings." In *Architectural Engineering Conf. 2008: Building Integration Solutions*, edited by M. Ettouney, 1–10. Reston, VA: ASCE.

Korman, T. M., L. Simonian, and E. Speidel. 2010. "How building information modelling has changed the MEP coordination process." In *Challenges, Opportunities and Solutions in Structural Engineering and Construction*, 959–963. London: Taylor & Francis.

Liu, W., H. Guo, H. Li, and Y. Li. 2014. "Retracted: Using BIM to improve the design and construction of bridge projects: A case study of a long-span steel-box arch bridge project." *Int. J. Adv. Rob. Syst.* 11: 125.

Longley, P. A., M. F. Goodchild, D. J. Maguire, and D. W. Rhind. 2005. *Geographic information systems and science*. Chichester, UK: Wiley.

Majcher, J. 2019. "Everything worth knowing about the IFC format." Accessed April 13, 2021. https://bimcorner.com/everything-worth-knowing-about-the-ifc-format/.

Marzouk, M., and M. Hisham. 2012. "Applications of building information modeling in cost estimation of infrastructure bridges." *Int. J. of 3-D Inf. Model.* 1 (2): 17–29.

Mehrbod, S., S. Staub-French, N. Mahyar, and M. Tory. 2019. "Beyond the clash: Investigating BIM-based building design coordination issue representation and resolution." *J. Inf. Technol. Construct.* 24: 33–57.

Muñoz, J. N., F. V. Manzanares, and M. M. Gonçalves. 2020. "BIM approach to modeling a sports pavilion for university use." *Appl. Sci.* 10: 8895.

NBS (National Bureau of Standards). 1987. "Codes for the identification of the states, the District of Columbia and the outlying areas of the United States, and associated areas." Accessed February 7, 2022. https://nvlpubs.nist.gov/nistpubs/Legacy/FIPS/fipspub5-2.pdf.

Polat, N., M. Uysal, and A. S. Toprak. 2015. "An investigation of DEM generation process based on LIDAR data filtering, decimation, and interpolation methods for an urban area." *Measurement* 75: 50–56.

Price, C. V., N. Nakagaki, K. J. Hitt, and R. M. Clawges. 2006. *Enhanced historical land-use and land-cover data sets of the US Geological Survey*. Data Series 240. Washington, DC: USGS, US Dept. of the Interior.

Puri, N., and Y. Turkan. 2020. "Bridge construction progress monitoring using lidar and 4D design models." *Autom. Constr.* 105: 102832.

Rishikeshan, C. A., S. K. Katiyar, and V. V. Mahesh. 2014. "Detailed evaluation of DEM interpolation methods in GIS using DGPS data." In *Proc., 2014 Int. Conf. on Computational Intelligence and Communication Networks*, 666–671. New York: IEEE.

Saul, L. 2021. "Light box." Accessed April 13, 2021. https://www.instructables.com/light-box/.

Schwendel, A. C., I. C. Fuller, and R. G. Death. 2012. "Assessing DEM interpolation methods for effective representation of upland stream morphology for rapid appraisal of bed stability." *River Res. Appl.* 28 (5): 567–584.

Shim, C. S., N. R. Yun, and H. H. Song. 2011. "Application of 3D bridge information modeling to design and construction of bridges." *Procedia Eng.* 14: 95–99.

Sierra Aparicio, M. V. 2020. "Conceptual framework for incorporating access for maintainability considerations in BIM coordination." M.S. thesis, Georgia Institute of Technology, School of Civil and Environmental Engineering.

Son, H., C. Kim, and Y. Turkan. 2015. "Scan-to-BIM—An overview of current state of the art and a look ahead." In *Proc., 32nd Int. Symp. on Automation and Robotics in Construction*. Oulu, Finland. https://www.iaarc.org/.

Tobiáš, P. 2015. "An investigation into the possibilities of BIM and GIS cooperation and utilization of GIS in the BIM process." *Geoinformatics FCE CTU* 14 (1): 65–78.

Turkan, Y., F. Bosché, C. T. Haas, and R. G. Haas. 2014. "Tracking of secondary and temporary objects in structural concrete work." *Constr. Innovation Inf. Process Manage.* 14 (2): 145–167.

Turkan, Y., F. Bosché, C. T. Haas, and R. G. Haas. 2012. "Automated progress tracking using 4D schedule and 3D sensing technologies." *Autom. Constr.* 22 (2): 414–421.

Turner, J. M. 2008. "Announcing approval of the withdrawal of ten Federal Information Processing Standards (FIPS)." *NIST*. Accessed February 7, 2022. https://www.govinfo.gov/content/pkg/FR-2008-09-02/pdf/E8-20306.pdf.

United BIM. 2021. "MEP coordination- everything you need to know | importance & bene." Accessed April 13, 2021. https://www.united-bim.com/mep-coordination-bim-importance-benefits-outsourcing/.

USCB (United States Census Bureau). 2020a. "American National Standards Institute (ANSI) and Federal Information Processing Series (FIPS) Codes." Accessed February 7, 2022. https://www.census.gov/library/reference/code-lists/ansi.html.

USCB. 2020b. "Census Interactive Gallery." Accessed February 7, 2022. https://www.census.gov/library/visualizations/interactive.2020.html.

Volk, R., J. Stengel, and F. Schultmann. 2014. "Building information modeling (BIM) for existing buildings—Literature review and future needs." *Autom. Constr.* 38: 109–127.

Xu, Y., and Y. Turkan. 2019. "BrIM and UAS for bridge inspections and management." *Eng. Constr. Archit. Manage.* 27 (3): 785–807.

Yarmohammadi, S., and B. Ashuri. 2015. "Exploring the approaches in the implementation of BIM-based MEP coordination in the USA." *J. Inf. Technol. Construct.* 20 (22): 347–363.

Wang, Q., J. Guo, and M.-K. Kim. 2019. "An application-oriented scan-to-BIM framework." *Remote Sens.* 11 (3): 365.

Zhu, J., G. Wright, J. Wang, and X. Wang. 2018. "A critical review of the integration of geographic information system and building information modelling at the data level." *ISPRS Int. J. Geo-Inf.* 7 (2): 66.

Zhu, J., P. Wu, M. Chen, M. J. Kim, X. Wang, and T. Fang. 2020. "Automatically processing IFC clipping representation for BIM and GIS integration at the process level." *Appl. Sci.* 10 (6): 2009.

CHAPTER 15

PROFESSIONAL SERVICES AND DESIGN PROFESSIONALS' AGREEMENTS

David E. Woolley, Lisa D. Herzog

INTRODUCTION

As stated in *Hydrotech Systems, Ltd., V. Oasis Waterpark* (1991), the licensed design professional is required to have a rudimentary understanding of contract administration. A design professional's contract with their client is perhaps the most important document in a construction project because it defines the responsibilities and rights of the parties in connection with the project, *A&E Contract Review Guide* [Beasley n.d., p. 3, citing *Hydrotech Systems, Ltd., V. Oasis Waterpark* (1991) 52 Cal. 3d 998]. The design professional can either incur or avoid some liabilities based on the contract language particularly regarding the indemnity of others [Informed consent is a person's agreement to allow something to happen after the person has been informed of all the risks involved and the alternatives (Black's Law Dictionary, 9th ed. 2009, p. 346, col. 1).]. The required standard of care is usually directly linked to the scope of work.

It is equally important for a design professional to contractually agree as to what services will be provided and what services will not be provided. A design professional must understand that it is never acceptable to use a contractual agreement to violate the law or perform a negligent service that can harm an individual or the public. The protection that is afforded to a design professional through a written contract with a client is known as *privity of contract*. [Privity of contract is defined as "[t]he relationship between the parties to a contract, allowing them to sue each other but preventing a third party from doing so" (Black's Law Dictionary, 9th ed. 2009, p. 1,320, col. 1).]

This chapter discusses the basic elements of a legal contract, followed by special considerations related to design professional contracts, some basic contract clauses that should be included in a design professional's agreement, an example of a standard design professional's agreement [see Appendix: Exhibit A, the Engineers Joint Contract Documents Committee (EJCDC) Agreement between Engineer and Subconsultant for Professional Services 2020], and a specific clause to include in a design professional's employment agreement with their employer.

CONTRACTS 101—THE BASIC LEGAL PRINCIPLES

Individuals make casual promises all the time that may or may not be enforceable (Mallor 2013, *Introduction to Contracts*). Contract law helps us to determine which agreements are "worthy" of legal enforcement (Mallor 2013). In general, a person or a business is free to make whatever agreements they choose, but this freedom is not limitless. A contract is a legally enforceable agreement (Mallor 2013). A contract, in general, is defined as a legally enforceable exchange of promises or an exchange of a promise for an act that assures the parties to the agreement that their promises will be enforceable (Kubasek et al. 2017). Parties form a contract only if one side makes an offer and the other side accepts the offer. For a contract to be enforceable by a civil court of law, the following seven characteristics must be present (Mallor 2013): offer, acceptance, consideration, content, capacity, legality, and writing.

Offer

Offers include proposals to make a deal for buying something, selling something, providing services, or anything else. Only proposals made in certain ways amount to a legally recognized offer. The person making the offer is the "offeror," and the person receiving the offer is the "offeree." (Mallor 2013, p. 334). With an offer, the offeree should ask two questions: (1) Do the offeror's words and actions indicate a present intention to make a contract? and (2) Are the terms of the offer reasonably definite so that the offeree can understand what is being offered? (Mallor 2013, pp. 335–339).

For a valid contract to be formed, an offer must show the objective intent to enter into the contract (Kubasek et al. 2017, p. 257). Courts will look at the words, conduct, writing, and, in some cases, deliberate omissions of the offer (Kubasek et al. 2017). The offer must also be definite—there must be some reference to the subject matter, the number of items being offered, and the price of these items (Kubasek et al. 2017). The offer must also be communicated to the offeree intended by the offeror (Kubasek et al. 2017).

Offers are terminated by (1) revocation of the offer by the offeror, (2) rejection of the offer by the offeree, (3) a counter offer from the offeree (rejecting the first offer and making a new counter offer), (4) expiration of time (if specified in the offer or a "reasonable time" based on custom and practice between the parties or in the industry), (5) operation of law such as if the offeror dies or becomes incapacitated or the subject matter of the contract is destroyed (Mallor 2013, pp. 343–346). See also Kubasek et al. 2017, p. 259.

Acceptance

Acceptance is "a manifestation of assent to the terms of the offer made by the offeree in the manner invited or required by the offer." [Mallor 2013, p. 352, quoting Restatement (Second) of Contracts § 50(1) (1981)]. In determining if an offeree accepted an offer and created a contract, a court will look for evidence of three factors: (1) the offeree intended to enter the contract, (2) the offeree accepted on the terms proposed by the offeror, and (3) the offeree communicated his acceptance to the offeror (Mallor 2013, pp. 352–366). When a valid offer is outstanding, it remains effective until it is terminated or accepted (Mallor 2013).

Consideration

One of the conditions that separates a contract from an unenforceable social promise is that a contract requires voluntary agreement by two or more parties (Mallor 2013, p. 371). The concept of consideration distinguishes agreements that the law will enforce (contracts) from gratuitous promises (gifts), which are normally unenforceable (Mallor 2013). A common definition of consideration is "legal value, bargained for and given in exchange for an act or a promise." (Mallor 2013, p. 372). See also Kubasek et al. (2017, p. 263). A promise typically cannot be enforced against the person who made it (the "promisor") unless the person to whom the promise was made (the "promisee") has given up something of legal value in exchange for the promise (Mallor 2013).

For a valid contract to be formed, there must be a bargaining that leads to an exchange of value between the parties (the "offeror" and the "offeree") (Mallor 2013, p. 374). This value can be money for goods or services or an exchange of goods/services for other goods/services. Contracts cannot be a one-way street—both sides must receive some measurable benefit. If only one side receives a benefit but gives nothing up, a gift is given, but no contract is formed. Consideration is proof that the parties intended to be bound by their promises. There are several categories of agreements that do not constitute contracts, such as *illusory promises, preexisting duties,* and *past consideration* (Mallor 2013, pp. 375–381). Further discussion of these legal concepts is beyond the scope of this chapter.

Consent

The parties to a contract must give their consent to contract freely. A party cannot lack essential information that will be required to give informed and free consent (Mallor 2013, pp. 388–389). A contracting party who is subject to misrepresentation [*Misrepresentation* is defined as "the act of making a false or misleading assertion about something usually with the intent to deceive." An innocent misrepresentation is "a false statement that the speaker does not know is false; a misrepresentation that, though false, was not made fraudulently" (Black's Law Dictionary, 9th ed. 2009, p. 1,091, cols. 1 and 2)], fraud [*Fraud* consists of the following elements: (1) misrepresentation of a material (important) fact, (2) made with the intent to deceive the other party, (3) who reasonably relies on the misrepresentation, and (4) as a result is injured (Kubasek et al. 2017, p. 265)], mistake, duress [*Duress* is defined as any wrongful act or threat that prevents a party from exercising free will when executing a contract (Kubasek et al. 2017, p. 265)], or undue influence [If a party exerts mental coercion over another party, there is undue influence and no genuine assent. There are two requirements for undue influence: (1) There must be a dominant-subservient relationship between the contracting parties, and (2) This dominant-subservient relationship must allow one party to influence the other in a mentally coercive way (Kubasek et al. 2017, p. 267).] by the other contracting party can avoid and cancel (rescind) the contract because their consent was not real (Mallor 2013, pp. 388–389).

Capacity

A person must have the ability to give consent before they can be legally bound to an agreement (Mallor 2013, p. 402). A party must be competent (the ability to understand the nature of the transaction and the consequences of entering into it at the time the contract was entered into) (Kubasek et al. 2017, p. 267). For truly voluntary agreements to exist, the ability to give consent must involve a person being mature (adult) enough and having the mental ability to be capable of representing their own interests (Mallor 2013, p. 402). Capacity means the ability to incur legal obligations and acquire legal rights (Mallor 2013).

All parties to the contract must be of sound mind and must not be minors (age dictated by state law) (Mallor 2013, pp. 402–403). "Sound mind" means the person entering the contract is not mentally ill (meaning they do not understand the nature of their actions or may be adjudicated incompetent) or so intoxicated that they do not understand what they are doing when they enter the contract (this is a very tough standard to meet. because courts do not want a contracting party to claim intoxication to get out of a contract) (Mallor 2013). Adults of sound mind have the capacity to enter into contracts.

Legality

The contract's subject matter must be for a lawful purpose (Kubasek et al. 2017, p. 269). When an agreement involves an act or promise that violates some legislative or court-made rule, the public interests threatened by the agreement outweigh any interests that favor the agreement's enforcement (Mallor 2013, p. 416). The court will not help you enforce a contract that violates a statute or an important public policy. For example, a court will not help an individual enforce a contract for the sale of illegal drugs.

Writing

Many verbal agreements are enforceable (Mallor 2013, p. 434); however, attorneys always prefer that all contracts are in writing so that there is a clear understanding of the agreement between the parties, and it is memorialized—verbal contract disputes often lead to memory loss or ambiguity. In situations relating to claims, verbal agreements quickly deteriorate into a "he said, she said" fight, and the design professional has no written document to rely on (Beazley, *A&E Contract Review Guide* n.d., p. 3).

Certain types of contracts must be in writing to be enforceable. These contracts include the following: (1) contracts for an interest in land (mortgage, easement, sale, lease), (2) contracts for services that cannot be performed within one (1) year, (3) contracts to pay the debt of someone else, (4) contracts made by the executor of an estate to pay a debt of the estate, (5) contracts made in consideration of marriage, (6) contracts for the sale of goods for amounts of $500 or more (Mallor 2013, *Introduction to Contracts*, p. 435).

These six elements of a basic contract are simplified for the purposes of this chapter. To obtain more information about the elements of a valid contract, please refer any college-level business law textbook chapter(s) on contracts. Excellent references include Mallor (2013), Kubasek et al. (2017), and Beatty and Samuelson (2016).

KEY PROVISIONS FOR DESIGN PROFESSIONALS' CONTRACTS

Design professionals should be identified as such in written contracts rather than as *contractor* or *subcontractor* (Beazley n.d., p. 4). Acceptable alternatives to design professionals are the terms consultant or subconsultant (Beazley n.d., p. 4). These designations accurately reflect design professionals' status and their role in providing professional services, as opposed to the unacceptable designations that connotate the role of general contractors (Beazley n.d.) Once parties to the contract are defined, make sure the same designations are applied throughout the agreement and exhibits (Beazley n.d.).

California statutes provide a good basic framework of what must be included in a basic design professional's contract. According to the California Professional Land Surveyors Act, California Business & Professions Code § 8759, provisions that must be included in a professional land surveyor's written contract include the following:

> (a) A licensed land surveyor or registered civil engineer authorized to practice land surveying shall use a written contract when contracting to provide professional services to a client pursuant to this chapter. The written contract shall be executed by the licensed land surveyor or registered civil engineer and the client, or his or her representative, prior to the licensed land surveyor or registered civil engineer commencing work, unless the client knowingly states in writing that work may commence before the contract is executed. The written contract shall include, but not be limited to, all of the following:

1. A description of the services to be provided to the client by the licensed land surveyor or registered civil engineer.
2. A description of any basis of compensation applicable to the contract, and the method of payment agreed upon by the parties.
3. The name, address and license or certificate number of the licensed land surveyor or registered civil engineer, and the name and address of the client.
4. A description of the procedure that the licensed land surveyor or registered civil engineer and the client will use to accommodate additional services.
5. A description of the procedure used by any party to terminate the contract.

Certifications, Guarantees, and Warranties

"Design professionals should not give warranties, guarantees or certifications of their services" (Beazley n.d., p. 5). Design professionals provide services rather than goods and should be held to a negligence-based standard of care based on what a reasonable design professional (of the same type, i.e., architect, engineer, land surveyor) would do under same or similar circumstances (Beazley n.d.). Language that requires a design professional to guarantee or warranty services may elevate the standard of care to a perfection standard—much higher than the negligence standard (Beazley n.d.).

If a client requires the design professional to provide any certifications for a project, the certification should be to the usually accepted professional standard of care, and the language in the contract should certify "to the best of the design professional's information, knowledge and belief" (Beazley n.d.).

Practice Tips:

- Delete guarantees and warranties.
- Avoid words such as "all," "every," "insure," "ensure," "assure."
- Do not certify construction that was built by a contractor in compliance with code or in strict accordance with plans and specifications.
- Do not certify that the project will achieve any particular Leadership in Energy and Environmental Design (LEED) standard or sustainable design goal (Beazley n.d., p. 5).

Incorporation by Reference of Another Contract or Document

Commonly, a design professional's contract will incorporate by reference a *prime agreement* between the design professional's client and the owner (Beazley n.d., p. 7). Incorporating a document by reference makes the document part of the design professional's agreement with the same force and effect as if the document had been part of the design professional's contract (Beazley n.d.).

Incorporating any document by reference can create a significant risk management and liability issue for the design professional (Beazley n.d.). When this happens, the design professional must obtain and review the incorporated document to make sure that it does not contain inappropriate language that may increase the design professional's standard of care or include provisions that create insurability issues (inappropriate indemnity provisions or warranties and guarantees) (Beazley n.d.). In addition, the design professional should add language to their professional services agreement stating the following (Beazley n.d.):

> In the event of any discrepancy between the terms and conditions of the consultant agreement and the prime contract (or other document incorporated by reference), the consultant agreement will control.

Practice Tips:

- If the client incorporates any document by reference, review and confirm that the terms and conditions of that document are acceptable and consistent with the design professional's agreement.
- If a client refuses to allow a design professional's agreement to govern, list in detail within the design professional's agreement which provisions of the referenced document are not incorporated, or list in detail the provisions that the design professional agrees are incorporated (Beazley n.d., p. 8).

A Design Professional's Indemnity Obligation Must Be Negligence Based

"Indemnity is an agreement whereby one party agrees to assume the liability of another in the event of loss." (Beazley n.d., p. 8). California Civil Code § 2772 defines *indemnity* as a contract by which one engaged to save another from a legal consequence of the conduct of one of the parties, or of some other person. See also Black's Law Dictionary, 9th ed. (2009, p. 837, col. 2 (*indemnity clause*—contractual provision in which one party agrees to answer for any specified or unspecified liability or harm that the other party might incur; also termed *hold harmless clause* or *save harmless clause*). The concept of reimbursing a client for losses caused by the design professional seems fair; however, too often, indemnity provisions are too one-sided in favor of the client and require the design professional to assume the liability of the client regardless of fault (Beazley n.d.). Design professionals should reject these broadly written indemnity provisions and negotiate revisions so that their indemnity exposure is limited to the extent damages are caused by their negligent performance of services under the agreement (Beazley n.d.). Otherwise, if indemnity provisions are not appropriately negligence based, design professionals may be exposed to liability beyond the scope of their insurance coverage (Beazley n.d.).

Including language requiring the design professional to indemnify for their "recklessness, wrongful acts, intentional misconduct, willful misconduct" (or similar terms) can potentially expose them to liability beyond insurance coverage (Beazley n.d.).

Clients may attempt to include "breach of contract" among the laundry list of circumstances under which the design professional is required to contractually assume a duty to indemnify (Beazley n.d., p. 9). From a risk management and professional liability view, language requiring a design professional to indemnify a client for their breach of contract creates a huge additional exposure (Beazley n.d.). Most third-party claims will arise out of a performance-based issue by the design professional and will boil down to the allegations that, because the design professional breached the contract, they failed to meet the requisite standard of care and were negligent, thereby triggering their liability insurance (Beazley n.d.).

In addition, if a breach of contract claim does not arise out of the design professional's performance, the design professional's contractual obligation to indemnify the client for breach of contract may be beyond the coverage provided in a negligence-based professional liability insurance policy (Beazley n.d.).

Standard of Care in Negligence

The *standard of care* is the standard by which a design professional's performance is judged, and it is not a standard of perfection (Beazley n.d., p. 13). Instead, the standard for negligence by the design professional is "what a reasonable, practicing design professional would do under similar circumstances" (Beazley n.d.). Appropriate language that the design professional should include to clarify this point is as follows (Beazley n.d.):

> The Design Professional's services shall be performed in a manner consistent with that degree of skill and care ordinarily exercised by practicing design professionals performing similar services in the same locality, at the same site and under the same or similar circumstances and conditions. The Design Professional makes no other representations or warranties, whether expressed or implied, with respect to the services rendered hereunder.

Practice Tips:

- Make sure standard of care is reasonable and negligence based.
- Standard of care should be limited to the skill, care, and judgment ordinarily exercised by similarly situated design professionals performing the same services.
- Delete language that elevates the standard of care beyond ordinary, reasonable standards. Examples of words not to use are "highest, best, first-class and first rate."
- Delete "to the satisfaction of the client, in the client's judgment or non-negligent manner." (Beazley n.d., p. 14).

Indemnity—Duty to Defend

Design professionals should not extend the *duty to defend* in their professional services agreements. According to the California Court of Appeals (Court of Appeal of California, Sixth District. 2010) in *UDC-Universal Development, L.P. v. CH2M Hill*, the "duty to defend" should be specifically excluded from these agreements (*UDC-Universal Development, L.P., v. CH2M Hill* 181 Cal. App. 4th 10.).

The duty to defend is much broader than the duty to indemnify and can arise immediately, regardless of any actual or even alleged negligence by the design professional (email from Mike Dewitz-Cryan, RLI Corporation to Alicia Ingram of IOA USA dated December 29, 2020; Beazley n.d., p. 10). The word *defend* raises significant insurability issues, regardless of the insurance company involved (Beazley n.d., p. 10). The duty to defend does not depend on the outcome of the claim, whereas the duty to indemnify does not arise unless the outcome of a claim is adverse (ACEC Risk Management Committee n.d., p. 6). Because the duty to defend and the duty to indemnify are distinct obligations, the contract may impose a duty to defend the underlying claim even in the absence of a duty to indemnify [ACEC Risk Management Committee n.d., citing *Hollingsworth, v. Chrysler Corp.* (Del. 1965) 208 A. 2d 61)].

According to Beazley,

> Accordingly, when a design professional has a duty to defend, the design professional may be required to defend a claim based upon a mere allegation of negligence, unlike a duty to indemnify which is triggered by actual negligence. The duty to defend a client may be interpreted as a contractual obligation rather than an obligation triggered by adjudication of the design professional's negligence. As a contractual obligation, the duty to defend would not be covered by the design professional's professional liability insurance policy (Beazley n.d., p. 10). (See also *UDC-Universal Development, L.P., v. CH2M Hill* 181 Cal. App. 4th 10, Court of Appeal of California, Sixth District. 2010.)

Practice Tips:

- Delete the duty to defend.
- Delete "claims, suits, causes of action, actions, demands, allegations" because these words suggest a duty to defend (Beazley n.d., p. 10).

Indemnity—Limit Indemnitees

Similarly, indemnitees (those insured by the design professional) should be limited to the client, their affiliates (only if specified by name), and their respective officers, directors, and employees. Other entities or individuals such as agents, volunteers, representatives, successors and assigns, attorneys, lenders, consultants, subconsultants, contractors, subcontractors, servants, divisions, subsidiaries, parent companies, and related and affiliated companies are too broad (unpublished letter from Barbara Sable, Professional Enterprise Risk Solutions, RLI Professional Services Group to Alicia Igram, IOA Insurance Services, March 22, 2017). Often, a design professional's insurance carrier will not provide liability insurance coverage for these additional entities because they are too broad and not in privity of contract with the design professional (see, in general, Beazley n.d., p. 9).

Instead, parties to a contract, indemnitees, and those requesting additional insured status should be specific (unpublished letter from Barbara Sable Professional Enterprise Risk Solutions, RLI Professional Services Group to Alicia Igram, IOA Insurance Services, March 22, 2017). The extension of indemnity and additional insured status should be limited to parties in the contractual chain between the owner/client and the design professional's firm providing services. Clients and design professionals should be able to identify the name of any entity or individual within the contractual chain and to whom the clients deem it necessary for the policyholder (design professional) to indemnify and provide additional insured status. Other terms to consider are "partners" (when contracting with a partnership or LLP) or "member" (when contracting with an association or an LLC).

Practice Tips:

- Limit the indemnity obligation "to the extent damages are caused by the design professional's negligence."
- Avoid broad definitions of indemnitees (Beazley n.d., p. 10).

Liability to Owners Based on Claims by Third Parties

According to Beazley's *A&E Contract Review Guide*, by specifically including third parties in the "indemnitees" definition, these parties may be able to establish a third-party beneficiary status, at least with respect to the indemnity provision (Beazley n.d., p. 8). Some contracts include a "third-party beneficiaries" provision that has language along the lines of "Except with respect to the indemnity obligations, there are no third-party beneficiaries to this agreement" (Beazley n.d., pp. 8–9).

Practice Tips:

- Delete any third-party beneficiaries.
- Design professional's contract should explicitly disclaim third parties.
- Design professionals cannot name additional insureds on workers' compensation or professional liability insurance.
- Professional liability insurance is *per claim* and not *per occurrence.*
- Delete the requirement that a professional liability insurer must provide notice to the client if the aggregate insurance limits available for claims decrease or erode (Beazley n.d., pp. 21, 25).

Indemnity—Joint and Several Liability

Design professionals should be careful not to agree to overly broad liability coverage. For example, a design professional should not agree to joint and several liability coverages in a contract. Black's Dictionary (Black's Law Dictionary, 9th ed., 2009, p. 997, col. 2). defines *joint and several liability* as follows:

> Liability that may be apportioned either among two or more parties or to only one or a few selected members of the group, at the adversary's discretion. Thus, each liable party is individually responsible for the entire obligation, but a paying party may have a right to contribution and indemnity from nonpaying parties.

Accepting joint and several liability means the design professional may be stuck paying damages for the actions of another contractor with the client. To obtain contribution from another contractor, the design professional will need to sue that contractor in court. This is too much of a burden for a design professional's firm to absorb.

Indemnity—Limits on Liability to Policy Limits

Typically, a design professional is not in a financial position to take on liability above insurance policy limits. Therefore, it is important to limit liability by using one of the following clauses:

"Article 1 – Limitations of Liability

1.01 Limitation of Design Professional's Liability:

A. Design Professional's Liability Limited to Stated Amount, or Amount of Design Professional's Compensation: *To the fullest extent permitted by Laws and Regulations, and notwithstanding any other provision of this Agreement, the total liability, in the aggregate, of Design Professional and Design Professional's Subcontractors, officers, directors, members, partners, agents, employees and Subconsultants, to Owner and anyone else claiming by, through or under Owner for any and all injuries, claims, losses, expenses, costs, or damages whatsoever (including but not limited to direct, indirect, special, incidental, punitive, exemplary, or consequential damages) arising out of, resulting from, or in any way related to the Project, to Design Professional's or its Subconsultants' or Design Professional's Subcontractor's services, or to this Agreement, from any cause or causes whatsoever, including but not limited to the negligence, professional errors or omissions, strict liability, breach of contract, indemnity obligations, or warranty express or implied, of Design Professional or Design Professional's officers, directors, members, partners, agents, employees or Subconsultants, or Design Professional's Subcontractors, will not exceed the total amount of $___________ or the total compensation received by the Design Professional under this Agreement, whichever is greater. Higher limits are available for an additional fee.*

B. Design Professional's Liability Limited to Amount of Design Professional's Compensation: *To the fullest extent permitted by Laws and Regulations, and notwithstanding any other provision of this Agreement, the total liability, in the aggregate, of Design Professional and Design Professional's officers, directors, members, partners, agents, employees, Subconsultants, and Design Professional's Subcontractors, to Owner and anyone claiming by, through or under Owner for any and all claims, losses, costs, or damages whatsoever (including but not limited to direct, indirect, special, incidental, punitive, exemplary, or consequential damages) arising out of, resulting from, or in any way related to the Project or the Agreement from any cause or causes, including but not limited to the negligence, professional errors or omissions, strict liability, breach of contract, indemnity obligations, or warranty express or implied of the Design Professional or the Design Professional's officers, directors, members, partners, agents, employees, Subconsultants, or Design Professional's Subcontractors, will not exceed the total compensation received by Design Professional under this Agreement.*

C. Design Professional's Liability Limited to Amount of Insurance Proceeds: *Design Professional shall procure and maintain insurance as required by and set forth in* ***Exhibit*** ____ *to this*

Agreement. Notwithstanding any other provision of this Agreement, and to the fullest extent permitted by Laws and Regulations, the total liability, in the aggregate, of Design Professional and Design Professional's officers, directors, members, partners, agents, employees, Subconsultants, and Design Professional's Subcontractors to Owner and anyone claiming by, through or under Owner for any and all claims, losses, costs, or damages whatsoever (including but not limited to direct, indirect, special, incidental, punitive, exemplary, or consequential damages) arising out of, resulting from, or in any way related to the Project or the Agreement from any cause or causes, including but not limited to the negligence, professional errors and omissions, strict liability, breach of contract, indemnity obligations, or warranty express or implied, of Design Professional or Design Professional's officers, directors, members, partners, agents, employees, Subconsultants, or Design Professional's Subcontractors (hereinafter "Owner's Claims"), will be limited to (1) responsibility for payment of all or the applicable portion of any deductibles, either directly to the Design Professional's insurers or in settlement or satisfaction, in whole or in part, of Owner's Claims and (2) total insurance proceeds paid on behalf of or to Design Professional by Design Professional's insurers in settlement or satisfaction of Owner's Claims under the terms and conditions of Design Professional's applicable insurance policies up to the amount of insurance required under this Agreement.

(1) Such limitation will not be reduced, increased, or adjusted on account of legal fees paid, or costs and expenses of investigation, claims adjustment, defense, or appeal.

(2) If no such insurance coverage is provided with respect to Owner's Claims, then the total liability, in the aggregate, of the Design Professional and the Design Professional's officers, directors, members, partners, agents, employees, Subconsultants, and Design Professional's Subcontractors, to Owner and anyone claiming by, through, or under Owner, for any and all such uninsured Owner's Claims will not exceed $___________." (Language copied from Engineers Joint Contract Documents Committee available at edcdc.org.)

Choosing one of these form clauses will help the design professional limit their liability to the owner of a project. Without one of these protective clauses, the design professional may be liable for damages far beyond what is available with their insurance coverage. Huge damage awards with no cap stated in an agreement can put a design professional out of business or force them into bankruptcy.

Ownership and Use of Design Professional's Work Including Copyright

Typically, the design professional keeps ownership of their work product except for those documents required to be filed with public agencies. With regard to copyrights, a good clause to include in a design professional's contract (Woolley and Associates Inc. 2021) is as follows:

> All work product identified in this Agreement as within the scope of Services of Consultant, shall be deemed protected as if such work product was within the protections against third-party use and disclosure of general copyright law in the United States as well as [state where contract is to be enforced], including common law and statutory law, whether or not such work product actually is so copyrighted and without regard to whether or not such copyright law actually applies to such work product.

A design professional's contract should also limit a client's use of work product. For example, the contract (Woolley and Associates Inc. 2021) should state the following:

> **Use of Work Product.** Client agrees not to use or permit any other person to use final maps, exhibits, legal descriptions, surveys or other work product prepared by Consultant, which work product is not final and which is not signed, and stamped or

> sealed by Consultant. Client agrees that Consultant is not responsible for any such use of non-final work product and waives any right to claim liability against Consultant therefor.
>
> Client further agrees that final work product is for the sole use of Client for the specific purpose described in this Agreement. Such final work product may not be altered or reproduced in any way nor used on any other project or for any other purposes than as specifically authorized by Consultant in writing prior to such use, alteration, or reproduction.

A design professional's contract should also specify what happens when a client attempts to change a design professional's work product. The following (Woolley and Associates Inc. 2021) is a good clause:

> **Changes in Work Product.** In the event the Client agrees to permit or authorizes changes in the documents prepared by the Consultant pursuant to this Agreement, to which changes Consultant has not previously consented to in writing. Client acknowledges that such changes and the effects thereof are not the responsibility of Consultant and Client agrees that Consultant is automatically released from any and all liability arising therefrom and further agrees to defend, indemnify and hold harmless Consultant, its officers, directors, principals, agents and employees from and against all claims, demands, damages or costs arising therefrom unless caused by the sole negligence or willful misconduct of Consultant.

Increasingly, clients are seeking to transfer the copyright of the design professional's work product into their own names. A design professional should insist on language clarifying that any such transfer of copyrights does not impair the design professional's future use of the product's standard design details (Beazley n.d., p. 12).

If a client seeks copyright of the design professional's work product, the latter should consider the following language (Beazley n.d.):

> Notwithstanding any other provision in this Agreement, the Design Professional staff shall not be in violation of this Agreement if the Design Professional utilizes any standard details that may be incorporated into the work product generated by the Design Professional in connection with this Project. The Client understands that regardless of any transfer of ownership or copyright rights granted to the Client pursuant to the terms of this Agreement, the Design Professional shall in no way be restricted or prohibited from future use of any such standard details.

Insurance coverages may not extend to damages related to alleged patents, copyrights, trade secrets, or intellectual property rights that go beyond the damages caused by the design professional's negligence (email from Mike Dewitz-Cryan, RLI Corporation to Alicia Ingram of IOA USA, December 29, 2020).

Practice Tips:

- Try to maintain copyright and ownership of work.
- Any reuse of the design professionals' work without their consent in other projects or modifications done to current projects should be at the client's risk without liability or legal exposure to the design professional.
- Include document defense and indemnity protection in favor of the design professional for reuse and modifications of instruments of service, regardless of whether the design professional transfers ownership rights to the client (Beazley n.d., p. 13).

Include a Hold Harmless Clause in the Design Professional's Contract

If a client will agree, it is always good to include a *hold harmless* provision in a design professional's agreement. The following (Woolley and Associates Inc. 2021) is an example of this clause:

> **Hold Harmless.** Client agrees to be solely and completely responsible for job-site conditions during the course of Consultant's performance, including safety of all persons and property; that this requirement shall apply continuously and not be limited to normal work hours; and Client further agrees to defend, indemnify and hold Consultant harmless from any and all liability, real or alleged, in connection therewith, except liability arising out of the sole negligence or willful misconduct of Consultant.

Scope of Work

"It is important that the design professional's agreement include a sufficiently detailed and defined scope of services so the parties fully understand what each party will, and will not, do during the course of the project." (Beazley n.d., p. 23). Almost all claims result from disappointed expectations by one of the parties (Beazley n.d.). A precisely drafted scope of services is a proactive way to manage a client's expectations and avoid disappointments regarding what a client can anticipate from the design professional on any given project (Beazley n.d.). A sufficient scope of work provision should emphasize clarity, definition, and competency (Beazley n.d.).

Practice Tips:

- Ensure that the scope of work is sufficiently detailed and defined.
- Delete broad language requiring *any and all services necessary*, *complete design services*, and *adequate to meet the needs of the project* (Beazley n.d., p. 23).

Other Important Areas to Cover in a Design Professional's Agreement

1. Alternative dispute resolution provisions (mediation or arbitration) (Beware of "*binding*" arbitration clauses. You are agreeing to give up your rights to a civil trial. You are agreeing to be bound by the arbitrator's award absent misconduct by the arbitrator.). In general, a design professional's agreement should include a dispute resolution provision requiring nonbinding mediation as a condition precedent to litigation (or to binding arbitration) (Beazley n.d., p. 6). Mediation allows the parties an opportunity to engage in a creative dispute resolution procedure that is confidential, voluntary, and nonbinding (Beazley n.d.).

(a) A sample mediation clause is as follows:

"Dispute Resolution

A. Unless otherwise required by Exhibit H, Engineer and Subconsultant shall resolve all disputes in the following manner:

1. Engineer and Subconsultant agree to negotiate all disputes between them in good faith for a period of 30 days from the date of notice, prior to invoking mediation.

2. Engineer and Subconsultant agree that they shall first submit any and all unsettled claims, counterclaims, disputes, and other matters in question between them arising out of or relating to this Agreement or the breach thereof ("Disputes") to mediation. Engineer and Subconsultant agree to participate in the mediation process in good faith. The process will be conducted on a confidential basis, and must be completed within 120 days.

3. If the parties fail to resolve a Dispute through negotiations under Paragraph 6.07.A.1 or mediation under Paragraph 6.07.A.2, then:

A. either or both may invoke the applicable procedures of Exhibit H for final resolution of Disputes.

B. If Exhibit H is not included, or if no final dispute resolution method is specified in Exhibit H, then the parties may exercise their rights at law." (See Appendix for Exhibit A, copy of the Engineers Joint Contract Documents Committee, Agreement Between Engineer and Subconsultant for Professional Services 2020.)

(b) A sample binding arbitration clause is as follows (Woolley & Associates, Inc. 2021):

> **Arbitration of Disputes.** Any dispute arising out of or related to this Agreement shall be resolved by binding arbitration and not in a court of law. The dispute will be settled in accordance with the Rules of the American Arbitration Association, and judgment will be entered on the award. The arbitrator will award reasonable attorney's fees to the prevailing party. If a party, after due notice, fails to appear at or participate in the proceedings, the arbitrator will make an award based on the evidence presented by the party who does participate.

2. Attorney's fees provisions including the term "reasonable."

A sample attorney's fees provision is (Woolley and Associates 2021)

> **Costs of Dispute Resolution.** In the event that Client institutes a proceeding against Consultant, either directly or by way of cross-complaint, including a claim for indemnity, for alleged negligence, error, omission, or other failure to perform, wherein: (a) Client fails to obtain a judgment or award in Client's favor, (b) the action is dismissed, or (c) judgment or award is rendered for Consultant, Client agrees to pay Consultant immediately following the proceedings all costs of defense, including, but without limitation, reasonable attorney's fees, expert witness fees, court costs and any and all other reasonable expenses of defense.

3. Provisions clearly outlining the responsibility of the design professional and the owner.

4. Clearly written payment/compensation provisions for the design professional along with a reasonable time frame for payment.

5. Choice of law provision indicating where a civil case will be heard—state or federal court (depending on which court(s) have jurisdiction), which state and what county within the state will the case be filed and tried.

6. Client's approval—if this is included in a contract, this clause should also include *"such approval shall not be unreasonably withheld."*

7. Performance of others—a client should acknowledge in the agreement that the design professional is not responsible for the performance of services by third parties, including, but not limited to, design professionals in other disciplines (e.g., engineers, architects, surveyors, subcontractors, or suppliers).

8. Integration Clause/Entire Agreement—a statement that the written agreement is the final expression of the agreement between the parties. The following is an example of this clause:

> **Entire Agreement**. This Agreement contains the entire agreement between Client and Consultant relating to the project and the provision of services to the project. Any prior agreements, promises, negotiations or representations not expressly set forth in this Agreement are of no force or effect. Subsequent modifications to this Agreement shall be in writing and signed by both Client and Consultant.

9. Clear termination provisions if termination is by design consultant or by owner.

10. Severability—a severability provision provides that if a certain provision is deemed by a court to be illegal or unenforceable, the offending provision will be severed from the

agreement (Beazley n.d., p. 23). A survival provision ensures that in the event a provision is severed from the agreement, the remainder of the agreement remains in full force and effect as it is written (Beazley n.d.).

EXAMPLES OF DESIGN PROFESSIONALS' AGREEMENTS

Design professionals may wish to consider the sample language provided by the Engineers Joint Contract Documents Committee (EJCDC) Agreement between Engineer and Subconsultant for Professional Services (2020) as this design professional agreement contains many key clauses applicable to all design professionals (see Exhibit A).

DESIGN PROFESSIONAL'S EMPLOYMENT AGREEMENT

One item that should be included in a design professional's employment contract when the design professional works for an employer is the duty of the employer to indemnify the design professional employee after they leave the design firm. A sample clause is as follows (Woolley and Associates Inc. Standard Employment Agreement 2021):

> In addition to any other separate agreement with Employer concerning indemnification, to the fullest extent permitted by applicable law and the bylaws of Employer, as form time to time in effect, Employer shall indemnify Employee and hold Employee harmless for any acts or decisions made non-negligently, legally, and in good faith and within the scope of Employee's employment while Employee is performing services for Employer and Employer shall use its best efforts to obtain reasonable insurance coverage for Employee (provided the same may be obtained at reasonable cost) under any liability insurance policy or policies now in force or hereafter obtained during the term of this Agreement that cover other employees of Employer having comparable or lesser status and responsibility. To the same extent, Employer will pay and, subject to any legal limitations, advance all expenses, including reasonable attorney's fees and costs of court approved settlements, actually and necessarily incurred by Employee in connection with the defense of any action, suit or proceeding and in connection with any appeal thereon, which has been brought against Employee by reason of Employee's service as an agent of Employer while working legally, in good faith, non-negligently and within the scope of Employee's employment.

APPENDIX

This document has important legal consequences; consultation with an attorney is encouraged with respect to its use or modification. This document should be adapted to the particular circumstances of the contemplated Project and the controlling Laws and Regulations.

EJCDC® E-570, Agreement between Engineer and Subconsultant for Professional Services, is comprised of two parts: (1) the E-570 Agreement form, and (2) this Exhibits document, consisting of the standard exhibits used with E-570. The first part contains a Guidelines for Use section that pertains to both parts of E-570.

EXHIBITS TO AGREEMENT BETWEEN ENGINEER AND SUBCONSULTANT FOR PROFESSIONAL SERVICES

Prepared by

EJCDC®

ENGINEERS JOINT CONTRACT DOCUMENTS COMMITTEE

ACEC

American Council of Engineering Companies

ASCE

AMERICAN SOCIETY OF CIVIL ENGINEERS

NSPE®

NATIONAL SOCIETY OF PROFESSIONAL ENGINEERS

NOTE: EJCDC publications may be purchased at www.ejcdc.org, or from any of the sponsoring organizations above.

EXHIBIT A—SUBCONSULTANT'S SERVICES

Exhibit A Table of Contents

Guidance Notes—Preparing Exhibit A for the Specific Project

1. Exhibit A, Subconsultant's Services, as published contains a comprehensive scope of services, starting with Study and Report Phase services and continuing through Preliminary Design, Final Design, Construction Phase services, Post-Construction services, and potential Additional Services. However, EJCDC acknowledges that the scope of services for many Engineer-Subconsultant agreements will be much more limited in range. It is also common for Subconsultant services to be defined in a short form or attachment, often incorporating by reference portions of the Engineer's scope of services from the Prime Agreement. In some cases, the parties will elect to use a detailed scope of services document that is customized to the Subconsultant's area of expertise. Revise (or if appropriate replace) Exhibit A as published to reflect the needs of the specific subconsultant engagement.
2. To the extent that Exhibit A as published is used as the starting point for defining the Subconsultant's scope of services, Engineer and Subconsultant should carefully review and modify the published text. In some cases, this may entail supplementing the published provisions with more specific or detailed descriptions of services. In other cases, the parties may substantially trim the scope of services, for example by eliminating the Study and Report Phase, or by reducing the Subconsultant's role during construction.
3. Even when most of Exhibit A as published is retained, the user will note that there are many places in which it is necessary to furnish Project-specific information, as indicated by notes and prompts.
4. Exhibit A as published is coordinated with the other sections of the Agreement; and also with the terms and provisions of related EJCDC documents, most notably with the content of EJCDC® E-500 2020 (Owner-Engineer Agreement) and EJCDC® C-700 2018, Standard General Conditions of the Construction Contract, and the other 2018 Construction Series documents. When supplementing or revising Exhibit A for the specific Project, the user should take care to preserve this coordination, by using consistent terminology, and by noting and addressing any divergence between Subconsultant's

duties in Exhibit A (as modified) and in other sections of the Agreement, and Engineer's related duties in documents such as E-500 and C-700.

5. When a form or a customized scope of services is used (as an alternative to basing the scope on Exhibit A as published), the user should take additional precautions to assure coordination of duties, requirements, and terminology within the Agreement, and with respect to related documents.

Article 1 of the Agreement, Services of Subconsultant, is supplemented to include the following provisions:

Guidance Notes—Baseline Information

1. Inherent in establishing the scope of Subconsultant's services is clearly defining the nature of the facility or facilities to be studied, planned, designed, or built, to the extent known at the time this Agreement is drafted. Providing such information creates a mutual understanding by the parties as to the services that will be needed, as well as the level of effort and time (and thus cost) required to deliver those services. To accurately establish the scope of services, it is almost always in the best interests of Engineer and Subconsultant when the Engineer provides to the Subconsultant relevant, detailed baseline information regarding the facilities, to the extent available. In many cases the primary source of the baseline information will be the Owner.
2. If known, the baseline information regarding the physical improvements to be designed should include important controlling features, assumptions, and limitations that may affect Subconsultant's services. The information may be expressly stated below or by reference to external sources, such as requests for proposal or actual proposals, both of which often contain project descriptions, or by reference to baseline information included in the Prime Agreement.

Baseline Information: Engineer has furnished the following Project information to Subconsultant as of the Effective Date. Subconsultant's scope of services has been developed based on this information. As the Project moves forward, some of the information may change or be refined, and additional information will become known, resulting in the possible need to change, refine, or supplement the scope of services.

Notes to User

1. Here insert relevant information about the Project, either directly or by reference, to the extent known and necessary to give context to the Subconsultant's services that are specified in the remainder of Exhibit A.
2. Below are some suggested or sample categories of information that may have been furnished to Subconsultant as baseline information about the Project. Other categories may be relevant to the specific engagement.
3. Note that the term "Project" is broadly defined; in most cases it will be sufficient here to describe information relevant to those portions of the Project within Subconsultant's scope of services.

Project Title:
Type and Size of Facility:
Description of Improvements:

Expected Construction Start:
Prior Studies, Reports, Plans:
Facility Location(s):
Current Project Budget:
Funding Sources:
Known Design Standards:
Known Project Limitations:
Project Assumptions:
Other Pertinent Information:

Subconsultant shall provide Basic and Additional Services as set forth below.

Guidance Notes—Quantification of Certain Services

1. This Exhibit A includes numerous services that, depending on the nature of the Project, will likely be provided on multiple occasions, for repetitive activities, or recurring periods or intervals. Examples are progress meetings, responding to requests for information or interpretation (RFI), reviewing Bidder or Contractor "or equals" or substitution requests, and processing Change Proposals.
   a. Because the quantity of such repetitive activities is wholly dependent upon the needs of the Project, no single or "standard" specified quantity is offered here.
   b. The general approach of this Exhibit is that Subconsultant's services include a reasonable number of such repetitive activities, unless a bracketed selection or option is provided.
   c. Excessive quantities of repetitive activities are typically handled as Additional Services.
   d. The parties may desire specificity as to the quantity of any repetitive service to be provided, both to assure the Engineer as to the number of such activities to be included and to facilitate budgeting for Basic Services. Anywhere in this Exhibit that a repetitive service is stated, the parties may modify the wording to define a specific quantity for the service to be included as a Basic Service.

ARTICLE 1—BASIC SERVICES

Guidance Notes—Management of Engineering Services

1. Proper management of the Subconsultant's services includes certain administrative, scheduling, personnel coordination, reporting, and other tasks, directly and specifically related to the specific engagement. These project management tasks are necessary to effectively and successfully complete the required professional and technical services that are more commonly the focus of attention when drafting a subconsultant scope of services.
2. The following Exhibit A Paragraph 1.01 lists typical management tasks that will almost always apply, in some form, to each Phase of the Subconsultant's services, whether separately budgeted and tracked or simply performed as part of the required professional and technical services.
3. The listed management services are intended to include only those that are directly and solely applicable to the particular Subconsultant scope of services, and should not include general business management ("overhead") tasks that apply to or are performed as part of a core business

function (finance, accounting, human resources, legal, corporate, facilities, etc.) or to those tasks that are performed in the course of Engineer's management of multiple projects. Some owners, engineering firms, and agencies have specific rules and guidance defining what they will allow to be treated as billable project management, and the following scope should be modified for compliance with such applicable rules and guidance.

1.01 Management of Subconsultant's Services

A. As pertinent to those services and portions of the Project assigned to the Subconsultant, all phases of Subconsultant's services will include management of Subconsultant's Project-specific responsibilities, including but not limited to the following management tasks, whether separately tracked and itemized or included as being incidental to other phase and scope task items.

1. Develop and submit a Subconsultant's Services Schedule. The Subconsultant's Services Schedule will:

a. be consistent with and serve as a supplement to the Schedule of Deliverables set forth in Exhibit B.

b. be updated on a regular basis, and as required to reflect any programmatic decisions by Engineer or Owner.

c. include, but be not limited to, an anticipated sequence of tasks; estimates of task duration; interrelationships among tasks; milestone meetings and submittals; anticipated schedule of construction; and other pertinent Project events.

2. Develop and submit detailed work plans from Exhibit A tasks.

3. Coordinate services within Subconsultant's internal team and with Sub-subconsultants.

4. Prepare for and participate in meetings with consultants and contractors working on other parts of the Project that may affect, or be affected by, Subconsultant's services or resulting construction.

Notes to User

1. Paragraph 4.01 of the Agreement states that Subconsultant shall prepare invoices in accordance with its standard invoicing practices, and with the progress reporting and special invoicing requirements (if any) in Exhibit A Paragraph 1.01.A. Paragraphs 1.01.A.5 and 6 immediately below are the locations for the specific follow-up to Paragraph 4.01.
2. The progress reporting addressed in Exhibit A Paragraph 1.01.A.5 is typically submitted in conjunction with Subconsultant's periodic invoices. Reporting and invoicing obligations will usually be coordinated.
3. If Engineer or Owner have progress reporting or invoicing requirements that differ from Subconsultant's standard and routine reporting and invoicing procedures, modify the requirements in Exhibit A Paragraphs 1.01.A.5 and 6, as appropriate.
4. Relative to the standards set forth in Exhibit A Paragraph 1.01.B or elsewhere in Exhibit A, see Exhibit F, Electronic Documents Protocol, for additional information regarding the exchange of Electronic Documents by Electronic Means between Engineer and Subconsultant. The notes to

Exhibit F provide guidance for addressing exchange of Electronic Documents exclusively applicable between Engineer and Subconsultant, either by modifying or supplementing Attachment F-1 of Exhibit F or relying on sufficiently detailed descriptions of the Subconsultant's Electronic Documents deliverables in Exhibit A.

5. Prepare and submit **[monthly] [other periodic interval]** progress reports to the Engineer. Include summary of services performed in period, expected progress in next period, percent completion of current tasks, and a description of major issues or concerns.

6. Special Invoicing: In addition to, or as a substitute for, Subconsultant's standard invoicing, provide the specified additional information or documentation, following the invoicing procedures indicated: **[indicate Not Applicable; or specify required information, documentation, or special invoicing procedures]**.

7. Conduct ongoing management tasks, including:

   a. Maintaining communications records and files pertaining to or arising from Subconsultant's services;

   b. With respect to Subconsultant's services and other directly relevant parts of the Project, prepare for and participate in periodic progress meetings with Engineer and Owner to discuss progress, schedule, budget, issues, potential problems and their resolution; and

   c. Preparing agendas prior to and minutes following all Subconsultant-led meetings.

B. In all phases of Subconsultant's services, Subconsultant shall prepare draft and final Drawings in accordance with **[Engineer's CAD standards] [Owner's CAD standards] [other CAD standard (specify)]**, using [___] version [___] software.

C. The source documents for the draft and final Specifications in all phases of Subconsultant's services will be **[Subconsultant's standard specifications] [Engineer's standard specifications] [Owner's standard specifications]**, unless otherwise mutually agreed upon by the parties.

1.02 Study and Report Phase

A. For those portions of the Project assigned to the Subconsultant, or to the extent that a specific task is delegated to Subconsultant, Subconsultant shall:

1. Consult with Engineer to define and clarify Owner's requirements for the Project, including design objectives and constraints, space, capacity and performance requirements, flexibility, and expandability, and any budgetary limitations, and identify available data, information, reports, facilities plans, and site evaluations.

2. Identify potential solution(s) to meet Owner's Project requirements, as needed.

3. Study and evaluate the potential solution(s) to meet Owner's Project requirements.

4. Visit the Site, or potential Project sites, to review existing conditions and facilities, unless such visits are not necessary or applicable to meeting the objectives of the Study and Report Phase.

5. Assess initially available Project information and data, including the Baseline Information set forth at the beginning of this Exhibit A.

6. Advise Engineer of any need for Owner to obtain, furnish, or otherwise make available to Subconsultant additional Project-related information and data, for Subconsultant's use in the study and evaluation of potential solution(s) to Owner's Project requirements, and preparation of a related report.

7. After consultation with Engineer, recommend the solution(s) which in Subconsultant's judgment meet Owner's and Engineer's requirements for the Project.

8. Identify, consult with, and analyze requirements of authorities having jurisdiction to permit or approve construction or operation of the portions of the Project to be designed or specified by Subconsultant, including but not limited to impacts and mitigating measures identified in previously prepared environmental assessments for the Project provided to the Subconsultant or being concurrently prepared for Owner by others.

9. Advise the Engineer of any need for Owner or Engineer to provide data or services of the types described in Article 2, for use in Project design, or in preparation for Contractor selection and construction.

10. Assist Engineer in evaluating the possible use of building information modeling; civil integrated management; geotechnical baselining of subsurface conditions at the Site; innovative design, contracting, or procurement strategies; project delivery method; or other strategies, technologies, or techniques for assisting in the design, construction, and operation of Owner's facilities. The subject matter of this paragraph will be referred to in Exhibit A as "Project Strategies, Technologies, and Techniques."

11. Assist Engineer in identifying opportunities for enhancing the sustainability of the Project, and pursuant to Owner's instructions to Engineer, plan for the inclusion of sustainable features in the design.

12. Review with Engineer the thresholds established in applicable codes, standards, and design criteria specifically governing the ability of the proposed facilities or improvements to perform, and to absorb or avoid damage without suffering complete or substantial failure. As part of the review, identify additional risk assessment studies or tools that are available to evaluate the susceptibility of the facilities or improvements to natural and man-made events beyond the applicable established thresholds. Upon Engineer request, as an additional service, perform additional risk assessment studies or tools to further evaluate system resiliency beyond the applicable established thresholds.

13. Utilities, including Underground Facilities

    a. Review any utility mapping and surveys and other utilities documentation made available by Owner through Engineer. Take note of observable utilities during Site visit.

    b. Identify, in a preliminary manner and to the extent determinable by such mapping or other information provided by Owner through Engineer, and by observations at the Site, those utilities (whether above-ground utilities of any type, or

Underground Facilities) likely to be affected by the Project construction and additional utility facilities or extensions that will be needed to serve the Project.

c. If the impact on existing utilities or the need for additional utility facilities or extensions cannot reasonably be determined in a preliminary manner from mapping or other information provided by Owner through Engineer, or such information was not available from Engineer, then assist Engineer in evaluating the need to either obtain additional utility mapping and utility documentation during the Study and Report Phase, or undertake other alternative approaches and contingencies to account for utility uncertainties in this phase.

d. Advise Engineer of additional utility documentation and coordination needed during the design and construction phases to adequately assess, mitigate, and manage the impact of the Project (including any additional utility facilities or extensions needed to serve the Project) on existing utilities.

e. In advising Engineer, consider ASCE 38, "Standard Guideline for the Collection and Depiction of Existing Subsurface Utility Data" regarding the extent and identification and mapping of existing Underground Facilities during the design and construction phases.

   1) If Owner or Engineer has retained a land surveyor, utility engineer, or utility consultant, collaborate with such individuals or entities regarding the application of ASCE 38.

14. Develop a scope of work and survey limits for any topographic and other surveys necessary for design.

15. Prepare a report (the "Report") which will, as appropriate, contain schematic layouts, sketches, and conceptual design criteria with appropriate exhibits to indicate the agreed-to requirements, considerations involved, and Subconsultant's recommended solution(s).

    a. For each recommended solution, Subconsultant will separately tabulate Total Project Cost, itemizing those items and services included within the definition of Total Project Costs.

    b. Subconsultant will meet with Engineer (together with Owner at Engineer's option) to discuss the draft Report and receive Engineer's and Owner's comments.

16. Perform or provide the following other Study and Report Phase tasks or deliverables:

    a. **[List any such tasks or deliverables here]**.

17. Furnish the Report and any other Study and Report Phase deliverables to Engineer pursuant to the requirements of the Deliverables Schedule in Exhibit B, and review the deliverables with Engineer (together with Owner at Engineer's option).

18. Revise the Report and any other Study and Report Phase deliverables in response to Engineer's and Owner's comments, as appropriate, and submit revised deliverables pursuant to the Deliverables Schedule.

B. Subconsultant's services under the Study and Report Phase will be considered complete on the date when Subconsultant has delivered to Engineer the final Report (as revised) and any other Study and Report Phase deliverables.

Guidance Notes—Preliminary Design Phase

1. In the Preliminary Design Phase, all major decisions on the Project's scope, extent, design criteria, and quality should be finalized, and a preliminary design (usually including preliminary drawings and a preliminary list of expected specifications) established. The Preliminary Design Phase should fully set the stage for development of the final Drawings, Specifications, and other Bidding Documents.
2. The suggested report content included in Exhibit A for the Study and Report Phase (Exhibit A Paragraph 1.02 above) and the Preliminary Design Phase (Exhibit A Paragraph 1.03 below) are general guides to elements and subject matter often contained in such reports, and may be either more or less comprehensive than what is needed to establish the scope of Subconsultant's services for the specific Project.
3. As with all of Exhibit A, the scope for report content is intended to be customized for each particular project.
4. Often in engineering projects, the interim design submittals are labeled as 30%, 60%, 90%, and 100%, or other percentages based on owner usage, to indicate level of completeness of the overall design. These can be misleading labels, as level of completeness is not easily measured in objective metrics, and, on multidiscipline projects (the rule, not the exception), not all disciplines are at the same level of completeness at every design submittal. EJCDC does not use the percentage system to define design completion status, but as an approximate reference for users who are accustomed to percentages, completion of Subconsultant's services in the Preliminary Design Phase roughly corresponds to 30% design completion.
5. The Preliminary Design Phase services that follow include specific obligations with respect to utilities at the Site. These should be carefully reviewed and modified as necessary to conform to practices and Laws and Regulations applicable to the specific Project.

1.03 Preliminary Design Phase

A. After acceptance by Owner of the Report and any other Study and Report Phase deliverables (if Subconsultant's services under this Agreement included Study and Report Phase services); selection by Owner of a recommended solution; issuance by Owner of any instructions for use of Project Strategies, Technologies, and Techniques, or for inclusion of sustainable features in the design, or enhanced resiliency of the design; indication by Owner of any specific modifications or changes in the scope, extent, character, or design requirements of the Project desired by Owner; and any necessary changes, refinements, and supplementation of the Baseline Information set forth at the beginning of this Exhibit A, Subconsultant and Engineer shall discuss, resolve, and document in writing any necessary revisions to Subconsultant's scope of services, compensation (through application of the provisions regarding Additional Services, or otherwise), and the time for completion of Subconsultant's services, resulting from the selected solution, related Project Strategies, Technologies, or Techniques, sustainable design and resiliency instructions, specific modifications to the Project, or changes, refinements, or supplementation of the Baseline Information.

B. Upon written authorization from Engineer, for those portions of the Project assigned to the Subconsultant, or to the extent that a specific task has been delegated to Subconsultant, Subconsultant shall:

1. Review and assess all available Project information and data, including any pertinent reports or studies (whether prepared by Subconsultant or others) and any related instructions from Owner and Engineer.

2. Based on the threshold review and assessment of available information and data, advise Engineer of any need for Engineer or Owner to obtain, furnish, or otherwise make available to Subconsultant any additional information and data, for Subconsultant's use in the preparation of a Preliminary Design Phase Report.

3. Prepare a Preliminary Design Phase Report in the following format: **[specify the format that is appropriate to the scope of the design and sufficient for Engineer's purposes—ranging from a consolidated and comprehensive narrative report to an assemblage of required documents]**.

4. The Preliminary Design Phase Report will consist of final design criteria, preliminary drawings, a preliminary list of expected specifications, and written descriptions of the Project. The Preliminary Design Phase Report will consider the following matters to the extent applicable to the Project and as necessary to establish the basis of design for proceeding to final design and construction:

   a. The Project concept, intent, performance criteria, desired outcomes, Owner's standards and Owner directed improvements and facility elements as established in the Study and Report Phase and as expressly set forth in the Baseline Information section of this Exhibit A (collectively the "Project Goals").

   b. Recommended appropriate design criteria for each primary portion and significant discipline of the design necessary to address the Project Goals.

   c. Site conditions and characterization as known at the time of, or to be determined during, the Preliminary Design Phase, including topography; subsurface information; Constituents of Concern; cultural, historical, and archaeological resources at the Site; wetlands information; and evaluations of flora and fauna that may be affected by the Project.

   d. The time schedule for completion of the Project in accordance with Project Goals, including any recommended changes to the time required to complete the Final Design Phase (as set forth in Exhibit B, Deliverables Schedule) and estimated schedule(s) for construction.

   e. Identification of major items of materials and equipment, rationale for selection with consideration of quality, suitability, pricing, sourcing, regulatory, and bidding issues affecting recommended selection.

   f. Revised opinions of probable Construction Cost.

g. The impact of Project Strategies, Technologies, and Techniques, sustainable features, and enhanced resiliency selected by Owner for inclusion in the Project on the Project Goals, schedule and probable Construction Cost, including impact of multiple prime construction contracts, separate procurement of materials or equipment, and other alternate project delivery methods when the Project Goals necessitate and Owner authorizes.

h. Construction Phase quality assurance and quality control needs affecting development of Drawings and Specifications and other Final Design and Bidding Phase documents.

i. The effect of permits and authorizations by other entities and utility coordination needs on the Project.

j. Other matters and information pertinent to addressing the Project Goals.

5. In preparing the Preliminary Design Phase Report, use any specific applicable Project Strategies, Technologies, and Techniques authorized by Owner during or following the Study and Report Phase, and include sustainable features and enhanced resiliency, as appropriate, pursuant to Owner's instructions to Engineer.

6. Visit the Site as needed to prepare the Preliminary Design Phase Report.

7. If at any point in the Preliminary Design Phase it becomes apparent to Subconsultant that additional reports, data, information, or services of the types described in Article 2 are necessary, then so advise Engineer, and assist Engineer in obtaining such reports, data, information, or services.

8. Above-Ground Utilities

   a. Review above-ground utilities information obtained from Owner through Engineer and from observations at the Site.

   b. Make recommendations regarding any further identification, investigation, and mapping of above-ground utilities at or adjacent to the Site, for Subconsultant's design purposes or otherwise.

   c. Account for above-ground utilities, based on available information, when advancing design during the Preliminary Design Phase.

9. Underground Facilities

   a. Review Underground Facilities data furnished by Owner through Engineer. Assist Owner in reducing and managing risks associated with Underground Facilities by working together with Engineer and Owner to jointly establish a procedure ("Underground Facilities Procedure") for the further identification, investigation, and mapping of Underground Facilities at or adjacent to the Site, using ASCE 38, "Standard Guideline for the Collection and Depiction of Existing Subsurface Utility Data," as a basis for establishing such Underground Facilities Procedure.

   b. Such Underground Facilities Procedure must take into account the Site and the nature of the Project.

c. Use the Underground Facilities Procedure to aid in the performance of design services:

1) Account for Underground Facilities, based on available information, when advancing the design during the Preliminary Design Phase.

2) The Underground Facilities Procedure will include a plan to keep Underground Facilities information current as Engineer proceeds with the provision of design services, and to add new or relocated Underground Facilities information to the base utility or Site drawings.

3) To manage the potential impact of design changes on Underground Facilities, Engineer shall work together with Owner to modify or reapply the Underground Facilities Procedure as the design progresses and changes.

10. Mitigation of Utilities Conflicts

a. Identify potential conflicts between the Project (including existing and new facilities and structures) and above-ground utilities and Underground Facilities as reviewed in Exhibit A Paragraphs 1.03.B.8 and 9 above, and advise Engineer regarding the need for resolution of such conflicts with utility and Underground Facilities owners and permit agencies. Identify the potential need for the relocation of existing above-ground utilities and Underground Facilities.

b. Update the Underground Facilities Procedure as necessary for any Underground Utilities conflicts and relocations.

c. Provide assistance to Engineer in working together with Owner to jointly identify which specific parties or other entities will be responsible for implementation of the various specific parts of the Underground Facilities Procedure (including those parts that address resolution of Underground Facilities conflicts), and for resolution of above-ground utilities conflicts. Such identification will take into account Owner's authority and standing, as owner of the Site, with respect to Underground Facilities and above-ground utilities.

1) To the extent that Owner and Engineer agree that in addition to performing the design-related obligations set forth in Exhibit A Paragraphs 1.03.B.8 and 9, Engineer will also implement any non-design part of the Underground Facilities Procedure (including resolution of Underground Facilities conflicts), or undertake resolution of above-ground utilities conflicts, assist Engineer in carrying out such additional duties, as Additional Services under Article 2 of this Exhibit A.

11. Surveys, Topographic Mapping, and Utility Documentation

a. Coordinate with Owner's or Engineer's utility engineer, utility consultant, or land surveyor for the necessary field surveys, topographic mapping, and utility documentation required for Subconsultant's design purposes, or by the Underground Facilities Procedure.

b. If no scope of work and procedure for utility mapping has been established, selected, or authorized, then at a minimum Subconsultant will, for those portions of the project assigned to the Subconsultant, acting in coordination with Engineer, contact utility owners and obtain available information. Except as otherwise provided in this Agreement, Engineer agrees to seek from Owner acknowledgment that the information gathered from utility owners may be incorrect, incomplete, outdated, or otherwise flawed, and acceptance by Owner of all associated risks. To the extent Engineer obtains such acknowledgment and acceptance from the Owner, Engineer provides an equivalent acknowledgment and acceptance to Subconsultant on behalf of Subconsultant. Owner reserves all associated rights as to recourse against the sources of such flawed information and against third parties.

12. Prepare initial draft of a permit document for those portions of the Project assigned to the Subconsultant that identifies Owner's permit duties, Engineer's permit duties (and if applicable any proposed delegations of such duties to Subconsultant), and Contractor's permit duties, and the schedule for permitting activities.

13. Assist Engineer with Project Strategies, Technologies, and Techniques that Owner has chosen to implement in Exhibit A Paragraph 1.03.A.

Notes to User—If Subconsultant will have limited or no involvement in preparing Bidding/Proposal Documents or Front-End Construction Contract Documents, modify the following paragraph. See Guidance Note 2 at Exhibit A Paragraph 1.04.

14. Obtain from Engineer the Owner's instructions regarding Owner's procurement of construction services (including instructions regarding advertisements for bids, instructions to bidders, and requests for proposals, as applicable), Owner's construction contract practices and requirements, insurance and bonding requirements, electronic transmittals during construction, and other information necessary for the finalization of Owner's Bidding/Proposal Documents and Front-End Construction Contract Documents.

    a. Also obtain copies of Owner's standard Bidding/Proposal Documents and Front-End Construction Contract Documents (if other than the EJCDC 2018 Construction Series documents), and any other related documents or content for Subconsultant to include in drafts of the bidding-related documents (or requests for proposals or other construction procurement documents), and in drafts of the Project-specific Bidding/Proposal Documents and Front-End Construction Contract Documents, when applicable.

    b. Review Owner's instructions regarding procurement, bidding and contracting of construction services with respect to effects on the Project design, schedule and construction and address as needed in the Preliminary Design Phase deliverables.

15. Prepare the Preliminary Design Phase Report. This Report will consist of, as appropriate, separate or combined submittals in whole or summary, the Preliminary Design Phase documents listed in Exhibit A Paragraph 1.03.B.4, Subconsultant's findings (including those, if any, relating to permitting, utilities and Underground Facilities), and recommendations for advancing the Project to the Final Design Phase. The submittal will be in the format of a report, or otherwise organized and assembled for ease and practicality of use.

a. Based on the information contained in the Preliminary Design Phase documents, prepare a revised opinion of probable Construction Cost, and on the basis of information furnished by Engineer and Owner, assist Engineer in tabulating the various cost categories which comprise Total Project Costs.

b. Subconsultant will meet with Engineer (together with Owner at Engineer's option) to discuss the draft Preliminary Design Phase submittal and receive Engineer's and Owner's comments.

16. Perform or provide the following other Preliminary Design Phase tasks or deliverables:

a. **[List any such tasks or deliverables here]**.

17. Furnish the Preliminary Design Phase Report, opinion of probable Construction Cost, and any other Preliminary Design Phase deliverables to Engineer pursuant to the requirements of the Deliverables Schedule in Exhibit B, and review the deliverables with Engineer (together with Owner at Engineer's option).

18. Revise the Report and any other deliverables in response to Engineer's and Owner's comments, as appropriate, and submit revised deliverables pursuant to the Deliverables Schedule.

C. Subconsultant's services under the Preliminary Design Phase will be considered complete on the date when Subconsultant has delivered to Engineer the final Preliminary Design Phase Report (as revised) and associated documents, revised opinion of probable Construction Cost, and any other Preliminary Design Phase deliverables.

1.04 Final Design Phase

Guidance Notes

1. As noted above with respect to the Preliminary Design Phase, in Guidance Note 4, immediately preceding Exhibit A Paragraph 1.03 above, it is common to refer to interim design submittals as 30%, 60%, 90%, and 100%, to indicate the level of completeness of the overall design. Although EJCDC does not use these percentage labels (for the reasons stated in the cited Guidance Note 4 above), for reference purposes the Subconsultant's completion of the Preliminary Design Phase should bring the design to roughly 30% completion. The first Final Design Phase draft of the Drawings and Specifications (Exhibit A Paragraph 1.04.B.1) is a rough equivalent of 60% completion; the second draft (1.04.B.2) may be thought of as roughly 90% completion; and the third and final draft (1.04.B.3) as 100% completion.

2. This exhibit has been coordinated for use with the 2018 EJCDC Construction Documents. When other contract forms and bidding requirements will be used on the Project, it is important for the Owner and Subconsultant to carefully coordinate the Subconsultant's authority, responsibilities, and scope of services (especially during construction) under such other forms with the provisions of Exhibit A. If the final bidding and contract requirements are not known at the time of the execution of the agreement, some modifications to the agreement may be needed during the review by the parties under the Construction Phase services in Exhibit A Paragraph 1.06.A.

A. After acceptance by Owner of the Preliminary Design Phase Report and any other Preliminary Design Phase deliverables; issuance by Owner of any instructions for specific modifications or changes in the scope, extent, character, or design requirements of the Project desired by Owner; and any necessary changes, refinements, and supplementation of the Baseline Information set forth at the beginning of this Exhibit A, Subconsultant and Engineer shall discuss, resolve, and document in writing any necessary revisions to Subconsultant's scope of services, compensation (through application of the provisions regarding Additional Services, or otherwise), and the time for completion of Subconsultant's services, resulting from specific modifications to the Project, or changes, refinements, or supplementation of the Baseline Information.

1. The number of prime contracts for Work designed or specified by Subconsultant upon which the Subconsultant's compensation has been established under this Agreement is **[Enter number of contracts]**. If more prime contracts are awarded, Subconsultant shall be entitled to an equitable increase in its compensation under this Agreement.

B. Upon written authorization from Engineer, for those portions of the Project assigned to the Subconsultant, and to the extent that each specific task is delegated to Subconsultant, Subconsultant shall prepare final Drawings and Specifications indicating the scope, extent, and character of the Work to be performed and furnished by Contractor, in accordance with the Preliminary Design Phase Report (as revised) and other Preliminary Design Phase deliverables. As part of the preparation of the Drawings and Specifications, Subconsultant shall prepare interim design drafts as follows, pursuant to the Deliverables Schedule in Exhibit B, but not less than **[insert number of days]** days prior to each design review meeting between Owner and Engineer:

Notes to User—The number of drafts will vary, depending on project needs and complexity. Edit the following paragraph and the associated Exhibit B, Deliverables Schedule, accordingly.

1. First Final Design Phase draft of all Drawings and Specifications required by Subconsultant's scope of services.
2. Second Final Design Phase draft of Drawings and Specifications, addressing Engineer's and Owner's comments (if Owner's comments have been provided to Subconsultant by Engineer), and including appropriate design advancement.
3. Final Drawings and Specifications that address Engineer's and Owner's comments (if provided); complete the pertinent portions of the design; are suitable for estimating and pricing by prospective Contractors; and are complete and ready for construction.

C. In preparing the Specifications (and any bidding, contract, or other documents that are part of Subconsultant's scope of services), Subconsultant shall obtain from Engineer any relevant Owner constraints such as requirements for use of domestic steel and iron, other domestic purchasing requirements, statutory restrictions on utilizing proprietary specifying methods, and the like, and comply with or account for such constraints in drafting Specifications and other Project documents.

D. Subconsultant shall prepare or assemble draft Bidding/Proposal Documents and Front-End Construction Contract Documents as follows:

1. Such documents will be based on the 2018 EJCDC Construction Documents, or on the specific bidding or procurement-related instructions and forms, text, or content received by Engineer from Owner.

2. When Subconsultant is required to use other than the 2018 EJCDC Construction Documents, then Engineer will furnish to Engineer a copy of the required documents to be used for the Project's Bidding/Proposal Documents and Front-End Construction Contract Documents. Prior to the first final design submittal, Subconsultant will review the bidding and contracting requirements documents furnished by Engineer and furnish comments to Engineer for Engineer to discuss with Owner.

3. Subconsultant will furnish to Engineer the draft Bidding/Proposal Documents and Front-End Construction Contract Documents. Engineer will provide Subconsultant with Engineer's comments and comments and revisions to the draft documents, if any, received from Owner and Owner's legal counsel.

E. During the Final Design Phase the Subconsultant shall continue to account for above-ground utilities and Underground Facilities as the design advances and is finalized. This may include:

1. performing the services assigned to Engineer under the Underground Facilities Procedure described in Exhibit A Paragraph 1.03 above, including but not limited to the design-related tasks in Exhibit A Paragraph 1.03.B.9, to the extent delegated by Engineer to Subconsultant.

2. addressing required and proposed activities or mitigations identified in the analysis of utilities and by the Underground Facilities Procedure as having an impact on the final design, and considering such in preparing the Drawings and Specifications.

F. Subconsultant shall perform or furnish the following other Final Design Phase services:

Notes to User—It may be prudent to specify the number of site visits, especially if field verification of existing facilities is necessary.

1. Visit the Site as needed to assist in preparing the final Drawings and Specifications.

2. Assist with or prepare applications for permits and approvals, as follows:

   a. Update comprehensive permit document with Final Design detail.

   b. Prepare or assist the Engineer with preparation of the following applications for Owner's submittal to authorities having jurisdiction over the construction or operation of the Project:

      1) **[identify specific permit or approval applications]**.

   c. Confer with Engineer regarding revisions, if any, to the applications, and make appropriate revisions to the applications for resubmittal to the authority having jurisdiction.

   d. Provide technical criteria, written descriptions, and design data to Engineer for Owner's use in filing applications for permits from or approvals of the authorities having jurisdiction listed above, including applications for review or approval of the final design.

e. Identify and indicate in the Construction Contract Documents the permits and approvals for which Contractor will be responsible, including work permits, building permits, and other permits and approvals that will be Contractor's responsibility; and, in addition, indicate those permits initially obtained by Owner for which Contractor will be a co-permittee, together with associated requirements.

f. Unless expressly indicated otherwise, Subconsultant's scope and budget includes attending one meeting, or conference call, with each permit and approval-issuing agency to discuss the Project and receive the agency's comments on the application.

g. Subconsultant does not guarantee issuance of any required permit or approval.

h. Fees charged by authorities having jurisdiction for such permits or approvals are the responsibility of Owner.

3. Advise Engineer of any recommended adjustments to the opinion of probable Construction Cost. Furnish to Engineer an updated opinion of probable Construction Cost with the interim and final deliverables of the Drawings and Specifications.

4. After consultation with Engineer, include in the Front-End Construction Contract Documents any Electronic Document Protocol addressing specific protocols for the transmittal of Project-related correspondence, documents, text, data, drawings, information, and graphics, in electronic media or digital format, either directly, or through access to a secure Project website.

5. Upon direction of Engineer, assist Owner in assembling known reports and drawings of Site conditions, and in identifying the technical data contained in such reports and drawings upon which bidders or other prospective contractors may rely.

6. Review the preliminary schedule for the Construction Phase and advise Engineer when initial understanding of the Construction Contract Times must or should be revised, and furnish Engineer with recommendations on revisions to the proposed Construction Contract Times.

7. Subconsultant's project manager and other appropriate staff will participate in the following meetings and conference calls:

a. First draft design review meeting at Owner's office.

b. Second draft design review meeting at Owner's office.

c. **[Indicate others as appropriate for the Project]**.

d. Subconsultant will prepare and distribute minutes of each such Subconsultant-led meeting and conference call, indicating attendees, topics discussed, decisions made, and action items for follow-up.

8. Perform or provide the following other Final Design Phase activities or deliverables:

a. **[List any such tasks or deliverables here]**.

G. Subconsultant shall complete the Final Design Phase as follows:

1. Pursuant to the requirements of the Deliverables Schedule in Exhibit B, furnish for review by Engineer the final Drawings and Specifications (as set forth in Exhibit A

Paragraph 1.04.B.3 above), assembled drafts of other Construction Contract Documents, including the draft Front-End Construction Contract Documents; the draft Bidding/Proposal Documents; and any other Final Design Phase deliverables, pursuant to the requirements of the Deliverables Schedule in Exhibit B, and review the deliverables with Engineer (together with Owner at Engineer's option).

2. Revise the final Design Phase deliverables in response to Engineer's and Owner's comments, as appropriate, and submit revised deliverables pursuant to the Deliverables Schedule.

3. Subconsultant's services under the Final Design Phase will be considered complete on the date when Subconsultant has delivered to Engineer the final Drawings and Specifications; assembled drafts of the Front-End Construction Contract Documents; the draft Bidding/Proposal Documents; and any other Final Design Phase deliverables, as revised.

1.05 Bidding/Proposal Phase

A. After acceptance by Owner of the final Drawings and Specifications; assembled drafts of other Construction Contract Documents, including the draft Front-End Construction Contract Documents; the draft Bidding/Proposal Documents; the most recent opinion of probable Construction Cost as determined in the Final Design Phase, and any other Final Design Phase deliverables, and upon written authorization by Engineer to proceed, for those portions of the Project assigned to the Subconsultant, and to the extent that each specific task is delegated to Subconsultant, Subconsultant shall:

1. Assist Engineer and the Owner in advertising for and obtaining bids or proposals for the Work; issuing assembled contract, and bidding-related documents (or requests for proposals or other construction procurement documents) to prospective contractors; if applicable, maintaining a record of prospective contractors to which documents have been issued; attending pre-bid conferences, if any; and receiving and processing contractor deposits or charges, if any, for the issued documents.

   a. **[If applicable to Subconsultant's services, list method (such as Owner's procurement website, commercial plan room or web service, distribution by Engineer, etc.) that Owner will use for distributing Bidding Documents here.]**

2. Prepare addenda as appropriate to clarify, correct, or change the issued documents.

3. If the issued documents require, the Subconsultant shall evaluate and determine the acceptability of "or equals" and substitute materials and equipment proposed by prospective contractors, provided that such proposals are allowed by the bidding-related documents (or requests for proposals or other construction procurement documents) prior to award of contracts for the Work. Services under this paragraph are subject to the provisions of Exhibit A Paragraph 2.01.

4. Attend the bid opening; prepare bid tabulation sheets; and assist in evaluating bids or proposals, assembling final Construction Contracts for the Work for execution by Owner and Contractor, and in preparing notices of award to be issued by Owner for such contracts.

5. Provide information or assistance needed in the course of any review of any bids, proposals or negotiations with prospective contractors.

6. Consult with Engineer as to the qualifications of prospective contractors.

7. Consult with Engineer and Owner as to the qualifications of subcontractors, suppliers, and other individuals and entities proposed by prospective contractors, for those portions of the Work as to which review of qualifications is required by the issued documents.

8. If Owner engages in negotiations with bidders or proposers, assist Engineer with respect to technical and engineering issues that arise during the negotiations.

9. Perform or provide the following other Bidding or Negotiating Phase tasks or deliverables:

   a. **[List any such tasks or deliverables here]**.

B. The Bidding or Negotiating Phase will be considered complete upon award of Construction Contracts for the Work and commencement of the Construction Phase, or upon cessation of negotiations with prospective contractors.

1.06 Construction Phase

A. After completion of the Final Design Phase and concurrent with the Bidding/Proposal Phase, and after issuance by Owner of any instructions for specific modifications or changes in the scope, extent, character, design, schedule, number of prime construction contracts, and other construction requirements of the Project during the Construction Phase desired by Owner, the Subconsultant and Engineer shall discuss, resolve, and document in writing any necessary revisions to Subconsultant's scope of services or compensation (through application of the provisions regarding Additional Services, or otherwise), or the time for completion of Subconsultant's services, resulting from specific modifications to the Project.

1. Subconsultant shall be responsible only for those Construction Phase services expressly required of Subconsultant in Exhibit A Paragraph 1.06, as duly modified. With the exception of such expressly required services, Subconsultant shall have no design, or Submittal (including Shop Drawing) review, or other obligations during construction, and Engineer assumes the responsibility, if any, for providing or arranging for all other necessary Construction Phase administrative, engineering, and professional services that are required of Engineer by the Prime Agreement.

2. Notwithstanding the above, Subconsultant shall be responsible for any professional opinions and interpretations provided by the Subconsultant to Engineer, Owner or others during the Construction Phase or Post-Construction Phase, including interpretations or clarifications of the Construction Contract Documents, and shall remain responsible for all other services performed or furnished under this Agreement.

B. Upon successful completion of the Bidding and Negotiating Phase, and upon written authorization from Engineer, for those portions of the Project assigned to the Subconsultant, and to the extent that each specific task is delegated to Subconsultant, Subconsultant shall provide the following services:

1. General Administration of Construction Contract: Consult with Engineer in Engineer's role as Owner's representative as provided in the Prime Agreement and Construction Contract. Unless otherwise set forth in the scope of Basic Services (as duly modified), the extent and limitations of the duties, responsibilities, and authority of Engineer shall be as assigned in EJCDC® C-700, Standard General Conditions of the Construction Contract (2018) or other construction general conditions specified in this Agreement. Except as otherwise required by this Agreement, Subconsultant's Construction Phase communications to Owner and Contractor will be issued through Engineer.

   a. Subconsultant shall not be required to furnish or perform services contrary to Subconsultant's responsibilities as a licensed professional.

2. Subconsultant's Resident Representative (SRR): Provide the services of an SRR at the Site to assist Subconsultant and to provide more extensive observation of Contractor's Work. Duties, responsibilities, and authority of the SRR are as set forth in Exhibit D. The furnishing of such SRR's services will not limit, extend, or modify Subconsultant's responsibilities or authority except as expressly set forth in Exhibit D. SRR will be based in Engineer's field office at the Site. **[If Subconsultant will not be providing the services of an SRR, then delete this paragraph by inserting the word "DELETED" after the paragraph title, and do not include Exhibit D as part of this Agreement.]**

3. Selection of Independent Testing Laboratory: Assist Engineer in Owner's selection of an independent testing laboratory to perform required testing services.

4. Pre-Construction Conference: Participate in a pre-construction conference prior to commencement of Work at the Site.

5. Schedules: As pertinent to Subconsultant's scope, review, and, subject to the criteria of the Construction Contract, determine the acceptability of any pertinent schedule that Contractor is required to submit to Engineer, including the progress schedule, schedule of submittals, and schedule of values. Advise Engineer of Subconsultant's comments regarding acceptability of schedules.

   a. Schedules will be acceptable to Engineer as to form and substance:

      1) Progress Schedule: if it provides an orderly progression of the Work to completion within the Contract Times. Such acceptance will not impose on Engineer responsibility for the Progress Schedule, for sequencing, scheduling, or progress of the Work, nor interfere with or relieve Contractor from Contractor's full responsibility therefor.

      2) Contractor's Schedule of Submittals: if it provides a workable arrangement for reviewing and processing the required Submittals.

      3) Contractor's Schedule of Values: if it provides a reasonable allocation of the Contract Price to the component parts of the Work.

6. Baselines and Benchmarks: As appropriate, provide information to Engineer to establish baselines and benchmarks for locating portions of the Work designed by Subconsultant to the extent that such baselines and benchmarks are necessary (in Subconsultant's judgment) to enable Contractor to proceed.

7. Permits: Provide Engineer with copies of technical information and data previously obtained or developed by Subconsultant for Owner's use, or for Owner to provide to Contractor, in obtaining required permits and licenses delegated to Contractor by Owner.

8. Visits to Site and Observation of Construction: In connection with observations of Contractor's Work while it is in progress:

Notes to User—To the extent that the parties may want to better quantify the number of actual visits for budgeting or quality assurance purposes, it may be advisable to insert at the beginning of Paragraph 8.a an agreed upon frequency of Subconsultant's site visits, e.g. "at approximately weekly intervals during periods of active construction" in lieu of the more general "intervals appropriate to the various stages of construction."

a. Make visits to the Site at intervals appropriate to the various stages of the Work, as Subconsultant deems necessary, to observe as an experienced and qualified design professional, the progress of Contractor's executed Work. Such visits and observations by Subconsultant, including its SRR, if any, are not intended to be exhaustive or to extend to every aspect of the Work or to involve detailed inspections of the Work beyond the responsibilities specifically assigned to Subconsultant in this Agreement and the Construction Contract Documents, but rather are to be limited to spot checking, selective sampling, and similar methods of general observation of the Work based on Subconsultant's exercise of professional judgment, as assisted by the SRR, if any. Based on information obtained during such visits and observations, Subconsultant will determine in general if the Work is proceeding in accordance with the Construction Contract Documents, and Subconsultant shall keep Engineer informed of the progress of the Work. Subconsultant will make a report of Subconsultant's visit, summarizing Subconsultant's general observations and any significant findings.

b. The purpose of Subconsultant's visits to the Site, and representation by the SRR, if any, at the Site, will be to enable Subconsultant to better carry out the duties and responsibilities assigned by this Agreement and undertaken by Subconsultant during the Construction Phase, and, in addition, by the exercise of Subconsultant's efforts as an experienced and qualified design professional, to provide for Owner a greater degree of confidence that the completed Work will conform in general to the Construction Contract Documents and that Contractor has implemented and maintained the integrity of the design concept of the completed Project as a functioning whole as indicated in the Construction Contract Documents. Subconsultant will not, during such visits or as a result of such observations of the Work, supervise, direct, or have control over the Work, nor will Subconsultant have authority over or responsibility for the means, methods, techniques, sequences, or procedures of construction selected or used by any Constructor, for security or safety at the Site, for safety precautions and programs incident to any Constructor's work in progress, for the coordination of the Constructors' work or schedules, nor for any failure of any Constructor to comply with Laws and Regulations applicable to furnishing and performing of its work. Accordingly, Subconsultant neither guarantees the performance of any Constructor nor assumes responsibility for any Constructor's failure to furnish or perform the Work, or any portion of the Work, in accordance with the Construction Contract Documents.

9. Defective Work: If, on the basis of Subconsultant's observations or as indicated in documentation available to Subconsultant, Subconsultant believes that any part of the Work is defective under the terms and standards set forth in the Construction Contract Documents, Subconsultant will promptly notify Engineer of such defective Work. Such notice will communicate the scope, extent (to Subconsultant's understanding) of defect, and associated provisions of the Construction Contract Documents.

    a. Provide recommendations to Engineer regarding whether Contractor should correct such Work or remove and replace such Work, or whether Owner should consider accepting the defective Work in accordance with the provisions of the Construction Contract Documents.

    b. However, Subconsultant's authority to make such recommendation or Subconsultant's decision to exercise or not exercise such authority will not give rise to a duty or responsibility of the Subconsultant to Contractors, Subcontractors, material and equipment suppliers, their agents or employees, or any other person(s) or entities performing any of the Work, including but not limited to any duty or responsibility for Contractors' or Subcontractors' safety precautions and programs incident to the Work.

10. Compatibility with Design Concept: If Subconsultant has express knowledge that a specific part of the Work that is not defective under the terms and standards set forth in the Construction Contract Documents is nonetheless not compatible with the design concept of the completed Project as a functioning whole, then inform Engineer of such incompatibility, and provide recommendations for addressing such Work.

11. Clarifications and Interpretations: Accept from Engineer submittals that Engineer has received from Contractor and Owner concerning the requirements of the Construction Contract Documents (sometimes referred to as requests for information or interpretation—RFIs), or relating to the acceptability of the Work under the Construction Contract Documents, to the extent that any such submittal is pertinent to Subconsultant's scope of services. With reasonable promptness, recommend to Engineer a written clarification, interpretation, or decision on the issue submitted, or initiate an amendment or supplement to the Construction Contract Documents.

12. Non-reviewable Matters: If a submitted matter in question concerns the Subconsultant's performance of its duties and obligations, or terms and conditions of the Construction Contract Documents that do not involve (1) the performance or acceptability of the Work under the Construction Contract Documents, (2) the design (as set forth in the Drawings, Specifications, or otherwise), or (3) other engineering or technical matters, then Subconsultant will promptly give written notice to Engineer that Subconsultant will not provide a decision or interpretation.

13. Field Orders: Subject to any limitations in the Construction Contract Documents, recommend to Engineer any Field Orders requiring minor changes in the Work.

14. Change Orders and Work Change Directives: Recommend Change Orders and Work Change Directives to Engineer, as appropriate, and prepare Change Orders and Work Change Directives as directed by Engineer.

15. Change Proposals and Claims

    a. Review and promptly respond to Change Proposals submitted by Engineer to Subconsultant for comment. If the Change Proposal does not involve the design (as set forth in the Drawings, Specifications, or otherwise), the acceptability of the Work, or other engineering or technical matters pertinent to Subconsultant's services, then Subconsultant will so advise the Engineer.

    b. Provide information or data to Engineer regarding engineering or technical matters pertaining to Claims.

16. Differing Site Conditions: Upon request by Engineer, provide information to assist Engineer in responding to any notice from Contractor of differing site conditions, including conditions relating to Underground Facilities such as utilities, and hazardous environmental conditions.

17. Contractor's Submittals: Review and approve or take other appropriate action with respect to required Contractor Submittals, but only to determine if the items covered by the Submittals will, after installation or incorporation in the Work, comply with the requirements of the Construction Contract Documents and compatibility with the design concept of the completed Project as a functioning whole as indicated by the Construction Contract Documents. Such reviews and approvals or other action will not extend to means, methods, techniques, sequences, or procedures of construction or to safety precautions and programs incident thereto. Subconsultant shall meet any Contractor's Submittal schedule that Engineer has accepted.

18. Substitutes and "Or-equals": Consult with and advise Engineer concerning and the acceptability of substitute or "or-equal" materials and equipment proposed by Contractor, but subject to the provisions of Exhibit A Paragraph 2.01.

19. Inspections and Tests

    a. In consultation with Engineer, receive and review all certificates of inspections, tests, and approvals required by Laws and Regulations or the Construction Contract Documents. Subconsultant's review of such certificates will be for the purpose of determining whether or not the results certified indicate compliance with the Construction Contract Documents and will not constitute an independent evaluation that the content or procedures of such inspections, tests, or approvals comply with the requirements of the Construction Contract Documents. Subconsultant and Engineer shall be entitled to rely on the results of such inspections and tests.

    b. Through the Engineer, reply to Contractor requests for written concurrence that specific portions of the Work that are to be inspected, tested, or approved may be covered.

    c. Prepare for Engineer's issuance written requests to Contractor that specific portions of the Work remain uncovered.

d. As deemed reasonably necessary and through Engineer, request that Contractor uncover Work that is to be inspected, tested, or approved.

e. Pursuant to the terms of the Construction Contract, require additional inspections or testing of the Work, whether or not the Work is fabricated, installed, or completed.

20. Contractor's Applications for Payment: Based on Subconsultant's observations as an experienced and qualified design professional and upon Engineer's request review of Applications for Payment and accompanying supporting documentation:

a. Determine the amounts that Subconsultant recommends Contractor be paid. Recommend reductions in payment (set-offs) based on the provisions for set-offs stated in the Construction Contract. Such recommendations of payment will be in writing and will constitute Subconsultant's representation to Engineer, based on such observations and review, that, within the limits of Subconsultant's knowledge, information and belief, Contractor's Work has progressed to the point indicated, the Work is generally in accordance with the Construction Contract Documents (subject to an evaluation of the Work as a functioning whole prior to or upon Substantial Completion, to the results of any subsequent tests called for in the Construction Contract Documents, and to any other qualifications stated in the recommendation), and the conditions precedent to Contractor's being entitled to such payment appear to have been fulfilled in so far as it is Subconsultant's responsibility to observe the Work. In the case of unit price Work, Subconsultant's recommendations of payment will include final determinations of quantities and classifications of the Work (subject to any subsequent adjustments allowed by the Construction Contract Documents).

b. By recommending payment, Subconsultant shall not thereby be deemed to have represented that observations made by Subconsultant to check the quality or quantity of Contractor's Work as it is performed and furnished have been exhaustive, extended to every aspect of Contractor's Work in progress, or involved detailed inspections of the Work beyond the responsibilities specifically assigned to Subconsultant in this Agreement. Neither Subconsultant's review of Contractor's Work for the purposes of recommending payments nor Subconsultant's recommendation of any payment including final payment will impose on Subconsultant responsibility to supervise, direct, or control the Work, or for the means, methods, techniques, sequences, or procedures of construction or safety precautions or programs incident thereto, or Contractor's compliance with Laws and Regulations applicable to Contractor's furnishing and performing the Work. It will also not impose responsibility on Subconsultant to make any examination to ascertain how or for what purposes Contractor has used the money paid to Contractor by Owner; to determine that title to any portion of the Work, including materials or equipment, has passed to Owner free and clear of any liens, claims, security interests, or encumbrances; or that there may not be other matters at issue between Owner and Contractor that might affect the amount that should be paid.

21. Contractor's Completion Documents: Upon request from Engineer, assist Engineer in review of Contractor-provided maintenance and operating instructions, schedules, guarantees, bonds, certificates or other evidence of insurance required by the Construction Contract Documents, certificates of inspection, tests and approvals, and Shop Drawings, Samples, and other data approved as provided under Exhibit A Paragraph 1.06.B.17; and annotated record documents which are to be provided to Engineer and assembled by Contractor in accordance with the Construction Contract Documents. The extent of Subconsultant's review of these documents will be to check that Contractor has submitted a complete set of those documents that Contractor is required to submit.

22. Substantial Completion: Promptly after notice from Engineer that Contractor considers the entire Work ready for its intended use, visit the Site in company with Engineer, Owner, and Contractor, to review the Work and determine the status of completion. Assist the Engineer in following the procedures in the Construction Contract regarding the preliminary certificate of Substantial Completion, punch list of items to be completed, Owner's objections, notice to Contractor, and issuance of a final certificate of Substantial Completion. Provide support to Engineer regarding any remaining engineering or technical matters affecting Owner's use or occupancy of the Work following Substantial Completion.

23. Other Tasks: Perform or provide the following other Construction Phase tasks or deliverables:

    a. **[List any such tasks or deliverables here]**.

24. Subconsultant's Notice of Acceptability of the Work: Assist Engineer in conducting a final visit to the Site to determine if the portion of the Work designed or specified by Subconsultant, together with any additional portion of the Work specifically designated by Engineer for observation by Subconsultant, is complete and acceptable so that Subconsultant may recommend to Engineer, in writing, that Owner may be advised to make final payment to Contractor with respect to such portion of the Work. If Subconsultant is satisfied that such portion of the Work is complete and acceptable, provide a notice to Engineer using the form provided in Exhibit E to this Agreement (Subconsultant's Notice of Acceptability of Work) stating that such portion of the Work is acceptable, subject to the provisions of the Notice, to the best of Subconsultant's knowledge, information, and belief, and based on the extent of the services provided by Subconsultant under this Agreement.

25. Standards for Certain Construction-Phase Decisions: Subconsultant, when delegated by Engineer to assist Engineer or fulfill Engineer's decision making duties, will render decisions regarding the requirements of the Construction Contract Documents, and judge the acceptability of the Work, pursuant to the specific procedures set forth in the Construction Contract for initial interpretations, Change Proposals, and acceptance of the Work. In rendering such decisions and judgments, Subconsultant will not show partiality to Owner or Contractor, and in such decision-making duties will not be liable to Owner, Contractor, or others in connection with any proceedings, interpretations, decisions, or judgments conducted or rendered in good faith.

C. Duration of Construction Phase: The Construction Phase will commence with the execution of the first Construction Contract for the Project or any part thereof and will terminate upon written recommendation by Engineer for final payment to Contractor. If the Project involves more than one prime contract as indicated in Exhibit A Paragraph 1.04.A.1, then Construction Phase services may be rendered at different times in respect to the separate contracts. Subject to the provisions of Article 3 of the Agreement, Subconsultant shall be entitled to an equitable increase in compensation if Construction Phase services (including SRR services, if any) are required after the original date for completion and readiness for final payment of Contractor as set forth in the Construction Contract.

1.07 Post-Construction Phase

A. Upon written authorization from Engineer during the Post-Construction Phase, Subconsultant shall:

1. Together with Owner and Engineer, visit the Project to observe any apparent defects in the Work, make recommendations as to replacement or correction of defective Work, if any, or the need to repair of any damage to the Site or adjacent areas, and assist Engineer in consultations and discussions with Contractor concerning correction of any such defective Work and any needed repairs.

2. Together with Owner and Engineer, visit the Project within one month before the end of the Construction Contract's correction period to ascertain whether any portion of the Work or the repair of any damage to the Site or adjacent areas is defective and therefore subject to correction by Contractor.

3. Perform or provide the following other Post-Construction Phase tasks or deliverables:

a. **[List any such tasks or deliverables here]**.

B. The Post-Construction Phase services may commence during the Construction Phase and, if not otherwise modified in this Exhibit A, will terminate 12 months after the commencement of the Construction Contract's correction period.

ARTICLE 2—ADDITIONAL SERVICES

2.01 Additional Services Requiring Engineer's Written Authorization

A. If authorized in writing by Engineer, the Subconsultant shall provide Additional Services of the types listed below. These services are not included as part of Basic Services and will be paid for by Engineer as indicated in Exhibit J.

1. Obtain or provide specified additional Project-related information and data to enable Subconsultant to complete its Basic and Additional Services.

2. Preparation of special and customized reporting, invoicing, and related support documentation in addition to that identified to be provided under Basic Services.

3. Preparation of applications and supporting documents (in addition to those furnished under Basic Services) for private or governmental grants, loans, or advances in connection with the Project; preparation or review of environmental assessments and impact statements; review and evaluation of the effects on the design requirements for the Project of any such statements and documents prepared by others; and assistance

in obtaining approvals of authorities having jurisdiction over the anticipated environmental impact of the Project.

4. Services to make measured drawings of existing conditions or facilities, to conduct tests or investigations of existing conditions or facilities, or to verify the accuracy of drawings or other information furnished by Owner or others.

5. Services resulting from significant changes in the scope, extent, or character of the portions of the Project designed or specified by Subconsultant , or the Project's design requirements, including, but not limited to, changes in size, complexity, Owner's schedule, character of construction, or method of financing; and revising previously accepted studies, reports, Drawings, Specifications, or Construction Contract Documents when such revisions are required by changes in Laws and Regulations enacted subsequent to the Effective Date or are due to any other causes beyond Subconsultant's control.

6. Services resulting from Owner's request to evaluate additional Study and Report Phase alternative solutions beyond those agreed to in Exhibit A Paragraph 1.02.A.

7. Services required as a result of Engineer or Owner providing incomplete or incorrect Project information to Engineer.

8. Providing renderings or models for Engineer's or Owner's use, including development, management, and other services in support of building information modeling or civil integrated management.

9. Undertaking investigations and studies including, but not limited to:

    a. All-hazards risk assessments and other studies to evaluate the feasibility of enhancing the resiliency of the design;

    b. detailed consideration of operations, maintenance, and overhead expenses;

    c. the preparation of feasibility studies (such as those that include projections of output capacity, utility project rates, project market demand, or project revenues) and cash flow analyses, provided that such services are based on the engineering and technical aspects of the Project, and do not include rendering advice regarding municipal financial products or the issuance of municipal securities;

    d. preparation of appraisals;

    e. with respect to proprietary systems or processes requiring licensing, assisting Engineer and Owner in obtaining such licensing;

    f. detailed quantity surveys of materials, equipment, and labor; and

    g. audits or inventories required in connection with construction performed or furnished by Owner.

10. Furnishing services of Subconsultants for other than Basic Services.

11. Providing data or services of the types described in Article 2 of the Agreement, when Owner or Engineer retains Subconsultant to provide such data or services instead of Owner furnishing the same.

12. Providing the following services:

a. Services attributable to more prime construction contracts than specified in Exhibit A Paragraph 1.04.A.1.

b. Services to arrange for performance of construction services for Owner by contractors other than the principal prime Contractor, and administering Owner's contract for such services.

13. Services during out-of-town travel required of Subconsultant, other than for visits to the Site, Engineer's or Owner's office as required in Basic Services (Article 1 of Exhibit A).

14. Preparing for, coordinating with, participating in and responding to structured independent review processes, including, but not limited to, construction management, cost estimating, project peer review, value engineering, and constructability review requested by Engineer or Owner; and performing or furnishing services required to revise studies, reports, Drawings, Specifications, or other documents as a result of such review processes.

15. Preparing additional bidding-related documents (or requests for proposals or other construction procurement documents); preparing pre-qualification procedures and documents, and participating in pre-qualifying prospective bidders; and preparing Construction Contract Documents for alternate bids.

16. Assistance in connection with bid protests, rebidding, or renegotiating contracts for construction, materials, equipment, or services.

17. Preparing conformed Construction Contract Documents that incorporate and integrate the content of all addenda and any amendments negotiated by Owner and Contractor.

18. Services to assist Engineer or Owner in developing or modifying protocols for transmittal of Electronic Documents by Electronic Means after the effective date of this Agreement, either by revising or adapting Exhibit F (the EDP) to the Project or implementing other Electronic Documents protocols among Project participants.

19. Any services by Subconsultant in connection with Owner, Engineer or Subconsultant providing a Document to a Requesting Party under Exhibit F Paragraph 1.01.D, Electronic Documents Protocol, or any other distribution of a Document to a third party. Such services may include but are not limited to preparing the data contained in the requested Document in a manner deemed appropriate by Engineer or Subconsultant; and creating or otherwise preparing and distributing the Document in a format necessary to respond to Owner's direction or decision to provide the Document to a requesting party, including Contractor, in a format other than that required for deliverables from Subconsultant to Engineer or Owner; Compensation for these Additional Services is not contingent upon Owner's reimbursement from the requesting party.

20. Providing Construction Phase services beyond the original date for completion and readiness for final payment of Contractor, but only if such services increase the total quantity of services to be performed in the Construction Phase, rather than merely shifting performance of such services to a later date.

21. Preparing Record Drawings, and furnishing such Record Drawings to Owner.

22. Supplementing Record Drawings with information regarding the completed Project, Site, and immediately adjacent areas obtained from field observations, Engineer, Owner, utility companies, and other reliable sources.

23. Conducting surveys, investigations, and field measurements to verify the accuracy of Record Drawing content obtained from Contractor, Owner, utility companies, and other sources; revise and supplement Record Drawings as needed.

24. Preparation of operation, maintenance, and staffing manuals.

25. Protracted or extensive assistance in refining and adjusting of Project equipment and systems (such as initial startup, testing, and balancing).

26. Assistance to Owner in training Owner's staff to operate and maintain Project equipment and systems.

27. Assistance to Engineer or Owner in developing systems and procedures for (a) control of the operation and maintenance of Project equipment and systems, and (b) related recordkeeping.

28. Preparing to serve or serving as a consultant or witness for, or producing documents for or on behalf of, Owner in any litigation, arbitration, mediation, lien or bond claim, or other legal or administrative proceeding involving the Project (but not including disputes between Owner and Engineer).

29. Overtime work requiring higher than regular rates.

30. Providing construction surveys and staking to enable Contractor to perform its work other than as required under Exhibit A Paragraph 1.06.B.6; any type of property surveys or related engineering services needed for the transfer of interests in real property; providing construction and property surveys to replace reference points or property monuments lost or destroyed during construction; and providing other special field surveys.

31. Providing more extensive services required to enable Engineer to issue notices or certifications requested by Owner.

32. Extensive services required during any correction period, or with respect to monitoring Contractor's compliance with warranties and guarantees called for in the Construction Contract (except as agreed to under Basic Services).

33. Substantive design and other technical services in connection with Work Change Directives, Change Proposals, and Change Orders to reflect changes requested by Owner.

34. Services essential to the orderly progress of the Bidding and Construction Phases and not wholly quantifiable prior to those Phases or otherwise dependent on the actions of prospective individual bidders or contractors and including:

    a. making revisions to Drawings and Specifications occasioned by the acceptance of substitute materials or equipment other than "or equal" items;

b. services after the award of the Construction Contract in evaluating and determining the acceptability of a proposed "or equal" or substitution which is found to be inappropriate for the Project;

c. evaluation and determination of an excessive number of proposed "or equals" or substitutions, whether proposed before or after award of the Construction Contract; and

d. providing to the Engineer, Contractor or Owner additional or new information not previously prepared or developed by the Subconsultant for their use in applying for or obtaining required permits and licenses, in responding to agency comments on such applications, or in the administration of any such permits or licenses.

35. Services resulting from significant delays, changes, or price increases occurring as a direct or indirect result of materials, equipment, or energy shortages.

36. Additional or extended services arising from (a) the presence at the Site of any Constituent of Concern or items of historical or cultural significance, (b) emergencies or acts of God endangering the Work, (c) damage to the Work by fire or other causes during construction, (d) a significant amount of defective, neglected, or delayed Work, (e) acceleration of the progress schedule involving services beyond normal working hours, or (f) default by Contractor.

37. Implement coordination of Subconsultant's services with other parts of the Project.

38. Implement the specific parts of an Underground Facilities Procedure that are assigned to Subconsultant (but not including the design-related services already assigned to Subconsultant as a Basic Service in Exhibit A Paragraph 1.03.B.9).

39. As directed by Engineer, observe a specific portion of the Work (in addition to those designed or specified by Subconsultant) to assist in determining Substantial Completion and readiness of the completed Work for final payment;

40. Attend additional meetings, as directed by Engineer, including but not limited to design progress, pre-bid conference, bid opening, pre-construction conferences, construction progress, and other job-related meetings.

41. Services (other than Basic Services during the Post-Construction Phase) in connection with any partial utilization of the Work by Owner prior to Substantial Completion.

42. Evaluating unreasonable or frivolous requests for interpretation or information (RFIs), Change Proposals, or other demands from Contractor or others in connection with the Work, or an excessive number of RFIs, Change Proposals, or demands.

43. Reviewing a Shop Drawing or other Contractor Submittal more than three times, as a result of repeated inadequate submissions by Contractor.

44. While at the Site, compliance by Subconsultant and its staff with those terms of Engineer's, Owner's or Contractor's safety program provided to Subconsultant subsequent to the Effective Date that exceed those normally required of engineering personnel by federal, State, or local safety authorities for similar construction sites.

45. To the extent the Project is subject to Laws and Regulations governing public or government records disclosure or non-disclosure, Subconsultant will comply with

provisions applicable to Engineer, and, to extent Owner will compensate Engineer as Additional Services for Engineer's costs to comply with any disclosure or non-disclosure obligations beyond those identified in the Basic Services, the Engineer will compensate the Subconsultant as Additional Services.

Other additional services performed or furnished by Engineer not otherwise provided for in this Agreement.

REFERENCES

ACEC (American Council of Engineering Companies) Risk Management Committee. n.d. *An overview of indemnification and the duty to defend*. Washington, DC: ACEC.

Beatty, J. F., and S. S. Samuelson. 2016. *Business law and the legal environment*. 7th ed. Boston: Cengage Learning.

Beazley, n.d. A&E Contract Review Guide (citing *Hydrotech Systems, Ltd., v. Oasis Waterpark* (1991) 52 Cal. 3d 998), 3.

Black's Law Dictionary, 9th ed. 2009. 346, col. 1.

Court of Appeal of California, Sixth District. 2010. UDC-Universal Development, L.P., v. CH2M Hill. 2010. 181 Cal. App. 4th 10. Accessed January 11, 2022. https://casetext.com/case/udc-universal-dev-v-ch2m-hill.

EJCDC (Engineers Joint Contract Documents Committee), ACEC (American Council of Engineering Companies), and ASCE. 2020. *Agreement between engineer and subconsultant for professional services*. Washington, DC: ACEC.

Kubasek, N. K., B. A. Brennan, and M. Browne. 2017. *The legal environment of business a critical thinking approach*. 8th ed. London: Pearson.

Mallor, J. 2013. "Introduction to contracts." In *Business law: The ethical, global, and E-commerce environment*. 15th ed., 316. New York: McGraw Hill.

Woolley, D., and Associates Inc. 2021. "Agreement for professional services between client and consultant. language." Engineers Joint Contract Documents Committee. Accessed January 11, 2022. https://www.ejcdc.org/ejcdc/.

INDEX

Note: Page numbers followed by *f* and *t* indicate figures and tables.

Printed and bound by CPI Group (UK) Ltd, Croydon, CR0 4YY
16/09/2026
14963439-0001